Laser Interaction
and Related Plasma Phenomena
Volume 10

Laser Interaction
and Related Plasma Phenomena

Volume 10

Edited by

George H. Miley

*Fusion Studies Laboratory
University of Illinois at Urbana–Champaign
Urbana, Illinois*

and

Heinrich Hora

*CERN
Geneva, Switzerland
and University of New South Wales
Kensington, New South Wales, Australia*

SPRINGER SCIENCE+BUSINESS MEDIA, LLC

Proceedings of the Tenth International Workshop on Laser Interaction
and Related Plasma Phenomena, held November 11–15, 1991, at the
Naval Postgraduate School, Monterey, California

Library of Congress Catalog Card Number 86-651543 (ISSN 0148-0987)

ISBN 978-1-4613-6464-1   ISBN 978-1-4615-3324-5 (eBook)
DOI 10.1007/978-1-4615-3324-5

© 1992 Springer Science+Business Media New York
Originally published by Plenum Press, New York in 1992
Softcover reprint of the hardcover 1st edition 1992

PREFACE

The Tenth International Workshop on "Laser Interaction and Related Plasma Phenomena" was held November 11-15, 1991, at the Naval Postgraduate School, Monterey, California. This conference joined physicists from 11 countries (Australia, Canada, China, France, Israel, Italy, Spain, Switzerland, United Kingdom, USA, and the USSR). This meeting was marked by the inauguration of the **EDWARD TELLER MEDAL FOR ACHIEVEMENTS IN FUSION ENERGY.** This medal served as a celebration of the tenth conference in the 22-year series and as an opportunity to honor one of the world's greatest physicists and a leading pioneer in this field: Edward Teller. Four medals were awarded in the inaugural ceremony.

The first recipient of the medal was Nobel Laureate **Nikolai G. Basov**, who served for many years as Director of the Lebedev Physical Institute of the Academy of Sciences of the USSR. In his address to Edward Teller, Dr. Basov underlined that Dr. Teller was the first in history to produce an exothermal nuclear fusion reaction, the mechanism that may now lead to an inexhaustive, environmentally clean, and low cost energy source in the future. This goal, he stressed, becomes more crucial as the greenhouse effect may not permit burning of fossil fuels for much longer. Basov also reviewed events leading the International Quantum Electronics Conferences of 1963 where he disclosed the first publication on laser fusion and that of 1968 where he reported the first observation of fusion neutrons using a laser-irradiated target.

The second recipient was **John H. Nuckolls**, the director of the Lawrence Livermore National Laboratory in Livermore, California, who praised Edward Teller as "the father of inertial confinement fusion." In the late 1950s Nuckolls began to pursue the ignition of contained thermonuclear explosions of the smallest possible size, using the radiation implosion scheme developed by Teller. He was immediately inspired by the invention of the laser in 1960 to make supercomputer calculations of small radiation implosions driven by high-power lasers. Temporally shaped (tailored) multi-megajoule laser pulses were used in these calculations to implode "bare drop" DT targets to super-high densities with ignition at a central hot spot. Unshaped laser pulses were also used to implode thin hollow shell fusion targets to similar conditions. The leading paper on supercomputer calculations of the analogous electron implosion approach to fusion with multi-kilojoule lasers was written by John Nuckolls and co-workers [Nature **239**, 192 (1972)]. The rapid advances in laser technologies led to the initiation of the leading laser fusion program at Livermore as summarized by John Nuckolls when receiving the James Clark Maxwell Prize of the American Physical Society [Physics Today **35** (No. 9), 24 (1982)]. Through the early 1980s, he headed the thermonulcear target design group at Livermore. In this period he generalized the basic target design to couple to heavy ion beam drivers. Throughout this 25-year period from the late 1950s to the early 1980s, Nuckolls also had a leading role in designing nuclear explosive experiments that demonstrated--as underlined in the address by Edward Teller as a point of Nuckolls' special initiative--the feasibility of using relatively small energies to ignite small masses of DT fuel.

The third recipient was **Chiyoe Yamanaka**, founder and first director of the Institute of Laser Technology in Osaka, Japan. He presented a remarkable historical review about laser fusion work and how his laboratory built the second largest glass laser in the world for laser fusion research. He went on to review recent achievements such as smoothing techniques for direct drive. In this way it was possible to reach compressions of polyethylene to densities of up to 1000 times the solid state (about 1 kg/cm^3). The highest neutron yields achieved at this laboratory used direct drive with an almost adiabatic compression without stagnation or shock waves. These experiments verified the effectiveness of maintaining an also ideal adiabatic volume compression.

The fourth recipient, **Heinrich Hora** from CERN in Geneva, Switzerland, was cited as the initiator and promoter of the international conference series. In his acceptance speech, he reported on the new physics--in view also of Edward Teller's first initiatives--which had been developed by the studies of inertial confinement fusion with lasers. While the approach of Nuckolls and colleagues was to use a radiation drive to provide a uniform compression, the complexity of very high intensity laser interactions with a dense plasma has only recently become understood. Dr. Hora also outlined how the complicated pulsation (period of several picoseconds) can be avoided by beam-smoothing techniques. He proposed that this provides the physics solution for a direct drive laser fusion reactor with a naturally simple design that should lead to low cost energy production. To understand this problem, the macroscopic plasma theory had to be extended by the exploration of nonlinear mechanisms, arriving at a complete formulation of the hydrodynamic equations involved. A consequence of this was the treatment of double layers and surface energy for plasmas. That concept has now been successfully generalized to metals and to nuclei.

In his after-dinner speech, **Edward Teller** addressed the scientific merits that were achieved with inertial confinement fusion and the optimistic and realistic aspects of energy production. He underlined everyone's strong desire for international cooperation but acknowledged the implications of proliferation of technologies for defense application cannot be ignored. The recent disclosures from Saddam Hussein's research laboratories serve as a clear example of this problem. Still, he stated, important new discoveries cannot be held secret long. There are always ways that the knowledge gets out or is rediscovered. Thus, he proposed a sunset clause whereby any classified materials would be declassified in one year.

A special highlight of the conference was the first disclosure about the big Russian chemical explosive lasers by N. G. Basov. An explosive driven shock wave in XeF gas in 10-meter long and 1-meter wide tubes has produced megajoule laser pulses of 70-microsecond duration. The possibility of pulse shortening exists such that low cost and efficient lasers of this type may become available for laser fusion. Also such an arrangement might be used for igniting a self-sustained fusion reactor system of the type proposed in earlier workshops of this series by G. Miley. In that case the laser driver is directly pumped by fusion neutrons from the target implosion.

M. Sluyter, the director of the Office of Inertial Fusion within the U.S. Department of Energy presented a review of the activities in the U.S. ICF program. The main question facing the community is the demonstration that pulses in the megajoule range will produce ignition of a fusion target. This, he stated, will be like "getting the first man to the moon."

The laser fusion concept with the Nova upgrade at Livermore for megajoule laser pulses was reviewed by H. Lowdermilk. He described a new amplifier chain with a quadratic cross section that serves as the module for the big laser system. Another important innovation is the use of Raman scattering to produce a broad-band laser beam for smoothing. He suggested that despite the rather modest funding level at the present, the ICF program should lead to a power plant in 2025. However, he and others noted other possible strategies with earlier dates for the reactor.

The various new developments presented in the 61 papers at the conference are too numerous to mention or discuss in detail here. Some of special interest include the new diagnostic method of laser-produced plasmas that involved analyzing the emission of neutral atoms (A. S. Shikanov, Lebedev Institute), a new and very successful x-ray focusing optics technology (A. Rode, Australian National University), the presentation of light-ion-beam compression of targets with black-body radiation temperatures of about 60 eV (T. Mehlhorn, Sandia National Labs.), the Chinese result of achieving 150 eV radiation temperature in Hohlraum experiments (Peng Hansheng), and the use of circularly polarized laser light to produce a secondary confinement of laser-produced plasmas by magnetic fields of several megagauss (S. Eliezer, SOREQ, Israel). The first report on the demonstration of a nuclear flashlamp-pumped laser was given (W. H. Williams and G. H. Miley, University of Illinois) with atomic iodine laser emission and XeBr as flashlamp medium.

A special discussion meeting was arranged by G. Velarde (Madrid), the President of the International Society for Inertial Fusion Energy with the members of the board of directors of the society present at the conference. At this meeting, views were exchanged on how a more objective judgment between Inertial Fusion Energy (IFE) and the powerful magnetic confinement fusion (MCF) can be achieved. As was indicated before in several lectures and statements, IFE has, as a goal, achievement of a laser fusion reactor within about 15 to 25 years. This assumes a crash program will be started with total support of about $30 billion. In contrast, MCF needs more than 30 years and $150 billion. Further, the cost of energy from the MCF reactor may be much higher due to such problems as rapid erosion of the first wall and neutron damage to structures. The use of a flowing liquid-metal first wall in ICF or the Cascade reactor design avoid such programs.

Another discussion was organized by J. Mark (Livermore) on the problems of the greenhouse effect and the role for ICF power in future energy scenarios. In this context, issues of an even wider scope for the environment were articulated by various speakers. For example, N. G. Basov mentioned the application of laser-produced neutron sources for transmuting radioactive waste. He also noted that one important advantage of fusion energy is that the volume of the fusion fuel that must be transported is reduced by a factor of one million compared to fossil fuel. Relative to industry and power plants in general, he stressed that one should put high priority on studies of the toxicity of chemical substances and what damage has been done to human bodies. Finally, Dr. Basov underlined again the need for international cooperation, not only in ICF but in all areas of science related to the environment and mankind's mutual interests.

The articles contained in this volume represent a majority of the presentations at the International Workshop. However, for various reasons, full papers were not submitted for publication in connection with thirteen presentations. Persons interested in reviewing subjects in that category should consult the abstract booklet distributed at the workshop and available through either of the Co-editors of this proceedings.

The organizers of the workshop wish to acknowledge the generous support of the following organizations: the Fusion Studies Laboratory of the University of Illionis, the Naval Postgraduate School, and the United States Department of Energy. Special thanks go to the Fusion Power Associates (Dr. S. Dean) for support of student scholarships, and to UNESCO (Dr. S. Raither) for travel assistance for Professor Basov, one of the Edward Teller Medal recipients.

George H. Miley
U. of Illinois
USA

Heinrich Hora
U. of New South Wales
Australia

CONTENTS

II.   INTERACTION MECHANISMS

VI.   BASIC PHENOMENA

EDWARD TELLER

EDWARD TELLER MEDAL PRESENTATIONS

The articles submitted by the four recipients of the Edward Teller Medal for 1991 represent a written version of the remarks made by the awardees when they received the medals.

The citations presented with the award to each individual were as follows:

1. **Nikolai G. Basov:** Dr. Basov was a co-recipient of the 1964 Nobel Prize for pioneering work leading to the discovery of the laser. We now honor him for subsequent work, namely for his leadership and for pioneering research on ICF at the Lebedev Institute, leading to the 1963 disclosure of laser fusion concepts at the International QE Conference, and the following disclosure in 1968 of the first detection of fusion neutrons from laser irradiated lithium hydride targets. This led to the development of the key research laser projects Kalmar and Delfin, which have enabled pioneering research in ICF.

2. **John H. Nuckolls:** As the Director of LLNL, John Nuckolls has an international reputation for his many contributions during three decades at LLNL. We honor him tonight for his contribution to the pioneering calculations, disclosed in 1972, that suggested laser fusion fuel pellets could be ignited with much less energy than predicted there-to-fore, and his subsequent contributions to the understanding of hydrodynamic and instability phenonema associated with the compression of dense plasmas. This insight has provided the guidance behind the LLNL ICF program, one of the leading programs in the world.

3. **Chiyoe Yamanaka:** We are honoring Dr. Yamanaka for his inspiration and leadership in the establishment of the Institute of Laser Engineering at Osaka University and, more recently, the formation of the Institute for Laser Technology. Under his leadership, the scientific team at Osaka is internationally recognized as a leader in the field. Dr. Yamanaka's many contributions to ICF theory and experiment culminated in 1987 with the remarkable achievement of record compression densities approaching a kg/cm^3. The momentum achieved under Chiyoe lives on today as ILE remains a leader among the world's largest ICF laboratories.

4. **Heinrich Hora:** Internationally recognized for his pioneering research on non-linear forces, electronic double layers, volume ignition, and smoothing of direct drive beams for ICF, Dr. Hora is recognized here for his leadership in the community, including initiation in 1969 of the International Workshop series on Laser Interaction and Related Plasma Phenomena and for starting and serving as the first editor-in-chief of the international journal, **Laser and Particle Beams**.

LECTURE IN CONNECTION WITH THE

EDWARD TELLER MEDAL AWARD

Dr. Edward Teller

Lawrence Livermore National Laboratory
P. O. Box 808
Livermore, CA 94550

Let me start by thanking Heinrich Hora for his valuable work on international cooperation and in particular for getting all of us together here. Tonight, we are giving awards to three outstanding men from Russia, the United States and Japan. All this is also a great honor to one man present here who comes from Hungary.

The emphasis in this particular award is on the use of high intensity lasers in producing controlled fusion. I want first to mention in connection with this, Professor Basov, whose work produced high intensity lasers by starting from no laser at all. The very realization of lasers is to a great extent his own work. The only one who may have priority of many years over him is no less than Albert Einstein who came up with the basic idea of lasers one-half century before they became real. It is a pity that we no longer can ask Einstein himself about the all important connection between particles obeying Bose statistics and the instabilities which play such an important role when these particles are produced in the highest numbers.

It was just in connection with such instabilities that the work of my good friend, John Nuckolls, is particularly important. He calculated these instabilities and made predictions about them and also pursued them in connection with very specific experiments at Livermore. Of course, he made me blush with embarrassment by the things he said about me when accepting the medal, but I have to forgive him for the very specific reason that I suspect him of meaning what he said!

Laser Interaction and Related Plasma Phenomena, Vol. 10
Edited by G.H. Miley and H. Hora, Plenum Press, New York, 1992

1

The recent great accomplishments of Professor Yamanaka have brought the expectation of success on controlled fusion much closer. It is fantastic, but a fact that he managed to compress hydrocarbons to six hundred times their natural density. What next? To expect practical success in controlled fusion in less than ten years may require an excess of optimism but to make predictions of scientific developments for the next century far exceeds my confidence in my ability to see into the future. All of the recipients and many of those present have contributed and will contribute to the realization of controlled fusion.

Now before descending to practical issues, I want to elaborate on the details of my blackboard to which our Director, John Nuckolls referred. Of course, my blackboard contains my thoughts going back a year or so which I was too lazy to erase, but above my blackboard, you will find the conspicuous presence of the picture of a centipede installed there ten years ago by Sandy Guntrum, one of our charming secretaries. I told her that a centipede is the heraldic animal of any big establishment and should be presented with the motto, *Never let your right front foot know what your left hind foot is doing*; nor . . . . She promptly came back and asked me when I expected delivery of the important object. I told her that imitating the Pope's comment to Michelangelo concerning the Sistine Chapel, I will not insist on any time limitations. She delivered the illustration in a prompt and beautiful fashion without the motto, but with little red shoes on each foot of the centipede emphasizing their divergent actions. One of these shoes she explained to me having a hole in the big toe was to remind us of our parent institution, the University of California. I repeat all this because even if the physical feasibility of controlled fusion is solved, the question of economic feasibility will be up to the establishments of whose existence the centipede keeps reminding me.

Having mentioned these important facts, I would now like briefly to mention a point which is both a requirement for our success and may benefit from our success. This is the connection between international cooperation and secrecy. Let me correct myself at once. I should not talk about secrecy but rather about openness.

The first important sign of real changes in the Soviet Union was, indeed, *Glasnost*. The influence of openness in our scientific work and our cooperation and our general behavior can hardly be overestimated. It is now to a considerable extent up to the United States to help establish openness in many ways and, in particular, in connection with the research on controlled fusion. In a specific way, I would like to mention the use of nuclear explosives in connection with experiments which are usually described by the not-sufficiently-descriptive name of tests. By practicing openness, tests could give most valuable scientific results concerning the effects of high energy concentrations which may be difficult to reach even with lasers. In addition to controlled fusion, these high energy densities could lead to the study of states of matter not otherwise accessible to our experience. It would be, for instance, particularly interesting if high compressions could be produced at lower temperatures. It would be certainly interesting to see what the Curie point would do if ferromagnets are several fold compressed. One might even want to speculate what would happen under such conditions to superconductivity.

But returning to the general question of secrecy, I would like to be practical and therefore refrain from a flat statement which I dare not make that secrecy should be abolished. I want to propose, indeed, that secrecy in technical matters involving the collaboration of many people should be limited to a relatively short period, let us say, for instance, the period of one year. We have in the United States tried to keep secrets for a long period of time. I believe the result was that we managed to confuse our own people but did not manage to keep secrets from our competitors.

There are many who argue that secrecy is necessary to prevent the spreading of nuclear weapons. I claim that secrecy is actually of little or no help in delaying proliferation. But openness would be of great help in preventing secret proliferation.

At this meeting, where we are discussing international efforts toward realizing controlled fusion, we have a valid opportunity to speak up for the principle and practice of openness. What I have mentioned and what I will proceed to some little extent to elaborate, I cannot call a solution to the problem. But my words may serve the purpose of stimulating some thought toward practical measures so that we can strengthen international cooperation by the practice of openness.

Of course, the connection between the proliferation of nuclear weapons and secrecy brings us in direct touch with the hard core of our problem. In this regard, we have learned a lot in recent months in connection with the efforts of Saddam Hussein toward producing nuclear weapons. This problem is now under investigation by an international committee whose findings are both unanimous and striking. We now know that Saddam Hussein has spent billions of dollars on the development of nuclear weapons. His approaches were based on valid information. Indeed, secrecy had little to do with delaying his success.

These fortunate delays appear to be due to the technical difficulties which are particularly important in a country where there is a scarcity of people of sufficient know-how in work requiring high technology. A policy of openness would make it obvious whenever massive work on a secret project is undertaken for a substantial period of time. This would give us a better chance to deal with problems of proliferation before they become acute and most dangerous.

I should like to conclude by repeating the great satisfaction I feel for having my name connected with the international effort toward controlled fusion. I am particularly grateful for the opportunity to speak for collaboration and for openness in the company of those who can support constructive action on the specific subject of this conference and also the more general subject of progess and openness of knowledge.

LASER FUSION RESEARCH IN 30 YEARS:
LECTURE OF EDWARD TELLER AWARDEE

C. Yamanaka

Institute for Laser Technology and
Institute of Laser Engineering
Osaka University
2-6 Yamada-oka Suita Osaka 565
Japan

INTRODUCTION

The laser fusion has made a great progress in the last 30 years. Similarly there has been an active international collaboration to develop the magnetic confinement fusion for power research. Recent inertial fusion experiments on the direct driven fusion at Osaka have successfully got the high fusion neutron yield 10^{13} and the high density compression of 600 times normal fuel density. The electron degeneracy of core plasma is also observed. The recent U. S. Halite/Centurion program informed us of indirect driven fusion news that may give high confidence that high gain inertial fusion will be attainable for less than about 10 MJ of driver. However, the data base is not yet clear to determine the details for high gain. This question can only be solved by a large enough laser facility. The U. S. policy on indirect driven fusion program has discouraged international cooperation in ICF for a long time. However, experimental and theoretical progress in ICF in the international community has suggested that the time has come to eliminate unnecessary restrictions on information relevant to the energy applications of ICF. Now ICF is in the second stage of the development. The ignition and breakeven are in a scope of our program. The international collaboration shall be initiated as soon as possible.

It is the most encouraging event for the international cooperation that the Edward Teller Awards are given on the occasion of the 20th anniversary of the Workshop on Laser Interaction and related plasma phenomena which was initiated by Helmut J. Schwarz and Heinrich Hora in 1969 to the four members Nikolai Basov, John Nuckolls, Chiyoe Yamanaka and Heinrich Hora who have been earnestly involved in the laser fusion research for the last 30 years.

WORLD PROGRESS OF ICF RESEARCH

The benefits from achieving high gain ICF in the laboratory are too important for individual countries to try to solve the physics and technology issues in isolation.

Laser Interaction and Related Plasma Phenomena, Vol. 10
Edited by G.H. Miley and H. Hora, Plenum Press, New York, 1992

5

Since 1974, our continuous efforts have been performed to organize the international ICF community concerning the IAEA activity and other authorities. In 1990, the US Secretary of Energy James D. Watkins addressed at the thirteenth IAEA Conference on Plasma Physics and Controlled Nuclear Fusion Research according to the reports by Koonin's committee of NAS and Stever's FPAC and announced if inertial fusion has promise as an energy source - and I believe that it does - we should pursue that promise with sort of cost-effective international collaboration that marks magnetic fusion efforts such as the International Thermonuclear Experimental Reactor (ITER). This address was welcomed by all the participants. A fundamental change is expected. Now, to show the importance of the international collaboration, the world progress of inertial fusion is briefly reviewed setting particular remarks on the Japanese efforts.

At the Levedev Institute in USSR, subsequent pioneering research on ICF has been performed. It let to the disclosure of laser fusion concepts at the International Quantum Electronics Conference in 1963, followed by a presentation in 1968 of the first detection of fusion neutrons from laser irradiated lithium hydride targets. These progresses led to the development of the first world class research lasers, Kalmar and Delfin, which have enabled the Levedev scientists to carry out forefront research in ICF.

At the Lawrence Livermore National Laboratory the pioneering theoretical and experimental works have been performed in the last three decades. Especially a disclosure in 1972 of the implosion physics at the International Quantum Electronics Conference, Montreal by Edward Teller showed that the laser fusion targets could be ignited with much less energy than predicted there-to-fore if the fuel was compressed up to 100 or 1000 times of the normal density. And also the understanding of hydrodynamic and instability phenomena associated with the strong compression ICF targets were prevailed. Series of the Nd glass lasers, Augus, Shiva and Nova were developed to perform the laser fusion experiments. The major glass suppliers for the large Nd glass lasers are in Japan and Germany. LLNL and also LANL have been performed a lot of interesting works in the ICF research which provided essential guidance for the ICF program in the world. However the US classification policy of inertial fusion especially indirect driven fusion is a crucial problem. It hurts the morale of the US scientists who are unable to take credit for their creative work and often must endure the vexation of seeing nearly identical work published in the open literature by workers in Japan, Europe or the Soviet Union. Classification impedes progress by restricting the flow of information, and does not allow all ICF work to benefit from open scientific scrutiny. According to the patient efforts of international movement of several countries, the world situation for cooperation is changing to promote.

In France, the CEA laboratory at Lemeil built a Nd glass laser, phebus, 20kJ which is now open to the public use. Smaller Nd glass laser facility exist at the University of Rochester, Ecole polytechinique Palaiseau, the Shanghai Institute for Optics and Fine Mechanics and at several other places around the world significant progress has been made over the last five years. Since there is in principle no physics obstacle to achieving sufficient control of illumination uniformity and hydrodymanic instability to achieve high convergence, more intense international efforts should be made for exploring abundant and affordable energy.

ICF PROGRESS AT OSAKA

The scientific works at Osaka is internationally recognized. In particular our contributions to ICF theory and experiment culminated in 1987 with remarkable achievement of record compression densities approaching a kg/cm^3. The Institute of Laser Engineering, Osaka University has been a leader among the world's largest ICF laboratories. This is the statement developed by the Awards Committee for the Edward Teller Medal.

In 1972, the Institute of Laser Engineering, Osaka University was established in accordance with the Edward Teller's special lecture on "New Internal Combustion Engine" at Montreal. And also we had timely the first Japan-US Scientific Seminar at

Kyoto by the Japan Society for Promotion of Science in this year which was an origin of the international collaboration on the inertial fusion research where 30 scientists from the U.S., Germany, Britain, Soviet Union and Japan gathered together. At the meeting, K. Hushimi the President of Science Council of Japan attended and encouraged the participants as saying that some person said the ICF is a dark horse, but contrary to it , the ICF is now at the focus of illumination and of attention, you may even speak about a white horse .... In this way the international race of inertial fusion started. Our research on the laser plasma initiated in 1963 using ruby lasers and Nd glass lasers. The first issue of research was the laser-plasma coupling. The absorption mechanisms were thoroughly investigated a result of which was to propose the anomalous absorption caused by the plasma parametric instability. Nonlinear plasma instability due to laser drive became a worldwide popular subject. We investigated the self phase modulation of laser light by plasmas and also the nuclear excitation by electronic transition due to lasers.

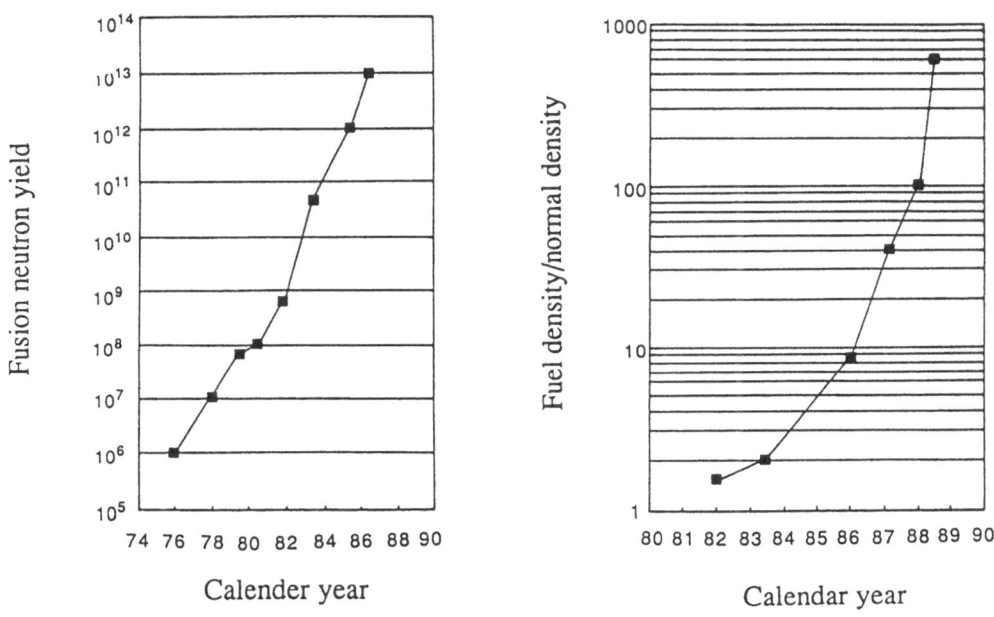

Fig. 1 Recent progress of ICF, neutron yield and implosion density

In 1975, we invented the so called indirect driven fusion concept "Cannonball Target" at our Daisen Summer Seminar which became later the Institute very popular in the world. At the age of oil crisis, the importance of new energy sources was well understood throughout the country. The Prime Minister T. Fukuda who was very favor to fusion sent us a message; "Resources are limited but Human wisdom is unlimited". In a fair wind to fusion research we set LEKKO CO_2 lasers to the Los Alamos group and competed with Livermore people by GEKKO glass lasers and compared the ideas to the Sandia team by REDEN beam machines. In the laboratory different kinds of these three drivers enabled us to compare the results of various plasma experiments and to review the qualification of drivers.

In 1983 the world largest of glass laser GEKKO XII was completed by the cooperation of the NEC and we got a fine estimation of the world leading laser fusion laboratory. As for the direct driven ICF, it is potentially more efficient but has significantly more stringent requirements on driver beam uniformity and the control of hydrodynamic instabilities. We had significant progress in this field using a novel type of uniform shell target and a random phasing smooth laser beam.

In 1985, the new idea of LHART (Large High Aspect Ratio Target) was devised by using an implosion simulation code. It could record a super shot of DT fusion neutron yield 10^{12} which was hurriedly after traced by the LLNL group. In 1986 the Fusion Power Associates gave a Leadership Award to this laser fusion achievement.

In 1987, The green light random phasing 12 beam of GEKKO XII glass laser irradiated a plastic shell target of nearly perfect sphericity to attain the 600 times normal density. The D-T fuel density reached $120gr/cm^3$ in absolute. The plasma is some what Fermi degenerated. These details were reported at the IAEA conference in Nice at 1988. Ablative pressure generation and hydrodynamic behavior of compressed fuel were experimentally and theoretically investigated. The implosion performance was optimized by using an appropriate aspect ratio of the target. The uniformity of laser irradiation as well as the pellet structure were essentially important to avoid the growth of instability.

Since the Edward Teller's lecture at Montreal, it has passed 20 years to attain the high compression densities of fuel predicted. These results give us high confidence that the ignition and burn of ICF will be attainable with a 100 kJ laser driver, such as GEKKO XII laser up grade.

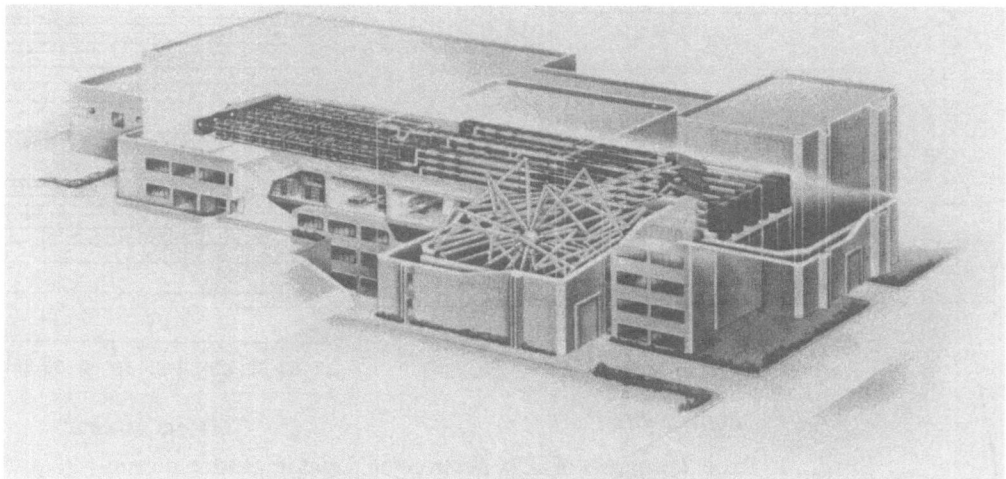

Fig. 2   GEKKO XII laser upgrade

PROSPECT OF ICF

At the Edward Teller Award ceremony, Teller predicted again the prospect of ICF research that the ignition and breakeven will be attained in the next 10 years before the 21st century following the recent results of high compression. He is content with the fulfillment of his first prediction of high density compression by Osaka. We shall say the Prospect of ICF seems to be very bright.

Applications of inertial confinement fusion include not only civil energy production but also physics at the laser-atomic frontier, nuclear matter under extreme conditions, cosmology, special isotope separation, food preservation, hydrogen production and advanced space propulsion. The pursuit of ICF will contribute substantially to overall scientific strength in several areas.

In the international collaboration, the essential research and technology development for fusion engineering and technology should be carried out in the following items,

    (1) High-average-power fusion drivers, lasers as well as heavy ion beams. They are also to produce the significant new applications in industry and science.

    (2) Power-plant use ICF target and fueling technology including cryogenic methods.

    (3) Confinement-chamber materials and energy conversion technology including studies of neutron damage and tritium handling.

The ICF reactor driver development is essential and adding to it, material research including target, tritium and structural materials and energy conversion systems need an intense technology development effort for fusion power plant to be a reality by 2025. The total integrated cost will be about $20 billion to get the first prototype ICF demonstrate power plant by 2025. Thus, after the year of 2000, when the breakeven might be attained, works on fusion energy technology would become heavy and strong. The international center for integrating a demonstration power plant of ICF should be contemplated. No other alternate energy source holds the bright promise of fusion and none has ever presented such formidable scientific and engineering challenges.

NEW BASIC PHYSICS DERIVED FROM LASER PLASMA INTERACTION*

Heinrich Hora

CERN, CH1211 Geneva 23, Switzerland**

INTRODUCTION

When Einstein's discovery of the stimulated emission was realised in the laser in 1960, a new chapter in physics was opened up. The ability to concentrate electromagnetic energy spatially into the range of the wave lengths, and temporally to the oscillation time of light (including x-rays), has permitted the concentration of light to an intensity 24 orders of magnitude higher than that of sunlight falling on earth and to energy densities of MeV/atom in solid state densities.

It was fully expected that the physics of the interaction of such radiation with matter would reveal completely unpredictable phenomena. The following examples should be considered only as an overture to how physics needs to be generalized and explored further in order to discover the many new phenomena which can be expected in the future.

These examples relate to high intensity interactions of laser radiation with condensed or gaseous materials where the radiation converts the atoms into electrons and ions of a high temperature plasma within a very short time. The interaction conditions are orders of magnitude different from those of the preceeding plasma physics and it is no surprise that a new area of plasma physics had to be developed and explored. All the phenomena considered here are highly nonlinear. Although these types of phenomena were well known in theory, continued experience of them has revealed as a general principle of nonlinear physics, that much higher accuracy is needed in specifying the initial presumptions of any model or theory than was necessary in the case of the earlier linear physics. One consequence of this is that it is now much more difficult to correctly predict results in theoretical physics, however, it is likely that perserverence will lead to the prediction of phenomena which could not now be imagined.

---

* Edward Teller Lecture delivered at the occasion of the award of the EDWARD TELLER MEDAL FOR ACHIEVEMENTS IN FUSION ENERGY, Monterey Cf.12 November 1991
** also from Department Theoretical Physics, University of NSW, Kensington 2033, Australia and Department Applied Physics, University of Central Queensland, Rockhampton 4702, Australia

Laser Interaction and Related Plasma Phenomena, Vol. 10
Edited by G.H. Miley and H. Hora, Plenum Press, New York, 1992

11

VARIOUS PHENOMENA

Studies of laser irradiation of gaseous or solid targets initially revealed classical behaviour for heating, ionization and gas dynamics with temperatures of up to 100,000 degrees (10 eV temperature). For slightly higher laser powers (above MW), however, the plasmas generated suddenly revealed ions with energies 1000 times higher than gas dynamic ones. The optical constant (i.e. dielectric response, including absorption) corresponding to conditions, had to be generalized from the classical values to intensity dependent nonlinear formulations which could include relativistic effects. The theory of self focusing of laser beams by ponderomotive forces and the relativistic self focusing was one consequence of such generalization.

The most ambitious application of a laser-plasma interaction is the generation of clean, low cost and inexhaustable nuclear fusion energy. The aim here is to use laser irradiation to both heat and compress a pellet or capsule of high density deuterium and tritium in order to ignite exothermal fusion reactions such that the fusion energy gained is much higher than the laser energy applied. There were two main problems to be solved (a) the interaction of the laser radiation appeared to be extremely complex, with unexpected and rarely understood instabilities, pulsations and anomalies, and (b) the achievement of sufficiently high gains required the ignition of a small core of the pellet (e.g. by a central spark) to initiate the reaction of the surrounding material by a fusion detonation wave.

Many laboratories contributed to providing solutions, but it was due to the triumphant achievements of the Lawrence Livermore National Laboratory in general, and of John Nuckolls [1] in particular that a solution based on an understanding of the physics can now be offered for the establishment of an economically competitive inertial fusion energy reactor at the beginning of the next century. Nuckolls solved the complexity of the interaction by using indirect drive and also by introducing the above mentioned spark ignition to produce sufficiently high gains.

The task of providing a physics solution for initiating a large scale developmental program for laser fusion reactors is now complete. It is now in the hands of the politicians to decide how it should be developed.

In support for this physics solution the following alternative scheme can also be considered [see Chapter 13 of Ref.2] as providing a safe solution which has been experimentally confirmed by underground nuclear explosions during the Centurion-Halite project. The alternative scheme solved the complexity of the interaction by clarifying the main problem of the interaction process (i.e. the pulsation in the 10 to 30 psec range), via a numerical modelling of this process and by the application of smoothing techniques which can suppress the pulsation by the use of short time coherence and the superposition of fields. These methods initially aimed at a smoothing of the laser irradiation in the lateral direction to achieve uniform intensity profiles and to avoid self-focusing and other instabilities, but owing to the experimental studies of pulsation by Maddever and Luther-Davies in Canberra, and our numerical understanding the concept of direct drive with smooth interaction and low reflectivity can be considered.

The achievement of sufficiently high gains (the second problem) is possible without the rather complex and very sensitive parameters required when using spark ignition. The earlier very inefficient volume compression and burn, led us to the numerical observation of a volume ignition mechanism such as in a Diesel engine (1978) with a decrease of the optimum ignition temperature to 4 keV by self heating of the reaction products and to temperatures even of only 1.5 keV, due to self-absorption of bremsstrahlung. The full equivalence of the results of this volume ignition with that of the spark ignition has been demonstrated and some interlinking of the processes have been possible.

According to this view, the fully established fusion reactor concept of Livermore can be supported by the possibility that the otherwise unclear properties of "mix" problems and sensitivity of parameters can be avoided completely in favour of the very simple and safe volume ignition process. The earlier question was whether laser ablation of pellets could produce this ideally adiabatic compression free of stagnation and shock waves. This problem was solved experimentally in 1985 by C Yamanaka, S Nakai, T Yamanaka and co-workers [3] where high fusion gains were measured.

12

COMPLETION OF THE EQUATION OF MOTION BY NONLINEAR FORCES

One central problem in laser plasma interactions was to find the correct equation of motion for the plasma. The problems were caused from the beginning by the very strong inhomogenities in density and temperature in laser irradiated plasmas. Most earlier theories of plasmas were based on the homogeneous collisionless plasma. Inclusion of collisions was one of the crucial problems at laser interaction. We have shown how dispersion functions from a pole changed from minus infinity for neglection of the collisions into values of nearly plus infinity when collisons were included. The poles of the optical constants near the critical densities especially caused very high gradients for the quantities in the near neighbourhood of these poles. The neglect of collisions led to fundamentally different results in resonances and other phenomena. A quantum modification of the collision frequency at high electron temperatures appeared to be necessary, although all the consequences of this fact have by no means been exhausted.

Prior to 1966, the problems of inhomogenities and collisions in the theory of the equation of motion, were not so dramatic. Spitzer's derivation for the equation of motion from the Boltzmann equation arrived at a force density in a plasma with an ion mass m_i and ion density n_i and a net velocity $\underline{v}$ of the space charge free and thermally equilibrated plasma of:

$$(1) \qquad f = m_i n_i (\frac{\partial}{\partial t} v + v \,.\, \nabla v) = -\nabla p + \frac{1}{c} j \times H + F$$

which is determined by the thermokinetic force f_{th}, given by the negative gradient of the gasdynamic pressure P, and by the Lorentz force due to electric currents in the plasma and the magnetic fields $\underline{B}$, and additional forces $\underline{F}$ like gravitation etc..

If the currents and magnetic fields are due to the high frequency field (e.g. of a laser), the Lorentz term appeared to be the pondermotive force

$$(2) \qquad \frac{1}{c} j \times H = \frac{1}{4\pi} \nabla . E^2$$

This force was derived first in 1846 as electroscriction by Kelvin and later by Helmholtz as a ponderomotive force where the elastomechanical analogues were used to understand the electric field $\underline{E}$. It was to the credit of Erich Weibel (1957) that he could show that electrons in high frequency fields of microwaves obey the same field gradient forces (2) as evaluated by Kibble (1966) from Lagrangeans.

When looking at the inhomogeneous plasmas the gradient of the density is essential, as expressed by the gradient of the optical refractive index ñ. The forces (2) in a plasma with one dimensional geometry were then (Hora, Pfirsch and Schluter 1967 [4])

$$(3) \qquad f_{nl} = - i_x \frac{E_v^2}{16\pi} \frac{\omega_p^2}{\omega^2} \frac{\partial}{\partial x} \frac{1}{\tilde{n}(x)}$$

given as the time averaged nonlinear force in the plasma due to the propagation of light of frequency ω with an amplitude E_v of the electric field and using the plasma frequency ω_p. This force is identical with the ponderomotive forces (2) as seen from the WKB approximation and stands for the Lorentz terms in (1), expressing now the dielectric explosion (Fig. 1) of a plasma density profile resulting from laser irradiation. This predicted dielectric explosion was discovered numerically as the generation of a caviton by Shearer, Kidder and Zink 1969 [5], based on the nonlinear force formula (3) driving thick blocks of plasma to the observed keV ion energies, and producing the charactertistic density minima (cavitons) (see Fig.2).

Difficulties appeared when the one dimensional geometry was extended to more dimensions (e.g. in the case of laser radiation obliquely incident on a plasma). In

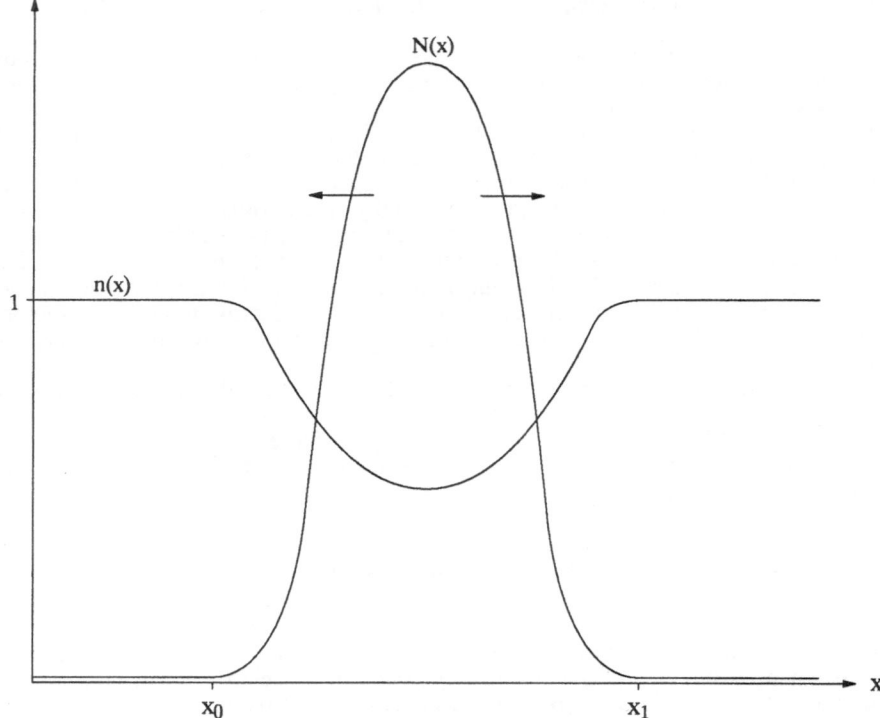

Fig. 1 Plasma with an electron density N(x) and subsequent real part of refractive index n(x) as irradiated by a laser from the l.h.s.. Arrows show the nonlinear forces by the negative gradient of 1/n(x), Eq. (3) [4].

this case the equation of motion had to be generalized such that the nonlinear force became [6]

$$(4) \quad f_{nl} = \frac{1}{c} j \times H + \frac{1}{4\pi} E \nabla . E + \frac{1}{4\pi} EE . \nabla(\tilde{n}^2 - 1) + \frac{\tilde{n}^2 - 1}{4\pi} E \nabla . E + \frac{\tilde{n}^2 - 1}{4\pi} E . \nabla E$$

where all vectors on the r.h.s. correspond to the oscillation of the high frequency field

of the laser and ñ is the complex refractive index, including the intensity dependent nonlinear modifications of the dielectric response and of the absorption. When

differentiating the last term, three terms appear and it should be noted that Schlüter's derivation of the equation of motion in 1950 reproduced one of the three nonlinear terms which did not appear in Spitzer's derivation from the kinetic theory. We showed from momentum conservation that we must use all the terms in our solution (4) to describe the force density based on non-transient collisionless plasma properties. When including collisions, the non-pondermotive terms appeared, determining the ordinary radiation pressure by absorption (e.g. perpendicularly to the density gradient of the plasma).

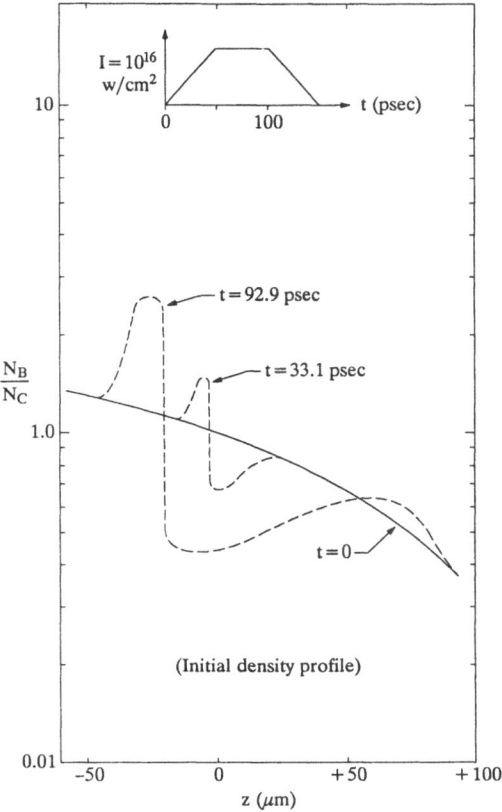

Fig. 2 Hydrodynamic computation of the electron density N_e per critical electron density $N_c = 10^{21}$ cm^{-3} at different times at dependence of the intensity I up to 10^{16} W/cm^3 is given by the upper insert. The monotoneous initial density changes at later times t to profiles with minima (discovery of the caviton) following Shearer, Kidder and Zink [5].

While our Eq.(4) was the final formulation for the time-independent solution, for the transient case several years of controversy occurred before it was decided which terms are needed. Beyond the former approximate solutions, we finally found the following solution (1985) [7]

$$(5) \qquad f_{nl} = \frac{1}{c} j \times H + \frac{1}{4\pi} E \nabla \cdot E + \frac{1}{4\pi}(1 - \frac{1}{\omega} \frac{\partial}{\partial t}) \nabla \cdot EE(\tilde{n}^2 - 1)$$

which is algebraically identical with the formulation using the Maxwellian stress tensor for the vacuum T,

$$(5a) \qquad f_{nl} = \nabla \cdot \left[T + (1 - \frac{1}{\omega} \frac{\partial}{\partial t}) \frac{\tilde{n}^2 - 1}{4\pi} EE \right] + \frac{1}{4\pi c} \frac{\partial}{\partial t} E \times H$$

The correctness and final generality of this nonapproximate equation of motion was first derived by the algebraic structure of the terms (1985) and was later confirmed by T Rowlands from Lorentz and gauge invariance.

NONLINEAR PRINCIPLE

On one hand, it was a basic achievement to derive the complete and general formulation of the force density in plasma theory, on the other hand one had to understand how in laser plasma interaction, the dielectric gradients according to our formula (3) produce the high plasma velocities by dielectric explosion. As seen from the WKB approximation, the electric laser field amplitude $E = E_v/|\tilde{n}|^{\frac{1}{2}}$ in a nearly collisionless plasma increases to very high values. For example, if the absolute value of the refractive index $\tilde{n}$ takes values of 1/10 or 1/100 (or much less in the plasma), the dielectric explosion is then due to the negative gradients of the increased E^2 values, and therefore the ordinary radiation pressure is increased dielectrically by a factor 10, or 100 (or more) respectively.

This could be seen immediately numerically and in experiments by several groups due to the generation of the density minimum near the critical plasma density at very high intensity laser interaction with plasmas. The action of the nonlinear force was checked also in the Boreham experiment [8] where a laser beam was focused into a low density gas and the generated electrons were emitted radially from the beam with energies in the range of 100 eV to 1 keV, as given by half of the maximum quiver energy of the electrons which the nonlinear force converts into translational energy.

While this conversion could be understood immediately from the nonlinear force by global considerations, the analysis of the single particle motion in the laser fields produced discrepancies. The reason was very simple: in the analysis, as usual, the transversal electric and magnetic field of the laser was used. These fields, however, do not exactly fulfill the Maxwellian equations if one considers a laser beam of finite diameter. When we derived the missing field components - it was the discovery of the first exact longitudinal (sic!) components of light in vacuum - their inclusion in the force produced agreement with the measurements.

This taught us that only with the Maxwellian exact solutions could we obtain the correct description, and that neglecting the very small longitudinal components led to a completely wrong result (i.e. a change from yes into no).

This was an example - and others were developed later - of how nonlinear physics cannot be done simply by using a next higher approximation of a second order extension, but one has to use the fully exact linear model or theory in order to arrive at correct predictions.

The old fashioned method of using theoretical physics to predict phenomena is certainly possible in nonlinear physics, but it is much more difficult to keep the correct linear ingredients and to avoid approximations. In this way, however, phenomena can be derived or predicted which cannot be thought of at all in physics.

For example, ion beam fusion is absolutely impossible according to Spitzer (1951) and magnetic confinement was considered as the way to progress. Spitzer was completely correct in his mathematics, physics and logic, but his result was nevertheless wrong: it was linear only. Nonlinear physics, however, does provide a solution for ion beam fusion, at least theoretically. Therefore, realizing the principle of accuracy in nonlinear physics will open completely new dimensions of physical research for the future as a recipe for elaboration and for solving very difficult mathematical problems.

The Boreham experiment [9] was an example for the application of the correspondence principle of electromagnetic interaction [10]. Contrary to Bohr's correspondence principle for the electronic states in atoms of very high fundamental quantum number, there is an easy way for an electromagnetic interaction to continously change from the quantum interaction to the classical interaction [10] just by continuously varying the laser intensity for the interaction. While the lower intensity results in quantum interaction as seen, for example, from multiphoton ionization [9], the higher intensities result in point-mechanical behaviour [8], as seen from the appearance of the Keldysh quasi-classical tunnel ionization [10].

16

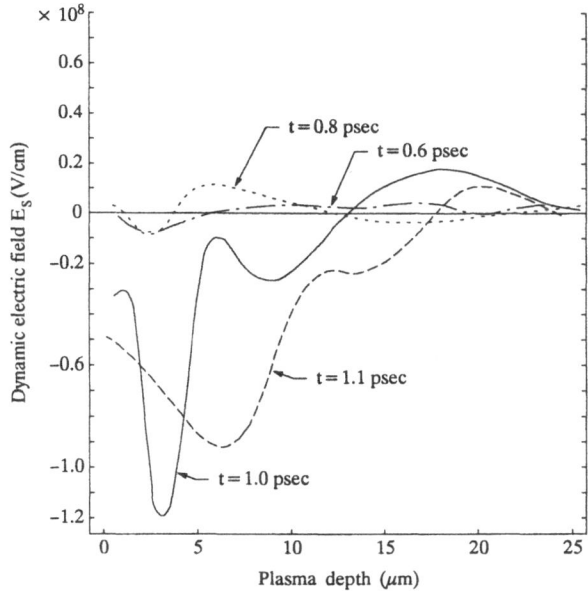

DOUBLE LAYERS AND SURFACE TENSION

Fig. 3 Amplitude $\underline{E}_s$ of longitudinal oscillations of the electric field driven by 10^{16} W/cm² Ng glass laser pulse incident from l.h.s. on a plasma slab generating an inverted double layer at the caviton near 5 micrometer [12]

Surface tension of liquids or solids is mostly related to dipoles of the molecules which are not saturated at the surface. Since high temperature, full ionized plasmas consist of electrons and nuclei only, no dipoles of the kind mentioned could be expected, and it would be strange to ask about surface tension of plasmas. The way in which this, nevertheless, is possible, became obvious from laser plasma interactions for studying nonlinear force.

Contrary to the space charge quasineutral theory of plasmas (which is correct only for homogeneous plasma), on which assumption the old two-fluid theory of the nonlinear force of section 3 is based, we knew from a semi-microscopic derivation of the nonlinear force that the light is pushing or pulling the whole column of the electron gas within the space charge neutral ion background, and the ions follow by electrostatic attraction, giving them the inertia for plasma motion. It was therefore necessary to use a genuine two-fluid model in contrast to the earlier two-fluid theory of

Schlüter and Spitzer. This genuine two-fluid model uses separate electron and ion fluids coupled only by Maxwell's equations. This reduces to the Poisson equation only in one dimensional geometry, but generally it results in the spontaneous magnetic fields of Megagauss values in laser produced plasmas in a three dimensional treatment of the genuine two-fluid model.

The electric fields $\underline{E}_s$ from the Poisson equation in plasmas are well known as ambipolar electric fields given by the gradient of any pressure. The electromagnetic field driven ponderomotion may also produce a pressure which causes electric fields in the same way, but generally the nonlinear force is not conservative and (so called "ponderomotive") potentials appear only in very special simplified conditions. Our genuine two-fluid model revealed the general electrostatic oscillations including collisional damping in any inhomogeneous plasma which contains the ambipolar field, but which is of very complex generality, including then the nonlinear force effects and oscillation given by the locally varying Langmuir (plasma) frequency and collisional damping.

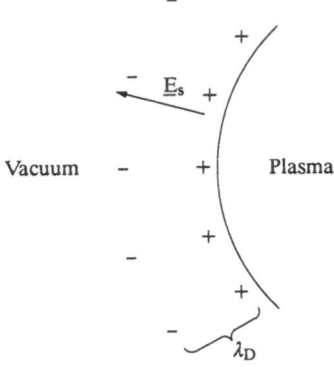

Fig. 4 Plasma Expanding into vacuum. The faster electrons move ahead of the slower (equithermal) ions establishing a double layer of the thickness of the Debye length D with an internal electric field $\underline{E}_s$

With laser irradiation of plasma, a new type of resonance was derived for perpendicular incidence of laser radiation. This was sought for a long time because resonance absorption (Forsterling-Denisov) works at oblique incidence only and is polarization dependent. We further derived strong second harmonic emission in very low density plasma with laser irradiation in agreement with observations.

This all resulted in a very general understanding of internal electric fields and double layers in laser produced plasmas and even the hydrodynamic derivation of the laser driven, very large amplitude, Langmuir oscillation with the Langmuir pseudo-waves was possible. The double layer results were in agreement with significant measurements at the SOREQ Nuclear Research Centre in Israel.

One could immediately see numerically, and as confirmed experimentally, how the cavitons produced by the nonlinear force resulted [11] in inverted double layers (Fig.3). The simple double layer at the surface of a laser-produced plasma expanding against vacuum is seen in Fig. 4. The electrons due to their smaller mass leave the surface where the heavier ions remain. The charge separation causes an electric field and the potential corresponds to the plasma temperature T or - at very intense laser irradiation - to the much higher potential produced by the nonlinear force.

The thickness of the double layer is of the order λ_D of the Debye length. If one integrates the electrostatic energy within the surface area one arrives at surface tension given by this energy per surface area as we derived together with Shalom Eliezer [12]

(6) $$\alpha = \frac{4\pi R^2 \int (E_s^2/8\pi)dR}{4\pi R^2} = \frac{(gkT)^2}{e^2 8\pi} = 4.75 \times 10^{-16}\ g^2 n^{\frac{1}{2}} T^{3/2}\ \frac{erg}{cm^2};\ [T] = {}^{\circ}K$$

where R is the radius of the sphere in Fig. 4, $\underline{E}_s$ is the "electrostatic" field in the plasma surface, k is the Boltzmann constant, and e is the electron charge.

This surface tension can have values of Joules/cm² in laser produced plasmas. It acts against the Rayleigh-Taylor instability in a way similar to the non-disintegrating water droplets. This causes the smooth surface of laser produced plasma plumes. Surface waves of a length shorter than about 100 times the Debye length are stabilized.

This can all also be applied to the degenerate electron gas in a metal. These electrons, similar to those of the plasma, tend to leave the ion lattice in this case not by thermal energy but with their Fermi energy of some eV. They are then stopped by the electric field that they generate as a surface double layer, resulting in the potential given as a work function of some eV, and a surface tension results in the same way as in Eq. (6). In cooperation with R S Pease [p176 of Ref.2], we derived values of the surface tension of metals that give good agreement with the experimental values. These surface tensions were all positive according to the model of the generation of the swimming electron layer above the lattice ions, in contrast to the jellium model of surface tension of metals which can give negative surface tensions contrary to measurement.

CONTAINMENT FORCE OF HADRONS IN NUCLEI AND PHASE TRANSITION INTO QUARK GLUON PLASMA

The surface tension for metals as given by this plasma model (expression (6)) for a degenerate plasma, can be extended to the case where a plasma is no longer defined for compensating charges, namely for a nucleus [13], just by substituting for the temperature T with the Fermi energy. Charges are present and Hofstadter's experiments showed how the charge distribution in a nucleus is constant in the interior, and how it decays over a quite long distance of about 3.5 fm from the constant value to zero at the surface of the nucleus.

We can now use the Fermi energy simply to define a similar "plasma like" surface tension and surface energy for the nucleus in conjunction with the hadron mass (that of protons or neutrons) and the Compton wave length λ_C for the nucleons

$$
(7) \qquad E_F = \frac{(3/\pi)^{2/3}}{4} \frac{h^2 n^{2/3}}{2m} \frac{1}{(\lambda_c/2)\,[n + 1/\,(\lambda_c/2)^3]^{1/3}}
$$

$$
(7a) \qquad E_F = \begin{cases} \dfrac{(3/\pi)^{2/3}}{4} \dfrac{h^2 n^{2/3}}{2m} & \text{(subrelativistic)} \\[3em] (3/\pi)^{2/3} \dfrac{hcn^{1/3}}{4} & \text{(relativistic)} \end{cases}
$$

$$
(7b)
$$

which changes from the subrelativistic branch into the relativistic one at a nucleon density of $(\lambda_C/2)^{-3}$. We note that the relativistic Fermi energy is not dependent on the particle mass and therefore is the same whether the mass is that of nucleons or quarks etc..

The internal energy E_i of the bismuth nucleus is mainly determined by the Fermi energy and to an extent by Coulomb repulsion, a dipole surface energy and a volume energy, to arrive at 4.17 GeV. The surface energy E_s of the bismuth nucleus using the Fermi energy as mentioned in an Eq. (6) produces a value of 4.14 GeV, confirming that our plasma surface tension model leads to a stable nucleus at a nucleon density as measured. We also derived a Debye length for the Hofstadter decay of the charge density of 3.64 fm in good agreement with the measurements.

What is interesting is that the ratio

$$
(8) \qquad E_s/E_i = \text{const } n^{1/6}
$$

becomes one at the nucleon density n of the stable nucleus. We see that for lower hadron densities there is indeed a surface tension and surface energy, but this does not compensate for the internal energy of the nucleus (Fig.5) [13]. We find further that our surface energy just corresponds to the tangling bond energy of the nucleons given by the Yukawa potential at the surface.

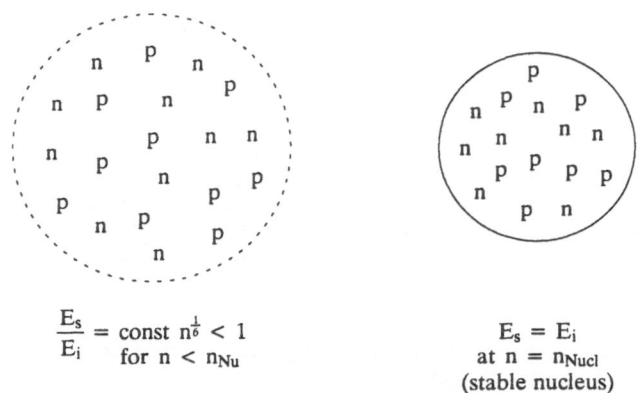

$$\frac{E_s}{E_i} = \text{const } n^{\frac{1}{6}} < 1$$
$$\text{for } n < n_{Nu}$$

$$E_s = E_i$$
$$\text{at } n = n_{Nucl}$$
$$\text{(stable nucleus)}$$

Fig. 5 Hadrons (protons p and neutrons n) at less than nuclear density n_{Nu} where the surface energy E_s is less than the internal energy E_i. Equality is reached at the known density of the nuclei explaining then how the surface energy is exactly compensating the internal energy for producing a stable nucleus.

What is even more interesting is what happens if the nucleus is compressed as in stars or for example, by the heavy nuclear collision where shock waves produce up to six times the nuclear density (Scheid, Muller and Greiner, 1970). The increase of the nuclear density will then just surpass the relativistic threshold in Eq. (7) and will reach a ratio given by Eq. (8) which does not depend on the nucleon density. The formation of nuclei makes then no sense at all. Since - as mentioned - there is not a mass defined for the particles, these can well be a quark gluon plasma as expected in the interior of dense stars. Only at an expansion surpassing the well known nucleon density of stable nuclei, will we have the formation of nuclei with hadron matter in their interior. In the reverse case, the generation of all kinds of nuclei may occur if a hadron ensemble is compressed from very low densities. Surpassing the well known stable nuclear density, the nuclei will be formed by the surface energy. A dependence on the nucleon number shows that nuclei with a number of nucleons above a limit (about 400) cannot be held together by the above plasma-Fermi surface tension, excluding the possibilities of obtaining superheavy nuclei.

We see how a very applied classical physics field such as laser produced plasmas can be generalized to describe such phenomena as the phase transition of hadron to quark matter when forming nuclei, understood from surface tension in agreement with the Yukawa potentials.

Acknowledgements

I am very grateful to Dr C S Taylor (CERN) for valuable comments when reading this text. Support by Dr Kurt Hubner, Dr Helmut Haseroth and Dr E J N Wilson (CERN) is gratefully acknowledged.

References

[1] J.H. Nuckolls. "Physics Today", 35 (No 9), 24 (1982); E. Strom et al LLNL Report No 47312, August 1988
[2] H. Hora. "Plasmas at High Temperatures and Densities", (Springer, Heidelberg 1991)
[3] C. Yamanaka, S. Nakai, T. Yamanaka, Y. Izawa, K. Mima, K. Nishihara, Y. Kato, T. Mochizuki, M. Yamanaka, M. Nakatsuka and T. Yabe. "Pellet Implosion and Interaction Studies by GEKKO XII Green Laser", Laser Interaction and Related Plasma Phenomena, H. Hora and G. H. Miley eds. (Plenum, New York 1986) vol 7, p395

[4] H. Hora, D. Pfirsch, and A. Schluter. "Z. Naturforsch" <u>22A</u>, 278 (1967)

[5] J. W. Shearer. R. E. Kidder and J. W. Zink, " Bull. Amer. Phys. Soc." <u>15</u>, 1483 (1970)

[6] H. Hora. "Phys. Fluids" <u>12</u>, 182 (1969)

[7] H. Hora. "Phys. Fluids" <u>28</u>, 3706 (1985)

[8] B. W. Boreham and H. Hora. "Phys. Rev. Letters" <u>42</u>, 776 (1979)

[9] S. Augst, D. Strickland, D. D. Mayerhofer, S. L. Chin and J. H. Eberly. "Phys. Rev. Letters" <u>66</u>, 1247 (1991)

[10] H. Hora and P. H. Handel. "New Experiments and Theoretical Development of the Quantum Modulation of Electrons (Schwarz-Hora Effect)" in Advances in Electronics and Electron Physics, P. W. Hawkes ed. (Acad. Press, New York, 1987) vol 69, p55

[11] H. Hora, P. Lalousis and S. Eliezer. "Phys. Rev. Letters" <u>53</u>, 1659 (1984)

[12] S. Eliezer and H. Hora. "Phys. Repts." <u>172</u>, 339 (1989)

[13] H. Hora. "Plasma Model for Surface Tension of Nuclei and the Phase Transition to the Quark Plasma" CERN-PS/DL-Note-91/05, August 1991

EDWARD TELLER MEDAL: ACCEPTANCE REMARKS*

John H. Nuckolls
Director
Lawrence Livermore National Laboratory

Tenth International Workshop on
Laser Interaction and Related Plasma Phenomena
Monterey, California

I am honored to receive this award. It is especially significant because Edward Teller is the father of inertial fusion. Teller's pioneering work in the extreme compressibility of matter, the radiation implosion concept, and the physics of thermonuclear burn are fundamental to the creation of very small scale inertially confined fusion explosions.   Edward also made key contributions by fighting to reduce secrecy and by promoting international collaborations.

In the late 1950's, I began to address the challenge of creating the smallest possible fusion explosions. My first supercomputer calculations of micro implosions/fusion explosions of DT masses as small as one milligram were completed in the spring of 1960, some months before the laser was invented. I believed that very small radiation implosions driven by a beam of energy (e.g., a charged particle beam) projected across an explosion chamber would be the best approach to ignition of small fusion explosions in the laboratory. This "indirect drive" approach minimized the energy beam symmetry requirements and maximally decoupled the implosion from the coupling of the energy beam to the target. In these early calculations, I used a spherical target with a very thin shell to relax the beam power requirement. When the laser was invented in late 1960, we immediately recognized its utility for inertial fusion. LLNL physicists Stirling Colgate, Ray Kidder and I independently calculated various methods of using high power lasers to implode and ignite various fusion target designs. Colgate and I calculated implosions in laser driven hohlraums.   Kidder applied a spherically symmetric pulse of laser light to the target without use of a radiation implosion.

---

*Performed under the auspices of the U.S. Department of Energy for the Lawrence Livermore National Laboratory under contract W-7405-ENG-48.

Laser Interaction and Related Plasma Phenomena, Vol. 10
Edited by G.H. Miley and H. Hora, Plenum Press, New York, 1992

23

In supercomputer calculations carried out in 1961, I utilized extreme pulse shaping to compress a bare drop of thermonuclear fuel in a hohlraum to super high densities and to achieve thermonuclear ignition and propagation from a small central hot spot. This spherical droplet minimized the target fabrication cost and made possible laser fusion power plants and spaceships (which I proposed at that time).

In the early 1960's, Ray Kidder organized and led the Laboratory's (and the world's) first experimental laser fusion program. This few million dollars/year program continued for ten years—when a greatly expanded laser fusion program was launched at LLNL.

In the 1960's and 1970's, my colleagues and I conducted nuclear tests of similar concepts.

In the early 1970's, my colleagues Lowell Wood, Ron Thiessen, George Zimmerman and I produced advanced computer calculations which suggested that the feasibility of laser fusion might be tested by igniting breakeven fusion targets with multi-kilojoule lasers—much smaller than the megajoule-scale lasers required for practical applications such as power plants. These limiting calculations assumed moderately symmetric illumination symmetry, and near perfect pulse shaping and laser target coupling. Symmetry and fluid stability were enhanced by utilizing an electron conduction alternative to indirect drive. This scheme made direct drive laser fusion targets feasible. These calculations were conducted with the LASNEX code newly developed by George Zimmerman.

In the early seventies I proposed declassification of laser fusion implosions of simple spherical droplets. This proposal initiated the laser fusion declassification process

I proposed exploding pusher targets which made possible experiments with 10 to 100 joule lasers. Variations of these targets yielded predictable and detectable numbers of thermonuclear neutrons in the first successful laser implosion experiments at KMS Fusion and at LLNL.

These developments provided the scientific basis for launching a greatly expanded laser fusion program at LLNL and for initiating the construction of a series of large solid state lasers, including the Shiva laser in the 1970's and culminating in the 1980's with Nova, the world's most powerful laser. This expanded laser fusion program was initiated in the early 1970's by Carl Haussmann and led by John Emmett.

In the seventies, the target design group which I recruited and led, proposed high gain indirect drive targets matched to heavy ion accelerators, developed the first successful laser driver indirect drive targets, and achieved compression of DT to 100 times liquid density.

I believe inertial fusion will continue to play a major role not only in basic physics and weapons research, but also as a future world source of energy capable of raising the standard of living on a global scale, and potentially capable of limiting global greenhouse warming by replacing fossil fuels. Just as the ignition of thermonuclear fusion on a micro-scale revolutionized weapons technology, in the 20th century the ignition of thermonuclear fusion energy on very small scales will revolutionize energy production in the 21st century.

COMMENTS ON THE HISTORY AND PROSPECTS FOR
INERTIAL CONFINEMENT FUSION

Nikolai G. Basov

Lebedev Physical Institute
of the Academy of Sciences
Moscow, Russia

It is a special favour to be here at a celebration for Edward Teller who was the very first in history to demonstrate a man-made exothermic nuclear fusion reaction. This represented the process of inertial confinement fusion (ICF) on a large scale. Now it is a most important aim for mankind to develop this process into a smaller controllable scale for production of energy.

I am very glad to present today some thoughts about fusion and especially on inertial confinement fusion.

Apparently, one of the most important applications of lasers is for ICF. "Laser fusion fever" started in 1962, and very quickly has become an independent scientific trend in physics and technology of thermonuclear fusion. Now one can speak separately on the history of ICF (see Table I).

The present short review of the ICF history does not involve many of the important theoretical and experimental aspects. Areas omitted for lack of space include: the generation of fast electrons and ions, the discovery of numerous laser-plasma effects, the evolution of the target stability problem, and the competition of long-wave and short-wave lasers, etc. While we will not discuss these issues, they have all been a vital part of ICF devlopment.

Such proportions of a laser light as high energy yield during short time and high flux density make it possible to attain specific energy deposition of 10^{18} W/g. This allows one to reach thermonuclear temperatures in the heated matter simultaneously with ultra high density compression. We paid attention to this fact (together with Prof. O. N. Krokhin) in 1962 (a report at the Executive Board of the Academy of Sciences of the USSR, March 1962), and in 1963 the first theoretical evaluations were reported at the III Conference on Quantum Electronics (Paris). Since that time, powerful laser interactions with matter have been studied extensively both

Laser Interaction and Related Plasma Phenomena, Vol. 10
Edited by G.H. Miley and H. Hora, Plenum Press, New York, 1992

25

Table I. Main Steps in the History of ICF

1963; N. G. Basov, O. N. Krokhin	Proposal to use lasers for controlled fusion
1968; N. G. Basov, P. G. Kryukov, O. N. Krokhin, Yu. V. Senatsky, S. D. Zakharov	Registration of thermonuclear neutrons in laser-produced plasma
1971; N. G. Basov, O. N. Krokhin, G. V. Sklizkov, S. I. Fedotov, A. S. Shikanov	First multi-base laser system "Kalmar" ("Russian Monster") for spherical target compression
1972; Livermore Lab., Los Alamos Lab. (USA)	Starting date for the financing of a National ICF Program in the USA
1974; Lebedev Physics Inst. (FIAN) Inst. of Applied Math.	Concept of low entropy compression of shell targets
1975-1978; FIAN	First experiments on low entropy shell target compression (deuterium densities reached 9 g/cm^3)
1970; Livermore Lab	Launching of 10 kJ Nd-laser "Shiva" (neutron yield reached 3×10^{10})
1978; Los Alamos Lab	Launching of 10 kJ CO_2-laser "Helios"
1979-1980 Inst. of Laser Tech. (Japan)	Concept of X-ray and "cannon-ball" targets
1979; Livermore Lab.	Density of compressed fuel reaches 20 g/cm^3
1981-1982 FIAN	Launching of 108-channel laser "Delfin" (stable compression of high-aspect targets and collapse time 200 km/hr)
1983; Livermore Lab.	Launching of a 20 kJ "Novetta" Nd-laser
1983; Inst. of Laser Tech. (Japan)	Launching of 30 kJ Nd-laser "Gekko"
1985-1989 Livermore Lab.	Launching of 130 kJ Nd-laser "Nova" (fuel density, 30 g/cm^3; neutron yield, 3×10^{13})
1990; Livermore Lab.	Appeal to the U.S. Congress on financing an upgrade laser
1990-1991 Inst. of Laser Tech. (Japan)	~ 600 g/cm^3 matter density reached with "Gekko-12" laser facility

theoretically and experimentally at various international labs including the Lebedev Physics Institute (LPI). Our experimental research resulted in the successful development of both ruby and Nd-glass lasers.

In autumn 1962, we put forward the idea of increasing the ruby-laser power by Q-modulation. Then, research on nanosecond pulse amplification using running-wave amplifiers were started at LPI. This work was aimed at development of laser sources of plasma. Experimental and theoretical investigations of the amplification processes in the saturation regime have allowed us to further reduce the laser pulse duration and to reach lasing powers of some GWs. Studies of the amplification processes in the pulses of a complicated multimode structure has resulted in the formulation of a model of solid-state lasers with passive mode synchronization.

At the same time there have been important developments in the methods for hot plasma diagnostics. These methods include the ability to achieve unique spatial-temporal resolution. Some of those methods, e.g. laser interferometry and Schlieren photography, have now come into routine use.

The observation of first thermonuclear neutrons in Nd-laser-produced plasma was an important early success in our experimental research. A year later that result has been reproduced in Limeil (France), and it proved the possibility of using a laser to drive thermonuclear reactions.

The early 70s can be viewed as the "period of laser target compression." In 1972, a 9-channel laser facility "Kalmar" for spherical target irradiation was launched at LPI. Pioneer experiments with spherical homogeneous targets have been carried out with "Kalmar." We first observed the generation of DD-neutrons and, later, the generation of secondary DT-neutrons. The secondary DT-neutrons provided conclusive evidence of a compressed nucleus. A group of scientists from Livermore, E. Teller's group, put forward an attractive and fruitful physical theory about supercompression using a time-profiled laser pulse at a Conference in Montreal the same year (1972).

In 1974, we proposed (together with the scientists from the Institute of Applied Mathematics) an alternative scheme for low-entropy compression by using a time homogeneous laser pulse and inhomogeneous high-aspect targets. During the next few years a great number of experiments on shell target compression were carried out with the "Kalmar" laser facility, and the results have completely proved our compression concepts.

We understood clearly that, when one uses thin shells, the basic problem arises from the possiblity of compression in stabilities. Thus, along with a theoretical-numerical study in the late 70s and early 80s, we carried out experiments on compressing the targets with the aspect ratio $\sim 10^2$ by using a 108-channel laser facility "Delfin" (launched in 1982). We reached compression of 3×10^3, proving the possibility of stable compression of such targets.

During the last years, the international effort in ICF

research has gained a new impetus. Thus, financing of ICF in the USA reached 200 million dollars per year. Likewise there is a special ICF National Program in Japan. A European Laser Center is planned to be opened, and one of the goals of this center will be a creation of a laser having the energy of hundreds of kilojoules. We hope that the scientists from our Institute will also collaborate in the international effort in this laser center.

Up to now, we have investigated physical processes, which take place during the target compression and burning. Numerous experimental techniques are used for this purpose, and they allow one to trace the various physical processes involved in the target compression (about 30 techniques have been worked out at LPI, they are different and very complicated). The scope of these modern experimental results confirms the practical feasibility of the next step in ICF: the attainment of 200 g/cm^3 density of compressed gas (for comparison note that the density of solid hydrogen is 0.1 g/cm^3) and target burn ignition at 200-kJ laser energy.

In our opinion, at the present time, the basic problem in the ICF is the further development of the physics of thermonuclear targets, especially in the compression range 10^4. Other issues include, the choice and creation of a suitable efficient driver for the ICF and the engineering-technological elaboration of a thermonuclear reactor design. All of these efforts have the goal of a project for a laser thermonuclear electric power plant which is practically realizable, economically profitable, and safe for people and nature.

DEMONSTRATION OF A NUCLEAR FLASHLAMP-PUMPED ATOMIC IODINE LASER

Wade H. Williams*
George H. Miley

University of Illinois
Fusion Studies Laboratory
Urbana, IL 61801

*presently at Lawrence Livermore
 National Laboratory

ABSTRACT

Experimental and theoretical results are presented on the demonstration of a nuclear flashlamp pumped atomic iodine laser.

The iodine laser, as a photolytically pumped system, is traditionally excited with electrical flashlamps. In the present approach, UV light from a nuclear pumped plasma is used as the excitation source. The excimer XeBr was chosen as the flashlamp medium, its 282 nm B-X emission being well within the photoabsorption band of common iodine lasants (e.g. C_3F_7I, CF_3I). Energetic products from the $^3He(n,p)T$ reaction were used to excite the gas. Parametric evaluations of XeBr under nuclear pumping allowed selection of an optimum gas mixture of 0.5 torr $CHBr_3$ and 200 torr Xe in 3 atm. ^3He, with a measured fluorescence efficiency of 1% (light out at 282 nm divided by energy deposited in the gas).

A laser cell was constructed consisting of two concentric cylinders, the outer, aluminum walled, containing the laser gas. Lasing was achieved with a peak power of approximately 20 mW using neutrons from a TRIGA pulse at the University of Illinois. The lasant was C_3F_7I at 15 torr. Flashlamp pump power into the lasant at lasing threshold agreed with literature value for this system, as did laser output dependance on C_3F_7I pressure.

This work demonstrates the feasibility of pumping the iodine laser with a nuclear flashlamp. The potential for scale-up to higher powers must be based on continuing studies concerned with the improvement of the fluorescence efficiency of XeBr, or on consideration of alternate fluorescers (e.g. KrF).

Laser Interaction and Related Plasma Phenomena, Vol. 10
Edited by G.H. Miley and H. Hora, Plenum Press, New York, 1992

29

BACKGROUND AND MOTIVATION

The work discussed was a study of a new approach to exciting the atomic iodine laser, namely the use of a nuclear pumped flashlamp. Experimental studies were conducted on a flashlamp gas under nuclear excitation, followed by construction and testing of a laser.

The conventional atomic iodine laser was first discovered in 1964 by Kasper and Pimentel [1,2]. In its most common pumping scheme, it is one of a large class of lasers, called photodissociation lasers, which use incoherent flashlamp light in one wavelength region to excite a laser gas, which then lases at a longer wavelength. The atomic iodine laser uses light in the ultraviolet region of the spectrum to cause photodissociation of a gaseous molecule containing iodine, typically C_3F_7I, or CF_3I. The free iodine atom formed from photodissociation is left in an excited electronic state in nearly 100% of dissociations. Lasing is achieved between this excited ($^2P_{1/2}$) state and the lower ($^2P_{3/2}$) ground state at a wavelength of 1.3152 micron. (Another important pumping scheme, not pertinent to this work, is through chemical generation of $O_2(^1\Delta)$, which reacts with I_2 to form excited atomic iodine.)

This laser has been well-studied since its discovery, principally for its application in Inertial Confinement Fusion (ICF) research. The largest iodine laser built to date is in the USSR, with an output of 100 TW and 30 kJ [3]. Considerable work on this laser has also been done in Garching [4]. The interested reader is referred to excellent references on the atomic iodine laser under usual pumping conditions [4,5].

EXCIMER FLASHLAMPS

The UV light source typically used to pump the atomic iodine laser is a xenon flashlamp. This lamp emits light as an electrical current is passed through a glass cell containing high pressure xenon, ionizing the gas. It was suggested by several workers [6,8] that this electrical flashlamp could be replaced by a nuclear flashlamp; that is, a fluorescing gas ionized by nuclear reaction products (i.e. fission fragments), rather than an electrical current. Wilson and Shapiro [6,7] performed kinetic modeling to predict flashlamp power output for a neutron flux as would be produced by a fast burst reactor. The present work was initiated as an experimental examination of these various predictions.

Three different gas mixtures, all excimers, were suggested as flashlamp gases by Wilson and Shapiro [6,7] due to the good overlap of the emissions of these gases with the photoabsorption bands of the common iodine lasants. Two important characteristics of excimers are 1) emission in a few, relatively narrow wavelength bands, often in the UV; and 2) an extraordinary efficiency (termed "fluorescence efficiency") at converting energy deposited in the gas into light emitted into the strongest of these bands. This efficiency is >50% for some gases.

The economic benefit in reduced size of having an efficient fluorescer in a large laser system is evident. The added advantage for the iodine laser of having a significantly monochromatic source is seen from the photodissociation cross-section of the three common atomic iodine lasants (see Figure 1). This photodissociation band is fairly narrow, located between approximately 230 and 330 nm. Thus a flashlamp source which has a broad spectral output and good fluorescence efficiency may still be inefficient in pumping this laser. Only ~10% of the electrical xenon flashlamp output commonly and for large iodine lasers is in the spectral absorption range of the lasant. (The efficiency based on electrical energy required is 8% [4].)

With its good fluorescence efficiencies and fairly monochromatic nature, an excimer with output within the lasant absorption band of 230 to 330 nm would have a reasonably high overall efficiency (compared to the 8% given above). A few excimers have been identified [6] with strong emission in this wavelength range, including KrF* at 249 nm, XeI* at 252 nm, XeBr* at 282 nm, Ar_2F* at 284 nm, and XeCl* at 308 nm. For these experiments, XeBr was chosen because of its better spectral overlap with the iodine lasants than KrF, XeI or XeCl, and because it was expected to have better fluorescence efficiency than Ar_2F.

Non-nuclear pumped excimers have previously been used to excite the iodine laser, including electrically-pumped KrF and XeCl. In those cases, the excimer was stimulated to lase, then this laser light was used to stimulate the iodine [4]. More closely related to this work was the use of e-beam-induced spontaneous emission from XeBr to achieve

FSL-91-63

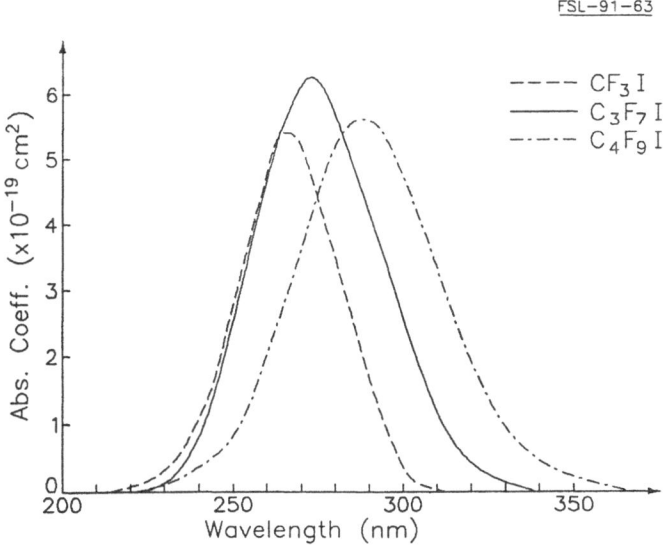

Figure 1. Photodissociation cross-sections for three common atomic iodine lasants.

lasing in iodine [9]. Those results are especially pertinent because there are important similarities between e-beam and nuclear pumping mechanisms.

NUCLEAR PUMPED LASERS

This approach of pumping with a nuclear flashlamp makes this laser one of a larger class of lasers termed Nuclear Pumped Lasers (NPLs). This work, however, differs from past research in NPLs in that all previous efforts have utilized nuclear reaction energy to directly excite the laser (no flashlamp) by, for example, having fission fragments slow down in the laser gas [10,11]. That approach is termed "direct pumping" of the laser, as opposed to the flashlamp pumping developed here. A review of past NPL work is given in references 10 and 11.

An interest in using nuclear reaction energy to pump any laser, including this flashlamp system, is motivated by two important characteristics. The first is that nuclear pumping uses neutrons reacting with an appropriate material, such as ^{235}U, to provide energy for the laser. Because neutrons have an ability to penetrate large distances through many materials before being absorbed, there is a potential to pump large gas volumes, necessary for developing high energy/high power systems. The second characteristic is the large amount of energy "stored" in relatively small amounts of nuclear fuel. This is an advantage in comparison with, for example, electrically pumped lasers, which utilize large (and expensive) capacitive storage systems to reach high powers.

While these are significant potential advantages, it should be mentioned that there are, of course, trade-offs. A significant challenge in developing a high-power nuclear pumped laser is the relatively low pump powers available in nuclear pumping (kW/cm^3 vs MW/cm^3 for e-beams), and slow rise times (10's of microsecs to msecs vs nsecs for e-beams). This restriction comes from the maximum neutron fluxes (approximately 10^{17} n/cm^2sec from a fast burst reactor) which can be readily generated over a large enough area for laser pumping. This limitation, for example, makes it difficult to nuclear pump excimer lasers, which require high pump powers. It is less restrictive with other systems.

NUCLEAR FLASHLAMP PUMPED LASERS

As mentioned above, this concept of flashlamp pumping differs from the direct pumping used in past nuclear pumped laser research. Although flashlamp laser pumping has not previously been successful, there has been significant study of the field in the past. Reference 8 gives a review of previous work.

There are several advantages to this approach of laser pumping with a nuclear flashlamp. The first is the possibility of exciting large fluorescer volumes at the modest pumping powers attainable in nuclear pumping (compared to e-beam pumping), and then coupling this fluorescence into a small laser volume to generate sizable laser pump powers, greater than would be attainable in direct pumping. This advantage utilizes well one major characteristic of nuclear pumping, the ability of neutrons to penetrate large distances in materials, and hence, excite large volumes.

A second advantage is one common to all flashlamp pumped lasers (nuclear or not) over direct pumped systems, and has to do with the spatial uniformity of excitation. With direct pumping, most of the deposited energy goes into heating the gas. Unless this energy can be distributed with good uniformity over the gas volume, optical inhomogeneities develop in the non-uniformly heated gas, spoiling the ability of a laser beam to propagate well in the volume. In nuclear pumped systems this is less of a problem when direct pumping with ^{3}He, where the energy deposition can be fairly uniform. In fission foil pumped systems, however, strong pump power gradients exist (if the foils are spaced more than a few cm apart), causing difficulties in laser cavity stability.

For flashlamp-pumped lasers, however, the energy going into the lasant is provided as light from the flashlamp. It is much easier to distribute this energy with good uniformity over the laser gas volume using mirrored walls, etc. In addition, if the flashlamp is fairly monochromatic, as with an excimer, most of this light energy is usable in pumping the lasant, so less goes into gas heating. Hence, large volumes can be pumped to high powers, as is done in present large atomic iodine lasers [4]. Optical inhomogeneities in the flashlamp are less important because it is only an incoherent radiation source. Hence, fairly steep energy deposition gradients in nuclear pumped flashlamps can be tolerated, allowing potential scaling to higher powers than may be possible in direct nuclear pumped systems.

A final advantage of nuclear flashlamp-pumped systems is the potential good efficiency of such lasers. Outside of exceptional results with the Ar/Xe laser [12,13] (1-3%), most systems have shown relatively low efficiency (<<1%) on direct nuclear pumping [10,11]. If nuclear pumped flashlamps, especially excimers, can be shown to be efficient, this coupled with the efficiency of converting fluorescence light to laser light, could mean an overall good system efficiency (1-5%).

APPLICATIONS

Such advantages of nuclear-pumped lasers would be especially important in large pulsed or steady-state laser systems (megajoule). Applications for such lasers are limited, but include Inertial Confinement Fusion, military defensive/offensive systems, and power transmission on earth or in space. Nuclear pumped lasers have been looked at for all these possibilities. To date, however, with the possible exception of the Ar/Xe laser [12,13], all lasers investigated have been too inefficient for current use. Other difficulties associated with nuclear pumping are the safety concerns of working around an intense nuclear radiation source, and activation of component materials. Also, an added complexity in scaling up to a large laser is the need to integrate a reactor and laser together. It is hoped that continued research in this field, such as that detailed in this work, will help overcome these obstacles, making nuclear pumped systems more attractive as high-power lasers.

Nuclear-pumped flashlamps could have numerous applications outside gas laser pumping. For example, intense UV light sources are an important tool in chemical processing. Prelas, et al. [14,15] have investigated such options and also discussed driving a solid state laser.

EXCIMERS, GENERAL

Excimers are a class of molecular compounds which have the common characteristic of only forming with the molecules being in an excited electronic state. In the ground state the molecules dissociate into their constituents atoms. The range of molecules covered by the term is broad, from simple diatomic excimers (e.g. Xe_2^*) to complex aromatic molecular systems. (The term "excimer" is short for "exited dimer." It strictly refers to homonuclear molecules, such as Xe_2^*; heteronuclear systems, such as XeBr*, are more accurately referred to as exciplexes, but "excimer" has come to be commonly used to refer to both classes.) We will concentrate on one class of excimers, the rare gas halides, of which XeBr is an example. References 16 and 17 provide excellent reviews.

EXCIMERS AS LASERS

The interest in excimer systems has increased greatly since the demonstration of their use in lasers [18]. Excimers meet the laser inversion criterion by definition, in that the molecules only exist in the excited state, and dissociate in the ground state, there are, then essentially no molecules in the ground state. Lasing has been demonstrated in numerous excimer systems, including potentially high powers applications [19].

FORMATION MECHANISMS

There are two important pathways for formation of the excimer state: the ion channel, and the metastable channel. Referencing Figure 2, the excimer is formed through the first channel when a positive rare gas ion and a negative halogen ion collide in the presence of a third body (to take up excess energy). The metastable channel occurs when a metastable rare gas atom collides with a halogen-bearing molecule. The rare gas atom attaches itself to the halogen atom, tearing it from the rest of the molecule (i.e. a "harpoon" reaction).

Both of these channels are present in a rare gas halide plasma. The extent to which one is dominant in the formation of the excimer depends on the pumping mechanism. For plasmas in which the energy starts from "the top" and dribbles down, such as in e-beam pumping, the ion channel tends to dominate. Conversely, for electrical discharges where excitation comes from relatively low energy electrons, the metastable channel is prominent [20]. These distinctions are approximate, however. In fact, XeBr does not follow this convention, with the metastable channel being dominant in e-beam plasmas [7,21,22].

The relative strengths of the two channels in nuclear pumping is not well known. Nuclear plasmas are typically considered to be very similar to e-beam generated plasmas for excimer laser pumping considerations [23], suggesting a strong ion channel for most excimers. Prior modeling work for nuclear pumped XeBr [7] suggests that the metastable channel dominates.

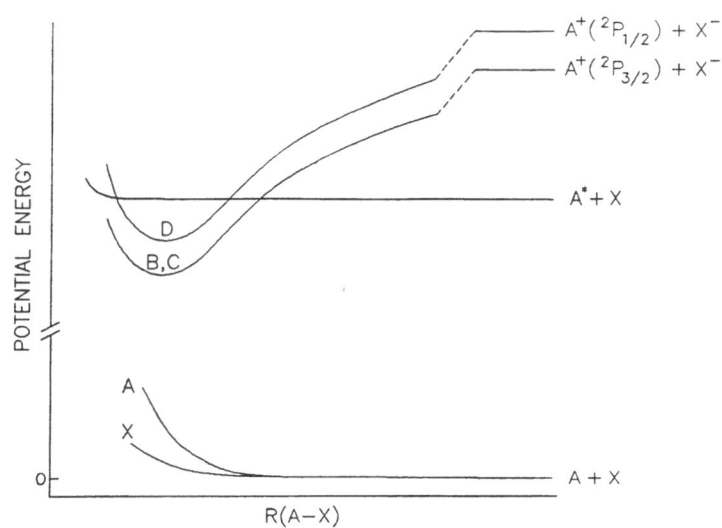

Figure 2.   Stylized potential energy curves for rare gas
             (A)- halogen (X) excimer systems (from ref. 16).

FLUORESCENCE EFFICIENCIES

An important characteristic of rare gas halides, and many other excimer systems, is the high efficiency at which energy deposited in the system can be converted into excimer fluorescence, principally on the B-X transition. This efficiency varies from system to system. As examples, efficiencies have been reported of approximately 50% for rare gas dimers [16, pg. 99; 17, pg. 52]; 28% [24], 8.3%

[25], and 12.5% [26] for KrF; 4.4% for ArF [27]; and 11 ± 5% for XeBr [9]. The kinetic reason behind such high efficiencies is the fast rates at which the excimers are formed in the mixtures combined with the lack of pathways out of the excimer state. From there, the short lifetime of the state, coupled with moderate quenching rates of the state by most of the gas species, yield a high fluorescence efficiency. This high efficiency is a key factor in the attractiveness of using an excimer as the nuclear flashlamp gas.

XEBR EXCIMER

As indicated above, XeBr is a rare gas halide excimer. It was chosen for these experiments because: 1) it was thought to have a good fluorescence efficiency (11% in e-beam pumped experiments [9]), and 2) its emission wavelength at 281.8 nm is well within the photoabsorption band of the common perfluoroalkyliodides (C_3F_7I and CF_3I) used as parent molecules in the atomic iodine laser (compare Figures 1 and 3).

XeBr is a fairly well-studied system, and was the first rare gas halide to lase [28]. Various methods have been used to attain lasing, including e-beams [22,28,29], e-beam controlled discharge [29], and electric discharges [30-32]. Stimulated emission has also been found on the Xe_2Br trimer

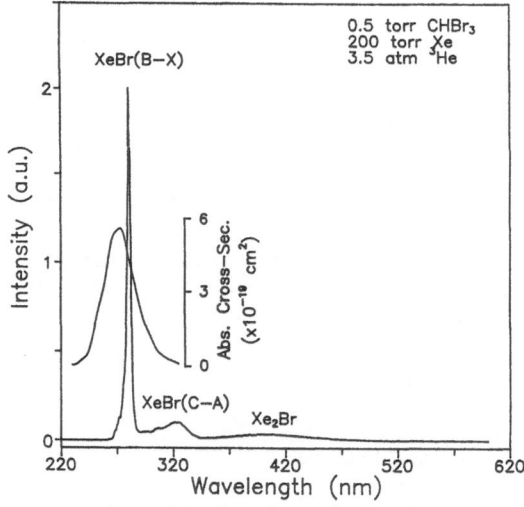

Figure 3. Emission spectrum of XeBr under nuclear pumping, as measured in these experiments (corrected for spectral sensitivity of the detector), with photo-dissociation cross-section of i-C_3F_7I superimposed.

using e-beams [33]. Development of this system as a laser, however, has been limited because of self-absorption of the laser line by gas constituents, particularly Br_2, and Xe_2+. These problems have been avoided to some extent by using alternate bromine donors, such as HBr, and by pumping with an electric discharge where few xenon dimer ions are formed [29,31]. It should be noted this self absorption is significant for a laser (perhaps a few percent per cavity length), but not large enough to effect performance of a flashlamp. (Reference 31 gives an estimated absorption of 0.017%/cm at 282 nm under laser pumping).

The only nuclear pumping of XeBr reported to date was a limited experimental study done at the University of Florida [34]. Pumping there was done with a single different gas mixture, somewhat different than found optimum in these results, and at neutron fluxes much lower than utilized in this work (10^{11} vs 10^{15} n/cm^2sec). A fluorescence efficiency (light out between 280 and 295 nm divided by energy deposited in the gas) of 0.37% was reported.

NUCLEAR PUMPING MECHANISMS

Nuclear pumping of the gases in a flashlamp utilizes the energy of fast ions produced in neutron-induced nuclear reactions and the associated secondary electrons. This makes the pumping efficiencies relatively insensitive to the type of ion doing the excitation. Examples of the reactions typically used are:

$$n + {}^{235}U \quad \text{-->} \quad ff + 200 \text{ MeV } (\sigma = 582 \text{ b}) \tag{1}$$

$$n + {}^{10}B \quad \text{-->} \quad {}^4He + {}^7Li + 2.3 \text{ MeV } (\sigma = 3837 \text{ b}) \tag{2}$$

$$n + {}^3He \quad \text{-->} \quad T + p + 0.8 \text{ MeV } (\sigma = 5327 \text{ b}) \tag{3}$$

The energies indicated are the kinetic energies of the reaction products. The cross-sections (σ) in these reactions are for thermal neutron energies, in units of barns (b) = 10^{-24} cm^2.

TRIGA REACTOR

The TRIGA reactor at the University of Illinois at Urbana-Champaign was used for this research. The TRIGA reactor is a light water research reactor. It is capable of both steady state (1.5 MW) and pulsed (6000 MW peak) operation. Some characteristics of the facility are summarized in Table 1.

A plan view of the reactor is shown in Figure 4. One of the irradiation facilities of the reactor which make it especially convenient for laser research is the "throughport." This 5 3/8"-diameter tube runs through the shielding, directly adjacent to the core. Experiments, such as a laser cell, can be mounted in a "carriage" and loaded in the port right next to the core. This allows the cell to

Table 1. TRIGA reactor characteristics for the throughport [35].

Steady-State Operation

Power	1.5 MW^{th}
Thermal Flux[a]	2.91e12 $n/cm^2 sec$
Fast Flux[b]	3.30e11 "

Pulsed Operation

Pulse Size	$3.00
Peak Thermal Flux	3.10e15 $n/cm^2 sec$
Peak Fast Flux	3.52e14 "
Peak Dose Rate (SiO_2)	7 MRad/sec
Pulse Duration, FWHM (msec)	12.5

[a]Thermal flux: 2200 m/s flux. Uncertainty: $\pm 2.72\%$.
[b]Fast flux: >2.8 MeV. Uncertainty: $\pm 1.86\%$.

receive a high peak neutron flux during a reactor pulse, while maintaining a reasonably uniform flux over a meter length. (The flux drops by approximately half about 1 foot on either side of the point of peak flux.) This throughport was used in all experiments described herein.

All experiments were done with $3.00 pulses from the reactor, giving a peak power of 1600 MW and peak flux of 3.1e15 $n/cm^2 sec$ for FWHM 12.5 msec. ("$3.00" is a measure of the reactivity insertion into the reactor core used to initiate the pulse.) As will be discussed below these pulses provided pump energies of 0.7 J/cm^3 in 2.5 atm. ^3He and 0.14 J/cm^3 in a 1-inch diameter ^{10}B coated cell at 1 atm. Ar. Respective peak powers were 56 W/cm^3 and 11 W/cm^3.

EXPERIMENTAL SET-UP

Also shown in Figure 4 is the set-up used for fluorescence experiments. Two different cells were constructed for fluorescence measurements, one for ^{10}B pumping, and the other for ^3He pumping. Details of the design of these cells will be given later. Both cells were placed on a carriage in the throughport, with vacuum/gas fill lines attached, running out to a gas handling station on one end of the port. Pneumatic valves on the carriage allowed remote control of gas inlet to the cell. Use of these lines allowed the cell to be evacuated and refilled with various gas pressures and mixtures without removing the cell from the reactor. This was necessary because the carriage and cell, though made of materials to minimize activation (mostly aluminum), still become sufficiently "hot" following a pulse of neutrons that they could not be handled without a long decay period (hours).

DIAGNOSTIC EQUIPMENT

Several diagnostics were implemented in different phases of the experiments. These included two different

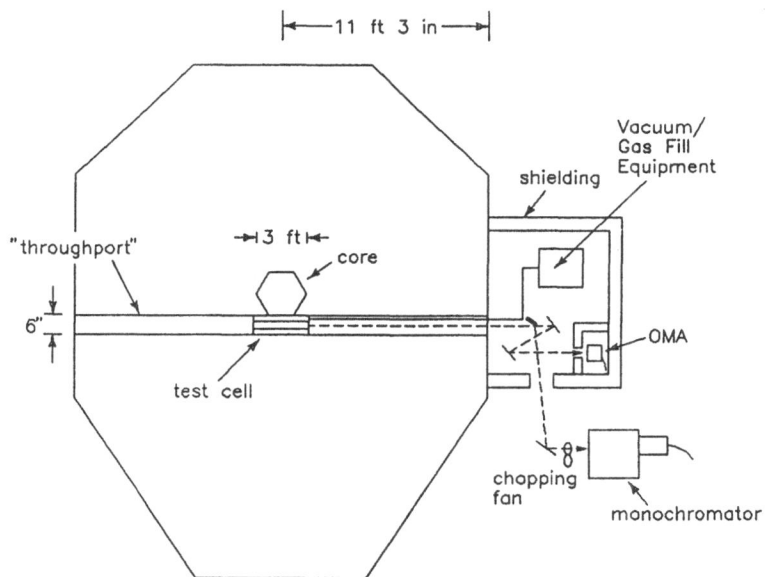

Figure 4. Experimental set-up for fluorescence experiments
(not to scale).

Optical Multichannel Analyzers (OMA's) to measure
fluorescence spectra (intensity versus wavelength): a
Princeton Applied Research[1] model 1450/1452, and a model OMA
III 1460/1456 from the same company. The spectrographs
attached to both OMA's were Thermo Jarrel Ash Inc. model
Monospec 18, with a 150 gv/mm, 450 nm blaze spectral grating
(Aires, Inc.) in the OMA model 1450/1452; and a 300 gv/mm,
300 nm blaze grating in the OMA 1460/1456. Both OMA's were
intensity calibrated using a D_2 light source: Gates Inc.
model DCR-30/60 G power supply with Nassau Inc. model D102F
lamp. The lamp was intensity calibrated by EG&G Gamma
Scientific Inc.

A monochromator/photomultiplier tube (PMT) was used to
collect information on the time dependance of the XeBr
fluorescence at the wavelength of interest (282 nm). The
monochromator was a CGA-McPherson model EU-700 with an RCA
31034 PMT.

An infrared sensitive InAs photodiode (EG&G Judson,
model J12 room temperature) with pre-amp was used to measure
laser output. A 1.31 micron bandpass filter was used in
front of the detector to allow measurement of only the laser
line.

A fast response (20 microsec) pressure transducer used
for energy deposition measurements in fluorescence
experiments was a Kulite, Inc. model IPT1100 with home-made

---

1   Indication of manufacturer is not intended as an
     endorsement of product.

pre-amp. As explained in the following section, this transducer was attached to the fluorescence and laser cells in the reactor throughport. It was found to operate surprisingly well in the intense radiation field during a pulse, with only a small radiation-induced current in the signal. The transducer lifetime was found to be approximately 6 dozen pulses (7.5 MRad dose) before performance started to degrade, presumably due to radiation-induced damage to the silicon diaphragm in the transducer. This degradation was observed as a discrepancy between fill gas pressures in the cell as registered on filling equipment gauges, as opposed to the readings indicated by the transducer.

Determination of the temporal shape of the neutron pulse was made using an uncompensated ion chamber located near the reactor core. The temporal shape of the neutron pulse as seen by the laser was assumed to be identical (for these purposes) to that recorded by the ion chamber, based on nearness of the ion chamber to the core. Ion current from the chamber was amplified through a picoammeter (Keithley model 415).

A digital oscilloscope (Nicolet 4094/4562) was used for data collection from the ion chamber, monochromator, and photodiode. An AT&T personal computer was used for some fluorescence data storage.

Correlation of timing between reactor pulse and data collection equipment was done using an adjustable-delay timing circuit constructed by the department electronics shop.

PRESSURE TRANSDUCER MEASUREMENTS

A piezoresistive pressure transducer was attached to experimental cells to measure energy deposited in the gas. The operation of this transducer is based on a measurement of the change in resistivity of a silicon diaphragm under stress. As nuclear pumping proceeds during a pulse, the gas heats and, due to the closed cell volume, the pressure increases a readily measurable amount. The energy deposited in the gas was determined using an ideal gas heat capacity for the gas mixtures of $C_v = 3/2R = 12.5J/(mol\ K)$, and neglecting losses due to optical emission.

This technique has been used for other applications [36,37]. For validity, the pulse length must be shorter than the time it takes for the gas to cool convectively/radiatively against the cell walls. Conversely, if the pulse is short compared to the speed of sound in the gas, pressure waves interfere with the measurement. Since the 12-msec reactor pulse is long compared to the transit time for a sound wave across the cell (60 microsec), only cooling of the gas during the pulse was of concern in these measurements, and this was corrected for as explained below.

^{10}B PUMPING

The pressure rise as measured in the fluorescence cell using ^{10}B pumping is shown in Figure 5, with the neutron pulse superimposed. This cell was 1-inch inside diameter by 12 inches long, with a thin ^{10}B coating (approximately 1 micron) on the inside. It can be seen that the pressure increased during the pulse to about 11.5 psi over fill pressure (800 torr), but then begins to decrease somewhat before the end of the pulse. This decrease is due to convective and radiative cooling of the gas against the cool cell walls.

A correction for the fall-off was done by curve-fitting the decrease in pressure to a single exponential, to represent the convective and radiative cooling. The entire rise and fall of the pressure was then modeled as a balance between a Gaussian heat input (the reactor pulse shape), and the exponential heat loss term. This calculation indicated a pressure rise of 12.6 psi without losses vs the 11.6 psi shown in the figure. This fractional correction (12.6/11.6) was assumed to be representative and used to adjust all subsequent measurements.

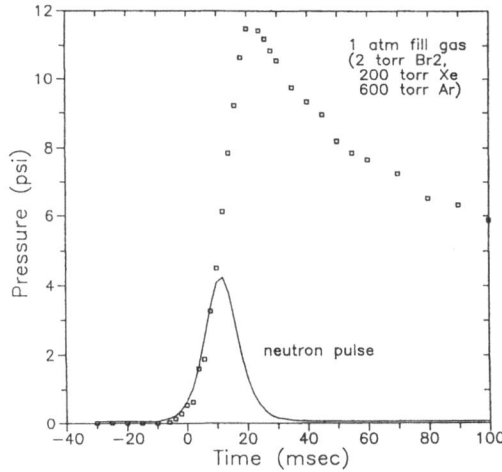

Figure 5. Measured pressure rise during a pulse with the ^{10}B-coated fluorescence cell.

A significant radiation-induced spike in the pressure transducer signal was observed (approximately a 5 psi equivalent voltage). This effect, however, peaks and dies away well before the peak of the pressure rise, so it does not interfer with the measurement.

The pressure rise of 12.6 psi was used to calculate a volume averaged energy deposition of 0.14 J/cm^3. Note though, that the alpha particles produced in the n-^{10}B reactions have a finite range (approximately 1.4 cm in 1 atm

He). Thus, the energy deposition in the cell is spatially non-uniform.

^{3}HE PUMPING

Experiments were also conducted measuring pressure rise and fluorescence with ^{3}He pumping in a cell with similar dimensions to the ^{10}B cell. Figure 6 shows results from several pressure rise measurements on fills of 0.5 torr CHBr$_3$, 200 torr Xe, and the balance ^{3}He up to a pressure of 3.7 atm. Also shown in the figure are several values for the theoretically predicted energy deposition for pure He, based on a knowledge of the geometry, neutron flux, energy released per absorption, etc. These predicted results were taken from Wilson and DeYoung [38], who did a detailed analysis of energy deposition in ^{3}He pumped lasers for various pressures and cell diameters. Scaling to neutron fluxes in the reactor was based on information given previously in Table 1.

A surprisingly close agreement is shown between the measurements and theory in Figure 6. Similar to the discussion above for ^{10}B pumping, however, one may expect some difference between the theoretical predictions for energy deposition and measured values because the predictions assumed pure ^{3}He, whereas these experiments included 200 torr Xe in up to 3.5 atm ^{3}He. The range of an n-^{3}He reaction proton in 3 atm. He is 1.5 cm, compared to

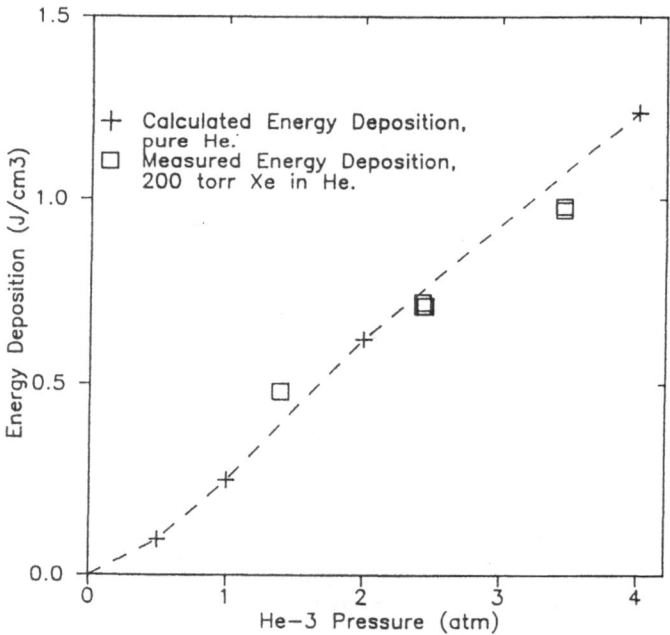

Figure 6. Comparison of energy deposition measured in He+200 torr Xe vs calculated (from ref. 38) for pure He for the 1-inch diameter test cell.

2.7 cm in 200 torr Xe. Hence, for the 2 cm diameter cell, the energy deposition should be somewhat higher with the addition of Xe. The fact that this does not show in the figure may be due to the fact that, as in ^{10}B pumping, the neutron flux dips at the ends of the cell, which was not taken into account in the calculations. (Depression of the neutron flux by the significant neutron absorption in the 3 atm. ^{3}He is taken into account in the calculations. This absorption reduces the flux at the cell location by approximately 35% [38].) Taking this dip into account would place the energy deposition in the He/Xe mixture above that for pure He, as expected.

GAS HEATING

The 0.98 J/cm^3 measured in ^{3}He experiments is a significant amount of energy, and generates a peak (spatially averaged) gas temperature of approximately 780°C. This is significant temperature increase for many gas phase reactions, and has the potential of affecting, either positively or negatively, the fluorescence efficiency of the excimer. Some evidence was found that, for these experiments, the fluorescence efficiency was greater on the falling side of the pulse than on rising side, possibly attributable to a change in gas temperature.

XEBR FLUORESCENCE UNDER ^{10}B PUMPING

The motivation behind these experiments was to determine the optimum Xe/bromine donor mixture to maximize XeBr fluorescence at 282 nm, which mixture could then be used in a laser configuration. ^{10}B was used because of its low cost, as compared to ^{3}He, even though it was not planned for use in a final laser design.

Measurements were made using three different bromine donor molecules and with varying pressures of these donors and of Xe. Ar was used as a buffer gas to maximize pump power, as it has a higher stopping power for alphas than He, so less alphas were lost to the wall of the cell. (The principal purpose of the buffer gas is to stop the alpha particles. It has a weak effect on gas kinetics, as buffer atoms act as third bodies in 3-body reactions. The buffer also increases the heat capacity of the mix, lessening temperature rise.)

Care was taken in cell design, alignment, and diagnostics for these experiments, as well as in ^{3}He fluorescence experiments, so that the entire fluorescence cell volume was viewed by optical diagnostics. This was necessary for a determination of absolute fluorescence output.

BROMINE DONORS

Three bromine-containing compounds were used in these experiments, Br_2, $CHBr_3$, and BBr_3. All three are liquids at room temperature. Only one data point was taken with BBr_3 because its reactive nature caused significant handling difficulties. All three materials have low personnel exposure limits, requiring care in experiments.

Choice of these compounds was aided by work reported [21] on the trimer, Xe_2Br. Several bromine donors were utilized in that research. The best, in terms of XeBr formation rate, was CBr_4. This material requires system heating, however, to achieve a sufficient vapor pressure, so was not used for this work. $CHBr_3$ was found to work almost as well.

^{10}B FLUORESCENCE CELL

A schematic of the cell constructed for these experiments is shown in Figure 7. The tube was 1 mm wall aluminum (1100). The boron coating was approximately 1 micron [39,40]. Fused silica windows, chosen because they were known from previous experiments [41] to experience minimal darkening under irradiation.

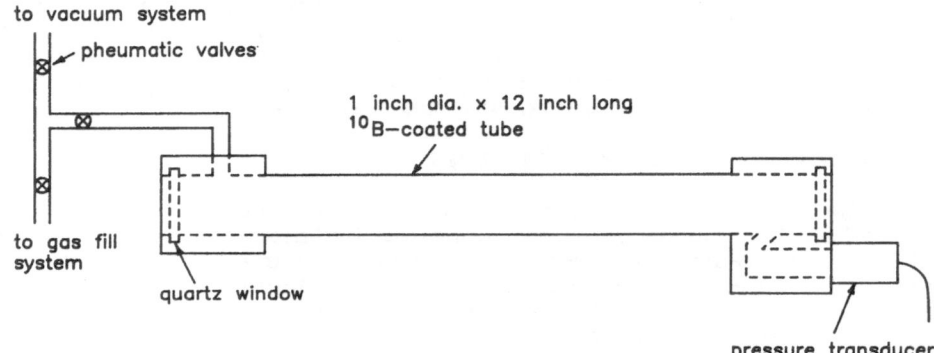

Figure 7. Schematic of ^{10}B-coated test cell (not to scale).

FLUORESCENCE RESULTS

A fluorescence spectrum of XeBr under ^{10}B excitation, as recorded with the OMA 1450/1452 system, is shown in Figure 8. The strong peak of interest for this work is seen at 282 nm. Two other peaks at 320 nm and 415 nm are also indicated. Other finer features in the spectrum are thought to be noise, since the fluorescence was relatively weak, just on the edge of the detector sensitivity. The spectrum has not been corrected for OMA spectral sensitivity. The lines observed with spectra taken of XeBr by other workers under different pumping mechanisms [9,21,30,42], indicating the three peaks seen can be correctly attributed to XeBr and Xe_2Br transitions. The measured linewidth of approximately 3 nm for the 282 nm peak was similar to that reported by other workers (1.6-2.8 nm in ref. 31).

As the line at 282 nm was of principle interest for this work, results presented below will focus on the dependance of the intensity of this peak as a function of gas mixture.

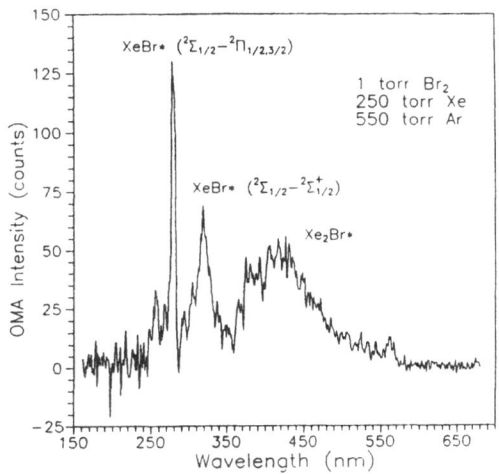

Figure 8. Fluorescence spectrum of XeBr under ^{10}B pumping (OMA 1450/1452).

Figure 9 shows measurements done using $CHBr_3$ as a bromine donor, varying the Xe pressure to optimize output. Light detection was done using the OMA 1460/1456 system. One data point using BBr_3 as a donor is also shown. Because its performance was similar to $CHBr_3$, and because its corrosive nature made it difficult to work with, no further experiments were done with this material.

These measurements indicated a weak dependance on Xe pressure. Other measurements (not shown) showed similar Xe-pressure dependance with the use of Br_2 as a donor The fluorescence at zero Xe pressure in Figure 9 was not XeBr fluorescence, as it was shifted in wavelength from 282 nm to approximately 286 nm. The origin of this fluorescence was unclear. It is close enough in wavelength, however, to the XeBr transition to be useful for iodine pumping, so is included in the figure.

The letters (A-D) in Figure 9 represent sequential pulses on the same fill of gas. That is, letter A represent the first pulse, letter B, the second, etc. (Data from some pulses were lost due to radiation effects on the OMA, so some sequences are incomplete; e.g., there is no pulse A at 200 torr Xe in Figure 9. Also, two fills were done at 200 torr Xe: hence, the primed notation.) Other data had seemed to suggest a decreasing output with added pulsing, which may have been due to the effect of impurities being driven from the boron coating into the gas on each pulse, contaminating the mixture. These data, however, seemed to indicate the variation was simply scattered in the data, due to a difficulty found in making reproducible measurements with OMA's in the high radiation field around the reactor.

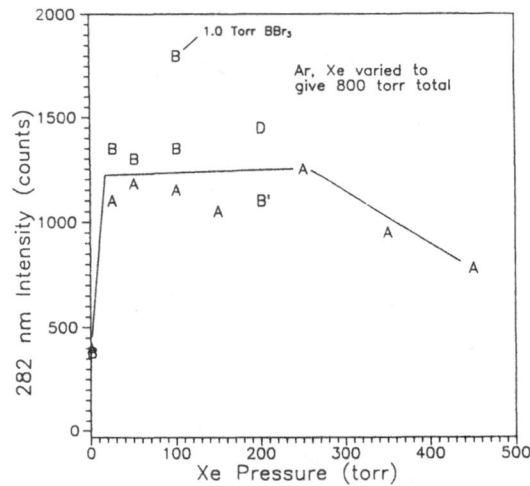

Figure 9.   XeBr emission at 282 nm as a function of Xe
             pressure under ^{10}B pumping (OMA 1460-1456).
             Bromine donor: 0.5 torr CHBr$_3$, except as noted.
             Letters represent sequential pulses on the same
             gas fill.   The unidentified emission at zero torr
             Xe had a wavelength of approximately 286 nm.

CHBR$_3$ FLUORESCENCE RESULTS

     The dependence on fluorescence output as a function of
CHBr$_3$ pressure is shown in Figure 10.   The optimum pressure
found here of 0.1-0.2 torr is somewhat lower than the 1 torr
reported in ref. 21 for investigations of Xe$_2$Br.   This may
be because these experiments were done at the higher
pressures first.   Consequently, there may have CHBr$_3$
adsorbed on the cell walls (on the boron coating) which
contributed to the fluorescence at the low partial pressure
readings.   The 282 nm fluorescence seen at zero CHBr$_3$
pressure is attributed to this effect.   Because of the
difficulty in measuring pressures below 0.5, this pressure
was used in laser experiments, even though a lower pressure
may have been preferable.

     Other measurements were also done with variable Br$_2$ and
Ar pressures.   These indicated little dependance on Br$_2$
pressure over the range 0 to 2 torr in 175 torr Xe/600 torr
Ar (though little data was taken).   Similarly, the effect on
282 nm fluorescence of variable Ar pressure over the range
150 to 650 torr in 1 torr Br$_2$/75 torr Xe was measured,
indicating no dependancy.   Based on all of this data, an

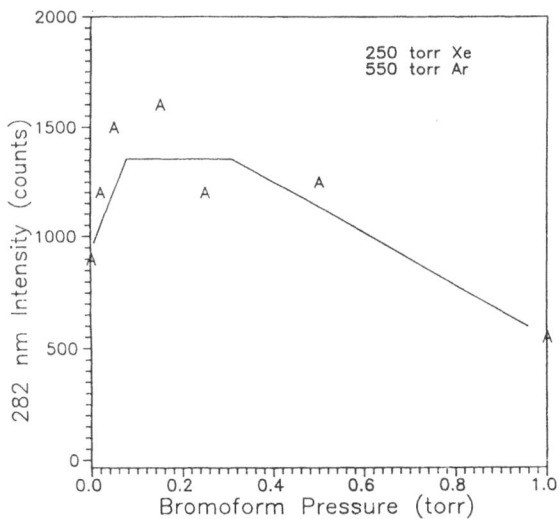

Figure 10. XeBr emission at 282 nm as a function of $CHBr_3$ pressure under ^{10}B pumping (OMA 1460/1456). Letters represnt sequential pulses on the same fill of gas. The emission at zero torr $CHBr_3$ is attributed to $CHBr_3$ adsorbed on the cell wall.

"optimum" Xe pressure of 200 torr is indicated for 0.5 torr $CHBr_3$.

The conclusion of these experiments with ^{10}B pumping was a determination of an XeBr mixture for use in a laser of 0.5 torr $CHBr_3$ and 200 torr Xe. This mixture was considered approximately optimum, although the fluorescence output dependance on pressure for both of these gases was not strong.

XeBr* FLUORESCENCE UNDER ^{3}HE PUMPING

Following the fluorescence experiments utilizing ^{10}B pumping described above, measurements were taken as the fluorescence of the "optimum" gas mixture determined of 0.5 torr $CHBr_3$ in 200 torr Xe, mixed with various pressures of ^{3}He.

^{3}HE FLUORESCENCE CELL

A schematic of the cell used in fluorescence experiments with ^{3}He is shown in Figure 11. The cell was aluminum with fused silica windows sealed with o-rings. Vacuum and gas filling of $CHBr_3$ and Xe were done through one port on the cell, and ^{3}He filling through another. The cell was constructed so the inside diameter tapered from a diameter of 0.93" on one end to 0.652" on the other. This facilitated viewing of the entire cell volume when in the throughport, necessary for accurate determinations of fluorescence efficiencies.

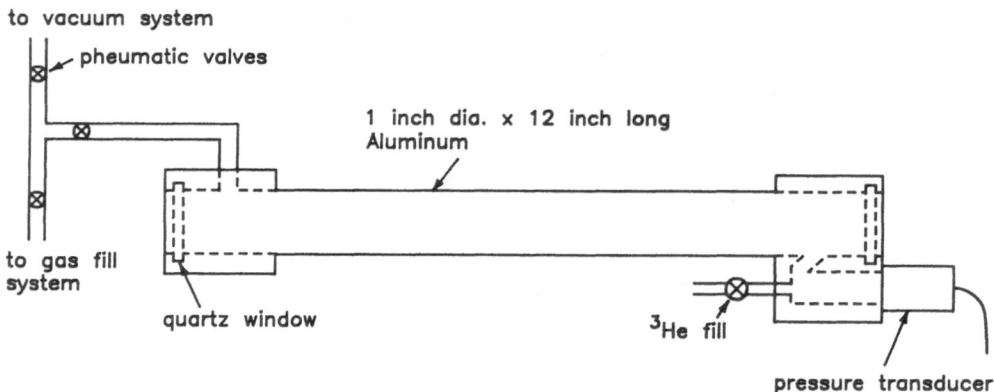

to vacuum system

pheumatic valves

1 inch dia. x 12 inch long
Aluminum

to gas fill
system

quartz window

^{3}He fill

pressure transducer

Figure 11.   Schematic of ^{3}He test cell (not to scale).

FLUORESCENCE RESULTS

All ^{3}He fluorescence measurements were taken with the
OMA 1460/1456 system. A typical spectrum not corrected for
spectral response is shown in Figure 12.   The XeBr
fluorescence at 282 nm is actually much stronger than the
other two emissions at 320 and 415 nm, so in order to make
good measurements of all peaks, a 1/16" piece of Pyrex glass
was used in front of the OMA as a filter. Spectrophotometer
measurements had been taken on the glass, showing a
transmission at 282 nm of 6.2%, but approaching 80% at 320
nm and 92% at 415 nm. When used in these measurements the
filter served to reduce the XeBr peak onto the scale of the
other two peaks.

Fluorescence measurements were taken with 0.5 torr
$CHBr_3$ and 200 torr Xe with ^{3}He to total pressures of 1.6 atm
to 3.7 atm. Figure 13 shows results at three pressures. A
new gas fill was used for each pulse shown. A linear output
with ^{3}He pressure in this regime is observed. (Changing ^{3}He
pressure affects both the pumping power and the gas
kinetics, so a linear dependence with pressure was not
necessarily anticipated.)

FLUORESCENCE TEMPORAL DEPENDENCE

In most fluorescence experiments, both with ^{3}He
pumping, as well as ^{10}B pumping, a measure of the temporal
dependence of the fluorescence at 282 nm was taken using a
monochromator/PMT.   A trace of the fluorescence, with
neutron pulse superimposed, is shown in Figure 14.   The
chopping in the signal is due to a chopping fan placed in
front of the monochromator to allow subtraction of
radiation-induced noise in the PMT. This noise is seen in
the figure as the small increase in the chopped signal above
zero during the 10-20 msec around the peak of the pulse.

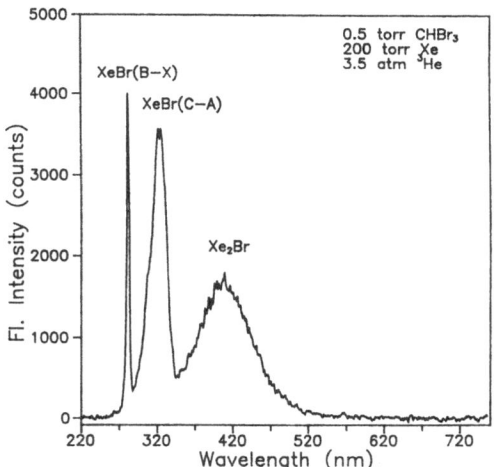

Figure 12. Spectrum of XeBr fluorescence under ^{3}He pumping,
not corrected for wavelength sensitivities of
OMA and filters (OMA 1460/1456).

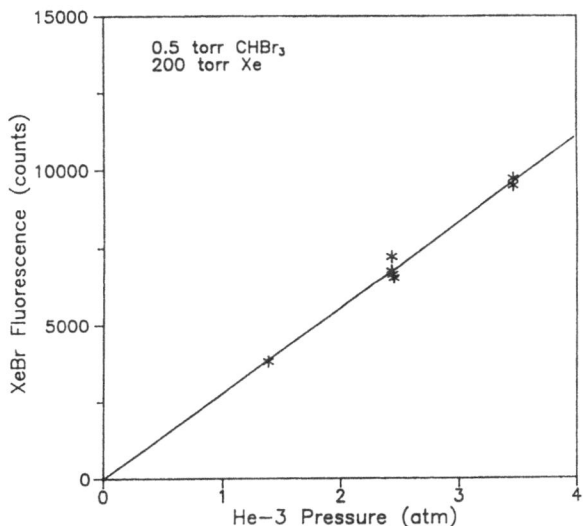

Figure 13. XeBr emission at 282 nm as a function of ^{3}He
pressure (OMA 1460/1456).

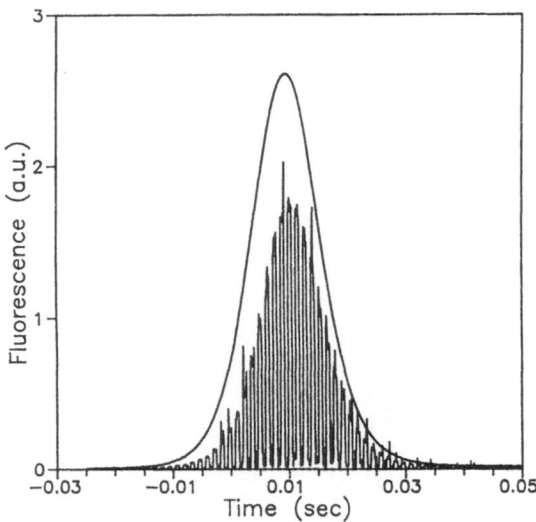

Figure 14.   XeBr emission at 282 nm during a pulse from a
$3.00 pulse (chopped signal) with reactor pulse
shape superimposed.

A time shift approximately 1 msec is apparent in the
data shown in the figure, with the neutron flux peaking
ahead of the fluorescence. No adequate explanation has been
found for this effect, though several possibilities were
considered from experimental artifacts to kinetic processes.

In calculations explained below comparing the
fluorescence with the reactor power in time, this time shift
was subtracted out, since the time shift, real or
artificial, would not have a direct bearing on system
efficiency.

Aside from this time shift, the fluorescence was found
to be generally in phase and symmetrical with the neutron
pulse. This is evidence that little bromine donor depletion
occurs. Depletion had been a concern in planning the
experiments due to the relatively large pulse energies
involved. In a few experiments with 1 torr Br_2 in 75 torr
Xe with little to no Ar buffer, a definite asymmetry in XeBr
fluorescence occurred. The fluorescence was seen to peak
well before the reactor power, and the temporal shape was
distorted, with the fluorescence having a fast rise time,
and then a much longer decay. This was very likely due to
Br_2 depletion, since the recombination of Br atoms into Br_2
requires a third body, usually a buffer gas atom.

FLUORESCENCE LINEARITY WITH POWER

An important consideration in the scaling for
application in a large laser system is the linearity of the
fluorescence intensity with pumping power. Because the
reactor neutron pulse is essentially Gaussian in time, a
measure of linearity can be ascertained from the ratio of

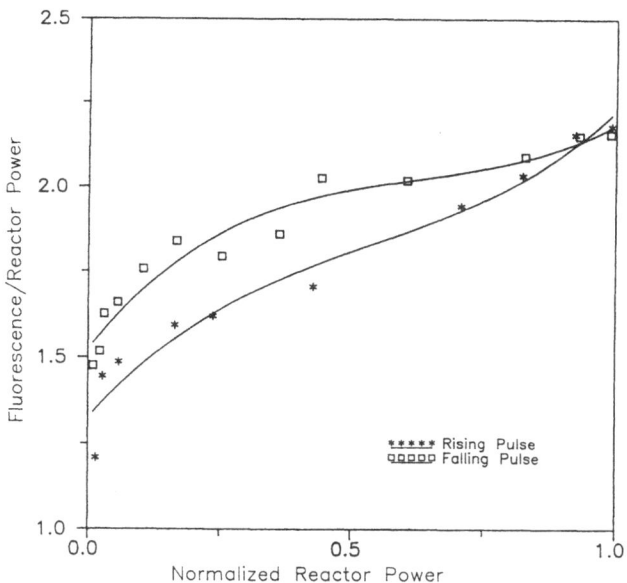

Figure 15. Ratio of instantaneous fluorescence intensity at
282 nm divided by reactor power, plotted versus
normalized reactor power (1.0 = pulse peak).
The solid curves are included as visual aids
only. (Data from Figure 13)

fluorescence intensity to reactor power at different times
(powers) during the pulse. The data shown in Figure 14 was
analyzed in this fashion. The results are shown in Figure
15. Data is shown for times during both increasing and
falling reactor power.

Two interesting points are seen. First, the
fluorescence efficiency appears to increase with pumping
power. Second, the ratio seems somewhat higher on the
falling side of the pulse than on the rising side. A
possible explanation is that the fluorescing gas heats
significantly during the pulse, so this change may be due to
a difference in gas temperature.

CORRECTED SPECTRA: FLUORESCENCE EFFICIENCY DETERMINATION

Utilizing the fluorescence spectra, as shown in Figure
12, the absolute fluorescing gas emission in J/cm^3-nm was
determined. This involved several factors, including the
distance from the cell to the detector, cell volume,
transmission characteristics of the Pyrex and neutral
density filters, and, most importantly, the wavelength-
dependant sensitivity of the OMA. This sensitivity was
determined by performing measurements on a deuterium lamp of
calibrated emission characteristics, which covered the UV
wavelength region of interest. The resulting corrected
spectrum is shown in Figure 16.

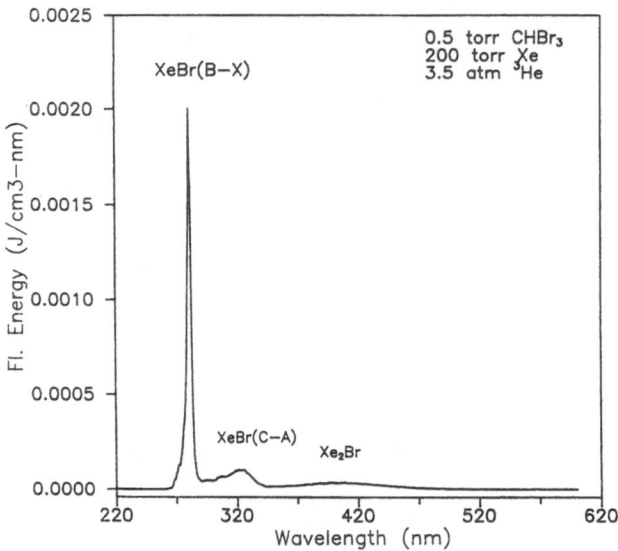

Figure 16. Fluorescence spectrum under ^{3}He pumping as shown
in Figure 12, corrected for spectral sensitivi-
ties of the OMA and transmitting/reflecting
optics.

FLUORESCENCE EFFICIENCY DETERMINATION

Integration under the 282 nm peak, as in Figure 16,
allowed a determination of the fluorescence energy/unit
volume available for laser pumping. This energy was
combined with the energy deposition results described above
to yield a 282 nm fluorescence efficiency, defined as the UV
light emitted around 282 nm, divided by energy deposited in
the gas.

^{10}B AND ^{3}HE PUMPING RESULTS

Efficiency calculations were made for experiments on
^{10}B pumping, ^{3}He test cell pumping, and the pumping of XeBr
with ^{3}He as the flashlamp in the laser cell. Table 2 lists
efficiencies for several fluorescence measurements. In all
fluorescence efficiency measurements, the uncertainty is
estimated as $\pm50\%$.

The efficiency in the ^{10}B pumping experiments was
somewhat higher than that in the ^{3}He pumping results. This
may be associated with the use of Ar as the buffer gas,
instead of ^{3}He. Less energy is required to generate an ion
pair in Ar/Xe mixtures than in the He/Xe mixtures (24.2 eV

Table 2. Summary of Fluorescence Efficiency Results

Pumping Mechanism	Fluorescing Mixture[a]	OMA System	Energy Depos. (J/cm^3)	Fluor. Energy (J/cm^3)	Fluor. Eff. (%)
^{10}B Test Cell	1 Br_2 75 Xe 725 Ar	1450/ 1452	0.14	0.0014	1.0
^{10}B Test Cell	0.5 $CHBr_3$ 200 Xe 600 Ar	1450/ 1452	0.14	0.002	1.4
^{10}B Test Cell	0.5 $CHBr_3$ 200 Xe 600 Ar	1460/ 1456	0.14	0.0016	1.1
^3He Test Cell	0.5 $CHBr_3$ 200 Xe 2.4 atm ^3He	1460/ 1456	0.71	0.0062	0.87
^3He Laser Cell	0.5 $CHBr_3$ 200 Xe 3 atm ^3He	1460/ 1456	0.54	$(0.0019)^b$	$(0.35)^b$

[a]pressure in torr, unless otherwise noted.
[b]cell view was partially eclipsed, so these values may be low.

in 200 torr Xe/600 torr Ar vs 28.1 eV in 200 torr Xe/2000 torr He), so more of the energy can end up in the excimer. The higher concentration of He also speeds 3-body reactions, one of which de-excites the excimer state.

^{3}HE PRESSURE DEPENDANCE

Consideration was also made of the effect of ^{3}He pressure on fluorescence efficiency. Efficiencies for the ^{3}He experiments are shown in Figure 17. The efficiency possibly exhibits a slight increase with pressure, though the limited data makes a definitive conclusion difficult.

Two factors are presented simultaneously in these results, the effect of changing pump power (due to higher ^{3}He pressure) on the efficiency, and the effect of more He atoms on the kinetics (which effects efficiency). It is not possible to distinguish between these two effects. From Figure 17 it can be inferred, given the positive effect of increasing pump power indicated in Figure 15, that increasing He atom concentration must have only a small (positive or negative) effect on fluorescence efficiency.

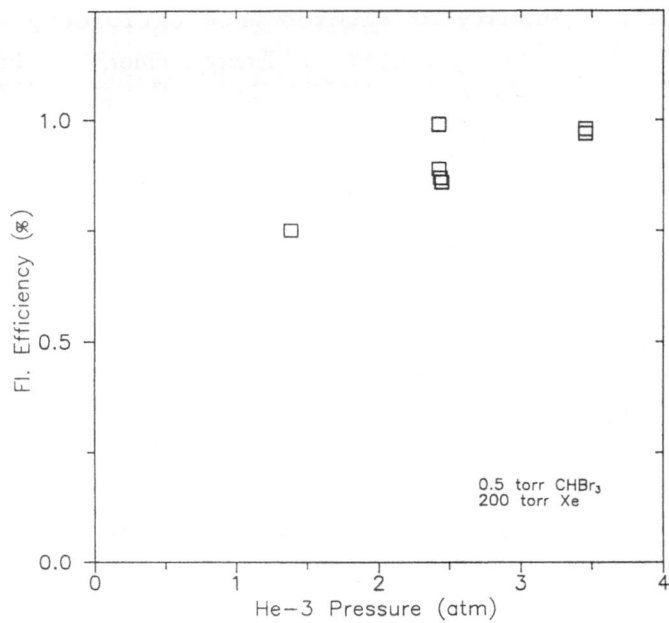

Figure 17. XeBr* fluorescence efficiency as a function of
^{3}He pressure. (Data from Figure 12)

COUPLING EFFICIENCY CALCULATIONS

One fundamental advantage of a nuclear flashlamp pumped laser is that a large flashlamp volume can be excited, and then the light from that volume concentrated into a small lasant volume. A corresponding difficulty involves coupling from a large volume, isotropic source into a small volume with good efficiency. (In the case of the laser used here the flashlamp volume was 2550 cm^3, and the lasant volume 153 cm^3.) The term "coupling efficiency" used here is defined as number of photons absorbed in the lasant, divided by the number born in the fluorescer.

In past conceptual studies of nuclear flashlamp pumped lasers, some attention has been paid to this problem [8,43,44].

Results are a strong function of cell geometry, however, and published results were viewed to be insufficient to design these experiments. Thus, a computer code was written for this application. The code was based on a ray tracing technique [45]. The basic approach was to model surfaces of objects in space with mathematical functions, and the light ray as a line in space. A simple geometry representing these experiments, is two concentric cylinders. The inner cylinder contains the lasant gas, and the annular region contains the fluorescer. The wall between the two is glass, and the inner surface of the outer cylinder is reflective. Photons are born in the fluorescing volume and bounce around inside the annular cavity; some intersect the inner cylinder and are fully, or partially, adsorbed by the lasant. If partially adsorbed, the ray continues on with decreased intensity. The ray also loses intensity at each bounce off the outer cylinder wall due to the finite reflectivity (approximately 82% for aluminum at 282 nm). As the photons were bounced around inside the cavity, a tally was kept of the energy lost, either as reflective losses, absorption in the glass, absorption in the fluorescer, or absorption in the lasant.

A small loss was incurred as the photon passes through the glass wall. This is because the glass darkens during the pulse, but was found to be negligible [41,46]. Likewise, a small absorption coefficient was included for the fluorescing medium itself, as some medium species, especially Br_2 and Xe_2+ [29,31] have appreciable absorption cross-sections at 282 nm. However, this too proved negligible.

The code was two dimensional, restraining photons to a cross-sectional plane of the laser. A random angle in the third dimension was included for each photon. This allowed, for example, the surfaces of the inner and outer cylinders of the laser geometry given above to be modeled with equations as circles of given radii.

The simple design of two concentric cylinders is shown in Figure 18, along with the modification of adding 1 to 4 longitudinal fins tangential to the inner cylinder. These fins served the purpose of deflecting photons which, in the simple concentric cylinder geometry, would bounce around inside the annulus, never intersecting the inner cell. A corresponding improvement in coupling efficiency is seen. Another modification to use a reflective annulus wall, as above, but have a rough surface so photons are reflected off at random angles. This, again, keeps photons from endlessly looping around the center cell without intersecting it. A final modification changes the fluorescing cell shape to approximately concave. This also allows more photons to be directed toward the center of the geometry. Similar dimensions (for example, cell radii) were used for designs shown in the figure to allow comparison. Lasant pressure used in these calculations was 100 torr C_3F_7I.

Through calculations such as these it was decided to construct the laser cell of two concentric cylinders, with a smooth, reflective annulus surface. This is far from optimum, but was estimated to be adequate to achieve lasing. Most importantly it is simple and inexpensive to build. A coupling efficiency of 15% was calculated for this geometry with 15 torr of lasant gas, the approximate pressure used in laser experiments.

NUCLEAR RADIATION EFFECTS ON LASANT

In considering the possibility of pumping an iodine laser in the radiation field from a nuclear reactor, some

55

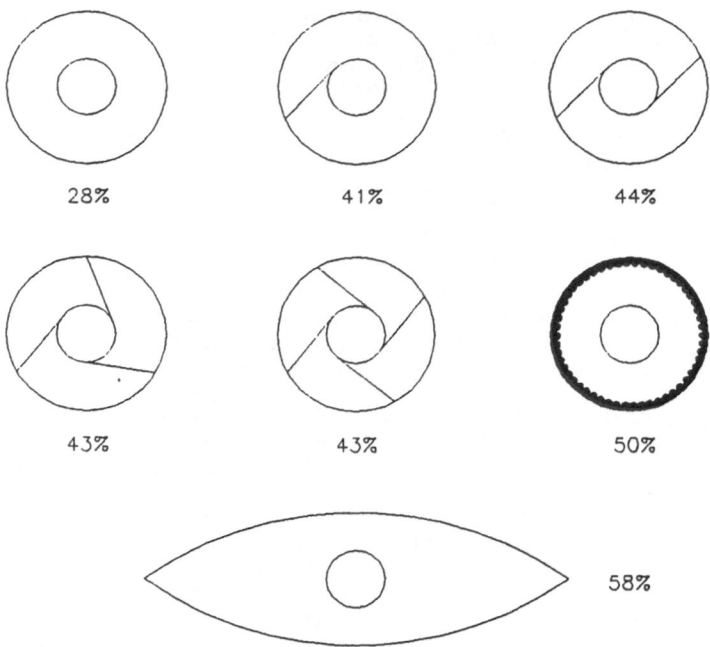

Figure 18. Calculated coupling efficiencies for various
geometries. Laser tube radii = 1 cm. Round
fluorescer tube radii = 2.5 cm. Parabolic-
shaped fluorescer tube cross-section: height
= 5 cm; length = 8 cm.

consideration was given to the question of whether the
radiation would damage the lasant. Such damage (radiolytic
decomposition) comes about as gammas and high energy neutrons break molecular bonds. An especially important
product occurs as atomic iodine atoms recombine with
themselves to form I_2. Molecular iodine is an especially
strong quencher of the upper laser state for the atomic
iodine laser, and cannot be tolerated in concentrations
above ~1%. A key concern, then, in the radiolytic decomposition of the lasant would be possible formation of I_2.

To evaluate this potential, the I_2 production rate in
CF_3I under Co-60 irradiation, as reported by McAlpine and
Sutcliffe [47], was applied to the total dose to the lasant
provided by the TRIGA in a pulse (0.1 MRad). This predicts
a 0.03% conversion of the lasant to I_2 from 50 reactor
pulses. It seems likely that I_2 formation from flashlamp-
induced photolysis of the lasant might represent a larger
problem, though there is little data on this. The
conclusion, then, is that radiolytic decomposition should
not be a problem in nuclear pumped laser applications
involving the present reactants.

LASER CELL DESIGN AND EXPERIMENTAL SET-UP

A schematic of the cell is shown in Figure 19. It consisted of two concentric cylinders, the inner to contain the lasant gas, and outer to contain the fluorescing $XeBr/^3He$ mix. The inner cylinder was 20 mm O.D. Suprasil fused silica tubing with 1 mm wall. Suprasil was selected because experiments [41] had shown the radiation-induced darkening of this material to be small. Also it has good transmission in the UV. The Brewster windows were made of Infrasil. Normal fused silica has a small water absorption band near the laser wavelength of 1.3 micron which, it was feared, would grow upon irradiation.

The outer cylinder was 3-1/16-inch I.D., 1/4-inch wall aluminum. The inside of the cylinder was polished on a lathe with fine alumina powder in deionized water to a good mirrored finish. This polishing allowed a prediction of the coupling efficiency through the use of the ray tracing calculations and also removed any aluminum oxide coating, increasing reflectivity. End plates of 5/8 inch aluminum were bolted on the cylinder with buna-n o-rings. The inner glass cylinder was also sealed to these end plates with the use of o-rings. Fittings were attached to the cell for gas fill and vacuum lines. A small fused silica window, 3/8 inch diameter, was installed in one end plate to allow viewing of the fluorescence chamber during a pulse.

The cell was constructed to allow cold fluorescence gas fills of ~3 atm. (about 9 atm. during a pulse due to gas heating) with pressures in the laser tube of up to several hundred torr.

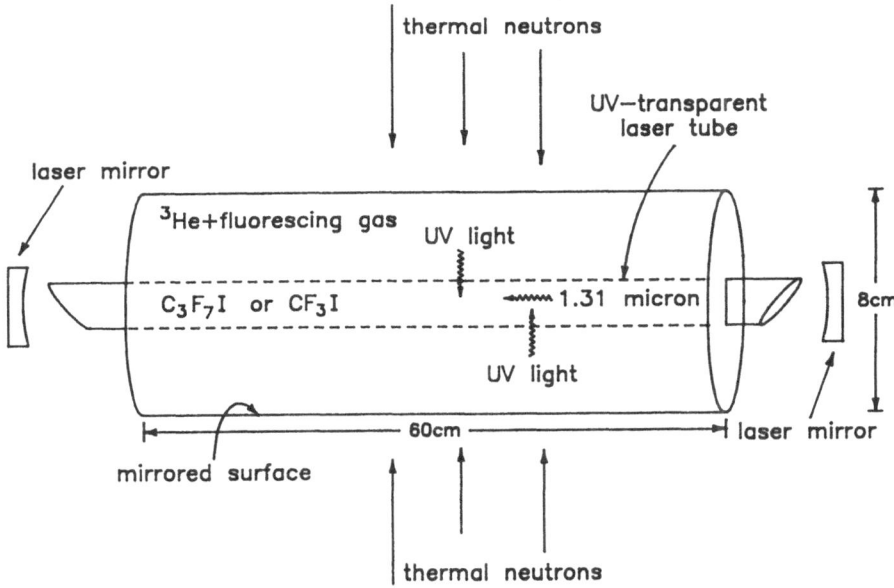

Figure 19.   Schematic of laser cell (not to scale).

Laser mirrors had reflectances of 99.7 and 98% at 1.31 micron, and were one inch in diameter. The output mirror was a fused silica substrate to minimize radiation darkening. The rear mirror had a Pyrex substrate.

A schematic of the experimental set-up is shown in Figure 20. The cell was viewed for two optical signals simultaneously, the laser line coming from the laser tube, and the fluorescence light coming through the small viewpoint in the flashlamp wall. Two sets of aluminum-coated mirrors were used to separate the two signals. The laser line was focussed onto a 2 mm diameter InAs photodiode, blocked with a bandpass filter which passed light only a few nm around 1.31 micron. The fluorescence signal was viewed with an OMA, as in fluorescence studies, and with a monochromator/PMT to look at the temporal dependance.

The attached gas handling lines allowed changing of both fluorescence and lasant gas fills. The fluorescer output was constant, however, and did not degrade with repeated pulses on the same fill. The lasant was replaced frequently because laser output was found to degrade under repeated pulses. The fluorescing gas mixture in all large experiments was 0.5 torr $CHBr_3$, 200 torr Xe, 3 atm ^{3}He.

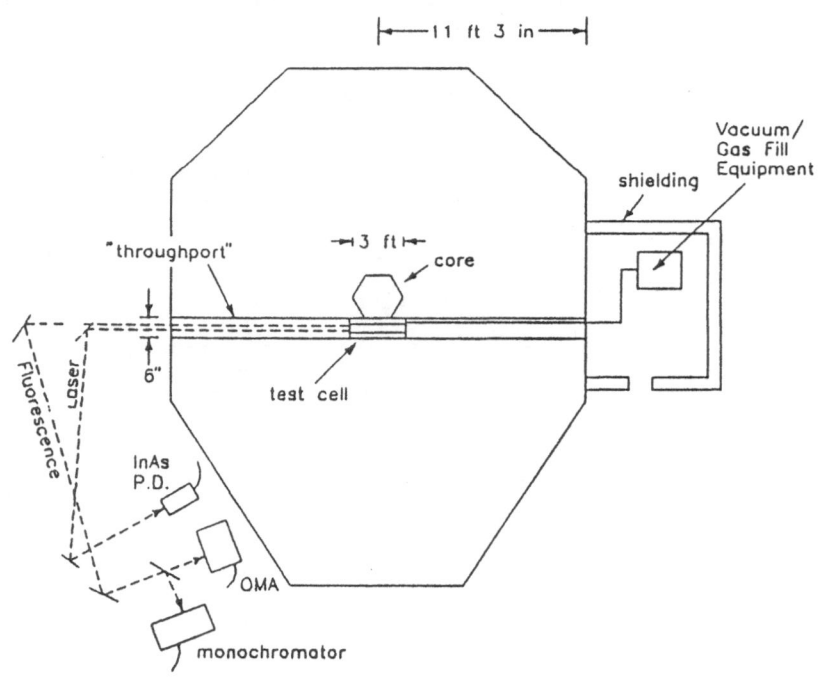

Figure 20. Setup for laser experiments in the TRIGA (not to scale).

Materials used as lasants were i-C$_3$F$_7$I and CF$_3$I, both common in atomic iodine lasers. Buildup of I$_2$ from photolytic and thermal decomposition in purchased bottles of C$_3$F$_7$I was found to be a problem. The I$_2$ is apparent upon visual inspection due to a purple tinge. To remove the iodine, the C$_3$F$_7$I was mixed with fine copper powder, which reacts with I$_2$ to form an insoluble compound. This procedure was repeated several times until the liquid showed no tinge. CF$_3$I, being a gas at room temperature, was not treated prior to use.

LASER SIGNAL

A trace of the laser output from the photodiode is shown in Figure 21, with the neutron signal superimposed. The laser signal jumps up suddenly as the pump power reaches threshold a characteristic of laser behavior. The lasing follows the shape of the neutron pulse until the pump power drops below threshold on the back side of the pulse, and the laser switches off.

The "noise" fluctuations in the laser signal are actually fast oscillations in the laser output termed "relaxation oscillators." They are common in atomic iodine lasers pumped at low pump rates [48,49]. They occur as the upper and lower laser states compete for atoms. These states have different lifetimes, depending on the atomic species and the gas kinetics. The resulting oscillations occur on a fast time scale (microsecond), faster than the resolution of the digital oscilloscope used here. For purposes of these experiments, these oscillations are unimportant, except as an evidence of lasing.

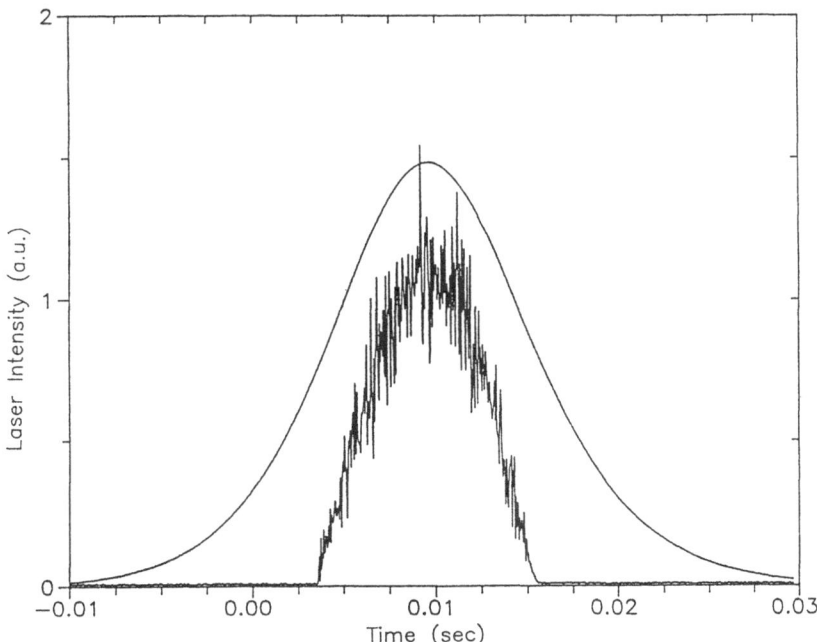

Figure 21. Trace of laser output from 15 torr C$_3$F$_7$I with reactor power superimposed.

Other evidences of lasing should also be noted. The photodiode used to register the laser signal was covered with a few nm wide, 1.31 micron bandpass filter. This minimized the possibility of other, non-laser light entering it. Radiation effects on the photodiode were too small to be measured, as indicated in experiments with an empty laser cell. Finally, the light registered by the detector could not have been spontaneous emission at 1.31 micron by the lasant. The measured signal (approximately 20 mW at peak) would have required a spontaneous emission power from the lasant in the cell of 8 kW, far larger than possible in this system. (The actual peak spontaneous emission, not measurable in these experiments, would have been <50 W.)

As mentioned, the peak laser power was determined approximately from the InAs photodiode as ~20 mW. When integrated over the pulse duration, the laser energy was approximately 150 microjoules.

The above data, and all laser data, were taken with C_3F_7I. An attempt was made to use CF_3I, but lasing was not achieved. This was due presumably to the somewhat lower photodissociation cross section of this material at 282 nm, or may have been due to I_2 contamination in the lasant.

Given the knowledge of the fluorescence intensity of the XeBr and coupling efficiency inside the laser cell, it is possible to calculate the flashlamp pumping power in the lasant at threshold. A value of approximately 0.4 W/cm^3 at threshold is found. This is reasonably close to the values of 0.06 and 0.07 W/cm^3 reported in the literature [48,50].

LASANT PRESSURE DEPENDENCE

Measurements were done to determine laser performance with changing lasant pressure. Results are shown in Figure 22. A wide scatter in the data was found, with poor repeatability. This is attributed to the fact that the laser was being pumped so close to threshold, and small variations in the system (e.g. impurities left from inadequate pumping on the cell between pulses) could effect performance significantly. Sufficient pulses were taken, however, to indicate lasing was only possible between approximately 5 and 25 torr, which agrees with other published data on the atomic iodine laser pumped at low powers [48,51]. At lower pressures there are insufficient lasant molecules to absorb the flashlamp light. At higher pressures, the lasant molecules quench the excited laser state. (In Figure 22 a curve has been drawn through the data only to help indicate this range, and is not meant to represent the actual dependance.)

LASER SCALE-UP

Based on these experiments, some comments can be made regarding scale-up to a large nuclear flashlamp pumped atomic iodine laser. An important conclusion is that these proof-of-principle results showed that the high radiation

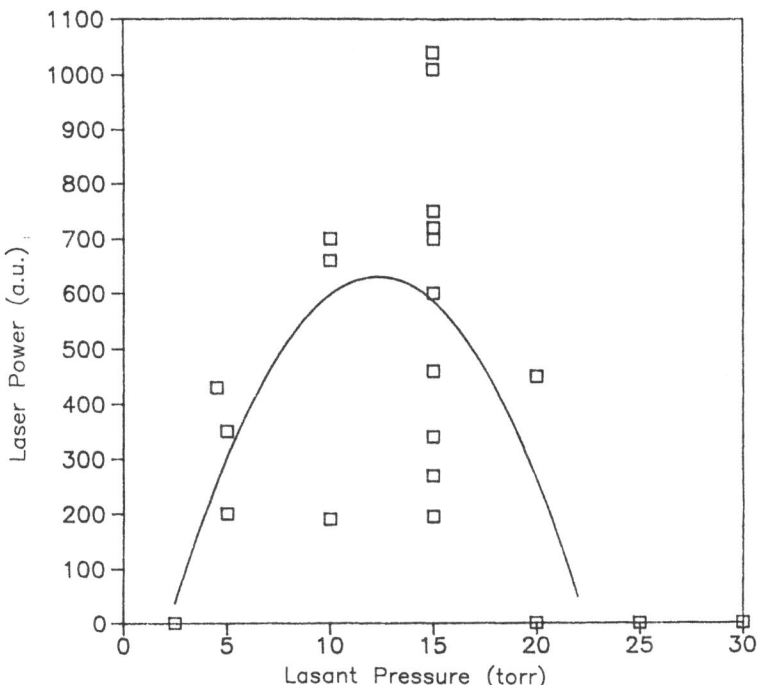

Figure 22. Lasant pressure dependence on laser output. The solid curve is included only as a visual aid, indicating that lasing was achievable between approximately 5 and 25 torr $i\text{-}C_3F_7I$.

field of nuclear pumping poses no special difficulties for this laser.

The parameter of principle interest in these experiments, however, was the fluorescence efficiency of an excimer under nuclear pumping, and measurements of <1% for this parameter were disappointing (compared to the 11 ± 5% determined in e-beam experiments [9]). If future experiments are able to increase this efficiency, the potential for a high power system becomes more promising. For example, with a flashlamp efficiency of, say, 10%, a coupling efficiency of 30%, and a lasant quantum efficiency of 20%, an overall system efficiency of ca. 0.6% is anticipated. This compares favorably with electrically pumped iodine laser efficiencies of 0.08-0.3% [4].

The fluorescence power which could be extracted from a nuclear flashlamp depends on the neutron flux, ^{3}He pressure, and fluorescence efficiency. For purposes of illustration, consider a box of dimensions 1 m by 1 m by 5 cm, containing 3 atm. ^{3}He in a 10^{17} n/cm^2sec neutron flux, and a fluorescence efficiency of 10%. This gives a flashlamp power of 11 MW/m^2, or 220 MW/m^3 of fluorescer. The energy emitted in the UV in a 150 microsec fast burst reactor pulse would be approximately 1.6 kJ.

As a specific example of scale-up, it is instructive to compare, to first order, how a nuclear flashlamp for pumping of a very large laser amplifier (ca. 1000 J) might compare to a capacitor-driven xenon flashlamp bank.

If we consider a nuclear source of 10^{18} n/pulse, such as from a fast burst reactor, then, given a 100% coupling efficiency of neutrons into the fluorescer, and of the fluorescence into the lasant, and a 10% conversion efficiency of ^{3}He reaction energy into UV light, we have a flashlamp providing around 10^4 J, approximately sufficient to pump a 1000 J amplifier. To generate this energy from xenon flashlamps, given an 8% efficiency of converting electrical energy into UV light [4], we need a 10^5 J capacitor bank. Such a bank, based on vendor data, would have a volume of approximately 60 ft^3. By contrast, a fast burst reactor is approximately 2 ft^3 in volume. In addition, the electricity generating equipment to charge the capacitors (presumably solar cells for space-based concerns) would be of extensive size. Hence, the potential for signficant advantage in volume (and weight) is indicated for the nuclear flashlamp over the xenon flashlamps.

SUMMARY AND CONCLUSIONS

This project was broken into two parts, studying nuclear pumped fluorescence of XeBr, and then using the resulting fluorescence to excite a laser. These results represent the first demonstration of the atomic iodine laser stimulated with nuclear-induced fluorescence. This is also the first demonstration of a nuclear flashlamp pumped laser of any kind.

While a number of detailed questions remain (e.g. the maximum efficiencies that can be obtained from a suitable nuclear-pumped excimer flashlamp, the maximum coupling efficiency, and the long term effect of radiation on laser materials and optics), these potentially open the path to a new approach to nuclear driven lasers. The use of a flashlamp makes it possible to pump a variety of lasers that might not be pumped directly by radiation. The use of a large flashlamp volume with its output focussed on a small laser volume also offers advantages in some applications. A next obvious step for evaluation of the use of this approach to ICF is to carry out a reactor design study. Also, simultaneous experimental studies at high power levels are essential to further explore scaling relations.

REFERENCES

1. Kasper, J. V. V., Pimentel, G. C., "Atomic Iodine Photodissociation Laser," Appl. Phys. Lett. 5, 231 (1964).
2. Kasper, J. V. V., Parker, J. H., Pimentel, G. C., "Iodine-Atom Laser Emission in Alkyl Iodine Photolysis," J. Chem. Phys., 43, 1827 (1965).
3. Kirillov, G. A., Murugov, V. M., Punin, V. T., Shemyakin, V. I., "High Power Laser System ISKRA V," Laser & Part. Beams 8 (4), 827 (1990).

4. Brederlow, G., Fill, E., Witte, K. J., <u>The High-Power Iodine Laser</u>, (Springer-Verlag, Berlin, 1983).
5. Hohla, K., Kompa, K. L., <u>Handbook of Chemical Lasers, ch. 12</u>, Gross, R. W. S. and Bott, J. F. ed., (Wiley & Sons, 1976).
6. Wilson, J. W., Shapiro, A., "Nuclear-Induced Excimer Fluorescence," <u>J. Appl. Phys.</u> <u>51</u> (5), 2387 (1980).
7. Wilson, J. W., "Nuclear-Induced XeBr* Photolytic Laser Model," <u>Appl. Phys. Lett.</u> <u>37</u> (8), 695 (1980).
8. Prelas, M. A., Boody, F. P., Miley, G. H., Kunze, J., "Nuclear Driven Flashlamps," <u>Laser & Part. Beams</u> <u>6</u> (1) 25 (1988).
9. Swingle, J. C., Turner, C. E., Jr., Murray, J. R., George, E. V., Krupke, W. F., "Photolytic Pumping of the Iodine Laser by XeBr*," <u>Appl. Phys. Lett.</u> <u>28</u> (7), 387 (1976).
10. Miley, G. H., McArthur, D., DeYoung, R., Prelas, M., "Fission Reactor Pumped Lasers: History and Prospects," Proc. Fifty Years with Nuclear Fission, ANS Conference (Gaithersburg, MD, April 26-28, 1989).
11. Schneider, R. T., Hohl, F., "Nuclear Pumped Lasers," in <u>Advances in Nuclear Science and Technology, Vol. 16</u>, 123 (Plenum, New York, 1984).
12. Alford, W. J., Hays, G. N., "Measured Laser Parameters for Reactor-Pumped He/Ar/Xe and Ar/Xe Lasers," <u>J. Appl. Phys.</u>, <u>65</u>, 3760 (1989).
13. Alford, W. J., Hays, G. N., Ohwa, M., Kushner, M. J., "The Effects of He Addition on the Performance of the Fission-Fragment Excited Ar/Xe Atomic Xenon Laser," <u>J. Appl. Phys.</u> <u>69</u> (4), 1843 (1991).
14. Prelas, M. A., "Nuclear-Driven Solid-State Lasers," Proc. Intl. Conf. on Lasers '89 (Soc. of Opt. and Quantum Elec., New Orleans, STS Press, 1990), 263.
15. Prelas, M., Kunze, J. F., "Direct Nuclear Energy Conversion Cycles Using Excimer Fluorescence," Proc. 6th Intl. Conf. on Emerging Nuclear Energy Systems (ICENES '91) (Monterey, CA, June 16-21, 1991) (to be published).
16. Hutchinson, M. H. R., "Excimers and Excimer Lasers," <u>Appl. Phys.</u> <u>21</u>, 95 (1980).
17. Rhodes, Ch.K., ed., <u>Excimer Lasers</u>, (Springer-Verlag, Berlin, 1979).
18. Basov, N. G., Danilychev, V. A., Popov, Yu. M., Khodkevich, D. D., "Laser Operating in the Vacuum Region of the Spectrum by Excitation of Liquid Xenon with an Electron Beam," <u>JETP Lett.</u> <u>12</u>, 329 (1970).
19. Cartwright, D. C., Figueira, J. F., McDonald, T. E., Harris, D. B. Hauer, A. A., "Status of Inertial Confinement Fusion Research at Los Alamos National Laboratory," <u>Laser Interaction and Related Plasma Phenomena</u>, <u>9</u>, 11 (Plenum Press, New York, 1991).
20. Verdeyen, J. T., <u>Laser Electronics, 2nd Ed.</u> (Prentice-Hall, New Jersey, 1989).
21. Wilson, W. L., Williams, R. A., Sauergrey, R., Tittel, F. K., Marowsky, G., "Formation and Quenching Kinetics of Electron Beam Excited Xe_2Br*," <u>J. Chem. Phys.</u>, <u>77</u>(4), 1830 (1982).

22. Bychkov, Yu. I., Mesyats, G. A., Tarasenko, V. F., "The-Short-Electron-Beam-Excited Excimer Lasers," Proc. Intl. Conf. on Lasers '80, 682, (Society of Optical and Quantum Elec., McLean, VA, 1981).

23. Moratz, T. J., Kushner, M. J., "A Comparison of Electron Beam and Heavy Ion Excitation of Rare Gas-Halogen Gas Mixtures," Proc. IEEE Conf. on Plasma Sci., 93 (Seattle, WA, May 1988).

24. Gerber, T., Luthy, W., Burkhard, P., "High Efficiency KrF Excimer Flashlamp," Opt. Comm. 35 (2), 242 (1980).

25. Kumagai, H., Obara, M., "New High-Efficiency Quasi-Continuous Operation of a KrF(B-X) Excimer Lamp Excited by Microwave Discharge," Appl. Phys. Lett., 54 (26), 2619 (1989).

26. Kumagai, H., Obara, M., "A High-Efficiency, High-Repetition-Rate KrF(B-X) Excimer Lamp Excited by a Microwave Discharge," Jpn. J. Appl. Phys. 28 (12), L2228 (1989).

27. Kumagai, H., Obara, M., "New High-Efficiency Quasi-Continuous Operation of a KrF(B-X) Excimer Lamp Excited by Microwave Discharge," Appl. Phys. Lett. 55 (15), 1583 (1989).

28. Searles, S. K., Hart, G. A., "Stimulated Emission at 281.8 nm from XeBr," Appl. Phys. Lett. 27 (4), 243 (1975).

29. Konovalov, I. N., Tarasenko, V. F., "E-Beam Pumped and E-Beam Controlled XeBr-Laser," Proc. Intl. Conf. on Lasers '80, 686, (Society of Optical and Quantum Elec., McLean, VA, 1981).

30. Shevera, V. S., Shuaibov, A. K., "Formation of Monohalides of Inert Gas in a Transverse AC Electric Discharge," Sov. Phys. Tech. Phys. 25 (4), 434 (1980).

31. Sze, R. C., Scott, P. B., "High-Energy Lasing of XeBr in an Electric Discharge," Appl. Phys. Lett. 32 (8), 479 (1978).

32. Fu, S. F., Chen, J. W., Liu, M. H., "Discharge-Pumped Multiwavelength Laser," Chin. Phys. 1 (3), 580 (1981).

33. Tittel, F. K., Wilson, W. L., Jr., Williams, R. A., "Spontaneous and Stimulated Emission Characteristics of the Excimer XeBr," Proc. 12th Intl. Quantum Elec. Conf., 126, (Muenchen, Germany, June 22-25, 1982).

34. Walters, R. A., "Spectral Emission of Nuclear Excited XeBr*," Proc. Workshop on Nuclear Pumped Lasers, NASA Conf. Publ. 2107, 33 (Hampton, VA, July 25-26, 1979).

35. Peach, R. O., M. S. Thesis, University of Illinois, 1988.

36. Brau, C. H., Ewing, J. J., "Spectroscopy, Kinetics and Performance of Rare Gas Halide Lasers," in Electronic Transition Lasers, ed. by J. I. Steinfeld, 195 (MIT Press, Cambridge, MA, 1976).

37. Torczynski, J. R., Gross, R. J., Hays, G. N., Harms, G. A., Neal, D. R., McArthur, D. A., Alford, W. J., "Fission-Fragment Energy Deposition in Argon," Nuc. Sci. Eng. 101 (3), 280 (1989).

38. Wilson, J. W., DeYoung, R. J., "Power Density in Direct Nuclear-Pumped ^{3}He Lasers," <u>J. Appl. Phys.</u> <u>49</u> (3), 980 (1978).

39. Zediker, M. S., <u>An Investigation of the Singlet Delta Oxygen and Ozone Yields from the Pulsed Radiolysis of Oxygen and Oxygen-Noble Gas Mixtures</u>, Ph.D. Thesis (University of Illinois, 1984).

40. Elsayed-Ali, H. E., Miley, G. H., "Ozone Dosimetry for Calibration of Power Deposition in Nuclear Pumped Lasers," <u>Rev. Sci. Instrum.</u> <u>55</u> (8), 1353 (1984).

41. Miley, G. H., Chapman, R., Nadler, J., Williams, W., "Radiation Effects on Nuclear Pumped Laser Optics," Proc. 10th Intl. Conf. on Lasers '87 (Lake Tahoe, NV, December 11-17, 1987).

42. Boivineau, M., LeCalve, J., "Formation of the XeBr* Excimer by Double Optical Excitation of the $XeBr_2$ van der Waals Complex," <u>J. Chem. Phys.</u> <u>84</u> (8), 4712 (1986).

43. Prelas, M. A., Jones, G. L., "Design Studies of Volume-Pumped Photolytic Systems using a Photon Transport Code," <u>J. Appl. Phys.</u> <u>53</u> (1), 165 (1982).

44. Javedani, J. B., Prelas, M. A., "Concentrating Properties of Simple Two-Dimensional Geometries for Isotopic Light," <u>Laser Interaction and Related Plasma Phenomena,</u> 7, 155 (Plenum Press, New York, 1986).

45. Glassner, A., <u>An Introduction to Ray Tracing</u> (Academic, London, 1989).

46. Palma, G. E., Gagosz, R. M., "Optical Absorption in Fused Silica During Irradiation: Radiation Annealing of the C-Band," <u>J. Phys. Chem. Solids</u>, <u>33</u>, 177 (1972).

47. McAlpine, A., Sutcliffe, H., "The Radiolysis of Trifuoroiodomethane in the Gas Phase," <u>J. Phys. Chem</u>. <u>73</u>, 3215 (1968).

48. DeYoung, R. J., "Low Threshold Solar-Pumped Iodine Laser," IEEE <u>J. Quant. Elec.</u> <u>QE-22</u> (7), 1019 (1986).

49. Yariv, A., <u>Quantum Electronics, 3rd ed.</u>, 560 (Wiley & Sons, New York, 1989).

50. Witte, K. J., Burkhard, P., Luthi, H. R., "Low-Pressure Mercury Lamp Pumped Atomic Iodine Laser of High Efficiency." <u>Opt. Comm.</u> <u>28</u> (2), 202 (1979).

51. DeYoung, R. J., Conway, E. J., "Progress in Solar-Pumped Laser Research." Proc. Intl. Conf. on Laser '85, 467 (Soc. of Opt. and Quant. Elec., Las Vegas, NV, STS Press, 1986).

A COMPARISON OF ELECTRICAL, FUSION-GENERATED-ION, AND FISSION-GENERATED-ION ICF DRIVERS

Mark A. Prelas[1], and Frederick P. Boody[2]

[1]Fusion Research Laboratory

Nuclear Engineering Program

University of Missouri-Columbia

Columbia, MO 65211

[2]Current Address: Nuclear-Pumped Laser Corp.

P.O. Box 641, Columbia, MO 65205-0641, USA

Introduction

In this paper, arguments are presented for various inertial confinement fusion laser systems (electrical driven laser, ion--generated by fusion reactions-- driven laser, and ion--generated by fission reactions--driven laser) which show that high plant efficiencies can be achieved with low efficiency fission-generated-ion driven lasers 0.5%. It is the unique properties of the fission-generated-ion driven system which provides it a distinct advantage over the electrical driven system and the fusion-generated-ion driven system.

Inertial confinement fusion systems are examined from the perspective of energetics. Three types of laser fusion drivers are examined; a laser driver pumped by conventional electrical means (i.e., electrical discharge, or electrical flashlamp), a laser driver pumped by the ions generated in the fusion reaction (i.e, direct ion excitation, or ion excited flashlamps), and a laser pumped by ions generated in fission reactions (i.e., direct ion excitation, or ion excited flashlamps).

Model

In order to compare the various methods of powering lasers (electrical drivers, fusion-generated-ion drivers, and fission-generated-ion drivers), the important variables must be determined for the energy flow of the system. In the analysis of energy flow, the variables which were examined are the pellet gain, the energy conversion system efficiency, the laser efficiency, the fraction of energy recirculated to the laser system from the energy conversion system, the fraction of charged particles recirculated from the fusion reaction to the laser system, the fusion to fission energy ratio, the fluorescer efficiency, the fluorescence absorption efficiency, the plant's rated electrical power output, and the thermal efficiency of the power plant. [Pellet gain (G) is defined as the fusion energy out of the pellet divided by the laser energy into the pellet. The energy conversion system efficiency (η_c) is defined as the electrical energy out of the system divided by the thermal energy into the system. Laser efficiency (η_{laser}) is defined as the laser energy divided by the energy into the laser media. The recirculation energy (f) is the fraction of electrical energy from the energy conversion system

Laser Interaction and Related Plasma Phenomena, Vol. 10
Edited by G.H. Miley and H. Hora, Plenum Press, New York, 1992

67

used to drive the laser system. The fusion charged particle fraction (f_{cp}) is the fraction of energy in charged particles used to drive the laser. The fusion to fission energy ratio (E_{fusion}/$E_{fission}$) is the fraction of fusion energy produced versus the fraction of fission energy produced. The fluorescer efficiency (η_f) assumes that the ion driven systems use flashlamps. The fluorescence absorption efficiency (η_a) is the efficiency of the fraction of fluorescence energy absorbed by the laser. The plant's rated electrical power output (E_{load}) is the electrical power output from the plant. The thermal efficiency of the power plant is the electrical power production from the plant versus the thermal energy.]

Electric Powered Driver

In Figure 1, the energy flow in an inertial confinement fusion power plant using an electrical powered driver is shown. Fusion energy flows into the energy conversion system and

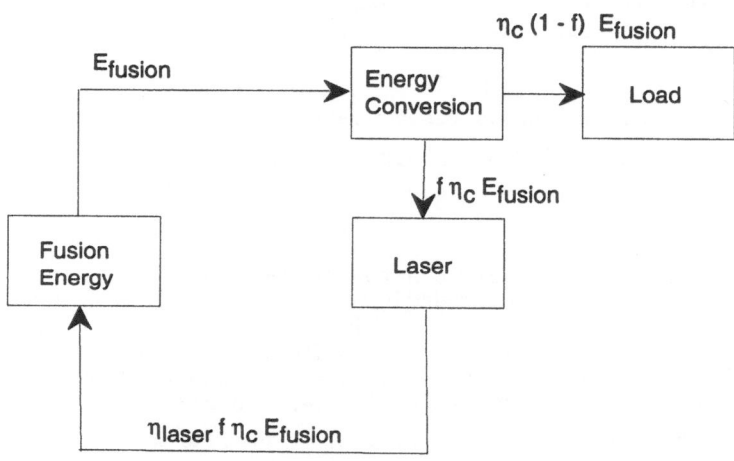

$$E_{laser} = \eta_{laser}\ ^f \eta_c\ E_{fusion}$$

$$E_{load} = \eta_c\ (1 - f)\ E_{fusion}$$

$$E_{fusion} = G\ E_{laser} = G\ \eta_{laser}\ ^f \eta_c\ E_{fusion}$$

$$f = 1/(G\ \eta_{laser}\ \eta_c)$$

$$\eta_{plant} = E_{load}/E_{fusion} = \eta_c\ \{(G\ \eta_{laser} - 1)/(G\ \eta_{laser})\}$$

E_i = Energy in i; f = fraction of fusion energy recycled to laser; η_{laser} = laser wall plug efficiency; η_{plant} = plant thermal efficiency; η_c = energy convertor efficiency; G is pellet gain.

FIGURE 1 Energy flow diagram of an inertial confinement fusion system using an electrical driven laser.

is converted into electricity. A fraction (f) of the electrical energy is then recirculated to the laser driver. In this approach, there are eight variables as shown in Figure 1. By choosing E_{load} to be 1 GigaWatt-s electric, and η_c to be the efficiency of a standard steam cycle (33%), there remains only two independent variables (G and η_{laser}). Thus, it is possible to determine E_{laser}, f, E_{fusion}, and η_{plant} as a function of G and η_{laser}.

Figure 2 is a plot of the laser energy as a function of the pellet gain and the laser efficiency. As expected, it can be seen in Figure 2 that as G and η_{laser} increases, the required laser energy decreases. Figure 2 is a very good summary of the dilemma of electrical drivers in inertial confinement fusion: Unfortunately, the best plant parameters require high gain pellets and high laser efficiencies which are very difficult to obtain in reality.

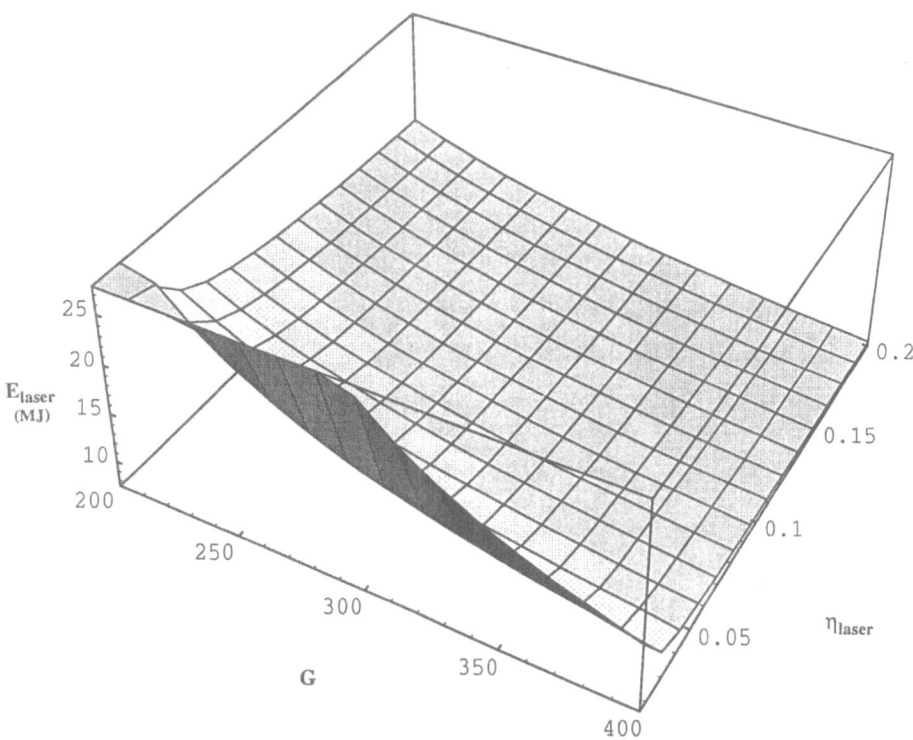

FIGURE 2   Estimated required laser energy as a function of pellet gain and laser efficiency for a 1 GigaWatt electric inertial confinement fusion power plant using a 33% efficient energy conversion system.

Another problem with electrical drivers is that the power plant efficiency is highly dependent upon the laser efficiency. As can be seen in Figure 3, the maximum power plant efficiency is equal to the energy conversion system efficiency. In the case of a standard steam cycle energy conversion system, this efficiency is 33%. As long as the laser efficiency is above 10%, the recirculation fraction remains relatively low (see Figure 4) and the plant efficiency is a significant fraction of the energy conversion system efficiency. It is very difficult however, to find a laser system with an efficiency of 10% or greater. (Ion beam drivers look attractive compared to lasers because it is relatively easy to build an ion beam driver with an efficiency greater than 10%.)

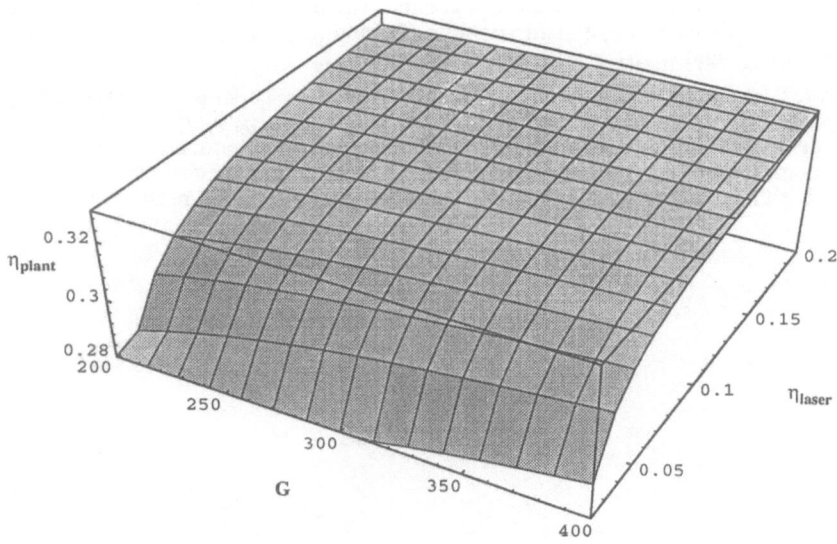

FIGURE 3 Estimated thermal plant efficiency as a function of pellet gain and laser
efficiency for a 1 GigaWatt electric inertial confinement fusion power plant
using a 33% efficient energy conversion system.

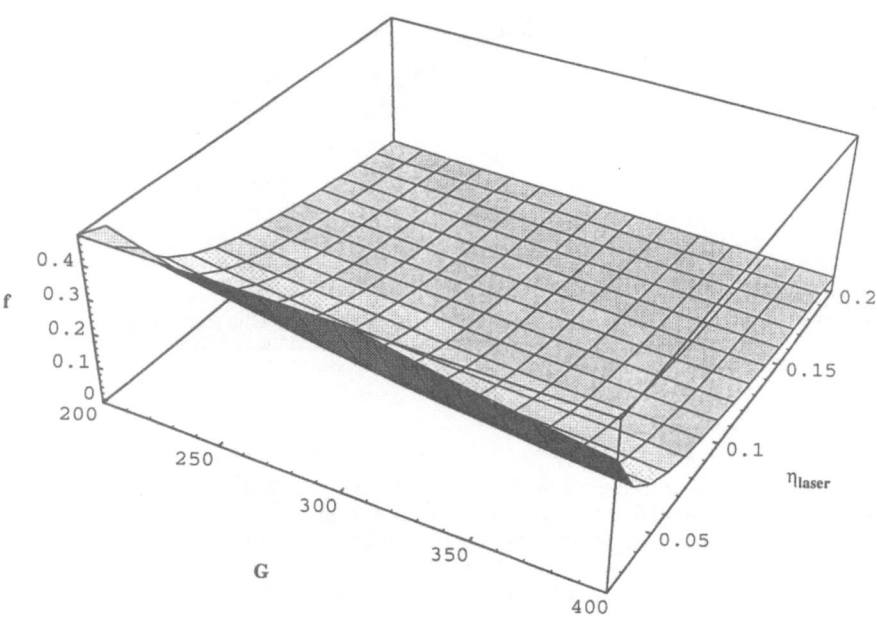

FIGURE 4   Estimated fraction of power recirculated to the laser system as a function of
pellet gain and laser efficiency for a 1 GigaWatt electric inertial
confinement fusion power plant using a 33% efficient energy conversion
system.

Fusion-Generated-Ion Drivers

The concept of using the ions generated in fusion reactions to power the laser driver was first proposed by Miley 1977. A subsequent study was performed using the long lived oxygen singlet delta molecule to drive an iodine laser (Wessol, Prelas, and Merril 1988).

Fusion reactions produce ions and neutrons as shown below:

Eq. 1 $D(T,n)He^4 + 17.6$ MeV {n(14.08 MeV) & He^4 (3.52 MeV)}

Eq. 2 $D(D,n)He^3 + 3.27$ MeV (50% branching) {n(2.45 MeV) & He^3 (0.82 MeV)}

Eq. 3 $D(D,p)T + 4.03$ MeV (50% branching) {p(3.02 MeV) & T (1.01 MeV)}

The products from fusion reactions can be used to excite laser media. This paper will focus on one of the many possible methods of exciting a laser with fusion products; solid-state lasers driven by excimer flashlamps (Prelas, Boody, Woodall, Speziale, and Wills, 1987, Prelas, and Charlson, 1987, and Prelas and Boody, 1991). Excimers which produce fluorescence within the absorption bands of advanced solid-state laser materials can be excited by charged particles (Prelas, 1981, Prelas, Boody, Kunze, and Miley 1988). Neutrons can also be used to transfer energy to liquid or solid media through elastic scattering, inelastic scattering, and neutron capture -with subsequent charged particle, gamma ray, or neutron emission- (Prelas, Romero, and Pierson). Elastic scattering was examined in detail by Pappas, 1988, as a method of exciting liquid xenon excimers.

The steps in the excimer energy conversion process to be considered are: 1) η_t the efficiency of the transport of the fission fragment energy from the fuel into the fluorescer gas, 2) η_f the fluorescence generation efficiency, 3) η_c the photon coupling efficiency to material used in the converter, 4) η_p the useful product (electricity, chemicals, or laser photons) generation efficiency from the converter material per photon absorbed, and 5) η_{ex} the extraction efficiency, i.e., the extractable electrical energy (or other desirable energy form). (Processes 1, 2 and 3 represent the nuclear driven flashlamp and processes 4 and 5 represent the photon energy conversion module.) Hence, the system efficiency will be (Prelas 1981, Prelas, and Charlson 1987, Prelas, Boody, Kunze, and Miley 1988, and Prelas, Boody, Charlson and Miley 1991):

Eq. 4 $\eta_s = \eta_t\, \eta_f\, \eta_c\, \eta_p\, \eta_{ex}$

and the power density deposited in the photon energy conversion material is:

Eq. 5 $Pd_p = Pd_f\, \eta_f\, \eta_c\, V_f/V_p$

where Pd_f is the fission fragment power deposition in the fluorescer region (W/cm^3), V_f is the volume of the fluorescer (cm^3), and V_p is the volume of the converter material (cm^3).

Similarly the fluorescence intensity on the surface of the photon energy converter material is:

Eq. 6 $I_p = Pd_f\, \eta_f\, \eta_c{'}\, V_f/A_p$

where $\eta_c{'}$ is the efficiency for coupling the fluorescence to the converter material, and A_p is the area of the converter material (cm^2). (Note, the volume coupling efficiency, η_c, and the surface coupling efficiency, $\eta_c{'}$, will differ, with $\eta_c > \eta_c{'}$.)

In the case of laser excitation, the fluorescence needs to be coupled by waveguide to the laser system. This aspect of the energy conversion process has been examined by (Boody, and Prelas 1991, and Boody 1991). From this work the concentration of fluorescence onto the laser crystals has been examined in detail from both a theoretical and experimental perspective.

The energy flow for a fusion-generated-ion driver is shown in Figure 6. Differences between the fusion-generated-ion driver and electrical driver are in the energy flow in the laser system. Since the fusion-generated-ion driver uses a fluorescer which can operate at high temperature, the waste heat in this system is used in the energy conversion system. In contrast, the waste heat in the electrical driver will be so low in quality (temperature) that it is simply dumped to a sink. The ability to recycle waste heat makes an enormous difference in the performance of the inertial confinement fusion system.

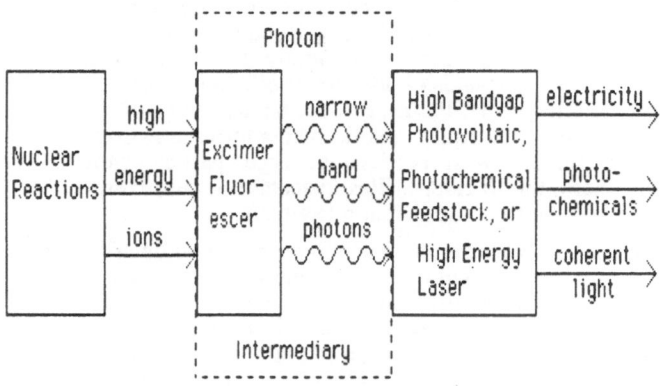

FIGURE 5   Illustration of the excimer energy conversion concept.

As discussed in the paper by Prelas, and Boody 1991, the efficiency of solid-state lasers driven by excimer lamps excited by fusion-generated-ions is intrinsically higher than electrical driven lasers (laser efficiencies between 12 and 20%). Despite the intrinsic higher efficiencies, it can be seen in Figure 7, that the effect of laser efficiency on the size of the laser is not as drastic as the electrical driver case shown in Figure 2. For example, examining a 5% efficient laser in both fusion-generated-ion drivers and electrical drivers, the fusion-generated-ion laser is 15 MJ and the electrical laser is 20 MJ (see Figures 2 and 7). In addition, as shown in Figure 8, the plant efficiency is a very weak function of laser efficiency which is in contrast to the electrical plant (see Figure 3).

Even though the recirculation energy fraction f and the charged particle fraction f_{cp} are different concepts, they serve similar roles in their respective plant concepts. For example, if the recirculation energy exceeds 1, then the power plant would need additional energy beyond what is being produced in fusion to drive the laser. Conversely, if the charged particle fraction exceeds 1, then the laser would require additional energy beyond what the fusion produces. In order for the laser to work with various fusion reactions the charged particle fraction must be less than the fraction of energy emitted in charged particles as shown in equations 1 and 2. Thus if the fusion reaction is D-T, the charged particle fraction must be less than 0.2.

Fission-Generated-Ion Driver

The fission-generated-ion driver concept is very similar to the fusion-generated-ion driver concept in that the charged particles generated in the respective nuclear reactions are used to drive the laser. In this case, a separate fission reactor is used to drive the laser system as shown in Figure 10. Thus, the fusion to fission energy ratio ($E_{fusion}/E_{fission}$) is used to describe the fraction of fusion energy generated per unit fission energy used. This ratio is very similar to the gain concept, in that the goal is to produce as much fusion energy as possible per unit of fission energy input. It is assumed that this laser will use the excimer driven lamp concept.

As can be seen in Figure 10, the energy flow for the fission-generated-ion driver concept depends upon two independent power systems; an inertial confinement fusion reactor; and a fission reactor. The fission reactor, like the fusion-generated-ion drive, can recirculate waste heat to the energy conversion system.

Feedback Fusion

$$\eta_c \{f_n + 1 - f_{cp} \eta_a \eta_f\} \; E_{fusion}$$

$f_n \; E_{fusion}$

Energy Conversion → Load

$f_{cp} \; (1 - 1 - f_{cp} \eta_a \eta_f) \; E_{fusion}$

Fusion Energy → $f_{cp} \; E_{fusion}$ → Laser

$f_{cp} \; \eta_{laser} \; \eta_a \; \eta_f \; E_{fusion}$

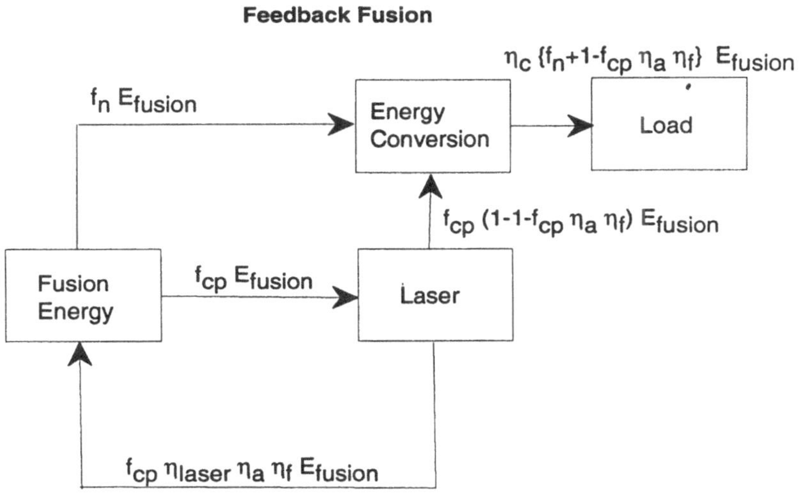

$$E_{laser} = f_{cp} \; \eta_{laser} \; \eta_a \; \eta_f \; E_{fusion}$$

$$E_{load} = \eta_c \{f_n + f_{cp} + 1 - f_{cp} \eta_a \eta_f\} \; E_{fusion}$$

$$E_{fusion} = G \; E_{laser} = G \; f_{cp} \; \eta_{laser} \; \eta_a \; \eta_f \; E_{fusion}$$

$$f_{cp} = 1/(G \; \eta_{laser} \; \eta_a \; \eta_f)$$

$$f_n = (1 - f_{cp})$$

$$\eta_{plant} = E_{load}/E_{fusion} = \eta_c \{f_n + 1 - f_{cp} \eta_a \eta_f\}$$

E_i = Energy in i; f_{cp} = fraction of fusion energy in charged particles f_n = fraction of fusion energy in neutrons; η_a = fraction of fluorescence absorbed by laser; η_f = fluorescence production efficiency; η_{laser} = laser energy/energy deposited; η_{plant} = plant thermal efficiency; η_c = energy convertor efficiency; G is pellet gain.

FIGURE 6   Energy flow diagram of the fusion-generated-ion driver. The plant power output is 1 GigaWatt electric, and the energy conversion system is 33% efficient.

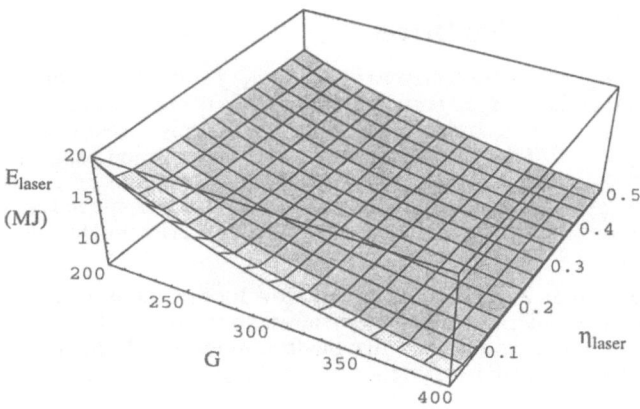

Figure 7 Laser energy for the fusion-generated-ion driver as a function of pellet gain and laser efficiency.

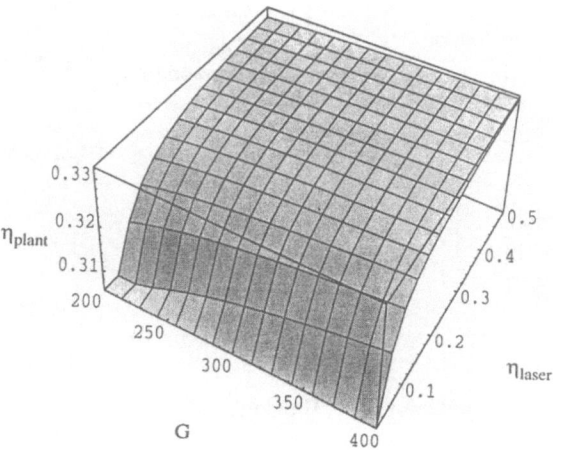

Figure 8 Power plant efficiency as a function of pellet gain and laser efficiency. The power plant is 1 Giga Watt and the energy conversion efficiency is 33%.

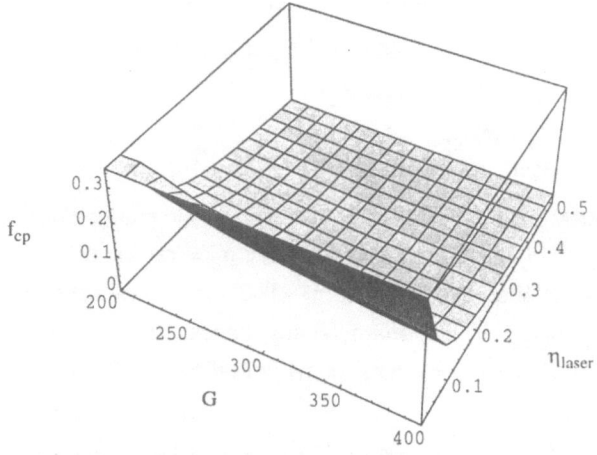

Figure 9 The required fraction of charged particle energy released in fusion as a function of pellet gain and laser efficiency. The plant power is assumed to be 1 Giga Watt electric with an energy conversion efficiency of 33%.

Fission Powered Driver

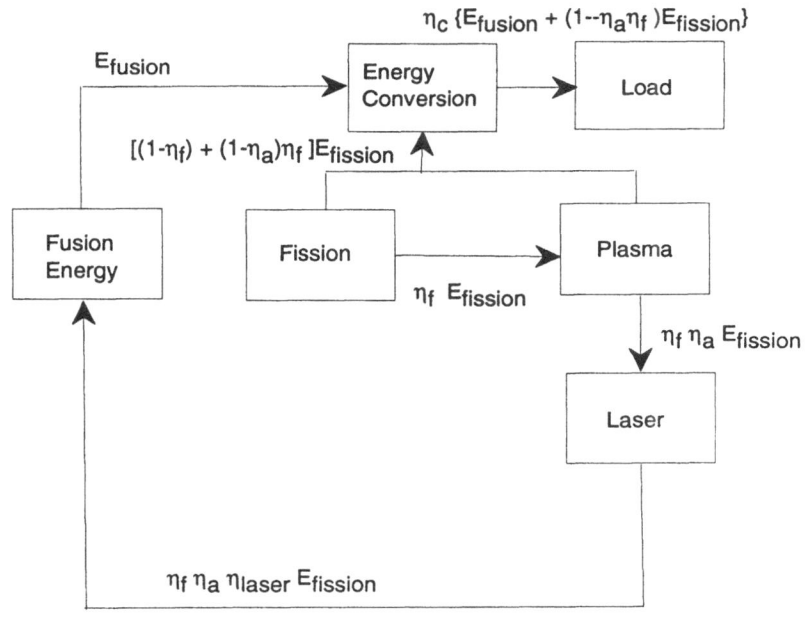

$$E_{laser} = \eta_{laser}\, \eta_a\, \eta_f\, E_{fission}$$

$$E_{load} = \eta_c\, (E_{fusion} + (1 - \eta_a\, \eta_f)\, E_{fission})$$

$$E_{fusion} = G\, E_{laser} = G\, \eta_a\, \eta_f\, \eta_{laser}\, E_{fission}$$

$$E_{fusion}/E_{fission} = G\, \eta_a\, \eta_f\, \eta_{laser}$$

$$\eta_{plant} = \eta_c\, \{(G\, \eta_a\, \eta_f\, \eta_{laser} + 1 - \eta_a\, \eta_f)/(G\, \eta_a\, \eta_f\, \eta_{laser} + 1)\}$$

E_i = Energy in i; η_a = fraction of fluorescence absorbed by laser;
η_f = fluorescence production efficiency; η_{laser} = laser
energy/energy deposited; η_{plant} = plant thermal efficiency; η_c =
energy convertor efficiency; G is pellet gain.

FIGURE 10 Energy flow in the fission-generated-ion driver concept.

In the fission-generated-ion concept, the laser energy is relatively low regardless of the laser efficiency (see Figure 11). The power plant efficiency is always near the energy conversion system efficiency (Figure 12). From Figure 11, it will be noted that the required laser power at low efficiencies is lower than the required laser power at high efficiencies. This may seem inconsistent but it should be noted that at lower laser efficiencies, a higher fraction of the plant's power output comes from fission power (Figure 13).

There are substantial advantages in using the fission-generated-ion driver over the fusion-generated-ion driver and the electrical driver. The laser energy is always substantially lower than the requirements for the fusion-generated-ion driver and the electrical driver. Furthermore, the plant efficiency is always very near the energy conversion system efficiency.

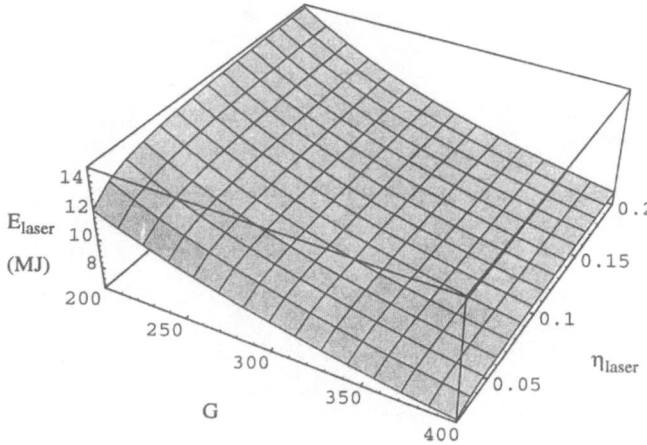

FIGURE 11 Laser energy as a function of pellet gain and laser efficiency for the fission-generated-ion system. The power from the plant is 1 Giga Watt electric and an energy conversion efficiency of 33%.

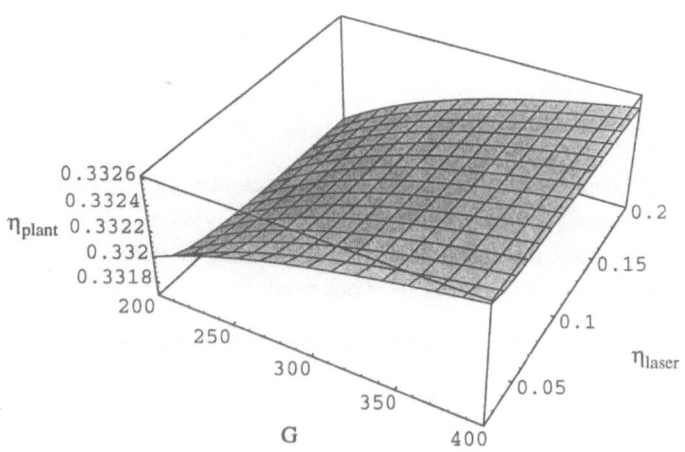

FIGURE 12 The plant efficiency as a function of pellet gain and laser efficiency for the fission-generated-ion system. The power output from the power plant is 1 Giga Watt electric and an energy conversion efficiency of 33%.

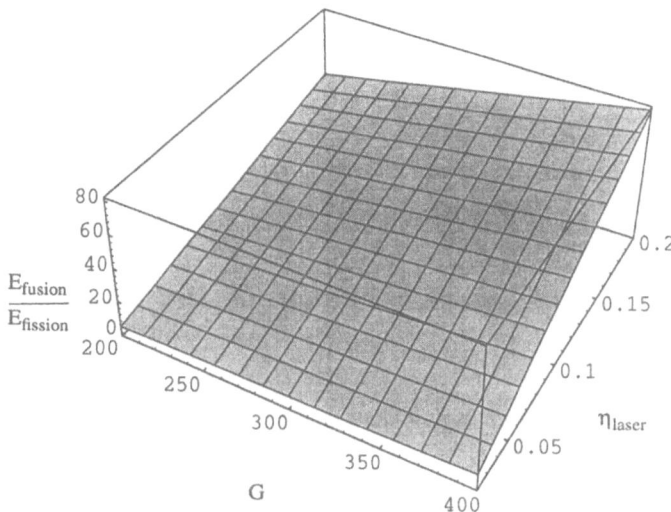

FIGURE 13 The fusion to fission energy ratio as a function of pellet gain and laser efficiency. The power from the plant is 1 Giga Watt electric and the energy conversion system efficiency is 33%.

Most intriguing, however, is that laser efficiency is not a significant problem. The major effect of low laser efficiencies is to reduce the fusion to fission energy ratio. For example, as shown in Figure 13, a laser efficiency of 1% with a pellet gain of 100 will still lead to a fusion to fission energy ratio of 2.5. This is still a very significant amount of fusion power.

In the paper by Prelas and Boody 1991, the efficiency of an excimer excited solid-state laser was discussed and shown to be between 12 to 20%. Thus, one would expect a fission-generated-ion driven concept to be capable of achieving very large fusion to fission energy ratios (see Figure 13).

Conclusions

Inertial confinement fusion systems based on electrical drivers has two significant problems which it must overcome; the driver efficiency, and the pellet gain. By looking at two relatively new concepts, a comparison between electrical drivers, fusion-generated-ion driver, and fission-generated-ion driver was made. The results of this study demonstrate some important points. First of all, the laser energy and plant efficiency is a strong function of the pellet gain and the laser efficiency for electrical drivers. Conversely, the laser energy and plant efficiency is a weak function of the pellet gain and the laser efficiency for fusion-generated-ion drivers. Perhaps the most interesting part of this study was the fission-generated-ion driver in which an independent fission reactor is used to drive a laser which in turn is used to drive an inertial confinement fusion event. In the fission-generated-ion concept, the laser efficiency had relatively little effect upon the laser energy and plant efficiency. The major effect was on the fusion to fission energy ratio.

It seems that the use of fission energy to drive a laser may be a beneficial approach to inertial confinement fusion. A low efficiency laser system could be used as a driver for an inertial confinement fusion system. Low efficiency laser systems can not be used in the electrical driver concept and the fusion-generated-ion driver concept.

Acknowledgments

The authors wish to thank their colleague Prof. George Miley of the University of Illinois for many fruitful discussions. Additionally, the author is grateful to the Department of Energy's Idaho Office for support of this work.

References

F. P. Boody, "Remote Pumping of Solid-State Lasers Pumped by Remotely-Located Nuclear-Driven Fluorescers," Ph.D. Dissertation, Nuclear Engineering Program, University of Missouri-Columbia, August 1991.

F. P. Boody, and M. A. Prelas, "Very High Average Power Solid-State Lasers Pumped by Remotely-Located Nuclear-Driven Fluorescers", *Advanced Solid-State Lasers*, Optical Society of America (To be Published 1991).

Gu G., Kunze J. F., Boody F. P., and Prelas M. A., "A UF_6 fueled Visible Nuclear-Pumped Flashlamp", *Space Nuclear Power Systems 1988*, M. S. El-Genk and M. Hoover, editors, Orbit Book Company, 153-160 (1989).

D. Pappas, Seminar on Fusion Neutrons As an Excitation source For a Xe Excimer Laser, US-Japan Seminar on Laser Fusion, Honolulu Hawaii, August, 1988.

M. A. Prelas, "A Potential Fusion Light Bulb For Energy Conversion," Poster Paper APS Meeting on Plasma Physics October 11-16, 1981, Bult. Am. Phys. Soc., 26(7), 1045 (1981); APS Press Release picked up by AP and story reported by various newspapers worldwide (See for example Washington Missourian, Washington, Missouri, Wednesday. January 6, 1982, page 7 or Inside R & D Vol. 10, Number 41, October 14, 1981).

M. A. Prelas, J. B. Romero, and E. F. Pearson, "A Critical Review of Fusion Systems For Radiolytic Conversion of inorganics to Gaseous Fuels," Nuclear Technology/Fusion, Vol. 2, 143-164 (1982)

M. A. Prelas, F. P. Boody, J. F. Kunze, and G. H. Miley, "Nuclear-Driven Flashlamps", Lasers and Particle Beams, 6(1), 25,(1988).

M. A. Prelas, E. J. Charlson, "Synergism in Inertial Confinement Fusion: A Total Direct Energy Conversion Package", *Lasers and Particle Beams*, Vol. 7 (3), 449-466 (Aug 1989).

M. A. Prelas, and F. P. Boody, "Nuclear-Driven Solid-State Lasers for Inertial Confinement Fusion", *Laser Interactions & Related Plasma Phenomena*, **Vol 9**, Plenum Press, New York, 197-210 (1991).

M. A. Prelas, E. J. Charlson, F. P. Boody, and G. H. Miley, "Advanced Nuclear Energy Conversion Using a Two Step Photon Intermediate Technique", *Prog. In Nuclear Energy*, **23** (3), pp. 223-240 (1990).

D. E. Wessol, M. A. Prelas, B. J. Merrill, and T. Speziale, "Feasibility Study of a Nuclear Driven $O_2(^1\Delta)$ Generator to Power an 18 MW Average Power Iodine Laser for Inertial Confinement Fusion," *Laser Interactions and Related Plasma Phenomena*, **Vol. 8**, Plenum Press, New York (1988).

COMPUTATION OF THE COMPRESSION OF FREQUENCY MODULATED PULSES IN LINEAR DISPERSIVE MEDIA USING A TIME-DOMAIN METHOD: EXAMPLES AND GUIDELINES*

G. E. Sieger, D. J. Mayhall, and J. H. Yee

Lawrence Livermore National Laboratory
University of California
Livermore, CA

INTRODUCTION

For the case of radio and microwave pulses propagating in the earth's ionosphere, which motivated this work and is the source of our examples, the dispersion process is nearly time-reversible, since collisional losses are very small. As a result, a long transmitted pulse of suitable amplitude and frequency modulation can be received as a compressed pulse of much higher amplitude at a chosen location in the ionosphere. The compression factor will be limited by source bandwidths and modulation control, distortion of broadband signals by antennas, and uncertainties in the propagating medium. In this paper, we present examples of compressed pulses for a variety of source waveforms. Our computer similations were performed by convolving the source signal with the dispersed waveform of a triangular element. We have obained a time-domain expansion, in which terms due to moments of the electron density, an earth's axial magnetic field, and refraction of oblique trajectories, can be identified.

Previous reports by others[1-4] on the compression of chirped pulses in linear dispersive media have mostly considered analytic properties or solutions for special cases. Felsen[1] gives an approximate description of dispersion in terms of time varying instantaneous frequency and amplitude, which cannot be used to accurately compute the amplitude or shape of compressed pulses. Felsen also discusses a variety of analytical methods, but he does not provide any examples of computed waveforms, presumably because the work predates the widespread use of computers.

El-Khamy and McIntosh[2] derive the optimum transmitted pulse envelope for a band-limited source with linearly swept frequency. They assume the phase shift in the medium to be a quadratic function of frequency. This form of dispersion could represent, for example, a second order Taylor expansion of a more general dispersion relation, which would be adequate for a sufficiently narrow bandwidth. In a further paper, El-Khamy[3] shows examples of chirping band-limited pulses both for compression and for maximum spreading, as a method of binary

* Work performed under the auspices of the U.S. Department of Energy by the Lawrence Livermore National Laboratory under contract number W-7405-ENG-48.

Laser Interaction and Related Plasma Phenomena, Vol. 10
Edited by G.H. Miley and H. Hora, Plenum Press, New York, 1992

79

signaling. He points out that this method of signaling may be suitable for security systems, since binary detection by a noncoherent detector would be possible only at a specified receiver location. For such signals, El-Khamy and Shaaban[4] calculate the performance (error rate vs signal-to-noise ratio) of noncoherent and partially coherent binary detectors.

TIME DOMAIN EXPRESSION FOR PULSE DISPERSION

Our computational method is based on the existence of an asymptotic expression for the dispersion a plane-wave pulse in a linear medium.[5,6] If a pulse has spread to many times its initial length after traveling a distance R through a medium with a real dielectric constant, $n(\omega,r)$, which varies slowly along the direction of propagation, then the dispersed pulse can be approximated as the following[7]

$$E(R,t) \approx 2\sqrt{2\pi}\ \mathrm{Re}\left\{\frac{F(\omega_s)}{\left[\phi''(\omega_s)\right]^{1/2}}\exp\left[\phi(\omega_s)\right]\right\} \tag{1}$$

where the phase is

$$\phi(\omega) = i\omega\left[\int_0^R n(\omega,r)\frac{dr}{c} - t\right] \tag{2}$$

and

$$F(\omega) = \frac{1}{2\pi}\int_0^\infty E(0,t)\,e^{i\omega t}\,dt \tag{3}$$

Also, $\phi''(\omega)$ is the second derivative of ϕ with respect to ω, and the stationary-phase frequency, ω_s, is given by

$$\phi'(\omega_s) = 0 \tag{4}$$

The results presented in this paper are for the case of a collisionless electron plasma with a static magnetic field parallel to the direction of propagation. The refractive index is[9]

$$n = \left[1 - \frac{\omega_p^2}{\omega^2}\left(1 \pm \frac{\omega_c}{\omega}\right)\right]^{1/2} \tag{5}$$

where $\omega_c = eB_0/mc$ is the electron-cyclotron frequency, and the plus (minus) sign refers to the ordinary (extraordinary) wave. The plasma frequency, ω_p, is given by

$$\omega_p^2 = \frac{4\pi e^2}{m} N_e \tag{6}$$

where e, m, and N_e are the electron charge, mass, and number density, respectively. N_e and B_0 can be slowly varying functions of r.

Substituting Eq. (6) into Eq. (2) and expanding in powers of $1/\omega$ gives

$$\phi(\omega) = i\omega\left\{\frac{R}{c}\left[1 - \frac{1}{2}\frac{\langle\omega_p^2\rangle}{\omega^2} \pm \frac{1}{2}\frac{\langle\omega_p^2\omega_c\rangle}{\omega^3} - \frac{1}{8}(1+A)\frac{\langle\omega_p^4\rangle}{\omega^4} + \cdots\right] - t\right\} \tag{7}$$

where <> indicates an average over the propagation path. An extra factor A has been included in the $1/\omega^4$ term to treat approximately the case of an oblique trajectory (propagation at an angle to the direction in which the medium varies). The phase advance of a refracted ray, obtained from an application of Fermat's principle, is derived in the appendix.

Also, from Eq. (7)

$$\phi''(\omega) = -i \frac{R}{c} \left[\frac{\langle \omega_p^2 \rangle}{\omega^2} \pm 3 \frac{\langle \omega_p^2 \omega_c \rangle}{\omega^3} + \frac{3}{2}(1+A) \frac{\langle \omega_p^4 \rangle}{\omega^4} + \cdots \right]$$ (8)

By using Eq. (7) in Eq. (4), the stationary-phase frequency can be related to the delayed time:

$$\tau = t - \frac{R}{c} = \frac{R}{c} \left[\frac{1}{2} \frac{\langle \omega_p^2 \rangle}{\omega_s^2} \pm \frac{\langle \omega_p^2 \omega_c \rangle}{\omega_s^3} + \frac{3}{8}(1+A) \frac{\langle \omega_p^4 \rangle}{\omega_s^4} + \cdots \right]$$ (9)

Formally, $\omega_s(\tau)$, defined implicitly by Eq. (9), can be used to express $\phi(\omega_s)$ and $\phi''(\omega_s)$ in Eq. (1) as functions of delayed time. To carry out this procedure explicitly, it is convenient to define the following dimensionless parameters:

$$\delta = \langle \omega_p^2 \rangle / \omega_s^2$$
$$\mu = (c\tau/R)^{1/2}$$ (10)

The required steps are somewhat laborious, but straightforward. First, from Eq. (9), δ can be obtained as an expansion in powers of μ:

$$\delta = 2\mu^2 \mp 4\sqrt{2}\, a\,\mu^3 + \left(24a^2 - 3b\right)\mu^4 + \cdots$$ (11)

where

$$a = \langle \omega_p^2 \omega_c \rangle / \langle \omega_p^2 \rangle^{3/2}$$
$$b = \left[\langle \omega_p^4 \rangle / \langle \omega_p^2 \rangle^2 + A \right]$$ (12)

Expanding in powers of δ or μ is useful, as noted in the Introduction, only for the case of a weakly dispersive medium, for which

$$\mu \cong \frac{1}{\sqrt{2}} \frac{\langle \omega_p^2 \rangle^{1/2}}{\omega_s} \ll 1$$ (13)

Next, Eq. (11) can be substituted into Eqs. (7) and (8) to obtain expansions in powers of μ:

$$\phi(\omega_s) = if(\tau)$$
$$= \sqrt{2} \frac{R}{c} \langle \omega_p^2 \rangle^{1/2} \mu \left[1 \pm \frac{a}{\sqrt{2}} \mu + \left(\frac{1}{4} b - a^2 \right) \mu^2 + \cdots \right]$$ (14)

and

$$\phi''(\omega_s) = ig(\tau)$$
$$= 2\sqrt{2} \frac{R}{c} \langle \omega_p^2 \rangle^{-1/2} \mu^3 \left[1 + 3 \left(\frac{1}{4} b - a^2 \right) \mu^2 + \cdots \right]$$ (15)

One final step is required to obtain a suitable time-domain formula for computing dispersion. The initial waveform can be expressed, in piecewise-linear approximation, as a sum of triangular elements. Each triangle has a width $2\Delta t$, where Δt is the sampling interval. The transform of $E(0,t)$ becomes

$$F(\omega) = \sum_k F_k(\omega)$$
$$= \frac{\Delta t}{2\pi} \sum_k E(0, k\Delta t) \left[\frac{\sin(\omega \Delta t/2)}{\omega \Delta t/2} \right]^2 \exp\left[i\omega k \Delta t \right]$$ (16)

Since Eq. (1) is valid for short pulses, it can be used to calculate the dispersed waveform of a triangular element. The dispersion of an arbitrary pulse can then be evaluated as the sum of dispersed triangular elements, each with an appropriate time-shift and amplitude. The result is the following:

$$E(R, \tau = n\Delta t) = \left(\frac{2}{\pi}\right)^{1/2} \Delta t \sum_k E(0, k\Delta t)$$

$$\times \left[\frac{\sin(\omega_{nk}\Delta t/2)}{\omega_{nk}\Delta t/2}\right]^2 \frac{\cos[f(t_{nk}) + \pi/4]}{|g(t_{nk})|^{1/2}} \tag{17}$$

where $t_{nk} = (n-k)\Delta t$, and $\omega_{nk} = \omega_s(t_{nk})$ is found from Eqs. (10) and (11).

EXAMPLES OF COMPRESSED PULSES

For the examples presented here, with one exception, the propagation path length is 300 km, through a uniform electron density of 10^6 cm^{-3} ($\omega_p/2\pi = 8.97$ MHz). The exception is a case of a linear-ramped electron density, as described below. Conceptually, the simplest way to obtain a signal which will be compressed is to time-invert a pulse which has been dispersed in the same medium. The first example, shown in Fig. 1a, is the dispersed waveform of a Gaussian-shaped video pulse of peak amplitude 7.72 and full-width-half-maximum (FWHM) 72 ps, which has been time-inverted. The waveform in

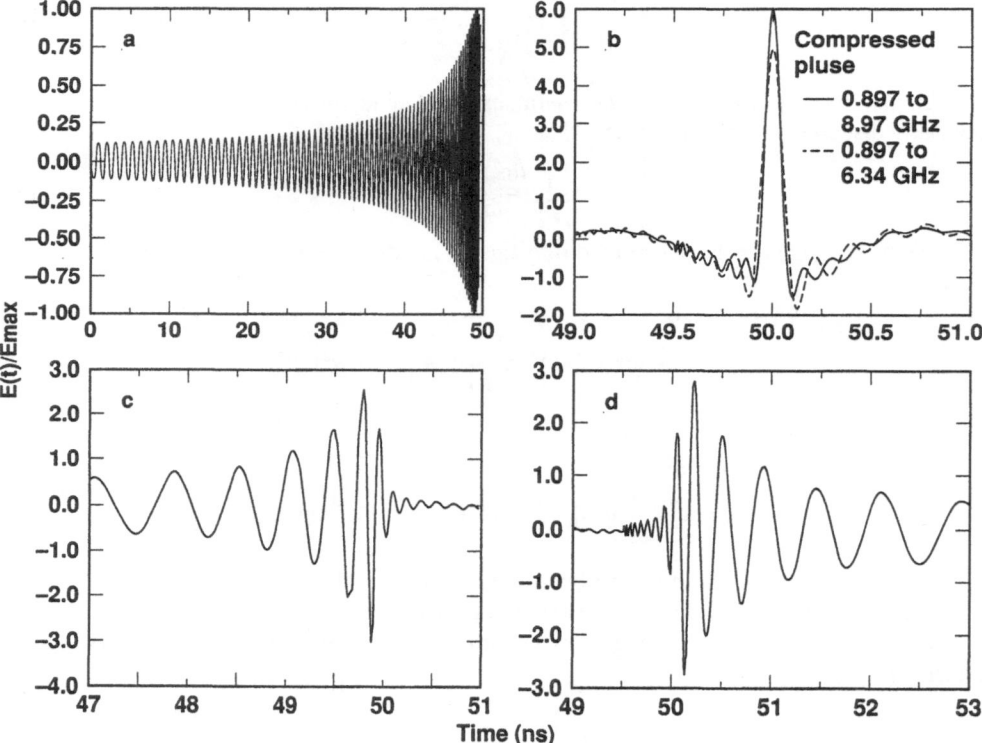

Fig. 1. Example of a chirped waveform (a), which is compressed in an electron plasma into a nearly monopolar video pulse (b). Waveforms (c) and (d) show the signal at 90 per cent and 110 per cent, respectively, of the optimum path length.

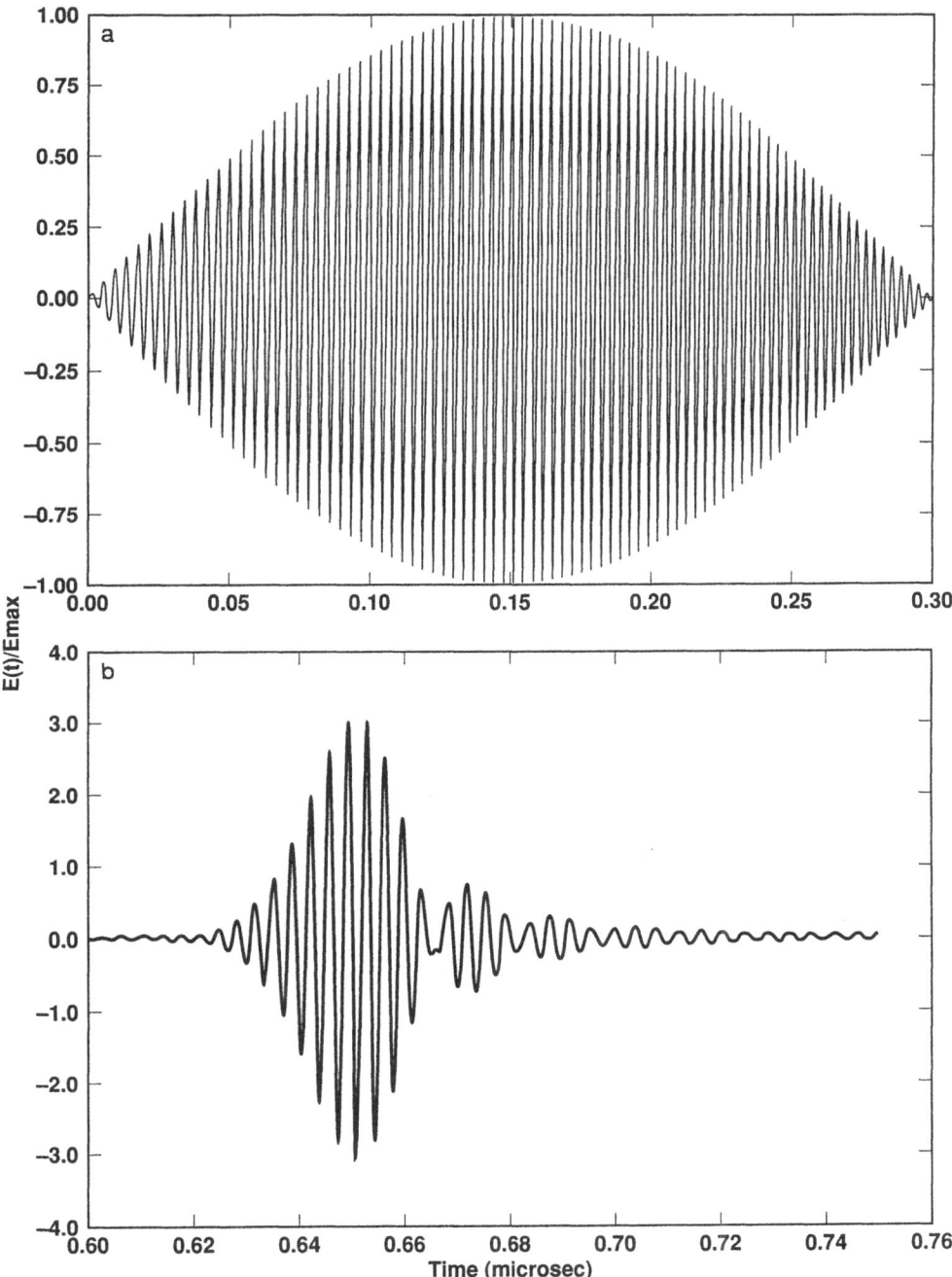

Fig. 2. Example of a linearly chirped signal with a sine envelope (a), and the resulting compressed pulse (b). The center frequency is 284.6 MHz and the frequency spread is 30 per cent.

Fig. 1a has also been truncated to include only modulation frequencies from 0.897 to 8.97 GHz. The signal arriving at the end of the propagation path is the solid curve in Fig. 1b, which has been compressed by a factor of about 40. The dashed curve in Fig. 1b is the result when an additional 0.5 ns is truncated from nose of the signal in Fig. 1a, which reduces

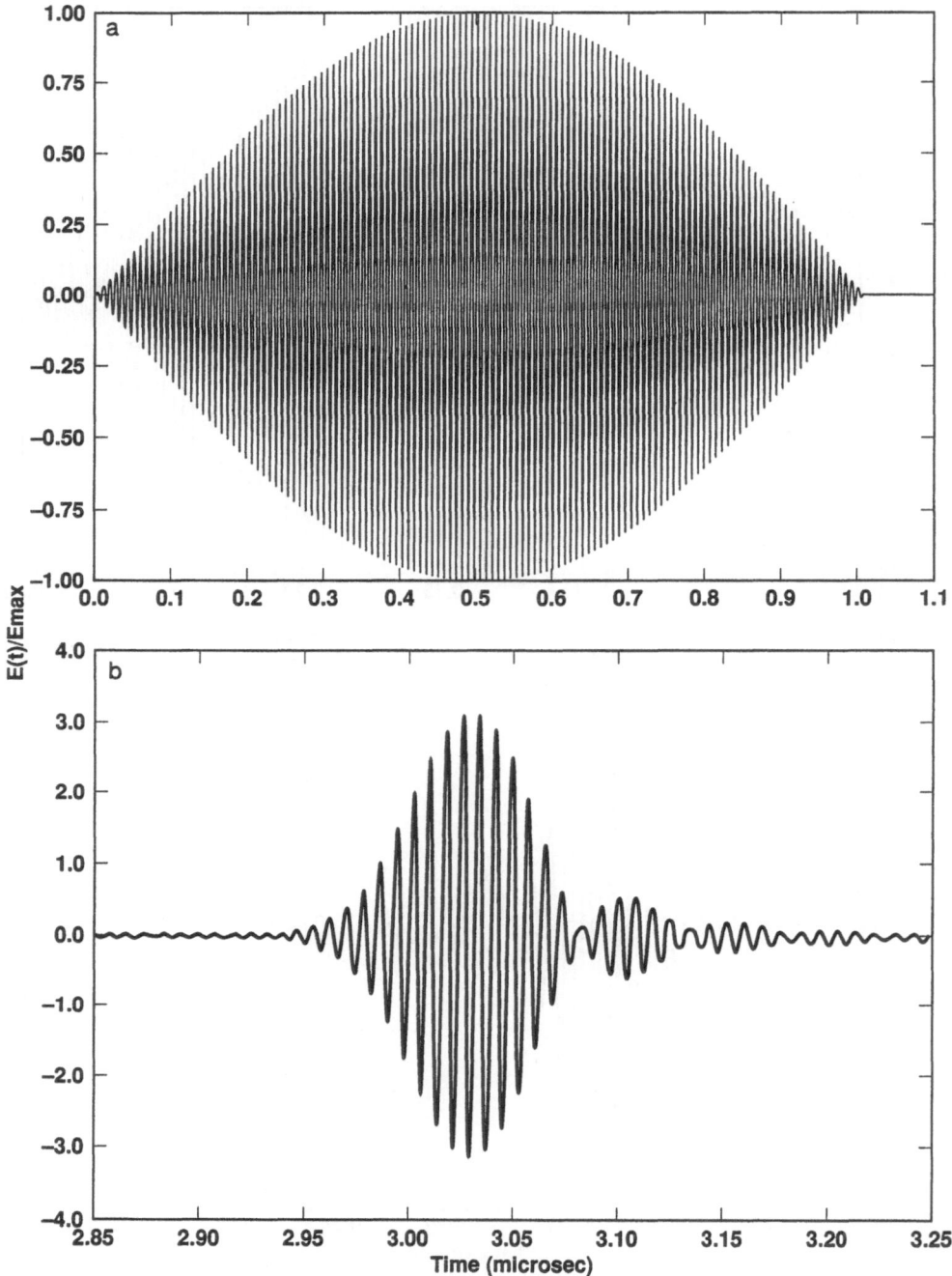

Fig. 3. Example of a linearly chirped signal with a sine envelope (a), and the resulting compressed pulse (b). The center frequency is 126.5 MHz and the frequency spread is 20 per cent.

the upper modulation frequency to 6.34 GHz. Fig. 1b shows the extent to which the amplitude of the compressed signal is reduced below the value of 7.72, and the presence of a bipolar component, due to different degrees of truncation. Figs. 1c and 1d show how the signal in 1b would appear, after traveling a distance of 90 per cent and 110 per cent of the optimum path, respectively.

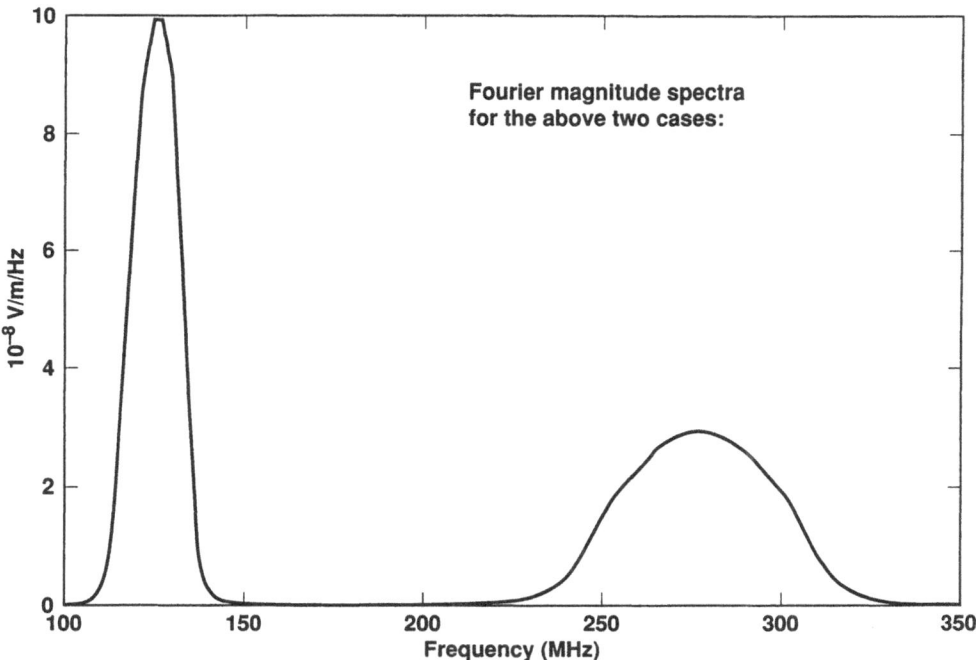

Fig. 4. Fourier magnitude spectra of the waveforms in Figs. 2 and 3.

An example of a simple waveform which is a good candidate for compression is a linearly chirped signal with a sine envelope. The signal in Fig. 2a has a center frequency of 284.6 MHz and a 30 per cent sweep in modulation frequency. These values were chosen to result in a compression factor in time of about 10, which is nearly achieved, as can be seen in Fig. 2b. The confinement of the compressed pulse is not perfect, as there is some tail structure. The amplitude gain is slightly greater than a factor of 3. Similarly, the waveform in Fig. 3a has a center frequency of 126.5 MHz and a 20 per cent sweep in modulation frequency. Again, the goal of a factor of 10 in compression in nearly achieved, as seen in Fig. 3b. The Fourier magnitude spectra of the signals in Figs. 2 and 3 are shown together in Fig. 4. For a given medium and compression ratio, one can choose between a longer, narrower-band, lower-frequency pulse or a shorter, broader-band, higher-frequency pulse.

Fig. 5 shows an example of attempting to compress a linearly chirped sine-envelope pulse by a factor of 20. The signal in Fig. 5a has the same center frequency as in Fig. 3 (126.5 MHz), but the frequency spread is increased to 28.3 per cent. Fig. 5b shows that that amplitude gain of the compressed pulse is increased to nearly a factor of 4, but more of the pulse energy is in the tail. Figs. 5c and 5d show how the signal in 5b would appear, after traveling a distance of 90 per cent and 110 per cent of the optimum path, respectively. It can be seen that a 10 per cent error in the path length results in only a slight reduction in the signal amplitude. It is interesting that although much of the energy in Figs. 5c and 5d appears in the tails, there are no significant precursors to the main pulses.

Our next example shows the effect of an axial earth's magnetic field. The signal in Fig. 6a is chirped from 80 MHz up to 120 MHz over a period of about 4 μs. After propagating in the absence of an axial magnetic field, this signal is compressed into the center pulse in Fig. 6b (labelled A). Waveforms B and C in Fig. 6b are the extraordinary wave and the ordinary wave, respectively, for the case of an axial field of 0.32 gauss ($\omega_c/\omega_p = 0.1$). The pulses B and C are circularly polarized, and it would be very difficult to predict their

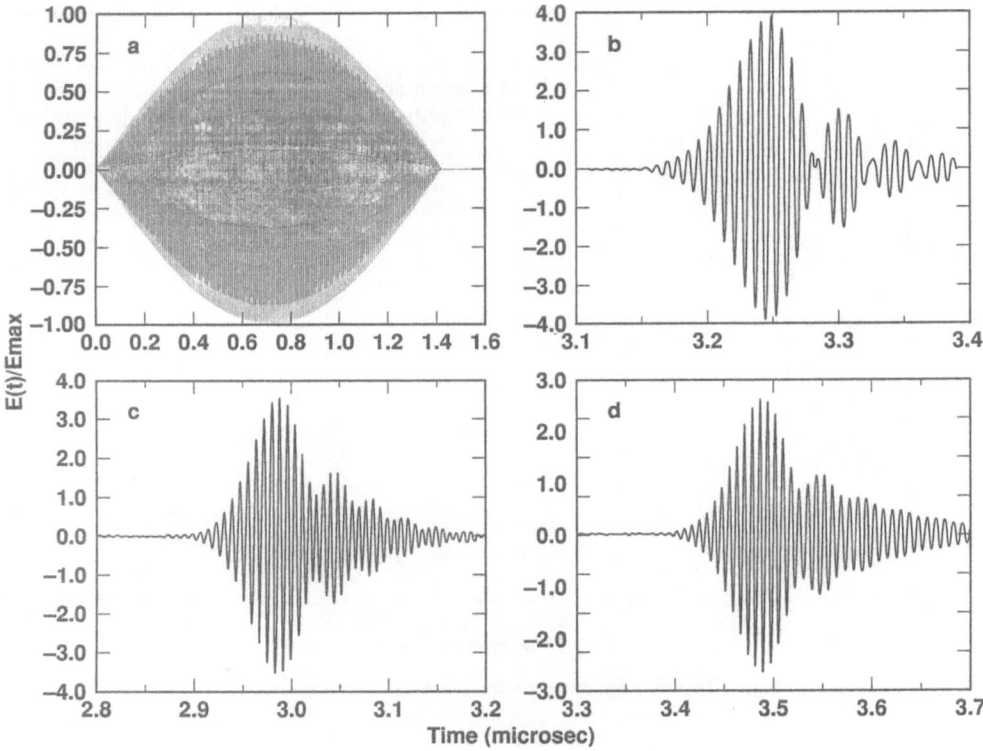

Fig. 5. Example of a linearly chirped signal with a sine envelope (a), and the resulting compressed pulse (b). The center frequency is 126.5 MHz and the frequency spread is 28.3 per cent. Waveforms (c) and (d) show the signal at 90 percent and 110 per cent, respectively, of the optimum path length.

relative orientation at any particular location. A low modulation frequency was chosen for this example to dramatize the effect of the axial field. In order to reduce this effect, a higher range of modulation frequencies should be chosen for pulse compression.

Fig. 7 is an example of the combined effects of transmitting from a focusing antenna and compression in the dispersive medium. The source waveform is again as in Fig. 1a. The focused and compressed signals at three angular locations near the central lobe are shown in Fig. 7. For this calculation, the FWHM of the central lobe is 1 degree for a 5 GHz harmonic wave. The shape of the signal changes with location angle, since the antenna pattern is frequency dependent. Note that the high-frequency structure appearing at the time near −0.5 ns in Fig. 7 is not due to numerical inaccuracy. This structure is due to the time-windowing (truncation) of the source waveform (Fig. 1a).

Fig. 8 is an example of oblique propagation through a plasma in which the electron density is proportional to the altitiude, z, increasing to 10^6 electrons/cm^3 at at a height of 300 km. The endpoints of the path are at $(x,z) = (0,0)$ and $(520,300)$ in km. Along this path $<\omega_p^4>/<\omega_p^2>^2 = 4/3$. The phase advance due to refraction, represented by the factor A in Eqs. (7) − (9), and (12), is derived in the appendix. For this case A = 0.75. Waveform A in Fig. 8 is the optimal compression of a 10 μs transmitted pulse with a modulation frequency range of 60 to 200 MHz, in which the electron density along the path is a uniform 5×10^5 cm^{-3}. Waveform B in Fig. 8, which has propagated through the ramped electron density, is reduced in amplitude by about a factor of 2. Of course, the transmitted signal could be optimized for any particular path through a non-uniform medium, but such detailed information may not be available.

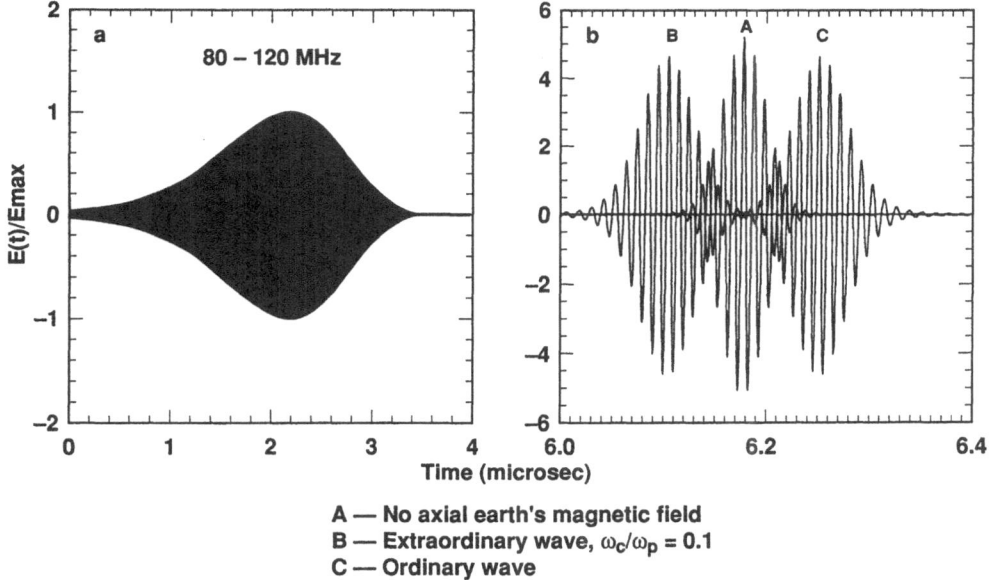

A — No axial earth's magnetic field
B — Extraordinary wave, $\omega_c/\omega_p = 0.1$
C — Ordinary wave

Fig. 6. Example of a waveform chirped from 80 MHz up to 120 MHz (a), and the resulting compressed pulses (b). In (b), signal A is in the absence of an axial magnetic field, and B and C are the extraordinary wave and the ordinary wave for the case $\omega_c/\omega_p = 0.1$.

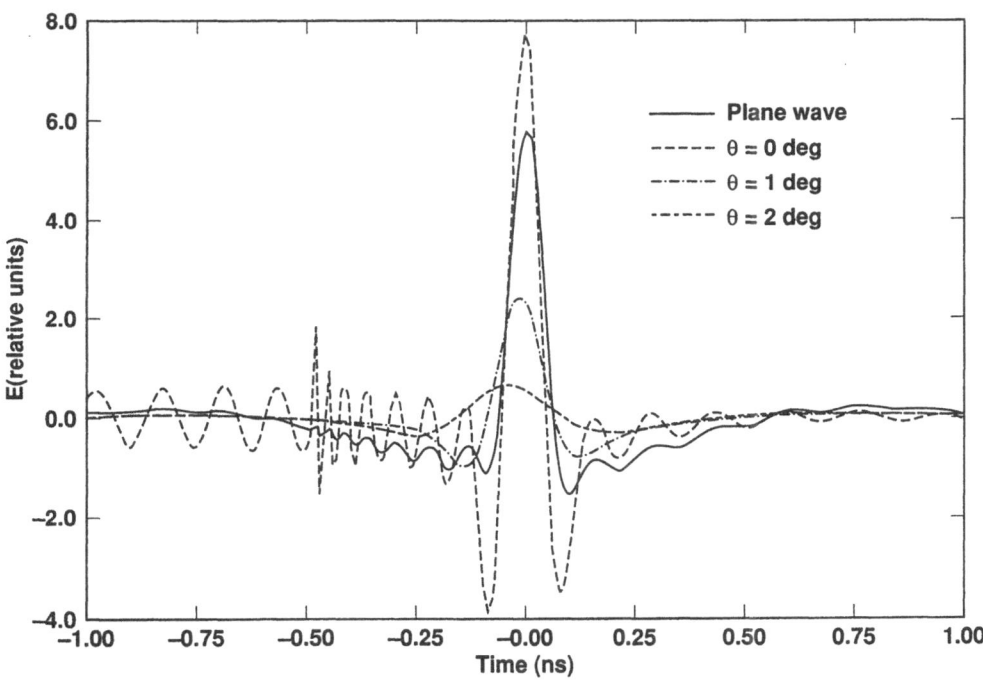

Fig. 7. Distortion of the compressed signal in Fig. 1 near the central lobe of a focusing antenna.

SUMMARY

We have investigated computationally the extent to which frequency-modulated pulses can be compressed in the earth's ionosphere. A time-domain expansion is presented, which can be convolved with a source waveform to calculate dispersion in a collisionless electron plasma. This expansion is useful for modulation frequencies sufficiently greater than both the electron plasma frequency and the electron cyclotron frequency. In this expansion, terms appear which represent moments of the electron density (or, equivalently, the square of the electron plasma frequency), birefringence due to an axial magnetic field, and the effect of refraction of oblique trajectories, when the plasma has a non-uniform density profile along one direction.

A correctly modulated pulse with a bandwidth of one decade can be compressed by a factor of about 40, with accurate knowledge of the dispersive medium. A linearly-chirped sine-envelope pulse can be compressed by a factor of about 15, even with a 10 per cent uncertainty in the propagation path length. Examples of birefringence due to an axial magnetic field and distortion due to refraction in a non-uniform plasma have been presented. These effects must be taken into account for modulation frequencies which are not much greater than the electron plasma frequency or the electron cyclotron frequency.

REFERENCES

1. L. B. Felsen, "Asymptotic Theory of Pulse Compression in Dispersive Media," *IEEE Trans. Antennas Propagat.* AP-19, 424 (1971).
2. S. E. El-Khamy and R. E. McIntosh, "Optimum transionospheric pulse transmission," *IEEE Trans. Antennas Propagat.* AP-21, 269 (1973).
3. S. E. El-Khamy, "Matched swept frequency digital modulation for binary signaling in inhomogeneous dispersive media," *IEEE Trans. Antennas Propagat.* AP-28, 29 (1980).
4. S. E. El-Khamy and S. E. Shaaban, "Matched Chirp Modulation: Detection and Performance in Dispersive Communication Channels," *IEEE Trans. Commun.* 36, 506 (1988).
5. G. C. Sherman and K. E. Oughstun, "Description of pulse dynamics in Lorentz media in terms of the energy velocity and attenuation of time harmonic waves," *Phys. Rev. Lett.* 47, 1451 (1981).
6. R. J. Vidmar, F. W. Crawford, and K. J. Harker, "Delta function excitation of waves in the earth's ionosphere," *Radio Sci.* 18, 1337 (1983).
7. H. Jeffreys and B. Jeffreys, *Methods of Mathematical Physics*, 3rd ed., Cambridge University Press, New York, N.Y., sections 17.04 and 17.05 (1956).
8. ibid., section 10.041.
9. I. Tolstoy, *Wave Propagation*, McGraw-Hill, New York, N.Y., section 6–3 (1973).

APPENDIX: REFRACTION OF OBLIQUE TRAJECTORIES FOR THE CASE OF A LINEAR-RAMP ELECTRON DENSITY

We have estimated the effects of refraction on pulse compression for the case of an electron density proportional to the altitude, z. Again assuming $\omega_p^2 \ll \omega^2$, from Eq. (5) the phase velocity is, in the absence of an axial magnetic field,

$$
\begin{aligned}
v_{ph}(\omega, z) = \frac{c}{n} &\cong c\left(1 + \frac{1}{2}\frac{\omega_p^2(z)}{\omega^2}\right) \\
&= c\left(1 + \frac{z}{z_0}\right)
\end{aligned}
\tag{A1}
$$

where $z_0 = 2\omega^2 z / \omega_p^2(z)$. Using Fermat's principle for a sinusoidal wave, it has been shown that each ray is a circular arc.[8] In particular, the ray which passes through the two points $(x,z) = (0,0)$ and (x_1, z_1), satisfies

$$
\left(x - z_0 \tan\theta\right)^2 + (z + z_0)^2 = z_0^2 \sec^2\theta
\tag{A2}
$$

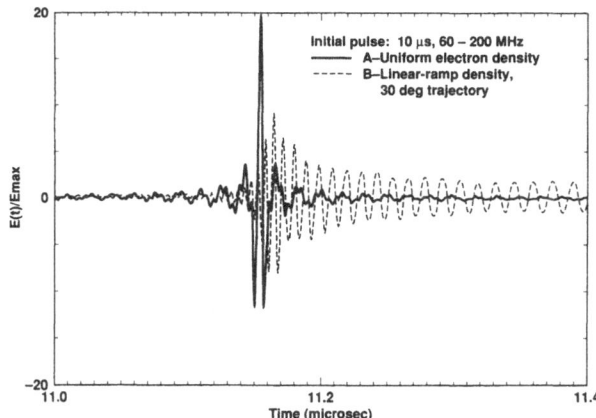

Fig. 8. Compression of a 10 μs pulse chirped from 60 MHz up to 200 MHz. Signal A is for the case of a uniform electron density, and signal B is for the case of an oblique trajectory through a linearly ramped electron density. The integrated electron density is the same for the two cases.

where θ, the initial angle of the ray from horizontal, is given by

$$\tan\theta = \frac{z_1}{x_1} + \frac{x_1^2 + z_1^2}{2x_1 z_0} \tag{A3}$$

Following Ref. 8, the time required for a phase front to travel from (0,0) to (x_1, z_1) is

$$
\begin{aligned}
t_{ph} &= \int_{path} \frac{dr}{v_{ph}} \\
&= \frac{z_0}{c}\left[\tanh^{-1}\left(\frac{\tan\theta}{\sec\theta}\right) - \tanh^{-1}\left(\frac{\tan\theta - x_1/z_0}{\sec\theta}\right)\right]
\end{aligned}
\tag{A4}
$$

The time that would be required for a phase front to travel in a straight line from (0,0) to (x_1, z_1) is

$$t_{ph,0} = \frac{z_0}{c}\left(1 + \frac{z_1^2}{x_1^2}\right)^{1/2} \ln\left(1 + \frac{z_1}{z_0}\right) \tag{A5}$$

Refraction of oblique trajectories can be treated approximately by adding a correction term to the phase, as in Eq. (7). The effect is that the lower frequency components arrive slightly earlier than they would without refraction. By expanding Eqs. (A4) and (A5) in powers of $1/\omega$, we find that, to leading order in $1/\omega$, the time advance of a phase front is

$$\Delta t = t_{ph,0} - t_{ph} \cong \frac{1}{32}\frac{R}{c}\left(\frac{x_1}{z_1}\right)^2 \frac{\langle\omega_p^4\rangle}{\omega^4} \tag{A6}$$

where, for a linearly-ramped electron density,

$$\langle\omega_p^4\rangle = \frac{4}{3}\langle\omega_p^2\rangle^2 \tag{A7}$$

For this case, the correction factor A in Eq. (7) becomes

$$A \cong \frac{1}{4}\left(\frac{x_1}{z_1}\right)^2 \tag{A8}$$

89

TENDING TO SATURATED GAIN OF SOFT X-RAY LASER
AT 23.2 AND 23.6NM*

Zhang Guoping,Sheng Jiatian,Yang Minglun,Zhang Tanxin
Shao Yunfeng, Peng Huimin, He Xiantu

Institute Of Applied Physics And Computational
Mathematics

Yu Min

China Academy of Engineering Physics

Wang Shiji, Gu Yuan, Zhou Guanlin, Ni Yuanlong
Yu Songyu, Fu Sizu, Mao Chusheng, Tao Zucong

Shanghai Institute of Laser Plasma

Chen Wannian, Lin Zunqi, Fan Dianyuan

High Power Laser and Physics Joint Laboratory,Shanghai
Institute of Optics and Fine Mechanic,Academia Sinica

ABSTRACT

We proposed a design named "Four-Target Series Coupling"
to work on soft x-ray laser in Ne-like Ge plasma. The total
length for four targets is up to 5.6cm. The gain length
product (GL) for small signal is up to about 18 for both lines
at 23.2 and 23.6nm ,and the effective GL is 16.4 and 15.7 for
these two lines respectively. These two lines are obviously
tending to saturation. The divergence of X-ray laser beam is
about $3 \sim 4$ mrad. Before experiments, we developed a series of
codes to simulate these experiments, got the parameters about
the distance and angle between the targets in order to get
best coupling.

Laser Interaction and Related Plasma Phenomena, Vol. 10
Edited by G.H. Miley and H. Hora, Plenum Press, New York, 1992

91

Recently,the aim of the X-ray laser research is focused on realizing amplification of spontaneous emission (ASE) at the "Water Window wavelength (2.3-4.4nm)" and getting saturated ASE output. In order to decrease the loss of laser line intensity by refractive deviation of the beam in plasma, we have proposed and carried out a novel design named "Double-Target Opposing Coupling" early in 1990. We have obtained high gain soft X-ray laser output in Ne-like Ge plasma[1], only using a one-terawatt laser facility named "Shen Guang". In this round of experiments for saturated gain, we developed the above idea with the design named "Multiple - Targert Series Coupling". We have got a tending saturated output of soft X-ray laser from Neon-like Ge plasma.

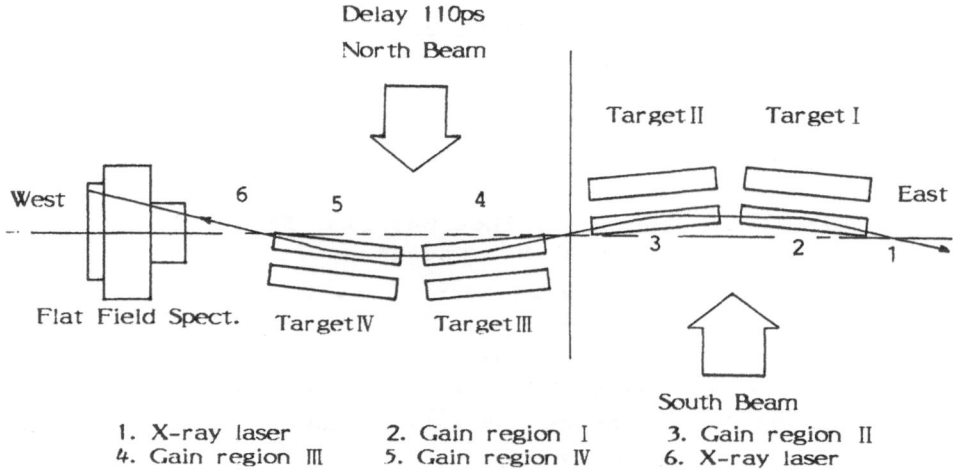

Fig.1. Principle of Four Targets Coupling

The principle of "Four-Target Series Coupling" is shown in Fig.1. From targets I to II or from targets III to IV ,the gradients of electronic density are almost the same. However, from target II to III they are opposite. For the first case, we call it "Following Coupling". For the second case, we call it "Opposing Coupling". For each single target, it is 0.2cm thick, 0.6cm wide and 1.4cm long or less, and the total length of four targets amounts to 5.6cm or less. In order to reduce the loss by the refractive deviation of the beam and obtain best effect for the series coupling, we design special holder for target. The distance and angle between targets can be accurately adjusted and measured, so that the match of plasma gain region for each target could be ensured.

These experiments were carried out on the "Shen Guang" laser facility at High Power and Physics Joint Laboratory. For each beam, the laser line focus is 3.06cm long and 120 μm wide.The intensity on the line focus is about $(0.8-1.0) \times 10^{13}$ w/cm^2 and the 1.05μm laser pulse duration is about 1 ns.

A flat field grazing incidence extreme ultraviolet (XUV) spectrometer with a 1200 lines/mm grating recorded the axial spectrum[2]. Time-averaged spectra with angular resolution are obtained using Kodak 101 XUV film in the focal plane. Variable attenuation is obtained with aluminium filters.

We developed a series of codes to simulate this experiments. We worked out one-dimensional and non-equilibrium radiation hydrodynamic JB19 code[3]. The physical processes, which are similar to be considered by LASNEX code[4], except the dielectronic processes and the collective absorption via resonance absorption, are also involved in JB19 code. For line focus experiments, we developed the pseudo-two-dimensional idea[5], proposed two-direction pseudo-two-dimension idea. The simulation of plasma condition is performed in one-dimensional column geometry, using a radius of curvature which is equal to twice of laser line width. In this way we simulate the two-dimensional effect for both exploding foil and slab target. The laser-produced plasma conditions, which are simulated by JB19 code, are fed into the ALPHA code[6] to get the gain and the spontaneous emissiom rate for Ne-like 3P-3S lasers.These conditions are fed into the XBY code[7], which simulates the soft x-ray laser line propagation and amplification in lasing medium, to get the intensity of laser lines on the film of photography with time resolution and angular resolution. Before these experiments, we adjusted some data in JB19 and ALPHA codes to match the results of the "Double-Target Opposing Coupling" expriments and got the information about the distance and angle between the targets for this round of experiments. In experiments we adjusted these parameters to get best effect from the series coupling. In fact the ALPHA code can correctly simulate the gain only for laser line at 23.2 nm.

According to the simulation, we suggested x_{ij} and θ_i values before this round of experiments. The x_{ij} is the distance between the target i and j in the normal direction. The θ_i is the angle between target i and focus line. For the "Two-Target Following Coupling" we suggested : θ_1, θ_2 is 4 ~ 5 mrad, $|x_{12}|$ is less than 5 μm. When $\theta_{1.2}$ is less than 2 mrad and more than 6 mrad , or $|x_{12}|$ is more than 20μm , the output of laser line will decrease obviously. For "Three-Target series Coupling",we suggested : θ_3 is also 4 ~ 5 mrad, and x_{23} is about 250 μm. When θ_3 changes 2 mrad or x_{23} changes ± 15μm , the output of laser line will decrease obviously. For the fourth target, θ_4 is also about 4 ~ 5

Tab. I   The peak of laser line intensity I (photons/μm)
at 19.6nm in the "Two-Target Series Coupling"

Exp. No.	Target data	I
083001	$L_{1+2}=28$ mm, $\theta_{1,2} \sim 4.8$mrad	7.00×10^5
083102	$L_{1+2}=28$ mm, $\theta_{1,2} \sim 6.0$mrad	2.90×10^5
090201	$L_{1+2}=28$ mm, $\theta_{1,2} \sim 1.9$mrad	3.08×10^5
090401	$L_1 = 28$ mm, $\theta_1 \sim 0$mrad *	1.61×10^5

* single target

Tab. II   The peak of laser line intensity I (photons/μm)
in the "Three-Target Series Coupling"

Exp. No.		092001	091801	091902	092401
Target data	θ_1	~ 4.5	~ 4.5	~ 4.5	~ 4.5
	θ_2	~ 4.5	~ 4.5	~ 4.5	~ 4.5
	θ_3	~ 4.5	~ 4.5	~ 4.5	~ 7
	x_{23} (μm)	~ 250	~ 275	~ 220	~ 235
λ (nm)	19.6	899×10^3	888×10^3	710×10^3	436×10^3
	23.2 I	227×10^5	105×10^5	133×10^5	149×10^5
	23.6	214×10^5	116×10^5	123×10^5	128×10^5

mrad and $|x_{34}|$ is less than 5 μm. When θ_4 changes 2 mrad or x_{34} changes $\pm 10\mu$m , the output of laser line will decrease obviously. Another method to choose $\theta_{3,4}$ is that the laser beam from the second or the third target incidence along the surface of the third or the fourth target with the angle about $-4 \sim -5$ mrad. In the following coupling, some rays are got in high amplification in the first one or three targets, and so are they in the next one. At the same time, other rays are got in low amplification in the first one or three targets, and so are they in the next one. In these cases, the divergence of laser beams decreases very quickly as the plasma length increasing. In the opposing coupling, it is different, some rays are got in high amplification in the first two targets, but they got in low amplification in the third one. At the same time, other rays are got in low amplification in the first two targets, but they are got in low amplification in the third one. In this case, the divergence of laser beams decreases very slowly.

At first, we worked on "Two-Target Following Coupling". We adjusted the $\theta_{1,2}$. The results from these experiments are given in Tab.I. When $\theta_{1,2}$ is about 4.5 mrad , the best coupling is achieved. Secondly, we worked on "Three-Target Series Coupling". The $\theta_{1,2}$ is about 4.5 mrad . We adjusted θ_3 and x_{23}. The results of it are given in Tab.II . When θ_3 is about 4.5 mrad and x_{23} is about 250 μm , the best coupling is observed. At last, we work on "Four-Target Series Coupling". We adjusted $\theta_{3,4}$ and x_{23} . The $\theta_{1,2}$ is also about 4.5 mrad. The results for it are put in the Tab.III . When θ_3 is about 4.5 mrad and θ_4 is about 0 mrad and x_{23} is about 250 μm, the best coupling is observed. The results of experiments show that the simulation is probably good.

Tab.III The peak of laser line intensity I (photons/μm)
in the "Four-Target Series Coupling"

	Exp. No.	092102	092402	092701
Target	θ_1 (mrad)	∼ 4.5	∼ 4.5	∼ 4.5
	θ_2 (mrad)	∼ 4.5	∼ 4.5	∼ 4.5
	θ_3 (mrad)	∼ 4.5	∼ 7	∼ 4.5
data	θ_4 (mrad)	∼ 0	∼ 4.5	∼ 2
	x_{23} (μm)	∼ 250	∼ 235	∼ 250
λ	19.6	2.26×10^7	7.95×10^6	1.52×10^7
(nm)	23.2 I	2.02×10^8	1.18×10^8	1.99×10^8
	23.6	1.48×10^8	7.96×10^7	1.36×10^8

The intensity data of these three laser lines at wavelength 19.6, 23.2 and 23.6 nm from the best coupling are plotted in Fig.2 against the plasma length. For two laser lines at 23.2 and 23.6 nm, the data in Fig.2 show a clear exponential increasing till plasma length reaches about 4.8cm. In this region, the gain(G) is 3.27 and 3.14 cm^{-1} respectively. With plasma length at 4.8cm, the GL reaches 15.7 and 15.1 respectively. Above 4.8 till 5.6cm, the intensity of these two lines increases continuously. The maximum of peak intensity for these two lines on the film of photography without absorbtion by aluminium filters reaches 2×10^8 and 1.5×10^8 photons/μm , respectively. However , their gains decreased obviously. The fourth target is only 1.4cm long or less, the intensity of laser line at 19.6 nm is still exponential increasing, and we adjusted $\theta_{3.4}$ and x_{23}, however, the intensity of these two laser lines is still increasing slowly , from above three reasons we think that the decrease of the gain is not due to the laser beams beeing refracted out of plasma gain region, so we are sure that the laser lines are tending to saturation. We believe this is the first unequivocal demonstration tending to saturation of an XUV laser. With 5.6cm length, the GL of small signal for both lines reaches about 18, the effective GL is 16.4 and 15.7, respectively.

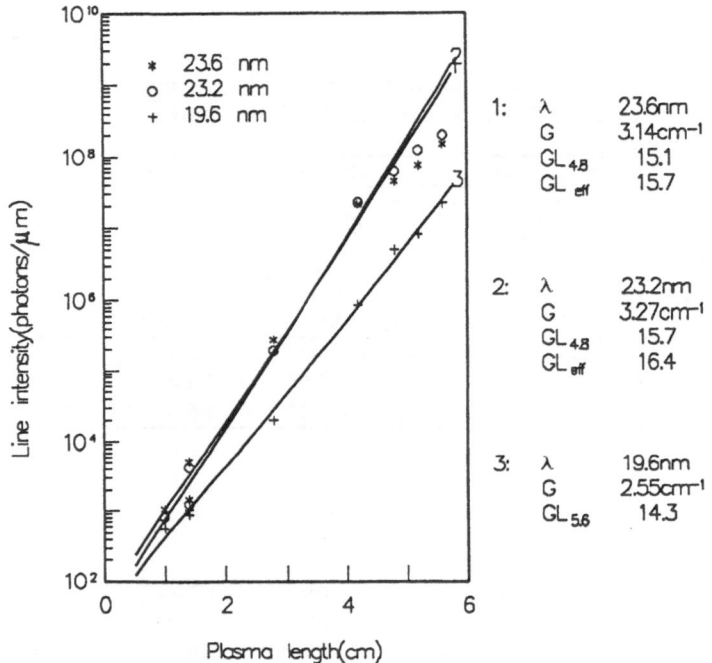

Fig. 2. Soft X-ray Laser Line Intensity Scale the Plasma Length

Differing from these two lines, the intensity of laser line at 19.6nm shows clear exponential increasing till the plasma length reaches 5.6cm. Its G is 2.55 cm^{-1}, maximum GL is 14.3, the maximum peak intensity is about 2×10^7 photons/μm.

According to these experiments the divergence of laser beam for these three lines obviously decreses as the plasma length(L) increasing. When L is 5.6cm the divergence for these three laser lines is about 3 ~ 4 mrad in the plane constructed by the normal of target and focus line . The main results are given in the Tab.Ⅳ . The data in Tab.Ⅳ show that the divergence of laser beam decreaces very quickly in the following coupling and it increases very slowly in the opposing coupling.

Tab.Ⅳ The divergence D(mrad) of laser line
scale the target length L(cm)

Exp. No.		91301	91201	92001	92002	92101	92102	92701
L(cm)		1.4	2.8	4.2	4.8	5.2	5.6	5.6
19.6		12.3	–	10.9	5.8	4.4	3.9	3.0
λ 23.2 D		13.5	9.3	8.5	8.1	5.5	3.7	2.9
23.6		15.1	10.0	9.4	8.3	5.7	3.7	2.7

A great deal of results from these experiments will improve our understanding for Ne-like Ge laser mechanism. We can use the following and opposing coupling to increase ASE intensity of laser lines from slab targets. We are going to realize saturated gain for these two lines using an XUV mirror in next round of experiments.

Acknowledgement

The authors would like to thank prof. Deng Ximing and prof. Du xiangwan.

Refrences

[1]Wang Shiji et al., Science in china, series A (1991)151 (in Chinese).
[2]Ni Yuanlong et al., High Power Laser and Particale Beams (HPLPB), vol.3(1991)242 (in Chinese).
[3]Zhang Guoping et al., HPLPB, vol.2(1990)298 (in Chinese).
[4]G.B.Zimmerman and W.L.Kruer, Comments Plasma Phys. Controlled Fusion 11(1975)51
[5]W.C.Mead et al, UCRL-84684 Rev I. (1981)
[6]Zhang Tanxin, private communication.
[7]Yang Minglun, private communication.

* Project supported by the Laser Domain of the Chinese National High Technology Plan

PROGRESS IN ICF AND X−RAY LASER

EXPERIMENTS AT CAEP

H. S. Peng, Z. J. Zheng, S. J. Wang, Y. Cun, J. G. Yang, and Z. C. Tao

China Academy of Engineering Physics

INTRODUCTION

The laser program at China Academy of Engineering Physics has made remarkable progress in ICF and x−ray laser research during the past few years. Hohlraum physics and imploding hydrodynamics have the priority in experimental investigations. Laser−plasma coupling, x−ray conversion and transport, nonlinear instabilities and suprathermal electrons, and fast ions for a variety of target configuration have been studied and neutron productions by indirect drive have been successfully demonstrated. Furthermore, x−ray laser experiments have been focused on pursueing large gain length products for Ne−like germanium 3p−3s transitions and a gl = 14.6 has been achieved with a proposed double−target coupling scheme using the LF−12 two beams irradiating oppositely and synchronously.

INDIRECT−DRIVE NEUTRON PRODUCTION IMPLOSIONS

The laser fusion programs conducted in many laboratories have made substantial progress in addressing the target physics and developing laser and target science and technology for both indirect−and direct−drive approaches to achieving ICF[1,2]. The encouraging accomplishments and the predictable potential of ICF in both civilian and military applications have attracted attention of not only scientists over the world but also governments of a quite number of countries. Consequently, a major step has been under consideration in order to demonstrate ignition and moderate gain by 2000[3].

The primary objective of the laser fusion program on the LF−12 and LF−11 laser facilities [4] at CAEP has been to address the basic physics of laser−plasma interactions, the hohlraum physics associated with creating the radiation environment for imploding ICF capsules and the imploding hydrodynamics. The majority of the experiments have focused on hohlraum laser plasma physics to study laser−plasma coupling, x−ray generation and transport, nonlinear instabilities, and suprathermal electron productions with $1.05\mu m$ driving light. Although the short wavelengh ($<0.35\mu m$) has been proved to be crucial due to the high efficiencies of laser−plasma energy coupling and x−ray conversion, and low suprathermal electron production yields, the physics of laser−plasma interaction for the fundamental frequency light must be understood well and the results are necessary for and applicable to future investigations with short wavelength light.

Here we report the results of neutron production implosion experiments by indirect driving at CAEP, which yielded detectable neutrons on a small laser facility for the first time to our knowledge. The experiments were carried out on the two−beam LF−12 Nd: glass laser facility, delivering each 600−700J in a 0.6−0.8ns Gaussian pulse with a risetime

Laser Interaction and Related Plasma Phenomena, Vol. 10
Edited by G.H. Miley and H. Hora, Plenum Press, New York, 1992

99

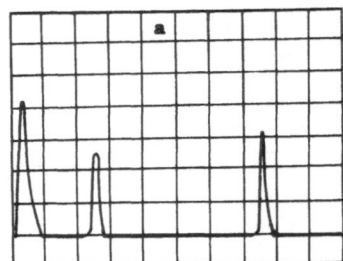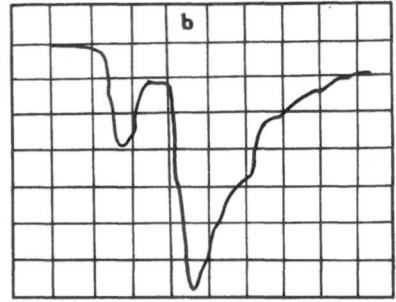

Fig. 1 Oscilloscope traces of fusion neutrons recorded with (a) BF₃ counters (1V/div, 20μs/div) and (b) photomultipliers (0.5V/div, 10ns/div). The first peak corresponds x rays, while the second neutrons in (b).

less than 0.2−0.3ns at 1.05μm. The energy imbalance of the two pulses was less than 10%, while the synchronization within 10ps. The targets used were consisted of a cylindrical gold case 500−800μm in diameter with a suspended glass capsule inside containing 10−to 30−atm equimolar DT gas. Diameters of the capsules were about 200μm with a wall thickness of 1.0−5.0μm. The diagnostics included a number of photodiodes, calorimeters, and flat response XRDs for energy coupling measurements, two Dante spectrometers and a filter−fluorescer spectrometer for radiation temperature and suprathermal electron temperature measurements, an x−ray streak camera and pinhole cameras to observe the temporal and spatial behaviours of x−ray emissions from case holes, Faraday cups for collecting ions, and two optical streak cameras for monitoring the time profiles of the incident laser pulses. Besides, two scintillator−photomultiplier detectors (sensitivity of 5×10^4 neutrons/shot) and three gated BF₃ counters with paraffin moderator (sensitivity of 2×10^5 neutrons/pulse) were employed to measure neutron yields. The neutron detectors were calibrated with both neutron tubes and a D. C. accelerator neatron source.

In the experiments, the two beams of the LF−12 facility illuminated the inner walls of the case through two injection holes, converting their part energy into subkev x rays, which then drove the implosions of the DT−filled capsules. For those targets in which capsules were not shielded by anything but the mounting Formvar film pieces 300 nm thick, the neutron yield was about 3×10^5 neutrons/shot. Figure 1 shows the typical oscilloscope signals of the fusion neutrons. To verify that these implosions were really driven by x rays, we used similar targets and irradiated the capsules directly with the two beams of the same parameters, obtaining only 5×10^4 neutrons/shot. This result indicated that the scattered light, in the previous regime, from the case walls could hardly make significant contribution to producing neutrons. In addition, another experiment was subsequently performed where only one beam was delivered onto the inner wall of the hohlraum, and 3×10^4 neutrons/shot were observed surprisingly. In this extremely asymmetrical illumination situation, the neutron yield of the same level as for the two−beam direct drive shot could only be attributed to radiation driving. Furthermore, by measuring hard x rays we inferred the temperature of suprathermal electrons to be 20KeV. The range of the 20KeV electrons was estimated 30μm in glass, and thus, the energy deposited by the electrons in the shell could be neglegible in comparison to that by x rays, even though the suprathermal electrons spent about 4−5% of the laser energy. Having found the evidence for indirect−drive neutron production, we tried new targets of specially designed configurations in which the capsules were shielded from being exposed to the scattered laser light, suprathermal electrons and fast ions, and produced 5×10^5 neutrons/shot. However, to generate this clean radiation environment for imploding capsules, most of the energy delivered by the laser beams was lost. According to the data given by the diagnostics, calculations showed that only 10−20J of subkev x rays were distributed to the capsule surface to generate the pressure for driving the implosions at a radiation temperature of 100−120eV. 1−D neutron yield was numerically simulated to be $10^7−10^9$ neutrons/shot and four orders of magnitude could be reduced by the 2−D effects of 2−beam asymmetry, mixing and breaking. The experiments also showed a large departure of the implosions from 1−D modeling. Apparently the targets were designed to work at high ion temperature, low fuel densities and small radial compression ratios and therefore could not scale to high gain.

Although we have successfully produced fusion neutrons by radiation driving on such a laser facility, which has given us experiences and confidence, we still have a long way to go in many aspects. To increase the energy efficiency and suppress suprathermal electrons, scattered light and fast ions, which are fatal to high gain implosions, we are working very hard to convert the 1.05μm light into 0.35μm light. Second, the output of the LF−12 laser facility is not powerful enough to conduct indirect−drive implosions for addressing hydrodymamics. Therefore, we are planning to upgrade the LF−12 facility to a larger output of 0.35μm light. Furthermore, improved target configurations and more sophisticated diagnostics have been designed and developed so that we will be able to gain a better understanding both in hohlraum physics and imploding hydrodynamics.

LARGE GL (14.6) GERMANIUM LASER EXPERIMENTS

Since the first demonstration of the Ne−like selenium lasers, large gain length products have been pursued in many laboratories in order to address the physics and develop the technology for producing applicable x−ray laser beams at short wavelength. LLNL first reached the gl value of 16 for the collisionally excited transitions at 20.6 and 20.9nm in selenium plasmas on the Nova two−beam facility [5], and subsequent efforts have not succeeded yet in setting a new record, except that the photoionized Cs vapor 96.9nm laser has been produced with an extrapolated small signal gain of exp(83) in a total length of 17cm [6].

As the laser power required to create plasmas for producing short wavelength lasing gains increases dramatically, scaling inversely as the lasing wavelength to the 9/2 power[7], attempts to achieve larger gl values for shorter wavelengths have been restricted by the laser facilities available at present. However, for longer wavelengths, target physics turns out to be major issues. To obtain larger gl values, one naturally lengthens the plasma medium, which, thus, brings two obstacles to the amplified spontaneous emission to overcome−plasma "ageing" and refraction. The lasing durations of the lines at 20.63 and 20.96nm with thin foil targets measured at LLNL are roughly 200ps[8], which indicates that the plasma conditions for lasing are transient both/either temporally and/or spatially. If the transmission time of light from one end of the plasma column to the other is comparable to the lasing duration, a ray originating first from fresh medium will meet, gradually, ageing medium along the amplifier, resulting in a limited effective length. Second, refractive losses are expected to be significant due to electron density gradients radially in the amplifier particularly for collisionally excited soft x−ray ASE with thick targets. To overcome these two problems many methods have been proposed and experimentally investigated. Saturated 108.9−nm Xe Ⅲ and 96.9−nm neutral Cs lasers have been demonstrated using travelling−wave laser−produced−plasma excitation with low density lasing media [6,8]. Hagelstein [9] has proposed that by alternately propagating the laser beam through plasmas expanding in opposite directions it should be possible to bend the beam first to the left and then to the right, and so forth, in order to keep it within the regions of gain for thick targets.

To suppress and compensate for the "ageing" and refraction of plasma simultaneously, we have designed an innovative "double−target coupling" amplification scheme and a gl of 14.6 for the Ne−like germanium 23.2 and 23.6−nm lines has been achieved.

The experiments were conducted on the two−beam LF−12 Nd: glass laser facility early in 1990. Based on the results obtained with germanium slab targets in 1989[4], two slabs were used and arranged as shown in Figure 2. The slabs were illuminated by two oppositely incident 1.05−μm, 1−ns laser pulses with an irradiance of 1.2×10^{13}W/cm² at 120−μm wide focal lines. The first slab was 2.2−cm long, while the second one ranged from 0.6 to 1.8cm. The beam to irradiate the second slab was delayed by about 90ps corresponding to the time needed for light to cover the first target and the separation between them axially. A specially designed mount was used to precisely adjust the targets positions and orientations for optimizing the coupling of the two lasing plasma columns. Numerical simulations showed that the lasing zone was less than 100−μm thick in the target normal direction about 50−100μm away from the surface and the refraction angle in the 2.2−cm long lasing medium was about 9 mrad.

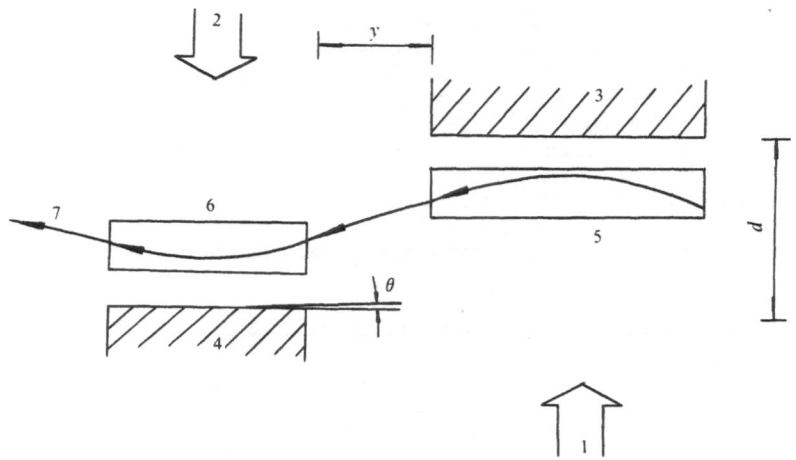

Fig. 2 Double – target coupling amplification scheme
1.laser beam I , $t=0$ 2. laser beam II , $t=(90\pm5)$ ps   3, 4. slab targets   5, 6. gain zones   7.x – ray laser

A flat field grating spectrometer was employed to measure either the lines intensities with Kodak 101 – 07 film or the time history of lasing with a soft x – ray streak camera. The slit of 15 μm in width was oriented in parallel with the target normal. Aluminium filters were placed between the slit and the grating to control the lines intensities to be recorded so that the film could work in its linear range. Besides, a grazing – incidence grating spectrometer, a crossed slit camera and KAP cyrstal spectrometers were installed to monitor plasma behaviors.

The results of temporally and spectrally integrated line gain coefficients are summarized in Table 1 and the curves of line intensity vs amplifier length for the transitions at 23.2

Table 1. Time – integrated gains of 3p – 3s transitions in Ne – like germanium plasmas

λ (nm)	Laser irradiance 1.2×10^{13} W/cm²			Laser irradiance 0.8×10^{13} W/cm²		
	L (cm)*	G (cm⁻¹)	GL	L (cm)	G (cm⁻¹)	GL
19.6	4.0	2.67	10.7	4.4	2.19	9.6
23.2	4.0	3.66	14.6	4.4	2.53	11.1
23.6	4.0	3.66	14.6	4.4	2.52	11.1
24.7	3.4	2.57	8.7			
28.6	3.4	3.52	12.0			

* Here L is the maximum target length in the experiments

and 23.6nm are depicted in Figure 3. The largest gl is 14.6 which seems to be on the brink of saturation as signed by the rollover at 4cm in the figure. The time histories of 19.6, 23.2, 23.6 and 28.6 – nm transitions were recorded as shown in Figure 4 without noticeable difference among themselves even though the intensities of the four lines spanned two orders of magnitude. The gain started 0.2 – 0.3ns after the onset of the plasma coutinuum and rose abruptly to a plateau lasting for 0.7 – 0.9ns. The time relation between the pumping laser and x – ray laser pulses was not able to be accurately determined due to the absence of an absolute timing fiducial for the optical and x – ray streak cameras. The long duration of the gains implied a very interesting feature of the slab targets, where the lasing medium was fairly stationary. Simulations have also showed that the dimentions and position of the gain regions of thick targets are substantially stable. This phenomenon can be easely understood by analyzing the physical picture of the laser interaction with target plasmas. Once the surface of a slab target is illuminated by a pumping beam, the plasma is produced and then expands outwards. If the pumping power is suitable, a zone with required electron density and temperature, and ions abundance for lasing will be formed somewhere close to the surface, for ex-

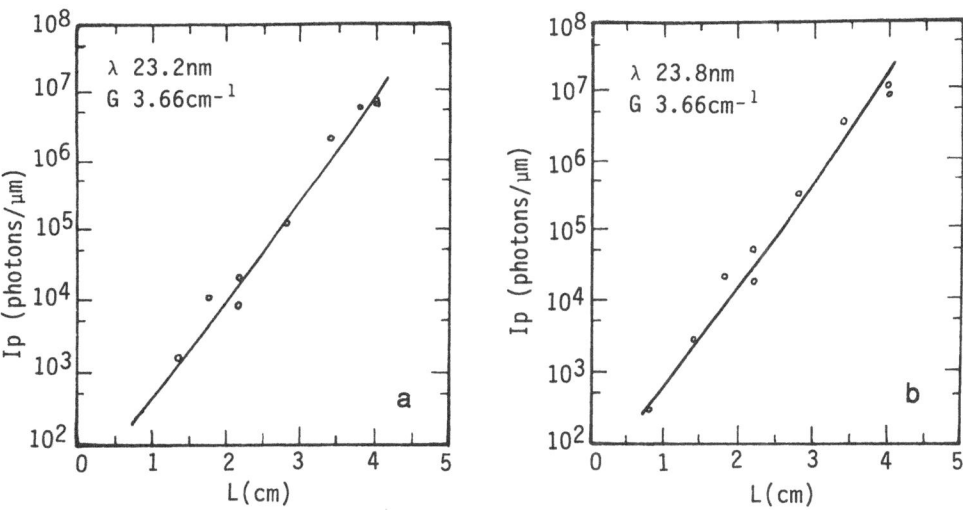

Fig. 3 Line intensities vs target length at (a) 23.2 and (b) 23.6 nm with an irradiance of 1.2×10^{13} W/cm²

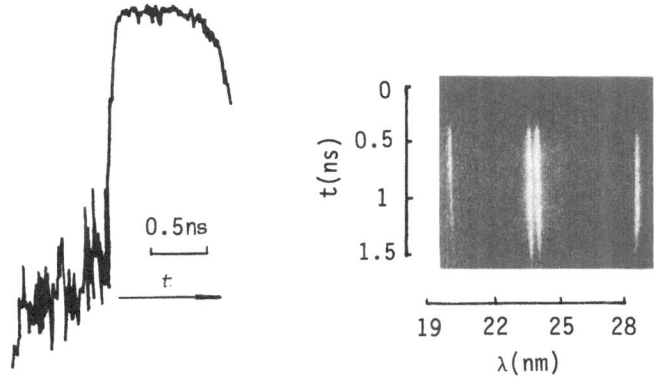

Fig. 4 time histories of the Ne−like germanium 3p−3s transitions

ample $50-100\mu m$ apart as mentioned above. The plasma flow keeps moving at the ion acoustic speed into and then out of the zone, where lasing takes place. Being pumped by a long laser pulse, a slab target thick enough for burning can thus provide a "stable" plasma flow and, therefore, a long lasting lasing zone before having been damaged mechanically . This may be the advantage of slabs over thin exploding foils, where plasmas can only lase for a short period of time because of the absence of fresh plasmas to add on.

To take this advantage of long lasting durations of slabs for developing soft x−ray resonant cavities, we conducted another experiment to demonstrate double−pass amplification of amplified spontaneous emission and to study the survivability of multilayer mirrors. At the same time we checked the temporal behaviors of the lasing medium. The experiment set up and diagnostics were the same as described for the former experiment except that a single germanium slab target 2cm long was used and a flat mirror was placed. The Mo/Si mirror had 15 layers and a 12.6nm period (Mo 4.6nm, Si 8.0nm) with a calculated reflectivity of 15% at 23.2nm. The distance between the mirror and the end of the slab was adjustable from 1 to 4cm. The mirror normal could rotate from 4 to 10 mrad with respect to the target surface so that the plasma refraction angle would be matched properly.

The time−integrated double−pass signal was five times more intense than the single−pass ASE without apparent differences while the matching angle changed from 5 to 8 mrad. The peaks of time−resolved double−pass signals were ten times bigger. The gain lifetime of the single slab target without the mirror was measured to be about equal to the duration of the 1−ns optical driving pulse. The double−pass signals started 200−250ps later than the single−pass ones in consistence with the experiment arrangement of a 2−3cm separation between the mirror and the end of target 2cm long. The durations of the double−pass signals recordod for 2−cm and 3−cm separations were about 450ps and 700ps, respectively. While for a 1−cm separation, we could not see any double pass amplification. The dependence of the double−pass signal duration and intensity on the separation may be attributed to the mirror damage caused probably by scattered optical laser light, plasmas, x rays or ASE. It is apparent that the damage is later and weaker the farther the mirror from the target. However, this will, in turn, reduce the signal intensity reflected from the mirror back for double−pass amplification and waste more time. We have not got enough information for making a reasonable compromise yet from the experiment.

In addition, we measured the spatial distributions of the ASE from the 2−cm long lasing column in the directions both perpendicular and parallel to the target surface. The former was about 12 mrad with a refracted angle of 8 mrad, while the latter about 22 mrad with only one peak on the axis.

In conclusion we have shown that the novel double−target coupling with a time delay has been successfully employed to achieve a gain length product of 14.6 at 23.2 and 23.6nm with slab targets. To answer whether the gain rollover means a marginal saturation or not, more experiments and simulations should be conducted. Furthermore, solid targets have demonstrated a long gain lifetime which will be of great importance in developing x−ray laser resonant cavities, offering opportunity for many amplification passes, if plasma refraction affects are overcome.

ACKNOWLEDGEMENT

The authors wish to thank M. Yu, R. Y. Hu and Y. B. Fu for support and advice, T. X. Chang, X. T. He and their co−workers for fruitful collaboration in numerical simulations, the crew on the LF−11 and LF−12 laser facilities for excellent operations and our colleagues for completing the challenging experiments. The x−ray laser experiments were supported by the National Laser Program.

REFERENCES

[1]   E. Storm, et al., "Progress in laboratory high gain ICF: Prospects for the future," Eighth Session of The International Seminars on Nuclear War, Erice, Sicily, Italy, August 20−23, 1988.

[2]   E. Storm, et al., "Progress toward ignition in the laboratory" IAEA−CN−53/B−2−3, Thirteenth International Conference on Plasma Physics and Controlled Nuclear Fusion Research, 1990.

[3]   H. G. Stever, et al., "Final Report", Fusion Policy Advisory Committee, September, 1990.

[4]   H. S. Peng, et al., Laser Interaction and Related Plasma Phemonena, November 1989.

[5]   B. J. MacGowan, et al., Proc, SPIE, 688, 36 (1986).

[6]   C. P. J. Barty, et al., OSA Proc on Short Wavelength Coherent Radiation: Generation and Applications, Vol. 2, 13 (1988).

[7]   N. M. Ceglio, et al., Energy and Technology Review, LLNL, p7, December 1989.

[8]   M. H. Sher, et al., Opt. lett. 12, 891−893 November (1987).

[9]   P. L. Hagelstein, Plasma Physics 25, 1345 (1983).

DECREASE OF RADIATION TRAPPING IN A LASER-PRODUCED PLASMA WITH VOLUME-REDUCING DEVIATIONS FROM CYLINDRICAL GEOMETRY

W. Brunner and R.W. John

Institute of Nonlinear Optics and Short-Time Spectroscopy
Rudower Chaussee 6, O-1199 Berlin, Germany

1. Introduction

In optimizing X-ray laser schemes based on laser-produced plasmas, a precise prediction of the amount of radiation trapping and its influence on the gain is of current interest (cf. |1,2|). The re-absorption of resonance radiation from the lower laser level is known as limiting the gain, primarily by re-populating this level. Especially in experiments demonstrating X-ray lasing in the 3 --> 2 transition in H-like ions, the reduction of the Balmer-α gain by trapping of the Lyman-α line is to study.

In X-ray gain calculations the form of the laser-produced plasma arising from fiber or foil targets, with the length of the plasma much larger than its transversal extension, is usually approximated by a 1-dimensional cylindrical geometry. However, there are deviations of the plasma geometry from the assumed axial symmetry. To determine how the degree of radiation trapping does depend on such deviations, at first we model them. Irradiating the target with two line-focused laser beams from opposite sides, the form of the arising plasma may be approximated, instead by a full cylinder of radius R_1, refinedly by a cylinder longitudinally cut out, with sector angle Φ, up to a radius $R_2 < R_1$, on the two diametrically opposite sides between the laser beams where the target was less strongly illuminated (see. Fig. 1). The plasma arising from the irradiation of a target by two opposite laser beams is depicted as a longitudinally twofold symmetrically cut out cylinder, with major radius R_1, minor radius R_2, and sector angle Φ. The resonance line reabsorption is considered in a point P of the plasma; r, φ are the radial and the angular coordinates, respectively, of P.

Laser Interaction and Related Plasma Phenomena, Vol. 10
Edited by G.H. Miley and H. Hora, Plenum Press, New York, 1992

105

2. Basic equations and results

In general, the determination of the reabsorption - affected population density implies the simultaneous treatment of the evolution equations for the level population densities of the ions and the equation of radiative transport. The problem involves the plasma geometry and factors such as, e.g., line profiles and, in the differentially expanding plasma, the variation of the plasma flow velocity.

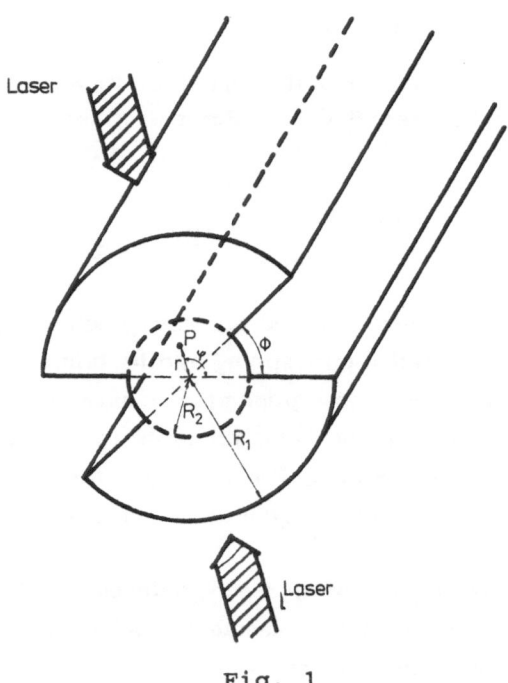

Fig. 1

Especially the latter was taken into consideration in a previous paper |3|, some of the results were used in |4|. A known approximate approach to the problem uses the concept of the escape probability - the probability for the photons to escape out of the plasma volume (|5|; cf. also |6|, p. 188, and, for summarizing results for several line profiles and geometries, |7|). Dating back to the treatment of radiation diffusion given in |8|, the influence of line trapping on the level population densities is approximately taken into account by using a reduced coefficient for spontaneous emission of the corresponding line that is generated by multiplying the Einstein A-coefficient with the escape probability.

The escape probability E in any point P of the plasma is represented by the integral ([5], cf. also [6], pp.184)

$$E(r) = 1 - \int_V dr' G(r, r')$$

(1)

with the kernel

$$G(r, r') = \frac{1}{4\pi |r-r'|^2} \int_{-\infty}^{\infty} d\omega \, L(\omega) \, k(\omega) \exp\{-k(\omega) |r-r'|\},$$

(2)

where r is the radius vector of the point P, and the spatial integration is taken over the plasma volume V; L (ω) is the normalized emission profile, $\int_{-\infty}^{\infty} d\omega \, L(\omega) = 1$, and k ($\omega$) the absorption profile, respectively, for a specific line.

To study the effect of the described plasma geometry on the escape probability, we assume at first, in the present paper, that in the integral (2) the absorption profile is very slowly varying with frequency in comparison with the emission profile. So, to a good approximation, under the integral sign we may take the absorption profile at its line center value k (ω_0), we write shortly k. Accordingly one obtains the kernel (cf. [6], p.187)

$$G(r, r') = \frac{k}{4\pi |r-r'|^2} \exp(-k|r-r'|)$$

(3)

which leads, via eq. (1), to the escape probability

$$E(r) = 1 - \frac{k}{4\pi} \int_V \frac{dr'}{|r-r'|^2} \exp(-k|r-r'|) .$$

(4)

We will evaluate the escape propability (4) for the specific non axisymmetric plasma geometry described in sect.1 (see Fig. 1). Introducing cylindrical coordinates, r', φ', z' with the origin on the cylinder axis, and assuming the plasma length L to be much larger than the major radius R_1, the escape probability in points P of the plasma with coordinates r, φ, o reads:

$$E(R_1, R_2, \Phi; r, \varphi) = 1 - \frac{k}{4\pi} \{ [\int_0^{R_1} dr' r' (\int_\Phi^\pi d\varphi' + \int_{\pi+\phi}^{2\pi} d\varphi') +$$

$$\int_0^{R_2} dr' r' (\int_0^\Phi d\varphi' + \int_\pi^{\pi+\phi} d\varphi')] \int_{-\infty}^{\infty} dz' \frac{\exp[-k(r^2+r'^2-2rr'\cos(\varphi-\varphi')+z'^2)^{\frac{1}{2}}]}{r^2+r'^2-2rr'\cos(\varphi-\varphi')+z'^2} \} .$$

(5)

Studying the escape probability in the vicinity of the cylinder axis,

$$E(R_1, R_2, \Phi; r, \varphi) \approx E_{(2)}(R_1, R_2, \Phi; r, \varphi) = (E)_{r=o} + r(\frac{\partial E}{\partial r})_{r=o} + \frac{r^2}{2}(\frac{\partial^2 E}{\partial r^2})_{r=o},$$

(6)

an explicit expression $E_{(2)}(\tau_1, \tau_2, \Phi; \tau, \varphi)$ is obtained ([9], [10]); here, the optical depths $\tau_1 = kR_1$, $\tau_2 = kR_2$; and $\tau = kr$ were introduced.

Averaging this escape probability over the angular coordinate of the plasma points,

$$\bar{E}(\tau_1,\tau_2,\Phi;\tau) := \frac{1}{2\pi}\int_o^{2\pi} d\varphi\, E_{(2)}(\tau_1,\tau_2,\Phi;\tau,\varphi)\,,$$

one is led to the formula (|10|)

$$\bar{E}(\tau_1,\tau_2,\Phi;\tau) = \tau_1\left(K_1(\tau_1) - \int_{\tau_1}^{\infty} dt\, K_o(t)\right) + \frac{\Phi}{\pi}\left[\tau_2\left(K_1(\tau_2) - \int_{\tau_2}^{\infty} dt\, K_o(t)\right) - \right.$$

$$\tau_1\left(K_1(\tau_1) - \int_{\tau_1}^{\infty} dt\, K_o(t)\right)\right] + \left(\frac{\tau}{2}\right)^2\{K_o(\tau_1) + \frac{1}{\tau_1}\int_{\tau_1}^{\infty} dt\, K_o(t) +$$

$$\frac{\Phi}{\pi}\left[K_o(\tau_2) + \frac{1}{\tau_2}\int_{\tau_2}^{\infty} dt\, K_o(t)\right.$$

$$\left.-K_o(\tau_1) - \frac{1}{\tau_1}\int_{\tau_1}^{\infty} dt\, K_o(t)\right]\},\tag{7}$$

$$0 < \tau << \tau_2 < \tau_1, 0 < \Phi < \pi\,.$$

$K_0(x)$, $K_1(x)$ are the modified Hankel functions (cf. |11|, pp. 374); the integral $\int_x^{\infty} dt\, K_0(t)$ has been also tabulated (see |11|, pp.479).

3. Conclusion

For a plasma in the form of a longitudinally cut out cylinder, with sector angle Φ (see Fig. 1), the length L of the plasma much larger than its major radius R_1, and by using the line center approximation for the absorption profile in the frequency integral, we have calculated the φ-averaged escape probability $\bar{E}$ for a specific line in any point P near the plasma axis, with coordinates r, φ. The obtained formula (7) shows explicitly the increase of the escape probability with increasing sector angle Φ. So, it is clearly seen how the described deviations of the form of the plasma from the cylindrical one lead to a diminution of the trapping of resonance radiation outgoing from a lower laser level what again implies a higher population inversion for the lasing transition in the ions. In a forthcoming paper |12|, for the same plasma geometry we will calculate the escape probability under inclusion of a considerable frequency variability of the absorption.

References

| 1| D.C. Eder, Phys. Fluids B1, 2462 (1989)

| 2| G.J. Pert, in: X-ray Lasers 1990. Ed. by G.J. Tallents.
Institute of Physics Conf. Ser. N. 117 (IOP Publishing Ltd,
Bristol 1991) p. 143

| 3| W. Brunner and R.W. John, Laser Part. Beams $\underline{9}$, 817 (1991)

| 4| W. Brunner, R.W. John and Th. Schlegel, Plasma Phys.
Controlled Fusion (1991, in press)

| 5| T. Holstein, Phys. Rev. $\underline{72}$, 1212 (1947)

| 6| H.R. Griem, Plasma Spectroscopy (McGraw-Hill, New York 1964)

| 7| F.E. Irons, J. Quant. Spectrosc. Radiat. Transfer $\underline{22}$, 1 (1979)

| 8| L.M. Biberman, Dokl. Akad. Nauk SSSR $\underline{59}$, 659 (1948)

| 9| W. Brunner and R.W. John, in: Proceedings book of the 21st
European Conference on Laser Interaction with Matter Held in Warsaw, Oct.
21-25, 1991 (to appear)

|10| W. Brunner and R.W. John, Laser Part. Beams 1992 (to be submitted)

|11| M. Abramowitz and I.A. Stegun, Handbook of Mathematical Functions
(Dover Publications, Inc., New York 1964)

|12| W. Brunner and R.W. John, contribution to the XVIII International
Quantum Electronics Conference (IQEC '92), Vienna, June 14-19, 1992

X-RAY OPTICS OF ARRAYS OF REFLECTIVE SURFACES

A.V. Rode[1], H.N. Chapman[2], K.A. Nugent[2], and S.W. Wilkinson[3]

[1]Laser Physics Centre, The Australian National University,
Canberra, ACT 2601, Australia
[2]School of Physics, The University of Melbourne, Parcville,
Victoria 3052, Australia
[3]CSIRO, Division of Materials Science and Technology, Clayton,
Victoria 3168, Australia

1. INTRODUCTION

We present a new way of focusing X-rays by capillary arrays, using total external reflection at grazing angles from the interior surfaces of hollow channels[1-4]. Thin Bragg-difraction curved crystals can also be used in a transmissive mode to focus X-rays[5] and the crystal netplanes can be considered as arrays of reflective surfaces.

Capillary arrays and focusing crystals can be grouped together in a class of devices we call *reflective arrays*.

Examples of refractive arrays are Micro-Channel Plates (MCP) as a capillary arrays, or Bragg crystals with the netplanes normal to the crystal surface, as in the Couchois X-ray spectrometer[6]. The imaging properties of these devices are very similar.

The reflective array is achromatic device. There are no chromatic aberrations, just as there are no chromatic aberrations in a mirror. Other properties are that it is spatially invariant, which means that the image is given by a point spread function convolved with the object function, and it exhibit translational invariance. That means that the device is very tolerant to misalignment errors.

The refractive index of most materials in the X-ray range is only slightly less than one, and so in MCP X-rays reflect by total external reflection at glancing angles up to the critical angle of reflection. For 1.54 Å x-rays incident on led glass, for example, the

Laser Interaction and Related Plasma Phenomena, Vol. 10
Edited by G.H. Miley and H. Hora, Plenum Press, New York, 1992

111

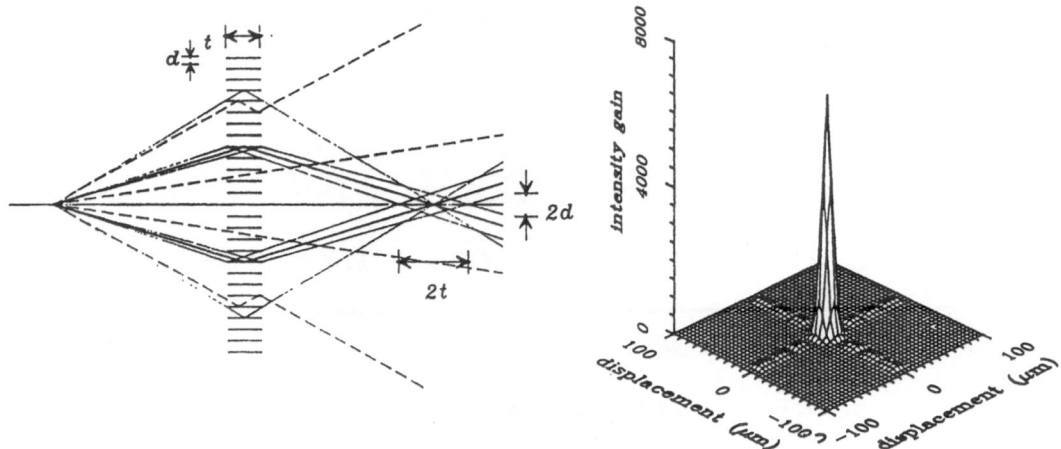

Figure 1. This diagram illustrates the principles of imaging with reflective arrays and the point spread function for a 2-D array of channels of square cross-section. Intensity gain profile is in relative units for MCP of d=10μm, t=2.82mm, open area 0.63 and critical angle 5 mrad.

critical angle is only about 5 mrad. For longer wavelength the critical angle increases, therefore it is seen, that this technique is well suited to the focusing of X-rays in a wide spectral range. We should point out that neutrons may also undergo total external reflection from surfaces at glazing angles, and so this method can be used to condense these particles too.

In this paper we will consider:
- properties of reflective arrays;
- ray aberration theory of reflective arrays;
- theoretical and experimental results for the focusing 1 - 8 keV X-rays
  using microchannel blanks, that is an array of cylindrical channels.

The motivation behind this work was to develop a fast condensing polychromatic element that would be cheap, robust, and compact, for use in X-ray microscopy and in X-ray analysis experiments, such as fluorescence or scattering experiments.

2. RAY OPTICS OF ARRAYS OF REFLECTIVE SURFACES

Let us consider a simple spherical mirror and a stack of plane mirrors aligned normal to the spherical surface with a radius of curvature *R*.

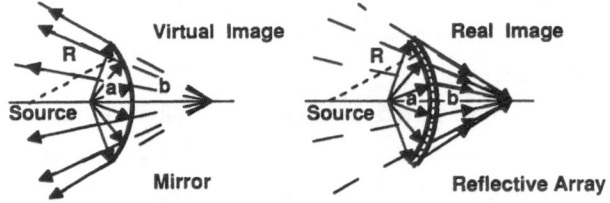

Figure 2. The real and the virtual images formed by mirror (left) and reflective array (right).

Due to the simple law of reflection and the shape of the mirror, paraxial rays form a virtual image when the source point is placed between the mirror and the focal point. Now, if we rotate a plane mirror by 90° then the reflected ray from that mirror would change orientation by 180° and it will be travelling in the exact opposite direction to the ray reflected from the unrotated mirror. Therefore, if we were locally rotate the mirror surface so that we are left with many small flat mirrors, each one radiating out from the spherical surface then rays would be reflected along these paths to the virtual image. Thus, we have a new type of geometric lens made from a stack of flat mirrors; one whose properties are almost equivalent to those of a spherical mirror, except that the **real** image formed by a reflective array corresponds to the **virtual** image of the mirror, and *vice versa*.

The paraxial rays from the source at distance *a* from the plate that are reflected in a meridional plane will be focused to a point distance *b* from the channel plate, where the lens equation is:

$$\frac{1}{a} + \frac{1}{b} = \frac{2}{R}$$

The sign conversion used here is the standard Cartesian convention where points located left of the origin are negative.

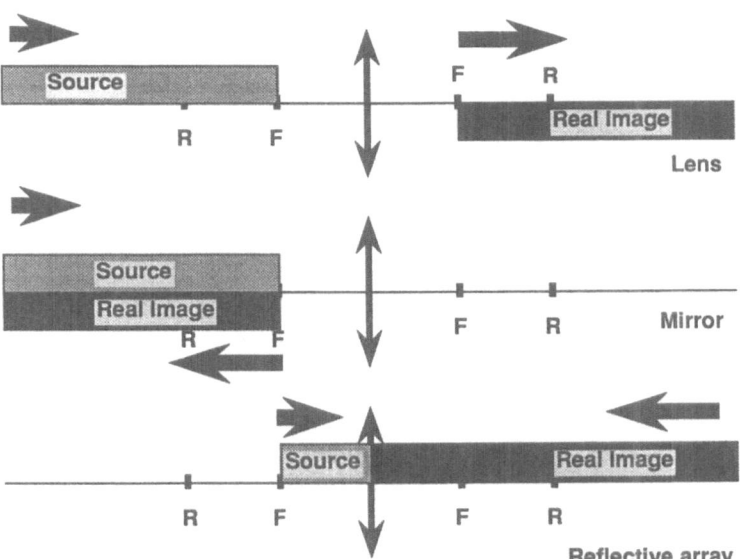

Figure 3. Areas of source and real image for lens, mirror and reflective array. Arrows indicate the directions in which the source and the real images move.

The two cases of transmitting (as in a capillary array) and reflecting rays (as in a simple mirror) can be combined into one by considering transmission through the array of reflecting surface as a refraction from the surface into a medium of refractive index n = -1 (it is assumed here that the reflective array is in vacuum).

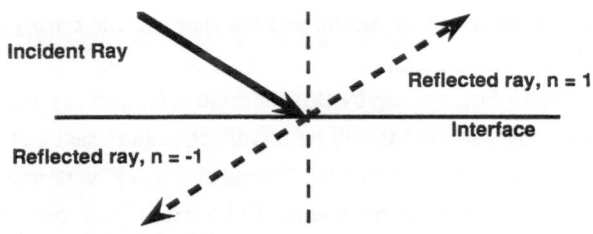

Figure 4. Reflecting rays in negative and positive refractive index case.

The difference between positive and negative reflection is that rays propagate in exactly opposite directions after reflection. Using a refractive index of -1 is often used to obtain equations regarding to reflective surfaces from equations derived for a refraction surface.

Reflective arrays possess a unique property; namely, the orientation of the reflectors may be completely independent of the surface profile of the array. That is, in comparison to simple optical elements, reflective array has an extra degree of freedom, similar to a variation in the refractive index with position in a lens. The deflection angle of a ray is determined by the orientation of the reflective surfaces and the point of the deflection depends of the substrate surface.

3. ABERRATION THEORY

We distinguish the orientation of reflective surfaces from that of the substrate by defining a "super-surface" which is normal to all reflectors and is not necessarily related in any way to the substrate[7].

For the purpose of aberration calculations we define a generalized reflective array:
- the channel width and length both approach zero;
- the broadening of a point spread function due to reflection curve or channel (crystal) width is ignored;
- the reflectivity is 100%;
- in the case of two dimensional focusing rays remains in a meridional plane after reflection.

The generalized reflective array is not an array at all but an interface, defined by the surface $z_1 = f_1(x,y)$, which may reflect or refract rays according to the orientation of another surface $z_2 = f_2(x,y)$ - a "super-surface", defined at each point of interface.

We calculate the aberrations for two distinct cases:

a) - where the centre of curvature of spherical super-surface C_{ss} is displaced from the centre of curvature of the spherical interface C_{int};

b) - where the interface is spherical and the super-surface is inclined at a constant angle Ψ to the interface at each point on the interface (this case does not exhibit rotational symmetry).

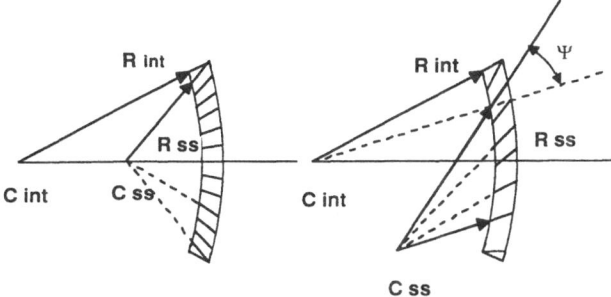

Figure 5. Two cases of supersurface. R_{int} is the radius of interface curvature; R_{ss} is a distance from C_{ss} to each point of supersurface.

The starting point is Fermat's principle and the analysis is based on variational calculus. The Hamilton optics[8] could be applied to a reflective array where the interface and super-surface are independent, but the mathematics becomes too complicated due to the fact that the super-surface is defined separately at each point of the interface. Therefore, instead of the powerful and general theory of Hamilton optics we will turn to the simple treatment, calculating the difference in optical path length between the extreme ray and the distance along the the optic axis from source to image points.

3.1. INTERFACE AND SUPER-SURFACE OF DIFFERENT CURVATURES

We consider a reflective array which has a spherical interface of radius R_{int} and spherical supersurfaces sharing a centre of curvature C_{ss}, a distance R_{ss} from the interface (Fig.6). The positions of the centres of curvature C_{int} and C_{ss} define an axis of rotational symmetry. Since the system exhibit rotational symmetry, only the distance of the source point from the optic axis is necessary for calculating aberrations.

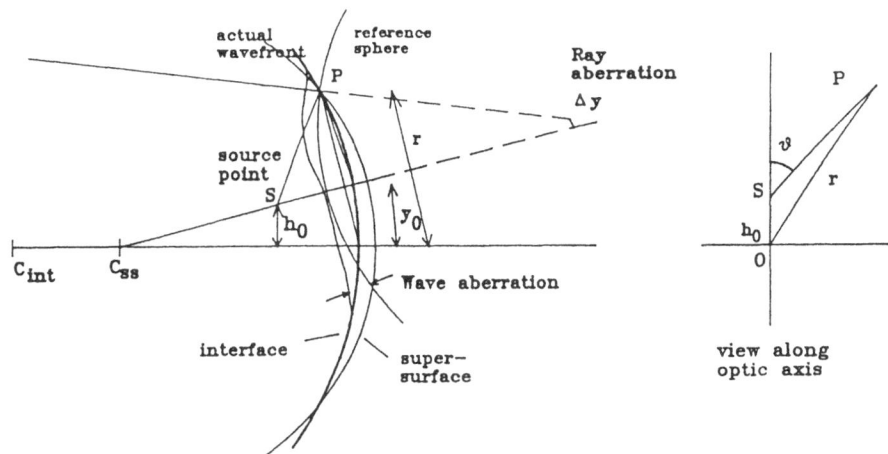

Figure 6. Geometric construction for the calculation of the wave aberration.

For systems with an axis of rotational symmetry a source point located at a distance $h_0 = R_{ss}(1/a - 1/R_{ss}) \cdot y_0$ from the axis will have ray aberrations given by:

$$\Delta x = -b \cdot (4 \, _0C_{40} r^3 \sin\theta + 2 \, _1C_{31} y_0 r^2 \sin\theta\cos\theta + 2 \, _2C_{20} r \sin\theta);$$

$$\Delta y = -b \cdot [4 \, _0C_{40} r^3 \cos\theta + _1C_{31} y_0 r^2 (1 + 2\cos^2\theta) + 2(_2C_{22} + _2C_{20}) y_0^2 r \cos\theta +$$

$$+ _3C_{11} y_0^3];$$

The coefficients in the above expression give the five Seidel aberrations (the subscripts in these coefficients refer to the powers of y_0, r, and respectively of terms in the expression of wave aberration).

$$_0C_{40} = \frac{1}{4}\left(\frac{2}{R_{int}} - \frac{3}{R_{ss}}\right) \cdot \left(\frac{1}{a} - \frac{1}{R_{ss}}\right)^2 \qquad \textbf{Spherical aberration}$$

$$_1C_{31} = -\left(\frac{1}{R_{int}} - \frac{2}{R_{ss}}\right) \cdot \left(\frac{1}{a} - \frac{1}{R_{ss}}\right)^2 \qquad \textbf{Coma}$$

$$_2C_{22} = -\frac{1}{R_{ss}}\left(\frac{1}{R_{ss}} - \frac{1}{a}\right)^2 \qquad \textbf{Astigmatism}$$

$$_2C_{20} = \frac{1}{2R_{ss}^2}\left(\frac{1}{R_{ss}} - \frac{1}{R_{int}}\right) \qquad \textbf{Curvature of field}$$

$$_3C_{11} = \left(\frac{1}{R_{ss}} - \frac{1}{R_{int}}\right) \cdot \left[\left(\frac{1}{R_{ss}} - \frac{1}{R_{int}}\right)^2 + \frac{1}{R_{ss}^2}\right] \qquad \textbf{Distorsion}$$

Plots of aberration curves (Δx, Δy) for three image positions for a reflective array with $R_{ss} = 500$ mm, $a = 200$ mm and $b = -1000$ mm for the distances from the optical axis $h_0 = 1$mm, 0.4 mm, and 0 mm (see Fig. 7). All curves are plotted to the same scale and the position of the paraxial image is at the origin of the axes; individual curves are plots of (Δx, Δy) for constant r (range from 0 to 10 mm, in steps 0.25 mm), varying θ.

The left-hand column shows the aberrations of an ordinary spherical mirror ($R_{int} = R_{ss}$); spherical aberrations are most noticeable.

The middle column shows the aberrations for the case where $R_{int} = R_{ss}/2$ and there is no coma.

The right-hand column shows the case $R_{int} = 2R_{ss}/3$, in which spherical aberrations is zero.

116

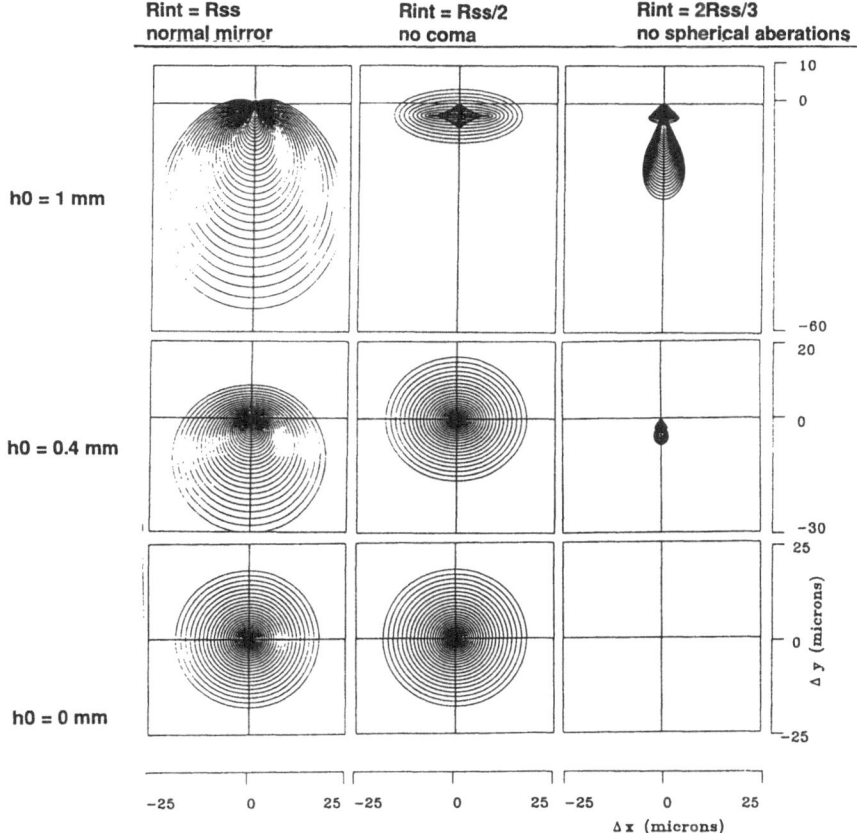

Figure 7. Aberration curves for the case where the extend of rays are limited by the angle of reflection, for three different curvatures of the interface: $R_{int} = R_{ss}$ (i.e. normal mirror); $R_{int} = (2/3)R_{ss}$; and $R_{int} = R_{ss}/2$. The image is located 25 mm from the optic axis.

X-ray astronomers have proposed that a flat capillary array ($R_{ss} = \infty$), in which channels lie normal to a spherical surface, be used as an X-ray telescope. It can be seen from the above expression that the magnitude of the spherical aberration of such an array has three times the spherical aberration of an ordinary mirror.

3.2. SUPER-SURFACE INCLINED TO THE INTERFACE BY A CONSTANT ANGLE

Since this system does not exhibit rotational symmetry there are non-Seidel aberrations. We assume the supersurface is spherical at each point, and the radius of curvature is such that the centre of curvature lies on the chief ray. The source point is located at S and is reflected at the point P, the centre of curvature of the array is C_1. A major difference between the previous case is that the position of the super-surface centre C_2 is not fixed for all super-surfaces but is a function of P.

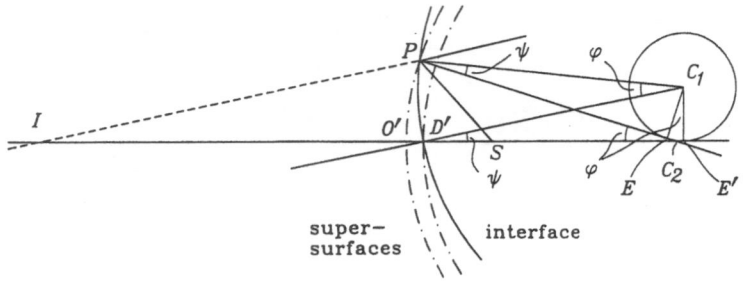

Figure 8. Geometric construction for the calculation of the wave aberration for the case where super-surfaces are inclined at a constant angle to the interface. Aberrations are calculated only for points on the axis C_2O'.

The aberration function can not be reduced to a sum of Seidel aberrations due to the absence of rotational symmetry, but for a small angle Ψ it is possible to write aberration function expanded to a second-order in Ψ:

$$A(P) = \frac{r^2 \cos^2\theta\ \Psi^2}{2R} - r^3 \cos\theta\ \Psi \left(\frac{1}{a^2} - \frac{2}{Ra} + \frac{3}{2R^2}\right) + \frac{r^4}{4R}\left(\frac{1}{a} - \frac{1}{R}\right)^2 + \cdots$$

The first three terms can be recognised as astigmatism, coma and spherical aberrations respectively. From this equation we can conclude:

- there is no dependence on the image height, since the aberration has only been calculated for points on optic axis;

- the polynomial in R and a in the second (coma) term has no real root, so there is no source position along the axis in which coma is zero, the minimum amount of coma occurs for $a=R$;

- coma is the only aberration for a flat array.

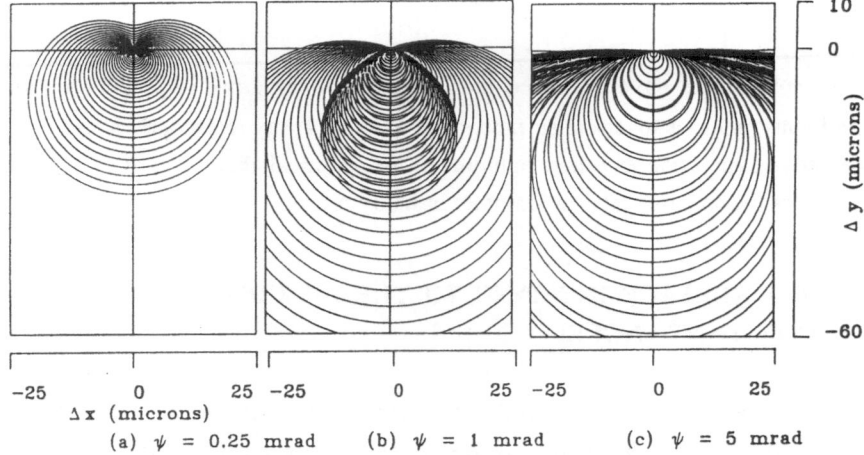

(a) $\psi = 0.25$ mrad   (b) $\psi = 1$ mrad   (c) $\psi = 5$ mrad

Figure 9. Aberration curves for the calculation of the wave aberration for the case where super-surface are inclined at a constant angle to the interface. Aberration curves are plotted for $\Psi = 0.5$ mrad (a), 2 mrad (b), and 5 mrad (c); the imaging parameters are $R = 500$ mm, $a = 200$ mm, $b = -1000$ mm, with r varying from 1 mm to 10 mm in 1 mm steps.

4. EXPERIMENTAL TEST: X-RAY FOCUSING USING MICRO-CHANNEL PLATES

Micro-Channel Plates consists of a small slab of lead glass in which there are hundreds of thousands of small hollow channels. They are used as image intensifiers, in which each channel acts as a photomultiplier when the faces of the plate are biased.

The ordinary flat MCP with cylindrical channels manufactured by Galileo Electro-Optics[9] was used in the experiments: d = 10 μm, t = 2 mm, open area 63%, the critical angle is 4.8 mrad. Rays can be reflected from the channel into all directions and only a small portion of each channel reflects rays into a "focus", so the focusing properties are somewhat different to square-pore arrays.

The experimental results[10] were recorded using two experimental set-ups:
- a microfocus X-ray source with a Cu-target, which produces 1.54 Å K_α-radiation;
- a laser-produced plasma with a spherically curved Silicon and Quartz crystals to monochromatize the radiation emitted by the plasma (6.2 Å and 8.4 Å).

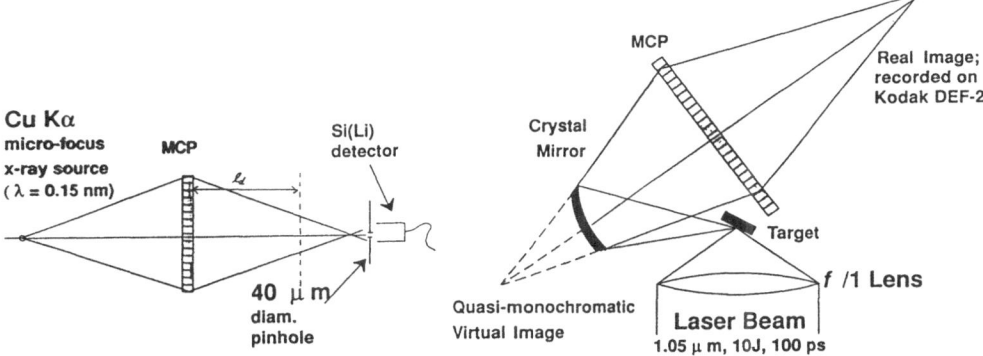

Figure 10. Schematic diagram showing the configuration used for experiments performed using a micro-focus X-ray tube and a laser-produced plasma.

The collector flux-gain for 1.54 Å is shown on Fig. 11. The simple model was not adequate in describing the focusing performance. To fit the data we needed to take into account:
- surface micro-roughness (wavelength dependant scattered factor);
- channel misalignment (misdirecting rays to broadening the profile and reduce the peak height);
- multiple reflections.

Measurements of intensity profiles of images formed by cylindrical-channel MCPs have been made at X-ray wavelength of 1.54 Å, 6.2 Å, and 8.4 Å. In the case of 1.54 Å X-rays, profiles were measured for various source to MCP distances. The model with RMS surface roughness of 36 Å, channel misalignment 0.3 mrad (1' of arc) were fitted 8 different profiles with various source-plate distances simultaneously. The parameters should be good for the long wavelength also.

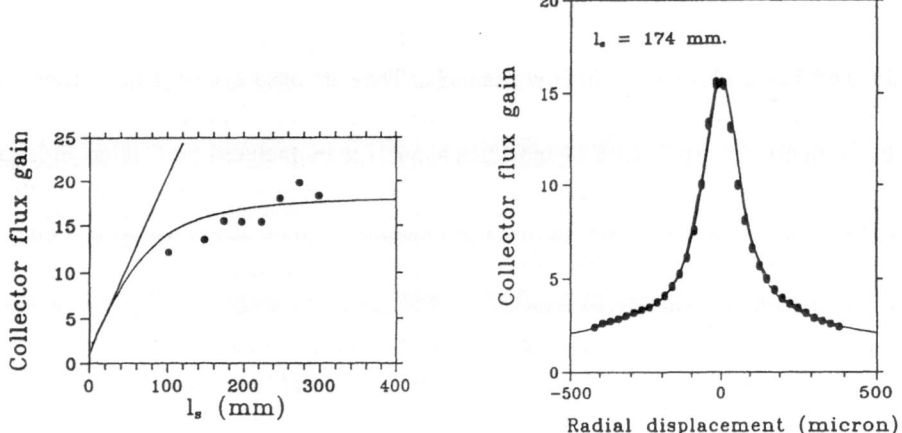

Figure 11. Experimental collector flux gains for various source-MCP distances and intensity gain profile (dots) and fitted profile calculated from the model of images of the micro-focus X-ray source (λ=1.54 Å) formed by the MCP.

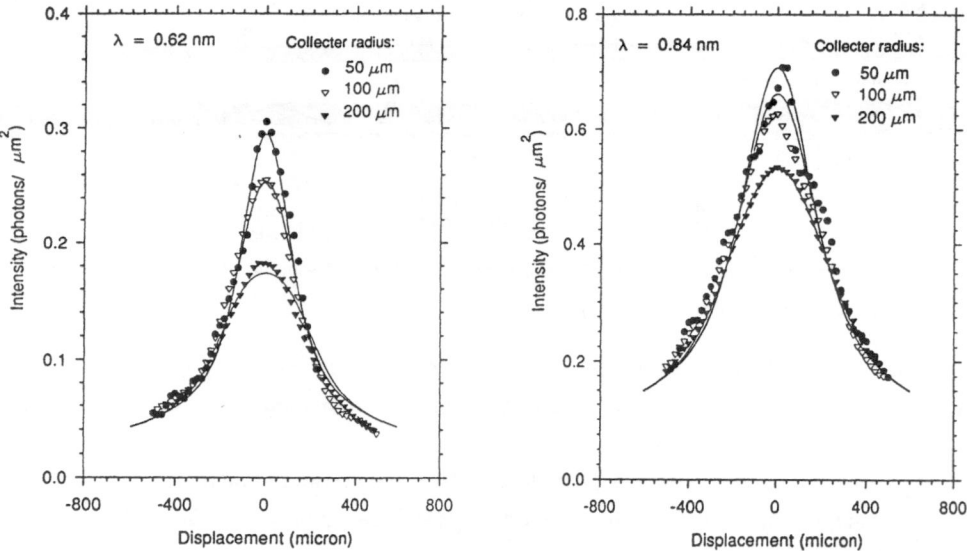

Figure 12. Experimental intensity profiles and fitted profiles calculated from the model (lines) of images of the plasma at 6.2 Å and 8.4 Å for a source-MCP distance of 210 mm. The data is shown for various collector radii: r_c= 50 μm (circles); r_c= 100 μm (hollow triangles), and r_c= 200 μm (fitted triangles). The flux-gain is 33±1 for 6.2 Å and 24±1 for 8.4 Å.

At the end we would like to present the image of a wire mesh (Fig.12) with a period 250 μm and 100 μm wire thickness, which was obtained with a laser-produced plasma and Q-crystal as a 8.4 Å X-ray source and MCP as a reflective array image device. The mesh was located at 10 mm from the laser-produced plasma (see Fig.10). The virtual image formed by the crystal converted to the real image by MCP. The resolution of the image is about 50 μm in horizontal direction (determined by crystal mirror astigmatism at the incident angle of 82°), and is about 20 μm in vertical direction (determined by the channel size of 10 μm).

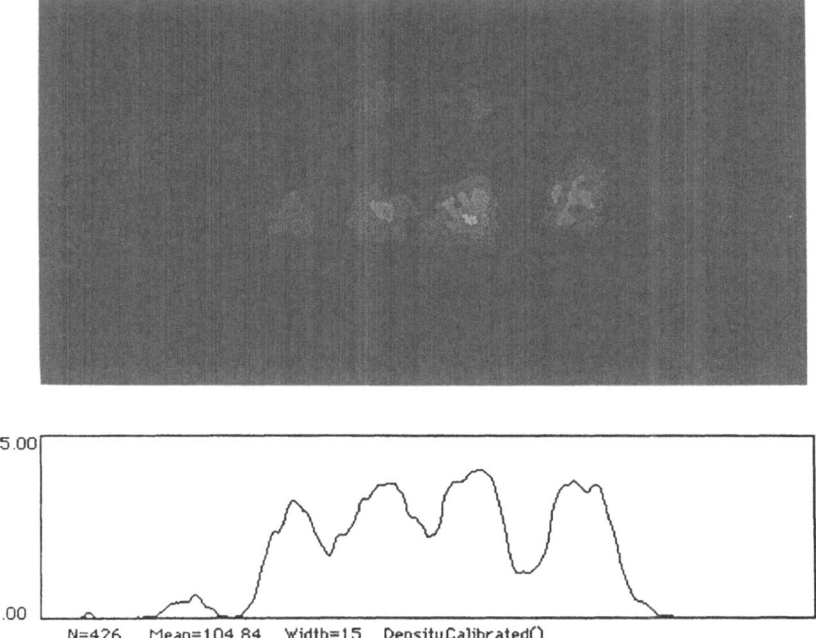

N=426   Mean=104.84   Width=15   DensityCalibrated()

Figure 13. X-ray image and a density trace of a wire mesh (period is 250 μm, wire thickness is 100 μm) obtained with MCP as a reflective array image device. The resolution is 50 μm in horizontal direction and 20 μm in vertical direction.

5. CONCLUSION

1. We have presented an analysis of the geometric optics of arrays of reflective surfaces. The two types of transmission and reflection arrays have been generalized into an array of supersurface, which are surfaces that either reflect as a mirror or refract as an interface between media of refractive index from 1 to -1, that means that transmitted rays (n=-1) propagate in exact opposite direction to reflected from a super-surface rays (n=1).

The supersurface may be completely independent of the interface of the array. The deflection angle of a ray is determined by the orientation of the reflective surfaces and the point of reflection depends on the substrate surface. The geometrical optics of supersurface has one more degree of freedom and includes the imaging properties of an ordinary spherical mirror as a particular case when a supersurface is coincident with an interface.

2. We have considered the aberrations of array of reflective surfaces in two general cases, one in which the supersurface shares a common centre of curvature different to the centre of curvature of the array interface, and another in which the supersurfaces were inclined at an angle to the interface. The optics were analysed by calculating the third-order (Siedel) aberrations of the images.

The aberrations calculations are justifiable to any system of arrays of reflective surfaces for focusing radiation or particles, where the ray optics is acceptable. It is expected, that these calculations will find use for optimising performance of reflective arrays or estimating the tolerances to which arrays must be manufactured.

3. It was experimentally shown that MCP can be used as an array of reflecting surfaces for X-ray focusing and imaging. The theoretical model have been found to explain all of the available data, taking into account multiple reflections of rays inside channels, scattering due to surface irregularities and channel misalignment.

We have developed a new approach to X-ray collimation and focusing in a wide spectral range, from γ-rays to VUV-radiation, - we have shown that the method works and we can accurately predict the performance of our devices.

6. REFERENCES

1. S.W. Wilkins, A.W. Stevenson, K.A. Nugent, H.N. Chapman, and S. Steenstrup, "On the Concentration, Focusing, and Collimation of X-rays and Neutrons Using Microchannel Plates and Configurations of Holes", Rev. Sci. Instrum. **60**, 1026 (1989).

2. H.N. Chapman, K.A. Nugent, S.W. Wilkins, and T.J. Davis, "Focusing and Collimation of X-rays Using Microchannel Plates: an Experimental Investigation", J. X-Ray Sci. Tech. **2**, 117 (1990).

3. P. Kaaret and P. Geissbühler, "Lobster Eye X-ray Optics using Microchannel Plates", Proc. SPIE **1546**, to be published.

4. G.V. Fraser, J.E. Lees, J.F. Pearson, M.R. Sims, and K. Roxburgh, "X-ray Focusing Using Microchannel Plates", Proc. SPIE **1546**, to be published.

5. M.A. Blokhin, "Methods of X-Ray Spectroscopic Research", ch.IV, Pergamon Press, Oxford (1965).

6. Y.J. Cauchois, Phys. et Rad., **3**(7), 320 (1932); **4**(2), 61 (1933)

7. H.N. Chapman, A.V. Rode, "Geometric Optics of Arrays of Reflective Surfaces", in preparation (1992).

8. M. Born and E. Wolf, Principles of Optics, Pergamon Press, Oxford (1980).

9. J. Ladilas Wiza, "Microchannel Plate Detectors", Nucl. Instrum. Meth. **162**, 587 (1979).

10. H.N. Chapman, A.V. Rode, K.A. Nugent, S.W. Wilkins, and B. Luther-Davies, "X-Ray Focusing Using Microchannel Plates II: Experiments", in preparation (1991).

SCALING LAW FOR THE TEMPERATURE AND ENERGY OF X–RAY IN LASER CAVITY TARGETS AND IN COMPARISON WITH EXPERIMENTAL MEASUREMENTS

Zhang Jun and Pei Wenbing

Institute of Applied Physics and Computational Mathematics
P.O. Bx 8009 Beijing, 100088. P.R. China

ABSTRACT

The dependence of the radiation temperature on the X–ray energy produced by the laser in the cavity targets is investigated in theory according to the information given by the numerical simulation of the laser X–ray conversion, and the corresponding scaling law of them is obtained. The method estimates the X–ray conversion efficiency of the cavity targets. According to the yielded scaling law, the X–ray conversion efficiency of the cavity targets can be determined by the laser absorption efficiency and the radiation temperature measured experimentally. By the method the X–ray conversion efficiency about 45–50%, is obtained for the cavity experiment conducted in 1989,1990, the calculated radiation temperatures about 140–150 ev are in agreement with ones measured for two beam targets, laser energy each beam ≈ 350–650J, pulse width ≈ 0.6–1.1ns.

I. INTRODUCTION

Laser–X–ray Conversion Efficiency (LXCE) is an important and interesting characteristic index in the cavity targets physics studying of the laser fusion, and the higher it is, the better. However it is difficult to measure directly the LXCE in the cavity targets case, and it often needs to be deduced by measuring the radiation temperature or the energy of X–ray. Therefore, to give the scaling law is of practical significance.

From our numerical simulation results, we know that almost all the X–ray produced by laser in the cavity would penetrate into the wall of the cavity and heat it except the energies escaped through the open holes. Therefore, if we know the dependence between the radiation temperature and the penetrated energy, and the laser energy absorbed by the cavity targets which can be directly measured, the LXCE can be determined by the radiation temperature measured experimentally.

The purpose of this paper is to deduce the scaling dependence of the radiation temperature on the penetrated energy. To do this we suppose the equilibrium between X–ray and wall matter, where the density is high, is reached, so we can use the radiation transfer equation of single temperature to investigate the radiation penetration problem.

Laser Interaction and Related Plasma Phenomena, Vol. 10
Edited by G.H. Miley and H. Hora, Plenum Press, New York, 1992

125

II. TEMPORAL CHARACTERISTIC OF X–RAY TEMPERATURE IN LASER CAVITY TARGETS

The variation of the radiation temperature with the time is drawn in fig.1 that is measured from the diagnostic hole of the cavity target on the LF–12 Laser in 1989[1]. The numerical simulations also give the similar curve. The results show that X–ray pulse is widened compared to the usual laser pulse and its width is about 1.5–2.0 times of the laser pulse. This fact has been verified by much experiments and numerical calculations.

The measured curve (fig.1) can be fitted into the following form

$$T = \begin{cases} T_{01} exp(\alpha_1 t) & if \quad 0 < t \leqslant 1.5ns \\ T_{02} exp(-\alpha_2 t) & if \quad 1.5 \leqslant t < 4ns \end{cases}$$

where $T_{01} = 62.25ev$, $\alpha_1 = 0.518 / ns$; $T_{02} = 189.85ev$, $\alpha_2 = 0.258 / ns$.

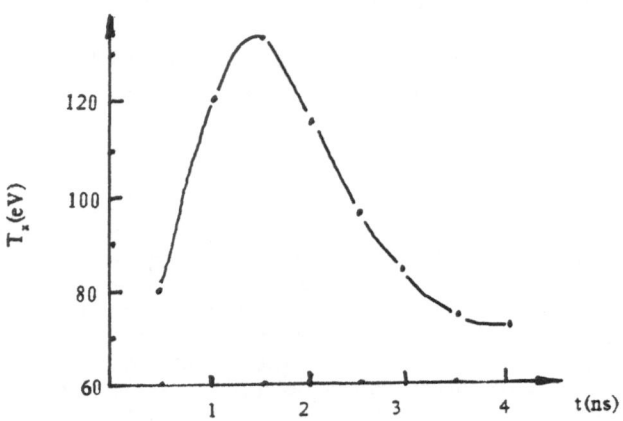

Fig.1. Variation of X–ray temperature in the cavity target with time

III. THE SELF–SIMILAR SOLUTION OF THE RADIATION THERMAL CONDUCTION EQUATION

Because in the interior of the cavity the plasma density is low and the radiation mean free path is large, the spase distribution of the X–ray temperature is smooth, and according to the above obtained results, in the following we assume the radiation temperature of the penetration region boundary increases exponentially with time.

The material of the wall is high Z element, Au, the thickness of which is much larger than the radiation mean free path. So the diffusion approximation can be used.

The radiation mean free path is taken as

$$l = l_0 \left(\frac{\rho_0}{\rho} \right)^m \left(\frac{T}{T_0} \right)^n , \tag{1}$$

and the equations of state are

$$\left. \begin{aligned} P &= P_0 \sigma' \left(\frac{T}{T_0} \right)^s \\ c_v &= c_{v0} \sigma'^{-1} \left(\frac{T}{T_0} \right)^{s-1} \end{aligned} \right\} \tag{2}$$

126

And we will use planar geometry since the cavity radius is much larger than the thickness of the wall.

With the above assumptions, we can obtain the self—similar solutions of the thermal conduction equation in two extreme conditions.

In Ref.2, a self—similar solution has been given with the boundary temperature increasing exponentially with time , but the equation of the state of perfect gas has been used. Ref.3 has shown another self—similar solution with constant boundary temperature with using also the same equation of state.

1. SELF—SIMILAR SOLUTION IN CONSTANT DENSITY APPROXIMATION

The radiation thermal conduction equation of 1—D plane geometry is

$$C_v \frac{\partial T}{\partial t} = \frac{ac}{3\rho_0} \frac{\partial}{\partial x} \left(l \frac{\rho}{\rho_0} \frac{\partial T^4}{\partial x} \right) \tag{3}$$

To compare with the solution in constant pressure approximation, which we will discuss latter, we take Lagrangian space coordinate, and dimensionless temperature T and density ρ :

$$\tau = \frac{T}{T_0}, \qquad \sigma = \frac{\rho}{\rho_0} \tag{4}$$

Substituting (1), (2) and (4) into (3), the equation becomes as

$$\frac{\partial \overline{T}}{\partial t} = K \frac{\partial^2 \overline{T}^{m_1}}{\partial x^2} \tag{5}$$

where

$$\left. \begin{aligned} \overline{T} &= \tau^s \\ m_1 &= \frac{n+4}{s} \\ K &= \frac{4acl_0 s T_0^3}{3(n+4)\rho_0 C_{v0} \sigma^{m+r-2}} \end{aligned} \right\} \tag{6}$$

Assuming

$$\overline{T}(x,t) = J_0 e^{2\alpha t} f(z) \tag{7}$$

and taking self—similarity parameter as

$$z = \left(\frac{\alpha}{KJ_0^{m_1-1}} \right)^{\frac{1}{2}} x \exp[-(m_1-1)\alpha t] \tag{8}$$

eq.(5) is reduced as

$$2f(z) - (m_1-1)z \frac{df}{dz} = \frac{d^2 f^{m_1}}{dz^2} \tag{9}$$

The asymptotic solution of eq.(9) can be obtained by means of power series expansion

$$f(z) = B\left(1 - \frac{z}{z_0}\right)^{n_1}\left(1 + b\left(1 - \frac{z}{z_0}\right)^{n_2} + \cdots\right) \tag{10}$$

Substituting eq.(10) into eq.(9) and comparing the coefficient of the same power of the two sides of the equation, we can get that

$$n_1 = \frac{1}{m_1 - 1}, \qquad\qquad n_2 = 1$$

$$B = \left[\frac{z_0(m_1 - 1)}{m_1^{\frac{1}{2}}}\right]^{\frac{2}{m_1 - 1}}, \qquad\qquad b = \frac{1}{2m_1(m_1 - 1)} \tag{11}$$

Using the boundary condition

$$f(0) = 1, \qquad\qquad at \qquad z = 0 \tag{12}$$

we can obtain the location of the radiation wave front in the similarity space

$$z_0 = \frac{m_1^{\frac{1}{2}}}{m_1 - 1} \tag{13}$$

and therefore the path of the radiation wave

$$X_R = z_0\sqrt{\frac{K}{\alpha}}\, \bar{T}_i^{\frac{m_1 - 1}{2}} = Z_0\sqrt{\frac{K}{\alpha}}\left(\frac{T_i}{T_0}\right)^{\frac{n+4-s}{2}} \tag{14}$$

where $\bar{T}_i = J_0 e^{2\alpha t}$ is $\bar{T}(0,t)$ at the beginning location of the radiation wave. Solution (11) and (12) show that if $n = 2$, $s = 1.5$, the error introduced by omitting the second term in the expanding expression (11) is never more than 4%. So the solution of eq.(9) can be written as

$$\bar{T}(x,t) = J_0 e^{2\alpha t}\left(1 - \frac{z}{z_0}\right)^{\frac{1}{m_1 - 1}} \tag{15}$$

In comparison with the numerical solution of eq.(9), the error of the analytical solution eq.(15) is less than 4%[4].

2. SELF–SIMILAR SOLUTION IN CONSTANT PRESSURE APPROXIMATION

We can also give a self–similar solution in constant pressure approximotion, in this case the equation to be solved is

$$C_v\frac{\partial T}{\partial t} - T\left(\frac{\partial P}{\partial T}\right)_{\rho}\frac{1}{\rho^2}\frac{\partial \rho}{\partial t} = \frac{ac}{3\rho_0}\frac{\partial}{\partial x}\left(l\frac{\rho}{\rho_0}\frac{\partial T^4}{\partial x}\right) \tag{16}$$

where we have taken Lagrangian coordinate.

The constant pressure has the form

$$P \propto \rho'.T^s = \rho_i'T_i^s \tag{17'}$$

the reduced density is

$$\sigma = \sigma_i \left(\frac{\tau_i}{\tau}\right)^{\frac{s}{r}} \tag{17}$$

where the parameters with subscript i show the quantities at beginning surface of radiation wave.

Using eqs.(1), (2) and (17′), eq.(16) is reduced as follows

$$\frac{\partial \tilde{T}}{\partial t} - K_2 \frac{\tilde{T}}{\tau_i} \frac{\partial \tau_i}{\partial t} = \overline{K} \tau_i^{(2-m-r)\frac{s}{r}} \frac{\partial^2 \tilde{T}^{m_2}}{\partial x^2} \tag{18}$$

where

$$
\left.
\begin{aligned}
&\tilde{T} = \tau^{\frac{s}{r}} \\
&m_2 = (n+4)\frac{r}{s} + m - 1 \\
&\overline{K} = \frac{4acl_0 T_0^{3\frac{s}{r}}}{3\left[n+4+(m-1)\frac{s}{r}\right]\sigma_i^{m+r-2}K_1} \\
&K_1 = C_{vo}\rho_0 + \frac{s^2}{r}\frac{P_0}{T_0} \\
&K_2 = \frac{s^3}{r^2}\frac{P_0}{T_0 K_1}
\end{aligned}
\right\} \tag{19}
$$

Assuming

$$
\left.
\begin{aligned}
&\tilde{T}(x,t) = J_0 e^{2\alpha t} f(z) \\
&\tilde{T}_i = J_0 e^{2\alpha t}, \qquad \tau_i = (J_0 e^{2\alpha t})^{\frac{r}{s}}
\end{aligned}
\right\} \tag{20}
$$

and taking the similarity parameter as

$$z = X\sqrt{\frac{\alpha}{k}}(J_0 e^{2\alpha t})^{-\frac{1}{2}(m_2-m-r+1)} \tag{21}$$

eq.(18) can be reduced to

$$2(1 - \frac{r}{s}K_2)f(z) - (m_2 - m - r + 1)z\frac{df}{dz} = \frac{d^2 f^{m_2}}{dz^2} \tag{22}$$

We also find a solution of the same form as eq.(10), with

$$
\left.
\begin{aligned}
&n_1 = \frac{1}{m_2 - 1} \\
&B = \left[z_0^2 \frac{(m_2-1)(m_2-m-r+1)}{m_2}\right]^{\frac{1}{m_2-1}}
\end{aligned}
\right\} \tag{23}
$$

The location of the radiation wave front in the similarity space is determined by

129

$$z_0 = \left(\frac{m_2}{(m_2 - 1)(m_2 - m - r + 1)} \right)^{\frac{1}{2}} \qquad (24)$$

and the path of the radiation wave in the Lagrangian frame is

$$X_R = z_0 \sqrt{\frac{K}{\alpha}} \tau_i^{\frac{s}{2r}(m_2 - m - r + 1)} = Z_0 \sqrt{\frac{K}{\alpha}} \left(\frac{T_i}{T_0} \right)^{\frac{n+4-s}{2}} \qquad (25)$$

The asymptotic solution of eq.(18) is

$$\tilde{T}(x,t) = J_0 e^{2\alpha t} \left(1 - \frac{z}{z_0} \right)^{\frac{1}{m_2 - 1}} \qquad (26)$$

IV. THE SCALING LAW OF THE TEMPERATURE AND THE ENERGY OF THE RADIATION

The scaling law can be easily found by the above results. When the radiation wave spreads to X_R, the energy that is lost due to heating the material of the wall is

$$E = \int_0^{X_R} \rho_0 dx \int_0^T C_V dT' = \int_0^{X_R} \frac{1}{s} C_V \rho_0 T dx \qquad (27)$$

According to the energy conservation, the energy should be equal to the radiation energy flux q_R at the boundary up to time t, e.g. $q_R = E$. Integrating q_R over the area of the interior surface of the cavity, we obtain the total radiation energy

$$F_R = q_R S = ES \qquad (28)$$

where S is the total area of the wall. Substituting the self–similar solution obtained in the above section into eq.(27), the scaling law can be got as follows.

1. Results of constant density approximation

$$\frac{F_R}{S} = E = C_{V0} T_0 \rho_0 \sigma^{r-1} \sqrt{\frac{k}{\alpha m_1}} \frac{1}{s} \tau_i^{\frac{n+4+s}{2}} \qquad (29)$$

or

$$\tau_i = \xi_1 \sigma^{\frac{m-r}{n+4+s}} \left(\frac{F_R}{S} \right)^{\frac{2}{n+4+s}} \qquad (30)$$

where

$$\left. \begin{aligned} \xi_1 &= \left(\frac{1}{T_0 \rho_0} \sqrt{\frac{(n+4)\alpha}{K'}} \right)^{\frac{2}{n+4+s}} \\ K' &= \frac{4acl_0 T_0^3 C_{V0}}{3(n+4)\rho_0} \end{aligned} \right\} \qquad (31)$$

If the dimensionless temperature is taken as $T_0 = 1$, then we have the relation

130

$$T_i = \xi_1 \sigma^{\frac{m-r}{n+4+s}} \left(\frac{F_R}{S}\right)^{\frac{2}{n+4+s}} \tag{32}$$

2. Results of constant pressure approximation

$$\frac{F_R}{S} = E = \frac{C_{v0} T_0 \rho_0}{s} \left[\frac{(m_2 - 1)\overline{K}}{m_2(m_2 - m - r + 1)\alpha}\right]^{\frac{1}{2}} \sigma^{r-1} \tau_i^{\frac{n+4+s}{2}} \tag{33}$$

or

$$\tau_i = \xi_2 \sigma_i^{\frac{m-r}{n+4+s}} \left(\frac{F_R}{S}\right)^{\frac{2}{n+4+s}} \tag{34}$$

where

$$\left.\begin{array}{l} \xi_2 = \left[\frac{1}{C_{v0} T_0 \rho_0} \left(\frac{\alpha m_2(m_2 - m - r + 1)}{K'(m_2 - 1)}\right)^{\frac{1}{2}}\right]^{\frac{2}{n+4+s}} \\[3mm] \overline{K'} = \dfrac{4ac l_0 T_0^3}{3[(n+4)rs + (m-1)s^2]K_1} \end{array}\right\} \tag{35}$$

and

$$T_i = \xi_2 \sigma_i^{\frac{m-r}{n+4+s}} \left(\frac{F_R}{S}\right)^{\frac{2}{n+4+s}} \tag{36}$$

From the above results we can see that these two scaling laws in the above approximations have the same form, in which the powers are the same and only the coefficients are different. In addition, we have also obtained the same power of the scaling law in case of constant temperature boundary condition. Thus we believe that the scaling dependence of radiation temperature on energy,especially the exponent power $\frac{2}{n+4+s}$, is quite reasonable and reliable,even so in more complicated situation.

If fundamental parameters are taken as

$$l = 3.42 \times 10^{-3} \sigma^{-1.45} \tau^{2.2}, P = 27.2\sigma^{0.9569} \tau^{1.482}, C_v = 23.4\sigma^{0.0431} \tau^{0.482}$$

that is, $l_0 = 3.42 \times 10^{-3}$, m = 1.45, n = 2.2, $p_0 = 27.2$, r = 0.9569, s = 1.482, $c_{v0} = 23.4$, then in the constant density approximation the scaling law becomes:

$$\left.\begin{array}{l} T_x = 195.3\sigma^{0.0642} \left(\dfrac{E_x}{S}\right)^{0.26} \\[3mm] \dfrac{E_x}{S} = \sigma^{-0.247} \left(\dfrac{T_x}{195.3}\right)^{3.846} \end{array}\right\} \tag{37}$$

In (37) T_i is replaced by T_x, F_R by E_x and E_x is in 10^4J, S in cm^2, and T_x in ev. The temperature T_x in (37) is its maximum value, and E_x is in the total radiation energy in the cavity. In (37) we have taken account of the time behavior of radiation temperature during its downhill stage by dividing a factor

$$\left(1 + \sqrt{\frac{\alpha_2}{\alpha_1}}\right)^{0.26} \tag{38}$$

Similarly, the results in the constant pressure approximation are

$$
\left.
\begin{aligned}
T_x &= 200\sigma_i^{0.0642}\left(\frac{E_x}{S}\right)^{0.26} \\[2mm]
\frac{E_x}{S} &= \sigma_i^{-0.247}\left(\frac{T_x}{200}\right)^{3.846}
\end{aligned}
\right\}
\tag{39}
$$

where $\sigma_i \equiv \rho_i/\rho_0$ is the reduced density at the inner boundary. The temperature denoted in eq.(39) is also its maximum value. Therefore, if we know the radiation temperature in the cavity from experiment, we can deduce the radiation energy penetrated into the wall of the cavity, and vice versa.

V. SCALING LAW OBTAINED FROM RADIATION HYDRODYNAMICS

Valuable scaling laws can be obtained by studying similarity of the equations of radiation hydrodynamics[5]. The penetration depth of radiation can be shown by the location of the front of radiation wave in the system of Lagrangian space coordinate:

$$
X_R \sim \tau^{\frac{1}{2}} V_i^{\frac{m+r-2}{2}} T_i^{\frac{n+4-s}{2}} .
\tag{40}
$$

In the duration $(0,t)$ the total radiation energy penetrated into the cold medium from the boundary at $x=0$ with the radiation source can be described by the following form

$$
\frac{E_R}{S} \sim \tau^{\frac{1}{2}} V_i^{\frac{m-r}{2}} T_i^{\frac{n+4+s}{2}} .
\tag{41}
$$

The defference between Eq.(41) and Eqs.(32) or (36) is only the time factor,but the coefficients on the right side of Eqs.(32) and (36) include the factor α,the dimension of which is inverse of the time.

It is verified by numerical simulations of radiation hydrodynamics that the feature of the radiation wave is mainly determined by the boundary temperature and pulse width of radiation source, and quite weakly by the hydrodynamic phenomena because the power exponents of the density in Eqs.(40) and (41) are small.

According to the power exponents given by the above equations we have fitted the numerical results of radiation hydrodynamics for the element Au and obtained the relation of radiation temperature as a function of both the areal energy and pulse width of radiation source:

$$
T_x = 13.2\left[\frac{(1-\eta_e)E_x}{S/cm^2}\right]^{0.26}\left(\frac{\tau}{ns}\right)^{-0.13} \quad (ev)
\tag{42}
$$

where η_e is the escaped fraction of radiation energy E_x produced in the cavity targets, $(1-\eta_e)E_x$ is the radiation flux penetrated into the cavity wall.

The laser absorption efficiency η_a and X-ray conversion efficiency η_x are defined as

$$
\eta_a = \frac{E_a}{E_L}
\tag{43}
$$

$$
\eta_x = \frac{E_x}{E_a}
\tag{44}
$$

where E_L is the laser energy entered the cavity target, and E_a is the laser energy absorbed by the cavity target.

Fitting numerical simulation results of η_a and η_x we have obtained the empirical formulas

$$\eta_a = 0.755 I_{14}^{-0.118}, \tag{45}$$

$$\eta_x = 0.450 I_{14}^{-0.116}, \tag{46}$$

or

$$\eta_a \eta_x = 0.340 I_{14}^{-0.23} \tag{47}$$

where I_{14} is laser intensity in $10^{14} W/Cm^2$. Substituting Eqs.(43)–(47) into Eq.(42) we have obtained semi–empiric and semi–theoretical formula of the radiation temperature,

$$T_x = 9.95 I_{14}^{-0.06} \left[\frac{(1 - \eta_e) E_L / J}{S / Cm^2} \right]^{0.26} \left(\frac{\tau}{ns} \right)^{-0.13} \quad (ev)$$

$$= 19.9 \left(\frac{S^*}{S} \right)^{0.06} (1 - \eta_e)^{0.26} \left(\frac{E_L / J}{S / cm^2} \right)^{0.20} \left(\frac{\tau}{ns} \right)^{-0.07} \quad (ev), \tag{48}$$

where S^* is the shot area of the laser inside cavity target. Obviously, the dependence of radiation temperature on the ratio (S^* / S) is insensitive. Letting (S^* / S) be 0.5, we have the following equation,

$$T_x = 19.1 (1 - \eta_e)^{0.26} \left(\frac{E_L / J}{S / cm^2} \right)^{0.20} \left(\frac{\tau}{ns} \right)^{-0.07} \quad (ev). \tag{49}$$

Scaling law (49) has an important application. It can be used to determine the LXCE[6]. Since the power exponents of the temperature in eq.(41) merely depend on those of the temperature in the equaton of state and opacity, and so do the power exponents of the density, these scaling laws can provide a method experimentally to determine the power exponents of the temperature in the equation of state and opacity.

VI .COMPARISON OF THEORETICAL SCALING LAW WITH EXPERIMENTAL RESULTS

The laser energy absorbed in the cavity and laser pulse width can be exactly measured experimentally. The total area of the interior surface of the cavity can also be known. The radiation temperature in the cavity can be measured through the diagnostic hole or the incident hole of laser, and can also be calculated by Eq.(49). The radiation temperature in the cavity have been measured experimentally on LF12 laser in 1989 and 1990, listed in table.1 and 2, respectively. It is seen that the theoretical results calculated by Eq.(49) are in well agreement with experimental ones.

133

Table 1. Comparison of radiation temperature calculated by Eq.(49) with experimental ones measured for the two beams targets on LF12 in 1989 , $(\eta_e = 0.4)$.

target number	E_L (10^2J)	τ (ns)	T_X (ev)$_{cal}$	$\overline{T}_X$ (ev)$_{cal}$	T_X (ev)$_{exp}$
89052503	3.69 4.56	0.960 0.998	140.2 145.9	143.0	142
89052601	4.68 5.10	0.820 1.350	148.7 146.1	147.4	145
89053001	3.71 3.55	1.100 1.120	139.1 137.7	138.4	145
89053002	3.68 3.60	0.620 0.750	144.5 141.9	143.2	145
89053101	4.27 4.15	0.820 1.000	146.0 143.2	144.6	148
89053102	4.52 4.64	0.690 0.915	149.4 147.3	148.3	146

Table 2. Comparison of radiation temperature calculated by Eq.(49) with experimental ones measured for the two beams targets on LF12 in 1990 , $(\eta_e = 0.5)$.

target number	E_L (10^2J)	τ (ns)	T_X (ev)$_{cal}$	T_X (ev)$_{exp}$
102702NB	5.27 5.27	1.02 0.95	143.0 143.7	145 135
102703NB	6.41 6.18	0.96 1.00	149.3 147.8	153 143
102801NB	6.18 5.86	1.02 0.93	147.6 147.0	146 130
112102NB	5.41 5.82	0.59 0.61	149.3 151.2	151 155
112202NB	5.11 5.58	0.62 0.69	147.1 148.6	146 145
112301NB	5.65 5.62	0.68 0.51	149.1 152.0	156 147
112602NB	6.57 5.85	0.62 0.69	154.7 150.0	153 154
112603NB	5.78 5.52	0.89 0.71	147.0 148.0	145 149

VII. SUMMARY

The scaling laws of radiation temperature by various analytical approaches are same. So it is sure that these scaling laws are reasonable and available to some real nonsimilarity problems. Because the cavity target usually has two—dimensional structure, the exact determination of the LXCE is a complicated problem. Here we have provided an alternative approximate method to determine the LXCE . The method is probablely feasible when the escaped energy due to the effects of 2—D is known.

For the laser of wave length 1.06μ, the LXCE of the cavity targets is about 60—70% by means of the 1—D numerical simulations, and the LXCE determined by the scaling law is about 45—50% for the two—beam targets. The results show that the 2—D effect makes the 1—D results reduce about 10—20%.

Both the theory and the experiment have verified that the LXCE of the cavity target is greater than that of the planar target obviously. It is because of the limited volume of the cavity, and hence the kinetic energy converted from the laser energy greatly decreases.

The best approach of raising radiation temperature is to increase laser energy. If the other factors are improved, the various effects in target design must summarily be considered.

ACKNOWLEDGMENTS

We wish to thank Z.J.Zheng,K.X.Sun et al of southwest Institute of Nuclear Physics and Chemistry for providing us with rich experiment data from the LF12 laser, We also thank our colleague for providing us with a lot of 1—D numerical results.

REFERENCE

1. Z.J.Zheng,K.X.Sun et al. private communications (1989)
2. R.E.Marshak,Phys. Fluids, 1 (1958) 24
3. 'A.G.Petschek,R.E.Williamson, LAMS—2421 (1960), Las Alamos National Laboretery.
4. P.J.Gu,private communications (1988)
5. Zhang Jun and Pei Wenbing, "Similarity Transformations of Radiation Hydrodynamics Equations and Investigation on Laser of Radiative Conduction", (1991), to be published in the Phys.Fluids B.
6. Zhang Jun et al, ACTA Phycica Sinica, 40, (1991)424 (in Chinese).

X-RAY LASING IN A MUON CATALYZED FUSION SYSTEM

S. Eliezer[1,2], Z. Henis[2]

[1]*Instituto de Fusión Nuclear, E.T.S. Ingenieros Industriales*
José Gutierrez Abascal, 2; 28006 Madrid, Spain
[2]*Soreq N.R.C. Yavne 70600, Israel*

ABSTRACT

When an energetic negative muon enters a dense DT target it initiates the chain of reactions: (a) stopping and capture of μ into atomic levels of deuterium and tritium. (b)μ⁻ transfer from the duterium to the tritium which can occur both from excited and from the ground state of dμ. (d) Nuclear fusion of dt (the capital letters D and T denote the deuterium and tritium atoms while the small letters denote the nuclei). The muon is either released or captured by the nuclear fusion products (sticking).

The possibility of an X-ray laser with photon energy of 370 e V induced by the deexcitation of dμ is suggested in this paper. A rod of dense deuterium-tritium mixture is irradiated by a muon beam. The stuck populations of the excited states during the cascade of dμ and tμ are used for lasing.

INTRODUCTION

In 1947 after the discovery of the pion (π) and the π decay into μ, it was realized that the muon is a heavy electron with a mass (m_μ)

$$m_\mu = 206.77\, m_e = 105.66\, \frac{MeV}{c^2} \qquad (1)$$

where m_e is the electron mass. μ and e also differ by their lepton quantum number. Moreover, the muon is an unstable particle with a lifetime (τ_μ) of

$$\tau_\mu = 2.2.10^{-6} sec \qquad (2)$$

The size of a "μ-atom" (e.g. the hydrogen atom where the electron is replaced by a muon) is of the order of the muon Bohr radius a_μ given by

$$a_\mu = \frac{m_e}{m_\mu} a_e = 2.6.10^{-11} cm \qquad (3)$$

where a_e, is the Bohr radius of an electron atom.

When an energetic negative muon (μ^-) enters a compressed hydrogen gas (e.g. hydrogen liquid density defined by $n_0 = 4.25.10^{22}$ atoms/cm^3), the following chain of reactions ocfccurs: (a) Slowing down of μ^-. (b) Capture of μ^- into atomic levels and the cascade to the atomic ground state (1s). (c) "μ-molecular" formation and the de-excitation to the molecular ground state. (d) Nuclear fusion of the nuclei (usually hydrogen isotopes) which are kept close together by the negative muon. The μ^- is either released or captured by the nuclear fusion products (μ^- sticking). If the lifetime of the muon is long compared to the time scale of the other processes ([a] to [d]), then many fusion reactions can occur during the lifetime of a muon. This chain of reactions, resulting in the nuclear fusion process, is usually called muon catalyzed fusion.

The idea that negative muons might be able to catalyzed proton-deutron (p-d) fusion was first considered by Frank[1]. Deuterium-deuterium and deuterium-tritium fusions catalyzed by muons were suggested by Sakharov & Leebedev[2] and further redicovedred by Zelodovich[3] and analyzed by Jackson[4]. The first experimental observation by Alvarez et al[5]. of (pdμ) fusion

$$(pd\mu) \rightarrow {}^3He + \mu^- (5.4\ MeV) \qquad (4)$$

was a rediscovery and for short period of time the Berkeley group thought that they had "solved all the fuel problems of mankind for the rest of the time"[6].

+After this discovery the muon catalyzed fusion was studied in England[7] through the reaction

$$(pd\mu) \rightarrow {}^3He + \gamma (5.4\ MeV) + \mu^-$$ (5)

These experiments were followed by a series of theoretical calculations, mainly by Soviet scientist[8]. The main research resulted in the prediction that one could not expect more than a few catalyzed fusions per muon during his lifetime. It thus appeared that muon-catalyzed fusion reactions were useless as an energy source. This conclusion was questioned after Dzhelepov et al. [9]. experiments which showed that the rate of dd molecule formation depends strongly on temperature. Following this experiment it was suggested by Vesman[10] that the muon-molecules can be formed resonantly if there are weakly bound states of these molecules (i.e. the binding energy is smaller than the dissociation energy of the hydrogen molecules~ few e V). This idea was already mentioned by Zeldovich[3]. However, Dzhelepov et al. experiment, Vesman´s paper and the following calculations[11, 12] and experimental measurements[13] of the energy levels of the muon molecules were crucial steps in renewing the interest in muon catalyzed fusion. Using the same resonance theory formation for the dtμ molecule, as for the ddμ molecule, a very large dtμ formation rate was predicted[11] which was confirmed experimentally.

The number of fusion X_μ that one μ^- catalyzes during its lifetime were measured experimentaly to be of the order

$$
\begin{aligned}
&X_\mu \sim 1 && \text{for P-D (hydrogen-deuterium)} \\
&X_\mu \sim 10 && \text{for D-D} \\
&X_\mu \sim 100 && \text{for D-T}
\end{aligned}
$$ (6)

Therefore, the deuterium-tritium mixture seems to be the most interesting system. When a negative muon enters a dense deuterium (D)-tritium (T) target it starts a chain of reactions: (we denote by capital letters D and T the deuterium and tritium atoms while the small letter denotes the nuclei).

(a) Stopping and capture of μ^- by d or t and the cascade to the ground state of the hydrogen like atom,

$$\lambda_a: \quad \mu^- + D \rightarrow (d\mu) + e^-$$ (7)

$$\lambda_a: \quad \mu^- + T \to (t\mu) + e^- \tag{8}$$

where e- denotes an electron an λ_a is the rate of the processes (7) and (8). In general, the rate λ (sec^{-1}) is given by

$$\lambda = n\sigma v \text{ [sec}^{-1}\text{]} \tag{9}$$

where n [cm^{-3}] is the density of the target, σ[cm^2] is the cross section describing the process under consideration and v [cm/sec] is the velocity of the projectile. The rate of the muonic atomic formation in equations (7) and (8) for the liquid hydrogen density (n = 4.25×10^{22} cm^{-3}) is 4×10^{12} sec^{-1}

(b) μ^- transfer from the deuterium to the tritium

$$\lambda_{dt}: \quad (d\mu) + t \to (t\mu) + d \tag{10}$$

The rate for this process at liquid density is estimated [14] to be 2×10^8 sec^{-1}. The 48 eV difference in the binding energies of (dμ) and (tμ) causes the transfer of μ in an irreversible process described by equation (10).

(c) At this stage, a (dtμ) molecule ion may be formed at the center of an H$_2$ tipe molecule. The formation of the relevant hydrogen type mumolecules are described by

$$\lambda_{dd\mu}: \quad (d\mu) + D_2 \to [(dd\mu)\, d2e] \tag{11}$$

$$\lambda_{dt\mu d}: \quad (t\mu) + D_2 \to [(dd\mu)\, d2e] \tag{12}$$

$$\lambda_{dt\mu t}: \quad (t\mu) + DT \to [(dt\mu)\, t2e] \tag{13}$$

$$\lambda_{tt\mu}: \quad (t\mu) + T_2 \to [(tt\mu)\, t2e] \tag{14}$$

where in general the mumolecule is in an excited state. The values of the rates describing equations (11) to (14) were measured [14] to be $\lambda_{dd\mu} \approx 3 \times 10^6$ sec^{-1}, $\lambda_{tt\mu} \approx 3 \times 10^6$ sec^{-1}, and $\lambda_{dt\mu} \geq 10^8$ sec^{-1}, where $\lambda_{dt\mu}$ is defined by

$$\lambda_{dt\mu} = \lambda_{dt\mu\text{-}d}C_d + \lambda_{dt\mu\text{-}t}C_t \tag{15}$$

C_d and C_t are the concentrations of the deuterium and tritium nuclei, so that if the presence of He and other impurities are neglected one has,

$$C_d + C_t = 1 \tag{16}$$

The value of the dtμ rate is almost two orders of magnitude larger than the ddμ and ttμ rates. This phenomena can be explained by the resonant formation of these molecules. A degeneracy in the excited state of the dtμ ion and the excited state of the electron molecular complex is causing a strong resonance effect. The rate $\lambda_{dt\mu}$ is found to be dependent on the temperature since the kinetic energy of the tμ atom, which forms the dtμ ion, is temperature dependent.

(d) Next in the chain of reactions one has the deexcitation of dtμ, ddμ and ttμ ions to their ground states and the occurrence of the nuclear fusion reactions

$$\lambda_f: \quad d + t \rightarrow {}^4He + n + 17.6 \, MeV \tag{17}$$

$$\tfrac{1}{2}\lambda_{fd}: \quad d + d \rightarrow {}^3He + n + 3.3 \, MeV \tag{18}$$

$$\tfrac{1}{2}\lambda_{fd}: \quad d + d \rightarrow t + p + 4 \, MeV \tag{19}$$

$$\lambda_{ft}: \quad t + t \rightarrow {}^3He + 2n + 10 \, MeV \tag{20}$$

The fusion rates are estimated[15] to be $\lambda_f \approx 10^{12}$ sec^{-1} for the dt nuclear reaction while $\lambda_{fd} \approx \lambda_{ft} \approx 10^{11}$ sec^{-1}.

(e) During the fusion reaction there is a possibility that the muon sticks to the ^{4}He by forming a muonic helium ion (^{4}Heμ). In this case, the muon stays bound to the He particle and it is lost for the chain of reactions described above [(a) to (d)]. The sticking propabilities for the reactions given by equations (17) - (20) were estimated to be [16] $W_s \approx 0.5 \times 10^{-2}$, $W_d \approx 0.13$ [for equation (18)], $W_d' \approx 0.003$ [for equation (19)] and $W_t \approx 0.05$, implying that the sticking fraction to the He particle in the dt fusion is about 0.5% of the events, in the dd fusion about 15% and in the tt fusion about 5%. For the most important dt case, the value of W_s implies that the moun sticks to the He particle after catalyzing at most $1/W_s \approx 200$ fusions, no matter how fast are the other processes leading to the mesomolecule formation and fusion. The sticking process may be the bottle neck of the muon catalyzed fusion idea.

POPULATION INVERSION

The muonic cascade in the atomic levels is determined by the competition between several processes [17-20]:

(1) The chemical reaction

$$(\mu d)_i + D \rightarrow (\mu d)_f + D + D \qquad (21)$$

(2) The external Auger effect:

$$(\mu d)_i + D_2 \rightarrow (\mu d)_f + D_2{}^+ + e^- \qquad (22)$$

The Auger transitions have a dipole nature ($\Delta l = l_i - l_f = \pm 1$) and transitions with the minimum change in the principal quantum numer n permitted for the ionization of the moleucle D_2 are predominat. The rate of the Auger transition is density dependent since they are caused by collisions of excited muon atoms. The Auger transitions rate is maximal at n = 7; transitions with $\Delta n = 1$ becoming possible from this level. With decreasing n the Auger transition rate falls rapidly [15]

(3) Radiative transitions

$$(\mu d)_i \rightarrow (\mu d)_f + \gamma \qquad (23)$$

at a rate [19] given by

$$\lambda_{if} = 2/3\alpha \left| R_{if} \right|^2 w_{if}^3 \qquad (24)$$

where α is the fine-structure constant, R_{if} and w_{if} are the dipole moment and the energy of the transition respectively. The radiative transitions are important for small values of n.

(4) Stark mixing of the different states nl with a given n of the dμ atom. The Stark transitions are important at high densities (of the order of the hydrogen liquid density). For example, deexcitation of the state 2S of the dμ atom can occur through radiative transitions to the state 1S in a Stark collision [17];

$$\lambda(2S \rightarrow 2P \rightarrow 1S) = 4 \times 10^9 \emptyset \ [s^{-1}] \tag{25}$$

where $\emptyset$ is the density measured in liquid hydrogen density units.

(5) μ-transfer to tritium whichdepends both on the density $\emptyset$ and the tritium concentration C_t.

Similar processes [(21) to (25)] occur for the muonic cascade in tμ atoms.

The kinetics of the muon cascade in a deuterium-tritium mixture is thus determined by the following rate equations[21]:

$$\frac{dQ_i}{dt} = -\lambda_i Q_i + \Sigma \lambda_{ji} Q_j, \ \ 1 \leq i \leq 5, \tag{26}$$

where Q is the population of the level i of the dμ atom λ_i is the sum of the rates of the processes of deexcitation of the level i including muon transfer and λ_{ji}, is the rate of the j →i transition. Similar equations describe the population of the level i of the tμ atom (P_i will denote the $(t\mu)_i$ population).

We solved numerically the rate equations (26) starting at N = 5, with initial conditions $Q_5 = 1$, $Q_i = 0$ for $1 \leq i \leq 4$ (and similarly $P_5 = 1$, $P_i = 0$ for $1 \leq i \leq 4$). The levels 2P and 2S are treated separately. The population of the levels 3S, 3P and 3D are assumed to be temperature dependent and given by the Boltzmann distribution:

$$\begin{aligned} Q_{3D} &= 5Q_{3S} \exp(-\Delta_{3D}/T) \\ Q_{3D} &= 3Q_{3S} \exp(-\Delta_{3P}/T) \end{aligned} \tag{27}$$

where $\Delta_{3D} = 72.1$ meV, $\Delta_{3P} = 66.5$ meV are the energy shifts between the 3S and 3D, and 3S and 3P states which are of order of the relevant thermal energies and T is the temperature. This assumption is based on the fact that the thermalization rate of excited muonic atoms is 10^{12} s^{-1} at $\emptyset = 1$, so that thermal equilibrium among the substates of the M shell is established.

The strenghs of the radiative and external Auger transitions and muon transfer from deuterium to tritium taken from refs 22, 23 are shown in tables 1 and 2. In our calculations we have assumed a suppresssion [24] in the muon transfer process for levels n≥3. For atom-molecule collisions the transfer process can be strongly influenced by a subtantial change of the final state density. For the levels n = 1 and n = 2 the muon transfer can be easily accompanied by dessociation of the target molecule, whereas those for n = 3 and higher levels

Table 1
$(d\mu)_{i \to j}$ Transition rates

Transition	Auger rate $(10^{11} s^{-1})$	Radiative rate $(10^{11} s^{-1})$
5→4	50∅	negligible
4→3	9.5∅	negligible
4→1	-	0.03
3D→2P	1.6∅	0.125
3P→2S	0.64∅	0.04
3S→2P	0.18∅	0.012
3P→1S	-	0.32
2P→1S	-	1.3

cannot. Thus the muon transfer rates for the levels n = 1, n = 2 probably apply, but for states with n≥3 a molecular supprression mechanism may be active. It must be emphasized that uninhibition of the muon transfer process from all excited states (x = 1, where x is the suppression factor) as originally suggested [23] predicts very small values of Q_{1S}, the probability for a muon, being in the dµ ground state. For example for $\varnothing = 1$, $c_t = 0.5$ the calculated value from eqs. (26) of Q_{1S} is 0.08. Muon catalyzed fusion experiments are not compatible with a small value of Q_{1S}. The extrapolated experimental value [14] of Q_{1S} for $\varnothing = 1$, $c_t = 0.5$ is 0.7 ± 0.2, which is consistent with the suppression of the muon transfer process from the levels n≥3 of the dµ atom. Solving our rate equations (26) starting at N = 5, with initial conditions $Q_5 = 1$, $Q_i = 0$ for 1≤i≤4 and assuming a suppression factor x = 0 yields $Q_{1S} = 0.6$.

Table 2
$(d\mu)_n \to (d\mu)_n$ transition rates

n	Rate $(10^{11} s^{-1})$
5	10∅c_tX
4	8∅c_tX
3	8.7∅c_tX
2P	2.5∅c_t
2S	17∅c_t
1	$3 \times 10^{-3}\varnothing c_t$

144

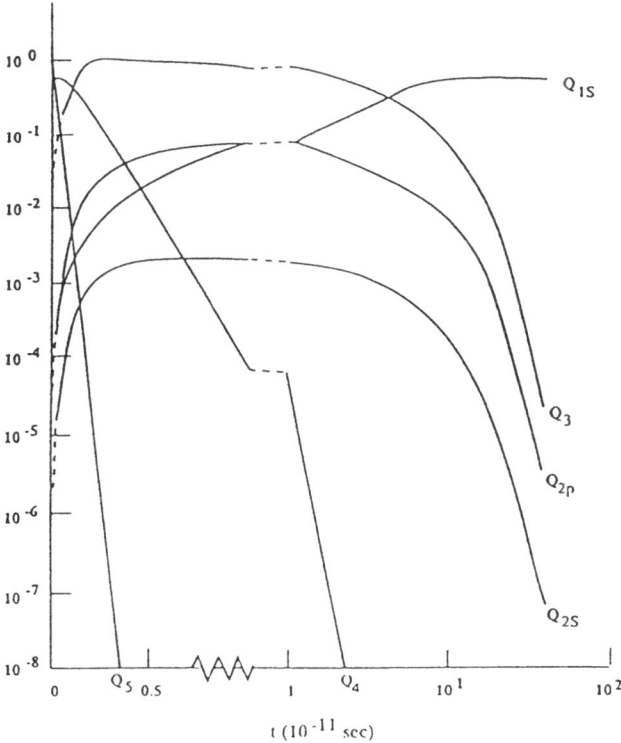

Fig. 1a. Time dependence of the populations of the muonic deuterium levels.

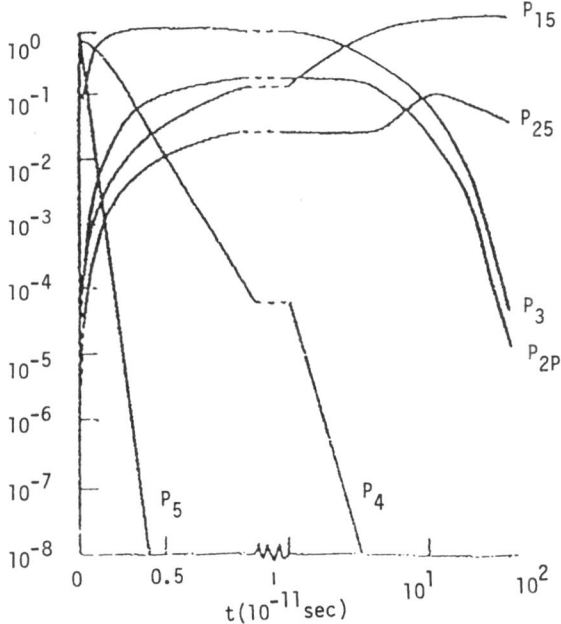

Fig. 1b. Time dependence of the populations of the mounic tritium levels.

The time dependence of the pupulations of the levels of dμ are shown in fig. 1a for dμ and in Fig. 1b for tμ. One can see the population inversion of muons between the levels n = 3, and n = 2. The large population of the level n = 3 occurs due to muon transfer suppression from this level. The above population inversion occurs naturally, requiring no pumping form a lower level, such as for conventional X-ray schemes. The duration of the population inversion occurs

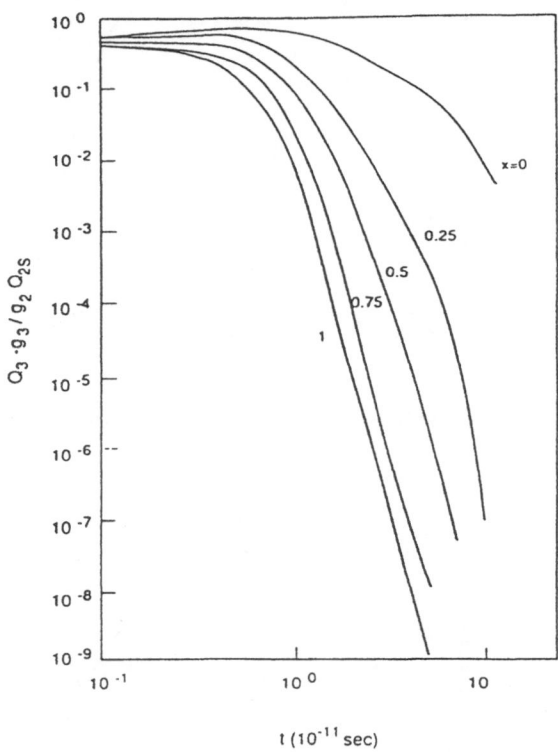

Fig. 2. Time dependence of Q3- (g3/g2) Q2S for different values of the suppression factor x.

naturally, requiring no pumping from a lower level, such as for conventional X-ray schemes. The duration of the population inversion depends on the suppression factor x. Fig. 2 shows the temporal dependence of the quantity $Q_3 - (g_3/g_2)Q_{2S}$, (where g_3 and g_2 are the degeneracy of the states n = 3 and n = 2) for different values of x.

THE POSSIBILITY OF X-RAY LASING

The intensity I of a laser system obeys the gain equation:

$$\frac{dI}{dZ} = \left(\gamma - \frac{1}{l_{op}}\right)I \tag{28}$$

where γ is the gain factor and l_{op} is the opacity. The gain factor is given by

$$\gamma = (Q_3 - Q_2\frac{g_3}{g_2}) N_\mu \lambda^2 g(v)/(8\pi^2 \tilde{n}^2 T_{sp}) \tag{29}$$

$\tilde{n} \simeq 1$ is the refraction index of the medium, $\lambda = 33.5$ Å for the laser under consideration (3S→2P). $T_{sp} = 2.3 \times 10^{-10}$ sec is the spontaneous decay lifetime. $g(v) = 4/(iT\Delta v)$ where Δv main contribution comes from Doppler broadening and is estimated to be $\Delta v \simeq 3.7 \times 10^{11}$ for a temperature of 1000^0K. Q_3-$Q_2 g_3/g_2 \simeq 0.7$ and N_μ is the density of the muons in the medium. In the Born approximation

$$l_{op}^{-1} \simeq 22 \varnothing \text{ cm}^{-1} \tag{30}$$

The lasing condition requires $\gamma \geq 1/l_{op}$ implying

$$N_\mu \geq 3 \times 10^{17}\varnothing \text{ (cm}^{-3}) \tag{31}$$

In order to minimize the heating problem we suggest the following preliminary scheme for producing an X-ray medium. The mean free path for elastic scattering of ground state $t\mu$'s and $d\mu$'s is about $10^{-4}\varnothing^{-1}$ cm and the range of the muon (captured in the target at an energy ~3keV) is less than $10^{-4}\varnothing^{-1}$cm. The range of the α particles, created by the fusion precess at an energy of 3.6 MeV is about $10^{-2}\varnothing^{-1}$cm. As long as the dimensions of the lasing medium are larger than the stopping length of μ and the mean free path of $d\mu$ and $t\mu$ but smaller than the ranges of α then the target will not overheat. So the lasing medium should be a thin rod with a width $10^{-2}\varnothing^{-1}$ cm. Using a target of dimensions 1cm (the length of the lasing medium), and a cross section $10^{-2}\varnothing^{-1} \times 10^{-2}\varnothing^{-1}$ cm^2, the appropriate volume of the target is $V \simeq 10^{-4}\varnothing^{-2}$ cm^3, so that the necessary total number of muons is

$$n_\mu^{tot} = N_\mu V \gtrsim 3 \times 10^{13}\varnothing^{-1} \text{ muons.} \tag{32}$$

For the above X-ray lasing scheme the muon must be cooled to be stopped in the target. Recently Nagamine suggested[25] a method to produce ultra slow negative muons. Futhermore, the muons are heating the target to a temperature of the order of $(N_\mu/N) \times 3$ keV $\sim 10^{-2}$ eV (N is the duterium-trititum density). The $\alpha\mu$ may also be absorbed in our medium. The $\alpha\mu$ heating is of the order of $(N_{\alpha\mu}/N) \times 10$ keV which is smaller by a factor of 10/3 $W_s \ll 1$ than the μ heating . However, if one muon catalyzes X_μ fusions then the temperature might increase accordingly. For $X_\mu \geq 10$ the target will heat and explode during the muon lifetime, and, in this case one would obtain short X-ray laser pulses.

CONCLUSION

We suggest to look experimentaly for 370 eV photons. The number n_γ of 370 eV noncoherent photons is estimated to be

$$n_\gamma = n_\mu{}^{tot} X_\mu q, \tag{33}$$

where $n_\mu{}^{tot}$ is the number of the absorbed muons in the d-t mixture, X_μ is the number of fusion catalyzed by one muon (maximun present experimental value is $X_\mu \approx 150$) .q is the ratio between the rates of the radiative and other decay modes of the level n = 3 and is given by

$$q = \frac{\lambda^{rad}}{\lambda^{Auger} + x\lambda^{transfer}} \tag{34}$$

For example, for the transition $3P \rightarrow 2S$, $q = 0.067$ assuming $x = 0$, while $q = 0.005$ for $x = 1$. The above experimental evidence could be very useful for the evaluation of Q_{1S}, which has proved to be very difficult to extract from experimental data. The opacity length for these photons is $I_{op} \approx 0.05 \emptyset^{-1}$cm. Therefore 370 eV photons might be deteced from a thin (with dimensions $\leqq$ 0.05 cm) DT target irradiated by a muon beam.

For a total number of muons $\sim 10^7$ and $X_\mu \sim 150$ one would expect $n_\gamma \sim 10^8$ for $x = 0$ and 10^7 for $x = 1$. This number of photons should be detectable (i.e. taking efficiency detection into account). From this measurement the value of x should be concluded. If x is of the order zero then an x-ray laser with 370 eV photon is promising.

Acknowledgement

One of the authors (S.E.) would like to thank Prof. G. Velarde, Prof. J.M. Martínez-Val and the staff of the *Instituto de Fusión Nuclear-Universidad Politécnica de Madrid* for their kind hospitality. The useful and enlightening discussions with the people of the Institute are very much appreciated.

REFERENCES

1. F.C. Frank, Nature **160**, 525 (1947).
2. A.D. Sakharov and P.N. Lebedev, Report of the Physics Institute Academy of Sciences of USSR (1948).
3. Ya. B. Zeldovich, Dokl. Akad. Nauk. SSSR **95**, 493 (1954).
4. J.D. Jackson, Phys. Rev. **106**, 330 (1957).
5. L.W. Alvarez et al., Phys. Rev. **105**, 1127 (1957).
6. L.W. Alvarez. 1968 Nobel prize acceptance lecture.
7. A. Ashmore et. al., Proc. Phys. Soc. (London) **71**, 161 (1958).
8. S.S. Gershtein and Ya. B. Zeldovich, Sov. Phys. Usp. **3**, 593 (1961).
9. V.P. Dzhelepov et. al, Zh. Eksper. Teor. Fiz. **50**, 1235 (1966) [Sov. Phys. JETP **23**, 820 (1966)].
10. E.A. Vesman, Zh. Eksper. Teor. Fiz. Pisma **5**, 113 (1967) [JETP Letters **5**, 91 (1967)].
11. S.S. Gershtein and L.I. Ponomarev, **Muon Physics Vol III**, Eds. V.W. Hughes and C.S. Wu, Academic Press, N.Y. p141 (1975).
12. L. Bracci and G. Fiorentini, Phys. Rep. **86**, 169 (1982).
13. W.H. Breunlich et. al., Phys. Rev. Lett. **53**, 1137 (1984).
14. S.E. Jones et. al., Phys. Rev. Lett. **51**, 1757 (1983); ibid. **56**, 588 (1986).
15. L.N. Bogdanova, V.E. Markushin and V.S. Melezhik, Zh. Eksp. Teor. Fiz. **81**, 829 (1981) [Sov. Phys. JETP **54**, 442 (1981)].
16. S.S. Gershtein et. al., Sov. Phys. JETP **53**, 782 (1981).
17. V. Markushin, Zh. Eksp. Teor. Fiz. **80**, 35 (1981) [Sov. Phys. JETP **53**, 16 (1981)].
18. M. Leon and H.A. Bethe, Phys. Rev. **127**, 637 (1962).
19. M. Leon, Phys. Lett B **35**, 413 (1971).
20. T.B. Day, G.A. Snow and J. Sucker, Phys. Rev. **118**, 864 (1960).
21. Z. Henis and S. Eliezer, Phys. Lett. A, **152** 472 (1991).
22. L.I. Menshikov and L.I. Ponomarev, J. Phys. **D2**, 1 (1986).
23. L.I. Menshikov and L.I. Ponomarev, JETP Lett. **39**, 663 (1984).
24. B. Müller et. al., **Muon-catalyzed Fusion**, Eds. S.E. Jones, J. Rafelski, G. Hendrik and J. Monkhorst (A.I.P. New York) p.105 (1989).
25. K. Nagamine, Proc. Japan Acad. **65B**, 225 (1989).

STATUS AND ISSUES IN THE DEVELOPMENT OF A GAMMA-RAY LASER

C. B. Collins, J. J. Carroll, K. N. Taylor,
T. W. Sinor, C. Hong, J. Standifird, and
D. G. Richmond

Center for Quantum Electronics
University of Texas at Dallas
P.O. Box 830688
Richardson, TX 75083-0688

ABSTRACT

A gamma-ray laser would stimulate the emission of radiation at wavelengths below 1 Å from excited states of nuclei. However, the difficulties in realizing such a device were considered insurmountable when the first cycle of study ended in 1981. Nevertheless, research on the feasibility of a gamma-ray laser has taken a completely new character since then. A nuclear analog of the ruby laser has been proposed and many of the component steps for pumping the nuclei have been demonstrated experimentally. A quantitative model based upon the new data and concepts of this decade shows the gamma-ray laser to be feasible if some real isotope has its properties sufficiently close to the ideals modeled.

INTRODUCTION

At the nuclear level, long-lived excited states are known as isomers. Populations of these nuclear metastables can store energies of tera-Joules (10^{12} J) per liter at solid densities for thousands of years. Such long storage times mean that it would not be necessary to pump a gamma-ray laser medium entirely in situ. For some cases the excited nuclei could be bred in a reactor from a parent material that captures a neutron or from a specific nuclear reaction acting upon precursive elements.

The problem of suddenly assembling a critical density of prepumped nuclei to reach the threshold for stimulated emission received much attention in a first cycle of study lasting from 1963-1981 and excellent reviews are available.[1] At the end of this period, it was generally concluded that such single photon, brute force approaches were essentially hopeless. In an encyclopedic review, Baldwin and coauthors[1] concluded the general impossibility of a gamma-ray laser based upon all techniques for pumping known in 1980. That review effectively terminated all traditional lines of approach to a gamma-ray laser. However, in the earlier work the greatest emphasis had been placed upon the use of intense particle fluxes for input energy.

Laser Interaction and Related Plasma Phenomena, Vol. 10
Edited by G.H. Miley and H. Hora, Plenum Press, New York, 1992

151

Toward the end of the first cycle of research the precursors of a new interdisciplinary approach began to appear.[2-8] These developed rapidly and launched a renaissance in the field. The basic theory[9-11] of upconversion at the nuclear level was in-place by 1982 for the two possible variants, coherent and incoherent upconversion. Involving either multiphoton processes or multiple electromagnetic transitions to release the energy stored in isomers, many of the difficulties encountered with more traditional pumping schemes did not arise.

In the decade since 1980, research on the feasibility of a gamma-ray laser has taken a completely new character. In the first half decade a series of experiments verified that the concepts of quantum electronics could be applied at the nuclear level.[12-16] By 1986 the blueprint for a gamma-ray laser[11] had been established and substantial effort was initiated toward the demonstration of feasibility. The purpose of this article is to review the major advances of this past five years that have significantly increased the likelihood of the feasibility of a gamma-ray laser.

CONCEPTS

At first approach it would seem that the prospects for all ultrashort wavelength lasers would be vitiated by a very fundamental factor.[1] The basic ν^3 dependence of electron transition probabilities so limits the storage of pump energies that even now some of the largest pulsed-power machines are able to excite only milliJoules of x-ray laser output and then only at soft photon energies. In contrast there are four unique advantages of a gamma-ray laser that would accrue from its operation upon electromagnetic transitions of nuclei:

1) The constant linking ν^3 with lifetime is more favorable by orders-of-magnitude because of the accessibility of a variety of transition moments. The effects pumped by an input pulse can be integrated up to larger values for longer times.

2) Nuclear metastables store keV and even MeV for years. With upconversion schemes most of the pump power is input long before the time of use and triggering requirements are small.

3) Nuclear transitions need not have thermal broadening and natural linewidths are routinely obtained. Without broadening electromagnetic cross sections are large and values for 1 Å transitions typically exceed the cross section for the stimulation of Nd in YAG.

4) Working metastables can be concentrated to solid densities.

The essential concept driving the renaissance in gamma-ray laser research was the "optical" pumping of nuclei. In this case optical meant x-rays, but the fundamentals were the same. Useful, resonant absorption of pump power would occur over short distances to produce high concentrations of excited nuclei while wasted wavelengths would be degraded to heat in much larger volumes. In the blueprint[11] of 1982 for upconversions, one of several possible types of photopumping was envisioned to transfer the stored population of an isomer to a state at the head of a cascade leading to the upper laser level. Of the cases considered, the nuclear analog of the ruby laser embodied the simplest concepts for a gamma-ray laser. Not surprisingly, the greatest rate of achievement in the last five years has been realized in that direction and this review will be limited to the results along that line.

For ruby, the identification and exploitation of a bandwidth funnel were the critical keys in the development of the first laser. There was a broad absorption band linked through efficient cascading to the narrow laser level. Our theory called for a nuclear analog of this structure which was unknown in 1986 when the first phase of intensive experiments was started for SDIO. Now, that theory has been confirmed.

Whether or not the initial state being pumped is isomeric, the principal figure of merit for bandwidth funneling is the partial width for the transfer, $b_a b_0 \Gamma$. Constituent parameters are identified in Fig. 1 where it can be seen that the branching ratios b_a and b_0 specify the probabilities that a population pumped by absorption into the j-th broad level will decay back into the initial or fluorescent levels, respectively. It is not often that the sum of branching ratios is unity, as channels of decay to other levels are likely. However, the maximum value of partial width for a particular level j occurs when $b_a = b_0 = 0.5$.

Classified as (γ, γ') reactions in the literature of nuclear physics, these "optical" pumping processes have been known for over 50 years[17,18] although relatively few results have been published since that time. Practical difficulties with the calibration and availability of sources of irradiation had limited the degree of reproducibility achieved in earlier work.

The most tractable (γ, γ') reactions for study are those for the photoexcitation of stable isotopes up to isomeric levels. In some cases the product can live long enough to be readily examined after termination of the input irradiation and lessons can be learned that can be applied to the excitation of shorter-lived levels more useful in a laser. The archetypical case for basic study has been the reaction $^{111}Cd(\gamma, \gamma')^{111}Cd^m$ exciting the 48.6 min level at 396 keV. Three of the most recent measurements of the fluorescence efficiency were conducted in 1979, 1982, and 1987 as reported in Refs. 19-21, respectively. Probable errors were quoted as varying only from 7 to 14%, and yet no two of the measurements were even within a factor of 2 of each other. This discrepancy led to serious contentions over the way in which the expected fluorescence yields were calculated.[20] One of our early challenges was to place the "optical" pumping of nuclei onto a firm quantitative basis.

For a sample which is optically thin at the pump wavelength, a computation of the number of nuclei pumped into a fluorescence level in the scheme of Fig. 1 is straightforward. Most intense x-ray sources emit continua, either because bremsstrahlung is initially produced or because spectral lines are degraded by Compton scattering in the immediate environment. The time-integrated yield of final-state nuclei, N_f obtained by irradiating N_i initial targets with a photon flux Φ_0 in photons cm^{-2} delivered in a continuum of intensities up to an endpoint energy E_0 is,

$$N_f = N_i \Phi_0 \int_0^{E_0} \sigma(E) F(E, E_0) dE \quad , \quad (1)$$

where $F(E, E_0)$ is the distribution of intensities within the input spectrum normalized so that

$$\int_0^{E_0} F(E, E_0) dE = 1 \quad , \quad (2)$$

and $\sigma(E)$ is the effective cross section for the excitation of the final state from the initial.

All (γ,γ') reactions occurring at energies below the threshold for particle evaporation excite discrete intermediate states of nuclei as shown in Fig. 1. Although only one intermediate state appears in Fig. 1, there could be more. Each would be excited at a different pump energy but all would branch to some extent into the same fluorescence level, f. The j-th intermediate is shown in Fig. 1 as typical.

Although the width of level j is broad on a nuclear scale, it is narrow in comparison to the scale of energies, E over which $F(E,E_0)$ varies. Then, the final-state yield, expressed as the normalized activation per unit photon flux, $A_f(E_0)$ produced with bremsstrahlung having an endpoint of E_0 can be written from Eq. (1) as,

$$A_f(E_0) \equiv \frac{N_f}{N_i\Phi_0} = \sum_j (\sigma\Gamma)_{fj} F(E_j,E_0) \quad . \tag{3a}$$

In this expression $(\sigma\Gamma)_{fj}$ is the integrated cross section for the production of final-state N_f as a result of the excitation of the intermediate state E_j with bremsstrahlung described by the spectral function $F(E,E_0)$, so that

$$(\sigma\Gamma)_{fj} = \int_{E_j-\Delta}^{E_j+\Delta} \sigma(E)dE \quad , \tag{3b}$$

where Δ is an energy small compared to the spacing between intermediate states and large in comparison to their widths. Levels of this type are sometimes called gateways or doorways.

It is straightforward to show that,

$$(\sigma\Gamma)_{fj} = (\pi b_a b_0 \Gamma\sigma_0/2)_{fj} \quad , \tag{4a}$$

where $\sigma_0/2$ is the peak of the Breit-Wigner cross section for the absorption step, and

$$\sigma_0 = \frac{\lambda^2}{2\pi} \frac{2I_e+1}{2I_g+1} \frac{1}{\alpha_p+1} \quad , \tag{4b}$$

where λ is the wavelength of the gamma ray at the resonant energy, E_j; I_e and I_g are the nuclear spins of the excited and ground states, respectively; and α_p is the total internal conversion coefficient for the system shown in Fig. 1.

PUMP CALIBRATION

From the perspective of laser physics the most unreliable sources of energetic photons used in early studies of (γ,γ') reactions seem to have been the nuclear sources. Although assumed to emit line spectra, in actual usage they produced intensities which were dominated by the continua resulting from multiple Compton scatterings of photons by the large amounts of shielding in the irradiation environment. Such multiple scatterings are difficult to calculate and still impossible to measure in practical laboratory configurations. In contrast, the spectral intensities of

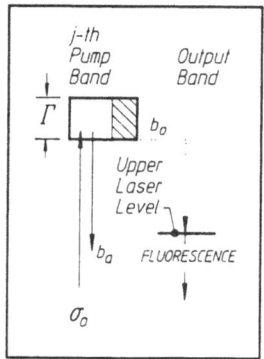

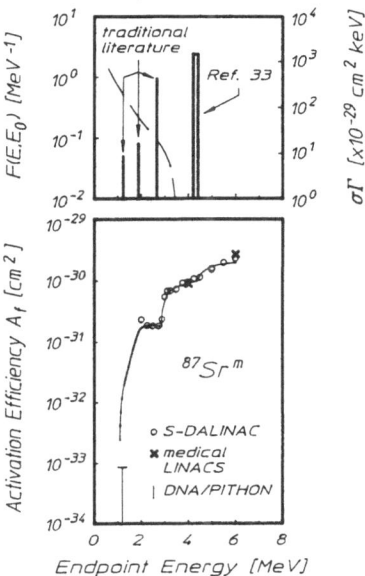

Figure 1. Schematic representation of the decay modes of a gateway state of width Γ sufficiently large to promote bandwidth funneling. The initial state from which population is excited with an absorption cross section σ_0 can be either ground or isomeric.

Figure 2. Activation efficiencies, A_f for the reaction $^{87}Sr(\gamma,\gamma')^{87}Sr^m$ are shown in Fig. 2b (lower) as functions of the endpoint, E_0 of the bremsstrahlung used for excitation. The solid curve in Fig. 2b plots values computed from Eq. (3a) using the gateway parameters found in the literature and plotted with the right axis in Fig. 2a (upper) together with calculated photon spectra like that given by the dashed line and plotted with the left axis in Fig. 2a. Figure 2b compares these A_f values with measurements obtained from four different accelerators.

bremsstrahlung are routinely calculated with high accuracy from measured accelerator currents and target geometries by well-established computer codes.[22]

In our experimental work of the last five years the bremsstrahlung from five accelerators in different experimental environments was used to verify the fluorescence model of Eqs. (1) - (4b) and to cross-check the accelerator intensities. The devices involved in this effort were DNA/PITHON at Physics International, DNA/Aurora at the Harry Diamond Laboratories, a 4 MeV and a 6 MeV medical linac at the University of Texas Health Sciences Center, and the superconducting injector to the storage ring at Darmstadt (S-DALINAC). Spectral intensities were calculated with the

Table I

Summary of nuclides, pump lines, and integrated cross sections for the excitation of delayed fluorescence suitable for use as calibration standards.

	PUMP LINE (keV)	$(\sigma\Gamma)_{fj}$ (10^{-29} cm^2 keV)
^{79}Br	761	6.2
^{77}Se	250	0.20
	480	0.87
	818	0.7
	1005	30
^{115}In	1078	20

EGS4 coupled electron/photon transport code[22] adapted for each individual configuration from closely monitored values of accelerator currents. In this way both $F(E,E_0)$ and Φ_0 were obtained. In some cases Φ_0 was separately verified by in-line dosimetry.

Of the many potential systems which might be used to confirm the formulations of Eqs. (1) through (4b), the literature[23] supports the calculation of integrated cross sections for very few. Table I summarizes those which are known with sufficient accuracy to serve as standards. In the convenient units of 10^{-29} cm^2 keV, values range from the order of unity to a few tens for bandwidth funnels that are sufficient for demonstrations of nuclear fluorescence from reasonable amounts of material at readily accessible levels of input.

In our experiments samples with typical masses of grams were exposed to the bremsstrahlung from the five accelerators for times ranging from seconds to hours for the continuously operating machines and to single flashes from the pulsed devices. The activations, A_f of Eq. (3a) were determined by counting the photons spontaneously emitted from the samples after transferring them to a quieter environment. Usual corrections were made for the isotopic abundance, for the loss of activity during irradiation and transit, for the counting geometry, for the self-absorption of the fluorescence, and for the tabulated efficiencies[23,24] for the emission of signature photons from the populations, N_f. The self-absorption correction required a calculation of photon transport which was verified in some cases by confirming that the same sample masses in different geometries with different correction factors gave the same final populations.

Results were in close agreement[25] with the predictions of Eq. (3a) used with the values of $(\sigma\Gamma)_{fj}$ shown in Table I. That work established a confidence level sufficient to support the use of nuclear activation as a means of selectively sampling spectral intensities of single pulses of intense continua to determine absolute intensities as functions of wavelength.[26] Having calibrated the spectral sources, the persisting uncertainties in the optical pumping of $^{115}In^m$ and $^{111}Cd^m$ were resolved.[27,28]

As part of the effort to establish better calibration standards, we reexamined the reaction $^{87}Sr(\gamma,\gamma')^{87}Sr^m$. Particularly valuable were the data obtained with the S-DALINAC because the endpoint of the bremsstrahlung could be varied. A change of the endpoint energy, E_0 of the bremsstrahlung, as well as altering Φ_0, modulates the spectral intensity function $F(E_j,E_0)$ at all of the important energies for resonant excitation E_j. The largest effect occurs when E_0 is increased from a value just below some intermediate state at E_j to one exceeding it so that $F(E_j,E_0)$ varies from zero to some finite value as shown in Fig. 2a.

Early work[29] on (γ,γ') reactions showed that a plot of activation, $A_f(E_0)$ as a function of bremsstrahlung endpoint energy displayed very pronounced activation edges at the energies, E_j corresponding to the resonant excitation of new intermediate states. Such edges enabled new gateways to be identified for a particular reaction.

The activation efficiencies, A_f calculated from Eq. (3a) using the literature values[29] plotted in Fig. 2a are shown in Fig. 2b together with the measurements obtained with four of the accelerators. As can be seen, agreement is very good between the different accelerators and between the experimental data and the model calculations. The units of A_f in Fig. 2b are those of area because they are a type of average cross section quite different from the σ_0 of Eq. (4a) that describe individual transitions. The small plotted values are the result of averaging the large σ_0 at the resonant E_j over the broad bandwidth of $F(E,E_0)$ in which most $E \neq E_j$.

These calibration studies served to confirm both the traditional model of nuclear activation summarized in Eqs. (1) - (4b) and to validate the EGS4 code for calculating bremsstrahlung intensities from measured accelerator parameters. *Now, there can be no reasonable doubt of procedures for quantitatively measuring fluorescence efficiencies if an experiment is carefully performed with a bremsstrahlung source of pump radiation.*

GIANT PUMPING RESONANCES

If expressed as partial widths the integrated cross sections for the excitation of $^{77}Se^m$, $^{79}Br^m$, and $^{115}In^m$ seen in Table I corresponded to 39, 5, and 94 μeV, respectively. While among the largest values reported prior to our studies, these results still left an aura of credibility to the traditional impressions that partial widths for exciting isomers would be limited to about 1 μeV.

Tempering expectations that integrated cross sections of even this size might be expected for the dumping of actual isomeric candidates for a gamma-ray laser was a concern for the conservation of various projections of the angular momenta of the nuclei. Many of the interesting isomers belong to the class of nuclei deformed from the normally spherical shape. For those systems there is a quantum number of dominant importance, K which is the projection of individual nucleonic angular momenta upon the axis of elongation. To this is added the collective rotation of the nucleus to obtain the total angular momentum J. The resulting system of energy levels resembles those of a diatomic molecule for which

$$E_n(K,J) = E_n(K) + B_nJ(J + 1) \quad , \tag{5}$$

where $J \geq K \geq 0$ and J takes values $|K|, |K| + 1, |K| + 2,....$ In this expression B_n is a rotational constant and $E_n(K)$ is the lowest value for any level in the resulting "band" of energies identified by other quantum numbers n. In such systems the selection rules for electromagnetic transitions require both $|\Delta J| \leq M$ and $|\Delta K| \leq M$, where M is the multipolarity of the transition.

In most cases of interest, the isomeric state has a large lifetime because its value of K differs considerably from those of lower levels to which it would otherwise be radiatively connected. As a consequence, bandwidth funneling processes such as shown in Fig. 1 must span substantial changes in ΔK and component transitions have been expected to have large, and hence unlikely, multipolarities. Initial expectations were that partial widths would decrease further as the values of ΔK needed for the transfer increased.

From this perspective the candidate isomer, $^{180}Ta^m$ was the most initially unattractive as it had the largest change of angular momentum between isomer and ground state, $8\hbar$. However, because it was the only isomer for which a macroscopic sample was readily available, $^{180}Ta^m$ became the first isomeric material to be optically pumped to a fluorescent level.

Table II

Recently measured values of integrated cross section, $(\sigma\Gamma)_{fj}$ for the reaction $^{180}Ta^m(\gamma,\gamma')^{180}Ta$. The gateway excitation energies, E_j for these levels are given at the centers of the ranges of energies that could be resolved experimentally.

Energy (MeV)	$\sigma\Gamma$ (10^{-29} cm^2 keV)
2.8 ± 0.1	12000 ± 2000
3.6 ± 0.1	35000 ± 5000

This particular isomer, $^{180}Ta^m$ carries a dual distinction. It is the rarest stable isotope occurring in nature and it is the only naturally occurring isomer. The actual ground state of ^{180}Ta is 1^+ with a halflife of 8.1 hours while the tantalum nucleus of mass 180 occurring with 0.012% natural abundance is the 9^- isomer, $^{180}Ta^m$. It has an adopted excitation energy of 75.3 keV and halflife in excess of 1.2×10^{15} years.[30]

In an experiment conducted in 1987 we exposed 1.2 mg of $^{180}Ta^m$ to the bremsstrahlung from the 6 MeV linac and obtained a large fluorescence yield.[31] This was the first time a (γ,γ') reaction had been excited from an isomeric target as needed for a gamma-ray laser and was the first evidence of the existence of giant pumping resonances. Simply the observation of fluorescence from a milligram sized target proved that an unexpected reaction channel had opened. Usually grams of material are required in this type of experiment. Analyses[31,32] of the data indicated that the partial width for the dumping of ^{180}Ta^m was around 0.5 eV.

To determine the transition energy, E_j from the ^{180}Ta^m isomer to the gateway level, a series of irradiations was made at the S-DALINAC facility using fourteen different endpoints in the range from 2.0 to 6.0 MeV.[33] The existence of an activation edge was clearly seen in the data shown in Fig. 3b. The fitting of such data to the expression of Eq. (3a) by adjusting trial values of $(\sigma\Gamma)_{fj}$ enabled us to determine the integrated cross sections for the dumping of ^{180}Ta^m isomeric populations into freely radiating states. Reported values[33] are summarized in Table II and shown schematicaly in Fig. 3a.

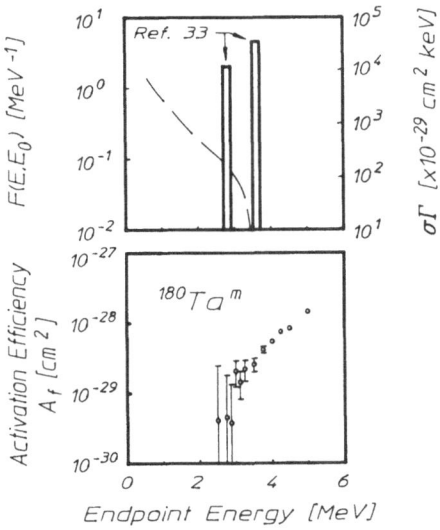

Figure 3. Activation efficiencies, A_f for the reaction ^{180}Ta$^m(\gamma,\gamma')^{180}$Ta are shown in Fig. 3b (lower) as functions of the endpoint, E_0 of the bremsstrahlung used for excitation. Gateway parameters obtained by fitting the data using Eq. (3a) are plotted with the right axis in Fig. 3a (upper). These parameters were determined using calculated photon spectra like that given by the dashed line and plotted with the left axis in Fig. 3a.

The integrated cross sections in Table II are enormous values exceeding anything previously reported for transfer through a bandwidth funnel by two orders of magnitude. In fact they are 10,000 times larger than the values usually measured for nuclei. With this result the restrictive guidelines customarily applied to optical pumping of nuclei are proven to be nearly 10^6 times too pessimistic.

While the width of the transfer process is difficult to interpret in a single-particle model, a puzzle of comparable complexity is found in the efficiency with which ΔK is transferred. It is an interesting speculation that at certain energies of excitation, collective oscillations of the core nucleons could break some of the symmetries upon which rest the identifications of the pure single-particle states. If single-particle states of differing K were mixed in this way, the possibility for transferring larger amounts of ΔK with greater partial widths might be enhanced. Some support for such a speculation was found in the unexpected enhancements measured very recently for the deexcitation of the ^{174}Hf^m isomer.[34] There also the decay of the isomer was found to occur primarily by transition through an intermediate state lying at 2685 keV in which K mixing occurred so that $\Delta K = 14$ was lost between isomer and the ground-state band. This is remarkably close to the energy of the K-mixing level at 2800 ± 100 keV for ^{180}Ta shown in Table II. The close similarity of the energetics shown in Fig. 4 for nuclei with such dissimilar single-particle structures does seem to support the identification of this K-mixing process with some type of core property varying only slowly among neighboring nuclei.

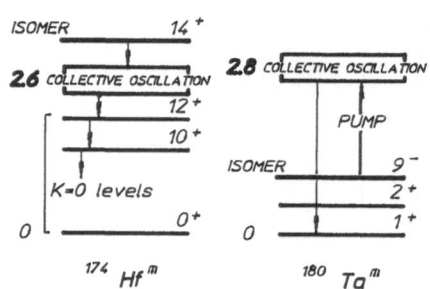

Figure 4. Energetics for the spontaneous decay of ^{174}Hf^m through an intermediate state providing $\Delta K=14$ compared with those for the dumping of ^{180}Ta^m through a similar intermediate state. In the case of ^{180}Ta^m, the dumping reaction provides $\Delta K=8$. Both gateways are expected to be admixtures of single-particle states, thereby producing significant K mixing.

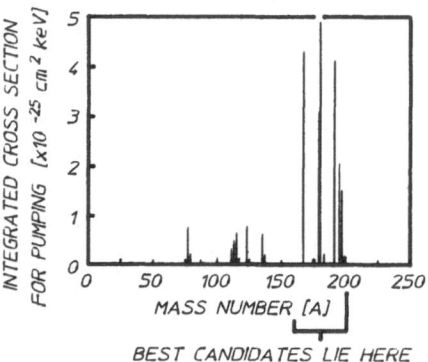

Figure 5. Integrated cross sections for pumping isomeric nuclei obtained in the survey of Ref. 35. The groupings of pumping strengths seen in the figure correspond to mass islands between magic numbers for neutrons and protons. The best candidates for a gamma-ray laser lie within the island containing the largest values of integrated cross sections corresponding to giant pumping resonances. Within each island the integrated cross sections vary only slowly with changing mass number, A.

A survey of 19 isotopes[35] conducted with the four U.S. accelerators over a fairly coarse mesh of bremsstrahlung endpoints confirmed the existence of giant resonances for breaking K in the region of masses near 180 as shown in Fig. 5.

Activation edges continued to support the identifications of integrated cross sections for pumping and dumping of isomers that were of the order of 10,000 times greater than usual values. A study with the higher resolution of the S-DALINAC[36] showed that the giant pumping resonances reappeared at lower masses near 120.

Whatever the mechanisms, the experimental fact remains that interband-transfer processes connecting to isomeric levels can be pumped through enormous partial widths reaching 0.5 eV, even when the transfer of angular momentum must be as great as $\Delta K = 8$, or even $\Delta K = 14$. It seems this is the nuclear analog of the giant resonance for pumping ruby at the atomic level. Elucidation of the process, together with identification of the gateways, has been propelled into a place of future importance.

CONCLUSIONS - A LASER MODEL

The model of a gamma-ray laser for the 1990's is not fundamentally different from the nuclear analog of the ruby laser described[11] in 1982. Envisioned as a thin film of diluent doped with isomeric nuclei and pumped with a flash of x rays in a slab geometry, the question of feasibility still rests on the degree to which the properties of some real nuclide approach those of the ideal being modeled. What has changed is that the discovery of giant pumping resonances enables some of the original constraints to be relaxed. The result is that the feasibility of a gamma-ray laser is orders-of-magnitude more probable than originally estimated in 1982. Because of this substantial improvement, it is useful to recompute the model in terms of the new data obtained in the past five years.

Since the better candidate isomers for a gamma-ray laser have never been fabricated in macroscopic amounts, the precise identity of the best nuclide to model is not known. Moreover, since feasibility is such a complex function of the nuclear parameters, the assumptions introduced into any model will critically affect the estimates of feasibility in strongly nonlinear ways. For the computation reported here the following parameters were assumed.

1) The pump band j in Fig. 1 is one of the newly discovered giant pumping resonances with a partial width of $b_a b_0 \Gamma = 1$ eV.

2) The pump transition is centered on an energy $E_j = 30$ keV.

3) The initial state is assumed to be isomeric with an excitation energy so high that 2) is possible.

4) The output transition is around 100 keV.

5) The nuclei are diluted in a thin film of diamond or Be.

6) The Borrmann effect contributes a factor of 10 to the enhancement of the ratio of cross sections for resonant to nonresonant transitions.

The most sensitive assumptions are those about the width and activation energy of the giant pumping resonance, statements 2) and 3). The range of excitation energies over which isomers can be found is very large. We have already shown that isomers can be dumped into the freely-radiating system, even through $\Delta K = 8$ or

ΔK = 14, so the only doubt here is a statistical one; whether or not a giant pump resonance can be found within 30 keV of an isomer.

Following our development[11] of 1982, under small signal conditions the midrange requirement of 10^{-4} is obtained for the pumped fraction,

$$\frac{N_f}{N_i} \geq 10^{-4} \quad . \tag{6}$$

This sets the pump intensity needed for threshold, and with it the amount of waste heat to dissipate.

The essential concept in the management of the thermal economy is that the mean free path (MFP) for a photon resonant with the nuclear transition is much shorter than the MFP for nonresonant, photoelectric absorption to produce heat. Also, the MFP for a photoelectron produced in the nonresonant channel is greater in the diluent than the MPF for the photons pumping the nuclear resonance. This means that a thin film of diamond can be doped to use most of the incident photons in the bandwidth of the giant pumping resonance while the majority of the nonresonant photons will pass through the film into the substrate which can be cooled by ablation or cryogenics. Moreover, primary photoelectrons produced by the small fraction of nonresonant events in the film can escape before their energy is degraded to heat.

The quantitative expression of this strategy is obtained by substituting Eq. (3a) into Eq. (6) and assuming a single giant resonance dominates so that the sum is unnecessary. Solving for $\phi_j = \Phi_0 F(E_j, E_0)$, and using the previous assumption that each pump photon carries 30 keV gives the spectral fluence, $F_j = E_j \phi_j$ at threshold,

$$F_j \geq 177 \ mJ \ cm^{-2} \ eV^{-1} \quad . \tag{7}$$

For the likely cases of rare earth or platinide elements, the 30 keV pump energy lies below the K edge and about 15 keV above the L edge. As a result, the primary photoelectrons resulting from the nonresonant absorption in the active medium should have energies of the order of 15 keV and ranges of 6.0 and 3.0 μm in Be and C, respectively.[37] Thus, only about 10% and 20% of the primaries, respectively, should be stopped in a 0.67 μm thick host film of Be or diamond. This is the thickness corresponding to the MFP for resonant absorption at a concentration of 10%. Then the fractions of the energy from the incident pump degraded into heat in the laser film because of nonresonant absorption become

$$f \ (Be) = 4.8 \times 10^{-4} \quad , \tag{8a}$$

$$f \ (C) = 2.4 \times 10^{-4} \quad . \tag{8b}$$

Considering that edge filters or ablation layers could reduce the bandwidth of the pump radiation to 3 keV before reaching the doped layer of active medium, the incident fluence lying outside the bandwidth for resonant absorption would be 3000 times greater than the value of Eq. (7). However, only the fractions of Eqs. (8a) and (8b) are capable of being degraded into heat in the sensitive layer. The resulting energy balance can be summarized at threshold by the first two lines of Table III.

Dividing those fluences by the 0.67 μm thickness gives the energy loading of the laser film shown in Table III. These values are quite significantly <u>below</u> the levels of heating required to degrade the recoil-free fractions in the case of the diamond lattice. Baldwin has summarized[1] the involved dependence of the recoil-free fraction of gamma transitions upon recoil energy, lattice parameters, and temperature. He shows that even at a temperature, T equal to the Debye temperature, Θ_D the recoil-free fraction is not significantly degraded (by more than a factor of 2) for a transition even as energetic as to give a classical recoil energy of 0.14 Θ_D. In diamond with Θ_D = 2230 K this means a transition of 100 keV is little affected by a temperature increase up to T = Θ_D.

It is a textbook computation[38] to estimate that the energy content of the phonons of a material with Θ_D = 2230 K at a temperature of T = Θ_D is about 11 kJ cm^{-3}. Comparing this with the estimated loading of 3.8 kJ cm^{-3} gives a "safety factor" of almost three. A comparable margin is obtained for the Be.

Table III

Summary of the thermal economy at threshold for a laser nuclide doped into a film of 0.67 μm thickness of the materials shown.

Lattice	Be	C(diamond)
Resonant input fluence	177 mJ cm^{-2}	177 mJ cm^{-2}
Fluence degraded to heat	127 mJ cm^{-2}	255 mJ cm^{-2}
Resonant energy density	2.6 kJ cm^{-3}	2.6 kJ cm^{-3}
Thermal loading	1.9 kJ cm^{-3}	3.8 kJ cm^{-3}

To summarize, it is convenient to recast the threshold fluence of Eq. (7) into more tangible terms. The spectral fluence of 177 mJ cm^{-2} eV^{-1} corresponds to 530 J cm^{-2} if the bandwidth of the pump x rays is arranged to be 3 keV, a practical separation which might be filtered between K edges. Even if pumped instantaneously, so that no waste heat were transported away, the thermal loading would reach only 1/3 of the limit for retaining the Mössbauer effect. If derived from an x-ray line of 30 eV width, the threshold fluence would be only 5.3 J cm^{-2}. In that case the thermal loading would reach only 1/300 of the critical limit for a diamond lattice.

Even beyond this point much can be done to reduce heating further. All calculations so far were done for the instantaneous generation of the waste heat. The time for the transport of a phonon across the 0.67 μm thickness of the working layer is of the order of only 100 psec so that the transport of significant amounts of heat from that layer into a diamond heat sink is possible on a nanosecond time scale. Yet most of the fluorescent levels of interest for inversion[11] have lifetimes of tens of nanoseconds to tens of microseconds. This is many times the period for the transport of phonons out of the inverting layer so that more orders of magnitude can be realized in reducing the thermal loading further below the limits specified so far. However, all these techniques require precise knowledge about the energy levels and absorption edges of the materials involved. Until the identity of the best candidate for a gamma-ray laser is known, the exact specifications of the solution to the disposal of the waste heat cannot be generally articulated. The examples considered here show that there are many orders-of-magnitude in the safety margin between likely amounts of heating and the much larger amounts which can be tolerated in stiff lattices such as Be and diamond.

The greatest significance is that the persistent tenets of theoretical dogma which have historically[1] inhibited the development of a gamma-ray laser are eliminated by studies of the past five years . There is no need to melt the host lattice in order to pump a nuclear system to the laser threshold. *There are no a priori obstacles to the realization of a gamma-ray laser. A gamma-ray laser is feasible if the right combination of energy levels occurs in some real material.* The overriding question to resolve is whether or not one of the better of the candidate nuclides has its isomeric level within a few tens or even hundreds of keV of one of the giant resonances for dumping angular momenta.

ACKNOWLEDGEMENT

This work was supported by SDIO/IST under direction of NRL.

REFERENCES

1. G. C. Baldwin, J. C. Solem, and V. I. Goldanskii, "Approaches to the development of gamma-ray lasers," Rev. Mod. Phys. 53:687 (1981).
2. V. S. Letokhov, Pumping of nuclear levels by x-ray radiation of a laser plasma, Sov. J. Quant. Electron. 3:360 (1974).
3. B. Arad, S. Eliezer and Y. Paiss, "Nuclear 'Anti-Stokes' transitions induced by laser radiation," Phys. Lett. 74A:395 (1979).
4. C. B. Collins, S. Olariu, M. Petrascu, and I. Popescu, "Enhancement of γ-Ray Absorption in the Radiation Field of a High Power Laser," Phys. Rev. Lett. 42:1397 (1979).
5. C. B. Collins, S. Olariu, M. Petrascu, and I. Popescu, "Laser-Induced Resonant Absorption of γ Radiation," Phys. Rev. C 20:1942 (1979).
6. S. Olariu, I. Popescu, and C. B. Collins, "Tuning of γ-Ray Processes with High Power Optical Radiation," Phys. Rev. C 23:50 (1981).
7. S. Olariu, I. Popescu, and C. B. Collins, "Multiphoton Generation of Optical Sidebands to Nuclear Transitions," Phys. Rev. C 23:1007 (1981).
8. C. B. Collins, "The Tuning and Stimulation of Gamma Radiation," Proceedings of the International Conference on Lasers '80, edited by C. B. Collins (STS Press, McLean, VA, 1981) pp. 524-531.
9. C. B. Collins, "Upconversion of Laser Radiation to γ-Ray Energies," Laser Technique for Extreme Ultraviolet Spectroscopy, edited by T. J. McIlrath and R. R. Freeman (AIP Conference Proceedings No. 90, New York, 1982) pp. 454-464.
10. C. B. Collins, "Upconversion of Laser Radiation to Gamma-Ray Energies," Proceedings of the International Conference on Lasers '81, edited by C. B. Collins (STS Press, McLean, VA, 1982) pp. 291-295.
11. C. B. Collins, F. W. Lee, D. M. Shemwell, B. D. DePaola, S. Olariu, and I. Popescu, "The Coherent and Incoherent Pumping of a Gamma Ray Laser with Intense Optical Radiation," J. Appl. Phys. 53:4645 (1982).

12. B. D. DePaola and C. B. Collins, "Tunability of radiation generated at wavelengths below 1 Å by anti-Stokes scattering from nuclear levels," J. Opt. Soc. Am. B 1:812 (1984).

13. C. B. Collins and B. D. Depaola, "Tunable Sub-Angstrom Radiation Generated by Anti-Stokes Scattering from Nuclear Levels," Laser Techniques in the Extreme Ultraviolet, edited by S. E. Harris and T. B. Lucatorto (AIP Conference Proceedings No. 119, New York, 1984) pp. 45-53.

14. C. B. Collins and B. D. DePaola, "Observation of coherent multiphoton process in nuclear states," Optics Lett. 10:25 (1985).

15. B. D. DePaola, S. S. Wagal, and C. B. Collins, "Nuclear Raman Spectroscopy," J. Opt. Soc. Am. B 2:541 (1985).

16. B. D. DePaola and C. B. Collins, "Tunability of radiation generated by wavelengths below 1 Å by anti-Stokes scattering from nuclear levels," J. Opt. Soc. Am. B 1:812 (1984).

17. B. Pontecorvo and A. Lazard, "Isomerie nucleaire produite par les rayons X du spectre continu," C. R. Acad. Sci. 208:99 (1939).

18. G. B. Collins, B. Waldman, E. M. Stubblefield, and M. Goldhaber, "Nuclear excitation of indium by x-rays," Phys. Rev. 55:507 (1939).

19. Y. Watanabe and T. Mukoyama, "Excitation of nuclear isomers by γ rays from ^{60}Co," Bull. Inst. Chem. Res., Kyoto Univ. 57:72 (1979).

20. M. Krcmar, A. Ljubicic, K. Pisk, B. Logan, and M. Vrtar, "Photoactivation of ^{111}Cd," Phys. Rev. C 25:2097 (1982).

21. I. Bikit, J. Slivka, I. V. Anicin, L. Marinkov, A. Ruydic, and W. D. Hamilton, "Photoactivation of ^{111}Cd^m without a 'nonresonant' contribution," Phys. Rev. C 35:1943 (1987).

22. "The EGS4 Code System," Walter R. Nelson, Hideo Hirayama, and David W. O. Rogers, Stanford Linear Accelerator Center Report No. SLAC 265, 1985 (unpublished).

23. Evaluated Nuclear Structure Data File (Brookhaven National Laboratory, Upton, New York, 1986).

24. E. Browne and R. B. Firestone, "Table of Radioactive Isotopes," edited by V. S. Shirley, Wiley, New York (1986) pp. 180-182.

25. J. A. Anderson and C. B. Collins, "Calibration of pulsed bremsstrahlung spectra with photonuclear reactions of ^{77}Se and ^{79}Br," Rev. Sci. Instrum. 58:2157 (1987).

26. J. A. Anderson and C. B. Collins, "Calibration of pulsed x-ray spectra," Rev. Sci. Instrum. 59:414 (1988).

27. C. B. Collins, J. A. Anderson, Y. Paiss, C. D. Eberhard, R. J. Peterson, and W. L. Hodge, "Activation of ^{115}In^m by single pulses of intense bremsstrahlung," Phys. Rev. C 38:1852 (1988).

28. J. A. Anderson, M. J. Byrd, and C. B. Collins, "Activation of ^{111}Cd^m by single pulses of intense bremsstrahlung," Phys. Rev. C 38:2833 (1988).

29. E. C. Booth and J. Brownson, "Electron and Photon Excitation of Nuclear Isomers," Nucl. Phys. A98:529 (1967).

30. E. Browne, "Nuclear data sheets for A = 180," Nucl. Data Sheets 52:127 (1987).

31 C. B. Collins, C. D. Eberhard, J. W. Glesener, and J. A. Anderson, "Depopulation of the isomeric state ^{180}Ta^m by the reaction ^{180}Ta$^m(\gamma,\gamma')^{180}$Ta," Phys. Rev. C 37:2267 (1988).

32. J. J. Carroll, J. A. Anderson, J. W. Glesener, C. D. Eberhard, and C. B. Collins, "Accelerated Decay of ^{180}Ta^m and ^{176}Lu in Stellar Interiors through (γ,γ') Reactions," Astrophys. J. 344:454 (1989).

33. C. B. Collins, J. J. Carroll, T. W. Sinor, M. J. Byrd, D. G. Richmond, K. N. Taylor, M. Huber, N. Huxel, P. von Neumann-Cosel, A. Richter, C. Spieler, and W. Ziegler, "Resonant excitation of the reaction $^{180}Ta^m(\gamma,\gamma')^{180}Ta$," Phys. Rev. C 42:R1813 (1990).

34. P. M. Walker, F. Sletten, N. L. Gjørup, M. A. Bentley, J. Borggreen, B. Fabricius, A. Holm, D. Howe, J. Pedersen, J. W. Roberts, and J. F. Sharpey-Schafer, "High-K Barrier Penetration in ^{174}Hf: A Challenge to K Selection," Phys. Rev. Lett. 65:416 (1990).

35. J. J. Carroll, M. J. Byrd, D. G. Richmond, T. W. Sinor, K. N. Taylor, W. L. Hodge, Y. Paiss, C. D. Eberhard, J. A. Anderson, C. B. Collins, E. C. Scarbrough, P. P. Antich, F. J. Agee, D. Davis, G. A. Huttlin, K. G. Kerris, M. S. Litz, and D. A. Whittaker, "Photoexcitation of nuclear isomers by (γ,γ') reactions," Phys. Rev. C 43:1238 (1991).

36. J. J. Carroll, T. W. Sinor, D. G. Richmond, K. N. Taylor, C. B. Collins, M. Huber, N. Huxel, P. von Neumann-Cosel, A. Richter, C. Spieler, and W. Ziegler, "Excitation of $^{123}Te^m$ and $^{125}Te^m$ through (γ,γ') reactions," Phys. Rev. C 43:879 (1991).

37. G. Knopf and W. Paul, "Alpha, Beta and Gamma-Ray Spectroscopy," edited by Kai Siegbahn, North-Holland Co., Amsterdam (1965) pp. 1 - 25.

38. C. Kittel, "Introduction to Solid State Physics, 6th Edition," Wiley, New York (1986) pp. 106.

GENERATION OF MAGNETIC FIELD BY CIRCULARLY POLARIZED LIGHT AND THERMOMAGNETIC INSTABILITY IN A LASER PLASMA

H.R. Strauss,[1] S. Eliezer,[2] Y. Paiss,[2] and A. Fruchtman[3]

[1]New York University, New York, N.Y., USA
[2]Soreq Nuclear Research Center, Yavne, Israel
[3]Weizmann Institute of Science, Rehovot, Israel

ABSTRACT

Circularly polarized laser light can be used to produce a poloidal magnetic field in laser produced inertially confined plasmas. The poloidal field combined with the naturally occurring toroidal field, gives a spheromak like magnetic field, which could be beneficial as a thermal insulator.

The thermomagnetic instability, driven by parallel gradients of the plasma density temperature, is studied in a magnetized plasma, in which the electron cyclotron frequency is larger than the electron collision frequency. The nature of the transverse heat conductivity in the magnetized plasma causes the instability to be driven by opposite density and temperature gradients.

CIRCULARLY POLARIZED LIGHT AND MAGNETIC FIELD GENERATION

Megagauss magnetic fields are routinely observed in laser produced plasmas. Many mechanisms have been proposed for this effect, but most of them ultimately rely on the thermoelectric e.m.f. produced by crossed temperature and density gradients.[1] This can produce megagauss fields with typical parameters. In an axially symmetric system, the magnetic field is in the tororoidal direction, that is, in the direction of symmetry. In some experiments, a poloidal field has also been seen. The mechanism has been attributed to the MHD dynamo effect.[2,3]

A poloidal field is important in laser fusion schemes which rely on the magnetic field for thermal insulation.[4] In recent experiments, a poloidal field has been produced

Laser Interaction and Related Plasma Phenomena, Vol. 10
Edited by G.H. Miley and H. Hora, Plenum Press, New York, 1992

167

by using a second laser to drive a current in an external coil.[5] In those experiments, the poloidal field led to improved plasma characteristics, such as longer lifetime and higher energy.

Here we propose a direct method for producing a poloidal field using circularly polarized laser light, which is capable of producing megagauss fields.

Consider a cold plasma fluid. The electrons move in an applied electric field according to the linearized law of motion

$$\frac{\partial \mathbf{v}_1}{\partial t} = -\frac{e}{m}\mathbf{E} \tag{1}$$

where $\mathbf{v}_1$ is the linearized electron velocity, and $\mathbf{E}$ is the applied electric field of the laser. The ions will be considered immobile. The electric field is incident in the z direction and circularly polarized in the x, y plane,

$$\mathbf{E} = E_1(\hat{x} + i\hat{y})\exp(-i\omega t + ikz) \tag{2}$$

We note that $| E |^2 = 2E_1^2$. The electrons also satisfy the continuity equation,

$$\frac{\partial n}{\partial t} = -\nabla \cdot (n\mathbf{v}) \tag{3}$$

The density is assumed to consist of a background and perturbed component, $n = n_0 + n_1$, with $n_1 \ll n_0$. The perturbed density satisfies,

$$-i\omega n_1 = -\mathbf{v}_1 \cdot \nabla n_0 \tag{4}$$

using (1) and assuming E_1 is spatially constant. We may then calculate the second order perturbed current

$$\mathbf{J} = -e\overline{n_1 \mathbf{v}_1}$$

and obtain

$$\mathbf{J} = \frac{e^3}{2m^2\omega^3} | E |^2 \nabla n \times \hat{z}, (\mathbf{k} \parallel \hat{z}) \tag{5}$$

which in is the toroidal direction. The current produces a poloidal magnetic field

$$\nabla \times \mathbf{B} = \frac{4\pi}{c}\mathbf{J},$$

which in order of magnitude is given by

$$B = \frac{\omega_p^2}{\omega^3}\frac{e}{mc} | E |^2 \tag{6}$$

where ω_p is the electron plasma frequency. This may be expressed in terms of the laser light intensity I_L by

$$I_L = \frac{c}{4\pi} | E |^2 .$$

Thus we find the poloidal magnetic field is

$$\frac{B}{\text{gauss}} = 2 \times 10^{-10}\frac{I_L}{\text{watt/cm}^2}\left(\frac{\lambda}{1\mu m}\right)^3\left(\frac{n}{10^{21}\text{cm}^{-3}}\right). \tag{7}$$

where λ is the laser wavelength. For example, taking $\lambda = 1\mu m$, as for a neodymium laser, and $n = 10^{21}cm^{-3}$, with $I_L = 10^{16}$ watts/cm², gives $B = 2 \times 10^6$ gauss. As a second example, suppose $\lambda = 10\mu m$, as in a CO_2 laser, $n = 10^{19}cm^{-3}$, and $I_L = 10^{14}$

watts/ cm². We have kept $I_L\lambda^2$ constant in the two examples, as this seems to be the case in practice. For this example, $B = 2 \times 10^5$ gauss.

THERMOMAGNETIC INSTABILITY

The generation of strong magnetic fields in laser-produced plasmas is a subject of major concern. One of the mechanisms for the spontaneous generation of magnetic fields is the thermomagnetic instability driven by parallel equilibrium density and temperature gradients.[6,7] The analyses of this instability assumed that the equilibrium plasma was unmagnetized. However, hot plasmas ($T = 10$ keV), even of density 10^{20}cm^{-3}, become magnetized ($\Omega\tau > 1$, Ω is the electron cyclotron frequency and τ is the electron collision time) already for a magnetic field of 300 G.

We assume that the process is so fast that the ions are immobile. The time scale is between the electron and the ion cyclotron periods and the spatial scale is between the electron and the ion skin depths. With the further assumption of quasi-neutrality, the electron density is fixed in time. The equations that govern the evolution of the electron thermal energy and of the magnetic field are the electron heat-balance equation

$$\frac{3}{2}\frac{\partial nT}{\partial t} + \nabla \cdot \left(\frac{3}{2}\,nT\mathbf{v}\right) + nT\nabla \cdot \mathbf{v} - \nabla \cdot K_{\wedge}\hat{h} \times \nabla T = 0 \;, \tag{8}$$

and Faraday's law combined with Ohm's Law

$$\frac{1}{c}\frac{\partial \mathbf{B}}{\partial t} = \nabla \times \left[\frac{\mathbf{v} \times \mathbf{B}}{c} + \frac{\nabla(nT)}{en}\right] = 0, \tag{9}$$

Ampere's law $(4\pi/c)\,\mathbf{j} = \nabla \times \mathbf{B}$, and the relation $\mathbf{j} = -en\mathbf{v}$. In these equations, n, $\mathbf{v}$, T are the electron density, flow velocity, and temperature , $\mathbf{B}$ is the magnetic field, $\mathbf{j}$ the current, $-e$ the electron charge, c the velocity of light in vacuum, and $\hat{h} \equiv \mathbf{B}/|B|$. These equations have been used to study fast magnetic evolution in plasma devices.[8] In a magnetized plasma ($\Omega\tau \gg 1$) the transverse thermal conductivity is[9] $K_{\wedge} \cong 2.5 \; cnT/(eB)$.

We study the stability of an equilibrium in which n, T depend only on x, and the magnetic field is constant in the z direction. We introduce perturbations $T_1, B_1 = B_{z1}$ which vary as $\exp(iky - i\omega t)$. The linearized equations (8), (9) become algebraic equations for the frequency ω. Expressing the gradients as $\partial T_0/\partial x = T_0/\ell_n$ and $\partial n_0/\partial x = n_0/\ell_n$, we obtain the dispersion relation,

$$\omega^2 - \sqrt{\frac{2}{3}}\frac{\lambda_D}{\ell_n}\left(1 + \frac{5}{6}\beta\right)\omega + \frac{\lambda_D^2}{\ell_n^2}\left[1 + \eta\left(1 - \frac{5}{6}\,\beta\right)\right] = 0 \;. \tag{10}$$

Here $\eta \equiv \ell_n/\ell_T$, $\beta = 8\pi nT/B^2$, and λ_D is the Debye length. For $\beta \gg 1$, the instability exists for negative $\eta < -5/12$. The growth rate is

$$Im\omega = kc\left(-\frac{5}{6}\frac{\lambda_D^2}{\ell_n\ell_T}\,\beta\right)^{1/2} \;. \tag{11}$$

The plasma magnetization reduces the growth rate from the unmagnetized case[6,7] by $c/(\omega_p\lambda_e)$, where λ_e is the electron mean free path. This is a very small number in laser produced plasmas. We conclude that the thermomagnetic instability is not likely to generate the megagauss fields observed in laser produced plasma, because of this reduction of the growth rate for $\Omega\tau > 1$.

169

Acknowledgement

H. Strauss acknowledges the hospitality of the Weizmann Institute of Science, where this work was performed, and the support of an Einstein Fellowship, the US National Science Foundation, and the US Department of Energy.

REFERENCES

[1] C. Max. "Laser-Plasma Interaction," R. Balian and J.-C. Adam, eds., North-Holland, Amsterdam (1982).

[2] J. Briand, V. Adrian, M. El Tamer, A. Gomes, Y. Quemener, J.P. Dinguirard, and J.C. Kieffer, *Phys. Rev. Lett.* 54:38 (1983).

[3] R. Dragila, *Phys. Fluids* 30:925 (1987).

[4] A. Hasegawa, K. Nishihara, H. Daido, M. Fujita, R. Ishizaki, F. Miki, K. Mima, M. Murakami, S. Nakai, K. Terai, and C. Yamanaka, *Nucl. Fusion* 28:369 (1988).

[5] H. Daido, F. Miki, M. Fujita, K. Sawai, H. Fujita, Y. Kitagawa, S. Nakai, and C. Yamanaka, *Phys. Rev.Lett.* 56:846 (1986).

[6] D.A. Tidman and R.A. Shanny, *Phys. Fluids* 17:1207 (1974).

[7] L.A. Bol'shov, Yu. A. Dreizin, and A.M. Dykhne, *JETP Lett.* 19:168 (1974).

[8] A. Fruchtman, *Phys. Fluids* B3:1908 (1991).

[9] S.I. Braginskii, Transport procession in plasmas, in "Reviews of Plasma Physics," M.A. Leontovich, ed., Consultants Bureau, New York (1965).

LASER DRIVEN PLASMA INSTABILITIES AT MODERATE LASER IRRADIANCES

L. A. Gizzi*, D. Batani, V. Biancalana, M. Borghesi, P. Chessa, I. Deha~,
A. Giulietti, D. Giulietti°, E. Schifano^, O. Willi#

Istituto di Fisica Atomica e Molecolare, Via del Giardino, 7 - 56100 Pisa, Italy

°Dipartimento di Fisica, Universita' di Pisa, Piazza Torricelli, 56100 Pisa, Italy
The Blackett Laboratory, Imperial College, London, U.K.
~ Universite' "H. Boumedienne", Algers, Algerie
^Ecole Polytechnique, Palaiseau, France
* also at Imperial College, London, U.K.

ABSTRACT

A systematic investigation is in progress on Second Harmonic (SH) and Three-Half Harmonic ($3\omega/2$) emission generated by the interaction of a 3 ns 1.064 μm Nd:YAG laser beam with underdense plasmas produced from thin plastic targets at irradiances up to 5×10^{13} Wcm^{-2}. 2-D time resolved imaging of the interaction region in SH light has been performed for a detailed characterization of SH radiation sources in the plasma. The onset of a SH emission regime dominated by Filamentation Instability is clearly evidenced by these images of the interaction region and confirmed by the anomalous scaling of SH power with incident laser power. The dependence of SH emission on target position has also been measured revealing an unexpected behaviour which we believe is linked to the effect of local beam geometry on Filamentation Instability growth rate mechanisms. A simple model of $3\omega/2$ emission is discussed which takes in account refraction of Two Plasmon Decay (TPD) produced electron waves. This model shows a strong dependence of $3\omega/2$ spectra on longitudinal density gradients. In contrast with 2ω, transverse density gradients do not contribute appreciably to $3\omega/2$ emission. We report on a new experiment performed to test the model and the effective reliability of $3\omega/2$ as a diagnostic tool to estimate electronic plasma temperature.

INTRODUCTION

Laser driven plasma instabilities are presently under extensive experimental and theoretical investigation world-wide due to the role they play in ICF experiments[1]. Great effort is being devoted in order to achieve control of these processes mainly by optimizing irradiation parameters through the introduction of laser beam smoothing techniques[2,3]. Among the large number of laser driven plasma instabilities, particular attention has been devoted to Filamentation Instability (FI). Incidentally it should be pointed out that the local increase of laser intensity produced by FI can give rise to interaction conditions suitable for instabilities to take place whose intensity threshold is higher than the average nominal intensity.

Our group is involved in collaborative experiments at Rutherford Appleton Laboratory (RAL), UK and at Ecole Polytechnique, France, to study some of these instabilities at intensity exceeding 10^{15} Wcm^{-2}. Particularly in the RAL experiment we study those effects

Laser Interaction and Related Plasma Phenomena, Vol. 10
Edited by G.H. Miley and H. Hora, Plenum Press, New York, 1992

171

in laser interaction with long-scalelength preformed coronal plasmas. In this experiment was proved that suitable beam smoothing techniques allow to control filamentation as well as stimulated Brillouin and Raman scattering[4]. A parallel study is performed at IFAM at lower laser intensity, up to 5×10^{13} Wcm-2. This intensity is still relevant to ICF and as been demonstrated to be able to excite several instabilities directly or via filamentation. Moreover at these levels of laser power a much larger number of shots can be produced during a single experiment, resulting in more accurate measurements.

In this paper we present recent experimental results obtained at moderate intensity. The experimental method simply consists in the direct irradiation of thin plastic foils with laser pulses whose duration and intensity on target allow the expanding plasma to become underdense early in the pulse. In particular the paper reports on the analysis of SH and $3\omega/2$ scattered radiation, which is one of the most direct approaches to plasma instabilities. A direct link between Filamentation Instability (FI) and 2w emission from underdense plasmas is not yet experimentally established strong evidence has been provided[5,6,7] which supports such correlation.

We discuss recent experimental investigation on forward 2ω emission from laser interaction with underdense plasmas. These measurements follow our previous analysis on side scattered SH radiation[7,8] and another interesting experiment[9] where filaments were artificially created into a plasma in order to study the mechanism of 2ω generation. Our measurements on forward emission clearly show that SH strongly emerges from the plasma self emission in presence of filamentation and filaments are the main source of 2w light. There is a definite confirmation that SH generation from underdense plasmas is very sensitive to the presence of transverse gradients (both density and field gradients). On the other hand, $3\omega/2$ spectra are mostly influenced by longitudinal density gradients as those produced by exploding foils. This observation was already considered in the interpretation of previous spectra[10] and is further supported by recent spectroscopic studies. They were performed using a new technique designed to detect in a single shot two separate spectra of $3\omega/2$ light emitted from both sides of the original target plane.

EXPERIMENTAL SET UP

A schematic diagram of the experimental set up with the main diagnostics for the study of SH and $3\omega/2$ emission is shown in Fig.1. A 3 ns (FWHM) 1.064 μm Nd:YAG (0.7 Å

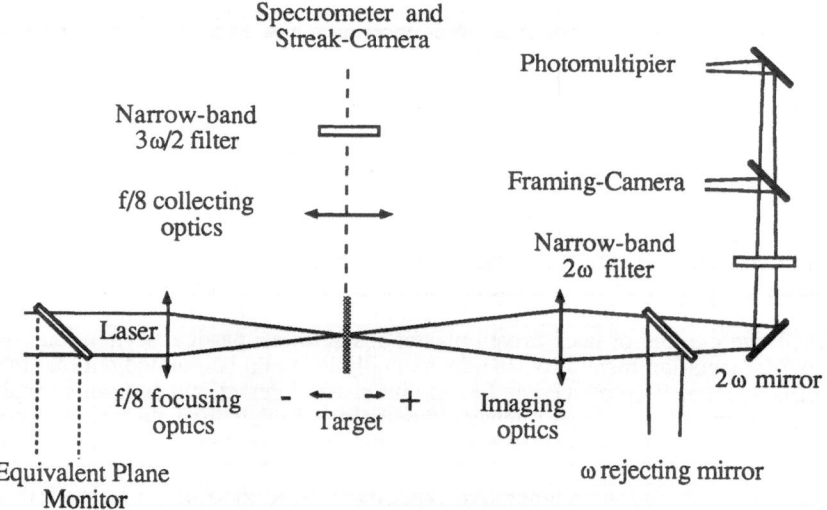

Fig.1 Schematic set up for 2ω and $3\omega/2$ measurements from laser produced plasma.

spectral band-width) laser beam was focused with an f/8 optics in a 60 μm focal spot on thin plastic (FORMVAR) targets whose thickness ranged from 0.1 to 1.8 μm. The laser intensity on target was varied between 5×10^{11} and 5×10^{13} Wcm^{-2}. The cross section of the laser beam in the focal region was monitored by an equivalent plane imaging optics.

A visible spectrometer coupled to a streak-camera was used for time resolved spectroscopy of 3ω/2 radiation. The temporal resolution was 30 ps while the overall spectral resolution was 3 Å. A 120 ps gate-time visible Framing-Camera (FC)[11] allowed 2-D time resolved imaging of the interaction region in SH light to be performed. A 80X magnified image of the plasma was produced on the input window of the FC whose spatial resolution is approximately 100 μm. Therefore the limit to the overall resolution in the plasma (≈ 2 μm) was finally set by optical aberration eventually present the imaging optics. Photomultiplier tubes and diode detectors coupled to suitable narrow-band interference filters were employed for energy measurements of scattered light.

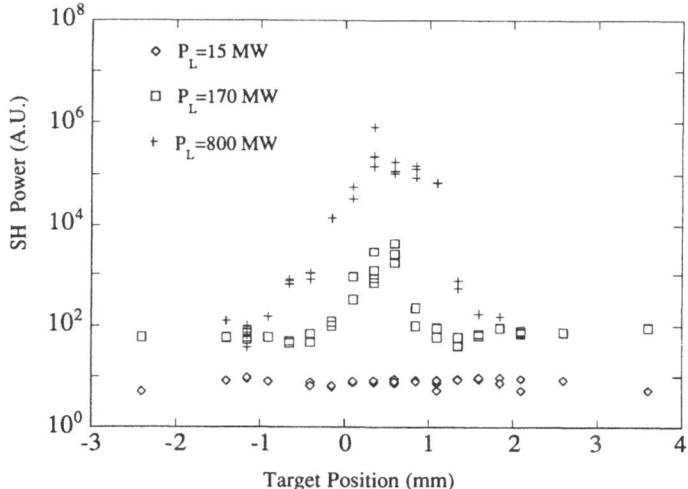

Fig.2 Forward Second Harmonic power as function of target position along the beam axis relative to the laser beam waist. Laser power levels were 15 MW, 170 MW and 800 MW and the target thickness was 1μm.

RESULTS AND DISCUSSION

Second Harmonic measurements

The graph of Fig.2 shows SH power as function of target position with respect to the laser beam waist for three different laser power levels. A spectral window of 30 Å centred on the exact $2\omega_L$ angular frequency, ω_L being the laser angular frequency, was selected by an interference filter. The transmitted light was detected by a photomultiplier tube. For laser power of 15 MW, corresponding to an intensity at the beam waist of approximately 5×10^{11} Wcm^{-2}, no dependence of the detected radiation was found on target position within the explored range. On the contrary an incident laser power of 170 MW resulted in a strong dependence of SH emission on target position with a maximum at approximately +600 μm from the nominal position of the laser beam waist.

Similar but enhanced behaviour was observed for laser power of 800 MW where the maximum emission was an order of magnitude higher than in the case of 170 MW laser power. Speculation on the shift of the maximum SH emission from the nominal focus requires a discussion on the methods used to determine the position of the focus itself. However this analysis is not relevant to the aim of this paper and is reported elsewhere[12].

It should be noted that the signal detected for marginal target positions, (typically for distances greater than 1.5 mm from the position of maximum SH emission), as well as the whole curve at the lowest laser power level, has to be attributed to bremsstrahlung radiation at the same frequency as the SH. Such conclusion is supported by comparing these measurements with analogous ones performed by tilting the interference filter. As a result of this tilting the spectral region allowed by the filter was shifted by the amount required to exclude narrow-band SH light from the detected spectral window. On the contrary the spectrum of bremsstrahlung radiation can be considered flat for a relatively large wavelength range compared to the SH band-width. Therefore, the contribution of bremsstrahlung radiation to the detected signal in both conditions is expected to remain fairly constant. In fact we observed that in condition of maximum SH emission, a tilt of the interference filter by 30° reduced the amplitude of the photomultipier signal by a factor of thousand. On the other hand a small signal variation was observed with tilted filter for marginal target positions or low laser power thus confirming the above assumption.

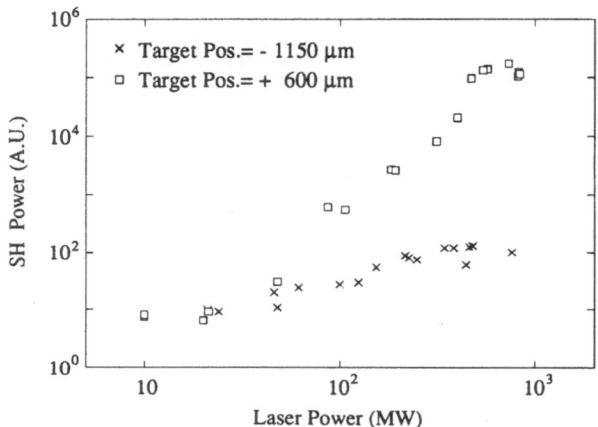

Fig.3. Forward Second Harmonic power as function of laser power for a 1μm target in the position of maximum emission and in a marginal position at -1.15 mm.

The most evident feature of the plots shown in Fig.2 is that SH emission clearly dominates over bremsstrahlung by up to three orders of magnitude. In addition we notice that an increase in the level of SH emission up to two orders of magnitude was detected for target displacement of approximately 200 μm. This value is definitely smaller than the overall depth of focus of our focusing optics which was measured to be approximately 600 μm. Therefore it is evident that some mechanism takes place which strongly modifies the laser intensity distribution generated by the focusing optics. This conclusion is supported by the dependence of SH emission on incident laser power. Fig.3 shows a plot of SH emission as function of laser power in a range from 10 MW to approximately 1GW with the target in position of maximum emission and, for comparison, in a marginal position.

In the range of laser power P_L between 0.1 and 1 GW, SH power P_{SH} scales approximately as the third power of P_L. On the other hand a sub-linear scaling of P_{SH} with laser power is measured for target in a marginal position. As stated above, most of the light collected at marginal target positions is due to bremsstrahlung radiation. This fact can explain the weak dependence of this curve on incident laser power. On the contrary, the more than quadratic scaling of P_{SH} with target in the position of maximum conversion efficiency can only be explained if extra processes are included which can "boost" the non-linear SH conversion efficiency. It is well known that ponderomotive and thermal effects can drive laser focusing processes in the laser-plasma interaction region. On the other hand another indirect experimental evidence of Filamentation Instability has already been established in our previous analysis of side scattered SH radiation in an underdense plasmas[7].

A more direct evidence of filamentary structures in our experiment was provided by the Framing Camera time resolved images. Fig.4 shows a time resolved 120 ps gate-time image of the interaction region in SH light at the time of the peak of the 3 ns laser pulse. The incident laser power was 800 MW and the target was 1μm thick and was set in the position of maximum SH conversion efficiency. The same timing was used to obtain a large number of images, including those shown in Figs. 5b and 6b.

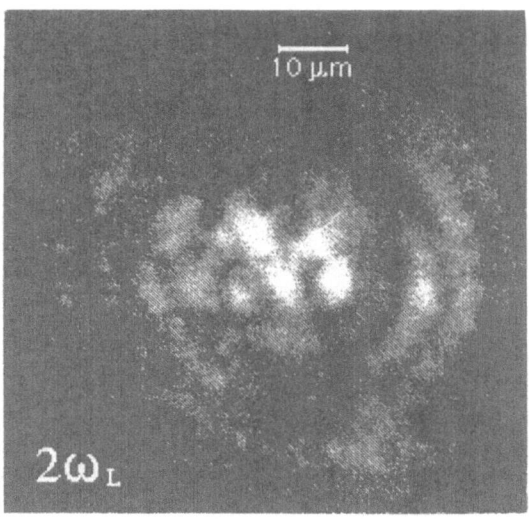

Fig.4   Time resolved Framing Camera image of the interaction region in Second
Harmonic light with target in the position of maximum SH emission.

This image clearly shows that SH emission comes from small circular regions, approximately 5 μm in diameter, which are well localized in the centre of the spot. The size of these structures is fairly reproducible shot by shot while their number and position change unpredictably. A simple question arising now is whether this pattern can simply be correlated to the intensity distribution in the laser beam or it is the result of more complex processes, probably initiated by laser beam inhomogeneities but ultimately due to plasma instabilities. A simple comparison between SH time resolved images and laser beam intensity distribution in the focal spot can give a valuable indication of the kind of coupling regime we are dealing with.

Equivalent Plane (EP) imaging was employed to study laser beam intensity distribution in the focal spot. An infrared (Kodak I.R. 4143) film was used to take EP images at different positions along the beam axis. Fig.5 shows a comparison between an EP image of the focal spot and another time-resolved FC image of the interaction region with the target in the same position of maximum SH conversion efficiency. Not only laser beam non-uniformities, but also optical aberrations of the focusing optics may contribute to the intensity distribution of Fig.5a. For this reason the clear difference between pictures of Fig.5a and Fig.5b cannot be considered as a conclusive evidence of the action of FI. Much more significant is that i) the size of the 2ω filamentary sources in Figs.4, 5b and several others pictures obtained in the same condition is close to the theoretical size expected for FI maximum growth; ii) 2ω images varied considerably shot by shot, while the ω equivalent plane images did not.

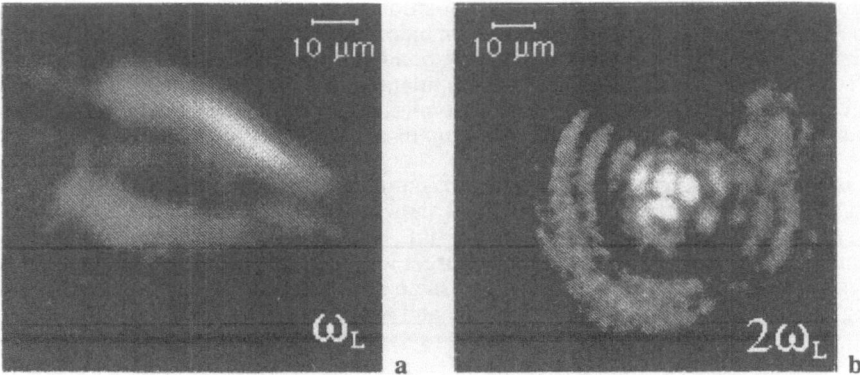

Fig.5 a)Equivalent Plane Image of laser focal spot at the waist of the f/8 main focusing optics and b)Framing Camera image of the interaction region in Second Harmonic light taken with the target at the same position of the equivalent plane image.

Therefore the laser intensity non-uniformities by itself cannot explain the SH source distribution nor the shot to shot variation evidenced by FC time resolved images. Filamentation process has to be included to explain the localized SH emission we observe in our experiment relatively to the target position of maximum of SH conversion efficiency. It should be emphasized that this filamentation dominated regime of SH emission seems to be confined in a narrow region of target positions around the maximum of Fig.2. as evinced from the following results. Fig.6 shows a set of images analogous of that of Fig.5 but taken with target at approximately -400 μm which, according to Fig.2 is still in the interval of high SH conversion efficiency but far from the maximum. In this case the laser intensity distribution shown by the EP image of Fig.6a is affected by the astigmatism of the beam. Although graphically attractive, the resemblance in shape of these two images should be carefully examined before drawing any conclusion as already pointed out in the case of Fig.5. In fact optical aberrations of the EP imaging system could introduce *ad hoc* changes in the laser focal spot intensity distribution. In addition the comparison of images is based on the corrispondence of the EP position in the case of Fig.6a and target position in the case of Fig.6b. Since large variations in the exact shape of the EP images is found for variation of the plane position on a 100μm scale, a comparison of the shape of the two images would require the knowledge of the exact location of 2ω sources in the plasma. And the accuracy on this measurement is limited by the depth of focus of the 2ω imaging channel.

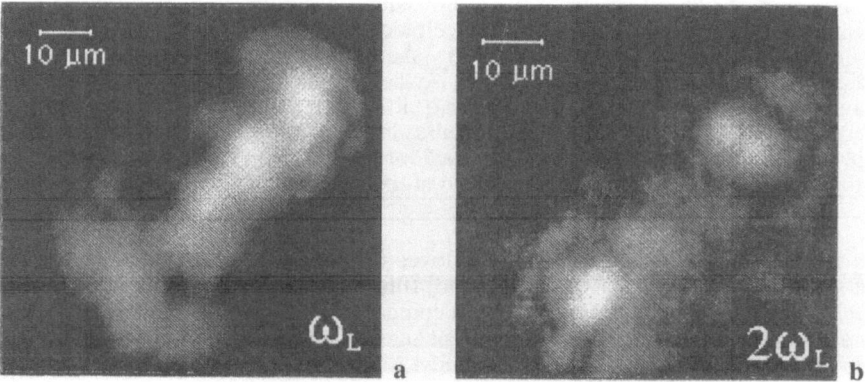

Fig.6 a)Equivalent Plane Image of laser focal spot 400 μm before the beam waist of the f/8 main focusing optics and b)Framing Camera image of the interaction region in Second Harmonic light taken with the target at approximately the same position of the equivalent plane image.

On the contrary it is important to point out that the pattern shown in this case by time resolved images was found to be highly reproducible shot by shot as already observed in the case of EP images.Therefore in this condition forward SH emission has to be attributed largely to intensity non-uniformities already present in the laser beam. Moving the target closer to the position of maximum SH emission there is a transition to a filamentation dominated regime.

The intermediate regime with SH emission but with low level of filamentation has been also investigated with measurements of SH power versus laser power. The different scaling of SH power with incident laser power typical of this regime is illustrated by Fig.7 compared with that relative to the filamentation dominated regime shown in Fig.3. Although the overall behaviour of these two plots is similar, the rate of growth of P_{SH} with P_L is different in the two cases. In contrast with the more than quadratic scaling of P_{SH} with P_L evidenced in the plot of Fig.3, the growth of P_{SH} in Fig.7 is closer to a quadratic regime.

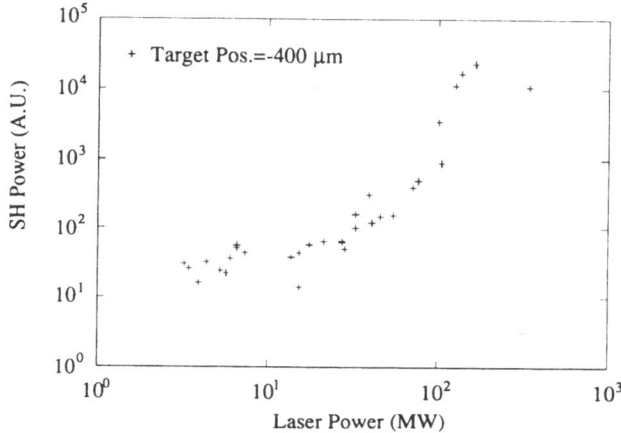

Fig.7 Forward scattered Second Harmonic power as function of laser power for a 1μm target in the same position as Fig.6.

This comparison suggests that relative position between laser beam focusing configuration and plasma density or temperature profile are not suitable in this intermediate regime, for an efficient growth of filamentation processes. Second Harmonic emission in this regime can therefore be attributed to transverse electric field gradients at the boundary of the laser beam waist as shown in Fig.6, and density gradients which originate from these intensity non-uniformities. We observe that the average laser intensity in the focal spot doesn't vary appreciably from Fig.5 to Fig.6. This suggests that it is possible to reduce the effect of FI at a given intensity on target by choosing appropriate target position in the region surrounding the beam waist.

Three-half harmonic measurements

3ω/2 harmonic emission consequent on Two Plasmon Decay (TPD) can be a useful tool to diagnose plasma temperature. According to the present understanding, 3/2 harmonic is generated by the coupling (Raman-like scattering) of laser photons with TPD plasmons excited in the $n_c/4$ layer by laser light itself. ω_L and 3ω/2 photons can propagate through this layer without strong refractive effects. For this reason spectral features of coupling plasmons can be directly inferred from the analysis of 3ω/2 light . TPD produced plasma waves are frequency shifted from $\omega_L/2$ value according to

$$\delta \omega = 3 k_L \chi v_e^2 /\omega_L$$

where k_L is the wave-vector of the laser photon, χ is the component of the plasmon in the direction of k_L reduced by $k_L/2$. Consequently 3ω/2 light should be frequency shifted by an amount proportional to the plasma temperature, the coupling coefficient being simply related to the matching condition in k-space. In terms of the detection angle θ with respect to the laser beam axis

$$\delta \omega = (3 k_L^2 v_e^2 /\omega_L) ((8/3)^{1/2} \cos \theta - 3/2).$$

Experimental measurements give indeed very broadened spectra, due to the strong refraction of plasmons before they couple. Plasmon frequency cannot be simply related to their wavevector because density gradients present in the plasma can affect the χ parameter[10,13]. The observed 3ω/2 spectrum is red or blue shifted with respect to the above formula whether density gradient component along the beam is negative or positive respectively.

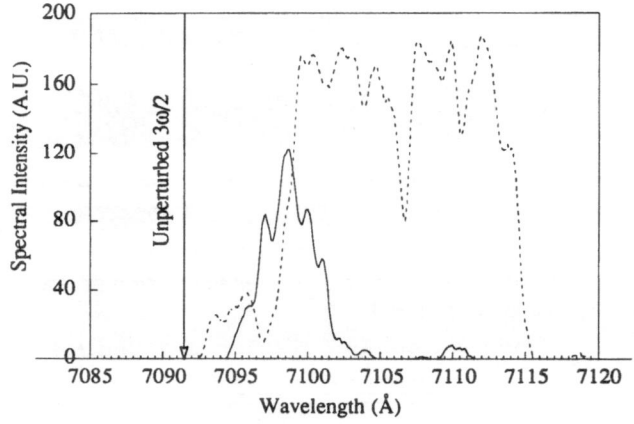

Fig.8 Densitometric traces of 3ω/2 spectra relative to the two plasma regions with opposite gradients along the beam axis. The plasma was produced from of 1.1 μm thick plastic target at an irradiance of 7×10^{12} Wcm-2.

Two plasma regions with opposite gradients along the beam axis are present in plasmas produced by irradiating thin solid targets. Spectra of 3ω/2 emitted normally to the laser beam axis by these two regions were detected in the same shot. Due to the spiky temporal emission of our laser spectra with both spatial and temporal resolution could be recorded. The laser-plasma interaction region was imaged out on the entrance slit of a spectrometer, with the laser axis parallel to the slit axis. The two different spectra resulting in the output plane were in turn imaged on the entrance slit of the streak camera which was fully open. In this configuration the temporal resolution was limited to the duration of a single spike. On the other hand a partial spatial resolution could be achieved. Pairs of spectra produced from different spikes were well separated by the streak.

In Fig.8 densitometric traces of two simultaneous spectra are reported. From the unperturbed $3\omega_L/2$ a shift of about 8 Å can be estimated for unpropagated plasmons. This value leads to an electron temperature of approximately 250 eV which is consistent with previous measurements[13]. This preliminary result is the first attempt to use spatial resolution in order to evaluate plasma temperature from broad $3\omega/2$ spectra. The technique, although promising, needs to be refined and to be supported by a more accurate modeling.

CONCLUSION

Our recent results on 2ω and $3\omega/2$ harmonics improve the capability of diagnose laser-plasma interactions in conditions of interest for ICF. In particular SH emission was found to be very suitable to detect and control Filamentation Instability. In this context the pure evaluation of FI intensity threshold seems not to be sufficient, because also the position of the target respect to the beam waist seems to be critical. In a range of position corresponding to few focal depth we observed three distinct regimes. At marginal position of the target respect to the waist the SH emission, if any, was merged into the plasma self-emission. At intermediate position, where laser intensity on target was close to its nominal maximum, SH level rose up to values definitely higher than continuum. In this latter condition the structure of the SH sources was found to be strictly related to the original non-uniformities of the laser beam. Closer to the beam waist SH jumped up more than three order of magnitude above the continuum, the SH source pattern was very unstable shot by shot and there was clear evidence for Filamentation Instability.

ACKNOWLEDGEMENTS

We are very grateful to Imperial College for supplying us with the 120 ps framing camera during the time of this experiment. We are indebted with L.Nocera and F.Bianconi for enlightening discussions on the theory of SH generation. The contribution of I.Deha was possible thanks to a grant from the International Centre for Theoretical Physics (Trieste, Italy). Two visits of O.Willi to IFAM-CNR in Pisa were supported by the British Council. The research program is fully funded by Consiglio Nazionale delle Ricerche.

REFERENCES

[1]William L. Kruer, Phys. Fluids B 3, 2356, (1991)

[2]Y. Kato, K. Mima, N. Miyanaga, S. Aringa, Y. Kittagawa, M. Nakatsuka and C. Yamanaka, Phys. Rev. Lett., 53, 1057 (1984)

[3]L. H. Lehmberg and S. P. Obenschain, Optics Comm. 46, 27 (1983)

[4]O. Willi, T. Afshar-Rad, S. Coe, A. Giulietti, Phys. Fluids B 2, 1318 (1990)

[5]J. A. Stamper, R. H. Lehmberg, A. Schmitt, M. J. Herbst, F. C. Young, J. H. Gardner and S. P. Obenschain,Phys. Fluids 28, 2563 (1985)

[6]J. Meyer and Y. Zhu, Phys. Fluids 30, 890 (1987)

[7]A. Giulietti, D. Giulietti, D. Batani, V. Biancalana, L. Gizzi, L. Nocera, and E. Schifano, Phys. Rev. Lett., 63, 524 (1989)

[8]A. Giulietti, D. Batani, V. Biancalana, D. Giulietti, L. Gizzi, L. Nocera, and E. Schifano, "Laser interaction and Related Plasma Phenomena", Vol.9, p.273 (Plenum Press, New York)

[9]P. E. Young, H. A. Baldis, T. W. Johnston, W. L. Kruer, and K. G. Estabrook, Phys. Rev. Lett., 63, 2812 (1989)

[10]D. Giulietti, V. Biancalana, D. Batani, A. Giulietti, L. Gizzi, L. Nocera, and E. Schifano, Il Nuovo Cimento, 13, 845 (1991)

[11]S. E. Coe, T. Afshar-Rad and O. Willi, Optics Comm. 73, 299 (1989)

[12]P. Chessa, "Interazione laser-plasma: studio della propagazione del fascio e della luce di seconda armonica emessa in avanti", Dipartimento di Fisica, P.za Torricelli, Pisa, Italy.

[13]L. Gizzi, D. Batani, V. Biancalana, A. Giulietti, and D. Giulietti, Laser Part. Beams, 10 (1991)

PULSATION AND STUTTERING OF LASER
INTERACTION WITH PLASMAS
AND ITS SUPPRESSION BY SMOOTHING

Meral Aydın[1] and Heinrich Hora[2]

[1]Department of Theoretical Physics,
  University of New South Wales, Kensington 2033, Australia
[2]CERN, CH1211 Geneva 23, Switzerland

Smoothing of laser-plasma interaction by ISI, RPP, SSD etc. was mainly directed to overcome lateral non-uniformity of irradiation. While these problems are in no way less important, we derived numerically the model of the Laue-rippling and hydro-relaxation model for explanation of the measured temporal pulsation in the 10 to 40 psec range, and how the smoothing schemes suppress these pulsations. The mechanism of pulsation appears from the numerical studies as the generation of a density ripple due to the partial standing wave in the corona where plasma is being pushed to the nodes. This self generated ideal Laue-Bragg grating causes a very high reflection of light at the very low density of the corona periphery. After a few psec, the density ripple within the corona is relaxed by hydrodynamics and the light can move again to the critical density with low net reflection until another density ripple is established, peripheric reflection occurs (as measured by Maddever and Luther-Davies) and the cycle closes to produce a sequence of pulsed plasma driving. Smoothing consists in the mechanism that the partial standing wave fields of the normally coherent laser irradiated plasma corona is then suppressed. The ISI smoothing uses 2 psec temporally incoherent radiation such that the standing wave laser field patterns are destroyed after 2 psec and the pushing of the plasma to the density ripple is avoided. A test of how the smoothing by a broad band laser beam works was simulated numerically where the suppression of the pulsation could be seen immediately compared to the monocromatic case. A further conclusion for testing this model may use the "question mark experiment". The result provides a physics solution of the laser interaction problem for direct drive inertial fusion energy.

INTRODUCTION

After the last 30 years of extensive studies of laser-plasma interaction for the aim of laser fusion it seems that the very complicated interaction mechanism has been studied in a completely wrong direction. The main emphasis was the study of instabilities and processes to avoid filamentation due to the lateral non-uniformity of the laser beams. It may have been

Laser Interaction and Related Plasma Phenomena, Vol. 10
Edited by G.H. Miley and H. Hora, Plenum Press, New York, 1992

181

overlooked that one basic difficulty is the temporal stochastic pulsation of the interaction with about 20 psec duration. Indications of this phenomenon were seen since 1973 and it may be clear now that these complications are now overcomed by smoothing, but smoothing -though experimentally successful- is interpreted and understood in a different way which discrepancy may now being solved in view of the pulsation and its suppression.

Twenty years ago it was known that the interaction of sufficiently high intensity lasers with targets produced on one hand purely thermal plasmas and on the other hand groups of plasmas with mostly superlinear and suprathermal properties. From laser irradiated aliminum spheres one could immediately see the thermally expanding core and the nonlinear cloud with keV ions[1]. The confusion about high and low temperature x-ray signals was clarified for the first time by Eidmann[2] showing that there are always both temperatures, a thermal one and a sprathermal temperature (due to "hot(?)" electrons).

When these plasmas were to be used of laser driven fusion, a most difficult situation appeared with highly varying reflectivity, low transfer of energy into the plasma, hot spots, self focusing, instabilities, anomalous kinds of scattering etc.. The situation was as crazy as it is with today's impossibility to understand what is going on in a tokamak. One aim was to achieve a smooth interaction. The first ingenious solution was to use "indirect drive"[3] where one option is the cannon ball about which the first publication was by Yabe et al[4] in 1975 converting the laser radiation into x-rays and let these then compress and heat a fusion fuel pellet.

The non-smooth interaction was considered to be caused by the non-uniform irradiation and smoothing was motivated by trying to spoil the otherwise ideal properties of the laser beam by artifically destroying the small spectral line width. When wide band irradiation is produced[5] where the most sophisticated method is the induced spatial incoherence (ISI)[6], a temporal coherence of 1 psec only is achieved. Another way of smoothing was introduced by varying the direction of beamlets within the beam by using the random phase plate (RPP)[7] or the "fly eye" lense array[8] or to combine diffraction and interference as "smoothing by spectral dispersion" (SSD)[9].

It seems evident that the motivation was mostly directed to suppress lateral non- uniformity of the laser intensity[10] when this type of smoothing was introduced. While these techniques proved an enormous success, e.g. the suppression of stimulated Raman scattering (SRS) from few percents to hundred times lower values[11], it seems that its essential influence had not at all been understood. The understanding of lateral uniformity in the laser intensity is indeed an important aim and its study[10] is at least a point of partial understanding.

The new message in understanding the recent progress in laser interaction with plasmas by smoothing, however, may be of a basically different kind than the earlier discussed[10] lateral beam uniformity, namely the most complex temporal pulsation of interaction with irregular sequences in the 10 to 40 psec range. A special view into this direction is presented here and a detailed numerical analysis may permit an understanding and an explanation why most conditions of smoothing are just the suppression of this pulsation apart from the non-uniformity problem.

PULSATION IN LASER-PLASMA INTERACTION

We mentioned the difficulty in lateral non-uniformity of irradiation in laser driving of plasmas. An obviously much more dangerous property is that of a strong pulsation of the interaction with a non-periodic sequence in the 20 psec range even if the incident laser intensity is completely constant in time. The first indication appeared from numerical studies. The Kinsinger code at ILE Rochester in 1973 described the one-dimensional interaction of a plane laser wave perpendicularly incident on a collisional plasma with some ramp of density profile.

Using the correct nonlinear optical constants with respect to the intensity dependence of the collision frequency and the ponderomotive and non-ponderomotive terms of the nonlinear force, the two-temperature model showed a very realistic response of the plasma as shown in figure 1. First the light penetrated to the critical density from where on the light decayed exponentially on the depth as expected and a strong mirror reflection produced a partial standing wave pattern with swelling and wave length stretching as known from the nonlinear force. The absorption caused that the standing wave was partial only. At steplike switching on and off the laser, within less than one picosecond, the laser pulse did no longer penetrate to the critical density but decayed to 0.1 percent at the critical density; the light was then reflected at the very low plasma density by the Bragg-von-Laue grating of the plasma density ripple which was produced by the nonlinear force in the standing waves.

While the net reflectivity was rather low when the light had a mirror reflection at the critical density due to the absorption of the light when moving back and forth in the plasma corona, the reflectivity was very high at the phase reflection of the Laue grating. This was a down casting result for laser fusion: the plasma did everything to prevent the light from being absorbed in the plasma. Motivated by this M. Lubin and his colleagues measured the reflectivity of a plasma when irradiated by a 100 psec laser pulse with nearly ideal rectangular time dependence and a rise and decay time of about 3 to 5 psec. The reflectivity showed then pulsation, changing between nearly 100 percent and a few percent within 10 to 20 psec irregularly up and down[12]. The reflected 3/2 harmonics had a similar pulsation but out of phase with the reflectivity[13] showing that only at low reflectivity the light went to the critical density where SRS produced the harmonics and was cut off when the reflectivity occured at the low plasma density by phase reflection.

In the recent years, similar pulsation or stuttering was observed. After discovering a completely irregular modulation of the backscattered spectra of the fundamental and the second harmonics of about 4 Angstrom width, it was clarified[14] that an otherwise smooth laser pulse pulls the plasma corona to a velocity of some 10^7 cm/sec within 10 (or so) psec and then the acceleration stops. After about 20 to 30 psec, another push is given to an additional velocity of the same kind. This gives just the modulated spectra while time resolved detection is unmodulated but just showing the shifts[14]. A pulsation of the reflectivity between few percent and 100 percent with an irregular sequence of about 20 psec was detected[14] similar to Lubin[12] and a likewise pulsation was measured for the H-alpha x-ray emission[15]. It was possible to clarify experimentally that the low reflection during the pulsation comes from the cut-off density while the high reflection is from the low density outermost corona[14].

Another pulsation in the 20 psec range was detected for the 3/2 harmonics[16] and what is most important: when adding a random phase plate (RPP)[7] for smoothing interaction, the harmonics emission is completely smooth without pulsation. The 20 psec pulsation process was directly identified also from measurement of generation of 25 keV ions of ionization up to $Z=18$ from tantalum targets irradiated by 30 psec or 3 nsec neodymium glass laser pulses[17].

NUMERICAL ANALYSIS OF THE STUTTERING PROCESS

Various models were discussed for explaining the mentioned pulsation. Motivated by the problems of lateral non-uniformity in connection with the RPP smoothing, filamentation and self focusing was considered as the reason for the pulsating 3/2 harmonics which simply are avoided by the random phase plate[16]. Futhermore, instabilities were discussed in this connection[14]. The fact that the pulsation is a rather stuttering mechanism with irregular random sequences, induced the view of chaos[18]. While all these and other mechanisms may be involved partially or cannot be excluded easily, the following numerical study may provide an explanation for the dominating mechanism of pulsation or stuttering and a consequence for interpreting the smoothing by ISI, RPP and other methods.

Knowing the interruption of the laser-plasma interaction as seen from the early computations (see above mentioned figures)[19,20] and the alternation from the mirror reflection at the critical density to the phase reflection by the Laue grating, we performed a computation with the real-time genuine two-fluid code of the laser-plasma interaction[20] for plane geometry including realistic nonlinear optical constants, nonlinear forces, equipartition time and thermal conduction with time steps 10 times below the shortest plasma oscillation time and with Maxwellian exact numerical computations of the temporally and spatially changing laser field in the corona[21].

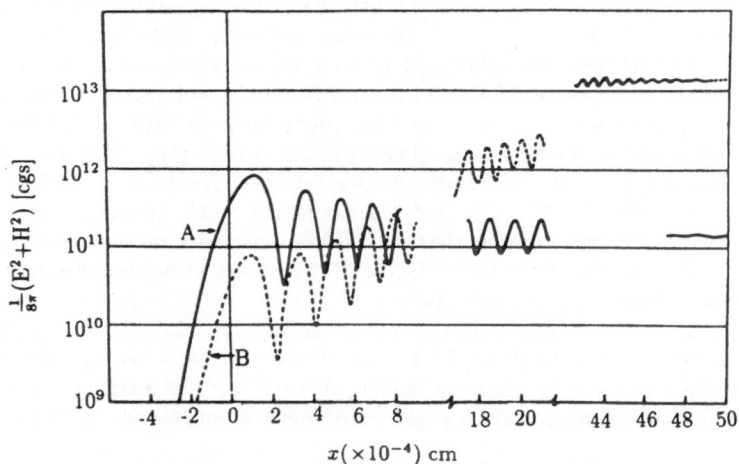

Figure 1. (a) A laser beam incident from the right side on a plasma of initial temperature of 100 eV and linear density increasing from zero at $x=50\mu m$ to the cut-off density at $x=0$ and then increasing more rapidly. The exact stationary (time-dependent) solution without retardation of the Maxwell equations with a nonlinear refractive index, based on an intensity dependent[23] colision frequency, results in an oscillation due to the standing wave and dielectric swelling of the amplitude (curve A). At a later time $(2 \times 10^{-12}$ sec), the laser intensity is 2×10^{16} W/cm^2 (curve B), where the relative swelling remains, but the intensity at $x=0$ is attenuated by dynamic absorption (Results from 1973 using the Kinsinger code[23].

The genuine two-fluid model[21] is based on the nonthermalized direct electromagnetic interaction between the laser light and the fully ionized plasma. In this model, electrons and ions are considered as seperate conducting fluids coupled by momentum exchange and the Coulomb interaction. To obtain the seven quantities which are electron and ion mass densities (ρ_e, ρ_i), temperatures (T_e, T_i), velocities (v_e, v_i) and the longitudinal electric field (E) (each depending on one spatial coordinate x and time t), it is necessary to solve seven equations, which are basically the equations of continuity for electrons and ions

$$\frac{\partial \rho_e}{\partial t} = -\frac{\partial(\rho_e v_e)}{\partial x} \tag{1}$$

and

$$\frac{\partial \rho_i}{\partial t} = -\frac{\partial(\rho_i v_i)}{\partial x}, \tag{2}$$

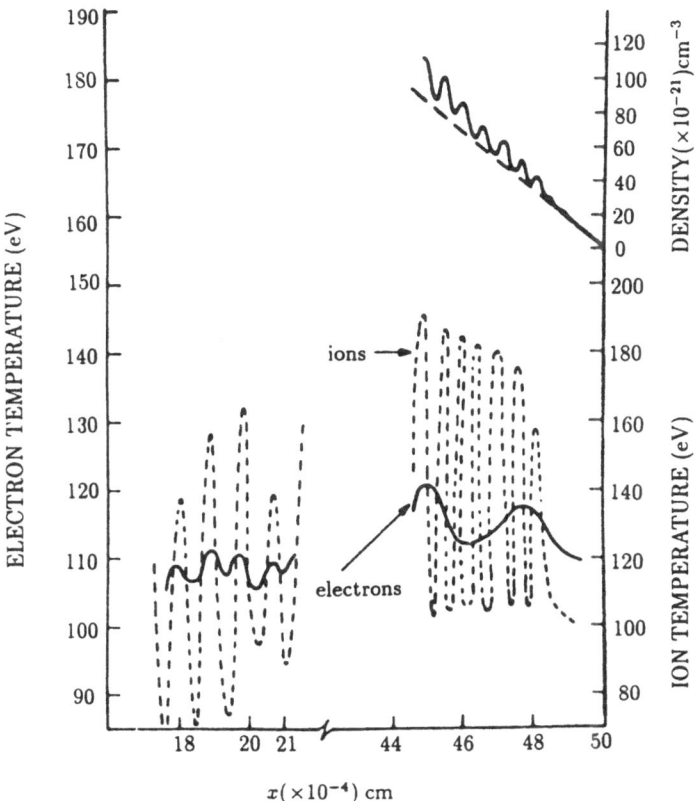

Figure 1. (b) The initial density (dashed line) and the density along curve B of 1(a), where a ripple is created by the nonlinear force pushing the plasma towards the nodes of the standing wave. The electron and ion temperatures are increased following the ripple by dynamic compression at conditions identical to curve B[23].

equations resulting from conservation of momentum for electrons and ions

$$\frac{\partial(\rho_e v_e)}{\partial t} = -\frac{\partial(\rho_e v_e^2)}{\partial x} - \frac{\partial P_e}{\partial x} - \frac{E}{4\pi}\frac{\partial E}{\partial x} - \rho_e \nu(v_e - v_i) + f_{nl} \tag{3}$$

and

$$\frac{\partial(\rho_i v_i)}{\partial t} = -\frac{\partial(\rho_i v_i^2)}{\partial x} - \frac{\partial P_i}{\partial x} - \frac{ZE}{4\pi}\frac{\partial E}{\partial x} + \rho_e \nu(v_e - v_i) + \frac{m_e}{m_i} f_{nl}, \tag{4}$$

equations resulting from conservation of energy for electrons and ions

$$\frac{\partial(\rho_e \epsilon_e)}{\partial t} = -\frac{\partial(\rho_e \epsilon_e v_e)}{\partial x} - P_e \frac{\partial v_e}{\partial x} - \frac{3kn_e}{2\tau}(T_e - T_i) + \frac{\partial}{\partial x}(\kappa_e \frac{\partial T_e}{\partial x}) + W_L \tag{5}$$

and

$$\frac{\partial(\rho_i \epsilon_i)}{\partial t} = -\frac{\partial(\rho_i \epsilon_i v_i)}{\partial x} - P_i \frac{\partial v_i}{\partial x} + \frac{3kn_e}{2\tau}(T_e - T_i) + \frac{\partial}{\partial x}(\kappa_i \frac{\partial T_i}{\partial x}), \tag{6}$$

and the Poisson equation

$$\frac{\partial E}{\partial t} = 4\pi e(n_e v_e - Z n_i v_i). \tag{7}$$

185

In these equations, P_e and P_i are the thermokinetic electron and ion pressures respectively ($P_e = n_e k T_e$, $P_i = n_i k T_i$, where k=Boltzmann constant, $n_e = \frac{\rho_e}{m_e}$, number density of electrons, $n_i = \frac{\rho_i}{m_i}$, number density of ions, m_e is the electron mass and m_i is the mass of an ion. Internal energy densities ϵ_e and ϵ_i are given as $\epsilon_e = \frac{3kT_e}{2m_e}$ and $\epsilon_i = \frac{3kT_i}{2m_i}$ respectively. The quantities ν, Z, W_L and τ are the collision frequency, the ion charge (which is unity for the hydrogen plasma), the density of electron heating power by laser absorption and the equipartition time respectively. The thermal conductivities (κ_e) of electrons and (κ_i) of ions are related by a relation[22]

$$\kappa_i = \kappa_e (\frac{m_e}{m_i})^{\frac{1}{2}}. \tag{8}$$

The nonlinear force f_{nl} can be obtained[23] as

$$f_{nl} = \frac{w_p^2}{8\pi w^2} sin^2(wt) \frac{\partial E_{Lx}^2}{\partial x}, \tag{9}$$

where w_p is the plasma frequency, w is the frequency of the laser electric field E_L which is expressed in the form

$$E_L = E_{Lx}(x,t) sin(wt). \tag{10}$$

The real part of the complex amplitude $E_{Lx}(x,t)$ (which is equal to $(\frac{8\pi I}{c})^{\frac{1}{2}}$, where I is the laser intensity and c is the speed of light) can be used to calculate nonlinear force f_{nl}.

The two step Lax-Wendroff method[21] is used to solve the continuity equations and the equations resulting from the momentum conservation (equations (1) to (4)). In this method these equations are in the form of finite difference equations which involve discrete quantities in time t and one-dimensional space variable x. The two steps are so called auxillary and main steps. At the main and auxillary steps, the conservative quantities ρ and ρv for electrons and ions are calculated at the grid and half grid points respectively. Under this scheme the continuity equations for electrons and ions (equations (1) and (2)) take the forms

$$\rho_{j+\frac{1}{2}}^{t+\frac{1}{2}} = \frac{1}{2}(\rho_{j+1}^t + \rho_j^t) - \frac{\Delta t}{2\Delta x}((\rho v)_{j+1}^t - (\rho v)_j^t) \tag{11}$$

at the auxillary step and

$$\rho_j^{t+1} = \rho_j^t - \frac{\Delta t}{\Delta x}((\rho v)_{j+\frac{1}{2}}^{t+\frac{1}{2}} - (\rho v)_{j-\frac{1}{2}}^{t+\frac{1}{2}}) \tag{12}$$

at the main step. Here Δx is the space interval to perform the differentiation in space variable x (number of grids=number of space intervals), and Δt is the integral time interval. Finite difference equations for the momentum equations (equations (3) and (4)) at the auxillary and main steps can be written in a similar fashion using the Lax-Wendroff scheme and they calculate the quantity ρv for electrons and ions.

The energy equations (equations (5) and (6)) and the Poisson equation (equation (7)) can be obtained in the same way at the two steps using the Lax-Wendroff method. The discrete quantities $\rho \epsilon$ and E are calculated implicitly using the "advanced" values (calculated at the next grid point at the previous time step) of densities and velocities.

At the boundaries, discrete quantities are calculated in different ways at the auxillary and the main steps. At the main step, calculations are done using the difference equation as usual and periodic boundary conditions are used when necessary (e.g. calculation of ρ using equation (12)). A second order extrapolation is applied to the boundaries at the auxillary step. Using this procedure, difference equations for the density can be written as

$$\rho_{N+\frac{1}{2}}^{t+\frac{1}{2}} = \frac{1}{2}(3\rho_N^t - \rho_{N-1}^t) - \frac{\Delta t}{2\Delta x}((\rho v)_N^t - (\rho v)_{N-1}^t) \tag{13}$$

and

$$\rho_{\frac{1}{2}}^{t+\frac{1}{2}} = \frac{1}{2}(3\rho_1^t - \rho_2^t) - \frac{\Delta t}{2\Delta x}((\rho v)_2^t - (\rho v)_1^t). \tag{14}$$

In equations (13) and (14) $N + \frac{1}{2}$ denotes the maximum number of grid points.

Time and space intervals are related to each other by a relation

$$(|v_f| + v_s)\frac{\Delta t}{\Delta x} < 1 \qquad (15)$$

which is known as the stability condition of the Lax-Wendroff method. Here v_f is the velocity of the fluid and v_s is the velocity of sound in the fluid. But the presence of the Coulomb interaction gives rise to an oscillating electric field which tries to maintain the charge neutrality. The period of these oscillations is in the order of the plasma period. In an early calculation[24], plasma frequency was calculated by linearizing the continuity equation of the electron fluid for a macroscopically neutral, cold, homogeneous plasma. As a result of this calculation, plasma frequency w_p was obtained as

$$w_p^2 = \frac{4\pi n_e e^2}{m_e}. \qquad (16)$$

More explanations about the plasma oscillations can be found in the work by Lalousis[21]. The time interval Δt is limited by the period of these oscillations and should be chosen sufficiently small to resolve these oscillations. The space interval is limited by the Debye length λ_D so that the condition $\Delta x > \lambda_D$ is necessary.

It is necessary to apply a filtering procedure[25] to filter some of the variables (ρ_e, ρ_i, v_e, v_i and E) with some frequency. To reduce any (negative) effect of filtering at the boundaries, filtered variables are extrapolated linearly at the boundaries. Although the number densities are introduced as input variables, the algorithm uses mass densities mostly in the difference equations as described above and in almost all cases, the calculations are stopped because the electron mass density becomes negative. To avoid this (and negative temperatures as well) temperature and mass density of electrons are recalculated using geometrical interpolation in some parts of the computer code.

RESULTS AND DISCUSSIONS

Using the Lax-Wendroff algorithm described in section II, number densities n_e and n_i, temperatures T_e and T_i, velocities v_e and v_i of electrons and ions, and the longitudinal electric field E are calculated for a laser intensity of $I = 10^{15}$ W/cm^2. Laser field energy density ε is obtained using the real part of the amplitude of the laser electric field E_L. As a result of computational experience, the time interval Δt is chosen to be 5×10^{-16} sec and the results are recorded at each picosecond[21]. Filtering at each ten time steps seems to give good results for this problem. The initial densities of electrons and ions are taken to be equal at each grid point of the plasma and the plasma is assumed to be at rest initially (i.e. $n_e = n_i$ and $v_e = v_i$ for all x at $t = 0$).

First set of results are obtained for the input values of $T_e = T_i = 3 \times 10^5$ K, the plasma width$=20\mu m$ and the number of grid points$= 150$[26,27]. The initial density profile is chosen to be linear with density values of 0.52×10^{21}cm^{-3} at $x = \Delta x$ (first grid point) and 1.3×10^{21}cm^{-3} at $x =$plasma width (last grid point). The results are given in figure 2.

When the laser light penetrates into the plasma, electron and ion densities start rippling up to the critical density (10^{21}cm^{-3}) and because of these ripples, laser light can no longer penetrate into the plasma and it is reflected back. Density ripples decay hydrodynamically within 10 psec, then laser light penetrates into the plasma, causing density ripples again and they also decay after some time. The rippling of electron and ion densities can be explained by the existence of the nonlinear force[23], as mentioned in the preceeding section, which causes the generation of partial standing waves and produces ideal Bragg reflection in the outermost part of the plasma. In figures 2(a) and 2(b), density ripples can be seen at times $t =$5-6, 13, 19-20, 25-26, 29 and 34 psec. The variables v_e, v_i, ε and E (figures 2(e), 2(f), 2(g) and 2(h)) follow

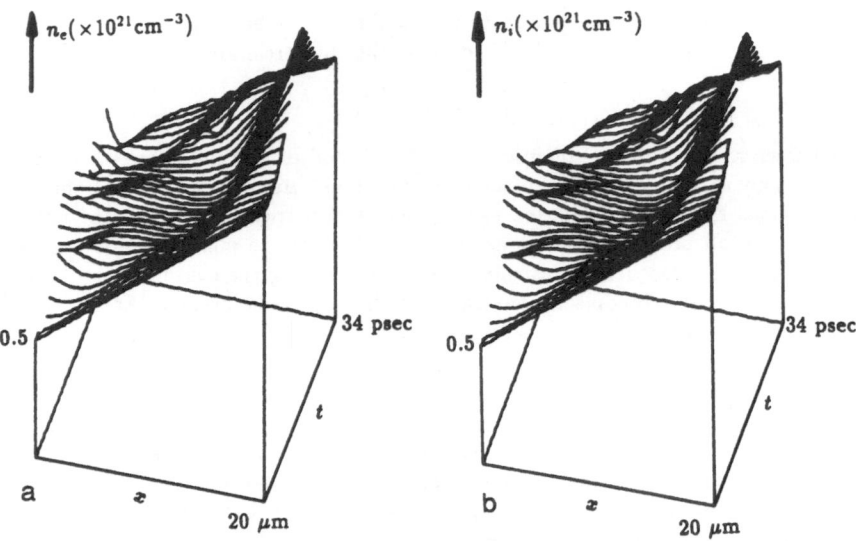

Figure 2. Time dependence of (a) the electron number density, (b) the ion number density when the plasma is irradiated by neodymium glass laser irradiation of 10^{15} W/cm^2 from the left hand side with a pulsating generation and relaxation of density ripples.

the density ripples (i.e. their ripples occur more or less at the same times that density ripples occur). Electron temperature T_e doesn't seem to oscillate and ion temperature T_i shows some rippling although it is not filtered very well (figures 2(c) and 2(d)). In figures 2(a) and 2(b), the density values at the lower boundary of the plasma seem to vary significantly as a function of time and increases unexpectedly at some time values. Depending on the values of the input variables, mostly on the density profile, one may see different behaviours at the lower boundary (for the other parts of the plasma, plots look very much similar in most of the cases). The possible sources of errors may be the use of the periodic boundary conditions and

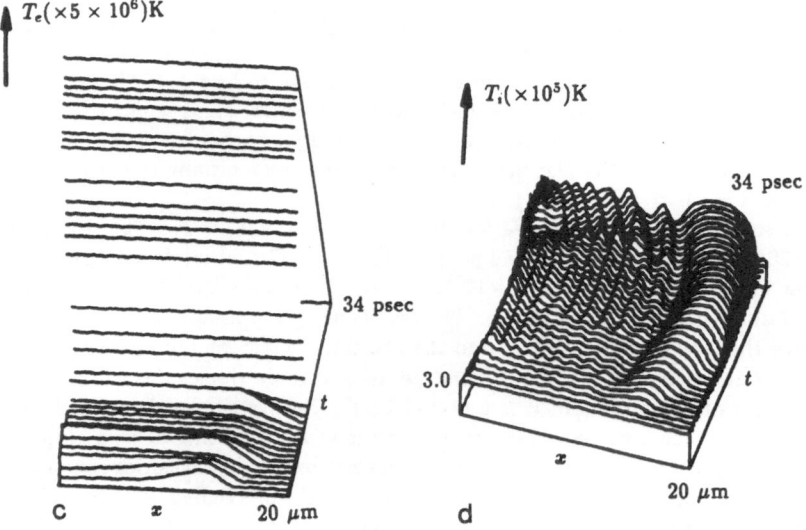

Figure 2. Time dependence of (c) the electron temperature, (d) the ion temperature for the same input parameters as in figure 2(a) and 2(b).

the filtering procedure. To see the results of these two effects, some "trial" runs are performed by removing periodic boundary conditions and then by not filtering ρ_e and ρ_i, using exactly the same input parameters used to obtain the first set of results given in figure 2.

First "trial" is to take the lower boundary value of the quantity ρv to be equal to its value at the first grid point. This leads to the assumption that at $x = \Delta x$ the density is assumed to be equal to its value at the previous time step (i.e. it is always equal to its initial value). Results for electron and ion densities show different behaviours at the lower boundary of the plasma and the difference between n_e and n_i increases as time increases. Unlike the case with periodic boundary conditions, the behaviour of n_e and n_i is very much similar for different input parameters. The other variables show similar behaviours to the results given in figure 2.

Second "trial" run is performed by not filtering ρ_e and ρ_i (v_e, v_i and E are filtered) using the original version of the computer code. Figure 3 shows density plots for this "trial" case.

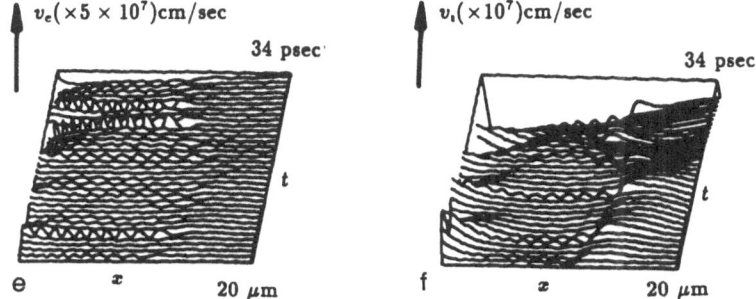

Figure 2. Time dependence of (e) the electron velocity, (f) the ion velocity for the same input parameters as in figure 2(a) and 2(b).

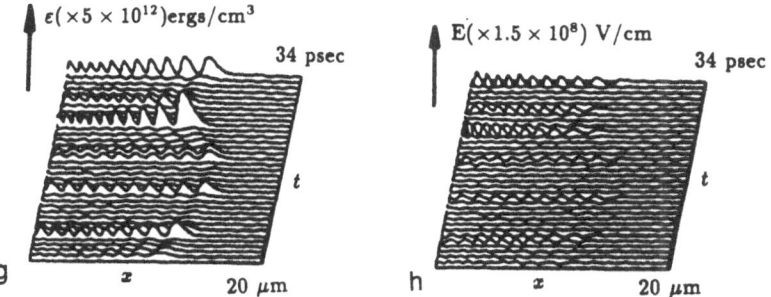

Figure 2. Time dependence of (g) the electromagnetic energy density, (h) the longitudinal electric field for the same input parameters as in figure 2(a) and 2(b).

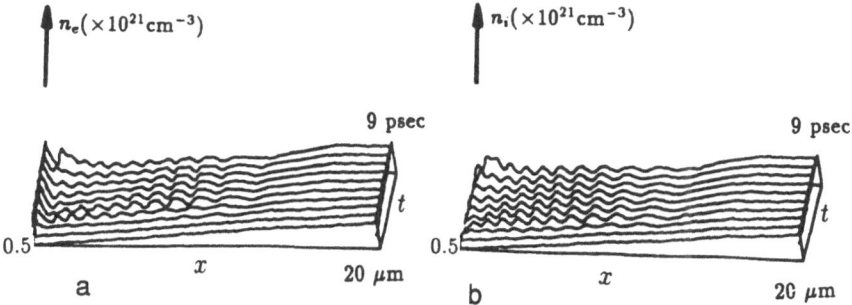

Figure 3. Time dependence of (a) the electron number density, (b) the ion number density using the same parameters as in figure 2 without filtering electron and ion mass densities.

While n_e and n_i are still different near $x = 0$ (and obviously density ripples can not be seen very well since they are not filtered), one can see the effect of filtering ρ_e and ρ_i clearly by comparing figure 2 and figure 3. The problem at the lower boundaries is almost completely solved by not filtering ρ_e and ρ_i. There is still a small difference, but one should always think the possibility of the effect of filtering other variables and some other problems.

Input values of the physical (density profile, ...) and technical (grid interval, filtering, boundary conditions, ...) parameters affect the behaviour at the boundaries, the "degree" of filtering and the time to run the code successfully and these may have different effects on different variables. As mentioned in the preceeding section, calculations are stopped mostly when numerical instabilities cause the appearence of the negative electron mass density. Because of the difference in electron and ion masses ($m_i = 1836m_e$), electron mass density is much smaller (around 10^{-7}gr/cm^{-3} $-$ 10^{-6} gr/cm^{-3}) than the ion mass density. This (and some factors because of the method generally, filtering, corrections, ...) may be the reason why the electron density becomes negative during calculations. After each picosecond, electron and ion mass densities are divided by the electron and ion masses respectively, to obtain number densities of electrons and ions. When the mass densities are filtered and the filtered values are used in further calculations at later times, any small negative effect may be enhanced (may even appear as a disastrous effect when small numbers like 10^{-7} are multiplied by a large number which is 1.098×10^{27}) and the difference between n_e and n_i may increase enormously as time increases. Another "trial" run could be performed by using number densities in the calculations rather than mass densities and by trying to reduce any negative effect of filtering or to avoid filtering, if possible.

By decreasing the initial value of the density at $x = \Delta x$, no results avoiding the mentioned discrepancy are obtained using the same set of input parameters. The only possibility is to increase the plasma width and the number of grid points, so that the grid interval Δx is more or less the same. The results for the set of input parameters $T_e = T_i = 10^5$ K, plasma width$= 45\mu m$, number of grids$= 338$, density values at $x = \Delta x$ and $x =$plasma width are about 0.024×10^{21} cm^{-3} (so it is 1 percent of the critical density at $x = 0$) and 1.1×10^{21}cm^{-3} respectively. We find then the discrepancy of the electron and ion densities beyond the expected double layer values[28] at the boundary ($x = 0$), but nevertheless the pulsation of generation and disappearence of the rippling can be seen. All the variables show ripples around $t = 1 - 2, 7, 15 - 16$ and 22 psec, difference to the first set of results (figure 2) being that the density ripples are weaker and can be seen only for fairly large x values (at which density versus x curves seem to be flattened).

Although calculations are mainly focused on the case for which intensity has a low value ($I = 10^{15}$ W/cm^2), some early results are obtained for $I = 10^{16}$W/cm^2 as well. These results are given in figure 4. To obtain these results, ρ_e and ρ_i are not filtered, only v_e, v_i and E are filtered. While one can see the effect of poor filtering or lack of filtering, it is good not to see any unexpected behaviour at the boundaries. A few (shorter) runs with higher temperatures up to 10^7K give similar results to those given in figure 4.

In almost all the results, the electron and ion velocities have values around 10^7 cm/sec and 10^6cm/sec respectively, they mostly fluctuate between positive and negative values when the density ripples occur (they are always positive for $I = 10^{16}$ W/cm^2). The laser field energy density ε has values around 10^{12}ergs/cm^3 for small x values and decreases as x increases, down to the values of 10^2 to 10^{-2}ergs/cm^{-3} at the maximum value of x which is equal to the plasma width (this point is located far inside the plasma with densities above the critical density and where the laser light decays exponentially beyond the critical density). The longitudinal electric field E is around 10^8 V/cm in magnitude, it fluctuates between positive and negative values when the density ripples occur and tends to decrease in magnitude as x increases, but the difference is no more than a few orders of ten when the large values of x is reached.

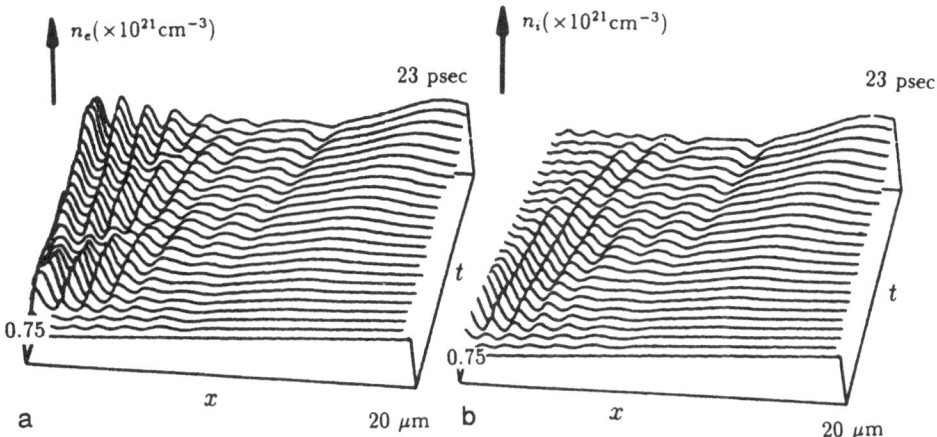

Figure 4. Time dependence of (a) the electron number density, (b) the ion number density when the plasma is irradiated by neodymium glass laser irradiation of 10^{16} W/cm^2 from the left hand side (without filtering electron and ion mass densities).

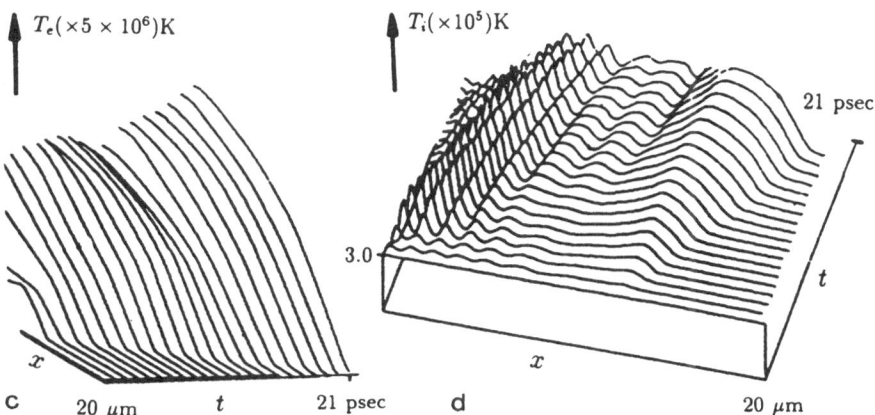

Figure 4. Time dependence of (c) the electron temperature, (d) the ion temperature for the same input parameters as in figure 4(a) and 4(b).

In all plots given in figures 2 to 4, x and t axes are scaled, the maximum values being the plasma width and the maximum time (that has been reached before the calculations are stopped when the electron mass density becomes negative) respectively. The axes showing calculated variables are not scaled since these variables are plotted as seperate curves as a function of x at each picosecond (with enough space between them to avoid mixing with each other). Each variable is divided by a number to see all the features clearly and this number is shown as a scale factor which is infact a measure of the order for the magnitudes of the variables plotted.

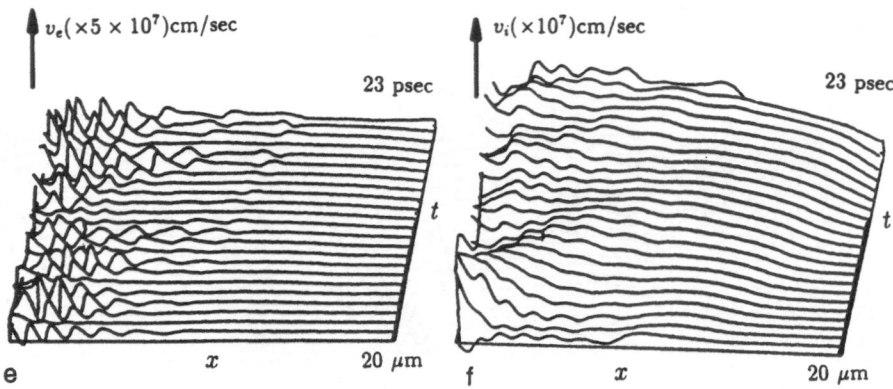

Figure 4. Time dependence of (e) the electron velocity, (f) the ion velocity for the same input parameters as in figure 4(a) and 4(b).

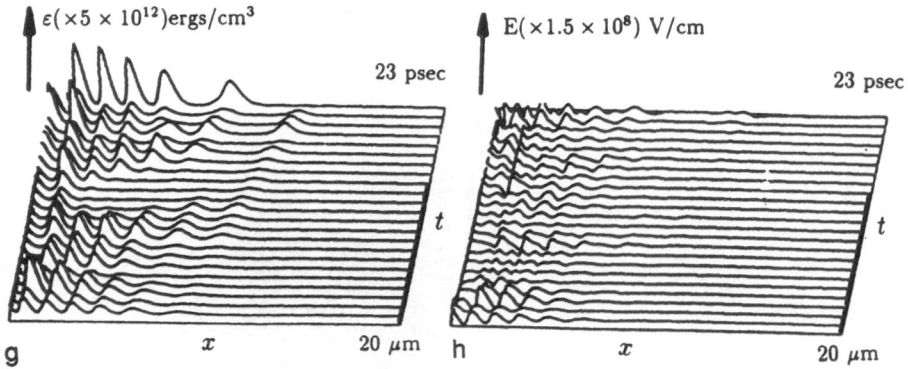

Figure 4.Time dependence of (g) the electromagnetic energy density, (h) the longitudinal electric field for the same input parameters as in figure 4(a) and 4(b).

NUMERICAL TEST OF SMOOTHING

The results of the numerical output of the preceeding section arrives at an unexpected clarification: it turns out that the laser light accelerates the plasma during few psec and then stops the interaction such that a plasma cloud of a velocity of about 10^7 cm/sec is reached. After about 10 psec the interaction begins again and pushes the plasma by a further additional velocity of about the same magnitude. The broad spectrum of the back scattered light is then simply a sequence of Doppler shifted light with a rather irregular pulsation in the range of 10 to 30 psec. Such a pulsation was seen also in the reflected light changing during this time up and down from few to nearly 100 percent which is similar to Lubin's early result[12,13].

It was possible to clarify experimentally that the laser light is alternately reflected at the critical density showing low reflectivity, while the other reflection after about 10 psec, each occurs in the outermost low electron density periphery of the plasma corona when the reflectivity is high.

The preceeding results permit an explanation of how the method of the induced spatial incoherence(ISI)[6] works (the coherence time of the laser light is about 2 psec). It is evident immediately from our hydrodynamic computation (figures 2 to 4) that if the standing wave profile then changes always after 2 psec, no density ripple can be produced. Laser irradiation is smooth with low reflectivity and high energy deposition in the plasma. Instabilities (SRS or SBS) are reduced by a factor of 100[6].

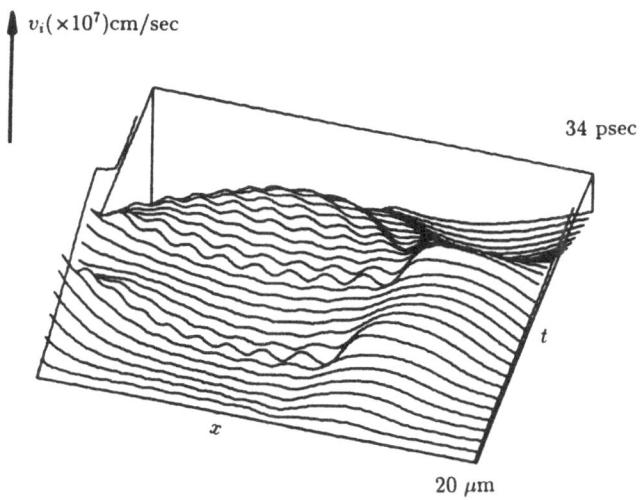

Figure 5. Ion velocity profiles similar to the case of figure 2, only for 0.6 times of the laser intensity when irradiated by one single laser frequency with a pulsating generation of ripples.

The action of smoothing by the random phase plate (RPP), introduced by Kato et al[7] consists in the generation of beamlets within the laser beam of which the phase is shifted forward and backward randomly. If a standing wave pattern is produced in one beamlet, the next beamlet has the standing wave pattern shifted. Any generation of a density ripple in one beamlet would then be laterally washed out hydrodynamically by the neighbour beamlet structure following our hydrodynamic result or due to the superposition of the diffraction spread fields of the beamlets. Again a good smoothing occurs and the achieved high deposition of laser energy into the irradiated plasma was one of the reasons for the famous production of high gain fusion neutron yields at direct laser drive[28].

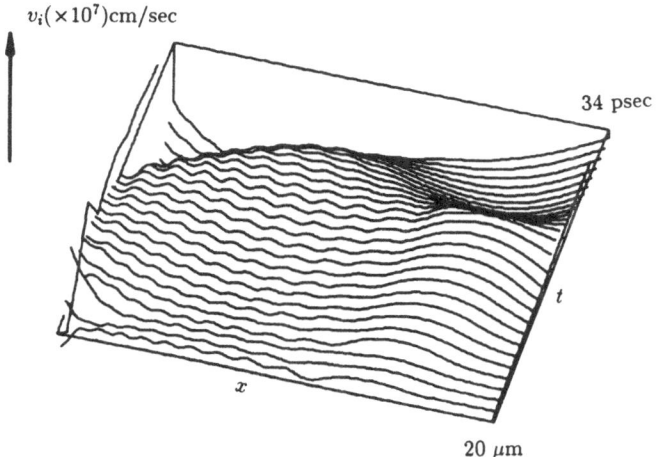

Figure 6. Same as figure 5 but with irradiation of a spectrum of 1 percent spectral width represented by three laser frequencies 0.5 percent spectrally apart.

Alternatively, smoothing was intended to be achieved by broad band laser irradiation or the "fly eye" optics[8]. A combination of this led to a very successful smoothing by spectral dispersion (SSD)[9]. The action of a broad spectrum of laser irradiation is immediately understandable from our model as this broad spectrum may not permit the standing wave patterns and the generation of density ripples as disscussed before.The numerical results for this test are presented in this section.

In order to perform the test mentioned above[26], a computation of the model similar to the case of figure 2 was performed for a Neodymium glass laser intensity of 6×10^{14} W/cm^2 and the same parameters as in the case of figure 2. If a single laser frequency is used, the velocity profile has a time dependence as shown in figure 5 with ripples at 8 and 15 psec (synchronous with ripples and pulsation in the density or the field properties). If a set of three laser frequencies each 0.5 percent different in wavelength are incident, the velocity profiles show the result of figure 6, where the pulsation is suppressed nearly completely.

These calculations are all done for a one-dimensional plane wave geometry. In reality the laser beam is of finite size, and divergent, and the plasma is in three dimensions. One extension of the plane wave geometry to beams may be justified by considering Sigel's question mark experiment[29]. Putting a question mark into the laser beam and looking for the picture produced by the reflected light after irradiating a plane target, it was found that the question mark was not turned around as expected from the reflection from an ideal mirror, but it appeared unchanged. This corresponds just to the reflection of the kind of a Bragg grating where the beamlets are reflected into themselves in full agreement with our[26] described density ripple result.

One modification may be the following suggestion: The Bragg reflection with high reflectivity from the very low plasma density occured only for some time. It may be asked whether the weak reflection for the short times occuring from the critical density do not have a turning of the question mark. Therefore an experiment with a time resolved question mark may have a pulsating up and down, or in time integrated experiments there may be a further weak picture of the question mark the other way around than reported[29].

CONCLUSION

The genuine two-fluid model[21] is used to calculate number densities (n_e, n_i), temperatures (T_e, T_i), velocities (v_e, v_i) of electrons and ions respectively, the longitudinal electric field (E) and the laser field energy density (ε) using a low temperature (10^5 K - 3×10^5 K) hydrogen plasma which is irradiated by a neodymium glass laser of intensity 10^{15} W/cm^2. Results for various density values near $x = 0$ (1 percent to 75 percent of the critical density which is 10^{21} cm^{-3}) with different technical parameters (filtering, boundary conditions, ...) show that electron and ion densities ripple up to the critical density at around 2, 6, 13, 20, 29 and 34 psec. This behaviour is explained by the effect of the nonlinear force[23] and the resultant partial standing waves in the plasma. Because of the density ripples, laser light can no longer penetrate into the plasma and it is reflected back. Density ripples decay hydrodynamically within 10 psec and the light goes into the plasma causing ripples again. These ripples produce ideal Bragg reflection in the outermost part of the plasma.

All the results with different input parameters (different density profiles, different intensities and the "trial" runs with different boundary conditions and filtering procedures) show similar features especially for the region which is sufficiently far away from the (real) lower boundary or the vacuum end of the plasma. The problems for small values of x seem to be mostly numerical/technical and can be solved by use of suitable parameters and/or procedures. The results using the input parameters which describe the whole plasma (for which the initial value of the density at $x = 0$ is 1 percent of the critical density) may need a careful consideration of the plasma near vacuum. By this means the well-known fact of the expansion of the plasma into vacuum can be observed clearly and any other possible mechanism can be investigated.

A further test is the use of a laser beam consisting of three frequencies with 1 percent wavelength spread. By comparing the results with those for the monochromatic case, one can see how the pulsation is suppressed when the beam with three waves is used as the laser source.

ACKNOWLEDGEMENTS

One of the authors (M.A.) would like to thank H. Szichman and P. Lalousis for their kind help to understand the Wax-Lendroff algorithm.

REFERENCES

1. A. G. Engelhardt et al, Phys. Fluids 13:212(1970)
2. K. Büchl, K. Eidmann, P. Mulser, H. Salzmann and R. Sigel, "Laser Interaction and Related Plasma Phenomena", H. Schwarz et al eds., Plenum, New York (1972) Vol. 2, p.503
3. J. H. Nuckolls, Physics Today 35:24(1982)
4. T. Yabe and K. Nishihara, Res. Rept. Inst. Plasma Phys., Nagoya University IPP-J-235 (1975)
5. X. Deng et al, Acta Optica Sinica 2:97(1983)
6. R. H. Lehmberg and S.P. Obenshain, Opt. Comm. 46:27(1983); R. H. Lehmberg, A J. Schmitt and S. E. Bodner, J. App. Phys. 62:2680(1987)
7. Y. Kato et al, Phys. Rev. Lett. 53:1057(1984)
8. X. Deng et al, Appl. Opt. 25:377(1986)
9. S. Skupsky et al, J. Appl. Phys. 66:3456(1989)
10. M. H. Emery, J. G. Gardner, R. H. Lehmberg and P. Obenschain, Phys. Fluids B3:2640(1991)
11. S. P. Obenschain, C. J. Pawley, A. M. Moistovych and J. A. Stamper, Phys. Rev. Lett. 62:768(1989)
12. M. Lubin, ECLIM'74 Garching, Abstracts p.34
13. S. Jackel, B. Barry and M. Lubin, Phys. Rev. Lett. 37:95(1976)
14. R. A. M. Maddever, Ph.D. thesis, Australian National University, 1988; R. A. M. Maddever, B. Luther-Davies et al, Phys. Rev. A41:2154(1990)
15. A. V. Rode et al, AINSE Plasma Conference, Feb. 1991, Lucas Heights, Australia
16. A. Guiletti et al, "Laser Interaction with Plasmas" G. Velarde et al eds., World Scientific, Singapore (1989) p.208
17. H. Hora, T. Henkelmann, H. Haseroth, C. E. Hill, G. Korschinek, R. Matulioniene, K. Langbein and C. S. Taylor, CERN-PS Report (1991)
18. S. Guskov et al, ECLIM'91 Warsaw Conference Paper P-66
19. H. Hora, "Laser Plasmas and Nuclear Energy" Plenum, New York (1975)
20. H. Hora, "Plasmas at High Temperature and Density" Springer, Heidelberg (1991)
21. P. Lalousis, Ph.D. thesis, University of New South Wales (1983); H. Hora, P. Lalousis and S. Eliezer, Phys. Rev. Lett. 53:1650(1984); H. Szichman, Phys. Fluids 31:1702(1988); Gu Min and H. Hora, J. Chin. Laser 16:656(1989); Gu Min and H. Hora, Laser and Particle Beams 9:381(1991); H. Hora, "Plasmas at High Temperatures and Density" Springer,Heidelberg (1991)
22. L. Spitzer, "Physics of Fully Ionized Gases" 2nd ed. Wiley, New York (1962)
23. H. Hora, Phys. Fluids 12:182(1969) ; H. Hora, "Physics of Laser Driven Plasmas" Wiley, New York (1981)
24. L. Tonks and I. Langmuir, Phys. Rev. 33:195(1929)
25. R. Shapiro, Rev. of Geophys. and Space Phys. 8:359(1970)
26. M. Aydın, Gu Min and H. Hora, Laser and Particle Beams 10:152(1992)
27. H. Hora and M. Aydın, Phys. Rev. A47 (1992)
28. S. Eliezer and H. Hora, Physics Reports 172:339(1989); H. Azechi et al, Laser and Particle Beams 9:193(1991)
29. K. Eidmann and R. Sigel, "Laser Interaction and Related Plasma Phenomena", H. Schwarz et al eds., Plenum, Newyork (1974) Vol. 3B, p.667; R. Sigel, K.Eidmann, C. H. Pant and P. Sachsenmeier, Phys. Rev. Lett. 36:1369(1976)

DISTRIBUTED ABSORPTION AND INHIBITED HEAT TRANSPORT

J. S. De Groot[1], K. G. Estabrook[2], W. L. Kruer[2], R. P. Drake[3], K. Mizuno[3], and S. M. Cameron[4]

[1]Department of Applied Science and Plasma Research Group, University of California, Davis, California 95616

[2]Y Program, Lawrence Livermore National Laboratory, Livermore, California 94550

[3]Plasma Physics Research Institute and Department of Applied Science, University of California, Davis and Lawrence Livermore National Laboratory, Livermore, California 94550

[4]Department of Applied Science, University of California, Davis, California 95616 and Lawrence Livermore National Laboratory, Livermore, California 94550

INTRODUCTION

High-gain laser-fusion pellets must be irradiated by[1] long (pulse width, $t_L \gtrsim$ 10ns) pulses of moderate to high intensity (incident laser energy flux, $I_0 \gtrsim 5 \times 10^{15}$ W/cm^2) laser light. These pulses produce large scale length plasmas that are cool enough that inverse bremsstrahlung is very strong, so that laser light absorption is not localized: rather, absorption is distributed[2,3] over densities from well below the critical density to the critical density. Distributed absorption has several important consequences for laser fusion. The temperature and density profiles impact the threshold for parametric instabilities that can degrade pellet performance. In addition, the ablation pressure is reduced due to distributed absorption, because laser light is absorbed at lower densities that are farther from the ablation surface. Hydrodynamic computer codes[4] such as LASNEX are used to study strongly absorbing plasmas, but it is difficult to use such complex codes to test alternative models for heat transport and laser light energy deposition. These considerations motivate the development of analytical models to improve our understanding. However, most previous models[5-9] are not applicable to strongly absorbing plasmas because the absorbed laser energy was deposited at one location in the plasma (usually the critical surface). Recently we presented[2] a model that incorporated distributed absorption, but it was assumed that the conduction region was in steady-state (as in previous models). As was shown in Ref. 2, these steady-state models only apply to low to moderate laser powers for typical laser pulse lengths.

We have recently presented[3] a model that applies to high laser powers. Typical spatial profiles of the temperature and density for a strongly absorbing plasma heated by a high power laser are shown in Fig. 1.

Laser Interaction and Related Plasma Phenomena, Vol. 10
Edited by G.H. Miley and H. Hora, Plenum Press, New York, 1992

197

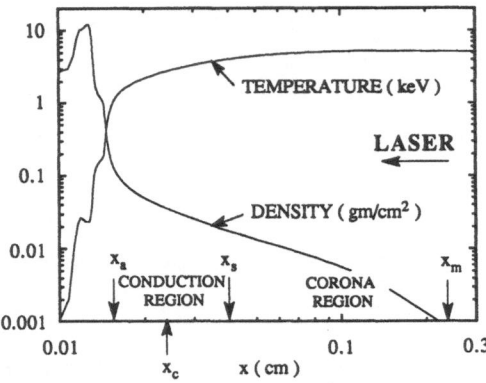

Figure 1. Temperature and density profiles for a strongly absorbing plasma. These results are from a LASNEX simulation of high power laser light (peak incident energy flux, $I_0 = 1 \times 10^{15}$ W/cm^2, pulse width, 2 ns (FWHM), and wavelength, $\lambda = 0.35$ μm) heating a thick aluminum target.

The blowoff plasma ($x \geq x_a$, the location of the ablation surface) is divided into two regions: The conduction region, where the absorbed laser energy is transported to the ablation surface and the slightly non-isothermal corona. We will show that the laser energy is primarily deposited in the corona region for laser parameters of interest to laser pellet fusion region. We take the sonic surface (x_s in Fig. 1) to be the boundary between the regions. The sonic point occurs at the critical surface (x_c in Fig. 1) for a weakly absorbing plasma[6] where the laser energy is deposited at the critical surface. The sonic surface is in general not located at the critical surface in strongly absorbing plasmas. The location of the temperature maximum (x_m in Fig. 1) is at the sonic surface in the steady state model. The energy deposited in the corona must be transported into the ablation region, so the temperature must increase in the corona. The location of the temperature maximum thus occurs at lower density for strongly absorbing plasmas.

We have presented a new model for a planar plasma heated by moderate to high laser powers. We use a self-similar model[3] for the conduction region. We have shown[3] that the well known self-similar solution for an isothermal plasma expanding into a vacuum is a reasonable model for the slightly non-isothermal corona region. Unlike previous models, our model explicitly includes the temporal evolution of the heat conduction region. It is shown that previous steady-state models apply only to a narrow range of the parameters (laser energy flux and pulse width) used in laser pellet fusion. The new model is shown to agree with flux-limited hydrodynamics simulations. The model and hydrodynamic simulations for classical heat transport show that the mass ablation rate and ablation pressure are essentially independent of laser wavelength for parameters relevant to laser fusion.

The absorbed laser energy is mainly deposited in the corona so the heat flux is a maximum in the corona near the sonic surface. Laser energy deposition in the corona and therefore heat transport in the corona has very important consequences for laser fusion. The point is that heat transport can drive ion acoustic turbulence which in turn can limit heat transport[10]. The energy transported down the temperature gradient is carried by suprathermal electrons. An electric field is generated to drive a return current of thermal electrons traveling down the temperature gradient. Ion acoustic waves are driven unstable if the return current is above a threshold so that ion turbulence is excited in the plasma. We apply the recently developed theory[10] of ion turbulence driven by heat transport to laser driven plasmas. We show that the threshold can be exceeded in high Z plasmas driven by high laser powers. The heat flux is only weakly limited if the isotropic part of the electron velocity distribution function is Maxwellian. However, the heat flux is strongly limited if the isotropic part of the electron velocity distribution function is strongly non-Maxwellian ($f \approx \exp(-v^5)$). Collisions in high Z plasmas drive the electron distribution non-Maxwellian for high laser powers. We show that ion acoustic turbulence can result is strong flux limiting in plasmas with $ZT_e/T_i \gg 1$, that are heated by high laser powers, i.e., $Zv_{os}^2 \gg v_e^2$.

HYDRODYNAMIC EQUATIONS

Consider a planar plasma driven by a laser energy flux ($I(t)$) that is suddenly switched-on and then remains constant in time, i. e., $I(t) = 0$, for $t < 0$, and $I(t) = I_0$ for $t \geq 0$. We have developed[3] self-similar solutions of the planar, one fluid hydrodynamic equations for each region. The equations for conservation of mass, momentum, and energy are

$$\frac{\partial \rho}{\partial t} + \frac{\partial (\rho v)}{\partial x} = 0 \tag{1}$$

$$\frac{\partial v}{\partial t} + v \frac{\partial v}{\partial x} = -\rho^{-1} \frac{\partial (\rho c^2)}{\partial x} \tag{2}$$

$$\frac{3}{2} \rho \frac{\partial c^2}{\partial t} + \frac{3}{2} \rho v \frac{\partial c^2}{\partial x} + \rho c^2 \frac{\partial v}{\partial x} = -\frac{\partial q}{\partial x} - \frac{\partial I}{\partial x} \tag{3}$$

where ρ is the mass density, v is the fluid velocity, c is the isothermal sound speed, $c \equiv (P/\rho)^{1/2}$, P is the pressure, q is the heat flux, $I(x)$ is the spatially dependent laser energy flux, $I(x) = I_0 \exp(-t(x))$, I_0 is the incident laser energy flux, and $\tau(x)$ is the inverse bremsstrahlung optical depth of the plasma from $x' = x$ to $x' = \infty$, i.e.,

$$t(x) = \int_x^\infty \kappa_{ib}(x') \, dx'. \tag{4}$$

The inverse bremsstrahlung absorption coefficient is given by[11]

$$\kappa_{ib}(x) \approx 3.4 \frac{(\rho/\rho_c)^2 Z \ln\Lambda_a}{\sqrt{1 - \rho/\rho_c} \, \lambda_\mu^2 \, T_{keV}^{3/2}} \tag{5}$$

where ρ_c is the critical mass density, $\ln\Lambda_a$ is the Coulomb logarithm for inverse bremsstrahlung absorption, λ_μ is the laser light wavelength (μm), and T_{keV} is the electron temperature (keV). The ponderomotive force is neglected since the radiation pressure due to the laser light is small compared to the plasma pressure for cases of interest. Radiation from the plasma and the inertial force due to ablative acceleration of the target have also been neglected. These terms are generally small[6] for parameters of interest to laser pellet fusion, especially for shorter wavelength lasers. The electron and ion temperatures are assumed to be equal, so that $c^2 = (Z+1)T/(A M_p)$, Z is the ionic charge state, A is the atomic weight, and M_p is the proton mass. The heat flux is given by the smaller of the classical[12] heat flux and a phenomenological model for limited[4] heat flux, i.e.,

$$q = - \text{Min}\left[K_0 T_e^{5/2} \left| \frac{\partial T_e}{\partial x} \right|, f n_e v_e T_e \right] \frac{\frac{\partial T_e}{\partial x}}{\left| \frac{\partial T_e}{\partial x} \right|} \tag{6}$$

where v_e is the electron thermal speed, $v_e = \sqrt{T_e/m}$, m is the electron mass, the heat flux coefficient is $K_0 = 1.8 \times 10^{29} / Z \ln \Lambda$, $\text{cm}^{-1} \text{ sec}^{-1} \text{ keV}^{-5/2}$ (from Ref. 12, with the terms slowly varying in Z evaluated for Z = 13), $\ln \Lambda$ is the Coulomb logarithm, and f is the flux limiter.

CONDUCTION REGION MODEL

We have shown[3] that the laser energy is deposited predominantly in the corona for moderate to high laser powers and typical pulse widths. We will therefore ignore laser deposition and develop a self-similar solution for the conduction region. We take the similarity variable to be, $s = (x - x_a) / \Delta_s(t)$, where, $\Delta_s(t)$ is the width of the conduction region, $\Delta_s = x_s - x_a$, to be determined. Guided by the LASNEX simulations, we assume that that the dependent variables can be written: $\rho(x,t) = \rho_s(t)R(s)$, $v(x,t) = c_s(t)V(s)$, and $c^2(x,t) = c_s^2 E(s)$, where ρ_s and c_s are the density and sound speed at the sonic surface. The boundary conditions are: $E(0) = V(0) = 1/R(0) = 0$, and $E(1) = V(1) = R(1) = 1$. The hydrodynamic equations become

$$rR - s(n+1)\frac{dR}{ds} = -\frac{c_s t}{\Delta_s}\frac{d(RV)}{ds} \tag{7}$$

$$(r+n)RV - s(n+1)\frac{d(RV)}{ds} = -\frac{c_s t}{\Delta_s}\frac{d(RV^2 + RE)}{ds} \tag{8}$$

$$\frac{3}{2}\left[2nE - s(n+1)\frac{dE}{ds} \right] + \frac{3}{2}\frac{c_s t}{\Delta_s}V\frac{dE}{ds} + \frac{c_s t}{\Delta_s}E\frac{dV}{dx} = \frac{c_s t}{\Delta_s}\frac{K_0 T_s^{7/2}}{\rho_s c_s^3 \Delta_s}\frac{1}{R}\frac{d\left[E^{5/2}\frac{dE}{ds} \right]}{ds} \tag{9}$$

where

$$\frac{\partial \rho_s}{\partial t} = \frac{r\rho_s}{t}, \tag{10}$$

and

$$\frac{\partial c_s}{\partial t} = \frac{nc_s}{t}. \tag{11}$$

Eqs. (7) - (9) are self-similar (i.e., not explicitly a function of time and space) if: The factor, $(c_s t)/\Delta_s$ and the factor, $K_0 T_s^{7/2}/(\rho_s c_s^3 \Delta_s)$ are not a function of time. Using Eqs. (10) and (11) the second condition results in, $r = 3n - 1$. Another condition results from requiring that the absorbed laser energy is the plasma energy source, i.e.,

$$I_0 = \beta \rho_s c_s^3 \tag{12}$$

where β is a constant that depends on the plasma spatial profile. This results in, $r = -3n$, or we find, $r = -1/2$ and $n = 1/6$.

The solution of Eqs. (7) - (9) require the solution of a set of four first order, nonlinear differential equations that must be solved numerically. However, guided by the LASNEX simulations we find the following approximate solutions: $E \approx s^{2/5}$, $V \approx s^{3/5}$, and $R \approx 1/V$. The approximate solutions for the conduction region are therefore

$$T \approx T_s [(x - x_a)/\Delta_s]^{2/5} \tag{13}$$

$$v \approx c_s [(x - x_a)/\Delta_s]^{3/5} \tag{14}$$

$$\rho \approx 1/v \tag{15}$$

Since the local sound speed is proportional to $T^{1/2}$, we see that one important prediction is that the Mach number ($M = v/c$) is equal to the normalized temperature. The predicted normalized temperature (from Eq. (13)) is in agreement with temperature (circles) and Mach number (diamonds) from LASNEX simulations as shown in Fig. 2.

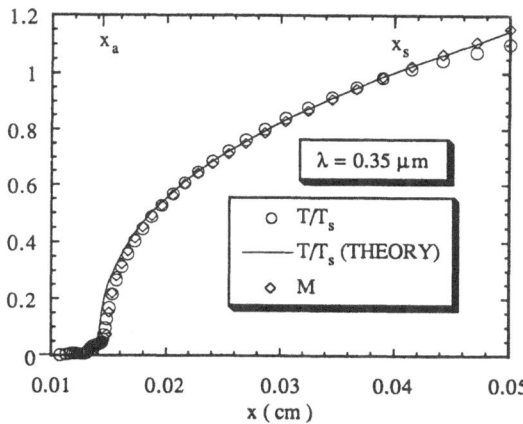

Figure 2. The spatial profile of the normalized temperature (circles) and Mach number (diamonds) from LASNEX calculations is in good agreement with model results (from Eq. (13), solid line). A thick aluminum slab was normally irradiated with a peak laser energy flux, $I_0 = 1 \times 10^{15}$ W/cm^2 and laser wavelength, $\lambda = 0.35\mu$m. Results are shown at the peak of the Gaussian laser pulse (FWHM = 2ns).

LASNEX simulations for $\lambda = 0.53$ and 1.059 μm are also in excellent agreement[3] with the theory. Classical heat transport was used in the calculations (the heat flux was below the flux limit, $f = 0.1$ for these calculations).

CORONA REGION MODEL

The well known similarity solution[4] for the expansion of an isothermal plasma into vacuum is used for the corona region, i.e.

$$\rho(x, t) = \rho_s \exp (- (x - x_s) / c_s t) \tag{16}$$

$$v(x, t) = c_s + (x - x_s) / t$$

The range of validity of this assumption is determined *a-posteriori*.. The isothermal rarefaction is maintained by a mass flux ($\rho_s c_s$) from the subsonic region and an energy flux[6] ($\rho_s c_s^3$) due to laser absorption.

Heat Flux in the Corona

The heat flux in the corona can be found by integrating Eq. (3) to obtain

$$q(x) = -\alpha(x)I_0 + \rho(x)c(x)^3 \tag{17}$$

where $\alpha(x)$ is the laser absorption in the plasma from x to ∞ , i.e.,

$\alpha(x) = 1 - \exp(-\tau (x))$. The terms with the derivative of the temperature (first and second terms in Eq. (3)) were neglected compared to the third term. It will be shown *a-posteriori* that these terms are small in the corona. We showed[2] that in the steady state conduction region model the temperature must be a maximum at the sonic surface. Thus, the heat flux is zero at the sonic surface and in the isothermal corona. In our present time dependent model the heat flux is not zero in the corona and the heat flux at the sonic surface is (Eq. (17) evaluated at $x = x_s$)

$$q_s = -\alpha_s I_0 + \rho_s c_s^3. \tag{18}$$

Eq. (18) can be reduced to the steady state model by setting $q_s = 0$, so that $\alpha_s = \rho_s c_s^3 / I_0$, which along with $I_0 = 4\rho_s c_s^3$, from the steady state model[6] gives $\alpha_s = 1/4$ in agreement with the steady state model.

WIDTH OF THE CONDUCTION REGION

The key features of our self-similar model are that the conduction region is subsonic and evolves in time. In our model, the sonic surface is the boundary between the conduction and corona regions. The width of the conduction region is found by requiring that the heat flux in the corona (Eq. (18)) and the heat flux in the conduction region are equal at the sonic surface. Eq. (13) for the temperature in the conduction region is used to find the heat flux at the sonic surface, $q_s = -(2/5) K_0 T_s^{7/2} / \Delta_s$. Substituting this equation into Eq. (18) and solving for Δ_s results in

$$\Delta_s = \frac{2}{5} \frac{K_0 T_s^{7/2}}{\alpha_s I_0 - \rho_s c_s^3}. \tag{19}$$

Another equation for the width is found by matching the spatial derivative of the velocity at the sonic surface. From the approximate solution for the subsonic region(Eq. (14)), the derivative of the velocity is

$$\frac{\partial v}{\partial x} (x \leq x_s) \approx 0.6 \frac{c_s}{\Delta_s} ,$$

and, $\frac{\partial v}{\partial x} (x \geq x_s) \approx \frac{1}{t}$ from the supersonic region (Eq. (16)). Another equation for the width of the subsonic region is therefore

$$\Delta_s \approx 0.6\, c_s\, t. \tag{20}$$

The sonic surface moves outward relative to the ablation surface at a velocity of $\approx 0.6\, c_s$.

PLASMA PROPERTIES AT THE SONIC AND ABLATION SURFACES

One condition on the plasma properties at the sonic surface is obtained by equating the absorbed laser flux, $I_0 t$, to the total plasma energy/area. The total plasma energy/area for the conduction region is obtained by using Eqs. (13) - (15) to form the energy density and then integrating over the conduction region. This energy/area is added to the energy/area for the corona ($4 \rho_s c_s^3$), resulting in

$$I_0 \approx 5.5 \, \rho_s c_s^3. \tag{21}$$

Another equation is obtained by equating Eq. (19) to Eq. (20), i.e.,

$$c_s t \approx 0.8 \, \frac{K_0 T_s^{7/2}}{I_0}. \tag{22}$$

In the spirit of our model, we assumed that the laser energy was deposited in the corona ($\alpha_s \approx 1$). The resulting density, temperature and mass ablation rate at the sonic surface are

$$\rho_s = 1.7 \times 10^{-9} \left[\frac{I_0}{Z \, t \, \ln\Lambda} \right]^{1/2} \left[\frac{A}{Z+1} \right]^{7/4} \text{gm/cm}^3, \tag{23}$$

$$T_s = 1.1 \times 10^{-5} \, [Z \, I_0 t \, \ln\Lambda]^{1/3} \left[\frac{Z+1}{A} \right]^{1/6} \text{keV}, \tag{24}$$

$$\rho_s c_s = 1.7 \times 10^{-4} \, \frac{I_0^{2/3}}{[\, Z \, t \, \ln\Lambda]^{1/3}} \left[\frac{A}{Z+1} \right]^{7/6} \frac{\text{gm}}{\text{cm}^2 \, \text{sec}}. \tag{25}$$

Here, the units are: I_0 (W/cm^2), λ (μm), and t (ns). The ablation pressure is

$$P_a = 2 \, \rho_s c_s^2 = 3.4 \times 10^{-11} \, \frac{I_0^{5/6}}{(Z \, t \, \ln\Lambda)^{1/6}} \frac{A^{7/12}}{(Z+1)^{7/12}} \text{Mbar}. \tag{26}$$

The key feature of these solutions is that the plasma properties at the sonic surface and the ablation pressure are not a function of laser wavelength. The mass density (Eq. (23)), temperature (Eq. (24)) at the sonic surface, and the ablation pressure (Eq. (26)) are compared to LASNEX simulations at the peak of the laser pulse in Fig. 3. The target is a thick aluminum slab (aluminum was taken as fully ionized in agreement with the LASNEX results). The width of the Gaussian laser pulse is, $t_L = 2$ ns (FWHM). The laser pulse width was reduced by a factor of 0.53 to compare our constant laser energy flux model to the Gaussian laser pulse used in the LASNEX simulations. (The factor is chosen so that the absorbed laser energy used in the model calculations is the same as in the LASNEX simulations.). The LASNEX results are in good agreement with the self-similar theory. The steady state theory gives results that disagree by a factor of 2 to 3.

VALIDITY OF THE SELF-SIMILAR CONDUCTION REGION MODEL

We have assumed that the laser light is strongly absorbed. The total absorption coefficient is

$$\alpha = 1 - \exp(-2\tau_c) \tag{27}$$

where τ_c is the optical depth of the plasma (Eq. (4) with $x = x_c$). The integral can be performed[13] using Eq. (16) and assuming that the temperature of the underdense plasma is

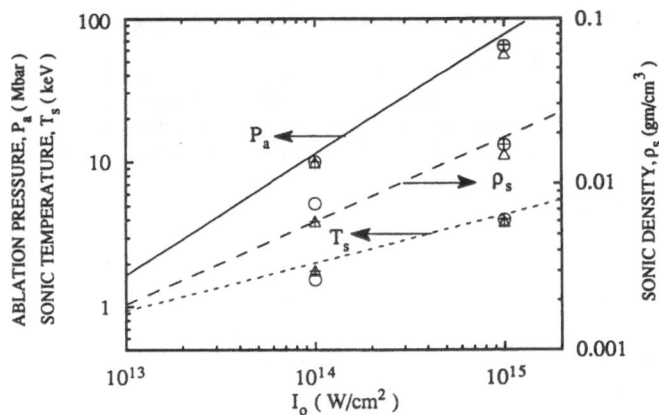

Figure 3. Model predictions for the sonic density (Eq. (23)), sonic temperature (Eq. (24)), and ablation pressure (Eq. (26)) are compared to LASNEX calculations. The temperature, T_S (keV) (dashes), density, r_S (gm/cm^2) (long dashes), and ablation pressure, P_a (Mbar) (solid line) are shown as a function of peak laser energy flux, I_O . Three wavelengths are shown, $l = 0.35$ (triangles), 0.53 (crosses), and 1.059 μm (circles). Results are shown at the peak of the Gaussian laser pulse (FWHM = 2ns).

approximately constant. Using Eqs. (23) and (24), the optical depth becomes

$$\tau_c \approx 1.1 \times 10^4 \frac{(Z \, t \, \ln\Lambda)^{2/3}}{I_0^{1/3} \lambda^2} \left[\frac{Z+1}{A}\right]^{1/3}. \tag{28}$$

Assuming that the the laser absorption is above 98%, results in, $\tau_c \geq 2$. The optical depth increases with time, thus, the plasma becomes strongly absorbing after a time, t_m (from Eq. (28) with $\tau_c \approx 2$). This minimum time is typically quite short, $t_m \lesssim 0.2$ ns.

As was shown in Ref. 2, the steady state conduction region model is applicable if $B \ll 1$, where, $B = \kappa_{ibs} K_0 T_s^{7/2} / I_0$. The parameter B is a the ratio of two of the three (the third is, $c_s t$) important characteristic lengths evaluated at the sonic surface, i.e., $B = $ the heat conduction scale length ($K_0 T^{7/2} / I_0$) divided by the inverse bremsstrahlung absorption length (κ_{ibs}^{-1}). We can understand the meaning of this parameter by relating B to the optical depth of the corona. Eq. (4) (with $x=x_s$ and using Eq. 16) can be integrated to obtain the optical depth of the approximately isothermal corona

$$\tau_s = \kappa_{ibs} c_s t / 2. \tag{29}$$

Using Eqs. (23) and (24), we find, $B \approx 2 \tau_s$. We see that the self-similar model is valid if $B > 1$, since then most of the laser absorption occurs in the corona. Thus, the parameter B indicates the region of validity of the steady state and the self-similar models. Using Eqs. (24) and (25) we find

$$B = 9.2 \times 10^{-9} \frac{I_0^{2/3} \lambda^2}{[t \, \ln\Lambda]^{1/3}} \frac{A^{7/6} Z^{5/3}}{[Z+1]^{19/6}}. \tag{30}$$

The parameter B decreases with time because the sonic surface moves outward into lower density plasma (ρ_s decreases with time, Eq. (23)). Thus, the steady state theory should become valid for very long pulses. For typical laser fusion parameters this time is quite long (~ 100 ns). The parameter B is a strong function of wavelength (~ λ^2). Thus, for longer wavelength laser light, lower laser powers and shorter pulse widths are required to satisfy the validity conditions for the self-similar model.

Since laser absorption occurs in the corona region, the temperature gradient must be non-zero in the corona. The term $\partial c^2/\partial x$ is neglected in the self-similar solution (Eq. (16)) for the corona. We can evaluate the effect of the non-isothermal corona by comparing this term to the term $v\,\partial v/\partial x$, that is retained in the self-similar solution. Using Eqs. (13), (14) and (16) we find that the two terms are comparable at the sonic surface and that $v\,\partial v/\partial x$ is larger over the rest of the corona. Thus, the isothermal solution is a reasonable model for the slightly non-isothermal corona.

The region of validity for the two models in laser pulse width - laser energy flux (t_L, I_0) space is shown in Fig. 4 with laser wavelength as a parameter.

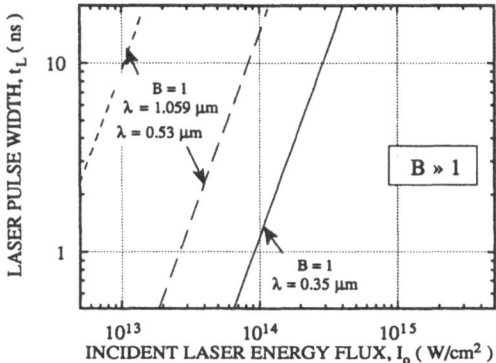

Figure 4. Region of validity of the steady state conduction region model and the self-similar model in laser pulse width - laser energy flux space. Lines for which, B = 1 (from Eq. (30)) are shown for three wavelengths: λ = 0.35 (solid line), 0.53 (long dashes), and 1.059 μm (dashes). The graph is drawn for fully ionized aluminum, but the results only change slowly with material (if A ≈ 2 Z).

The steady state model[2, 5-9] is valid for, B « 1, and the self-similar model is valid for, B » 1. The total absorption for the plasma is above 98% for Fig. 4 since the bremsstrahlung optical depth for the plasma, τ_c (Eq. (28)) is larger than 2. The graph is drawn for aluminum, but the results change slowly with material if (A ≈ 2 Z).

Thus, the following picture emerges for strongly absorbing plasmas. The plasma first evolves to a strongly absorbing state in a short time (t < t_m « 1ns). If the parameter B is small, B « 1, then the steady state theory is a reasonable approximation. However, if B > 1, then the self-similar theory applies. In this case, after a long time (when B (t) « 1) the steady state theory should again apply.

ION ACOUSTIC TURBULENCE AND INHIBITED HEAT TRANSPORT

The spatial profile of the normalized heat flux for a high laser power (I_0 = 1x10^{15} W/cm^2) is shown in Fig. 5.
The absorbed laser energy is mainly deposited in the corona so the heat flux is a maximum in the corona near the sonic surface. Classical heat transport theory[12] was used in these calculations. Suprathermal electrons carry the energy transported down the temperature gradient. An electric field is therefore generated to drive a return current of thermal electrons that travel up the temperature gradient. The effective drift velocity of the return current electrons is given by

$$u_{eff} = -\,\alpha_u \frac{q_e}{n_e v_e T_e}$$

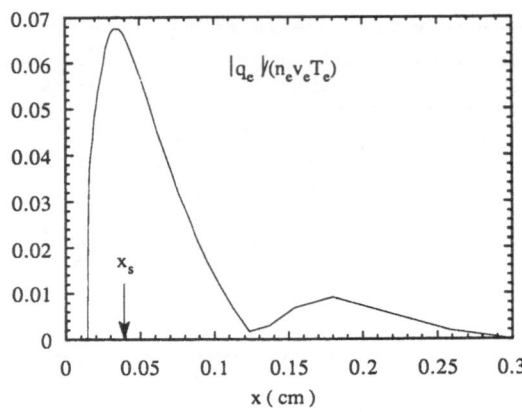

Figure 5. Heat flux for a strongly absorbing plasma. These results are from a LASNEX simulation of high power laser light (peak incident energy flux, $I_0 = 1 \times 10^{15}$ W/cm^2, pulse width, 2 ns (FWHM), and wavelength, $\lambda = 0.35$ μm) heating a thick aluminum target.

The coefficient, α_u, depends on the heat transport model. From classical transport theory[12], where the isotropic part of the electron velocity distribution is Maxwellian, $\alpha_u = 0.22$ for $Z = 1$ and $\alpha_u = 0.11$ for $Z \Rightarrow \infty$ (from Ref. 14). From Fig. 5, the maximum heat flux is

$$\frac{q_{max}}{n_e v_e T_e} \approx 0.068, \tag{31}$$

so ($Z = 13$ in the calculations) the maximum drift velocity is $u_{max} \approx 0.01 v_e \approx c_s$. This is a typical result for high power laser interactions. The importance of this condition is that the return current can excite ion acoustic waves in the plasma[15]. The threshold[15] for excitation of ion acoustic waves in a collisionless plasma with, $Z T_e/T_i \gtrsim 20$, is $u_{th} \sim c_s$ for $k\lambda_{De} \lesssim 1$ (we ignore a very slowly growing instability with a threshold[15] $u_{th} \lesssim c_s$ and $k\lambda_{De} \gtrsim 1$). The threshold is higher in collisional plasmas. The most important effect for laser fusion plasmas is ion-ion collisions[16]. The ratio of ion wave damping (due to ion Landau damping and ion-ion collisions) to electron Landau damping δ is shown in Fig. 6 (from Ref. 16). The ratio δ decreases for $k\lambda_{ii} < 0.1$ and agrees with fluid theory[17].

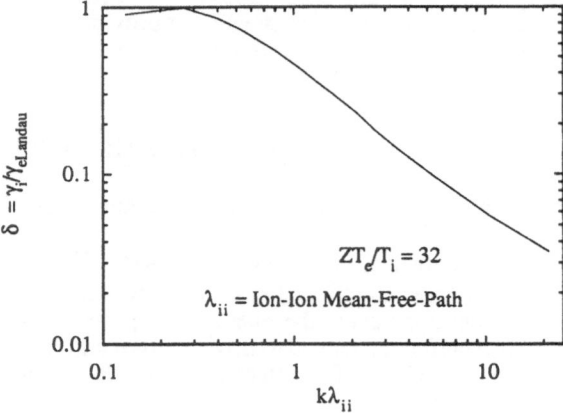

Figure 6. The ratio of ion wave damping (due to ion Landau damping and ion-ion collisions) to electron Landau damping (from Ref. 16).

Since the maximum value is $\delta_{max} \approx 1$ (this maximum is a slow function of ZT_e/T_i, see Ref. 16), we see that if the electron drift velocity is close to threshold, i.e. $1 < u_{eff}/c_s < 2$ then ion collisions stabilize a region of k-space.

The key result of the excitation of ion acoustic waves is the development of turbulence. Bychenkov, Silin, and Uryupin (Ref. 10) have developed a self-consistant theory of ion acoustic turbulence driven by heat transport. They find that the electron heat flux is limited, i.e.,

$$\frac{q_e}{n_e v_e T_e} \approx f_i \tag{32}$$

where f_i is the heat flux limit due to ion turbulence. For parameters of interest to laser fusion, the flux limit is $f_i \approx 0.18\sqrt{Z/A}$ - very weak flux limiting. A simple theory captures this result. As shown in Ref. 10, the effective collision frequency due to the ion turbulence is approximately

$$v_{eff} \approx \frac{v_e^2}{c_s\, L_T} \tag{33}$$

where L_T is the temperature gradient scale length, $L_T = (\nabla T_e/T_e)^{-1}$. The classical heat flux is given by[10] (for $Z \gg 1$)

$$\frac{q_e}{n_e v_e T_e} \approx 13\, \frac{v_e}{v_{ei}\, L_T}$$

where v_{ei} is the electron-ion collision frequency[10]. Using Eq. (33) for v_{eff}, we find

$$\frac{q_e}{n_e v_e T_e} \approx 13\, f(c_s, v_e) \approx 0.3\sqrt{Z/A}.$$

This compares with the value of the constant (0.2) obtained in Ref. 10.

Thus, we see that the heat transport is not strongly limited for laser fusion conditions if the isotropic part of the electron velocity distribution is Maxwellian. However, inverse bremsstrahlung absorption of the high power laser light in a plasma with $ZT_e/T_i \gg 1$ results in non-Maxwellian electron velocity distributions[18]. The point is that the electron heating rate (proportional to $v_{ei} E_0^2$ where v_{ei} is the electron-ion collision rate and E_0 is the laser electric field) can be much larger than the rate at which the electrons form a Maxwellian distribution ($v_{ee} v_e^2$ where v_{ee} is the electron-electron collision rate and $v_{ee} \approx v_{ei}/Z$). Thus if $Z v_{os}^2 \gg v_e^2$ the distribution becomes strongly non-Maxwellian and approaches the self-similar form $f_{eo} \sim \exp(- v^5)$. In this case[10] the heat flux with ion acoustic turbulence approaches $f_i \approx 0.06\sqrt{Z/A}$. Thus for $A = Z/2$, we find $f_i \approx 0.04$. Thus, ion acoustic turbulence can result is strong flux limiting in plasmas with $ZT_e/T_i \gg 1$, that are heated by high laser powers, i.e., $Z v_{os}^2 \gg v_e^2$.

SUMMARY

A new self-consistant model has been presented of the spatial structure of a laser heated planar, strongly absorbing plasma. Unlike previous models, this model explicitly includes the temporal evolution of the heat conduction region. It is shown that previous steady-state models apply only to a narrow range of the parameters (laser energy flux and laser pulse width) that are used in laser pellet fusion. The new model is shown to agree with flux-limited hydrodynamics simulations. The key feature of the new model is a self-similar solution for the spatial profile of the temperature, density, and fluid velocity in the conduction region. It is shown that the sonic surface moves outward at a velocity of about 0.6 of the sound velocity. The most important result of the new model for laser pellet fusion is that the plasma properties at the sonic surface and most importantly the ablation pressure are independent of laser wavelength.

We showed that the absorbed laser energy is mainly deposited in the corona so the heat flux is a maximum in the corona near the sonic surface. Laser energy deposition in the corona and therefore heat transport in the corona has very important consequences for laser fusion. The point is that heat transport can drive ion acoustic turbulence which in turn can limit heat transport. The energy transported down the temperature gradient is carried by suprathermal electrons. An electric field is generated to drive a return current of thermal electrons traveling down the temperature gradient. Ion acoustic waves are driven unstable if the return current is above a threshold so that ion turbulence is excited in the plasma. We applied a recently developed theory of ion turbulence driven by heat transport to laser plasmas driven by high laser powers. We showed that the threshold can be exceeded in high Z plasmas. The heat flux is only weakly limited if the isotropic part of the electron velocity distribution function is Maxwellian. However, the heat flux is strongly limited if the isotropic part of the electron velocity distribution function is strongly non-Maxwellian ($f \approx \exp(-v^5)$). Collisions in high Z plasmas drive the electron distribution non-Maxwellian for high laser powers. Therefore, ion acoustic turbulence can result is strong flux limiting in plasmas with $ZT_e/T_i \gg 1$, that are heated by high laser powers, i. e., $Zv_{os}^2 \gg v_e^2$.

Acknowledgements

We acknowledge useful discussions with Professor T. W. Johnston and Dr. J. P. Matte. This work was partially supported by the U. S. Department of Energy, the Plasma Physics Research Institute, U. C. Davis, and Lawrence Livermore National Laboratory, and partially performed under the auspices of the U. S. Department of Energy by the Lawrence Livermore National Laboratory under Contract No. W-7405-ENG-48.

REFERENCES

1. W.L. Kruer, *Comments on Plasma Physics and Controlled Fusion* 6, 167 (1981).
2. J. S. De Groot, S. M. Cameron, K. Mizuno, K. G. Estabrook, R. P. Drake, W. L. Kruer, and P. E. Young, *Phys. Fluids B* 3, 1241 (1991).
3. J. S. De Groot, K. G. Estabrook, W. L. Kruer, R. P. Drake, K. Mizuno and S. M. Cameron, To appear in *Physics of Fluids*, Feb. 1992.
4. G.B. Zimmerman and W.L. Kruer, *Comments on Plasma Physics and Controlled Fusion* 2, 85 (1975); J. Delettrez, *Can. J. Phys.* 64, 932 (1984).
5. F.S. Felber, *Phys. Rev. Lett.* 39, 84 (1977).
6. C.E. Max, C.F. McKee, and W.C. Mead, *Phys. Fluids* 23, 1620 (1980).
7. R. Fabbro , C.E. Max, and E. Fabre, *Phys. Fluids* 28, 1463 (1985).
8. P. Mora, *Phys. Fluids* 25, 1051 (1982).
9. Faiz Dahmani and Tahar Kerdja, *Phys. Fluids B* 3, 1232 (1991).
10. V. Yu. Bychenkov, V. P. Silin, and S. A. Uryupin, *Physics Reports* 164, 119 (1988); V. Yu. Bychenkov, V. P. Silin, and S. A. Uryupin, *Comments Plasma Phys. Controlled Fusion* 13, 239 (1990).
11. T. W. Johnston and J. M. Dawson, *Phys. Fluids* 16, 722 (1973).
12. L. Spitzer and R. Härm, Phys. Rev. 89, 977 (1953); S. I. Braginskii, in *Reviews of Plasma Physics*, Vol. 1, ed. M./ A. Leontovich (Consultants Bureau, New York, 1965).
13. V. L. Ginsberg, "Propagation of Electromagnetic Waves in Plasmas", (Pergamon, New York, 1965) Chap 6.
14. P. Monchicourt and P. A. Holstein, *Phys. Fluids* 23, 1475 (1980).
15. B. D. Fried and R. W. Gould, *Phys. Fluids* 4, 139 (1961).
16. C. J. Randall, *Phys. Fluids* 25, 2231 (1982).
17. A. F. Kuckes, *Phys. Fluids* 7, 511 (1964).
18. A. B. Langdon, *Phys. Rev. Lett.* 44, 575 (1980).

TWO-DIMENSIONAL CALCULATION OF ELECTROMAGNETIC PULSE GENERATION BY LASER-INITIATED AIR AVALANCHE SWITCHES[*]

D. J. Mayhall and J. H. Yee

Lawrence Livermore National Laboratory
University of California
P. O. Box 808, Mail Code L-156
Livermore, CA 94550

INTRODUCTION

The gas avalanche switch is a recently proposed, laser-initiated, high-voltage, picosecond-speed switch.[1,2] The basic conceptual switch consists of a set of pulse-charged electrodes, which is surrounded by a high-pressure (2-800 atm) gas. An avalanche discharge is initiated in the gas between the electrodes by ionization from a picosecond-scale laser pulse. The laser-induced initial electrons rapidly avalanche toward the anode in the applied electric field. The rapid avalanche is fueled by the immense number of electrons available in the high-pressure gas. Under the proper conditions, the rapid exponentiation of the electrons in the avalanche causes the applied voltage across the electrodes to collapse within picoseconds.

There are several simple versions of the basic switch. A parallel plate capacitor version consists of a gas between two parallel conducting plates. An applied voltage is impressed across the two plates. Dielectric spacers must confine the gas at ends and edges of the plates. Cylindrical coaxial capacitor structures are also possible. A simple, parallel plate, Blumlein type pulse generator geometry has a center electrode between two horizontal parallel conducting plates. In this structure, the center electrode is initially charged to a suitable positive high voltage while the two parallel plates remain at ground potential. The gas gap between the center electrode and one of the plates, usually the lower one, is rapidly illuminated with suitable laser light for initiation of an avalanche. If the ensuing voltage collapse occurs rapidly enough, it generates electromagnetic waves, which move outward from the center electrode toward the ends of the parallel plates. To investigate the generation and propagation of electromagnetic waves in such gas avalanche switches, we have developed a two-dimensional, electromagnetic, electron fluid computer code for avalanche ionization in gases.

[*] Work performed under the auspices of the U.S. Department of Energy by Lawrence Livermore National Laboratory under Contract W-7405-Eng-48.

Laser Interaction and Related Plasma Phenomena, Vol. 10
Edited by G.H. Miley and H. Hora, Plenum Press, New York, 1992

209

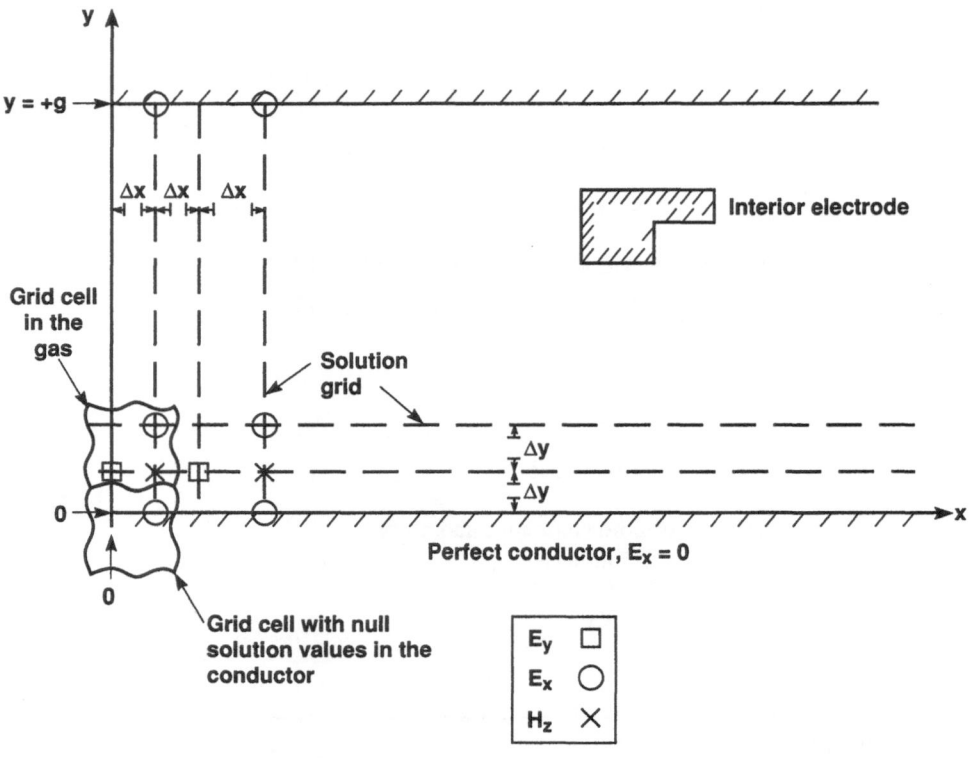

Figure 1. Solution geometry for two-dimensional, electromagnetic, electron fluid avalanche code.

THE TWO-DIMENSIONAL, ELECTROMAGNETIC, ELECTRON FLUID COMPUTER CODE

The two-dimensional, electromagnetic, electron fluid computer code solves Maxwell's curl equations for transverse electromagnetic (TM) modes in nonmagnetic gases between perfectly conducting parallel plates. At the same time, it also solves a set of electron fluid equations for the conservation of number density, momentum, and energy. The finite difference solution of these equations is simultaneous, self-consistent, and implicit. The coordinate system used is rectangular. A number of charged, rectilinear, perfectly conducting electrodes may be placed between the two parallel plates. Figure 1 shows a diagram of the geometry for the code. The x direction is the axial direction for the transmission line formed by the parallel plates. The y direction is the transverse direction.

Governing Equations

For a free space dielectric between the parallel plates, the relevant Maxwell's equations are

$$\frac{\partial E_x}{\partial t} = \varepsilon_0^{-1}\left(\frac{\partial H_z}{\partial y} - nev_x\right) , \tag{1}$$

$$\frac{\partial E_y}{\partial t} = -\varepsilon_0^{-1}\left(\frac{\partial H_z}{\partial x} + nev_y\right) \quad , \tag{2}$$

$$\frac{\partial H_z}{\partial t} = \mu_0^{-1}\left(\frac{\partial E_x}{\partial y} - \frac{\partial E_y}{\partial x}\right) \quad , \tag{3}$$

where E_x and E_y are the electric field components in the x and y directions, ε_0 and μ_0 are the gas permittivity and permeability, and H_z is the magnetic field component in the z direction. The quantity n is the electron density, v_x and v_y are the electron fluid velocity components in the x and y directions, and e is the electronic charge, -1.60×10^{-19} C.

The electron fluid conservation equations are

$$\frac{\partial n}{\partial t} = nv_i \quad , \tag{4}$$

$$\frac{\partial(nv_x)}{\partial t} = n\left[em^{-1}\left(E_x + \mu_0 v_y H_z\right) - v_m v_x\right] \quad , \tag{5}$$

$$\frac{\partial(nv_y)}{\partial t} = n\left[em^{-1}\left(E_y - \mu_0 v_x H_z\right) - v_m v_y\right] \quad , \tag{6}$$

$$\frac{\partial(nU)}{\partial t} = n\left[e\left(E_x v_x + E_y v_y\right) - v_u(U - U_0) - v_i \varepsilon_i\right] \quad , \tag{7}$$

where v_i is the ionization collision frequency, m is the electron mass, v_m and v_u are the electron-neutral molecule momentum and energy transfer collision frequencies, U is the electron fluid kinetic energy, U_0 is the average neutral energy, and ε_i is the average neutral ionization potential. To specialize the equations to air, we use collision frequencies for electrons in air. We also take U_0 as 0.025 eV and ε_i as 14 eV.

These four electron fluid equations constitute the first three moments of the Boltzmann equation with the assumption of a Maxwellian electron velocity distribution function. In addition, many other assumptions are also made. Convective terms are neglected. Pressure gradients and heat flows are also neglected. All electron number losses, from such processes as attachment, recombination, and diffusion, are ignored. Electron generation by photoionization is also neglected.

We further assume that the avalanche and electromagnetic wave generation are so rapid that ion motion is unimportant. Therefore, fluid equations for positive and negative ions are ignored. Similarly, neutral molecule fluid equations are also neglected. We thus assume that all the important events occur long before the formation of such ion current features as the hot channel in conventional spark gaps.

In addition, electrode pulse charging is assumed to take place instantaneously. Thus, the very real experimental problem of premature breakdown from field emission or insulator surface flashover is ignored. The generation of initial electrons by the laser pulse is also presently assumed to be instantaneous. Modeling of the laser deposition of electrons is thus neglected. These assumptions of instantaneous voltage application and initial electron induction mean that the calculational predictions are idealizations and therefore *"best case"* results. Due to technological limitations, these two idealizations may be difficult and expensive to achieve experimentally. Recent zero-dimensional circuit code calculations for electrode charging times of about 1 ns predict that, if the initial electron density at the start of charging is of the order of 10^7 cm^{-3} or greater, the voltage collapse will not be complete.[3] Residual voltage will remain across the electrodes after the rapid avalanche has ended. This incomplete

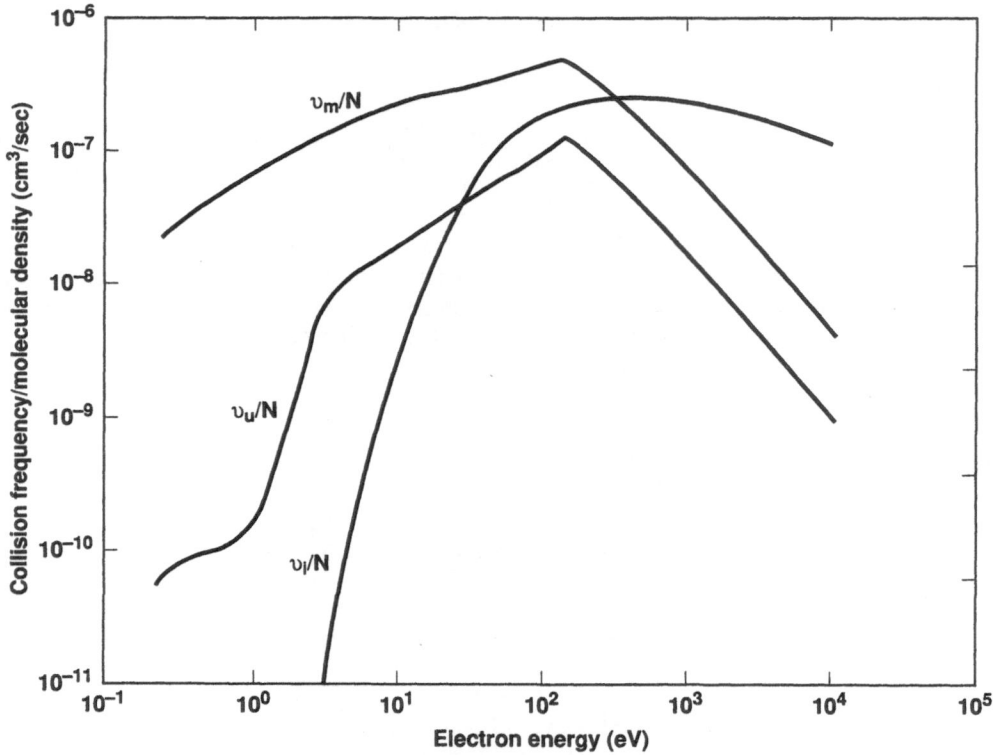

Density-normalized ionization frequency υ_i/N
momentum-transfer frequency υ_m/N,
and energy-transfer frequency υ_u/N,
used in the code calculations, vs average electron energy.

Figure 2. Collision frequencies for air.

voltage collapse will be accompanied by weak, long lasting electromagnetic pulses. These pulses represent an inefficient transfer of the initial electrostatic energy stored in the electrode structure into transmitted electromagnetic energy.

The collision frequencies the code uses for air are shown in Fig. 2 as functions of electron energy in eV. These frequencies are analytic fits to two types of data. Below electron energies of 150 eV, the data is from electron swarm measurements. Above 150 eV, the frequencies come from theoretical calculations with Maxwellian velocity distributions. These collision frequency fits were previously used for calculations of microwave pulse breakdown in low-pressure air[4-6] and microwave pulse propagation in the atmosphere.[7-9] These collision frequencies are assumed to scale directly with increasing neutral pressure.

Computational Geometry

Figure 1 shows the geometry for the finite difference computational solution of the governing equations. The coordinate direction z extends out of the plane of the figure. The space between the two parallel plates at y = 0 and y = +g, the plate spacing, is partitioned with a rectangular mesh. The mesh spacing in the x direction is Δx; that in the y direction is Δy. The two spacings are usually unequal and uniform. Solution points for the three electromagnetic components form a staggered solution grid. Similar grids are used in two-dimensional, explicit,

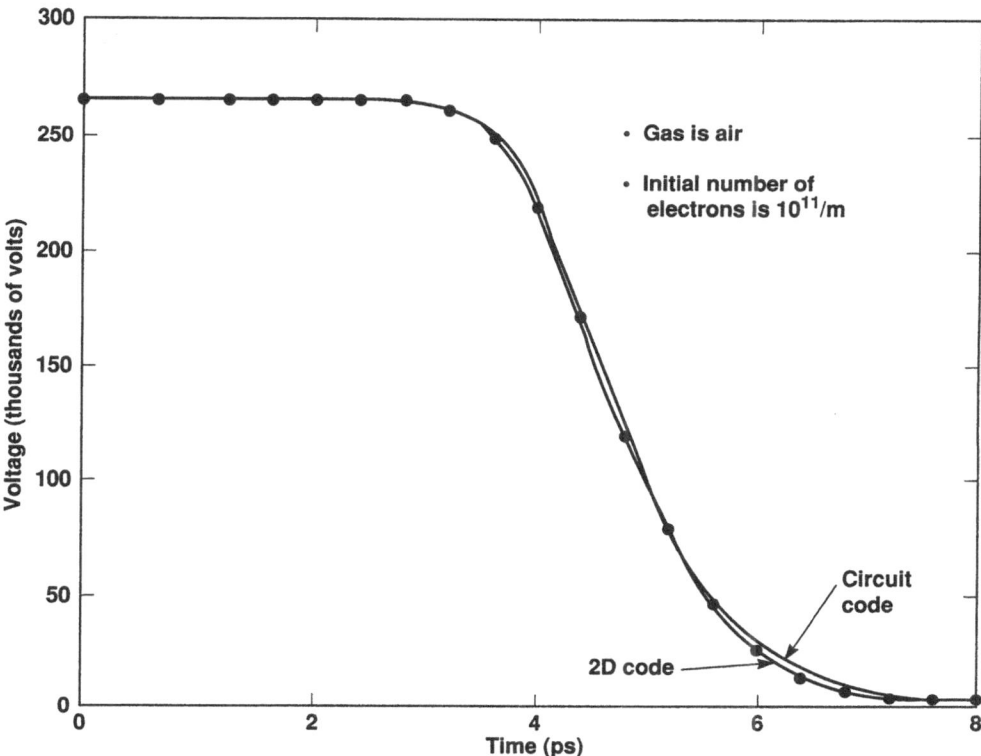

Figure 3. Comparison of parallel plate capacitor voltage waveforms at 266 kV and 350 atm. Alpha is the first Townsend coefficient used in the circuit code calculations and is equal to 0.67 times the reported value.

finite difference, electromagnetic scattering codes.[10] Such staggering of the field components results in convenient central differencing of the spatial derivatives in Maxwell's curl equations. As shown by the open circle in the inset legend box in Fig. 1, E_x is solved at even grid points on odd horizontal x-z planes. The lowermost of these planes coincides with the top surface of the lower conducting plate. The uppermost of these odd planes coincides with the bottom surface of the upper conducting plate. Since the plates are perfect conductors, E_x is set to zero along these two planes. E_y is calculated at odd grid points on even horizontal planes, and H_z is calculated at even grid points on these same planes. The four electron fluid variables are evaluated at the same grid points as H_z. The choice of these fluid variable locations means that averaged electric fields must be used in the two momentum equations. A rectangle of grid points, consisting of those for E_x, E_y, H_z, and the unused grid point above that for E_y, makes up a grid cell. An example of such a grid cell in the gas is shown in Fig. 1. The lower left grid cell in Fig. 1, which includes the coordinate system origin, is replicated along the lower plate in the x direction. These grid cells along the lower plate are special since all their solution variables have null values throughout the calculations. The grid cell for the gas shown in Fig. 1 is replicated in the x and y directions to complete the finite difference solution grid.

Implicit Numerical Solution

For numerical solution of Eqs. (1)–(7), central finite differences replace the spatial derivatives in Eqs. (1)–(3) in the interior of the computational mesh. The time derivatives of the products in Eqs. (5)–(7) are expanded and rewritten as time derivatives of v_x, v_y, and U. Seven ordinary differential equations (ODEs) in time result from this spatial discretization for each grid cell. The global number of ODEs for this type of grid is $i_{max} \cdot (j_{max} + 1) \cdot 7/4$, where j_{max} is the number of grid points in the y direction, and i_{max} is the number of grid points in the x direction. The global set of ODEs is solved by time integration with the block-iterative, optionally stiff, implicit solver GEARBI.[11] GEARBI is able to take dynamically variable time steps throughout the course of the integration.

GEARBI has attractive capabilities for problems involving both gas breakdown and electromagnetic wave propagation. The variable integration step allows reduction of the total solution time, compared to a fixed step integrator set at a small step to ensure good solution resolution. During regimes of rapid ionization or strong avalanching, a code with GEARBI can take very small steps, roughly of the order of the inverse of the ionization frequency. After the ionization slows significantly or stops, the code can speed up and use larger time steps. If electromagnetic wave propagation is the dominant time-varying process, these steps can exceed the Courant limit. The Courant limit for numerical stability of Maxwell's equations with explicit finite differencing is the time step limit given by $\Delta x/u$ in one dimension and $(\Delta x \cdot \Delta x + \Delta y \cdot \Delta y)^{1/2}/u$ in two dimensions, where u is the speed of light in the medium.[10,12] For a particular machine, the block-iterative solution method allows solution of problems with larger grids than is possible with other, more memory intensive methods because the memory requirement is less. This gain in problem size, however, is not free. More time, care, and effort must be invested to generate the required user-furnished subroutines. Another benefit of GEARBI is the stiff equation system option. This feature allows significant reduction of problem solution time. It allows the solver to largely ignore stiffness or the existence of vastly disparate time scales of variation among the various solution components. When this capability is used, the solver does not slow down to follow unimportant low-amplitude, high-frequency oscillations in some variable in some part of the grid.

VOLTAGE COLLAPSE IN A SIMPLE PARALLEL PLATE CAPACITOR SWITCH

The electron fluid computer code is first used to investigate voltage switching in a simple parallel plate capacitor. The magnetic forces in the momentum equations are ignored for this case. The geometry is that shown in Fig. 1 with the interior electrode removed. The plate spacing g is 100 microns. The plates extend 247.5 microns in the positive x direction from the coordinate origin. A laser pulse is assumed to generate 10^{11} electrons, which are uniformly distributed between the plates for distance of 1 m in the z direction. The top plate has a potential of 266 kV with respect to the lower plate. The right and left boundary conditions approximate the extension of the parallel plates and the intervening air gap to infinity in the positive and negative x directions. The air pressure between the plates is 350 atm.

After the simulation starts, the voltage across the plates remains essentially constant for about 3 ps and then collapses to close to zero over about 4 ps. The total voltage collapse or switching time is thus about 7 ps for the given initial conditions. Figure 3 shows the waveform of the voltage across the plates as the solid curve, labelled "*2D code*." The voltage collapse time from 90% of the initial voltage to 10% of that value is about 2 ps. The result of a zero-dimensional circuit code calculation for the same initial conditions by Cassell and Villa[2] is also shown in Fig. 3. This waveform is labelled "*Circuit code*." In the circuit code calculation, the first Townsend coefficient α was adjusted to 2/3 of its reported experimental value to make the two curves almost coincide.

Figure 4 compares the y direction electron fluid velocity from the electron fluid code simulation with the electron drift velocity in air measured by Bradbury and Nielsen.[13] These two velocities are plotted as functions of the reduced electric field E/p in V/(cm-torr) over the range, 0.1–16.5 V/(cm-torr). For the experimental data, E is the electric field in the direction of the electron drift and p is the neutral gas pressure. For the simulation, E is the electric field in the y direction. The agreement of the two curves over the specified range of reduced electric field is excellent.

An equivalent reduced first Townsend coefficient α_e/p may be defined in terms of the electron

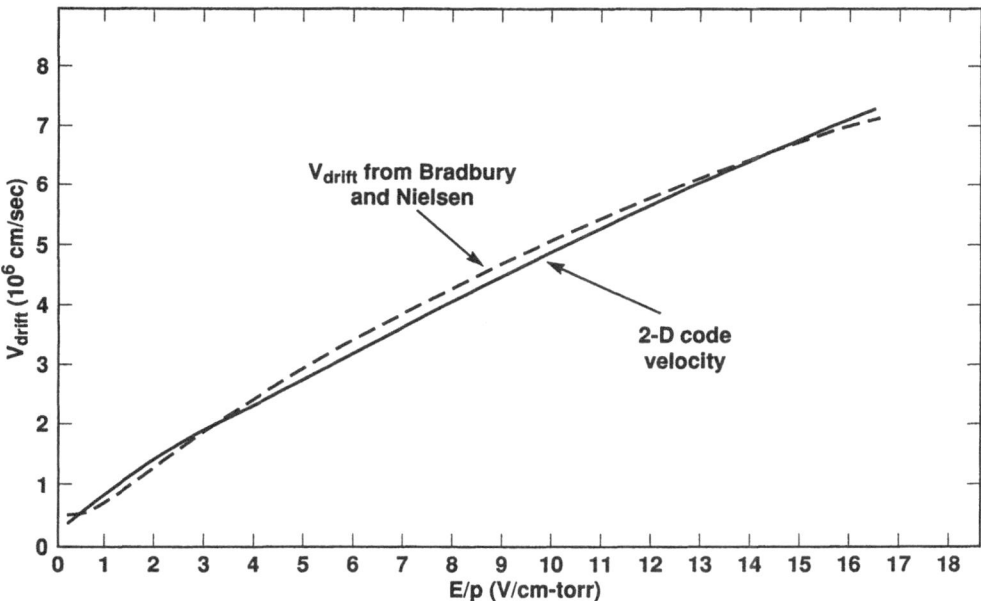

Figure 4. Comparison of electron drift velocities for parallel plate capacitor.

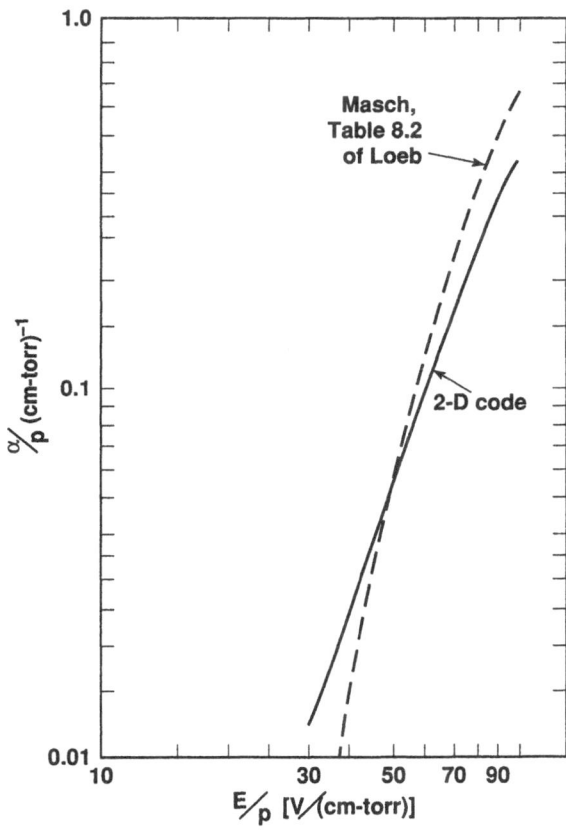

Figure 5. Comparison of first Townsend coefficients for parallel plate capacitor.

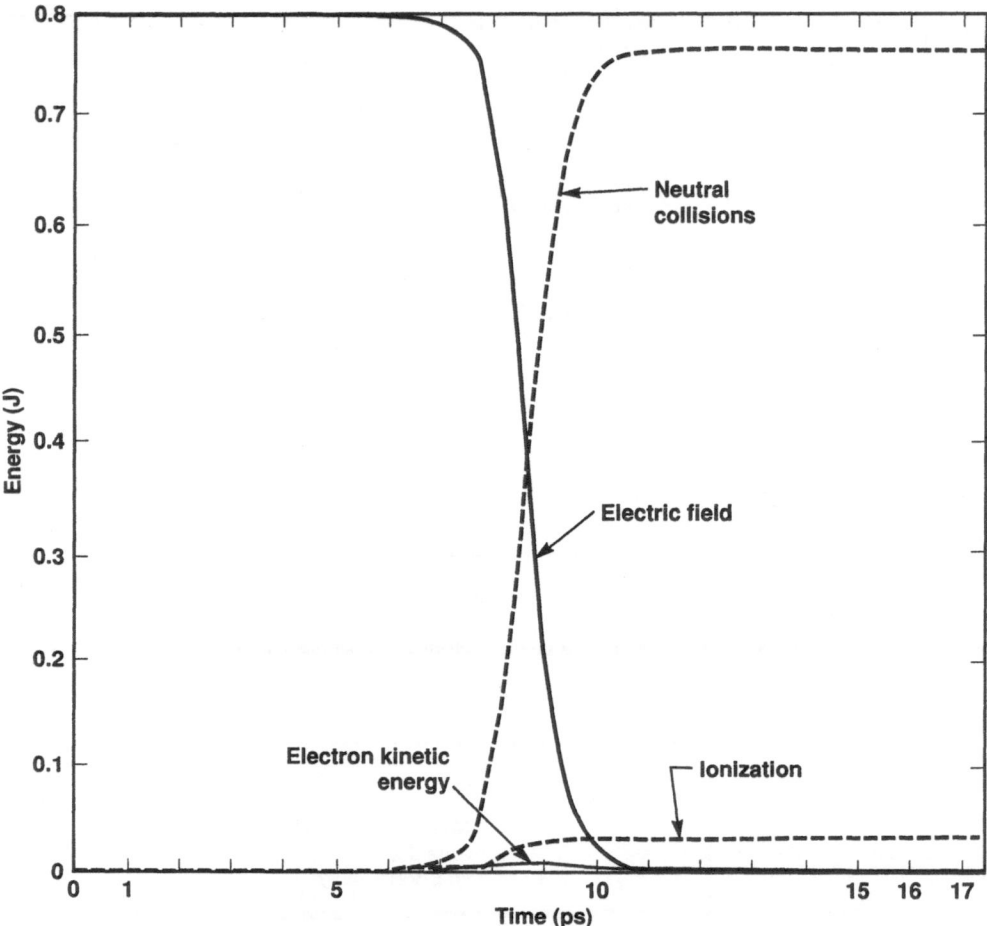

Figure 6. Energy balance for parallel plate capacitor.

fluid code variables by the relation $\alpha_e/p = \nu_i/(\nu_y p)$. The variation of this coefficient with the reduced electric field is shown in Fig. 5 as the solid curve. The reduced electric field varies from 30–100 V/ (cm-torr). The experimental data of Masch[14] is shown as the dashed curve. The agreement of the two curves is reasonably good for the range, $42 \le E/p \le 100$ V/(cm-torr). This is the range over which most of the electron avalanching and voltage collapse occur. At $E/p = 100$ V/(cm-torr), 98% of the electrostatic energy initially stored in the capacitor electric field remains in the electric field. At $E/p = 40$ V/(cm-torr), just 16% of the initial energy is still contained in the electric field. Below this latter value, the agreement between the two curves becomes worse and worse. Over most of the range of the reduced electric field from 52–100 V/(cm-torr), the reduced Townsend coefficient from the simulation is about 0.7 of the experimental value. This explains why the waveforms for voltage decay in Fig. 4 are in such close agreement. Part of the reason for the disagreement of the curves of α/p in Fig. 5 may be experimental uncertainty due to the measurement technique used in this older work. Part of the disagreement may also be due to inaccuracies in the collision frequencies used in the electron fluid computer code.

Figure 6 shows an energy balance as a function of time for this parallel plate capacitor simulation. About 96% of the initial electrostatic energy or 0.77 J is transferred to electron motion and dissipated through electron-neutral molecule collisions. About 4% of the initial electrostatic energy is lost through ionization of the neutrals. These processes result in a minute heating of the vast number of neutrals and

216

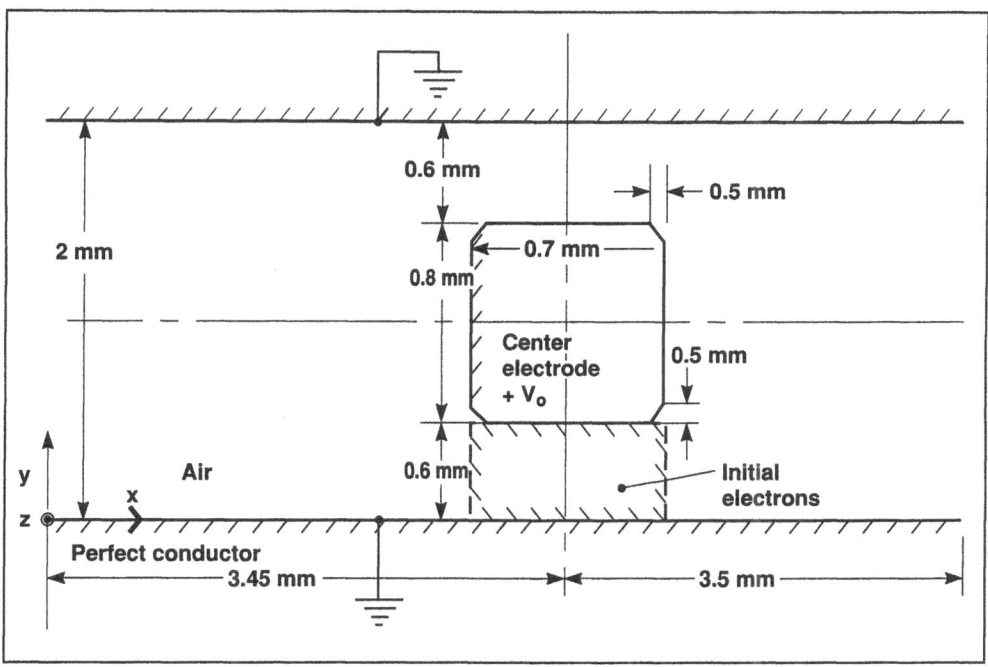

Figure 7. A simple pulse generator switch model. TEM boundary conditions at left and right.

a minute increase in the internal energy of the lesser number of ions. Virtually all the energy transfer is over by 11 ps into the simulation.

The agreement of the electron fluid code results with the circuit code results and the experimental data indicates that the electron fluid code can be used for prediction of gas avalanche switch behavior in more complex situations.

No electromagnetic waves are generated during this essentially zero-dimensional simulation with the two-dimensional code. Due to the completely uniform initial electric field and induced electron distribution, the seven variables in each grid cell are the same throughout the computational space at each output time of the code. Thus, the spatial derivatives of the electric field components are always zero everywhere. Therefore, from the right side of Eq. (3), no time varying H_z field occurs anywhere and no electromagnetic waves can develop. All the initial electrostatic energy eventually goes into heating and ionizing neutrals. Some nonuniformity in the electric field or the electron distribution must occur at some time for the generation of electromagnetic waves.

VOLTAGE COLLAPSE IN A SIMPLE PULSE GENERATOR SWITCH

A gas avalanche switch, which has nonuniformity in its applied electric field and is somewhat more geometrically sophisticated than the parallel plate switch, features a center electrode between the two parallel plates. This switch can function as a pulse generator. When a properly configured center electrode is charged to a suitable high voltage at the correct gas pressure, computations predict the generation of electromagnetic pulses with picosecond rise times and durations.

Figure 7 shows a rudimentary model of such a switch. The center electrode is a perfectly conducting rectangle with beveled corners. It is charged to the positive voltage V_0. The parallel perfectly conducting plates, above and below the center electrode, are initially at ground potential. Air surrounds the center electrode. The lower gap, the air gap between the center

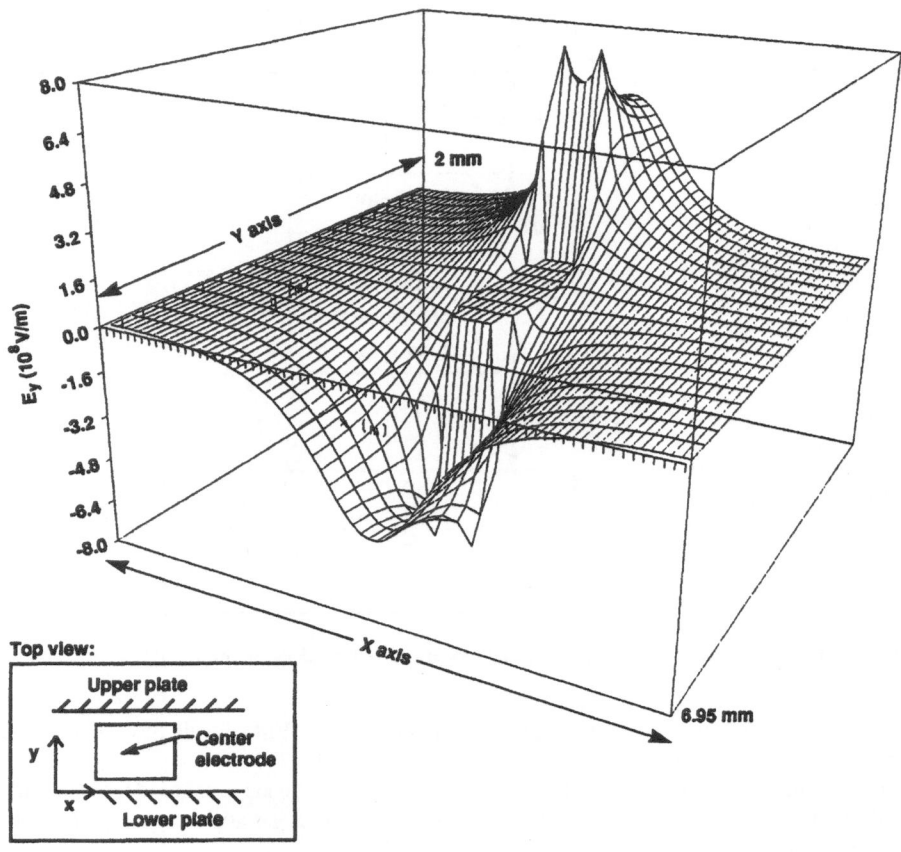

Figure 8. Initial distribution of E_y.

electrode and the lower plate, is uniformly filled with 1.7×10^{13} laser-generated electrons to an assumed depth of 1 m normal to the plane of the figure. The upper gap, the gap between the center electrode and the upper plate, contains no electrons. The modeled space extends from 3.45 mm to the left of the vertical center line of the center electrode to 3.50 mm to the right of that line. In this model, there are no dielectric windows at the left and right ends of the transmission line formed by the parallel plates. Instead, transverse electromagnetic (TEM) mode computational boundary conditions are specified. With these boundary conditions, outwardly traveling electromagnetic waves, whose predominant mode is transverse electromagnetic, pass across these boundaries with the generation of very little inwardly reflected component. This numerical treatment greatly reduces the contamination of the interior solution by spurious, nonphysical reflected waves from the two free space grid boundaries. At the left boundary, the condition is $E_y = \eta_0 H_z$, where η_0 is the free space intrinsic impedance, $(\mu_0/\varepsilon_0)^{1/2}$. At the right boundary the conditions are $H_z = E_y/\eta_0$ and $E_x = 0$. On the horizontal faces of the center electrode, $E_x = 0$; whereas, on the vertical faces, $E_y = 0$.

The finite element Poisson code STAT2D provides the initial electric field distribution for electron fluid code simulations. Figure 8 shows an initial distribution of E_y. The small top view at the lower left of the figure diagrammatically shows the relation of the switch electrodes to the field distribution. The x-z plane, which contains the coordinate origin, corresponds to the upper surface

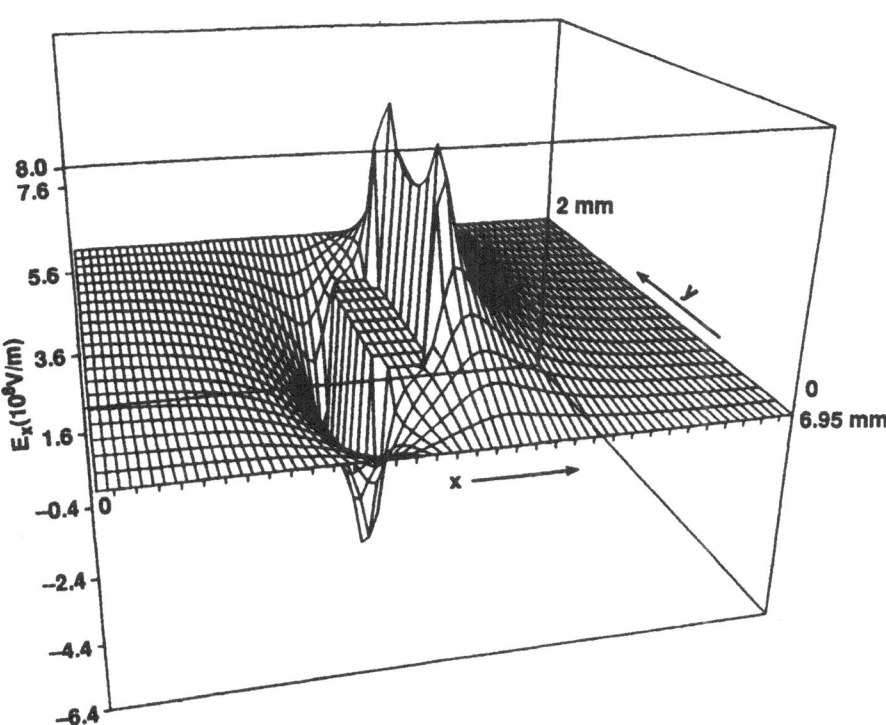

Figure 9. Initial distribution of E_x.

of the lower plate. This is the near x-z plane of the box enclosing the electric field in Fig. 8. The x-z plane at y = 2 mm corresponds to the lower surface of the upper plate. This plane is the far x-z plane of the box which encloses the electric field distribution. The region of zero field in the center of the figure corresponds to the center electrode. As required by the perfectly conducting boundary conditions, E_y is zero along the two vertical faces of the center electrode. The magnitude of E_y is greatest along horizontal lines just off the two horizontal surfaces. Along these lines, the magnitude of E_y peaks at the corners, where the greatest electric field enhancement is expected. Above the center electrode, the electric field is positive since it is directed along the positive y axis. Below the center electrode, the electric field is negative since it is directed along the negative y axis. Figure 9 shows the corresponding distribution of the E_x field.

Pulse Generation in the Kilovolt Applied Voltage Range

For most of the calculations in the kilovolt range of applied voltage, the magnetic forces in Eqs. (5) and (6) are neglected as being small compared to the electric forces. When the simulation starts with 292.4 kV on the center electrode at an air pressure of 27.2 atm, the voltage difference between the center electrode and the lower plate drops toward zero in roughly 3 ps. Figure 10 shows a waveform of this voltage difference for the voltage of the lower plate referred to the center electrode. This is the voltage difference close to the vertical center line in Fig. 7. After rebounding at 4 ps, this waveform slowly drops off to zero at 28 ps. The waveform ends slightly positive at the end of the simulation at 36.5 ps.

The rapid initial voltage collapse is due a rapid avalanche of electrons toward the center electrode. This displacement current neutralizes a large part of the applied positive charge on

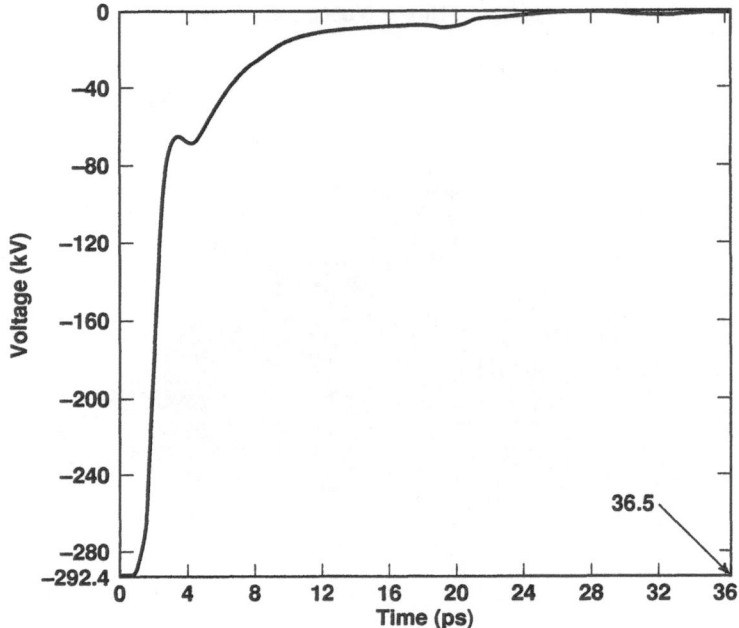

Figure 10. Waveform of the voltage between the center electrode and the bottom plate near the verticle center line at 292.4 kV and 27.2 atm.

Time = 1.1000 × 10⁻¹¹ sec

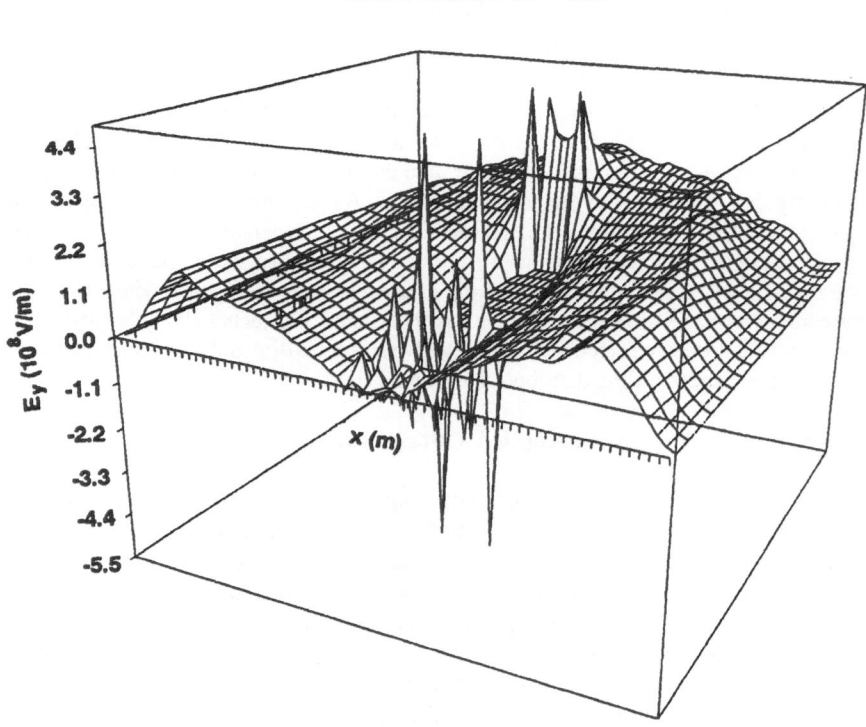

Figure 11. E_y at 11.0 ps into the simulation.

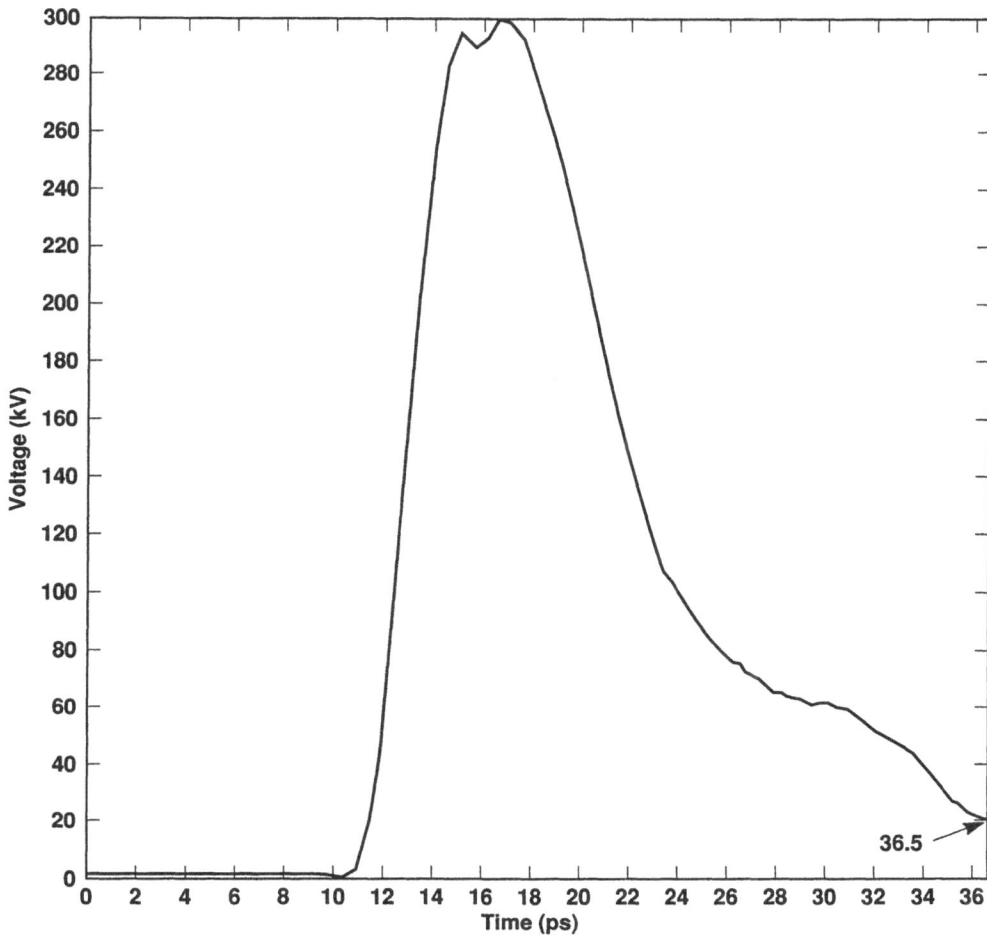

Figure 12. Voltage waveform at right boundary for 292.4 kV and 27.2 atm.

the center electrode and, in doing so, generates electromagnetic waves. The waves propagate outward from the center electrode toward the right and left computational boundaries. These waves also travel up the right and left sides of the center electrode, reflect at the upper gap, then move outward toward the left and right boundaries. This interaction with the upper gap causes delayed wave crests to move outward near the upper plate.

The E_y component of the electromagnetic waves is shown in Fig. 11 at 11.0 ps into the simulation. In the near x-E_y plane of the figure, which corresponds to the surface of the lower plate, two positive waves have moved out toward the boundaries. The leading edges of the waves have just impacted the boundaries. Figure 12 shows this incidence at the right boundary as the initial rise of the voltage waveform at 10.5–11.0 ps. This is the voltage of the upper plate with respect to the lower plate as calculated by the line integral of E_y across the right boundary. Two wave crests occur in Fig. 11 at the lower plate. It can be seen that as y increases from zero the wave crests lag farther and farther behind those at the lower plate. This lagging is believed to be due to the spreading of the wavefronts at the extremities of the lower gap and the delayed wave reflections at the upper gap.

The wave crests at the lower plate hit the boundaries at about 14.5 ps, slightly before the first peak of the voltage waveform in Fig. 12 at 15 ps. Discernible wave crests along the far x-E_y plane develop at about 12 ps. These lagging crests impact the open boundaries

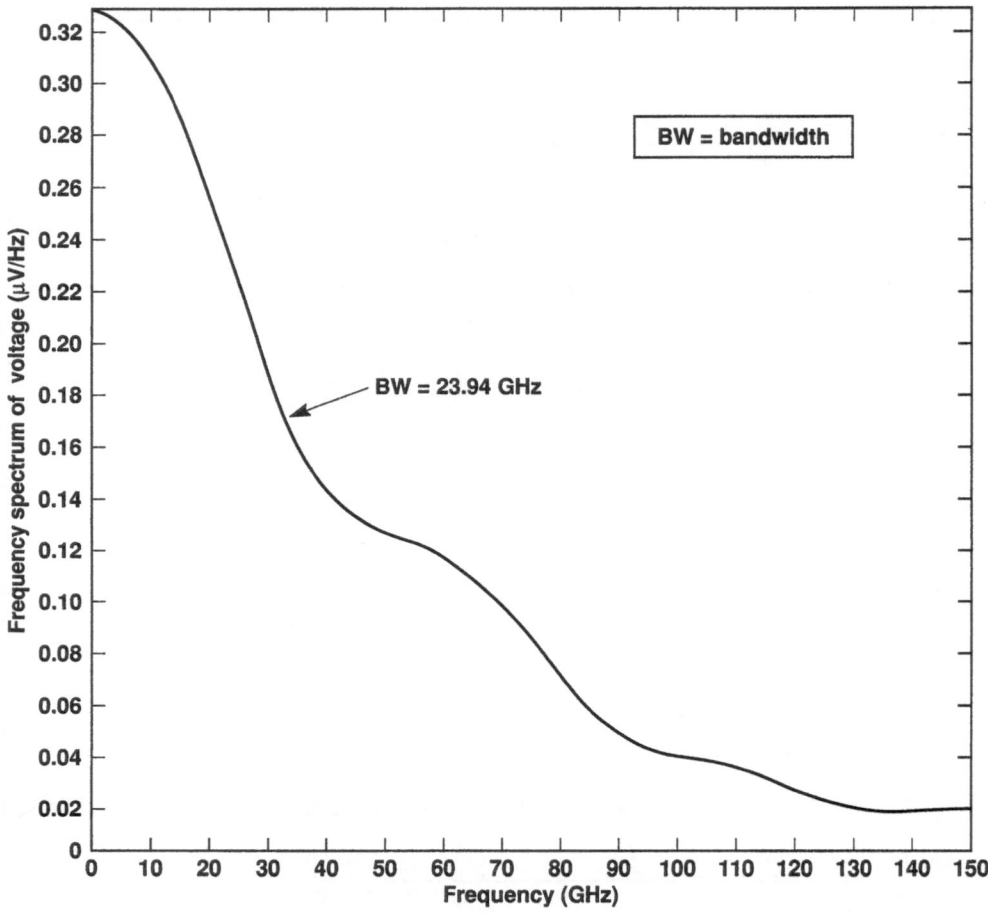

Figure 13. Frequency spectrum of voltage waveform at right boundary at 292.4 kV and 27.2 atm.

at about 16 ps, just before the second peak in Fig. 12 at 16.5 ps. After the second peak, the voltage drops to about 100 kV over about 6.5 ps. It then falls off more slowly to about 20 kV. This slower fall off seems to be correlated with a slow decay of the electric field in the upper gap.

The curious bipolar electric field spikes in the lower gap under the center electrode are presently believed to be nonphysical artifacts due to noise generated from the initial H_z component. They are reminiscent of two-dimensional, checkerboard noise from central finite differencing. The electric field components from STAT2D are thought to be inaccurate at the center electrode surface. These components are extrapolations of spatial derivatives of the scalar potential calculated some distance away at the finite element Gauss points. The use of spatial derivatives of these components in Eq. (3) is believed to cause aggravated inaccuracies, which result in imperfect cancellation of the righthand terms in Eq. (3). This causes an initial nonzero time derivative for H_z all around the perimeter of the center electrode. This noise extends well into the upper gap toward the upper plate. This disturbance appears in the upper gap within 0.5 ps, long before it could physically propagate from the lower gap, where the electron avalanching occurs. These bipolar spikes are one of a number of bothersome numerical problems, which can arise in such nonlinear simulations.

In addition to the strong waves in E_y shown in Fig. 11 and accompanying waves in

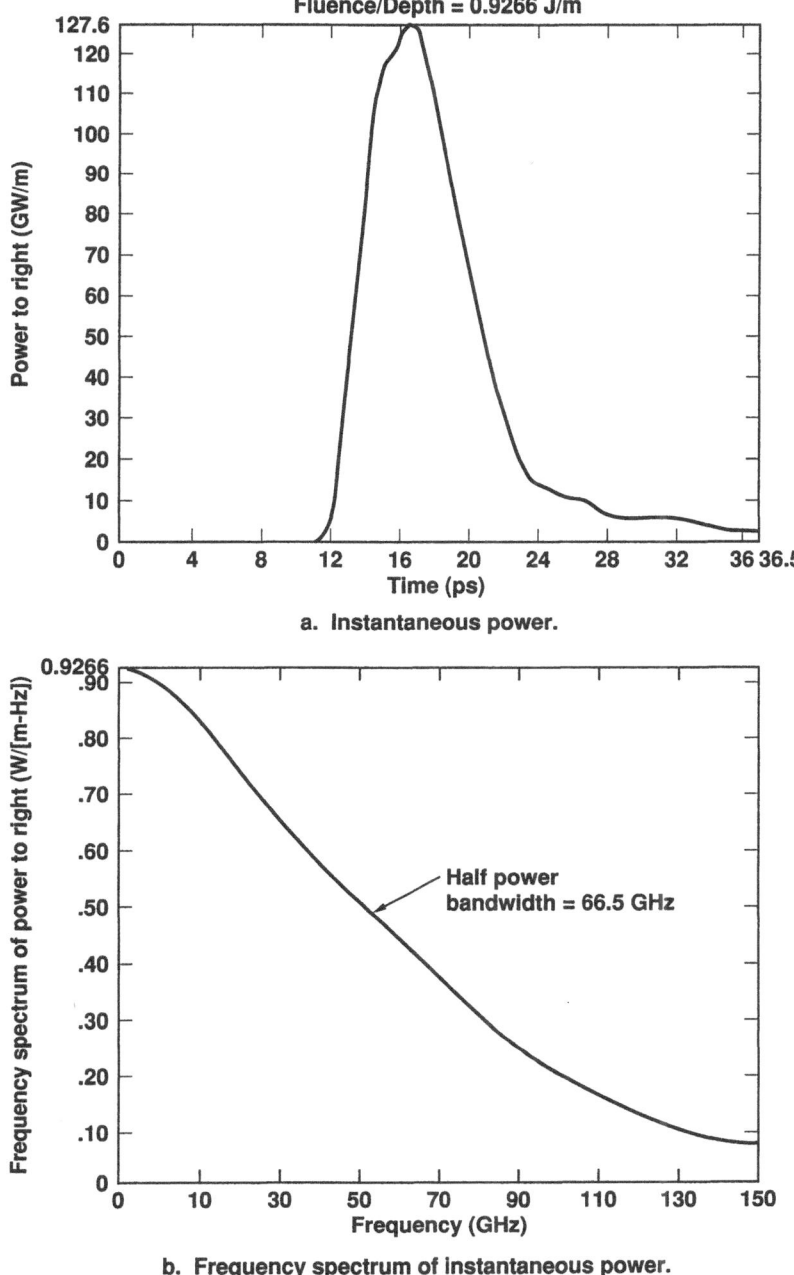

Figure 14. Transported power waveforms at the right boundary for 292.4 kV and 27.2 atm.

$\eta_0 H_z$ with amplitudes of about 2×10^8 V/m, weaker waves in E_x occur. These E_x waves have maximum values of about 5×10^7 V/m and propagate outward from the center electrode in complicated patterns. Thus, the generated mode structure is not purely TEM in nature.

Figure 12 has a peak induced voltage of 300 kV, a 10–90% rise time of 2.41 ps, and a full width at half maximum (FWHM) of 9.11 ps. The pulse duration is greater than 26 ps.

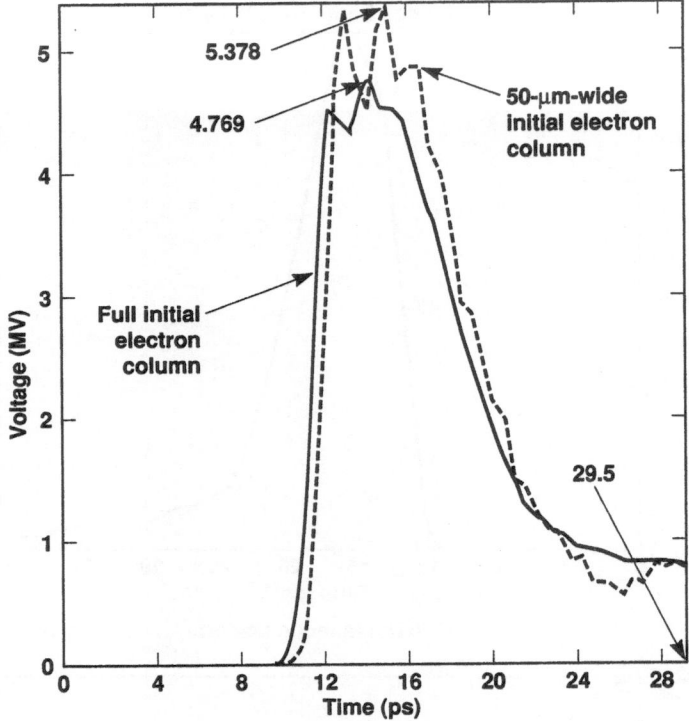

Figure 15. Voltage waveforms at the right boundary at 3.89 MV and 700 atm.

The peak pulse voltage is a 2.67% enhancement over the voltage initially applied to the center electrode. The initial electrostatic energy is 3.34 J/m. By 36.5 ps, 55.4% of this energy is transported across the open boundaries by electromagnetic waves.

Figure 13 shows the frequency spectrum of the voltage waveform in Fig. 12. It is assumed in generating the Fast Fourier Transform that the voltage drops abruptly to zero at 36.5 ps. The spectral peak is at dc. The 3 dB bandwidth is 23.9 GHz, which is the spectral width at values of 0.707 times spectral maximum.

Figure 14a shows the waveform of the outgoing electromagnetic power at the right boundary. As should be expected, it is more sharply peaked than the induced voltage waveform. The peak value is 128 GW per meter of transverse depth into the switch. By 36.5 ps, a fluence of 0.927 J/m has moved across the right boundary. A very similar amount has traversed the left boundary. This value is the time integral of the power waveform in Fig. 14a. The frequency spectrum of this instantaneous power waveform is shown in Fig. 14b. The 3 dB power bandwidth, which is defined as the spectral width at the half power points, is 66.5 GHz. This value is about 22% less than three times the corresponding induced voltage bandwidth.

Similar results occur with the center electrode voltage reduced to 227 kV and the pressure scaled linearly with this voltage. Such scaling keeps the initial nominal E/p at the center electrode constant, which makes the initial avalanches roughly the same. Reduction of the E/p at fixed voltage causes lengthening of the induced pulse rise time, FWHM, and duration, as well as a decrease in the peak pulse voltage. Increasing this E/p results in decreases in the induced voltage pulse rise time and FWHM, as well as an increase in the peak pulse voltage. Several induced voltage pulse characteristics for 227 kV on the center electrode are given by Table 1 for decrease of the pressure from 17 to 1 atm.

Table 1. Variation of right side induced voltage pulse characteristics with air pressure at 227 kV center electrode voltage.

Air pressure (atm)	Rise time (ps)	FWHM (ps)	Peak voltage (kV)	3 dB bandwidth (GHz)
17	2.56	8.98	236.6	26.9
15	2.28	8.38	240.5	23.9
7	1.87	7.57	264.2	39.6
1	1.51	7.12	277.7	42.5

It is also possible to lengthen the rise time and the FWHM of the induced voltage pulse by reducing the applied voltage with the pressure kept constant. This may be thought of as a pulse characteristic degradation by voltage detuning. Table 2 gives the variation of several of the characteristics of the induced voltage pulse as the applied voltage is decreased at roughly constant pressure. The rise time increases from 1.87 ps to 70.1 ps, and the FWHM increases from 7.57 ps to 167 ps. The peak voltage suffers a dramatic reduction from 264 kV to 1.71 kV.

Table 2. Variation of right side induced voltage pulse characteristics at roughly 8 atm air pressure.

Air pressure (atm)	Applied voltage (kV)	Rise time (ps)	FWHM (ps)	Duration (ps)	Peak voltage (kV)	Bandwidth (GHz)
7	227	1.87	7.57	>15.5	264.2	39.6
8	48.8	14.7	28.9	>109	18.5	9.28
8	24.4	70.1	167	>385	1.71	1.71

Pulse Generation in the Megavolt Applied Voltage Range

Computations with the same switch geometry and dimensions predict that induced voltage pulses with similar rise times and FWHMs occur with 3.89 MV on the center electrode and 700 atm of pressure. Figure 15 shows the right side induced voltage waveform as the solid curve for uniform laser-induced electron generation under the center electrode. Once again, the total number of induced electrons is 1.7×10^{13} electrons per 1 m depth of the switch. The rise time is 2.04 ps, the FWHM is 7.88 ps, and the duration is greater than 20 ps. The waveform may be expected to fall off beyond 29.5 ps in a manner similar to Fig. 12. The peak pulse voltage is 4.77 MV, which represents a 22.6% enhancement over the applied voltage. By 29.5 ps, 71.4% of the initial electrostatic energy of 422 J/m has crossed the open boundaries. Under the assumption of an abrupt drop of the voltage pulse to zero at 29.5 ps, the 3 dB bandwidth is 29.8 GHz. This is undoubtedly an overstatement since the pulse will not drop abruptly.

With the concentration of the initial, laser-induced electrons in a more narrow column, which bridges the center electrode and the lower plate, a higher peak induced voltage occurs. When 1.2×10^{12} electrons uniformly fill a 50 micron-wide column, centered under the center electrode, the dashed voltage waveform in Fig. 15 occurs. The rise time is 1.72 ps, and the FWHM is 7.39 ps. The peak pulse voltage is 5.38 MV, which is a 38.3% enhancement of the applied voltage. In this case, 82.9% of the initial 422 J/m of initial energy is transported out of the computational space by 29.5 ps. The bandwidth is 30.0 GHz.

The actual experimental realization of pulses under such megavolt conditions may be very difficult. Electrode charging to these voltages without premature breakdown may be extremely challenging.

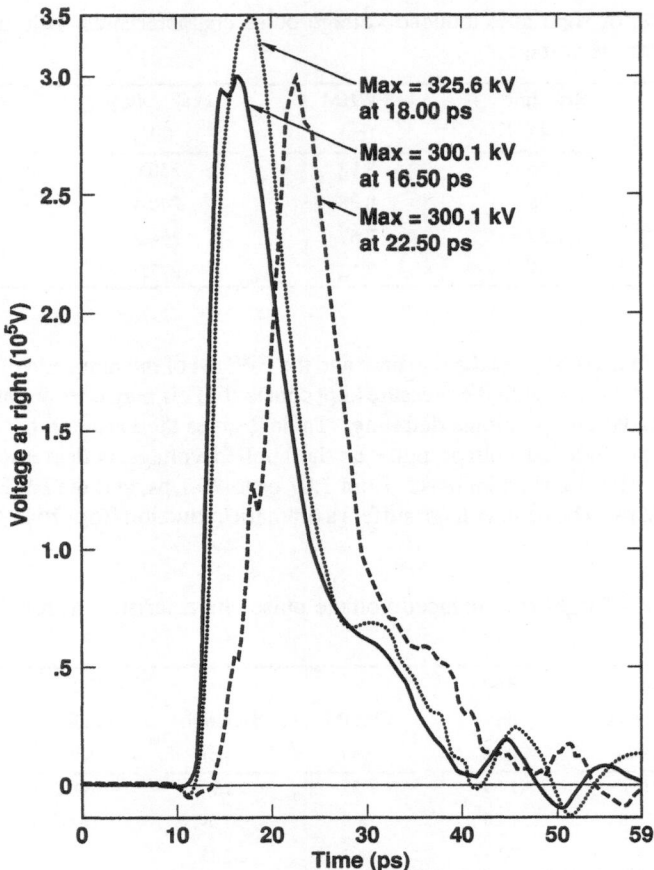

Figure 16. Voltage waveforms at the right boundary for 292.4 kV and 27.2 atm for variation of the initial electron number. The solid curve is for 1.2×10^{13} electrons/m uniformly distributed under the center electrode. The dashed curve is for 4.2×10^{2} electrons/m uniformly distributed under the center electrode. The dotted curve is for 1.2×10^{12} electrons/m concentrated into a 50 micron-wide column, centered under the center electrode.

Reduction of the Initial Electron Number

In the event of constraints on the ability to generate large numbers of initial electrons with a laser system, the effect of a reduction of the initial number of electrons in the gap between the center electrode and the lower plate becomes important. Simulations with uniform electron generation under the whole width of the center electrode and a reduction of the electron number from 1.7×10^{13}/m (4.04×10^{19} m^{-3}) to 4.20×10^{2}/m (10^{9} m^{-3}) show a surprisingly small difference in the induced voltage waveforms.

Figure 16 shows a waveform for 292 kV on the center electrode, 27.2 atm of pressure, and 1.7×10^{13} electrons/m as the solid curve. Superimposed on it as the dashed curve is the induced voltage waveform with the electron number reduced to 4.2×10^{2}. Compared to the solid waveform, the dashed waveform is delayed in reaching its peak at 22.5 ps. The solid waveform reaches its ultimate peak at 16.5 ps and has a cleaner rise to its first peak. The waveform for the lesser number of initial electrons is bumpier and has a more pronounced negative precursor. Surprisingly, it has a somewhat smaller FWHM at 8.67 ps versus 9.11 ps for the waveform for the larger number of initial electrons. Somewhat surprisingly, the pulse amplitudes are equal at 300 kV. The pulse duration from the precursor to the first zero crossing is 44.2 ps for the lesser electron number; whereas, it is 37.9 ps for the greater electron number. The 3 dB bandwidth for the lesser electron number is 21.0 GHz; whereas, for the greater electron number, it is 22.0 GHz.

Table 3 gives several of the relevant induced right side pulse characteristics for reduction of the initial electron number. The increase in delay to the peak of the induced voltage pulse with a decrease in the initial induced electron number is expected. With a lower initial electron number, the delay time or formative time for the electron density in the avalanche to affect the applied voltage becomes longer. Because the avalanche is somewhat slower for a lower initial electron number, the pulse rise time and duration should be longer. The bandwidth should therefore be more narrow. Because of the slower pulse, the peak voltage should be lower, but computationally, this is not the case. Apparently, the dynamics of the wave interaction around the center electrode cause the peak pulse voltage to vary slightly with a decrease in initial electron number. The same is true of the FWHM.

Table 3. Induced voltage pulse characteristics for reduction of initial electron number at 292 kV and 27.2 atm.

Initial electron number/m	Rise time (ps)	FWHM (ps)	Duration (ps)	Time to peak (ps)	3 dB bandwidth (GHz)	Peak value (kV)
1.7×10^{13}	2.41	9.11	37.9	16.5	22.0	300.1
4.2×10^{5}	5.07	8.60	42.3	20.0	21.0	301.8
4.2×10^{4}	5.47	8.13	42.8	20.5	21.0	295.7
4.2×10^{3}	5.87	8.13	43.3	22.0	21.0	298.9
4.2×10^{2}	6.00	8.67	44.2	22.5	21.0	300.1

Concentration of the Initial Electrons into a Narrow Column

When the initial electrons are concentrated uniformly into a 50 micron-wide column, which bridges the center electrode and the lower plate, an induced voltage pulse with a somewhat higher amplitude and slightly altered characteristics results. The simulations with a narrow initial electron column were all done with the magnetic forces included in the momentum equations. Comparisons of calculations at 292 kV and 27.2 atm with and without the magnetic forces, have shown no discernible differences in the induced voltage pulses. When 1.21×10^{12} electrons/m (4.04×10^{17} m^{-3}) were uniformly concentrated into 50 microns of width under the center electrode, a peak induced voltage of 325.6 kV occurred at the right side of the transmission line. This pulse had one peak, which occurred at 18.0 ps, as compared to the pulse for the wider initial electron distribution, which had two peaks and reached its maximum of 300.1 kV at 16.5 ps. The peak voltage for the narrow electron distribution is 8.5% higher than that for the wide distribution. This voltage waveform is shown in Fig. 16 as the dotted curve. It has a slight negative precursor, a more narrow FWHM at 8.75 ps, a bumpier tail, and a more rapidly occuring first zero crossing at 30.6 ps from the start of the negative precursor. This pulse has a bandwidth of 20.06 GHz under the assumption that the voltage drops abruptly to zero at 59.0 ps.

Table 4 shows a number of characteristics of the induced voltage pulse at the right side as the electron number per meter of switch depth decreases from 1.21×10^{12} to 30. The pulse rise time and the time to the voltage peak tend to increase with decreasing electron number. The pulse duration from the beginning of the negative precursor to the first zero crossing after the main pulse and the peak voltage increase. The FWHM decreases and the bandwidth increases as the initial electron number decreases.

Table 4. Induced voltage pulse characteristics for reduction of initial electron number in a 50 micron-wide centered column at 292 kV and 27.2 atm.

Initial electron number/m	Rise time (ps)	FWHM (ps)	Duration (ps)	Time to peak (ps)	3 dB bandwidth (GHz)	Peak value (kV)
1.2×10^{12}	3.00	8.75	30.6	18.0	20.0	325.6
3.0×10^{4}	3.13	8.50	33.4	22.0	21.0	336.7
3.0×10^{3}	3.25	8.31	33.8	22.5	21.0	336.3
3.0×10^{2}	3.50	8.25	34.4	21.5	21.0	337.6
3.0×10^{1}	3.44	8.13	34.8	22.0	22.0	341.6

Asymmetric Illumination of a Narrow Column

When the narrow column of initial electrons is centered under the center electrode, the induced voltage pulses at the left and right sides of the transmission are very similar. When the narrow column is moved to the extreme left side of the center electrode, the two induced voltage pulses are still generally similar, but differ more in detail. The greatest difference occurs in the peak pulse voltage. The pulse at the left has a lower peak value. For the cases of initial electron number considered, it is 10.6–19.9% lower than the pulse at the right. The difference is greatest for the greatest initial electron number, 1.21×10^{12} electrons/m. As expected due to the difference in transit time, the pulse at the left reaches its peak sooner than the pulse at the right.

Table 5 gives a number of characteristics of the induced voltage pulses at the left and right at 292 kV and 27.2 atm of pressure as the initial electron number decreases. The left side pulse always has a wider FWHM than the right side pulse and a shorter duration, as measured from the beginning of the negative precursor until the first zero crossing after the main positive pulse. As the initial electron number decreases, the rise time increases, the time to the peak value increases, and the main pulse duration increases. The length of the negative precursor increases, and the pulses become bumpier around the trailing part of the tail before the first zero crossing. It is noteworthy, that the gross pulse characteristics, such as the peak values, the rise times, and the FWHMs, do not greatly change as the initial electron number decreases by about eleven orders of magnitude.

The 3 dB bandwidth at the right side is calculated for the entire voltage waveform with assumption of an abrupt rise or drop to zero at 59 ps. Comparison with Table 4 shows that the asymmetric illumination of the lower gap causes the pulse bandwidth to be slightly degraded from the case of symmetric illumination. The trend toward a modest rise in bandwidth with decreasing initial electron number is, however, maintained.

GENERATION OF SHORTER PULSES BY DUAL GAP INITIATION

Reference to Table 3 shows that it is possible to reduce the induced voltage pulse FWHM by reducing the air pressure at fixed voltage on the center electrode. This reduction in pulse FWHM appears to be modest, at least, for 227 kV of applied voltage. When the pressure is reduced from 17 atm to 1 atm, the FWHM is reduced by only 20.7%. It seems possible to reduce the FWHM without changing the pressure or decreasing the critical dimensions of the switch electrodes.

The pulse FWHM can be strongly reduced by triggering the upper gap slightly after triggering the lower gap with a laser pulse. If the avalanches in the two gaps are roughly similar, a roughly equal amplitude, negative polarity electromagnetic wave arises from the upper gap. With no time delay between initiation of avalanches in the two gaps, the two opposite polarity

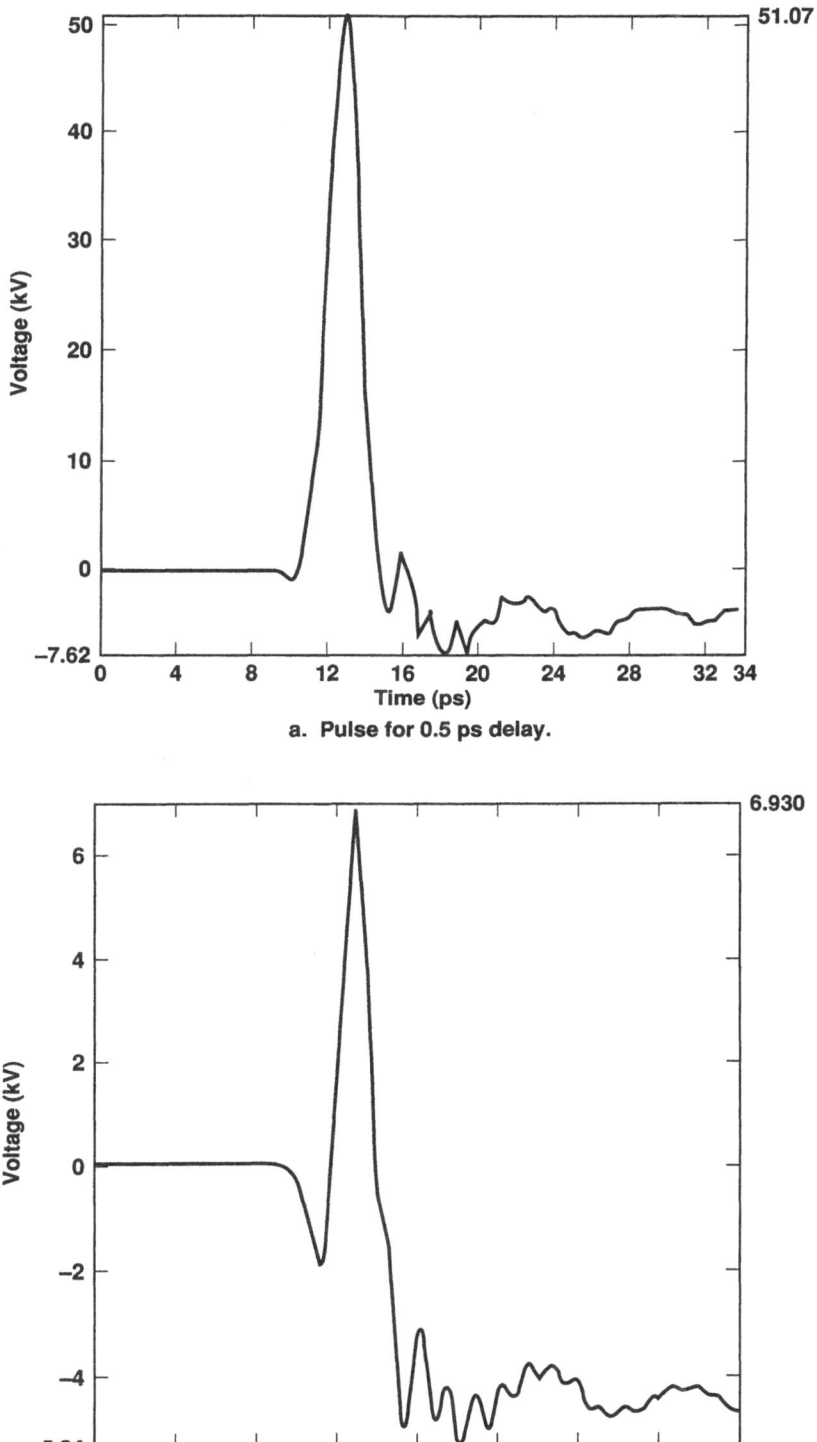

a. Pulse for 0.5 ps delay.

b. Pulse for 0.1 ps delay.

Figure 17. Voltage waveforms at the right boundary generated by wave interference at 227 kV and 15 atm.

229

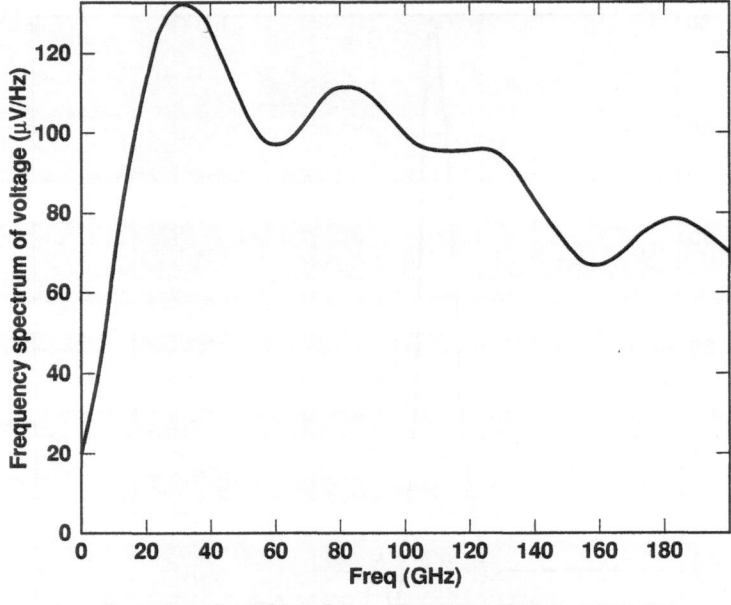

a. Spectrum for 0.5 ps delay. The bandwidth is 113.3 GHz.

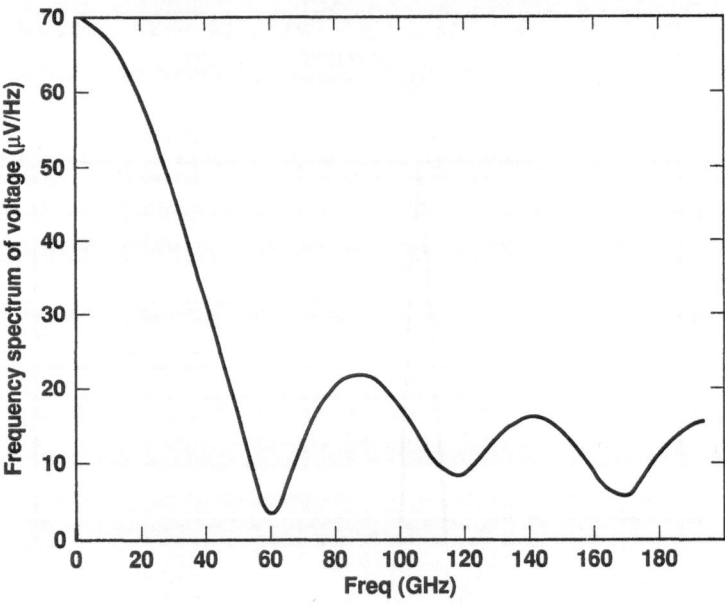

b. Spectrum for 0.1 ps delay. The bandwidth is 28.09 GHz.

Figure 18. Voltage frequency spectra at the right boundary generated by wave interference at 227 kV and 15 atm.

electromagnetic waves from the two gaps should almost completely cancel each other at the open boundaries. This cancellation or destructive interference should produce virtually no induced voltage pulses at the boundaries. With the avalanche initiation in the upper gap delayed somewhat from initiation in the lower gap, the wave from the upper gap should cancel only the tail of the wave from the lower gap. The width of the induced voltage should thus be reduced by a process of electromagnetic wave interference.

230

Table 5. Induced voltage pulse characteristics for reduction of initial electron number in a 50 micron-wide column under the extreme left side of the center electrode at 292 kV and 27.2 atm.

			Left Side			
						% Difference from right
Initial electron	Rise time	FWHM	Duration	Time to peak	Peak value	side peak voltage
number/m	(ps)	(ps)	(ps)	(ps)	(kV)	(%)
1.21×10^{12}	2.69	11.38	33.9	17.5	257.5	−19.9
3.00×10^{4}	4.69	9.50	36.7	19.5	291.9	−13.1
3.00×10^{3}	4.94	9.50	37.0	20.0	296.9	−12.2
3.00×10^{2}	5.13	9.50	37.6	20.5	300.0	−12.3
3.00×10^{1}	5.13	9.69	38.3	21.0	299.6	−10.6
			Right Side			
Initial electron	Rise time	FWHM	Duration	Time to peak	Peak value	3 dB bandwidth
number/m	(ps)	(ps)	(ps)	(ps)	(kV)	(GHz)
1.21×10^{12}	2.94	8.63	34.3	18.5	321.3	18.1
3.00×10^{4}	4.38	7.81	38.6	22.0	336.0	18.1
3.00×10^{3}	4.88	7.56	39.5	22.5	338.2	18.1
3.00×10^{2}	5.19	7.63	40.0	23.0	336.9	19.0
3.00×10^{1}	5.31	7.75	40.8	23.0	335.2	19.0

Figure 17 shows two induced voltage pulses calculated at the right side of the switch for two different delays in laser initiation of the upper and lower gaps. Equal initial electron numbers (1.7×10^{13} electrons/m) uniformly fill each gap. The pressure is 15 atm. The center electrode voltage is 227 kV. Figure 17a shows the voltage waveform when the delay is 0.5 ps. The narrow main pulse has an amplitude of 51.1 kV and a FWHM of 1.79 ps. This waveform has a slight negative precursor and a noisy tail with an amplitude of about 7.6 kV. This tail amplitude amounts to about 15% of the main pulse amplitude. The pulse amplitude is about 22.5% of the applied voltage. Thus, the price of a shorter pulse by this interference process appears to be a strong reduction in pulse amplitude. Figure 18a shows the frequency spectrum of the waveform in Fig. 17a. The 3 dB bandwidth is 113.3 GHz. This bandwidth may be fortuitously wide because the upper 0.707 spectral amplitude value falls just below the spectral valleys at about 60 and 100 GHz.

Figure 17b shows the waveform for a delay of 0.1 ps. The stronger interference of the two electromagnetic waves produces a main positive pulse with a lower peak value of 6.93 kV and a reduced FWHM of 1.14 ps. Unfortunately, in addition to the reduced pulse width, a much larger negative precursor and a relatively larger amount of trailing pulse occur. The trailing negative pulse has an amplitude of about 5.3 kV or 76.5% of the peak value of the desired narrow positive pulse. The peak value of the narrow pulse is 3.05% of the applied voltage. Figure 18b shows the frequency spectrum of this voltage waveform. The 3 dB bandwidth is 28.09 GHz. Notice that the spectral valleys at about 20, 60, and 160 GHz in Fig. 18a have been dramatically lowered in Fig. 18b. If the trailing signal in Fig. 17b were strongly attenuated, the pulse bandwidth should be greatly increased.

CONCLUSIONS

The simple, laser-initiated, air avalanche switch model computationally shows promise for the generation of high-voltage (50 kV–4 MV), broadband (~20 GHz) electromagnetic pulses of picosecond-scale rise time and duration. With the same physical structure, similar pulses occur for the ratio of the nominal peak electric field to pressure roughly constant. The induced voltage pulse characteristics may be varied by changing the applied voltage at constant pressure or by changing the pressure at constant voltage. The induced voltage pulse characteristics are not greatly affected by the reduction of the initial electron density over eleven orders of magnitude. The induced voltage pulse characteristics are likewise not grossly affected by asymmetric avalanche initiation in the lower air gap.

With the physical dimensions and pressure fixed, the FWHM of the induced voltage pulse can be shortened by a factor of 4.7 or greater by a wave interference technique. This technique involves dual initiation of the upper and lower air gaps with a slight time delay.

REFERENCES

1. F. Villa, "*High Gradient Linac Prototype: A Modest Proposal*," SLAC-PUB-3875, Jan. (1986).
2. R. E. Cassell and F. Villa, "*High Speed Switching in Gases*," SLAC-PUB-4858, Feb. (1989).
3. P. Pincosy and P. Poulsen, Lawrence Livermore National Laboratory, personal communication, Aug. (1991).
4. D. J. Mayhall, J. H. Yee, R. A. Alvarez, and D. P. Byrne, "Two-Dimensional Electron Fluid Computer Calculations of Nanosecond Pulse, Low Pressure Microwave Air Breakdown in a Rectangular Waveguide and their Verification by Experimental Measurement," *Laser Interaction and Related Plasma Phenomena*, H. Hora and G. H. Miley, eds., 8, Plenum Press, New York, 121–137 (1988).
5. G. E. Sieger, J. H. Yee, and D. J. Mayhall, "Computer Simulation of Nonlinear Coupling of High-Power Microwaves with Slots," *IEEE Trans. Plasma Science*, 17, no. 4, 616-621, Aug. (1989).
6. D. J. Mayhall, J. H. Yee, G. E. Sieger, and R. A. Alvarez, "Two-Dimensional Calculation of Sequential Electron Layer Formation by Crossed Microwave Beams in Air at Low Pressure," *Laser Interaction and Related Plasma Phenomena*, H. Hora and G. H. Miley, eds., 9, Plenum Press, New York, 515–524 (1991).
7. J. H. Yee, R. A. Alvarez, D. J. Mayhall, D. P. Byrne, and J. DeGroot, "Theory of Short, Intense Electromagnetic Pulse Propagation through the Atmosphere," *Phys. Fluids*, 29, no. 4, 1238–1244, April (1986).
8. J. H. Yee, D. J. Mayhall, and R. A. Alvarez, "Physical Phenomena Induced by Passage of Intense Electromagnetic Pulses (including CO_2 Lasers) through the Atmosphere," *Laser Interaction and Related Plasma Phenomena*, H. Hora and G. H. Miley, eds., 7, Plenum Press, New York, 901–913 (1986).
9. G. E. Sieger, D. J. Mayhall, and J. H. Yee, "Numerical Simulation of the Propagation and Absorption Due to Air Breakdown of Long Microwave Pulses," *Laser Interaction and Related Plasma Phenomena*, H. Hora and G. H. Miley, eds., 8, Plenum Press, New York, 139–148 (1988).
10. K. S. Yee, "Numerical Solution of Initial Boundary Value Problems Involving Maxwell's Equations in Isotropic Media," *IEEE Trans. Ant. and Propagat.*, AP-14, no. 3, 302–307, May (1966).
11. A. C. Hindmarsh, "*Preliminary Documentation of GEARBI: Solution of ODE Systems with Block-Iterative Treatment of the Jacobian*," UCID-30149, Preprint, Lawrence Livermore National Laboratory, Livermore, CA (1976).
12. R. D. Richtmyer and K. W. Morton, *Difference Methods for Initial-Value Problems*, 2nd Ed., Interscience Publishers, New York, 263 (1967).
13. L. B. Loeb, *Basic Processes of Gaseous Electronics*, University of California Press, Berkeley and Los Angeles, CA, 231 (1960).
14. L. B. Loeb, *Basic Processes of Gaseous Electronics*, University of California Press, Berkeley and Los Angles, CA, 676 (1960).

TWO-DIMENSIONAL CALCULATION OF HIGH-POWER
MICROWAVE BANDWIDTH BROADENING BY
LASER-INDUCED AIR BREAKDOWN[*]

D. J. Mayhall, J. H. Yee, and R. A. Alvarez

Lawrence Livermore National Laboratory
University of California
P. O. Box 808, Mail Code L-156
Livermore, CA 94550

INTRODUCTION

Wideband, high-power microwave pulses are expected to have a number of important future applications. One of these is wideband radar. The wide bandwidth should yield increased information for target characterization and identification.[1] The high power should yield increased target range for conventional objects. It should also make possible the detection of previously undetectable, reduced-signature objects.

One convenient way to obtain wideband, high-power microwave pulses with conventional technology is by the bandwidth broadening from air breakdown tail erosion. In this process, the tails of short (3–10 ns), high-amplitude (>1 MV/m) pulses are removed by passage through a low-pressure air breakdown cell. This tail erosion shortens the pulses and introduces high-frequency components, thereby producing transmitted pulses with broader bandwidths than the incident pulses. The air pressure of the cell must be matched to the characteristics of the incident pulse so that sufficient breakdown occurs to cause severe tail erosion. The relevant characteristics of the incident pulse are amplitude, frequency, pulse length, and pulse shape. The effects of these characteristics on tail erosion of microwave pulses in the earth's atmosphere have been examined in one-dimensional computer calculations.[2-4] Two-dimensional computer calculations of tail erosion in a rectangular waveguide have been experimentally verified.[5]

We have experimentally demonstrated that air breakdown tail erosion broadens the 3 dB bandwidths of 8 ns long, 2.8608 GHz pulses in a WR-284 rectangular waveguide at 3.5 torr pressure. The amplitude of the incident pulses varied over a modest range. The pulse bandwidth was broadened from 0.1466 GHz by 0.0097–0.039 GHz. The incident bandwidth, generated with a SLAC (Stanford Linear Accelerator Center) klystron and a pulse compression cavity,[6,7,9] was 5.12% relative to the incident carrier frequency of 2.8608 GHz. This relative bandwidth is already several times the 1–1.5% peak value, which is characteristic of most present integral

---

[*] Work performed under the auspices of the U.S. Department of Energy by Lawrence Livermore National Laboratory under Contract W-7405-Eng-48.

Laser Interaction and Related Plasma Phenomena, Vol. 10
Edited by G.H. Miley and H. Hora, Plenum Press, New York, 1992

233

high-power microwave sources.[8] The measured transmitted relative bandwidth from this modest tail erosion is 5.46–6.49%, an increase in relative bandwidth of 0.34–1.37% over the incident bandwidth.

The experiment was simulated with a two-dimensional, electromagnetic, electron fluid computer code for avalanche ionization in air. The simulation predicts 3 dB bandwidth broadening from 0.0293–0.127 GHz for a peak background electron density of 10 electrons/cm^3. The predicted relative bandwidth of the transmitted pulse is thus 6.15–9.56% or 12.5–47.4% greater than the measured value. Because the experimental detectors were not fast enough to resolve the individual rf cycles and hence the local frequency across the pulses, we presently believe the values from the simulation give better estimates of the true values than do the measured values.

In order to obtain estimates of the magnitude of the transmitted bandwidths possible, the computer code is also used to predict the bandwidth broadening of a specific, idealized incident pulse shape from the pulse compression cavity. Incident pulse amplitudes from 1–17.5 MV/m and peak background electron densities from 10–10^{11} electrons/cm^3 are considered. We assume that these electron densities can be rapidly generated with a properly configured focusing system for a pulsed laser beam. The incident bandwidth is 0.342 GHz or 12.0% relative. The predicted transmitted bandwidth varies from 0.352–3.214 GHz, that is, from 12.3–112% relative. The predicted transmitted spectral center frequency varies from 2.87–4.72 GHz. The transmitted pulse amplitude at 54.6 cm from the input port of the low-pressure waveguide section varies from 0.263–7.94 MV/m on the waveguide center line.

EXPERIMENTAL DEMONSTRATION OF BANDWIDTH BROADENING BY LOW-PRESSURE AIR BREAKDOWN

Experimental Configuration

The experimental demonstration of bandwidth broadening by low-pressure air breakdown was performed at the Lawrence Livermore National Laboratory 100-MeV linac, which has the nominal output pulse characteristics of 20 MW peak power, 2.856 GHz frequency, and 2 μs maximum width. At the beginning of the microwave pulse shaping, the output from a very stable, continuous wave master oscillator was gated to a pulse length of 100 ns with a Hewlett

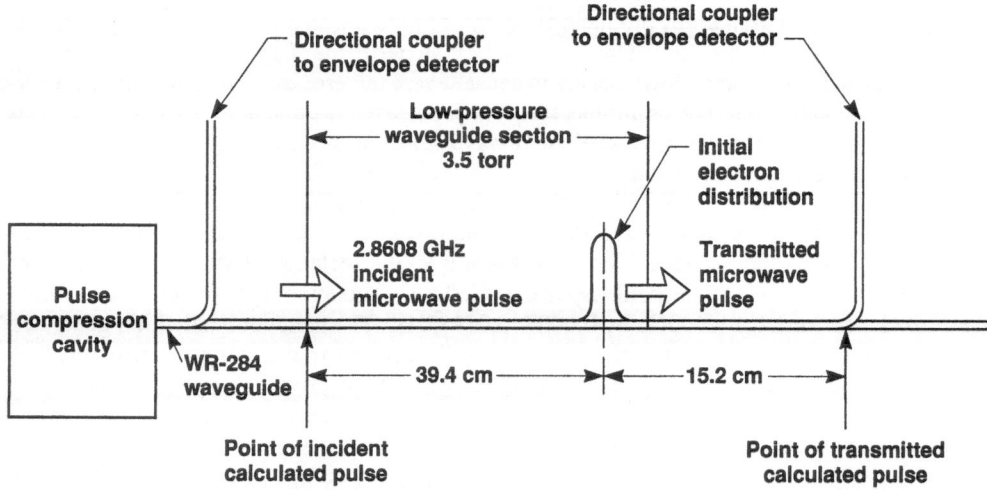

Figure 1. Simplified experimental arrangement.

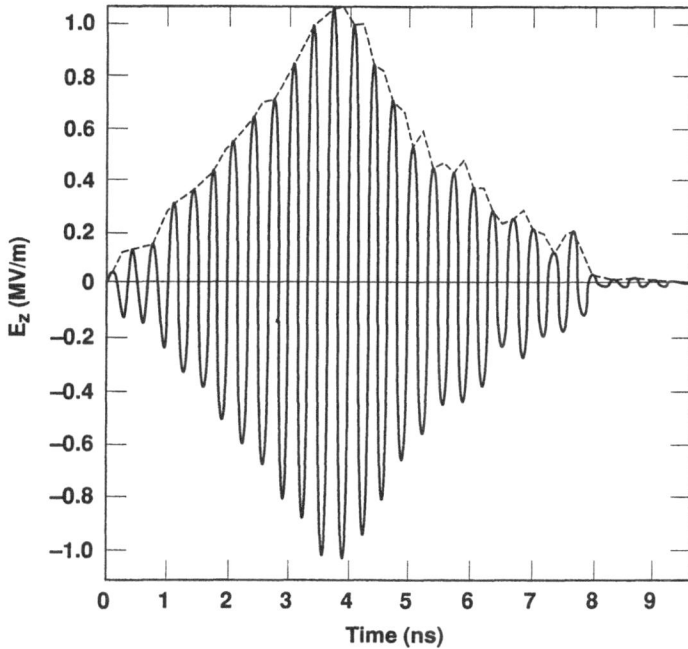

a. Measured incident envelope times sin $(2\pi \times 2.8608 \times 10^9 \ t)$.
Max input E_z field = 1.057 MV/m; min E_z field = −1.119 MV/m.

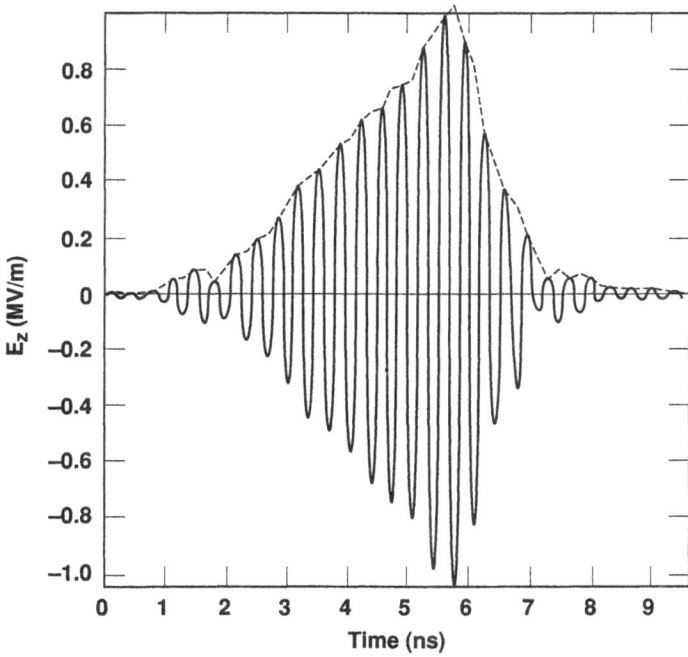

b. Measured transmitted envelope times sin $(2\pi \times 2.8608 \times 10^9 \ t)$.
Max transmitted E_z field = 0.9984 MV/m; min E_z field = −1.043 MV/m.

Figure 2. Estimated measured pulses. Electric field at 0.51 cm off waveguide center line.
1.164 MV/m incident amplitude on waveguide center line.
The dashed curves are the experimental envelopes.

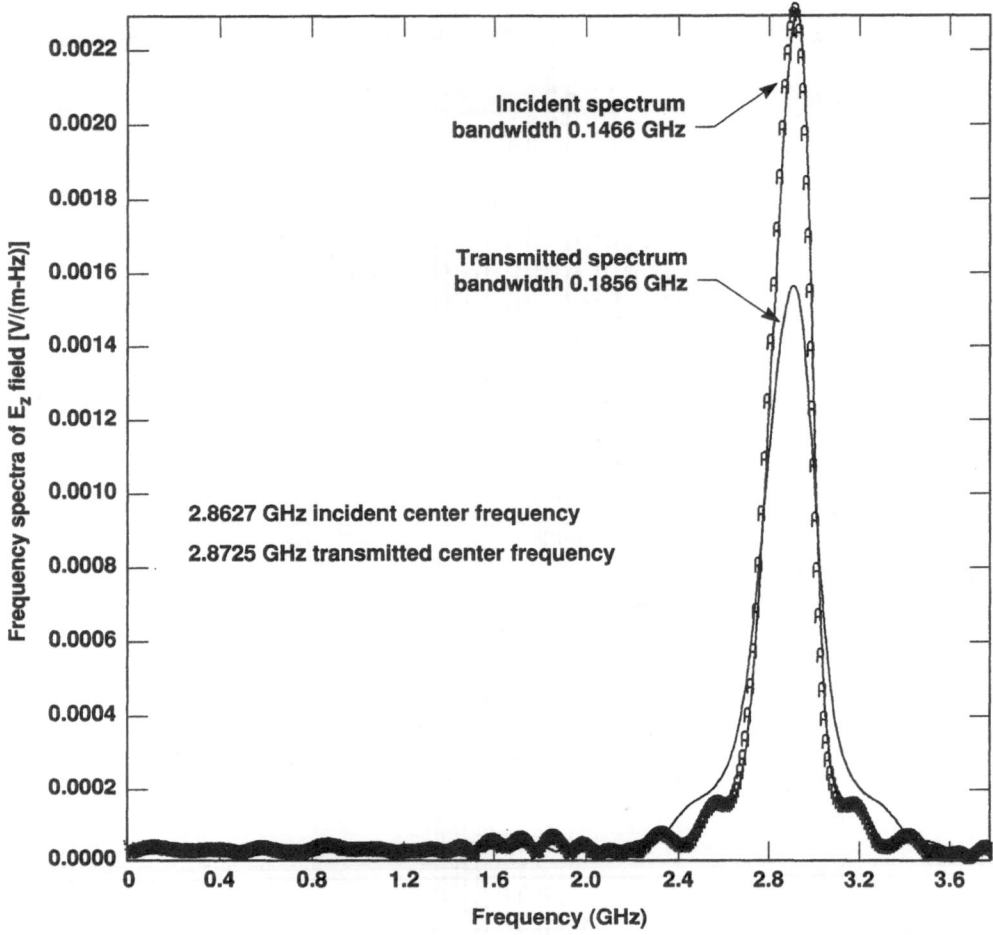

Figure 3. Estimated measured incident and transmitted frequency spectra. 1.164 MV/m incident amplitude on waveguide center line.

Packard 8732A PIN diode to supply a 500-mW drive pulse to a SLAC type driver klystron. This klystron generated a pulse at 200 ns and 20 kW for an RCA 8568 klystron. The power to the 8568 was controlled with a variable attenuator. The peak power to the 8568 was 15 MW. The output pulse from the 8568 passed through a vacuum to high-pressure waveguide window. The waveguide downstream from this window was pressurized to 30 psig with Freon-12 to prevent arcing. For protection of the 8568, the output pulse then traversed a circulator. Beyond this circulator, the output entered a pulse compression cavity.[6,7,9] This cavity delivered pulses with nominal widths of 5 ns, nominal rise times of 2 ns, and peak powers up to 100 MW. The output from the pulse compression cavity passed through another microwave window to a low-pressure, WR-284 rectangular waveguide test section. This waveguide test section was evacuated down to about 10^{-5} torr and backfilled with air at 3.5 torr.

A ^{60}Co gamma ray source was located to cause reproducible preionization of the air in the waveguide section. This generated a stable background electron density in the waveguide for initiation of air breakdown by incident microwave pulses of sufficient amplitude and pulse length. The presence of the background electrons made the microwave air breakdown highly reproducible. Directional couplers, which led to calibrated fast back diodes, were located upstream and downstream from the waveguide section. These detectors allowed measurement of the power envelopes of the incident, transmitted, and reflected pulses into and out of the waveguide section. A simplified diagram of the pulse compression cavity and the waveguide section is shown in Fig. 1. The initial electron distribution occurred at 39.4 cm downstream from the entrance to the waveguide section. At 15.2 cm beyond that point, a directional coupler sampled the transmitted pulse. This experimental arrangement is described in greater detail elsewhere.[5,9]

Experimental Electric Field Envelopes and Approximate Waveforms

Typical experimentally measured incident and transmitted electric field envelopes are shown in Fig. 2. Figure 2a shows an incident electric field envelope as the dashed curve. This envelope is for a position 0.51 cm off the waveguide center line along the waveguide "a" direction. This is in the direction of the long side of the waveguide. This envelope was measured with the directional coupler just upstream from the waveguide section. Since the detector was too slow to record each rf cycle, it recorded the envelope of the instantaneous average power. The electric field envelope along the waveguide center line was extracted from the average power envelope according to the peak electric field to power relation for the waveguide. The electric field envelope at 0.51 cm or one transverse computational grid spacing off the waveguide center line was then obtained by multiplication by the factor $\sin(6\Delta y\pi/14\Delta y) = 0.975$. The quantity Δy is the computational grid spacing in the transverse direction. This amplitude scaling was done so that inferred measured waveforms could be compared to simulation results directly from the computer code.

For a first order approximation to an instantaneous electric field pulse at the incident detector, the electric field envelope was multiplied by the function $f(t) = \sin(2\pi \cdot 2.8608 \cdot 10^9 t)$, where t is the time in seconds. The resulting waveform is not the true waveform since the pulse compression cavity imparts a chirp to the output pulse.[9] In addition, waveguide dispersion causes the high-frequency, high-velocity spectral components in the pulse to travel faster than the low-frequency, low-velocity spectral components as the pulse moves from the pulse compression cavity to the detector. This dispersion causes a sufficiently short pulse to spread out in time and thereby suffer a decrease in amplitude in travel down the waveguide. The local frequency across a pulse varies as the pulse travels down the waveguide. This frequency variation could not be determined because of the slowness of the detector response. It was therefore unknown. In addition, the phase angle of the first rf cycle was unknown and was assumed to be zero.

Figure 2b shows a transmitted experimental electric field envelope as the dashed curve. This envelope is for the electric field at 0.51 cm off the waveguide center line. The corresponding inferred electric field waveform is shown as the solid curve. The previously described extraction procedure was used to obtain this waveform from the experimental average power envelope. Figure 2b shows that some of the tail of the incident pulse has been removed or eroded by air breakdown. Also, the front of the transmitted pulse is longer than the front of the incident pulse. This lengthening is most likely a result of waveguide dispersion. The transmitted amplitude at 1.043 MV/m is lower than the incident amplitude at 1.119 MV/m. Some of this decrease in amplitude is due to waveguide dispersion, and some may be due to erosion of the pulse peak by air breakdown.

Experimental Bandwidth Broadening

The frequency spectra of the estimated measured electric field waveforms in Fig. 2 are shown Fig. 3. The top curve, labelled with the letter A, is the incident spectrum. It has a 3 dB bandwidth of 0.1466 GHz or 5.12% relative to the incident carrier frequency of 2.8608 GHz. The bandwidth is defined as the spectral width at 3 dB down from the spectral maximum value, that is, at 0.707 times the spectral maximum. Measurements of incident pulses of the type in Fig. 2a from the pulse compression cavity with a directional coupler and a spectrum analyzer give relative bandwidths of 1.3–2.3%. Measurements with a tuned YIG filter give relative bandwidths of about 6%.[10] The value computed from our estimated waveforms thus falls within the typical experimental bounds obtained from two other methods of measurement.

The spectral center frequency of a pulse is defined as the frequency at the spectral maximum. The incident pulse in Fig. 2a has a center frequency at 2.863 GHz.

The lower amplitude, slightly wider curve in Fig. 3 is the transmitted spectrum. It has a bandwidth of 0.1856 GHz or 6.49% relative to the incident carrier. The transmitted center frequency is 2.873 GHz. The air breakdown tail erosion thus causes an upward shift in center frequency of 0.010 GHz. An upward frequency shift is to be expected from tail erosion since the tail of the pulse is reduced and high-frequency spectral components are introduced into the pulse in the tail region.

The experimental results for the bandwidth broadening of three incident pulses are shown in the lower part of Table 1 under the heading, "Estimated measured transmitted bandwidths." These three pulses had very similar inferred electric field envelopes, but different incident amplitudes on the waveguide center line. These center line amplitudes are listed in the lefthand column. The pulses in Fig. 2 are for the incident amplitude of 1.164 MV/m. The transmitted bandwidths vary from 0.1563–0.1856 GHz or from 5.46–6.49% relative. These are modest increases of bandwidth over the incident bandwidth of 5.12%, but they demonstrate that tail erosion from air breakdown in the waveguide section does increase the pulse bandwidth. The experimentally measured transmitted bandwidths thus differ from the incident bandwidth of 0.1466 GHz by 0.0097–0.039 GHz or 0.34–1.4% relative.

Table 1. Summary of experimental 3 dB bandwidth broadening.

Estimated measured incident bandwidth		
Absolute bandwidth (GHz)		Relative bandwidth (%)
0.1466		5.12

Estimated measured transmitted bandwidths		
Amplitude (MV/m)	Absolute bandwidth (GHz)	Relative bandwidth (%)
0.6679	0.1563	5.46
1.022	0.1856	6.49
1.164	0.1856	6.49

COMPUTATIONAL DEMONSTRATION OF BANDWIDTH BROADENING

Computational Arrangement

A two-dimensional, finite difference, electromagnetic, electron fluid computer code for avalanche ionization of low-pressure air due to TE_{mo} modes in rectangular waveguides has been used to simulate bandwidth broadening due to tail erosion. The model for this code has been previously described in detail.[5,9] Three of the governing equations are the Maxwell curl equations for H_x, H_y, and E_z with x and y spatial derivatives. The coordinate x is directed along the waveguide axis, y is in the "a" or long transverse waveguide direction, and z is in the "b" or short transverse waveguide direction. The remaining three governing equations are continuity, momentum, and energy conservation equations for the electron fluid. These fluid equations are the first three moments of the Boltzmann equation under the assumption of a Maxwellian velocity distribution. A number of additional simplifying assumptions have been made. This computer code has been validated for low-pressure air breakdown by previously reported comparisons between measured and calculated electric field pulses.[5,9] The comparisons have varied from good to excellent.

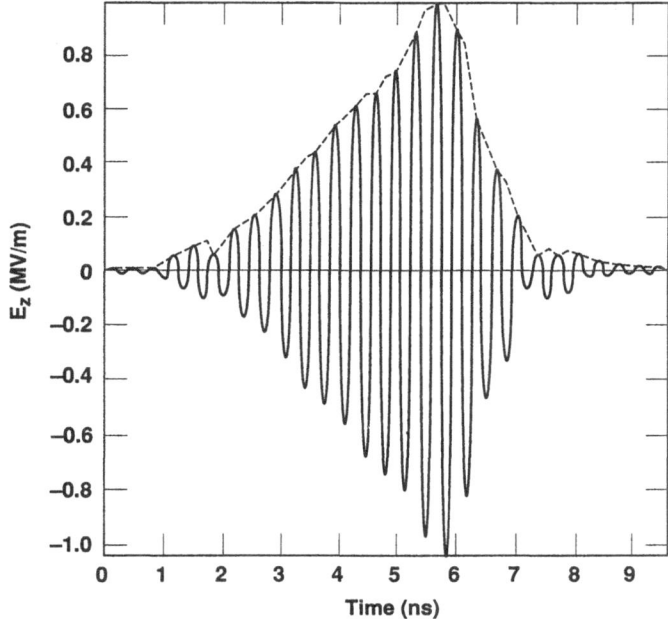

a. Measured transmitted envelope times sin (2 π × 2.8608 × 10⁹ t).
Max transmitted E_z field = 0.9984 MV/m; min E_z field = –1.043 MV/m.

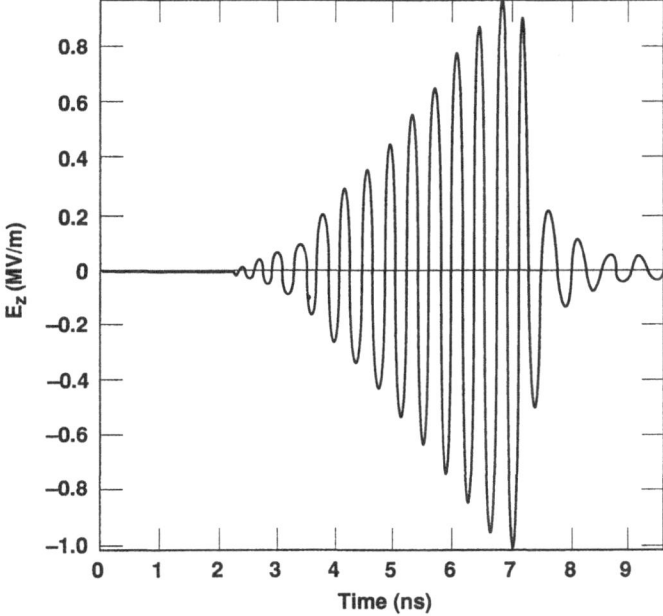

b. Calculated transmitted pulse; 54.6 cm from input.
Max transmitted E_z field = 0.9696 MV/m; min E_z field = –1.010 MV/m.

Figure 4. Transmitted pulses. Electric field at 0.51 cm off waveguide center line. 1.164 MV/m incident amplitude on waveguide center line. The dashed curve is the experimental envelope.

A typical experimental input pulse for simulation with this code is the pulse shown in Fig. 2a. This pulse was discussed in the section entitled *"Experimental Electric Field Envelopes and Approximate Waveforms."* A piecewise linear approximation to this pulse, referred to the waveguide center line by weighting with the function $\sin(y\pi/a)$, was fed into the left side of a two-dimensional, finite difference representation of the experimental, low-pressure, WR-284 waveguide section. The pressure was 3.5 torr. The section was 54.6 cm long. The low-pressure section was followed by a section of waveguide at 1 atm, which extended to 2.54 m from the left side of the grid. An assumed initial electron density distribution was centered at 39.4 cm into the low-pressure waveguide section. Based on photography of the luminous breakdown column,[9] this distribution was assumed to be Gaussian in shape in both the axial and transverse directions. The full width at half maximum (FWHM) was 3 cm in both directions. This value is close to the 2.13 cm diameter previously estimated to occur at 3 torr.[5] The peak initial electron density was assumed to be 10 electrons/cm^3. After encountering the initial electron density distribution, the incident computational pulse generated computational electron avalanching and air breakdown, which resulted in tail erosion. The eroded, transmitted pulse then travelled 15.2 cm on down the waveguide where it was sampled in time to yield the calculated transmitted pulse. Figure 1 indicates the locations of the calculated incident and transmitted pulses.

Computational Electric Field Waveforms

A calculated transmitted electric field pulse at 0.51 cm off the waveguide axis from the simulation is shown in Fig. 4b. The incident amplitude on the waveguide center line is 1.164 MV/m. The measured transmitted pulse for this incident amplitude is shown in Fig. 4a. The tail erosion is somewhat more severe for the calculated pulse than for the measured pulse. The calculated amplitude of 1.010 MV/m is 3.16% lower than the measured amplitude of 1.043 MV/m. This difference is probably due to the slightly stronger tail erosion in the simulation. A lower peak initial electron density in the simulation will probably give less calculated tail erosion and, hence, better agreement in amplitude. The measured pulse also shows more dispersive spreading at the front end than does the calculated pulse.

Computational Bandwidth Broadening

Figure 5 shows the frequency spectra of the calculated incident and transmitted pulses. The upper curve is the incident calculated spectrum, which is identical to the measured incident spectrum in Fig. 3. The lower curve in Fig. 5 is the calculated transmitted spectrum. It has a 3 dB bandwidth of 0.2736 GHz or 9.56% relative to the incident carrier. The calculated transmitted center frequency of 2.873 GHz is the same as the measured transmitted center frequency. The calculated transmitted bandwidth is 47.4% greater than the estimated measured transmitted bandwidth. The main reason for this difference is probably that the approximate analysis of the experimental average power envelope underestimates the transmitted bandwidth. The multiplication of the measured transmitted envelope by the sine function ignores the movement of the high-frequency components toward the front of the pulse during transit from the test section entrance port to the breakdown site and the introduction of more high-frequency components into the tail of the pulse by tail erosion. The multiplication also neglects the movement of these high-frequency components from tail erosion forward from the tail of the pulse in the transit from the breakdown site to the transmitted pulse detector. The multiplication thus neglects the expected, but unknown, variation of the local frequency across the transmitted pulse. Since the detector was too slow to resolve the rf cycles, the high frequencies generated by tail erosion are ignored. The experimental bandwidth is thus likely to be an underestimate. It is believed that the computer simulation picks up both the local frequency redistribution due to waveguide dispersion and the introduction of high-frequency components due to tail erosion. Previous calculations with the code have indicated the introduction of these high-frequency components at the tail of the pulse.[5] Calculated electric

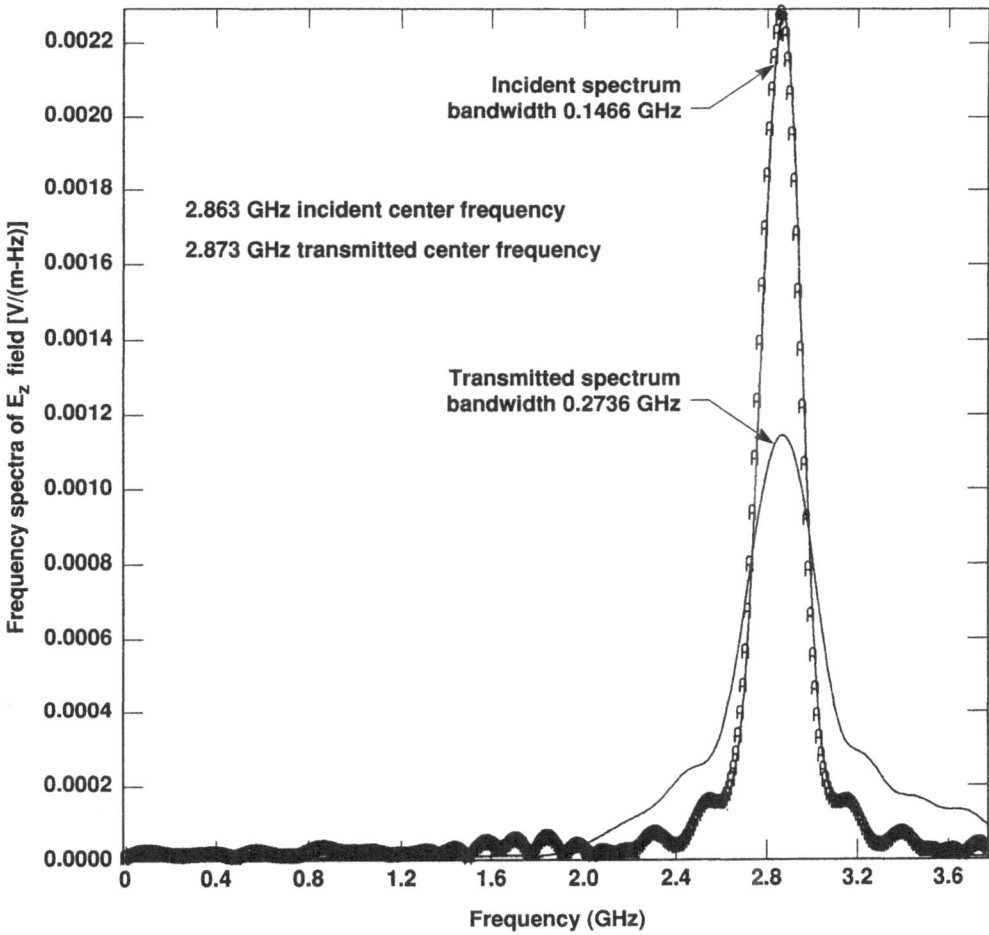

Figure 5. Calculated incident and transmitted frequency spectra. 1.164 MV/m incident amplitude on waveguide center line.

field waveforms for 0.8 torr show such components moving forward through the transmitted pulse and near the front of the reflected pulse. The movement of these high-frequency components causes a transient enhancement of the electric field amplitude over the incident pulse amplitude at 0.8 torr. Enhancements of up to 15% have been calculated.

Comparison with Experimental Bandwidth Broadening

The transmitted bandwidths from the simulation are compared to the estimated measured transmitted bandwidths in Table 2. The experimental bandwidths are given in the second and third columns. The calculated bandwidths are given in the fourth and fifth columns. The reference for the relative bandwidths is the incident carrier frequency. The percent difference between the calculated and measured values is shown in the sixth column. This percent difference becomes greater as the incident amplitude increases and, hence, as the amount of tail erosion increases. This trend of increasing disagreement with increasing incident amplitude constitutes further evidence that the lack of local frequency resolution in the measurements underestimates the bandwidth broadening. The increasing tail erosion with increasing incident amplitude should increasingly shorten the transmitted pulse and broaden the transmitted bandwidth. Because of the previously reported good agreement between measured and calculated electric field envelopes and the ability of the computer code to calculate waveguide dispersion and high-frequency injection, the code is believed to furnish better estimates of bandwidth broadening than is the approximate analysis of the experimental waveform envelopes.

Table 2. Comparison of experimental and computational bandwidth broadening.

Incident computational bandwidth			
Absolute (GHz)		Relative (%)	
0.1466		5.12	

Transmitted bandwidths					
Incident center line amplitude (MV/m)	Estimated measured transmitted bandwidth		Calculated transmitted bandwidth		Percent difference between calculated and measured bandwidths (%)
	Absolute (GHz)	Relative (%)	Absolute (GHz)	Relative (%)	
0.6679	0.1563	5.46	0.1759	6.15	12.5
1.022	0.1856	6.49	0.2540	8.88	36.9
1.164	0.1856	6.49	0.2736	9.56	47.4

PREDICTION OF BANDWIDTH BROADENING OF 3.5 NS PULSES

Computational Conditions

The experimental equipment can produce pulses with a 10–90% rise time of 0.4 ns. For estimation of the bandwidth broadening possible with pulses of this rise time, a number of electron fluid computer simulations have been done with incident pulses, which rise linearly to their peaks in 0.5 ns and then fall linearly in 3.0 ns. The present single pulse compression cavity can produce incident amplitudes to about 10 MV/m. The addition of a second cavity in parallel with the present one would boost the peak available amplitude to about 20 MV/m. Therefore, amplitudes up to 17.5 MV/m are considered. The computational conditions are, for the most part, the same as those for the previously discussed experimental simulations. The peak of the spatially Gaussian background electron density distribution varies from 10–10^{11} electrons/cm^3. In addition, transverse FWHMs of 10^4 cm are considered for several cases.

We assume that these background electron density distributions can be generated rapidly with a properly designed focusing system for pulsed laser beams. No effort has been made to design such a focusing system or to model the deposition of the electrons by the laser beam. The deposition of the background electrons is assumed to be instantaneous compared to the nanosecond time scale of the incident microwave pulse. The carrier frequency of the incident pulse is 2.8608 GHz. The incident 3 dB bandwidth is 0.342 GHz or 12.0% relative. An incident pulse is shown in Fig. 6a for an incident amplitude of 10.0 MV/m on the waveguide center line. The peak value of this pulse in Fig. 6a is 9.453 MV/m because the pulse is calculated at 0.51 cm transversely off the center line.

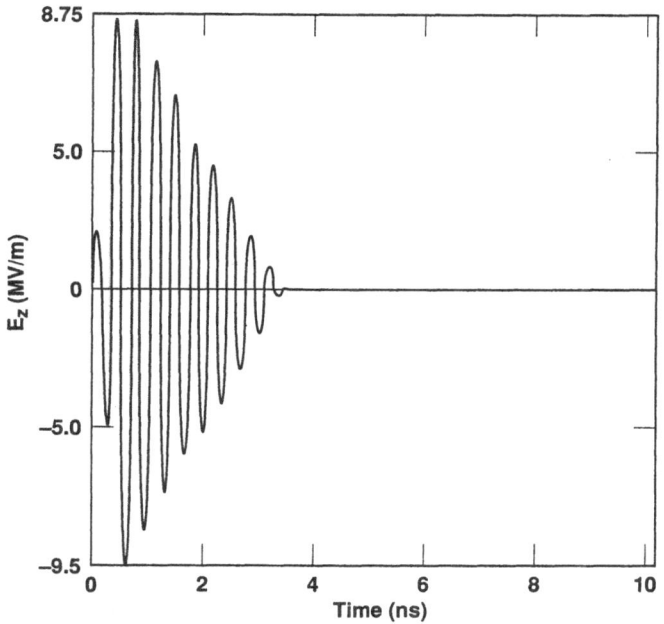

a. Incident pulse.
   Max input E_z field = 8.753 MV/m; min E_z field = –9.453 MV/m.

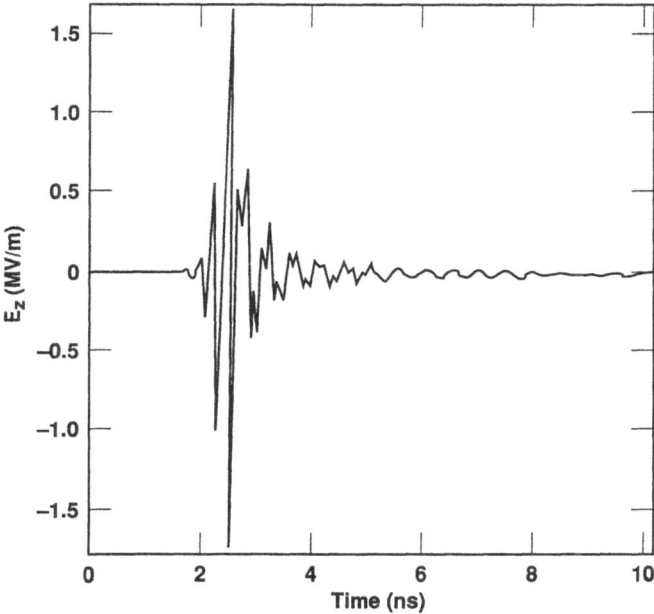

b. Transmitted pulse.
   Max transmitted E_z field = 1.671 MV/m; min E_z field = –1.774 MV/m.
   Pulse occurs at 54.6 cm from the input port.
   Initial electron density distribution: peak value = 10^9 electrons/cm^3,
   axial FWHM = 3 cm, transverse FWHM = 3 cm.

Figure 6. Calculated fast rise time pulses. Electric field at 0.51 cm off waveguide center line. 10 MV/m incident amplitude on waveguide center line.

Computational Results

The transmitted pulse for a center line amplitude of 10.0 MV/m, electron density distribution FHWMs of 3 cm in both the axial and transverse directions, and a peak background electron density of 10^9 electrons/cm^3 is shown in Fig. 6b. Strong tail erosion has occurred. Some dispersive spreading of the nose of the pulse is evident. It also appears that higher frequencies are present in the front of the pulse, while lower frequencies are present in the tail of this pulse. Low-amplitude, high-frequency oscillations are superimposed on the low frequencies in the tail of the pulse. The transmitted pulse amplitude is 1.774 MV/m at one transverse grid spacing or 0.51 cm off the waveguide center line. Multiplication of this amplitude by $0.975^{-1} = 1.026$ gives an amplitude of 1.820 MV/m on the waveguide center line at 54.6 cm downstream from the input port of the waveguide section.

Figure 7 shows a plot of the distribution of the local frequency across this pulse. This distribution is calculated from a digitization of the waveform in Fig. 6b. A linear interpolation algorithm is used to estimate the time of the zero crossings from the electric field-time data pairs. The high-frequency noise in the tail of the pulse is ignored. The frequency at the front of the pulse reaches 4.81 GHz. After about a nanosecond, the frequency falls to about 2 GHz and proceeds to decrease slowly in the long, dispersed tail of the pulse.

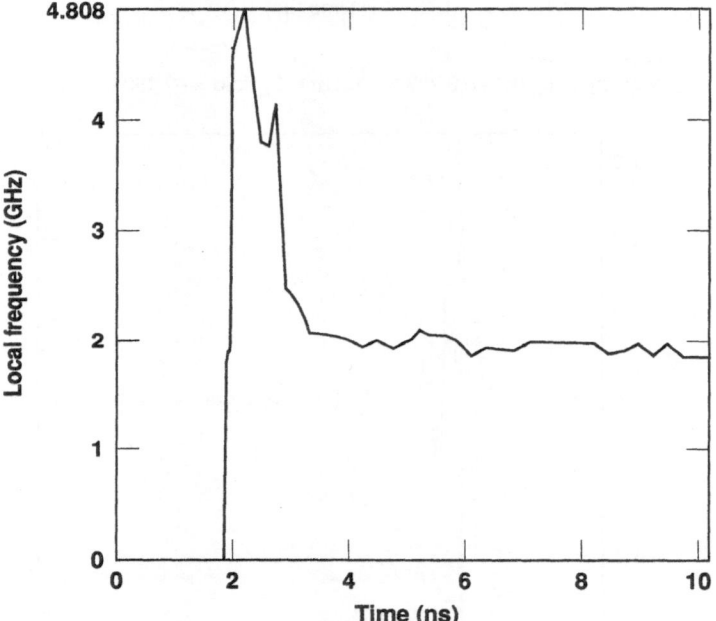

Figure 7. Local frequency distribution across the transmitted pulse.
The E_z waveform is shown in Figure 6b.
Initial electron density distribution:
peak value = 10^9 electron/cm^3, axial FWHM = 3 cm,
transverse FWHM = 3 cm.

The incident and transmitted frequency spectra for this simulation are shown in Fig. 8. The incident spectrum is the upper curve. The transmitted spectrum, which is the lower, much broader curve, has a bandwidth of 1.886 GHz or 65.9% relative. The transmitted bandwidth is thus 5.51 times the incident bandwidth. The transmitted center frequency is 3.742 GHz or 1.31 times the incident center frequency of 2.863 GHz.

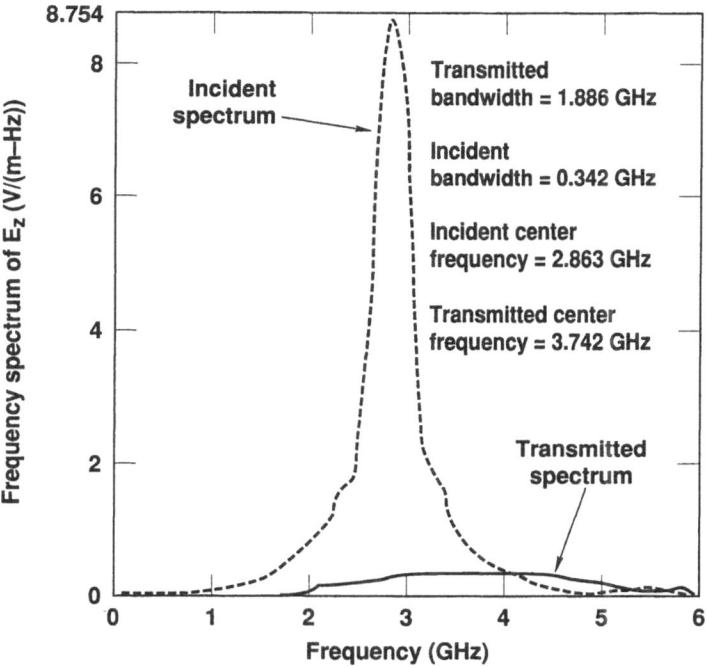

Figure 8. Frequency spectrum of the transmitted pulse.
The E_z waveform is shown in Figure 6b.
Initial electron density distribution: peak value = 10^9
electrons/cm^3, axial FWHM = 3 cm, transverse FWHM = 3 cm.

Figure 9 shows the variation of the transmitted bandwidth with the incident center line amplitude. The parameter for the curves is the peak background electron density, which varies from 10–10^{11} electrons/cm^3. The FWHM of the background electron density is 3 cm in the axial and transverse directions. The curve for 10^5 electrons/cm^3 ends at 17.5 MV/m. At 18.0 MV/m for this peak density, the computation stopped prematurely because the electron velocity at some grid point became mildly relativistic and the nonrelativistic governing equations became invalid. Computations at 18 MV/m and above require the solution of a set of relativistic electron fluid equations. Although we have developed a relativistic computer code, based on relativistic fluid equations, it has not been experimentally validated. For this reason, relativistic waveguide calculations have not yet been systematically conducted. The same relativistic invalidity occurs at 17.5 MV/m for 10^7 electrons/cm^3 and at 17.0 MV/m for 10^9 and 10^{11} electrons/cm^3. The curve for 10 electrons/cm^3 ends at 12.5 MV/m and the curve for 10^3 electrons/cm^3 ends at 9.5 MV/m because calculations have not yet been extended to greater incident amplitudes.

The peak transmitted bandwidth is 3.214 GHz at 11.5 MV/m of incident amplitude and 10^{11} electrons/cm^3 of peak initial electron density. This represents a relative bandwidth of 112%, which is quite large. It is 9.40 times the incident bandwidth.

All six curves rise fairly smoothly with roughly parabolic shapes to about 4–5 MV/m. Thereafter they develop bumps and wiggles of various widths and heights. Each curve reaches a peak and then falls off somewhat. This behavior is probably not unreasonable since the governing equations are highly nonlinear and coupled to various degrees. The general trends seem reasonable. For a fixed background electron density, the transmitted bandwidth increases with increasing incident amplitude. Then a saturation occurs. As the incident amplitude increases, the rate at which the electron density exponentiates increases. This causes stronger tail erosion, shorter transmitted pulses, and hence broader transmitted bandwidths. For a fixed incident amplitude, the transmitted bandwidth increases with increasing background electron

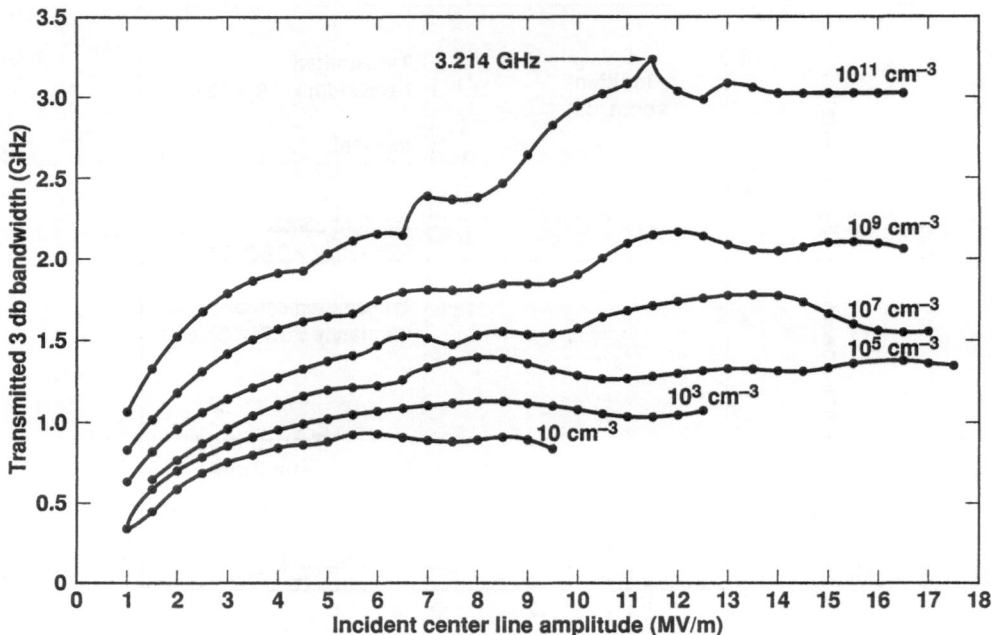

Figure 9. Transmitted 3 dB bandwidth versus incident waveguide center line amplitude.
The peak initial electron density is the parameter for the curves.
Initial electron density distribution: axial FWHM = 3 cm, transverse FWHM = 3 cm.

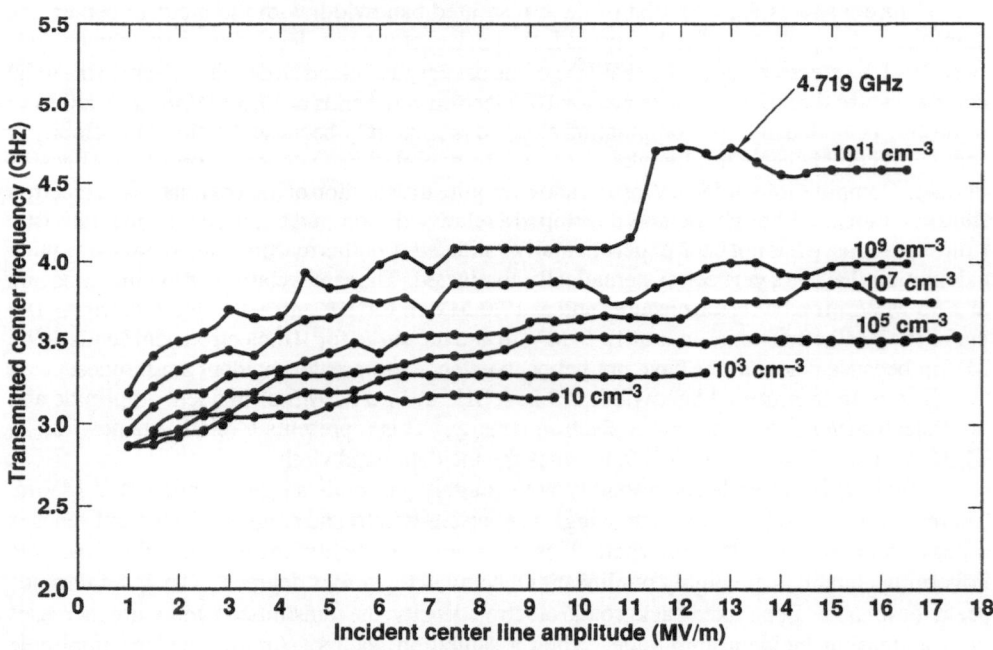

Figure 10. Transmitted center frequency versus incident waveguide center line amplitude.
The peak initial electron density is the parameter for the curves.
Initial electron density distribution: axial FWHM = 3 cm, transverse FWHM = 3 cm.

density. If the initial electron density is greater and that density increases at the same rate over the same period of time, the final density will be greater and the tail erosion stronger. The stronger tail erosion will cause a shorter transmitted pulse with a broader transmitted bandwidth. At the present time, all the bumps and wiggles in these curves cannot be explained. The curves may be smoother if the output time resolution in the calculations is increased from the present 0.05 ns of output sampling interval.

Figure 10 shows the variation of the transmitted center frequency with incident amplitude on the waveguide center line. Once again the parameter for the curves is the peak background electron density. The same general trends occur as for the transmitted bandwidth. The maximum transmitted center frequency is 4.719 GHz or 1.65 times the incident center frequency.

Figure 11 shows the variation of the transmitted waveguide center line pulse amplitude with incident pulse amplitude at 54.6 cm downstream from the input port of the waveguide section. For a fixed background electron density, the transmitted amplitude generally increases with increasing incident amplitude. This increase becomes progressively smaller as the peak electron density increases. Lower transmitted amplitudes result for greater background electron densities at fixed incident amplitudes.

The transmitted amplitudes for fixed incident amplitude and peak background electron density should drop as the pulse moves down the waveguide. The shape of the curves may be expected to change at greater axial positions due to the variation of waveguide dispersion with transmitted pulse length. The system linearity of the empty waveguide should cause the transmitted bandwidth to remain constant, although a spreading of the pulse and a redistribution of the frequency components in time across the pulse will occur.

Figure 12 shows the effect on the transmitted bandwidth of increasing the transverse FWHM of the background electron density distribution from 3 cm to 10^4 cm. The incident amplitude varies from about 1–17.5 MV/m, and the peak initial electron density varies from 10^3-10^5 electrons/cm^3. As might be expected, the transversely broader initial electron distribution generally causes a greater transmitted bandwidth than does the more narrow distribution. There are regions of incident amplitude where this is not true, and the transmitted bandwidth for the narrow transverse distribution exceeds that for the wide transverse distribution.

Figure 13 shows the effect of increasing the transverse FWHM of the background electron distribution on the transmitted center frequency for peak electron densities of 10^3 and 10^5 electrons/cm^3. The broader distribution generally creates a slightly greater transmitted center frequency. Note that this is not true for the range of 8.5–13.0 MV/m for the peak electron density of 10^5 electrons/cm^3.

Figure 14 shows the effect of the wider distribution on the transmitted pulse amplitude at the axial position of 54.6 cm. As can be seen, the effect is very small.

CONCLUSIONS

Approximate analysis of experimentally measured incident and transmitted air breakdown tail erosion envelopes for short pulse lengths at 3.5 torr shows bandwidth broadening. This bandwidth broadening increases with increasing incident amplitude. Electron fluid computer code simulations of these experiments also show bandwidth broadening. The simulations predict greater bandwidth broadening than the experimental analysis by up to 47%. The experimental results are believed to be underestimates of reality because of the lack of time resolution of the microwave cycles in the measurements. Experimental resolution of the individual microwave cycles is, at present, undoubtedly an expensive and challenging task. However, it should yield valuable information about the distribution of local frequency across the incident and transmitted pulses.

Predictions of the bandwidth broadening of an incident pulse with an 0.5 ns linear rise and a 3.0 ns linear fall are made with further code simulations. These simulations use a pressure of

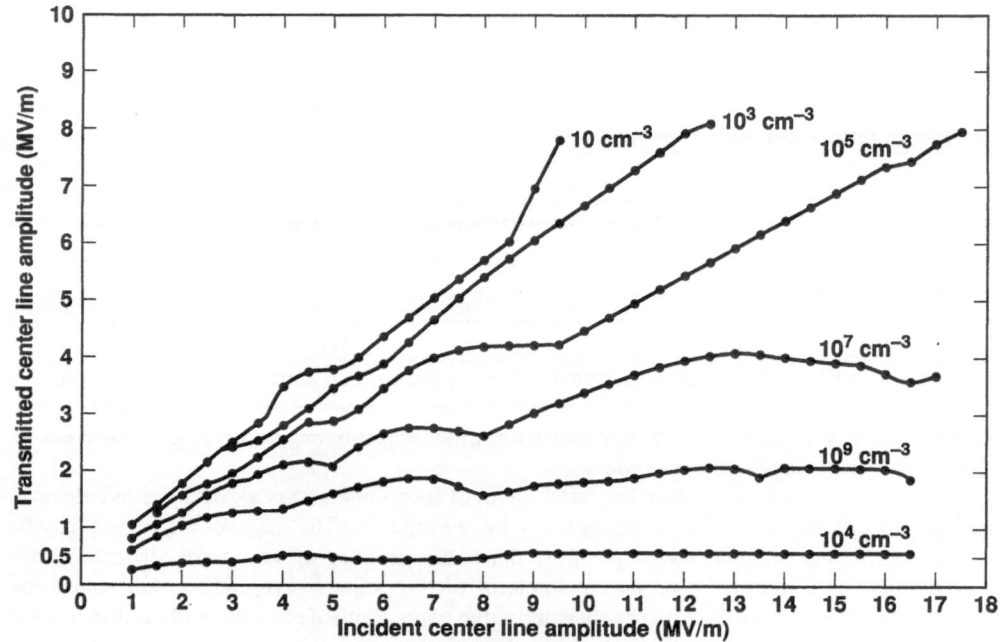

Figure 11. Transmitted center line amplitude versus incident waveguide center line amplitude at 54.6 cm from the input port.

The peak initial electron density is the parameter for the curves.

Initial electron density distribution: axial FWHM = 3 cm, transverse FWHM = 3 cm.

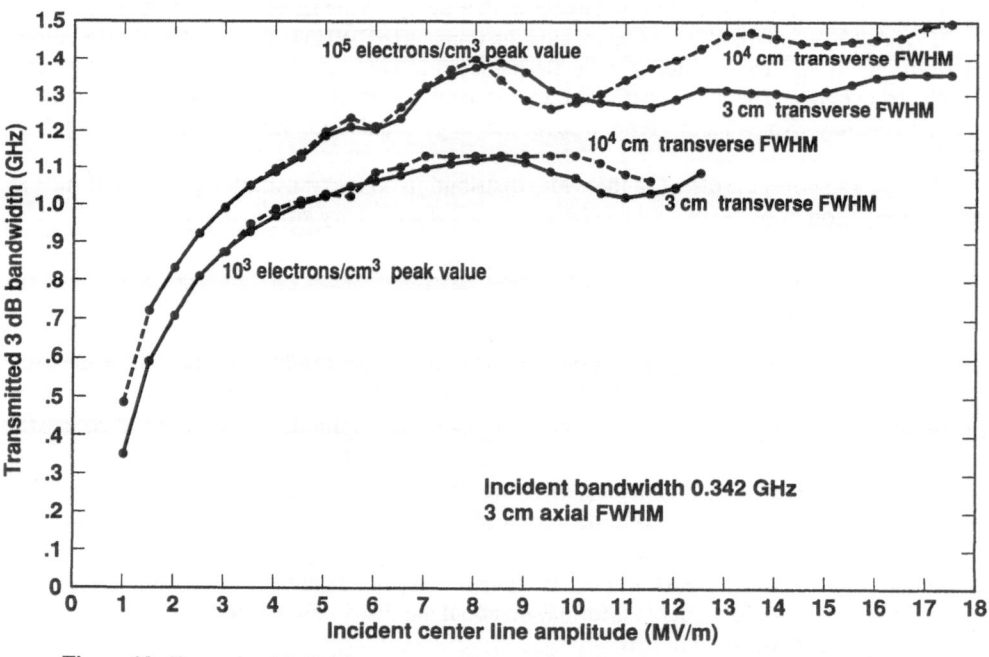

Figure 12. Transmitted 3 dB bandwidth for variation of initial electron density transverse FWHM.

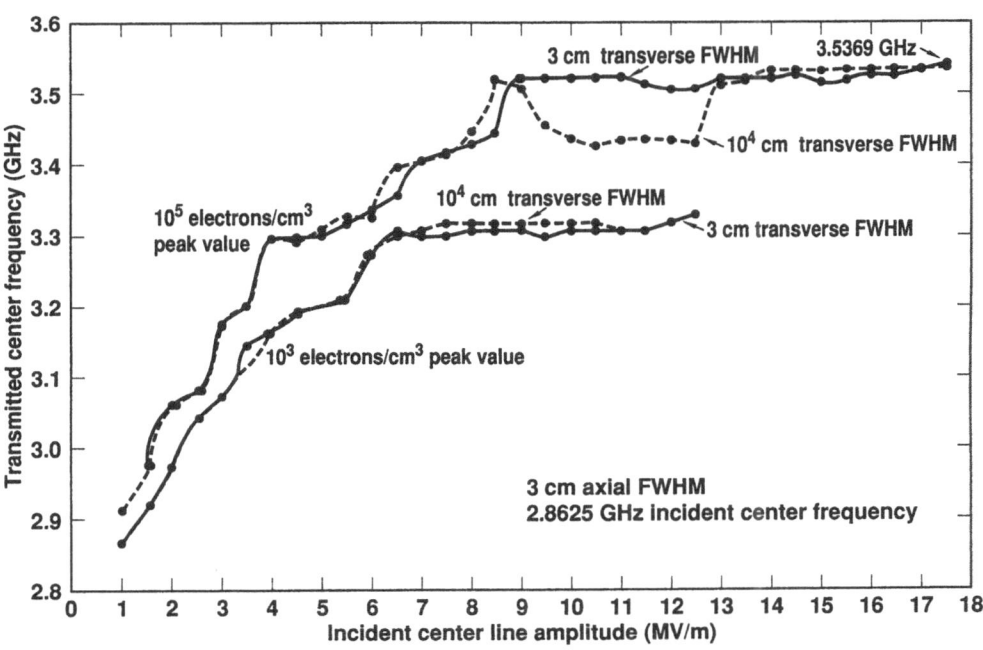

Figure 13. Transmitted center frequency for variation of initial electron density transverse FWHM.

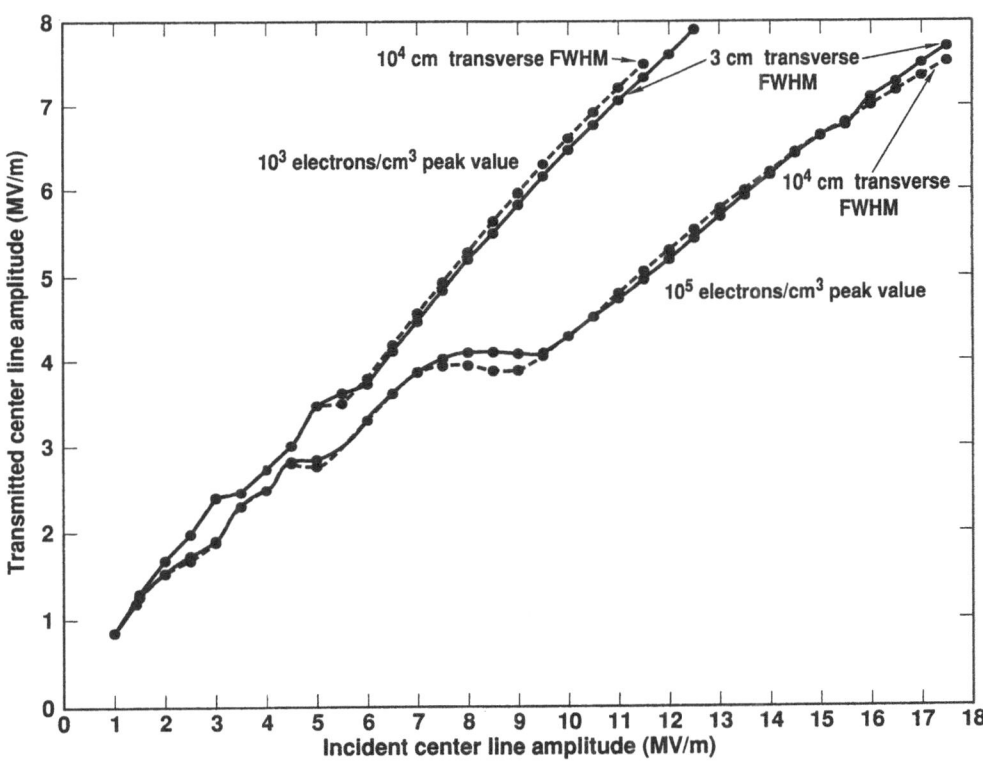

Figure 14. Transmitted center line amplitude at 54.6 cm from the input port for variation of initial electron density transverse FWHM.

3.5 torr, incident amplitudes from 1–17.5 MV/m, and peak background electron densities from $10\text{-}10^{11}$ electrons/cm^3. We assume that such electron densities can be rapidly generated with suitable focusing of a laser pulse. A peak transmitted 3 dB bandwidth of 3.21 GHz or 112% relative occurs for 10^{11} electrons/cm^3. Transmitted spectral center frequencies up to 4.72 GHz and transmitted pulse amplitudes up to 7.94 MV/m occur at 54.6 cm from the input port of the waveguide interaction section. Even greater transmitted bandwidths may result at greater peak background electron densities and other air pressures. Greater transmitted bandwidths should be possible with gases which ionize more rapidly than air, such as, hydrogen, argon, and azulene.

Laser-initiated pulse tail erosion presently appears to be a very promising, relatively easy way to generate very wideband, high-power microwave pulses with fairly conventional klystron technology. Furthermore, the electromagnetic, electron fluid computer code modeling should furnish good prediction of experimental results.

ACKNOWLEDGEMENTS

The authors thoroughly thank Dr. Paul Bolton of Lawrence Livermore National Laboratory for making available his experimental data on air breakdown tail erosion of microwave pulses.

REFERENCES

1. Mechanical and Electronic Engineering Division, Los Alamos National Laboratory, *"Ultra-Wideband Radar, Research, and Development Considerations,"* LA-UR-89-1420, Los Alamos National Laboratory, June 5 (1989).
2. J. H. Yee, R. A. Alvarez, D. J. Mayhall, D. P. Byrne, and J. DeGroot, "Theory of Intense Electromagnetic Pulse Propagation through the Atmosphere," *Phys. Fluids*, **29**, 1238–1244, April (1986).
3. J. H. Yee, R. A. Alvarez, and D. J. Mayhall, "Physical Phenomena Induced by Passage of Intense Electromagnetic Pulses (Including CO_2 Lasers) through the Atmosphere," *Laser Interaction and Related Plasma Phenomena*, **7**, H. Hora and G. H. Miley, eds., Plenum Press, New York, 901–913 (1986).
4. G. E. Sieger, D. J. Mayhall, and J. H. Yee, "Numerical Simulation of the Propagation and Absorption Due to Air Breakdown of Long Microwave Pulses," *Laser Interaction and Related Plasma Phenomena*, **8**, H. Hora and G. H. Miley, eds., Plenum Press, New York, 139–148 (1988).
5. D. J. Mayhall, J. H. Yee, R. A. Alvarez, and D. P. Byrne, "Two Dimensional Electron Fluid Computer Calculations of Nanosecond Pulse, Low Pressure Microwave Breakdown in a Rectangular Waveguide and Their Verification by Experimental Measurement," *Laser Interaction and Related Plasma Phenomena*, H. Hora and G. H. Miley, eds., **8**, New York, Plenum Press, 121–137 (1988).
6. R. A. Alvarez, D. L. Birx, D. P. Byrne, E. J. Lauer, and D. J. Scalapino, "Application of Microwave Energy Compression to Particle Accelerators," *Particle Accel.*, **11**, 125-130 (1981).
7. R. A. Alvarez, D. P. Byrne, and R. M. Johnson, "Prepulse Suppression in Microwave Pulse-Compression Cavities," *Rev. Sci. Instr.*, **57**, no. **10**, 2475-2480, Oct. (1986).
8. D. Giri, Pro-Tech, personal communication, Sept. 11 (1991).
9. D. P. Byrne, *"Intense Microwave Pulse Propagation through Gas Breakdown Plasmas in a Waveguide,"* Ph.D. Thesis, UCRL-53764, Lawrence Livermore National Laboratory, Livermore, CA, 94550, Oct. 8 (1986).
10. P. R. Bolton, Lawrence Livermore National Laboratory, personal communication, March 26 (1990).

PARTIALLY COHERENT LIGHT FOR IMPROVING IRRADIATION UNIFORMITY IN DIRECTLY DRIVEN LASER FUSION EXPERIMENTS

N. Miyanaga, H. Nakano, K. Tsubakimoto, K. Takahashi, M. Oshida,
H. Azechi, M. Nakatsuka, K. Nishihara, K. Mima, and S. Nakai

*Institute of Laser Engineering, Osaka University,
Yamada-oka 2-6, Suita, Osaka 565, Japan*

T. Kanabe and C. Yamanaka

*Institute for Laser Technology,
Yamada-oka 2-6, Suita, Osaka 565, Japan*

I. INTRODUCTION

In the recent directly driven laser fusion research, the feasibility of high density compression has been demonstrated experimentally [1-3]. To obtain such a compression, the improvement of laser irradiation uniformity was one of most important issues, and in these experiments the irradiation uniformity was greatly improved by installing random phase plates (RPPs) [4] in front of final focusing lenses. The highest density achieved using a random-phased frequency doubled Nd:glass laser [1,2] almost meets that required for the fuel ignition. However, the observed neutron yields were significantly smaller than predictions by one-dimensional hydrodynamics code, suggesting that a hot spot at the center of compressed fuel may be mixed with a surrounding cold part.

To ignite the target, DT fuel should be compressed to the density-radius product (ρR) greater than ~0.3 g/cm^2 with the temperature of ~5 keV. The driver energy (E_d) required to achieve this condition can be drastically relaxed with increasing the fuel compression density (ρ), which is scaled as $E_d \propto \rho^{-2}$. Therefore, a key subject relevant to the ignition is how to obtain the high density hot spot. Nevertheless, the compression performance would be limited by the hydrodynamic instability such as the Rayleigh-Taylor (R-T) instability caused from the imperfection of the target fabrication and the laser irradiation nonuniformity.

To date, several techniques such as the smoothing by spectral dispersion (SSD) [5], optical fiber oscillator (OFO) [6], induced spatial incoherence (ISI) [7] and echelon-free ISI [8] have been proposed for improving the irradiation uniformity. These are, in principle,

Laser Interaction and Related Plasma Phenomena, Vol. 10
Edited by G.H. Miley and H. Hora, Plenum Press, New York, 1992

251

based on the short coherence time which causes the rapid change of the focused beam pattern to smooth out small scale nonuniformities. Among them, the incoherency may be most important concept to obtain laser uniformity of $\sigma_{rms} \approx 1\%$ because the partially coherent light (PCL) does not gives rise to interference speckles if the beam pattern is averaged over a time interval longer than the coherence time. Moreover, the focused beam pattern consists of a number of independent beamlets because of the reduced spatial coherence, and it rapidly changes due to the reduced temporal coherence. In this paper we report recent activities and developments on the improvement of laser irradiation uniformity at Nd: glass laser Gekko XII, Osaka university.

II. GUIDELINE OF UNIFORM IRRADIATION AND REQUIREMENTS TO PARTIALLY COHERENT LIGHT SOURCE

2-1 Ideal beam pattern on the target

First of all, we consider a beam pattern with a smooth envelope to minimize the irradiation nonuniformity for the spherical target. The examination described below is important to reduce large scale nonuniformities which are determined by the number of laser beams and the focusing geometry. And here we ignore small scale nonuniformities which are discussed in following chapters. Usually the near field pattern (NFP) of laser beam is nearly flat top, thus long wavelength components appear without exception in the modal distribution of irradiation nonuniformity because of the overlap of neighboring beams. To avoid these nonuniformity components, the pattern of focused beam on the target plane is desired to have a bell-shaped intensity envelope.

In order to easily obtain the bell-shaped intensity distribution with a spot size corresponding to the target diameter, RPP is favorable because the far field pattern (FFP) is given by the square of sinc function $(sinc(x)=sin(\pi x)/\pi x)$. The spot size can be controlled by adjusting the F number of beamlets divided by RPP.

We calculated the spherical harmonics distribution of irradiation nonuniformity by changing a ratio of the target size to the beam spot size in order to obtain a minimum σ_{rms} value for the existing Gekko XII system, in which dominant nonuniformity modes are $l =$

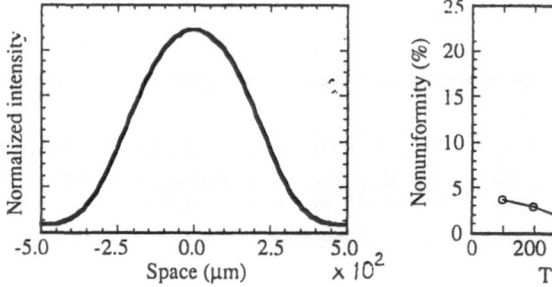 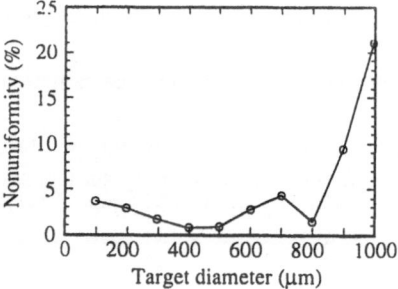

FIG. 1 Optimization of the intensity envelope on the target plane. A smooth bell-shaped intensity distribution reduces σ_{rms} down to ~1% level by optimizing the spot size corresponding to the target diameter (500μm).

6 and 12 [1]). Figure 1 shows typical nonuniformities calculated for the envelope which was composed from a huge number of shifted $sinc^2(x)$ functions. Here RPPs were used, and this envelope can be obtained by PCL of which coherence area is smaller than the element size of RPP. The simple $sinc^2(x)$ function does not give a minimum σ_{rms} because dominant nonuniformity modes remain as a result of beam overlap. As shown in this figure, we can basically reduce σ_{rms} down to 1% level by optimizing the envelope shape of intensity distribution even if we use only twelve beams. And the increase of the number of beams would give an enough tolerance to keep this uniformity even though the target shrinks during the compression [9]. Therefore the improvement of uniform irradiation results in the way to obtain a smooth and controlled bell-shaped intensity distribution.

2-2 Superposition of independent patterns

As described above, our interest is to avoid small scale speckles which appear in the FFP (or quasi FFP) of random-phased beam. So far as small scale nonuniformity, the beam smoothing can be achieved as a result of the incoherent (not electric field but intensity) superposition of many patterns. In order to treat the intensity distribution, it is convenient to use the statistical analysis of speckle pattern [10].

FFP of random-phased coherent beam can be expressed as a fully developed speckle pattern (FDSP). The probability density distribution function, $P(x)$, of intensity of FDSP obeys the well known negative exponential distribution as,

$$P(x) = \frac{1}{<I>} \exp(-x),\qquad(1)$$

where x $(= I / <I>)$ is the intensity normalized to the average. If we superpose a number of such patterns, the resultant probability density distributions become narrower as shown in Fig. 2, and these distributions can be expressed as,

$$P_N(x) = \frac{x^{N-1} N^N}{(N-1)! <I>} \exp(-Nx) .\qquad(2)$$

The variance, S^2, and the standard deviation, S, of $P_N(x)$ is easily given by,

$$S^2 = 1/N, \qquad S = 1/\sqrt{N}.\qquad(3)$$

Thus S^2 and S decrease as the number of pattern, N, increases. If we can measure N as the number of patterns independent each other, it can be approximated as,

$$N = t_{av} / \tau_c\qquad(4)$$

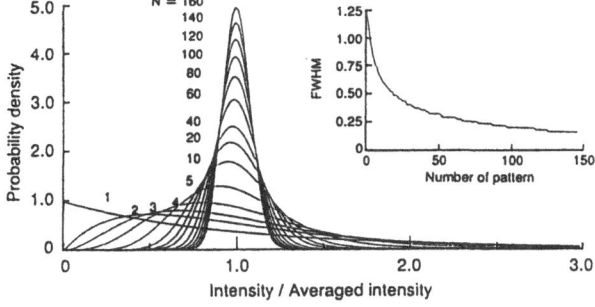

FIG. 2 Probability density distribution functions calculated by superposing fully developed speckle patterns. N is the number of patterns.

where t_{av} is the averaging time and τ_c is the coherence time. The computer analyses of SSD introduced into Nova system at Lawrence Livermore National Laboratory (LLNL) in USA showed that the initial smoothing by a factor of $\sqrt{2}$ occurred in a time scale of the coherence time[11]. Therefore Eq. (3) could be rewritten as,

$$S = 1/\sqrt{1+(t_{av}/\tau_c)} . \qquad (5)$$

In order to reduce S (i.e. to increase N), τ_c should be as short as a few ps. However, it takes too long time to relax S to an asymptotic level of a few %.

Explanation described above is one of possible mechanisms of beam smoothing. Now consider how we can apply the above treatment to PCL. Here, the beam consists of a number of small beamlets which are spatially incoherent each other. Therefore the intensity distribution of FFP, to which beamlets are focused and added incoherently, must have a probability density distribution (PDD) different from that of FDSP. The instantaneous PDD of PCL which may be obtain by averaging the intensity distribution for a time scale of τ_c. If this PDD can be expressed by Eq. (2) with $N=N^*$, Eq. (5) is replaced by

$$S^* \approx 1/\sqrt{1+N^*(t_{av}/\tau_c)} . \qquad (6)$$

Therefore the spatial coherence should be reduced enough to increase N^* ($N^* \gg 1$), so the required uniformity can be achieved in a hydrodynamic time scale of implosion.

2-3 Requirements to partially coherent light source for Gekko XII

If we divide the spherical harmonic components of the irradiation nonuniformity into two parts, long wavelength nonuniformities would cause the target deformation due to the forced R-T instability[1] during the acceleration, since the thermal smoothing would not be expected to reduce such nonuniformities. The intensity envelope of focused beam pattern at a target plane can be controlled by shaping the far-field pattern (FFP) at the front end of PCL source because FFP is relayed by spatial filters. On the other hand, shorter wavelength nonuniformities may give rise to the irregularity imprint on the target surface which would grow up with large growth rate even after ablation pressure nonuniformities disappear due to the thermal smoothing[12, 13].

As described in section 2-2, the beam smoothing to reduce small scale non-uniformities is determined by both temporal and spatial coherencies which characterize PCL. The coherence time, which is estimated as a reciprocal of the band width (Δv), is desired to be much less than 10 ps after the harmonic conversion. Then the overall spectral width ($\Delta \lambda$) of fundamental light ($\lambda = 1.053 \mu m$) should be much larger than ~0.2 nm. On the other hand, the spatial coherence is characterized by the number of spatial modes which has large beam divergence ($\Delta \theta$), i.e. a large spot size in the Fraunhofer region. In the main amplifier chain of existing Gekko XII system, the laser beam is transferred down to a target chamber by several spatial filters which can completely maintain the divergence corresponding to a FFP spot size of ~40-50$F\lambda$, where F is the F number of optics.

To generate PCL we adopted two schemes; one is the amplified spontaneous emission (ASE) from Nd:glass rod[14], and the other is the modal dispersion of a broad-band laser light in a multimode optical fiber[15]. For both PCL front end systems, the control of spectral width (especially the instantaneous spectrum) and the time dependent spatial frequency characteristics are important.

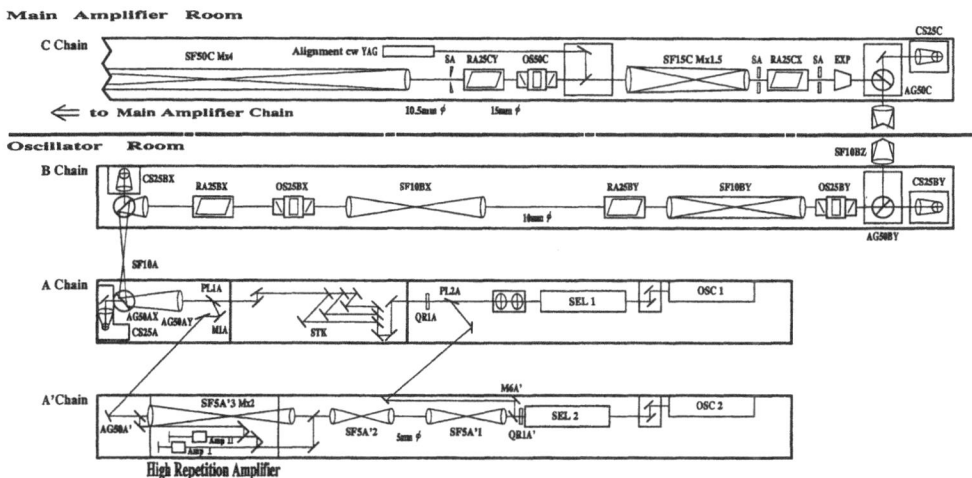

FIG. 3 Modified system arrangement of oscillators and preamplifier chain of Gekko XII. NFP at the output of SEL 2 in the B' chain is image relayed to a soft aperture SA in the C chain.

III. PARTIALLY COHERENT LIGHT SOURCES AND IMAGE RELAY SYSTEM

3-1 Modification of Gekko XII preamplifier chain

In order to keep the beam divergence of $40F\lambda$, a preamplifier chain of Gekko XII has been modified as shown in Fig. 3. All components (amplifiers and optical shutters) are set at optimum positions to be prevented from the damage due to the beam focusing, as the usual coherent laser is also used.

The near-field image is also relayed by spatial filters. An accumulated magnification from the end of PCL system (the output of optical shutter SEL 2 in Fig. 3) through a target chamber is nearly 90. A NFP at the output of PCL front end is completely relayed to an apodization aperture (soft aperture SA) in the C chain which is the starting point of image relay of main amplifier chains. Then the diameter of PCL beam is expanded from 5 mm to 320mm (both are nominal) including the apodization by a factor of 2/3 at the aperture (SA).

The beam divergence corresponding to $40F\lambda$ (note $\lambda = 1.053$ μm) is $\Delta\theta \approx 130$ μrad at the target chamber, while $\Delta\theta \approx 8.4$ mrad at the end of PCL front end. This beam divergence is not enough to smooth out the intensity distribution at the target of ~500 μm diameter which was used in the high density campaign of plastic shell. However, this divergence is too large to keep high efficiency of harmonic conversion. The harmonic conversion will be discussed later.

In modifying the preamplifier chain, other issues were considered as follows. 1) Two beams from the coherent laser (OSC 1) and PCL beam can be synchronized with an timing error less than ~100 ps. 2) We can flexibly switch the oscillator chain from one to the other. 3) The beam alignment (centering at components, pointing and focusing onto pinholes of spatial filters) accuracy should be high enough to keep uniform patterns at near field and far field regimes.

3-2 ASE front end and spectral selection

We used a Nd:glass rod of 25 mm in diameter to generate an ASE. The ASE front-end system consists of this ASE generator, a four-path amplifier (total gain of 10^6-10^7), Pockels cells and a spectrum slicer as shown in Fig. 4. A NFP at front surface of glass rod is reflected back by a rear side mirror, and is relayed to a hard aperture (HA) which is the starting point of NFP. An effective clear aperture of generator was chosen to be 20 mm. Finally, a central uniform part of this clear aperture is apodized at SA in the C chain (see Fig. 3. Beam diameter is 13 mm at this position.). The beam divergence was selected to be $40F\lambda$. (This will be improved in near future.) Spatial filters are installed to keep this large beam divergence and to relay the NFP.

The four-pass amplifier consists of a rod amplifier of 25-mm diameter, a Pockels cell (OS), a Faraday rotator (FR) and retardation plates. To suppress the parasitic oscillation, the opening time of OS was set at 10 ns, and appropriate apertures were installed at far-field positions of spatial filters.

The spectrum was sliced by a brazed grating (10 cm square, 1200 grooves/mm) coupled by a telescope with a variable iris set at the far-field image point. The spectral dispersion at the iris is 1.4 mrad/nm, here the grating was set at nearly the retro-reflection angle. By changing the size of iris, the spectral width can be continuously controlled in a range of 0.6-2.6 nm (full width at half maximum, FWHM). In this spectrum slicer the beam diameter was expanded for reducing the divergence down to ~500 μrad, so that we can obtain sharply cut spectra. The spatially dependent time delay (maximum 400 ps) due to the grating was not significant compared with the pulse width (~10 ns) at the grating.

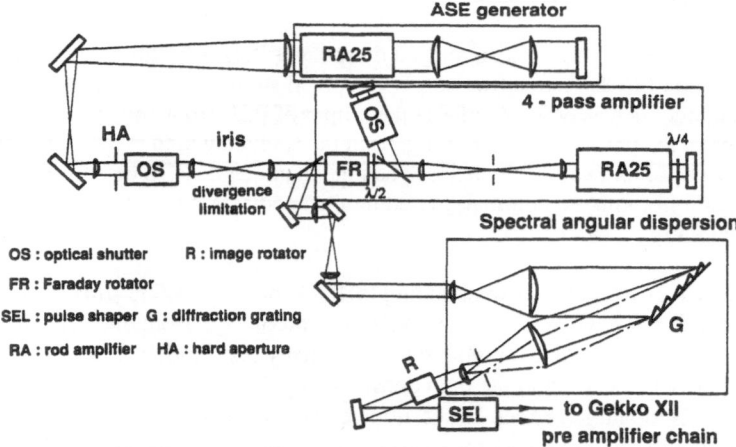

FIG. 4 Schematic arrangement of ASE front end. This system consists of four parts; a generator of glass rod, an amplifier including a Faraday rotator (FR) and a Pockels cell (OS), a spectrum slicer which give the angular dispersion of spectrum, and a pulse shaper (SEL 2 in Fig. 3). These parts are connected by spatial filters. The output energy of a pulse shaper is ≤30 μJ in 2.2 ns pulse width.

The output energy of this front end was ≤30 μJ which was limited by the parasitic oscillation, and the pulse shape was nearly Gaussian (2.2 ns FWHM) after passing through the pulse shaper. Here we should note that the grating makes roles of not only the spectral control but also the angular dispersion of spectrum. This angular dispersion is similar to SSD installed into Nova system at LLNL. The angular dispersion of PCL (ADPCL) which has the divergence in the direction perpendicular to the spectral dispersion was greatly effective not only for the beam smoothing as discussed later but also for improving the efficiency of frequency conversion.

3-3 Optical fiber system

A) <u>System construction</u>

In reference 6, they injected an output pulse from a broad-band Nd:glass oscillator into a multimode optical fiber to generate PCL. However, we used a mode-locked Nd:YLF oscillator (~10 μJ/pulse), since the mode-locked laser is very stable, and the number of photons included in a period of τ_c is large enough to assure good statistics, even after the beam is spectrally broadened and divided into many spatial modes.

An optical fiber system is schematically shown in Fig. 5. The spectral width of short (90-100 ps, FWHM) pulse is easily broadened by using the self phase modulation (SPM) effect in a single mode optical fiber. The single mode fiber here used was a polarization maintaining silica fiber of 9-μm core diameter. In broadening spectral width, the stimulated Raman scattering process competes with SPM, thus the maximum spectral width is limited to about 0.4 nm for ~100 ps Gaussian pulse. Therefore, once the chirped pulse of 0.4 nm band width was compressed to ~15 ps (FWHM) by a grating pair (1800 grooves/mm), so that τ_c becomes closer to the reciprocal of the time integrated band width Δv_{av}. As we showed in the next chapter, the time resolved spectra of ASE (also the usual broad band laser might have the similar characteristics) significantly differ from that time averaged.

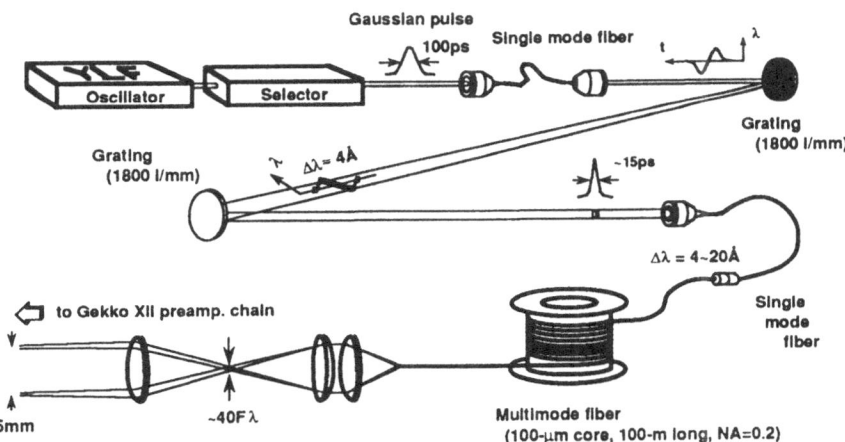

FIG. 5 Optical fiber system as an alternative generator of partially coherent light (PCL). A frequency chirped pulse of ~100 ps is compressed to ~15 ps, and next it is again spectrally broadened at the second single mode fiber. This broad-band laser becomes PCL after passing through a long multimode fiber because of spatial mode conversion.

Then this compressed pulse was again injected into the second chirping system. The spectral width was controlled in a range of 0.4-2.5 nm, by adjusting the incident energy into the second single mode fiber. The spectrally broadened short pulse was focused into a step index multimode fiber to reduce the spatial coherence. A typical core diameter (a_{core}) of the multi-mode fiber was 100 μm, and the numerical aperture (NA) was 0.2. This NA corresponds to $F \approx 3$, thus the output beam is simply expected to have the divergence of ~40 $F\lambda$.

B) Required length of multimode optical fiber

In order to induce the spatial incoherence using the multimode fiber, the output pulse must forget the temporal information about the incident one. So, the length of multimode fiber satisfies at least the next condition,

$$t_d = \frac{n_{core}L}{c}(\frac{1}{\cos\theta}-1) >> \tau_p \qquad (7)$$

where n_{core}, c and τ_p are the index of refraction at core, the speed of light and the incident pulse width, respectively. θ is the maximum reflection angle which is determined by NA to be $\theta = 7.66°$. Then the maximum transit time delay of multiply bouncing rays relative to the straight forward ray is given as $t_d = 0.45L$ (ps), where L is measured in a unit of cm.

Since $\tau_p \approx 15$ ps, a extremely long fiber may not be needed. However, the condition of Eq. (7) is not sufficient for the beam smoothing, because spatial modes transferring through the fiber interfere each other generating the speckle at the output plane of fiber. Therefore, a much more restrict criterion to avoid the interference between modes is expressed in terms of the transit time of M-th mode, τ_M,

$$\Delta\tau_M = \tau_M - \tau_{M-1} >> \tau_c \qquad (8)$$

where the time delay between adjacent two modes, $\Delta\tau_M$, decreases as the principal mode number M decreases. Here, τ_M can be calculated from the group velocity of M-th mode,

$$\tau_M = \frac{L}{c}\frac{\partial\beta}{\partial k_0}$$

$$= \frac{LN_l}{c}\frac{1+\delta(1+\frac{\varepsilon}{4})[b-1+\frac{d(Vb)}{dV}-1]}{\sqrt{1+2\delta(b-1)}} \qquad (9)$$

$$N_l = n_{core}\left[1-(\lambda/n_{core})(dn_{core}/d\lambda)|_{\lambda=\lambda_0}\right]$$

$$\delta = (n_{core}^2 - n_{clad}^2)/2n_{core}^2$$

$$V = k_0\frac{a_{core}}{2}n_{core}\sqrt{2\delta}$$

$$b = (\beta^2 - k_0^2n_{clad}^2)/(k_0^2n_{core}^2 - k_0^2n_{clad}^2)$$

$$\varepsilon = -(n_{core}/N_l)(2\lambda/\delta)(d\delta/d\lambda)$$

where n_{clad} is the refractive index of the cladding material. N_1, δ, and b are the group index, the specific index difference and the normalized propagation constant, respectively.

By calculating $\Delta\tau_M$ numerically, we obtained $\Delta\tau_{37} = 130$ ps, $\Delta\tau_{18} = 110$ ps and $\Delta\tau_2 = 20$ ps for $L = 100$ m. All of these transit time differences satisfy Eq. (8), then the spatial modes are recognized as independent patterns.

The maximum mode number, M_{max}, is so large as,

$$M_{max} = \frac{V}{2} - 1$$
$$= \frac{\pi a_{core}(NA)}{2\lambda} - 1 \qquad (10)$$

Therefore, a large number of spatial modes can be superposed incoherently at the end of fiber, which suggests that the beam smoothing would efficiently occur in a long multimode fiber. The total number of degenerated LP (linearly polarized) modes, N_{mode}, corresponding M_{max} is calculated as,

$$N_{mode} = V^2 / 8 \qquad (11)$$

here components with perpendicular polarization are excluded.

IV. CHARACTERISTICS OF PARTIALLY COHERENT LIGHT SOURCE

4-1 General concern

Characteristics of PCL relevant to the beam smoothing are, 1) spectral width and time resolved spectrum, 2) temporal and spatial coherencies, and 3) time resolved beam pattern and speed of smoothing. These characteristics were measured and analyzed with reasonable resolutions.

The time integrated spectral width should be controlled in a range of ~0.4-2 nm. The minimum width is given by the longest coherence time allowable (~10 ps at $\lambda = 1.053$ µm), and the maximum width may be limited by the spectral narrowing in the main amplifier chain. The primary concern to the spectrum is addressed in undesired structures in the shape and the temporal behavior which result in high coherence of PCL.

The spatial coherence can be measured according to Young's interference experiment. The coherent area of PCL beam is determined by the beam divergence, i.e. the source size, δ_s. The size of coherent area, A_c, at a beam diameter of D is given by,

$$\frac{D}{A_c} = \frac{\delta_s}{F\lambda} \ . \qquad (12)$$

Thus, A_c determined by the beam divergence of $\delta_s = \alpha \times F\lambda$ can be approximates as $A_c = D/\alpha$.

The time resolved beam pattern should be measured in order to estimate PDD of intensity, the spatial frequency characteristics and its time evolution. Then the smoothing time can be evaluated from this time resolved pattern. These measurements are carried out using a streak camera of 2 ps resolution.

4-2 ASE beam

The uncontrolled ASE spectrum showed a Gaussian shape of ~2 nm (FWHM). The spectral band width was controlled by changing the iris set at a far field position of a lens just after the grating in Fig. 4. Obtained spectra are shown in Fig. 6, and these have sharply cut edges. This sharp edge may be important to achieve the power balance. The usual Gaussian spectrum which has long tails is deformed by the gain narrowing and the gain saturation near the center of spectrum as is shown in the next chapter. The narrowing and saturation depend on the amplification characteristics of each chain. Thus output spectra of main amplifier chains might slightly differ from each other, causing differences in the time dependent harmonic conversion efficiency.

Figure 7 shows a temporal behavior of ASE spectrum before the grating pair. The time resolved spectra consists of several narrow spikes, while their envelope are nearly Gaussian. These spiky spectra were observed for a time duration of ~100 ps. This phenomenon may be due to the correlation of local field around seeds of emission. The duration of spectral spikes is much longer than the reciprocal of time integrated band width (Δv_{av}), so that the coherence would remain for these spectral components giving rise to the long time scale intensity perturbation [14].

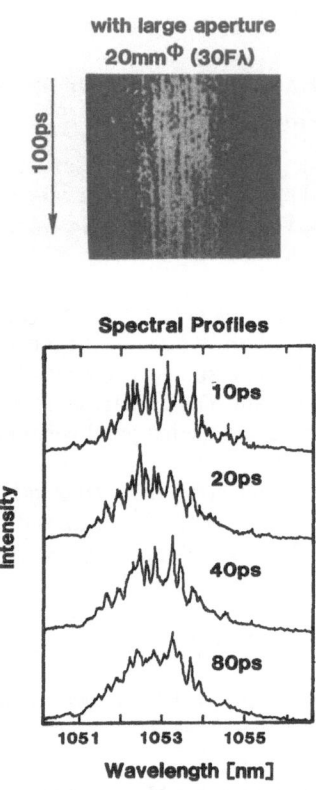

with large aperture
20mm$^{\Phi}$ (30Fλ)

100ps

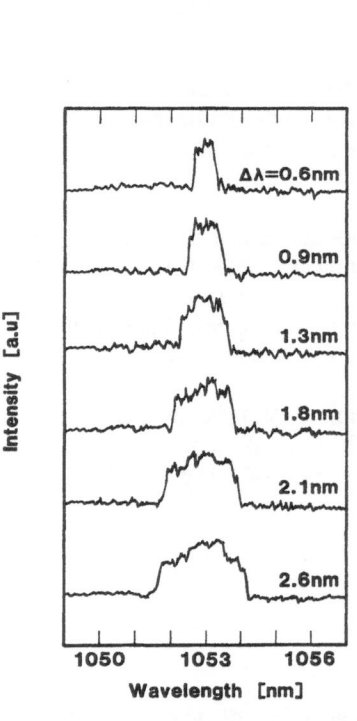

FIG. 6 ASE spectra controlled using a spectrum slicer (monochrometer) which consists of a beam expander, a grating and a beam minifier.

FIG. 7 Typical time resolved ASE spectrum which composed from narrow spikes with durations longer than the coherence time inferred from a time integrated spectral width.

4-3 Fiber-smoothed beam

A) Spectral width and time integrated far-field pattern

The spectral width determined by the SPM in the second single mode fiber varies as a function of the incident energy of compressed pulse (~15 ps, FWHM). Thus we can control the band width in a range of 0.4-2.5 nm. Although the spectral shape has in principle sharp edge because of SPM, there is of course a peak due to unchirped component at the center. This peak will be reduced in near future using a filtering technique.

This chirped 15 ps pulse was injected into a multimode fiber. The output beam from the multimode fiber was collimated using 10× microscope objective to have a diameter of ~5 mm. Figure 8 shows typical pictures of time integrated FFP (in this case, these correspond to the near field images at the output surface of fiber) measured by a cooled CCD camera. Inserted as a reference is a result for usual narrow band laser ($\Delta\lambda$ = 0.002 nm, transform limited band width of 100 ps pulse). With increasing the length of fiber (L) and the band width ($\Delta\lambda$), the beam pattern become smoother to vanish the speckle which is visible for the narrow band laser. This phenomenon can be clearly seen in comparing PDDs shown in Fig. 9. Double-peaked distributions are due to fringes in Fig. 8, and these fringes were not caused by the interference of the fiber-smoothed beam itself but by the interference inside the CCD camera. The standard deviation of intensity distribution can easily reach to a few % level. Furthermore we should note that PDD for narrow band laser also does not show a maximum probability density at zero intensity. This observation suggests that PDD no longer agrees with that of the fully developed speckle after passing through the long multimode fiber, if the transit time spread is much longer than the pulse width.

In fact, the low contrast of speckle for narrow band laser can be explained by the incoherent intensity addition of speckle patterns, here the pulse width was extended to a few ns in passing through a 100 m multi-mode fiber. The effect of intensity addition on the beam smoothing is more significant for broad band pulses. A quantitative relation between the beam smoothing and the coherence is investigated as follows.

B) Spatial and temporal coherencies

The spatially coherent area measured by Young's interferometer technique was about 1/30 of the beam diameter. An experimental setup is shown in Fig. 10. The fiber-smoothed beam of ~5 mm diameter was expanded to 18.1 mm by a telescope for observing interference patterns with a good spatial resolution. A double slit was set at an imaging point of a NFP at L_0, and a far field image of rays transmitted through slits was observed by a S-1 streak camera. Figure 11 shows typical interference patterns and their spatial traces obtained with different slit distances.

The degree of coherence is evaluated by the visibility defined as $(I_{max} - I_{min})/(I_{max} + I_{min})$, where I_{max} and I_{min} are intensities at a peak and neighboring valleys of interference pattern, respectively. In this experiment the beam divergence was limited by setting an aperture (see Fig. 10, diameter = δ_s) at a far field position of L_2.

Estimated degree of coherence is plotted in Fig. 12 as a function of slit separation

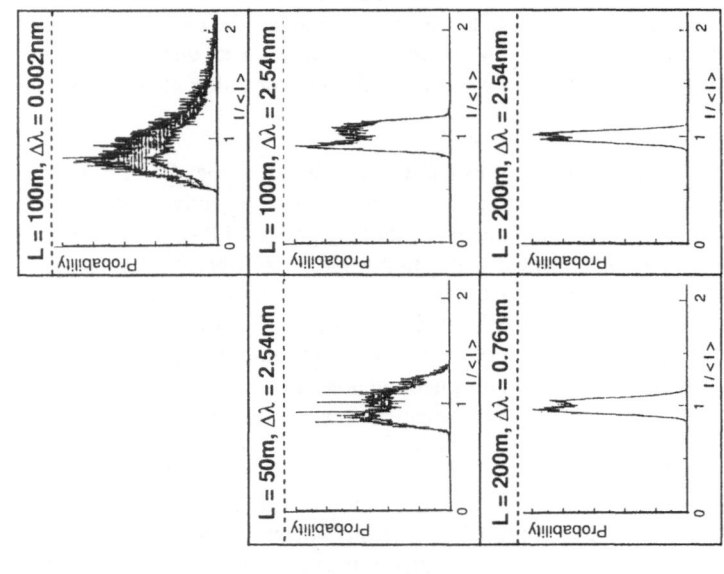

FIG. 9 Probability density distributions of time integrated far-field patterns shown in Fig. 8.

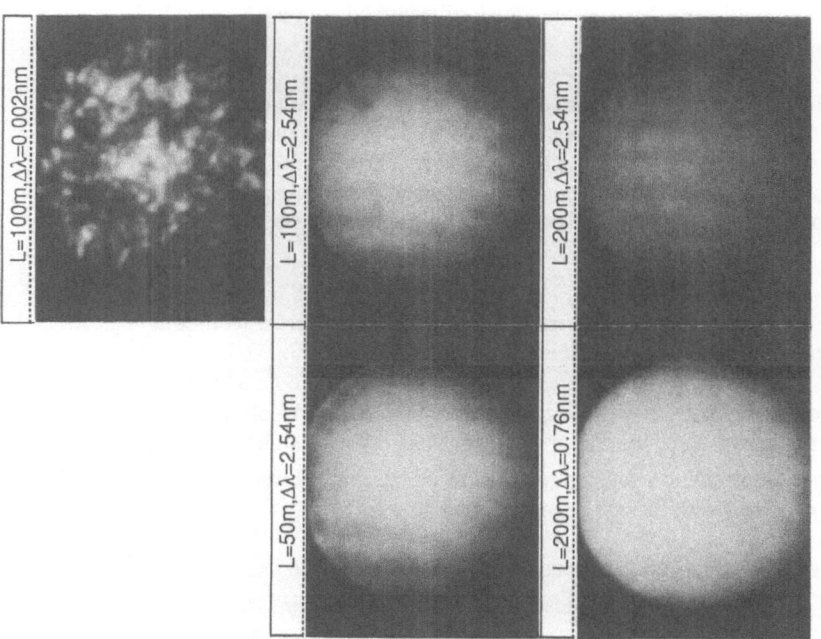

FIG. 8 Typical time integrated far-field patterns of fiber-smoothed beams. The smooth pattern was obtained with broader spectral width and longer multimode fiber. Also inserted is a speckle pattern taken with a narrow band laser. Interference fringes are due to CCD camera itself.

262

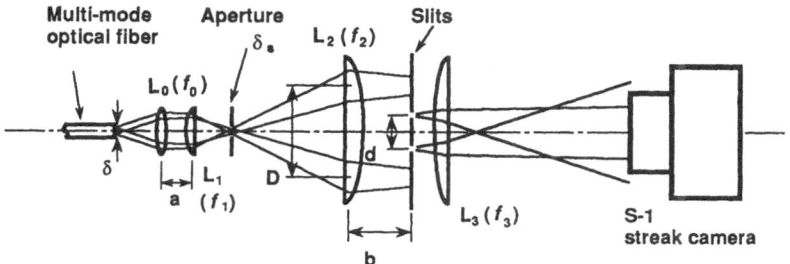

FIG. 10 Experimental setup for measuring the spatial coherence by use of Young's interferometer technique. Focal length of lenses were $f_0 = 13$ mm (L_0), 510 mm (L_1), 1750 mm (L_2) and 300 mm (L_3), respectively. Nominal beam diameters were 5 mm at L_1 and 18.1 mm at slits.

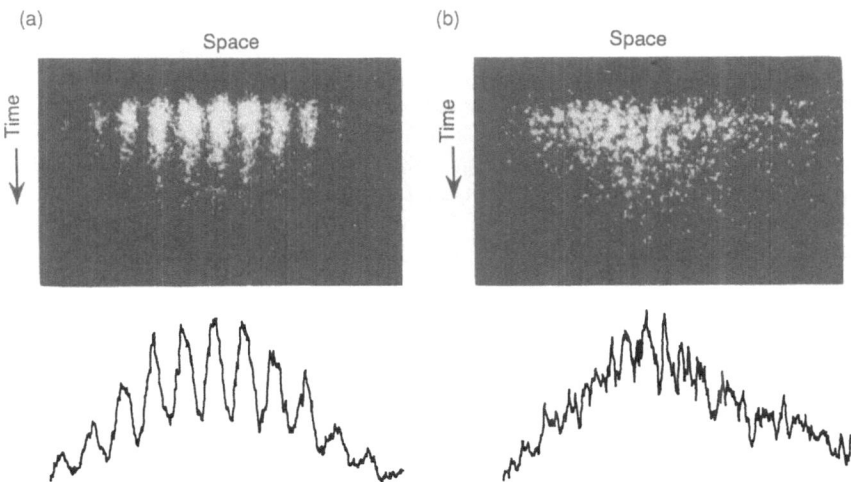

FIG. 11 Typical results of interference measurements obtained with different separation distance between slits. In this case the beam diameter was 6 mm at slits. Slit separation distances are 150 μm (a) and 230 μm (b), respectively.

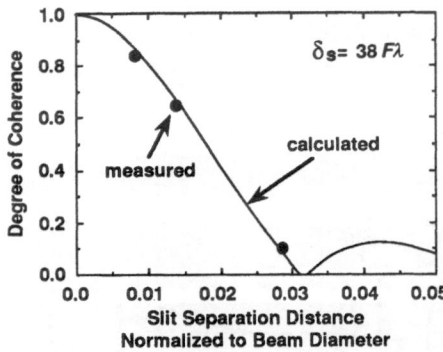

FIG. 12 Measured degree of coherence taken by changing slit separation distance (normalized to the beam diameter in the figure). Solid line represents a theoretical curve calculated for $\delta_s = 38F\lambda$ and $D = 18.1$ mm.

distance with a theoretical prediction [16] for the source size of $\delta_s = 38F\lambda$. A size of coherence area is defined as a slit separation distance at which the degree of coherence first becomes zero. Thus A_c was concluded to be about 1/30 of the beam diameter, as it was estimated from Eq. 12. If we amplify this PCL through the main amplifier chain and use RPP of 2-mm element size, the coherence area includes about 20 elements of RPP at which the beam diameter is nominally 320 mm. Therefore nonuniformities arising from the residual mutual coherence across the beam would be reduced at the target plane.

The coherence time was also inferred from time resolved observations of the Young's interference. This method is not an actual measurement of temporal coherence, but an auto correlation of temporal behavior of interference would give an information about the temporal coherence, if the slit separation distance is enough smaller than the size of coherence area so that two sources can be recognized to be almost the same. Figure 13 shows interference patterns and their traces along the time axis which were obtained by the similar arrangement illustrated in Fig. 10, but with a high resolution of a few ps. Here the time averaged spectral widths were 0.25 nm and 0.41 nm. Figure 14 shows the autocorrelation of the temporal traces of Fig. 13, and the coherence times were about 15-25 ps and ~10 ps, respectively. Similar data were obtained during the entire pulse. Therefore we can conclude that the coherence time of fiber-smoothed beam is given by the time averaged band width as $1/\Delta \nu_{av}$, even if the seed pulse is much shorter than the output pulse.

C) Time resolved far-field pattern

As described above, PCL generated from the optical fiber system showed good statistical characteristics, i.e. the time and spatial coherencies were determined by the time averaged band width and the beam divergence, and large scale structures were not observed in both time and spatial characteristics. Next we analyzed time resolved FFP data taken by a S-1 streak camera to estimate the smoothing time. Fig. 15 shows streaked FFPs and their time integrated spatial traces (image of the output surface of fiber) for the fiber length of $L = 100$ m. One can clearly see that the intensity profile is smoother for broader band width. This result can be explained by the fact that the number of independent spatial modes becomes larger with increasing band width (see section 3-3).

Similar streaked FFP data were obtained for a fixed band width ($\Delta\lambda = 1.59$ nm) by changing the fiber length as shown in Fig. 16. The fiber longer than 100 m is effective to smooth the output beam pattern.

These streaked data were reduced to the time evolution of power spectrum of the spatial intensity distribution. Fig. 17 shows nonuniformities which decreased as the beam pattern was time integrated. The time origin corresponds to the leading edge of output pulse for both cases of $\Delta\lambda = 0.76$ nm (a) and $\Delta\lambda = 1.59$ nm (b). Relatively large nonuniformities with lower modes would contain the sensitivity distribution of the streak camera. A time required for smoothing is shorter for the broader band width, and asymptotic nonuniformity levels for the broader band width are smaller than those for narrower band width. The nonuniformity of higher spatial frequency modes became halfway larger than that of medium modes, although it was initially smaller. This phenomenon would be explained by the transit time different for each spatial mode, so lower modes appear in the early part of the pulse, and higher modes appear later. However, the temporal change of mode structure was not so obvious because of the possible mode scrambling in the multimode fiber. Thus beam can be smoothed in a time scale shorter than 100 ps by using such the simple optical fiber system.

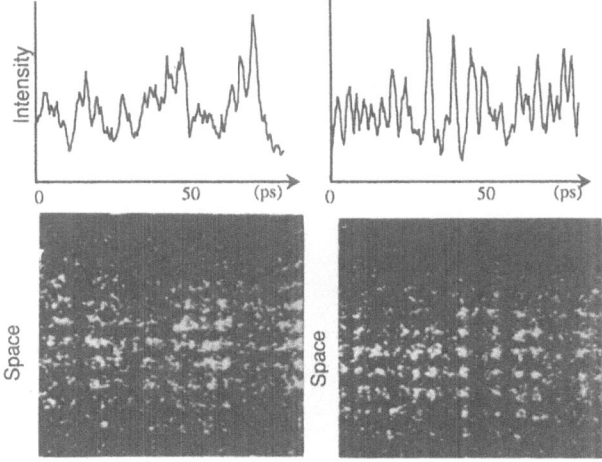

FIG. 13 Temporal behaviors of interference observed with double slit of which separation distance was enough smaller than the size of coherence zone.

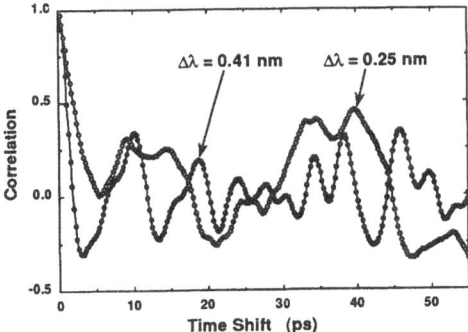

FIG. 14 Autocorrelation functions corresponding to the time resolved interference patterns of Fig. 13.

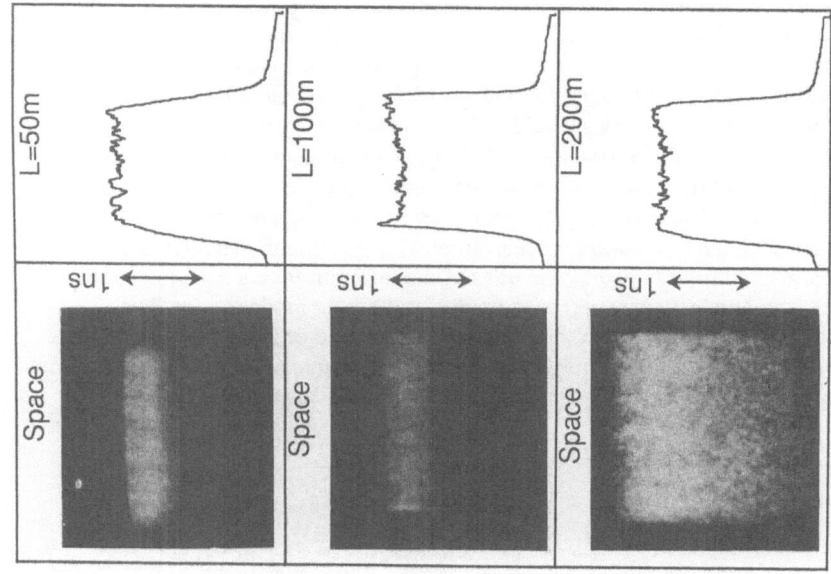

FIG. 16 Time resolved measurement of far-field patterns taken for a spectral width of 1.59 nm. Traces are time integrated intensity distributions.

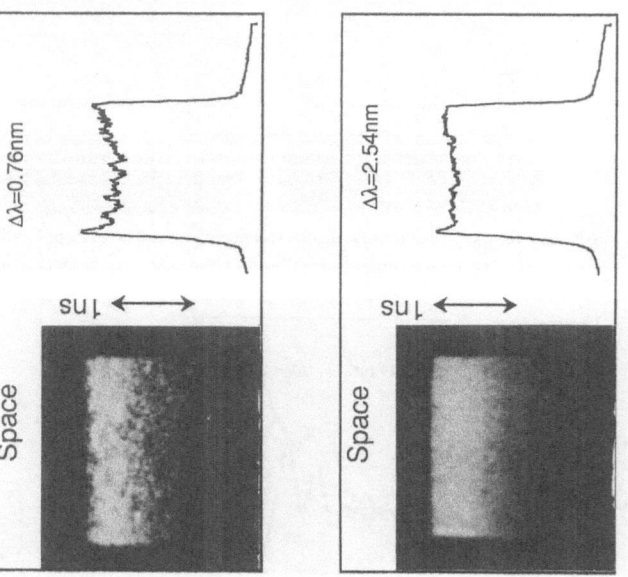

FIG. 15 Time resolved measurement of far-field patterns taken for a 100 m fiber. Traces are time integrated intensity distributions.

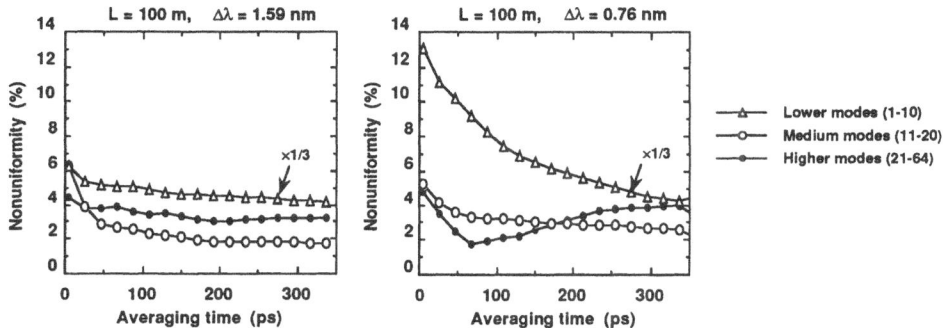

FIG. 17 Nonuniformity decreases as the spatial intensity distribution is time averaged. In the figure, nonuniformity is shown by three parts of the power spectrum, and the data for lower modes are scaled down by a factor of 3.

4-4 Instantaneous number of independent patterns, N*

To estimate the smoothing time qualitatively, the standard deviation of time resolved spatial intensity distribution was studied on the fiber-smoothed beam (~40$F\lambda$). In Fig. 18, the measured standard deviation is plotted as a function of averaging time. Dotted lines represent $S*$ calculated from Eq. (6) with $\tau_c=1/\Delta\nu_{av}$. It is Obvious that the measured standard deviation is quite smaller than that estimated by assuming the intensity addition of fully developed speckle patterns. Fitted curves indicate $N* = 3.5$ for $\Delta\lambda = 0.76$ nm and $N*$ = 8 for $\Delta\lambda = 1.59$ nm, where $N*$ corresponds to the number of independent patterns contained in a time scale of τ_c.

As mentioned above, the instantaneous pattern (during τ_c) of fiber-smoothed beam is expressed as the incoherent addition of many independent patterns. So the smoothing time of fiber-smoothed PCL would be remarkably shorter at the leading tail of pulse in comparison with that for ISI, because the number of beamlets of ISI is quite small in early part of the pulse. Here we defined the smoothing time of a single beam (τ_{sb}) as a time at which the normalized standard deviation of intensity distribution reaches to 10%. Then τ_{sb} is given by,

$$\tau_{sb} \approx 100\frac{\tau_c}{N*} . \tag{13}$$

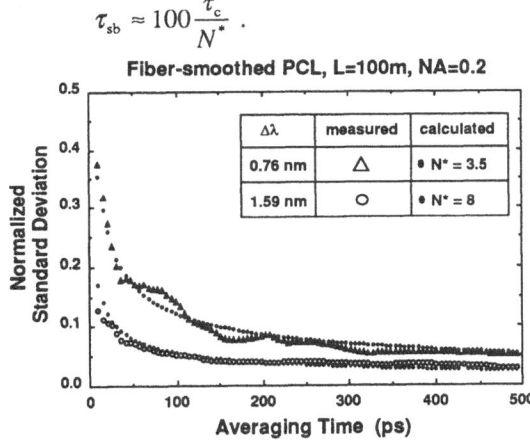

FIG. 18 Standard deviations of intensity distributions of the fiber-smoothed beam. Here the fiber length was 100m. Dotted lines are fitted curves obtained using Eq. (6) with $N* = 3.5$ for $\Delta\lambda = 0.76$ nm and $N* =8$ for $\Delta\lambda = 1.59$ nm.

267

From Fig. 18, we can estimate that $\tau_{sb} = 150$ ps for $\Delta\lambda = 0.76$ nm and $\tau_{sb} = 32$ ps for $\Delta\lambda = 1.59$ nm.

N* may be proportional to the fiber length and inversely proportional to the band width, which is expected from Eqs. (8) and (9). Therefore, even if the band width is limited by the frequency doubling efficiency, we can keep the smoothing time short using longer multimode fiber. But the optimum fiber length should exist because there is the limitation on the number of spatial modes which can transmit the fiber, and because the number of spatial modes included in a unit time decreases with increasing the fiber length. A method to solve this problem will be discussed in section 6-4.

4-5 Angular dispersion of PCL

A) Improvement of second harmonic conversion efficiency

The efficiency of second harmonic conversion can be improved by angularly dispersing the spectrum. The dispersion of index matching along the e-axis was measured to be 239 µrad/nm for type II KDP crystal. Because this angular dispersion should be satisfied at the final beam aperture, the corresponding dispersion at the output of PCL front end is 22.9 mrad/nm. To obtain this divergence using the grating of 1200 grooves/mm, the beam diameter was expanded to 80 mm at the grating (see section 3-1 and Fig. 4).

For the type II KDP, the polarization of the fundamental should be aligned to a direction 45° away from the e-axis, while the direction of angular dispersion should be parallel to the e-axis. So the direction of the spectral dispersion was rotated by 45° using Dove prism (R in Fig. 4).

B) Effect of angular dispersion on the beam smoothing

The focal spot diameter (FWHM) of PCL is evaluated from beam divergence as $\alpha F\lambda$. Substituting $\alpha = 40$, $F = 3.15$ and $\lambda = 1.053$ µm (α is defined for the fundamental light), the spot size is much smaller than the usual target sizes. Therefore RPP is effective to enlarge the spot size, and the bell-shaped intensity envelope of random phase beam is favorable to reduce low mode nonuniformity. Moreover RPP is needed for smearing out a central zero intensity region which is due to a beam stop set for an aim of optics protection from ghost focusing. Thus we should investigate the uniformity of PCL in conjunction with RPP.

Consider a situation where RPP is installed between the KDP crystal and the final focusing lens. Figure 19 schematically illustrates smoothing effects expected for the narrow band laser with RPP (a), PCL beam with RPP (c) and ADPCL with RPP (c). As is known very well, RPP creates small scale speckle structure. When PCL beam was used instead of the coherent laser, the substantial improvement occurs due to the reduced mutual coherence across the beam. The reduced mutual coherence causes many different speckle patterns giving rise to defused speckle peaks, although the broadening of speckle peak due to the interference between multiple wavelengths is not significant enough to smooth out the speckle. However, the very finite number of RPP elements contained in the coherence area cause an additional intensity perturbation of which wavelength become lager with increasing α. If the size of coherence zone can be matched to the RPP element size, the wavelength of this intensity perturbation reaches to the spot size, then this perturbation will disappear. Another small scale intensity perturbation determined by the residual mutual coherence would remain even if time averaged. This nonuniformity may be clearly revealed at quasi far-field positions.

On the other hand, much more smooth intensity profile could be obtained by using ADPCL. In this case, speckle patterns created by different wavelength components are continuously distributed along the dispersion direction. And the total shift of pattern is equal to or greater than the broadening due to beam divergence. Thus the remarkable smoothing occurs in the dispersion direction, whereas the smoothing in the perpendicular direction is the same as that for the simple PCL.

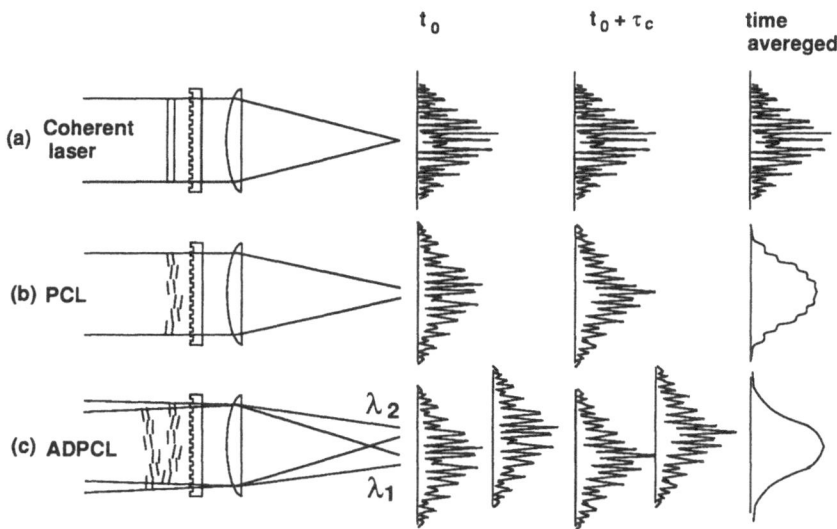

FIG. 19 Schematic illustration of beam smoothing for the coherent laser with RPP (a), PCL with RPP (b) and ADPCL with RPP (c), respectively.

C) Primary experiment at the fundamental wavelength

The additional smoothing due to the angular dispersion was examined at the output of PCL front end. Figure 20 shows a streaked pattern taken with a RPP for 1.053 μm wavelength. The ratio of the element size of this RPP to the diameter of ADPCL was the same as that at the target chamber. Data were taken at the quasi far-field position equivalent to the focusing condition of $d/R = -5$ with $R = 250$ μm. Also shown in Fig. 20 are spatial distributions for different averaging times. In this case the spectral width was limited to be $\Delta\lambda \approx 0.4$ nm which matched to the maximum beam divergence of determined by the acceptance of pulse shaping optical shutter (SEL 2). The intensity profile was rapidly smoothed, and this smoothing characteristics is shown in terms of the standard deviation of intensity in Fig. 21. Solid lines are fitted curves calculated using Eq. (6) with $N^* = 3.5$ for the beam divergence of $10F\lambda$, $N^* = 11.5$ for $20F\lambda$, and $N^* = 16.5$ for $40F\lambda$, respectively.

Single beam smoothing times estimated from Eq. (13) were $\tau_{sb} = 286$ ps for the beam divergence of $10F\lambda$, 87 ps for $20F\lambda$, and 61 ps for $40F\lambda$, respectively, as the spectral width was 0.4 nm. The smoothing time for $40F\lambda$ beam divergence (and $\Delta\lambda = 0.4$ nm) is substantially shorter than that obtained with the simple PCL generated by the optical fiber system ($\sim40F\lambda$, $\Delta\lambda = 0.76$ nm). N^* is nearly proportional to the beam divergence, so it is expected that the smoothing time is significantly shorter than that of SSD. Moreover, when we introduce RPP into this angularly dispersed PCL system, the target irradiation uniformity would be greatly improved in comparing with the usual PCL[17] (without angular dispersion) in conjunction with RPP.

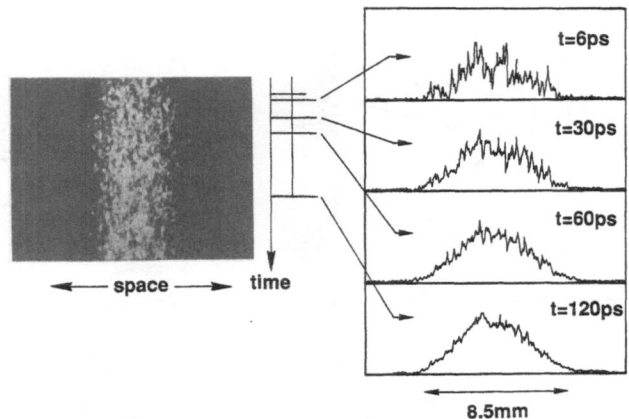

FIG. 20 Time resolved spatial intensity distribution of the angularly dispersed partially coherent light taken with a RPP. Time averaged intensity profile become rapidly smooth.

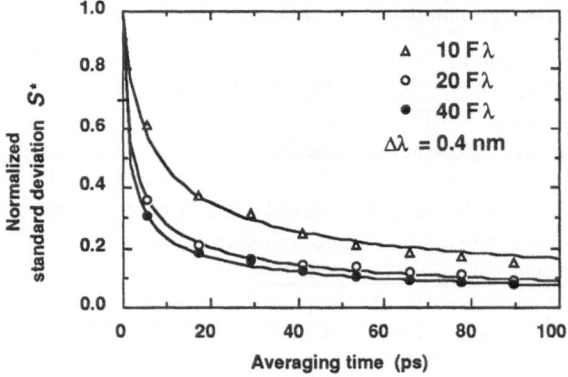

Beam divergence	N *
10 Fλ	3.5
20 Fλ	11.5
40 Fλ	16.5

FIG. 21 Standard deviations of spatial intensity distributions obtained with different beam divergences. Fitted curves indicate that the number of independent patterns (N^*) contained in a time scale of τ_c increases with increasing the beam divergence.

270

V. AMPLIFICATION AND SECOND HARMONIC CONVERSION CHARACTERISTICS

5-1 Amplification characteristics

PCL composed from small spatial spikes of which duration time is order of $1/\Delta v$. Therefore, our interests are in 1) energy gain characteristics, 2) the spectral narrowing, 3) the beam break-up and 4) any difference between the usual coherent laser and PCL in these amplification characteristics through main amplifier chain.

The gain characteristics observed by changing input energy was well predicted by an extension of the Frantz-Nodvik equation [18] for homogeneously broadened media, where the gain was modified as a function of a ratio of the incident spectral width to the gain width [14]. Figure 22 shows amplification characteristics of main chains. Data of two beams are plotted for different chain input energies by comparing with characteristics of the coherent laser beam (solid lines). There was not a significant difference between the PCL and the coherent laser.

In Fig. 23, the spectral width measured at the output of main amplifier chain is plotted as a function of chain output energy. The significant spectral narrowing was not observed at an energy range of 0.4-1.4 kJ/beam for ASE with the input spectral width of 1.7 nm (Gaussian). On the contrary, the spectral broadening occurred, probably due to the gain saturation near the center of spectrum, in the high power operation regime exceeding 1 kJ/beam. This characteristic is not favorable to the second harmonic generation (SHG), because the extended spectral wings depends on the saturation characteristics of each chain, thus it may degrade the power balance. For this reason, the spectral slicer (see Fig. 4) was installed into the PCL front end in order to cut wings sharply.

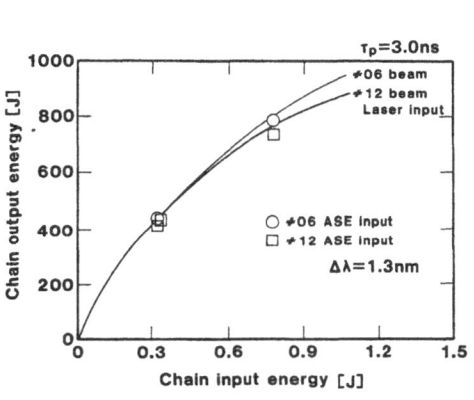

FIG. 22 Energy gain characteristics of main amplifier chains for PCL.

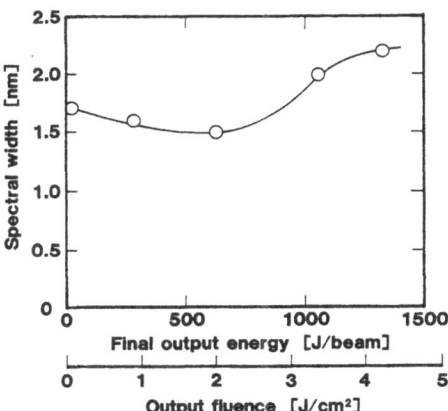

FIG. 23 Spectral width of chain output plotted as a function of the output energy. Here the spectral shape of incident ASE was Gaussian.

The significant beam break-up did not appear in NFP at output energies less than 1.4 kJ/beam. This would mean that high spatial frequency components does not build up due to the rapid change of beam pattern, suggesting no reduction of the maximum allowable output energy. These observed amplification/propagation characteristics can be applicable

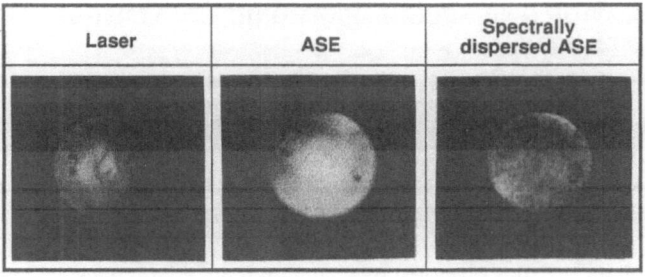

FIG. 24 Near-field patterns of the coherent laser (a), the simple PCL (b) and the ADPCL (c).

also to PCL generated by the optical fiber system. In addition, small scale intensity ripples in NFP of ADPCL is less sensitive to growth. NFPs taken for the coherent laser (a), PCL (b) and ADPCL (c) are compared in Fig. 24. Any characteristic is not so inferior to that obtained with the coherent laser. On the contrary, PCL is easy to treat in such a high power laser system since it is free from interference ripples arising from Fresnel diffraction.

5-2 Second harmonic conversion of high power PCL

Harmonic conversion is one of key issues addressing the efficient implosion. There are two subjects to be studied; one is the conversion efficiency, the other is the reproducibility of SHG among beam chains. The doubling efficiency depending on the spectral width of fundamental (ω) was measured for the type II KDP crystal of 20-mm thickness as shown in Fig. 25. In this measurement the incident power density was kept at ~0.3 GW/cm2. Low and scattered doubling efficiencies were observed at the wide ω spectral width of around 2 nm, and a little narrowing of 2ω spectrum was observed. If we require the efficient conversion with a reasonable reproducibility comparable to the narrow band laser, ω spectral width should be limited below ~1 nm.

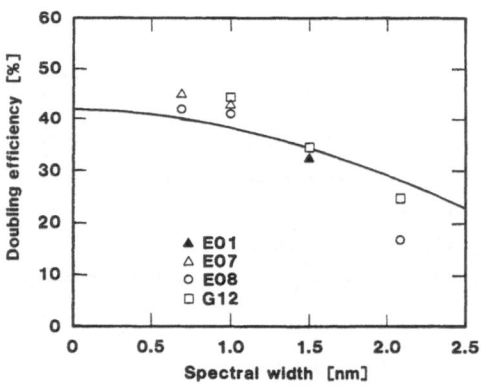

FIG. 25 Second harmonic conversion efficiency depending on the spectral width of ASE. Here the angular dispersion was not adopted.

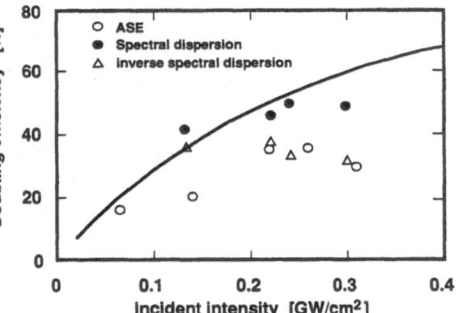

FIG. 26 Second harmonic conversion efficiency as a function of the incident fundamental power; dotted line: coherent laser, O: simple PCL, ●: ADPCL with phase matching, Δ: ADPCL with inverse dispersion direction.

Figure 26 shows doubling efficiencies measured by changing the incident power density. Here the beam divergence was kept to be $30F\lambda$ (~100 µrad at KDP crystal). A dotted line represents the mean efficiency for the narrow band laser. Plotted were taken with the simple PCL (open circles) and ADPCL (closed circles). Although the substantial improvement of doubling efficiency was observed by adjusting the angular dispersion of spectrum (phase matching angle) along the e-axis of KDP crystal, the significant reduction of efficiency occurred at higher intensity comparing with the narrow band laser. Also plotted (open triangles) are data obtained with the inverse dispersion of spectrum which degraded the conversion efficiency. The detailed SHG process of PCL are now under the study.

VI. TARGET IRRADIATION UNIFORMITY

6-1 Irradiation uniformity for spherical target without angular dispersion

The frequency-doubled PCL was focused through a RPP, and a time integrated pattern of the focused beam was measured with a magnification of 10. Figure 27 shows patterns of ASE ($\Delta\lambda\approx2$ nm, $\Delta\theta\approx30F\lambda$) and fiber-smoothed beam ($\Delta\lambda\approx1.6$ nm, $\Delta\theta\approx39F\lambda$. In this measurement, only the first single mode fiber and the multi mode fiber in Fig. 2 were used.). Quasi far-field patterns for the coherent laser beam with and without RPP are also shown for the comparison. The focusing conditions were equally chosen as $d/R=-5$ for the target of 500-µm diameter. A spot size determined by the beam divergence (~100 µrad) at the target plane is ~100 µm which is enough larger than the speckle size. Thus, it can be clearly seen that small scale speckles due to RPP is smoothed out for both ASE and fiber-smoothed beam. This smoothing effect appeared in the probability density distribution of intensity as a small standard deviation and small amplitudes of higher and lower intensity components.

Using these beam patterns, the irradiation uniformity was calculated by superposing twelve beams onto a spherical target of 500-µm diameter. Spherical harmonic components

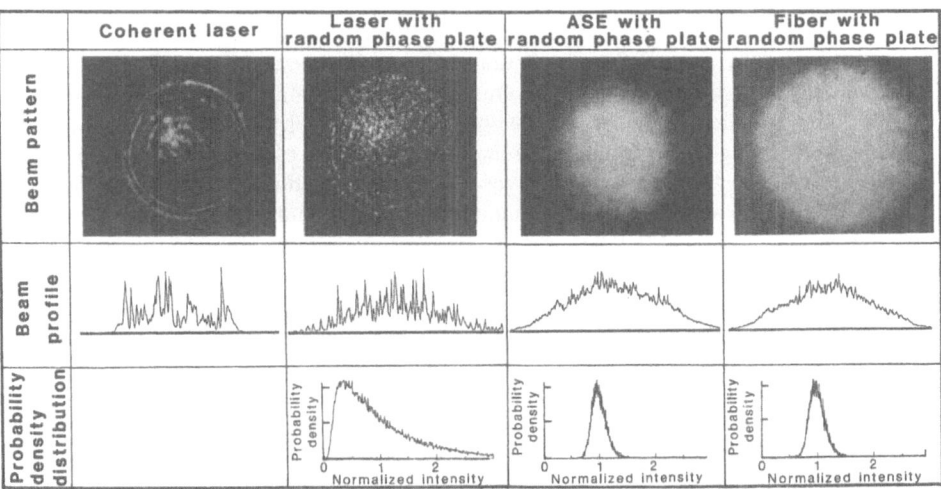

FIG. 27 Quasi far-field patterns for the different lasers. (a): coherent laser without RPP, (b): coherent laser with RPP, (c): ASE with RPP, and (d): fiber-smoothed beam with RPP. These patterns were observed at the plane equivalent to the focusing condition of $d/R = -5$ ($R = 250$ µm).

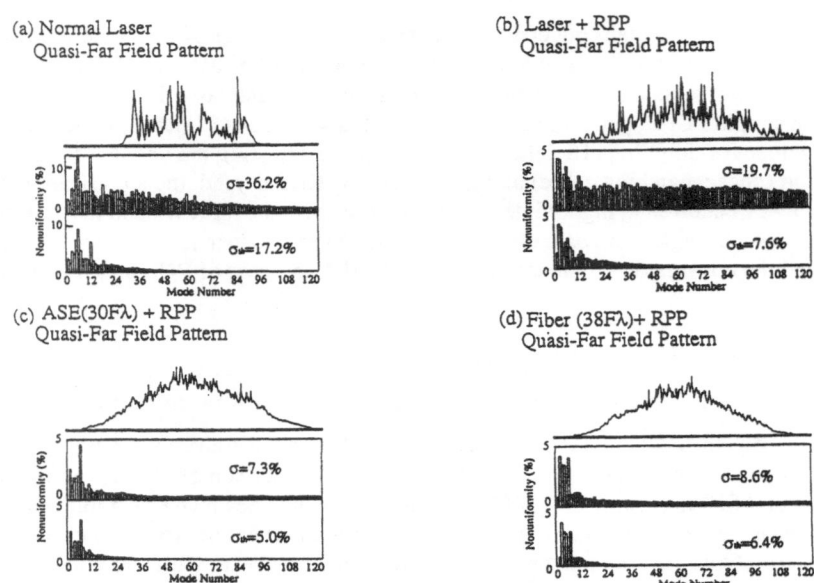

FIG. 28 Spherical harmonic modes of irradiation nonuniformity was calculated for each case of Fig. 27. Upper and lower histograms correspond to distributions without and with thermal smoothing.

of irradiation nonuniformity are shown in Fig. 28. The upper and lower histograms for each case correspond to distributions without and with thermal smoothing, respectively. Here we assumed 5% smoothing, i.e. an amplitude of l-th mode is reduced with a factor of $exp(-0.05l)$. Figure 28 suggests that R-T instability growth seeded by the shorter wavelength imprint would be greatly reduced in the case of PCL irradiation by a factor of ~5. A subject still remains is the control of the envelope shape of intensity distribution in order to suppress the growth of forced R-T instability with long wavelength.

6-2 Computer simulation of irradiation uniformity

In order to simulate the target irradiation, PCL was treated as follows. The frequency doubled PCL is composed from a large number of monochromatic components of which wavelengths are randomly chosen. The intensity of each wavelength component was determined according to an assumed time integrated spectrum of $\Delta\lambda(\omega)/2$ (FWHM). Here we assume that each component is plane wave with a propagation direction different with each other. Propagation directions are randomly distributed within an angle of $\delta\theta$. Although this situation does not realize PCL necessarily, these assumptions make it easy to treat PCL in a hydrodynamics simulation. Furthermore, these assumptions would be reasonable at least for ASE because it was composed from narrow spectral components of which time duration were ~100 ps (much longer than the coherence time evaluated from the time integrated band width).

PCL was divided into a huge number of beamlets by RPP, then focused by a lens of $F/3.15$. Simulation results showed that the focused beam pattern was improved with increasing $\delta\theta$, and almost independent on the number of wavelength components (N_w) when $\delta\theta$ is smaller than ~10 μrad. This calculation result may be explained by the fact that the maximum shift of speckle pattern is the same order of the distance between peaks of speckle. As the beam divergence became larger, the pattern was smoothed with increasing N_w because of the superposition of many shifted speckle patterns. Simulations showed that the time averaged intensity at a fixed position became closer to an asymptotic value by the time of 15-20 times τ_c, which is consistent with the observed smoothing time discussed in section 4-5 c). A new code treating the spatial incoherency is under the development which will be coupled with a hydrodynamics code, HISHO, in order to estimate the spatial and temporal incoherency required for future experiments at Gekko XII upgrade.

6-3 Irradiation uniformity with angular dispersion

The time integrated intensity distribution of the frequency-doubled ADPCL was measured by a cooled CCD camera. The image on a plane equivalent to that for $d/R=-6$ ($R=250$ μm, i.e. focusing condition of $d/R=-5$) is shown in Fig. 29. The measurement was carried out by changing the spectral width of fundamental and the beam divergence. Inserted is a diagram showing directions of spectral dispersion of ω and 2ω. Although the streak along the 2ω dispersion is visible when the beam divergence is small, this intensity modulation disappears for larger beam divergence.

The improvement on medium and higher mode nonuniformity is clearly seen in Fig. 30. This figure shows three dimensional displays and intensity contours of quasi far-field patterns ($\Delta\lambda=0.8$ nm) shown in the bottom of Fig. 29. Also shown is the same plots of data for the simple PCL which was measured with the same experimental setup and the same d/R condition. In these intensity profiles, the asymmetry in the horizontal direction was due to a damage in the KDP crystal of the diagnostic beam. Probability density

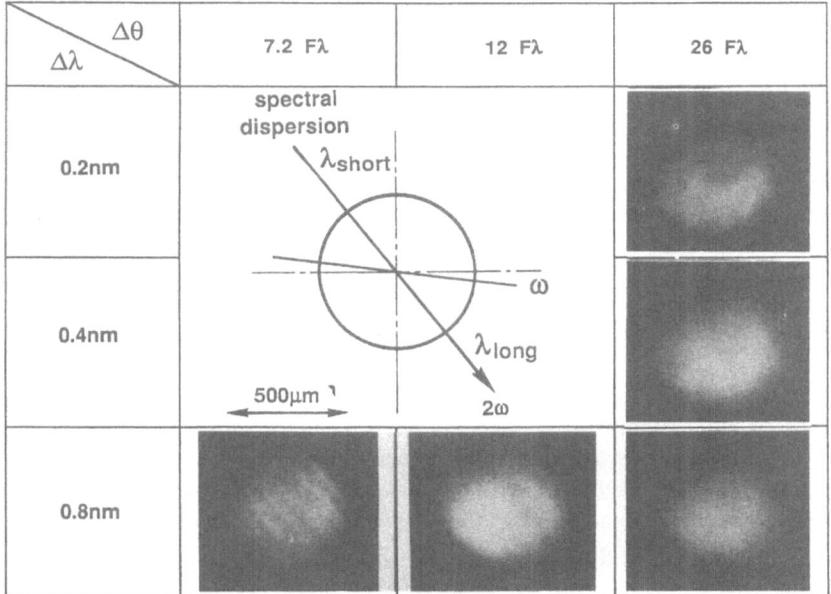

Δθ : beam divergence   Δλ : spectral width   DL : diffraction limit

FIG. 29 Quasi far-field patterns ($d/R = -5$) of ADPCL observed by changing the beam divergence and the spectral width.

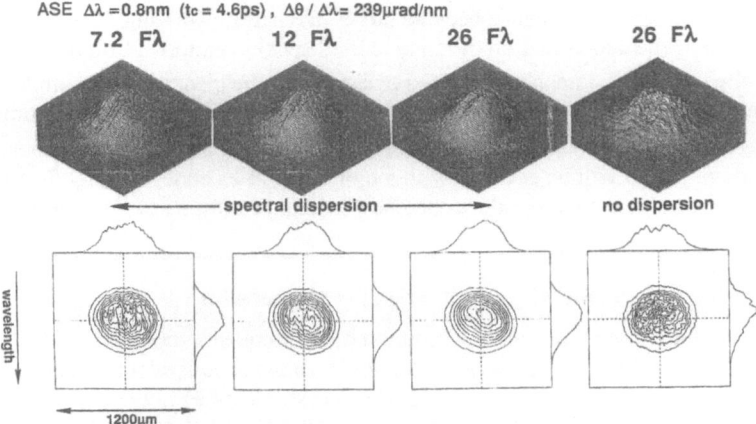

ASE $\Delta\lambda$ =0.8nm (t_c = 4.6ps) , $\Delta\theta$ / $\Delta\lambda$= 239µrad/nm

7.2 Fλ 12 Fλ 26 Fλ 26 Fλ

← spectral dispersion →   no dispersion

wavelength

1200µm

FIG. 30 Three dimensional displays and intensity contours of quasi far-field patterns of ADPCL taken with RPP ($\Delta\lambda$ = 0.8 nm). Also shown are the data for the simple PCL with RPP. The equivalent focusing condition is d/R = -5 (R = 250 µm).

distributions corresponding to patterns in Fig. 29 are shown in Fig. 31. A distribution for the narrow band laser is inserted for the comparison. The standard deviation S became smaller with increasing $\Delta\lambda$ and $\Delta\theta$, then reached to 3.3% at the condition of $\Delta\lambda$ = 0.8 nm and $\Delta\theta$ = 26$F\lambda$. Therefore the remarkable reduction is expected on the irradiation nonuniformity, especially for medium and higher spherical harmonic components.

The irradiation uniformity was estimated for the target of 500-µm diameter as shown in Fig. 32. Here we used the quasi FFP for $\Delta\lambda$ = 0.4 nm and $\Delta\theta$ = 26$F\lambda$. Nonuniformities with mode numbers larger than 20 decreased by a factor of greater than 8 or more in comparison with the coherent laser with RPP, suggesting the significant reduction of R-T growth seeded by the irradiation nonuniformity imprint. In regard to the lower modes, the control of intensity envelope was not satisfactory. An elliptical distortion of intensity contour arising from different characteristics along directions perpendicularly intersecting each other, thus it caused low mode nonuniformity. So the improvement of envelope shape may be the most important subject in applying ADPCL to the implosion experiment. The calculated rms nonuniformity was 6.3%, and the thermally smoothed nonuniformity was 5.1% assuming reduction factor of $exp(-0.05l)$. These values will be improved in near future by introducing additional smoothing techniques described later.

Time resolved intensity distributions were also measured by streaking the quasi near-field patterns shown in Fig. 29, and it was observed that the substantial smoothing occurred by <100 ps. Another issue which affects the uniformity is the intensity dependence of pattern due to the nonlinearity of SHG. Fig. 33 shows the difference in modal distributions of nonuniformity estimated at a high intensity operation (($\bigcirc$): fundamental energy of 1120 J/beam) and a low intensity (($\blacksquare$): 11 J/beam). Whereas a little enhancement of nonuniformity can be seen at around $l \approx$ 6-24, the significant degradation of σ_{rms} was not observed.

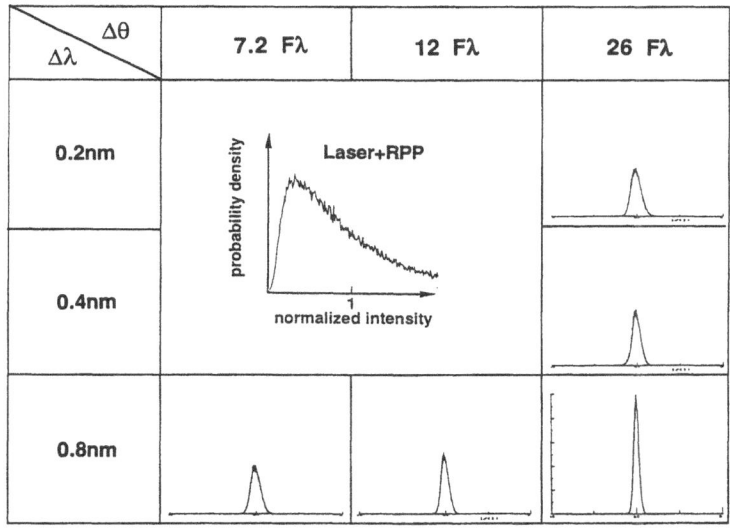

$\Delta\theta$ $\Delta\lambda$	7.2 Fλ	12 Fλ	26 Fλ
0.2nm			
0.4nm			
0.8nm			

FIG. 31 Probability density distributions of quasi far-field patterns shown in Fig. 29. Inserted is a distribution measured for the coherent laser with RPP.

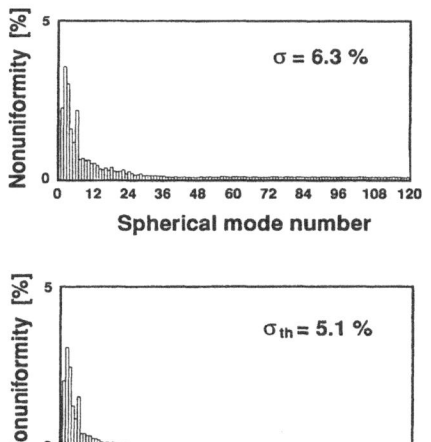

FIG. 32 Spherical harmonics distribution of irradiation non-uniformity estimated from a measured quasi far-field pattern of ADPCL with RPP. Upper and lower histograms correspond to distributions without and with thermal smoothing.

6-4 Other smoothing techniques

A) Polarization control by a distributed wave plate

Beam smoothing strongly depends on the instantaneous number of patterns (N^*) as discussed in chapter IV. A simple way to increase N^* is the addition of perpendicularly polarized two beams which do not interfere each other. Two polarization components can be distributed across the beam as the phase randomized by RPP. A distributed wave plate

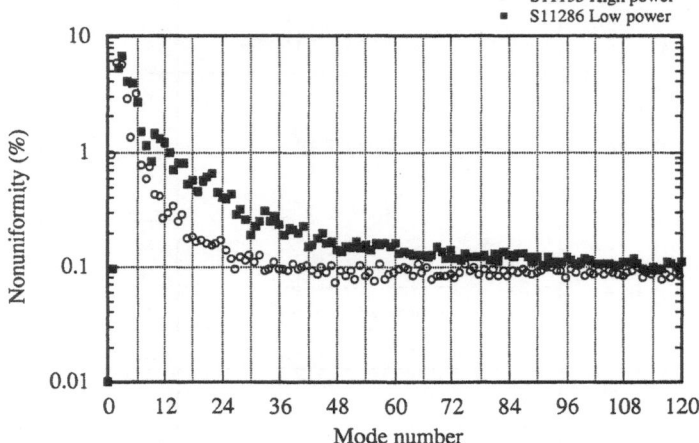

FIG. 33 Comparison of nonuniformities at different intensities of ADPCL. Chain output (ω) energies are 1120 J/beam ($\bigcirc$) and 11 J/beam ($\blacksquare$).

(DWP) can be made from segmented liquid crystal arrays. The computer calculation was carried out for the plane wave with DWP in front of RPP. The probability density distribution was improved in comparison with a case in which only RPP was used, so that the nonuniformity was reduced by a factor of $\sqrt{2}$. Here the number of polarization elements was 64, the element size of RPP was 2 mm, and the whole beam diameter was 320 mm. Calculations also showed that the smoothing did not so strongly depend on the arrangement of wave plate elements. A preliminary experiment using small array of wave plates has been done, and we observed the improvement of uniformity consistent with the theoretical estimation.

B) Two dimensional angular dispersion

In the same manner as DWP, two ADPCL beams of which dispersion directions are perpendicular each other can be added incoherently, if the time shift between them is larger than the coherence time. This method can be easily adopted at the output of PCL front end. In this case we need a large array of KTP crystal as a second harmonic generator, which relax the phase matching condition because of its large acceptance angle and large nonlinear coefficient. We have done preliminary experiments on SHG of broad band laser using KTP crystal. The growing technique of large KTP crystal is being developed at Osaka university.

C) Intensity addition of independent patterns using optical fiber coupler

The increase of instantaneous pattern number N^* could be achieved by using the optical fiber technology. Now we have prepared a multimode optical fiber divider (1×32) and coupler (32×1). The idea is as simple as 32 PCL beams are delayed with fibers of different lengths, and coupled again to one beam. The delay time difference will be optimized to obtain the maximum N^*. In addition, the optical fiber coupler will be useful to control the pulse shape in near future experiments.

D) Repetitively chirped pulse by cross phase modulation using different wavelength laser

To date, we are adopting the self phase modulation in order to obtain broad spectrum. Another technique is being planed in which two different wavelength lasers will be used. A promising approach may be to use the beat between a Nd:YLF ($\lambda = 1.053$ µm) laser and a frequency selected Nd:YAG ($\lambda = 1.052$ µm) laser. Using cross phase modulation in an optical fiber, we can obtain repetitively chirped long pulse of which chirping cycle is as short as 3.7 ps. This broad band chirped pulse can be applicable to a PCL generator using multimode optical fiber, and also useful for the modified SSD[11] which has been introduced into Nova system.

VII. CONCLUSION

A variety of efforts have been carried out to improve the laser irradiation uniformity by introducing the partially coherent light (PCL) into the glass laser system Gekko XII. Front-end generators of PCL have been constructed for the ASE generator and the optical fiber system. Both front ends have a capability of band width control. The coherence area was about 1/30 of the beam diameter which was consistent with the beam divergence of $30\text{-}40F\lambda$. This induced spatial incoherency give rise to the increase of independent patterns contained in the time scale of τ_c, thus the smoothing time is shortened comparing with the usual broad band laser without beam divergence. The coherence time was measured to be well below 10 ps. The overall smoothing time was enough shorter than 100 ps. The optimization of beam divergence and spectral width is now under the study.

The images (NFP and FFP) of PCL were relayed to the target chamber through amplifier chains. The gain characteristics can be explained by the modified Frantz-Nodvik equation, and no significant reduction of energy gain was observed with a slight change of spectrum. The amplification characteristics showed the capability of high power PCL system because the significant growth of intensity ripples was not observed.

However, the second harmonic conversion efficiency should be improved by a double KDP crystal technique [19] or new doubling crystals such as KTP. The substantial improvement of SHG was obtained by introducing the angular dispersion of spectrum into PCL system. Now, an approach using segmented KTP crystals is under the development.

The beam pattern on the target plane was greatly improved in comparison with that of normal laser beam with RPP. Moreover, the angularly dispersed PCL with RPP showed quite smooth intensity profile. Especially, higher mode nonuniformities were drastically suppressed, which is important to reduce the growth of R-T instability caused from the laser nonuniformity imprint. Although the envelope shape of the pattern has not been optimized yet, the irradiation nonuniformity (σ_{rms}) of Gekko XII will be reduced from 19.7% (estimated for the coherent laser with RPP) to 6.3%, and also the thermally smoothed nonuniformity (σ_{th}) will be reduced reaching to 5.1%. These values will be improved to a level of $\leq 3\%$ in near future.

ACKNOWLEDGEMENTS

The authors would like to thank K. Yagi for his contribution in carrying out this work, and greatly appreciate the useful discussions and the cooperation of T. Jitsuno, N. Nishi and the laser operation crew in doing the experiments.

REFERENCES

1) S. Nakai, et al., *Laser Interaction and Related Plasma Phenomena*, Vol.(19), ed. by H. Hora and G. H. Miley, Plenum Press, (1990) p. 25.

2) H. Azechi, et al., Laser and Particle Beams, **9**, 193 (1991).

3) R. L. MaCrory, et al., Nature, **335**, 225 (1988).

4) Y. Kato et al., Phys. Rev. Lett., **53**, 1057 (1984).

5) S. Skupsky, et al., J. Appl. Phys., **66**, 3456 (1989).

6) D. Veron, et al., Opt. Commun., **65**, 42 (1988).

7) R. H. Lehmberg, S. P. Obenschain, Opt. Commun., **46**, 27 (1983).

8) R. H. Lehmberg, A. J. Schmitt, S. E. Bodner, J. Appl. Phys., **62**, 2680 (1987).

9) S. Skupsky and K. Lee, J. Appl. Phys., **54**, 3662 (1983).

10) J. W. Goodman, *Laser Speckle and Related Phenomena*, ed. by J. C. Dainty, Springer, Heidelberg (1975) pp. 9-75.

11) S. N. Dixit, M. A. Henesian, et al., This calculation result was reported at Conference on Lasers and Electro-Optics, Baltimore, Maryland, USA, May 14, 1991.

12) K. A. Brueckner and S. Jorna, Rev. Mod. Phys., **46**, 325 (1974).

13) W. M. Manheimer, D. G. Colombant and J. H. Garder, Phys. Fluids, **25**, 1644 (1982).

14) H. Nakano, et al., Opt. Commun., **78**, 123 (1990).

15) H. Nakano, et al., proceeding of IAEA Technical Committee Meeting on Inertial Confinement Fusion Driver, April, 1991, Osaka, Japan (PI-06).

16) M. Born and E. Wolf, *Principles of Optics*, Pergamon Press, Oxford (1975).

17) N. Miyanaga, et al., proceeding of IAEA Technical Committee Meeting on Inertial Confinement Fusion Driver, April, 1991, Osaka, Japan (PI-07).

18) M. D. Rotter, et al., Opt. Commun., **71**, 311 (1989).

19) M. S. Pronko, et al., IEEE J. Quantum Electron., **QE-26**, 337 (1990).

TIME-RESOLVED MEASUREMENTS OF PLASMA CONDITIONS AND GRADIENTS IN RADIATIVELY-HEATED FOILS

J. Edwards, M. Dunne, D. Riley, R. Taylor, O. Willi, S.J. Rose[1]

Blackett Laboratory, Imperial College, London SW7 2BZ, U.K.
[1] Rutherford Appleton Laboratory, Chilton, Didcot, Oxon. OX11 0QX, U.K.

INTRODUCTION

It is now possible to produce intense approximately Planckian X-ray pulses in the laboratory with high-power laser systems. These are currently of considerable interest as possible drives in inertial confinement fusion experiments. Also, they can be used for the production of uniform, gradient-free plasmas which cannot be produced by other means in the laboratory and which are ideal for the testing of theoretical calculations (of e.g. opacities and radiative transfer) relevant for laboratory as well as astrophysics research.[1,2] Therefore, investigations of the interaction of intense soft-X-ray pulses with targets are important for the design of future experiments. In this paper we present the results of several experiments performed using the VULCAN laser at the Rutherford Appleton Laboratory, and designed to investigate the interaction of soft-X-ray pulses with low-Z planar targets.[3,4,5]

EXPERIMENTAL

Soft-X-ray pulses were generated from the rear of laser-irradiated, planar high-Z converters placed in close proximity to the target foil (figure 1).

Laser Interaction and Related Plasma Phenomena, Vol. 10
Edited by G.H. Miley and H. Hora, Plenum Press, New York, 1992

281

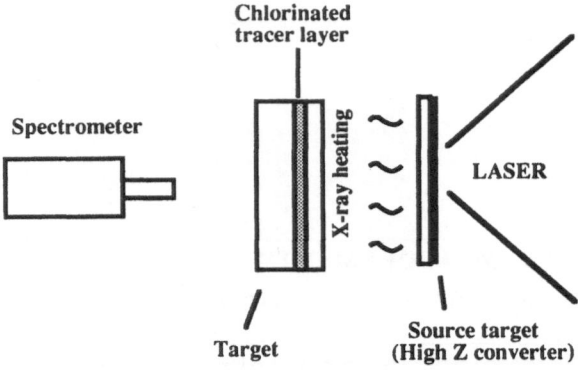

Figure 1. Experimental set-up for radiation heating experiments.

The laser wavelength was 0.53μm, the irradiance was around 10^{14}Wcm^{-2} and the pulse length was typically 1ns. The conversion of laser to X-ray energy was measured with time-resolving and time-integrating XUV photodiodes. The targets were made of plastic (CH) up to 6μm thick, with and without a thin 0.3μm tracer layer of chlorinated plastic ($C_8H_6Cl_2$) buried at different depths in the foils. The soft-X-rays transmitted through the targets were analysed using time resolved XUV spectroscopy in the 50A spectral wavelength region. The spectral and temporal resolutions of the instrument were 0.3A and 50ps respectively. A more detailed description of the experimental set-up is presented elsewhere.[3]

Absorption of the heating radiation by bound-bound (b-b) transitions of chlorine ions in the tracer layer was used to provide a time history of the layer temperature as a function of depth in the target. These measurements were compared with radiation-hydrocode calculations which could be brought into agreement with the observations by varying the total soft-X-ray energy incident on the targets. The model thus established was used to predict conditions in pure CH samples and to show with the aid of a spectral analysis package that these conditions were consistent with the formation of an edge-like structure close to the cold carbon K-edge and resulting from bound-bound transitions.

MODELLING

Radiation-hydrodynamics

The experimental conditions are simulated using a multi-group radiation transport model coupled to the 1D Lagrangian hydrodynamics code.[4] The radiation transport model uses group-averaged Planck-mean opacities calculated either from a screened-hydrogenic average-atom (AA) model, SHOP, or from a detailed configuration calculation, IMP.[6] Both models provide tabulated opacity data for the radiation package and SHOP can be used to provide opacities in-line with the hydrodynamics. The SHOP model is similar to the XSN model[7] but does not include b-b transitions while the IMP code uses a UTA approximation to account for b-b opacity. Both opacity models assume local thermodynamic equilibrium. This approximation was assessed using a time-dependent AA model and was found to be a good approximation throughout most of the simulations. The emission from the gold foil is not calculated in the simulations. Instead, the experimentally measured X-ray pulse profile is used in the code.

Spectral Analysis

Detailed spectral analysis of the experimental results has been performed using a spectral analysis package, SAP. The atomic data base used in SAP is generated from a multi-configuration Dirac-Fock code.[8] In the calculation of the transition energies the initial and final atomic states are optimised separately. The configuration probabilities are calculated using the Saha equation (i.e. assuming LTE). Absorption spectra are predicted by solving the 1D transfer equation.

EXPERIMENTAL RESULTS

Photodiode Measurements

Measurements were made on bismuth layers 2000A thick suported on 1μm of CH for laser irradiances between 1 and 5 10^{14}Wcm^{-2} with a time-resolved XUV photodiode. The diode was positioned at an angle of 50 degrees to the target normal. It was found that the soft-X-ray pulse emitted to the rear of the target closely followed the laser pulse. The conversion efficiency in the 100eV-1.2keV photon energy region was measured to be 5.5±2% with no obvious dependence on the laser irradiance in the experimental range. This figure was obtained assuming a radiation field isotropic in the half-plane to the rear of the target The conversion efficiency corresponds to radiation temperatures in the 100-130eV range at the rear of the target. These values are in reasonable agreement with time-integrated diode measurements taken at the same time[9] and with other measurements taken under very similar experimental conditions.[10] The 5.5% figure is calculated as a fraction of the total laser energy delivered by the laser and does not take into account losses due to the RPPs. A very similar conversion efficiency was found when the same ρR of gold was used in place of the bismuth.

XUV Spectra From Targets with no Tracer Layer

Figure 2 shows streak records obtained when pure CH foils of different thickness were used (0, 2 and 6μm).

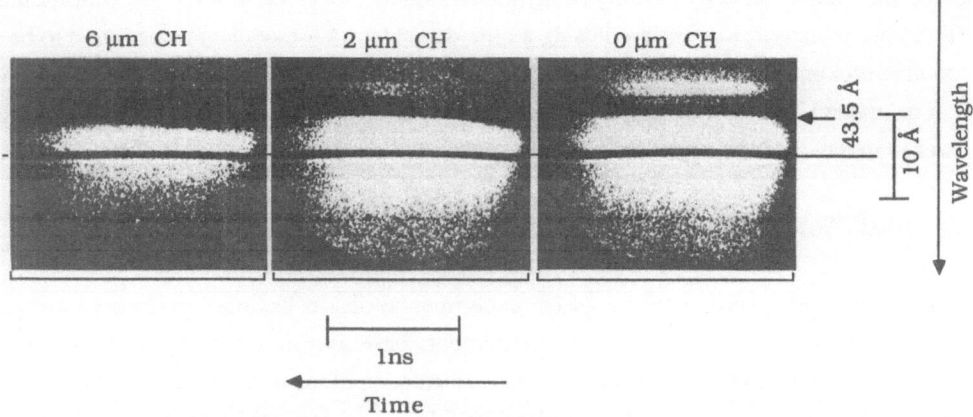

6 µm CH   2 µm CH   0 µm CH

43.5 Å

10 Å

Wavelength

1ns

Time

Figure 2. Streak records taken on the rear of CH foils

"Zero thickness" shows the streaked spectrum when no sample foil was present (source spectrum). The sharp feature at 43.5A in the source spectrum is due largely to the carbon K-edge in the instrumental filters. When a sample foil is present, it is very clear that an edge feature exists on the low energy side of the cold carbon K-edge and that this edge feature shifts towards higher energy, by up to approximately 10eV, (shorter wavelength) as the sample foil is heated. The shift is more rapid and pronounced for the thinner target. Spectra were also taken using sample CH foils of 3 and 4.5µm. These showed a corresponding delay and reduced shift of the edge feature compared to the 2µm case. It has been suggested that these edge features actually result from b-b transitions of partially ionized carbon in the sample foil.[4,5]

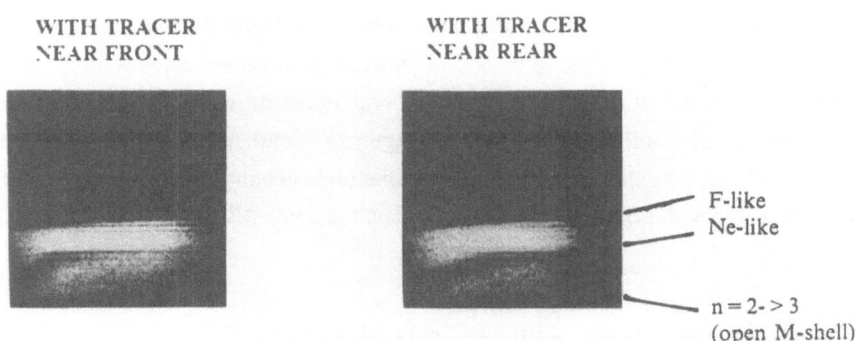

WITH TRACER
NEAR FRONT

WITH TRACER
NEAR REAR

F-like
Ne-like

n = 2-> 3
(open M-shell)

Figure 3. Streak records showing chlorine L-shell absorption features produced by a chlorinated tracer layer positioned near the front and rear surfaces of a radiatively heated CH foil.

XUV Spectra from Targets with Tracer Layers

Figure 3 shows streaked spectra transmitted through 2μm thick CH sample foils containing 0.3μm tracer layers 0.2μm below the surface (left) and 1.8μm below the surface (right).

The sharp feature at 43.5A is due to a combination of the instrumental filters and the sample foil. The clear absorption features observed in the two spectra above 43.5A are due to b-b chlorine L-shell transitions (mainly 2p-3d). The unresolved band of absorption beginning at around 60A and shifting rapidly towards shorter wavelength results from 2p-3d transitions of Cl to Mg like chlorine ions. The positions of 2p-3d transitions of Na- Ne- and F-like chlorine are also marked.

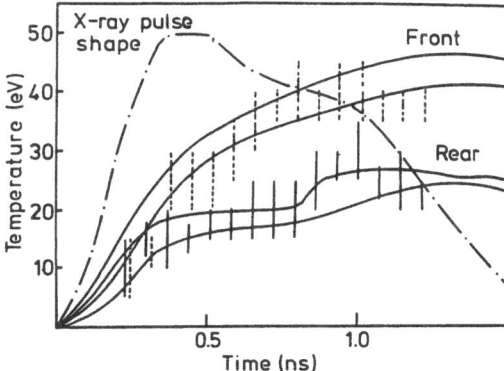

Figure 4. Comparison between measured and predicted temperature histories in chlorinated tracer layers buried at different depths in CH foils.

The two spectra are significantly different. When the layer is placed 0.2μm below the front surface of the sample, the unresolved band of absorption shifts towards shorter wavelength (higher ionization stages) noticeably more quickly than when the layer is placed 1.8μm below the sample surface. This indicates a more rapid ionization of the front layer. Also, absorption structure resulting from F-like chlorine can be seen when the layer is placed near the front of the target whereas no significant evidence for F-like chlorine can be found for the rear layer. The rear layer on the other hand exhibits strong absorption due to Ne- and Na-like chlorine later in the pulse. The higher degree of ionization occuring in the front layer clearly indicates that higher temperatures are attained than in the rear layer.

MODELLING

First, to analyse the chlorine absorption spectra, synthetic absorption spectra were generated using SAP for temperatures between 5 and 50eV at 5eV intervals and for densities between 1 and 0.005 g/cc with a ratio between successive values of approximately 2.

Temperature histories in the layers were inferred by comparison of the predicted and measured spectra. The density of the layer at each time was taken to be that predicted by the hydrocode. Figure 4 shows a comparison between the temperature histories inferred from the absorption spectra and those predicted by the hydrocode which required the assumption of a 2.5% soft-X-ray conversion efficiency from the gold source foil as opposed to the 5.5% measured value. (This discrepancy could be due to an incorrect assumption of an isotropic radiation field at the source target, variation in the conversion efficiency or the model opacities or the 1D assumption of the calculation.)

The solid curves show the range of values predicted by the hydrocode to exist in the layers, while the vertical bars show the experimental measurements. The uncertainty in the inferred

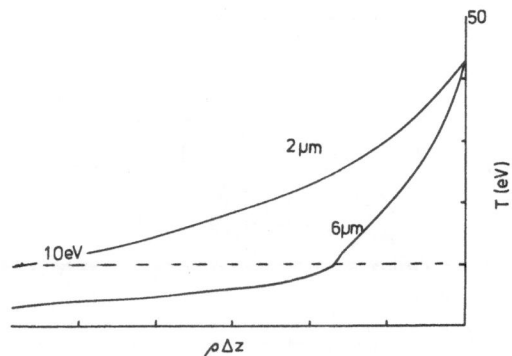

Figure 5. Predicted temperature profiles in CH foils at the peak of the pulse.

temperature histories arising from taking the layer densities to be those predicted by the hydrocode has been estimated by assuming a density error of a factor of 2 when comparing the experimental and synthetic spectra. This leads to an approximately 10% uncertainty in the inferred temperatures. The additional error resulting from an up to factor-of-2 change in the soft-X-ray conversion efficiency (see above) has also been included but is substantially less than the 10% resulting from the factor-of-2 assumed density error. The vertical error bars also include the estimated uncertainty both due to the discrete values of temperature and density at which the synthetic spectra were calculated and to gradients in the layers. A more detailed discussion of the experiment and analysis can be found elsewhere.[3]

Simulations for the pure CH targets were then performed using the hydrocode and the 2.5% conversion efficiency which gave agreement with the tracer targets. Figure 5 shows temperature predicted by the hydrocode to exist in the 2μm and 6μm CH targets near the peak of the pulse. The ordinate is a Lagrangian co-ordinate which is different for each foil. Note that the entire 2μm target is expected to have a temperature above 10eV by the peak of the pulse whereas the majority of the 6μm foil is expected to have a temperature well below 10eV at this time.

Figure 6 shows the transmission above a wavelength of 42A through a CH foil of initial thickness 3μm as calculated by SAP including only 1s-2p b-b transitions of carbon ions for three different plasma temperatures. The plasma density is taken to be that predicted by the hydrocode at the peak of the pulse, 0.1g/cc. Clearly, the b-b transitionsform an edge-like structure close to the cold carbon K-edge wavelength. Also, as the plasma temperature

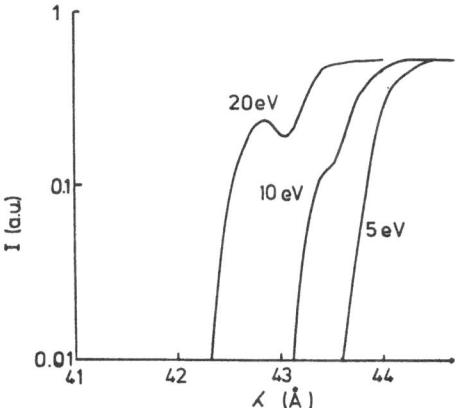

Figure 6. "Edge" structure resulting from carbon 1s-2p bound-bound transitions in a CH foil with column density $3\ 10^{-5}$ gcm^{-2} at three different temperatures.

rises, the position of the edge feature shifts towards shorter wavelength as a result of increased ionization. The maximum predicted "edge" position wavelengthin figure 3 is close to 44A, just 0.5A to the longer wavelength side of the cold carbon K-edge. In contast, the maximum measured red-shift of the edge feature is around 1.5A (for the 6μm foil) giving an edge wavelength of 45A. However, the position of the bound-bound "edge" depends not only on the accuracy of the atomic physics data but also on the plasma conditions, the lineshapes, the detail of the b-b structure included in the calculation and the sample thickness (which is larger for the 6μm foil than used for figure 6). Therefore, while further analysis is required to accurately predict the absolute position of the observed b-b "edge", there seems little doubt that the experimentally observed features result from b-b transitions of carbon ions.

CONCLUSIONS

In conclusion, experiments have been performed to measure the temperature histories in thin chlorinated tracer layers buried at different depths in planar CH targets using time-resolved XUV spectroscopy. Radiation-hydrocode simulations reproduced the measured temperature histories well but only when half the experimentally measured soft-X-ray flux was used in the calculation. This is probably due to the 1D assumption of the radiation transfer model and/or an incorrect assumption of an isotropic radiation field at the source. Spectra taken on pure CH foils exhibited an edge-like feature close to the cold carbon K-edge wavelength. This edge feature shifted by around 1A during the soft-X-ray heating and could be explained by absorption due to b-b transitions of carbon ions.

REFERENCES

1. S.J. Davidson et. al., Appl. Phys. Letts. **52**, 847, (1988).
2. J. Foster et. al., PRL **67**(23), 3255, (1991).
3. J. Edwards et. al., PRL **67**(27), 3780, (1991).
4. J. Edwards et. al., Europhys. Letts. **11**(7), 631, (1990).
5. J. Edwards et. al., "Proceedings of the International Workshop on Radiative Properties of Hot Dense Matter", World Scientific, Singapore, (1990).
6. S.J. Rose, accepted for publication in J. Phys. B, (1992).
7. W.A. Lokke and W.H. Grasberger, Lawrence Livermore National Laboratory Report No. UCRL-52276, (1977).
8. I.P. Grant et. al., Comput. Phys. Commun. **21**, 207, (1980).
9. C.A. Back et. al., "Proceedings of the International Workshop on Radiative Properties of Hot Dense Matter", World Scientific, Singapore, (1990).
10. J. Foster, private communication, (1989).

A SURVEY OF ION ACOUSTIC DECAY INSTABILITIES IN LASER PRODUCED PLASMA

Katsu Mizuno

Plasma Physics Research Institute
University of California, and Lawrence
Livermore National Laboratory, Livermore
California USA 94550

* In collaboration with R. P. Drake, W. Seka, R. Bahr, K. Estabrook, and J. S.DeGroot

INTRODUCTION

The ion acoustic decay instability (IADI) is a fundamentally important subject in plasma physics. The IADI is one of the concerns in laser fusion studies because of the production of hot electrons. Laser fusion requires a large scale plasma. We have for many years studied the IADI in microwave simulation experiments of laser-plasma interactions, and in laser produced plasma. The feature and the significance of the IADI in a large scale plasma are quite different from small scale plasma. The threshold is low, and the relatively weak intensity laser can heat significant hot electrons in a large scale plasma.

In the ion acoustic decay instability (IADI), an electromagnetic wave decays into two electrostatic waves: an electron plasma wave (epw) and an ion acoustic wave (iaw) near critical density (where electromagnetic wave frequency equals to plasma frequency). The instability occurs when the electromagnetic wave intensity exceeds a threshold. The ion acoustic decay instability has been studied by numerous authors by means of theory[1], microwave experiment[2], ionospheric experiment[3], and computer simulation[4]. There were evidences[5-7] that the IADI was excited in laser-plasma interactions. Microwave simulation experiments[8-9] of the laser-plasma interactions indicate that the epw of the IADI can produce significant quantities of hot electrons. Such hot electrons are of concern of laser fusion studies because they can preheat the target and degrade compression.

Laser Interaction and Related Plasma Phenomena, Vol. 10
Edited by G.H. Miley and H. Hora, Plenum Press, New York, 1992

289

On the other hand, it has been hoped that IADI would not constrain laser fusion target design, for two reasons. First, once the plasma has formed, the laser intensity will have been significantly attenuated by collisional absorption when it reaches the densities near the critical surface where IADI occurs. Second, it has been generally believed that the density profile near the critical surface would be quite steep even in a high-gain laser-fusion target, as the results of both ponderomotive and ablative steepening. As a result, it was anticipated that the threshold of the IADI may be substantially increased by convective stabilization.

A number of authors[5-7,10-14] have studied emissions near the second harmonic signals in laser produced plasmas. A Stokes sideband is often observed that is attributed[5-7] to the IADI. In the previous experiments, the use of very short (100 ps) laser pulses, very high laser intensities, or small laser spot sizes have produced complicated signals or high threshold values. In contrast, we have used relatively long laser pulses ($\tau_L \sim 1$ nsec) and large (up to D=900 μm) laser spot relevance to laser fusion. Our reports[15-16] show that the IADI threshold can be much lower than anticipated. In a large scale planar plasma, it reached homogeneous-plasma collisional values. Our experiments indicate that the IADI is excited on a shallow underdense shelf rather than on a steep density at the critical surface. The IADI can be excited on a large region. Hot electrons heat[4,9] when they travel the instability region, so hot electron heating increases with increasing the instability volume. An important point is that the hot electron heating can be significant even if the IADI is weakly excited, if the instability volume is large enough on a shallow density shelf. The heat flux of the hot electron increases proportionally to the product of the heating rate (anomalous collision frequency) of the epw and the instability volume. Our estimates based on the particle computer simulation indicate that a significant heating of hot electron can be possible.

The intensity of the laser light that penetrates to the critical density during the early phase of the laser fusion may be intense enough to exceed the low IADI threshold. The IADI excited on a large volume may cause a significant heating of hot electrons. These new results indicate that the IADI has potentially important impact on laser fusion.

EXPERIMENTAL DESIGN

(A) Experimental Scheme

Our purpose is to study the impact of the IADI on laser fusion. In our experiments, a moderate intensity laser ($I_L\lambda_L^2 = 3 \times 10^{12} - 1 \times 10^{14}$ W/cm^2-μm^2) normally incident on a large size plasma (Laser spot diameter D $\sim$ 600~900 μm), where I_L and λ_L are the intensity and the wavelength of laser. The moderate intensity laser ($I_L\lambda_L^2 = 10^{13}$-10^{14}) is relevant to the intensity of the early phase of the pulse in a proposed laser fusion, where the IADI is potentially important. We also have moderately long laser pulse length, $\tau_L \sim 1$ nsec. The choice of these parameters is important because (1) with a moderate intensity laser (typically 1×10^{13} W/cm^2-μm^2) irradiation, which is slightly above the IADI threshold, Stokes signal is simple, and it is relatively easy to analyze the results. On the other hand, for high power laser irradiation, the spectrum becomes broad and complicated. (2) For large size plasma ($D >> C_s\tau_L$, where C_s is sound speed), plasma expansion can be considered planar, so that we can use one dimensional theory to study the hydrodynamic properties of the plasma.

The Ion Acoustic Decay Instability can be excited when a laser is incident on a plasma with an intensity above a threshold value. The following energy and momentum conservations between laser light and plasma waves are satisfied.

$$\omega_L = \omega_e + \omega_i, \text{ and } k_L = k_e + k_i , \qquad (1)$$

where w_L, w_e, w_i, k_L, k_e, and k_i are frequencies of laser, epw, and iaw, and the wavenumber of laser, epw, and iaw, respectively. We obtain the wavenumber ratio $k_e/k_L = (c^2/3 v_e^2)$ near critical surface using Bohm-Gross dispersion relation (c is the speed of light, and v_e is electron thermal velocity). We have $k_e/k_L \sim 13$ for our typical experimental parameter of the electron temperature, $T_e = 1$ keV. Therefore, we can use dipole approximation, $k_e \sim k_i >> k_L$. Laser is normally incident on plasma, so that k_e and k_i are excited primarily perpendicular to the plasma density gradient by the IADI.

The two epws excited by the IADI couple to produce second harmonic emission. The second harmonic emission is electromagnetic wave. After the straight forward calculation, the wavenumber ratio of the k_e to the second harmonic $k_{2\omega}$ is written as $k_e/k_{2\omega} = [(\omega_L^2 - \omega_e^2)/(4\omega_L^2 - \omega_e^2) \times c^2/3v_e^2]^{1/2}$. It is shown later that the most unstable mode of the IADI is excited at the plasma density of $n_e/n_c \sim 0.86$, where n_c is critical density. Therefore, we obtain $k_e/k_{2\omega} \sim 2.8$ at the instability region. On the contrary to the coupling of Eq. 1, the $k_{2\omega}$ is not negligible for the k-matching condition.

The frequency of the epw signal (Stokes mode) is

$$2\omega = \omega_L + \omega_L' - (\omega_i + \omega_i'), \quad \omega_L \sim \omega_L', \text{ and } \omega_i \sim \omega_i' , \qquad (2)$$

so that the $\Delta\omega (\sim 2\omega_L - 2\omega)$ is twice of the ion acoustic wave frequency, or

$$\Delta\lambda = \Delta\omega \times \lambda_{2\omega}/2\omega_L \sim \omega_i \times \lambda_L^2/4\Pi c . \qquad (3)$$

The $\Delta\lambda$ is the wavelength difference between the $2\omega_L$ and the Stokes peak. We can estimate the ion acoustic wave frequency by measuring the $\Delta\lambda$.

(B) Experimental Arrangement

The experiments were conducted using the GDL laser system at the Laboratory for Laser Energetics (Fig. 1),

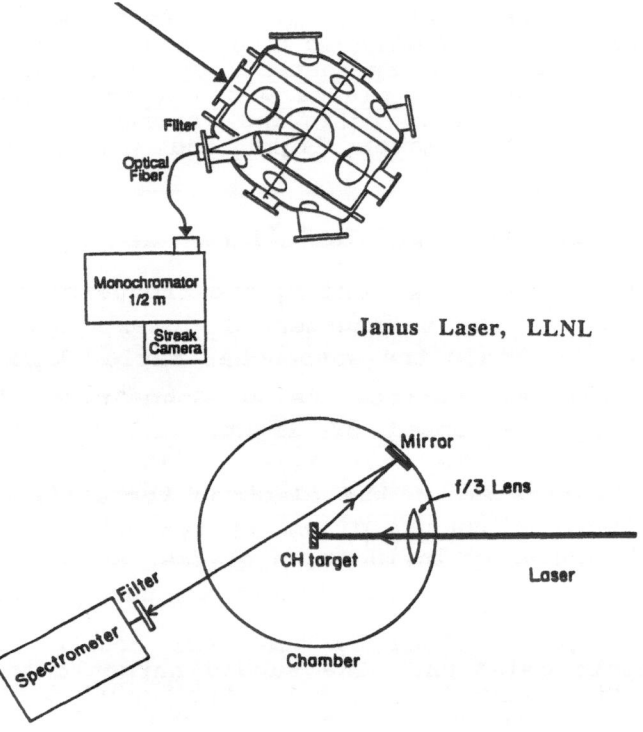

Janus Laser, LLNL

GDL, LLE, Univ. of Rochester

Fig. 1   The schematic of the experiments

University of Rochester, and Janus (Phoenix) Laser system at the Lawrence Livermore National Laboratory. GDL system and Janus laser system have similar parameters. Typical parameters are: laser wavelength λ_L = 1.06 μm and 0.53 μm., the laser pulse length τ_L = 0.8 nsec for GDL and 1.0 nsec for Janus laser, and the maximum laser energy is 200 J. The laser intensity I_L is varied from 10^{12} - 3×10^{15} W/cm^2 by controlling the laser spot size, and the laser energy, independently. The laser normally irradiates a planar CH target. The laser light was focused through an f/3 (GDL), or an f/2 (Janus laser) lens onto the target. The target was thick enough (50 μm) that no burnthrough was observed. We monitored the emission spectrum near the second harmonic ($2\omega_L$) of the incident laser, that was collected at 135^0 and 180^0 from the center axis of the incident laser, and in the plane of the laser electric field. A focusing mirror (GDL experiments) or a focusing lens (Janus experiments) was used to collect the emission near the second harmonic. The signal is fed into a spectrometer. The spectrum is streaked using a streak camera (resolution of 1 Å, 30 psec typical). The data are recorded on Kodak film. The data are digitized and processed using computers in LLNL and LLE.

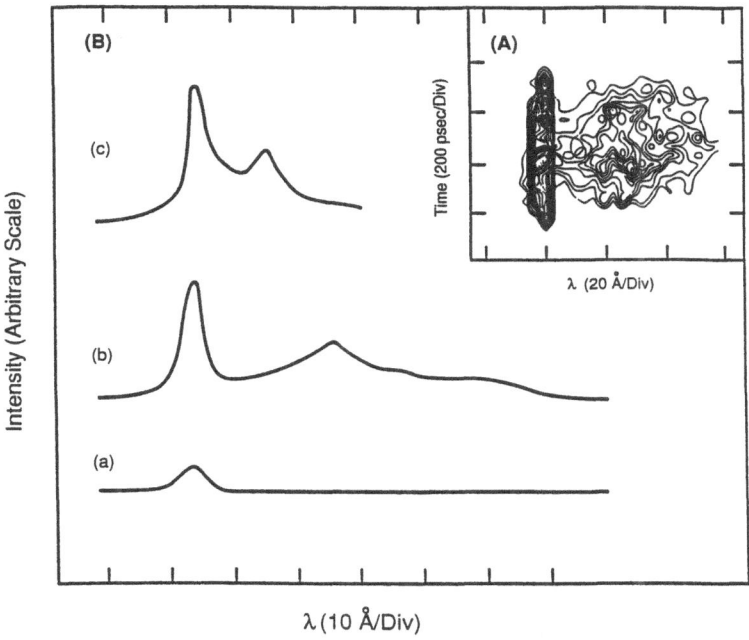

Fig. 2   (A) Two-dimensional contour plot of the IADI, (B) Time integrated spectrum of the second harmonic emission. (a) I_L=5.6x10^{12} W/cm^2, D=350 μm, λ=1.06μm; (b) 3x1013 W/cm^2, D=350 μm, λ_L=1.06 μm; and (c) 3x10^{13} W/cm^2, D=900μm, λ_L=0.53μm. The vertical scale for curve (c) is different from the other curves.

IADI IN LOW Z PLASMA

(A)   Second Harmonic Emissions

The inset of Fig. 2 is a typical time resolved spectrum of the second harmonic emission (the two dimensional intensity contour).

A narrow strong feature at $2\omega_L$ of the lower signal can be attributed[7,11] to the emission from the combined two epws excited by resonance absorption. The Doppler shift effect of the expanding critical surface is negligibly small compared to the Dl in our experiments. The Doppler shift effect of the expanding critical surface is negligibly small compared to Dl. The estimate of the Doppler effect assuming that the critical surface is expanding with a sound speed supports our observation that the Doppler shift is small. The Stokes structure attributed to the IADI is seen to the right (the red side) of the $2\omega_L$ signal. The Stokes signal is excited about 100 psec after the onset of the $2\omega_L$ signal. Notice that the Stokes intensity contour is wider and more complicated than that of the $2\omega_L$. Figure 2 shows the time-integrated second harmonic spectrum as a parameter of laser intensity. At a low intensity laser irradiation (curve (1)), the emission is seen only at $2\omega_L$. No Stokes sideband is observed. Above a well-defined threshold, the Stokes sideband appears with $\Delta\lambda \sim 23$ Å (curve (b)). A weak second Stokes peak is also seen with ~ 46 Å. No significant anti-Stokes signal is observed. The Stokes intensity increases strongly with the laser intensity. The spectrum becomes broad, and the Stokes peak is less clear.

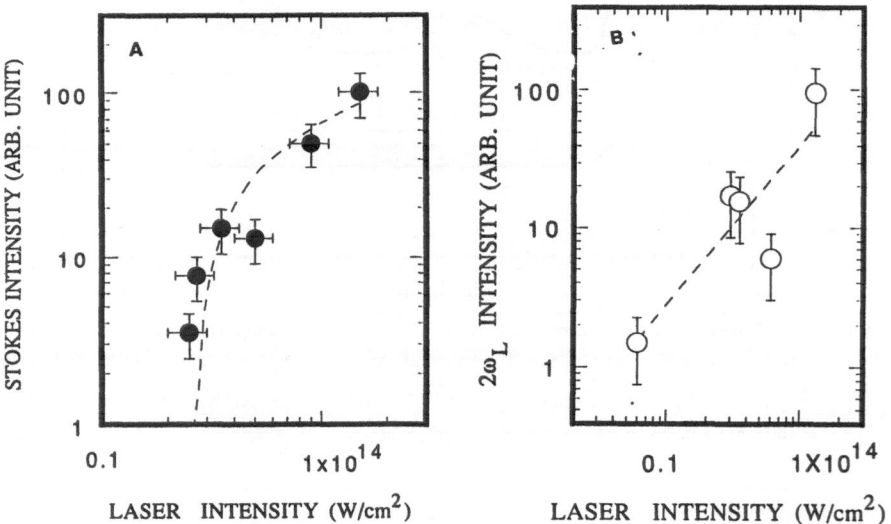

Fig. 3   Relative intensities of (A) Stokes mode, and (B) 2ω signal vs laser intensity for $\lambda_L = 1.06$ μm, and D=350 μm.

We have plotted Stokes (solid circles), and the $2\omega_L$ (empty circles) intensities vs. laser intensity in Fig. 3(A) and (B). The signal is integrated over the time and the wavelength of the each $2\omega_L$ and Stokes spectrum. The $2\omega_L$ and the Stokes spectra are slightly overlapping near the end of the spectrum, so that the minimum intensity point is taken as the boundary between the two spectra. No clear threshold behavior is observed in the $2\omega_L$ signals as is expected. Occasionally (for some laser shots), a significantly large fluctuation of the signal intensity (as high as factor 2) was seen from shot to shot. The Stokes mode is excited when laser intensity is above a threshold value. The Stokes intensity increases strongly with laser intensity near the threshold. No saturation of the Stokes intensity is seen with increasing laser intensity in our moderate power laser experiments ($I\lambda_L^2 - 10^{14}$). Because the laser quiver energy is much smaller than the plasma thermal energy (the ratio < 2×10^{-2}), a strong saturation mechanism of the epw such as trapping is not expected.

The IADI scales quite well from 1.06 μm laser results to short wavelength 0.53 μm laser results. A strong feature at $2\omega_L$, and a stokes spectrum at the red side are also observed with 0.53 μm laser experiments as shown in curve(c) of Fig. 2. The spectrum width of the Stokes mode is narrower than that (curve (b)) of 1.06 μm laser experiments. The $\Delta\lambda$ is about 12 Å, which is approximately the half of the value observed in 1.06 μm laser experiment. After a straight forward calculation, we obtain $\Delta\lambda - 2k_e\lambda_{De}$ x $(m/M(Z+3T_i/T_e))^{1/2}$ x λ_L, using Eq. 3 and ion acoustic wave dispersion relation. It is shown later that $k_e\lambda_{De}$ is about 0.23 at the Stokes peak, being independent of laser wavelength. Hence Dl is proportional to laser wavelength l_L. The measured Dl scales very well from the 1.06 μm laser to short wavelength 0.53 μm laser. This result gives an evidence that the measured Stokes signal can be attributed to the IADI.

We have measured the IADI emission simultaneously at 135^0 and 180^0 in order to investigate the k-spread effect and the plasma flow effect. The average Dl of 4 shots (laser intensity is between 1.1×10^{13}-2.6×10^{13}) is 22.9 Å for 135^0 emission and 23.1 Å for 180^0 emission. The Dl barely depends on the emission angle. The spectrum width of the Stokes mode is estimated from the ratio of the minimum and the peak values as is discussed in our paper[17]. The spectrum width of the 135^0 emission was about 20 % wider than that of the 180^0 emission.

(B) The IADI Instability Region

When laser spot size is much larger than plasma expansion distance, $D \gg C_s \tau_L$, the plasma may be considered as one dimensional. The first step is to analytically estimate the self-consistent plasma density profile using a simple one-dimensional theory[18,19]. When laser is normally incident on plasma, the plasma density is self-consistently modified by the ponderomotive force, and a step-plateau-like density profile (with a steep density gradient at critical surface and an underdense plateau) will be produced. After a straightforward calculation using the results given in Ref. 19, the average underdense plasma density n_{av} is written as

$$n_{av}/n_c = \frac{1}{2}(1 - 7.3 \times 10^{-9} \times (\frac{I_0 \lambda_0^2}{T_e})^{1/2}) \times (1 + \frac{1}{1 + 9 \times 10^{-9} \times (\frac{I_0 \lambda_0^2}{T_e})^{1/2}}) \quad (4)$$

, where I_0 (W/cm^2) is the local laser intensity near the instability region, λ_L (μm), and electron temperature T_e (keV). For the threshold laser intensity of 5×10^{12}, and $T_e = 0.5$ keV, the n_{av}/n_c is about 0.96. The plasma density modification due to the ponderomotive force is quite small.

It is straight forward to show that the plasma density, where the peak IADI is excited, has a fixed value being independent of plasma parameters. Using Eq. 1, Bohm-Gross dispersion relation for the epw, and the iaw dispersion relation, we obtain

$$(n_e/n_c)_{IADI} = 1 - 3M/m \times (\omega_i/\omega_L)^2 / (Z + 3T_i/T_e) \quad (5)$$

and

$$\omega_i/\omega_e = k\lambda_{De} / (m/M \cdot (Z + 3T_i/T_e))^{1/2} \quad (6)$$

, where m and M are the masses of electron and ion.

The Stokes peak appears at $k\lambda_{De} \sim 0.23$ due to the Landau damping barrier as discussed later. Therefore the plasma density of the IADI, $(n_e/n_c)_{IADI}$, is about 0.86 being nearly independent of plasma parameters. The important result is that the IADI density $(n_e/n_c)_{IADI}$ is less than the underdense plateau density n_{av}/n_c. Hence the IADI should be excited on the underdense plateau region. On the other hand, if $(n_e/n_c)_{IADI} > n_{av}/n_c$, the IADI will be excited at the steep plasma density near critical density.

(C) The IADI Threshold

We have shown that the convective loss of the epw is small, so the IADI can be described by the uniform plasma theory. Then, the dispersion relation of the IADI is given by a simple equation. The threshold laser intensity at the instability region is

$$I_0 \lambda_L^2 = 2.2 \times 10^{16} \cdot T_e(keV) \cdot \Gamma_e \Gamma_i / \omega_L k c_s \quad (7)$$

The Γ_e, and Γ_i are the energy damping rates of the epw (collisional and Landau damping), and iaw (Landau damping). The experimental results agree[16] well with the theory for 1.06 μm, and 0.53 μm lasers.

THE IADI IN HIGH Z PLASMA

(A) The IADI vs target material

We have plotted the time-integrated spectra as a parameter of target material in Fig. 4. Above well-defined threshold laser intensities for the IADI, Stokes sidebands are observed clearly in curves (a), (b), and (c). The shape of the Stokes mode depends strongly on the target material. For a low Z target such as CH, the spectrum is rather broad (curve (a)). The second Stokes mode is barely distinguishable. Clear, sharp, Stokes peaks are observed for higher Z targets such as Mo or Au. The 1st and 2Nd Stokes peaks are clearly distinguishable in curve (b). For the highest Z target (Au), no 2Nd Stokes peak is observed although the 1st Stokes mode has a well-defined sharp peak (curve (c)). The total area of the Stokes peak in curves (a,b,c) decreases with increasing the Z of target, indicating that the total intensity of the Stokes mode decreases strongly with higher Z.

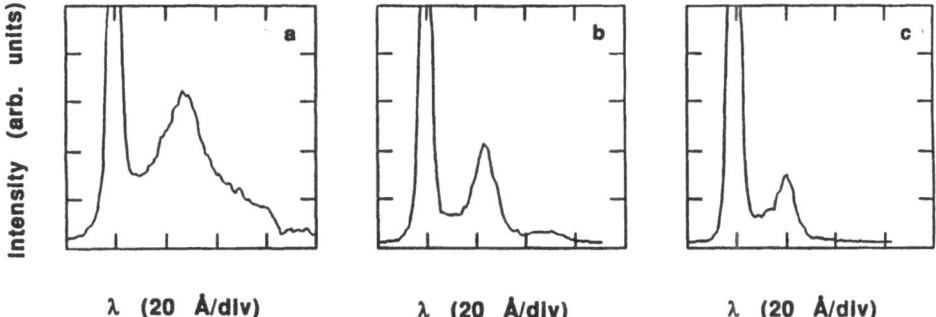

Fig. 4   (a) The time integrated IADI, and (b) the intensities of the Stokes mode (empty circles), and 2ω signal (triangle) vs atomic number.

Figure 5 (A) shows the measured threshold values vs the Atomic number of the target. We find that the threshold of the IADI is quite low (-2×10^{13} W/cm^2) even for the high Z target, and increases only weakly with increasing the atomic number. The change in the threshold value is less than 50 %, when the atomic number of the plasma varies from 13 (Al) to 79 (Au). This low threshold and the week Z scaling agree with the simple theory presented later.

The Stokes intensity depends strongly on atomic number of the target. The empty circles show the Stokes intensity vs. atomic number in Fig. 5 (B). The total Stokes intensity is measured for each atomic number at the laser intensity of $I_L = 3.6 \times 10^{13}$ W/cm^2, which is approximately twice the IADI threshold. The Stokes intensity decreases strongly with increasing atomic number of the target plasma. For comparison, we have plotted the intensity of $2\omega_0$ signal with empty triangle, which is measured simultaneously with each Stokes signal. The intensity of the $2\omega_0$ signal depends weakly on atomic number since it is not due to an instability. The error bars indicate the observed fluctuations of the intensity from one experiment to the next. We also have a few laser shot results that the measured intensity varied more than a factor 2. We attribute the observed Z scaling of the Stokes intensity to the decrease of the growth rate.

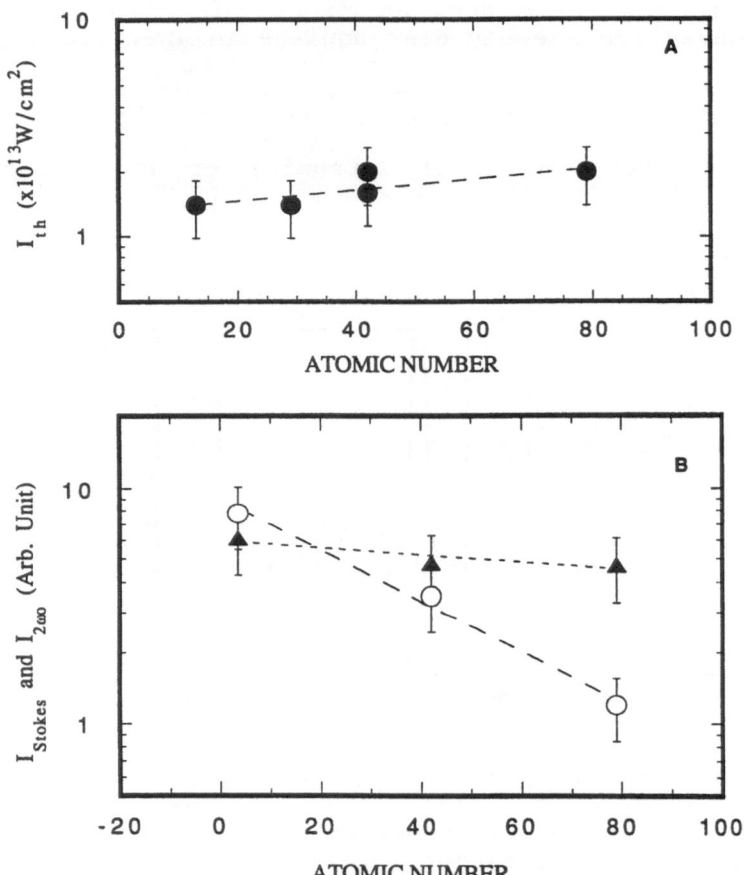

Fig. 5 (a) The threshold of the IADI, and (b) the intensities of the Stokes mode (empty circles), and 2ω signal (triangle) vs atomic number.

(B) The Z Scaling of the IADI

In this section, we discuss the Z scaling of the IADI assuming the flux limiter is $f = 0.1$ and thus that collisional effects dominate the threshold behavior, and that simple, homogeneous- plasma theory may describe the essential features of the experiment. In this case, We have

$$I_0 \quad \alpha \quad (\Gamma_e/\omega_0) \times (\Gamma_i/kC_s) \qquad\qquad (8).$$

On the other hand, the IADI growth rate, γ, is proportional to I_0 /Γ_e near the threshold ($\Gamma_i << \gamma << \Gamma_e$).

Equation 8 shows that the threshold laser intensity at the instability region is determined by the product of the damping rates of the electron plasma wave and the ion acoustic wave. The collisional damping of the epw, and also the attenuation of the laser light increase with Z. On the other hand, the damping of the iaw decreases with ZT_e/T_i. In the plasmas studied here, these opposite trends turn out to cancel each other, producing a weak dependence of the threshold on the target material. The detailed calculations indicate that the overall threshold value $I_{L,th}$ weakly depends on the atomic number in agreement with the experimental results.

The growth rate, on the other hand, is inversely proportional to Γ_e, and it is independent of ion wave damping if $\gamma > \Gamma_i$. Therefore, the growth rate decreases with Z. The growth rate is written as

$$\gamma \quad \alpha \quad (T_e^{1/2} / Z) \times (I_0 / I_L) \times I_L \qquad\qquad (9)$$

The growth rate is proportional to $(1/Z)$ and the laser attenuation (I_0 / I_L). Both values decrease with Z, so that the growth rate decreases strongly with atomic number. The growth rate scales as if the laser intensity is reduced by $(I_0/I_L)/Z$. As the result, the effective laser intensity for a Mo plasma is 20 times smaller than that for a CH plasma. Hence, it is quite reasonable to observe a weak IADI in high Z plasma, where the IADI is excited by a weak effective laser power.

HOT ELECTRON HEATING BY THE IADI

(A) Broad Spectrum Of the IADI

The spectrum shape depends strongly on the laser intensity. A sharp well-defined Stokes mode is excited with the weak intensity laser irradiation. The peak appears at the most unstable mode of the IADI, as is expected. The results agree well with the linear theory. When the incident laser intensity increases, the peak shifts from the original most unstable mode given by the linear theory, and the spectrum becomes broad. The original Stokes peak is now barely distinguishable. Hot electrons will heat when they travel in the broad-spectrum plasma wave by quasilinear diffusion[4].

(C)   Hot Electron Heating by 1 μm laser

The electron plasma waves with a broad spectrum, and excited in a large volume, may significantly heat hot electrons. Hot electrons can heat[2,8,9], when they traverse a short wavelength electron plasma wave excited by the IADI. The IADI can be excited[1] significantly in a region between critical density n_c and a density $n/n_c \sim 0.8$. We estimate the L_{IADI} from the plasma density profile obtained by the LASNEX calculations. We define the distance between the critical density and $n/n_c = 0.8$ as L_{IADI}. The length L_{IADI} is about 28 μm or $L_{IADI}/l_L \sim 28$, so the instability region can be a disk-shaped one with 28 μm width, and, for instance, 500 μm diameter. The electrons will enter the disk shaped instability region with random angles, and heat while they traverse the region.

The laser attenuates strongly due to the anomalous collision in the instability region. The attenuation length δ of the laser is about

$$\delta/\lambda_L \sim 1/2\pi (\omega_L/\omega_{pe})^2 \times 1/(\nu^*/\omega_L) \times (v_g/c) \sim 7 \times 10^{-2} \times (1/(\nu^*/\omega_L)) \qquad (10)$$

depending strongly on the anomalous collision frequency ν^*. The group velocity v_g of the laser is estimated at $n/n_c = 0.86$, where the IADI growth rate has a peak value. We can estimate the anomalous collision frequency using one dimensional PIC computer simulation results. For laser intensities in which we are interested, the anomalous collision frequency scales as

$$(\nu^*/\omega_L) = 3.86 \times 10^{-10} \times (\beta I_L \lambda_L^2/T_e)^{0.56} , \qquad (11)$$

where β is determined by the collisional attenuation of laser. The u* increases slowly when the normalized laser intensity $\beta I_L \lambda_L^2/T_e$ is larger than the turning point intensity 2×10^{14} W/cm^2-μm^2/keV. By combining Eqs. 10 and 11, we can estimate the laser attenuation length. The attenuation length is smaller than the instability width of 28 μm, indicating that most of the laser energy which reaches the instability region is absorbed there.

Let's estimate the approximate value of the hot electron heating. The electron plasma waves are excited by the IADI primarily along the electric field of the laser ($\vartheta \sim 0$, x-direction, where ϑ is measured from the axis perpendicular to the wave vector of the laser) because the wave number k_L of laser is much smaller than those of the plasma waves. Equating the energy flux of hot electrons leaving the heating region to the absorbed laser light, and using the anomalous collision frequency ν^* due to the electron plasma wave excited by the IADI, we have

$$n_H(\vartheta) v_H T_H \sin\vartheta = \nu^* \beta (c/v_g) L_{IADI} (E_L^2/8\pi) (n_e(\vartheta)/n_c) \qquad (12)$$

for electrons $n_e(\vartheta)$, which have an incident angle between ϑ and $\vartheta + \Delta\vartheta$. The parameters, n_H, and T_H is the hot-electron density, and temperature, and $v_H = (T_H/m)^{1/2}$. We assume that the $v*$ is uniform in space within the instability region for simplicity. The important result of the above equation is that the hot electron heat flux is proportional to L_{IADI}, so *hot electron can heat significantly even with a moderate intensity laser irradiation, if the instability width L_{IADI} is large.*

As is discussed previously in this section, the laser attenuation length δ is smaller than L_{IADI}, so that we can replace the L_{IADI}/λ_L in Eq. (12) with δ/λ_L of Eq. (10). The anomalous collision frequency is then cancelled out. Consequently, in a large scale plasma, hot electron heat flux is not a sensitive function of the anomalous collision frequency.

(D) Laser Wavelength Scaling of Hot Electron Heating

The hot electron heat flux is independent of laser wavelength in a large scale plasma, if the laser attenuation due to collision is small (β-1). The hot electron heat flux varies with laser wavelength through the value β. The b is the attenuation factor of the laser light due to the colisional damping, which is given by

$$\beta - \exp(-v_{ei}(n_c)L/c \times INT) , \tag{13}$$

where v_{ei} is electron-ion collision frequency, v_{ei}-3×10^{-6} $\ln\Lambda n_e Z/T_e^{2/3}$(eV), $\ln\Lambda$=7.8, and Z=3.5 for CH plasma. The INT is $x^2/(1-x^2)dx$ - 0.38 for linear density gradient. The laser attenuation is weak for 1.06 μm laser, and is significant for short wavelength 0.35 μm laser.

We have shown the estimated approximate values of the hot electron energy as a function of plasma scale length for several electron temperatures in Fig. 6. The laser wavelength is 1.06, and 0.35 μm, the intensity is $I_L l_L^2 = 10^{14}$ W/cm^2-μm^2, plasma is low Z, CH, and the plasma density profile is assumed to be linear. The 1.06 μm laser results is shown in Fig. 6(A). The strong hot electron heating is possible in a large scale plasma (L> 1 mm), if the electron temperature is larger than 1 keV. On the other hand, the laser attenuates significantly for short wavelength 0.35 μm laser. The estimated hot electron heat flux is shown in Fig. 6(B). A significant heating will be expected only when L is less than 1 mm. For low electron temperature (less than 1 keV), the hot electron will be heated only when the plasma size is small(L < 200 μm).

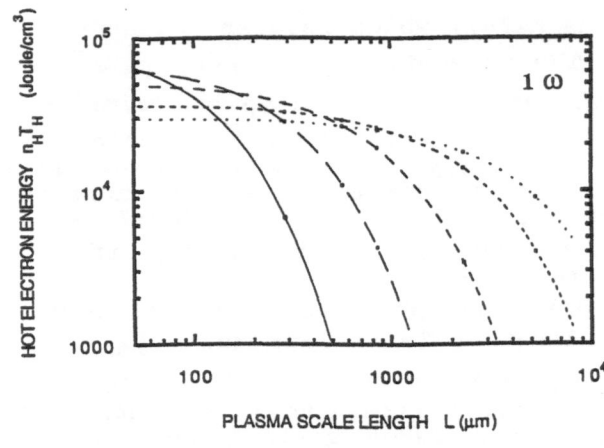

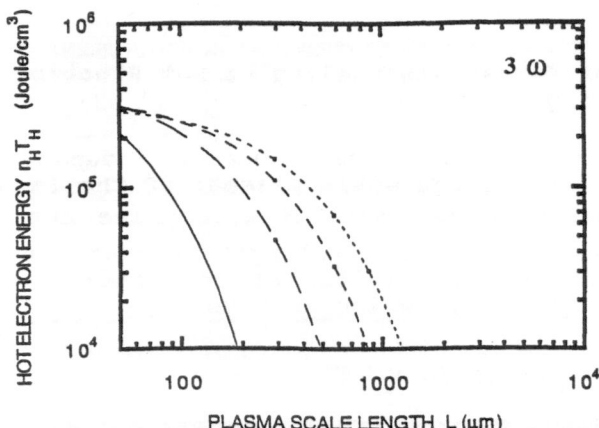

Fig. 6   Hot electron energy vs underdense plasma scale length for 1.06 μm(1ω), and 0.35 μm
(3ω)lasers. The electron temperature is 0.5, 1, 2, 4, 6 keV for 1ω, and 1, 2, 3, 4 keV for 3ω.

ACKNOWLEDGEMENTS

We would acknowledge useful discussions and
encouragements of E. M. Campbell, and J. Knauer.

This work was supported partially by the National Laser
Users Facility at the LLE, University of Rochester, with
financial support from the U.S. Department of Energy under
Cooperative Agreement. The work was Partially supported by
the Plasma Physics Research Institute, University of
California, and Lawrence Livermore National Laboratory.

REFERENCES

(1)   K. Nishikawa, J. Phys. Soc. Jpn. 24, 916; 1152 (1968); W. L.
      Kruer, The Physics of Laser Plasma Interactions (Addison- Wesley
      Publishing Company), 1988.

(2)   R. Stenzel, and A. Y. Wong, Phys. Rev. Lett. 28, 274 (1972); K. Mizuno, and J. S. DeGroot, Phys. Rev. Lett. 35, 219 (1975).

(3)   D. F. DuBois, H. A. Rose, and D. Russell, Phys. rev. Lett., 61, 2209 (1988); J. Geophys. Res. 95, 21, 221 (1990); LANL Report LA-UR-90-3463 (1991).

(4)   J. J. Thomson, R. F. Faehl, W. L. Kruer, and S. Bodner, Phys. Fluids 17, 973 (1974).

(5)   C. Yamanaka, T. Yamanaka, T. Sasaki, J. Mizui, and H. B. Kang, Phys. Rev. Lett. 32, 1038 (1974).

(6)   Xu Zhizhan, Xu Yuguang, Yin guangyu, Zhang Yanzhen, Yu Jiajin, and P. H. Lee, J. Appl. Phys. 54, 4902 (1983).

(7)   K. Tanaka, W. Seka, L. M. Goldman, M. C. Richardson, R. W. short, J. M. Soures, and E. A. Williams, Phys. Fluids 27, 2187 (1984).

(8)   K. Mizuno, J. S. DeGroot, K. G. Estabrook, Phys. Rev. Lett. 52, 271 (1984); Phys. Fluids 29, 568 (1986).

(9)   K. Mizuno, J. S. DeGroot, W. Woo, P. W. Rambo, K. G. Estabrook, Phys. Rev. A 38, 4333 (1988).

(10) J. L. Bobin, M. Decroisette, B. Meyer, and Y. Vital, Phys. Rev. Lett. 30, 594 (1973).

(11) K. Eidman and R. Sigel, Phys. Rev. Lett. 34,799 (1975).

(12) N. G. Basov, V. Yu. Byvhenkov, O. N. Krolieu, M. V. Osipov, A. A. Rapasov, V. P. Silin, G. V. Shliskov, A. V. Starodub, V. T. Tikhonchuk, and A. S. Shikanov, SOv. Phys. JETP 49, 1059 (1979).

(13) Y. Takeda, N. Nakano, and H. Kuroda, Phys. Fluids 31, 692 (1988).

(14) P. D. Carter, S. M. L. Sim, and T. P. Hughes, Opt. Commun. 27, 423 (1978).

(15) K. Mizuno, W. Seka, R. Bahr, R. P. Drake, P. E. Young, J. S. DeGroot, and K. Estabrook, in Laser Interaction and Related Plasma Phenomena, Vol, 9, edited by Hora, and G. H. Miley, 1990.

(16) K. Mizuno, P. E. Young, W. Seka, R. Bahr, J. S. DeGroot, R. P. Drake, and K. G. Estabrook, Phys. Rev. Lett. 65, 428 (1990).

(17) K. Mizuno, R. P. drake, P. E. Young, R. Bahr, W. Seka, and K. G. Estabrook, Phys. Fluids B3, 1983 (1991)

(18) K. Lee, D. W. Forslund, J. M. Kindel, and E. L. Lindman, Phys. Fluids 20, 51 (1977).

(19) K. Estabrook, and W. L. Kruer, Phys. Fluids 26, 1888 (1983).

CONVECTIVE GAIN OF PARAMETRIC INSTABILITIES: AN EXPERIMENTALIST'S VIEWPOINT

Steven H. Batha and Benjamin H. Batha

Plasma Physics Research Institute
University of California, Davis and
Lawrence Livermore National Laboratory
Livermore, CA   94550

I. MOTIVATION

The parametric instabilities inherent in laser-plasma interactions are a critical barrier to the realization of laser-driven inertial confinement fusion energy.[1] These instabilities couple the laser (pump) wave to two daughter waves which may be scattered-light (slw), ion-acoustic, or electron-plasma (epw) waves. Detrimental effects include the production of suprathermal electrons (two-plasmon decay,[2] ion-acoustic decay instability,[3] and stimulated Raman scattering [SRS][4]) and increased plasma reflectivity (stimulated Brillouin scattering [SBS] and SRS[5]). A basic understanding of the behavior of these instabilities, including instability intensity thresholds under various plasma conditions, intensity scaling, and saturation levels, is necessary to design efficient ignition capsules. An extensive theoretical and experimental program has been pursued since the mid-1970's to characterize these processes.[6,7]

At present, very few of the theoretical predictions about these instabilities have been confirmed by experiment. Because these instabilities were initially difficult to excite, much effort has been focussed on predicting and verifying intensity thresholds. These analyses required only (albeit difficult) linear instability analysis. An extension of the linear theory was to predict the convective, or spatial, gain of the instability in an inhomogeneous plasma. As the pump wave traverses the plasma, it excites the two daughter waves which join the pump wave in a feedback loop, thereby amplifying themselves.[8] The conditions needed to transfer energy from the pump to the daughter waves may only be satisfied for a small volume of the plasma because the three waves only retain the required phase relation for a short distance as they move through the plasma. The intensity scaling of a convective instability has only recently been confirmed for a single instability, namely stimulated Raman forward scattering (SRFS).[9] At high-enough intensities, however, nonlinear physics becomes important and the instability saturates.

A schematic of the intensity scaling of a convective instability is shown in figure 1. At low intensities, the gain of the instability is below a threshold value set either by wave

Laser Interaction and Related Plasma Phenomena, Vol. 10
Edited by G.H. Miley and H. Hora, Plenum Press, New York, 1992

305

damping or by plasma inhomogeneities, and the only observable emission is due to Thomson scattering and thermal emission from the thermal plasma. At a given intensity, the instability is amplified to a larger level than the background emission and becomes detectable. As the intensity is increased, the detected signal increases with a specific intensity scaling. At some point, assumptions made in the linear theory break down and the instability becomes nonlinear. The instability is said to have saturated.

An instability may also be considered to be absolute, in which case the waves undergo temporal growth at some point within the plasma. The intensity scaling of an absolute instability is also shown in figure 1. A linear calculation yields an intensity threshold at which the instability immediately reaches a saturated state because only nonlinear effects limit growth. Therefore, the absolute instability is in either of two states: below threshold or saturated.

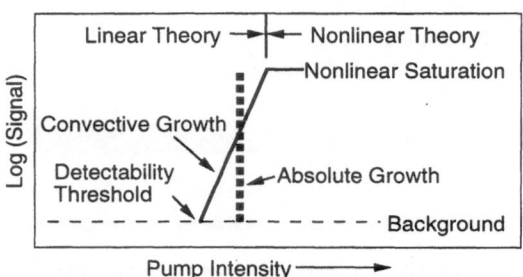

Figure 1. Schematic of a parametric instability defining some of the relevant terms.

Although a previous experiment[10] has shown the importance of absolute instabilities to laser-plasma coupling, the convective aspect of the parametric instabilities remains important for several reasons. With the higher energy, longer pulses, and larger targets of the next generation[11] of inertial-fusion laser systems, even low-gain convective instabilities may grow to large levels[12] with important consequences for coupling physics.[1] Even in present-day experiments the convective nature of the instabilities are important because the conditions for absolute instability can be satisfied for scattering in only a few directions, but must be convective in all other directions. A deeper understanding of convective instabilities may also lead to more stringent tests of linear theory (by examining intensity scaling), to the elucidation of the effects of a two-dimensional plasma on an instability, and to predictions of saturation levels.

Assumptions made in the analytic formulation of convective-gain calculations limit the applicability of previous results to actual target experiments. In this manuscript, work in progress to extend the well-known coupled-mode analysis of convective, parametric instabilities is presented. The primary assumption is that various plasma and wave attributes are allowed to vary throughout the plasma. Previous analytic calculations[13,14] assumed that these quantities did not vary from their value at the instability matching point. This is an appropriate assumption when the instability grows over a few laser wavelengths around the matching point, but is unreasonable when applied to plasmas which are thousands of laser wavelengths long. As will be seen, this assumption may give an incorrect value of the gain. The primary result of this work is derived in section III and is an expression for the gain coefficient of a convective instability assuming spatially-varying wave and plasma quantities (equations [34] and [35]). The effects of this result are illustrated by numerically evaluating equation (34) for stimulated Raman forward scattering in section IV. The emphasis of this manuscript is on the comparison to an experiment and the proposal of a linear saturation mechanism as presented in section V. The emphasis of a complementary work[15] is on the connection of the present numerical work with previous analytic results and comparing the predicted convective gain of forward and backward SRS. The analysis begins with a derivation of the coupled-wave equations which are chosen to model the evolution of a convective, parametric instability.

II. THE COUPLED-WAVE EQUATIONS

Both the traditional and new approaches to determine the convective gain coefficient begin with the coupled-mode equations which describe the interplay between the amplitudes of the two daughter waves. This set of equations has been derived by many authors,[13,16-22] but is repeated here in hopes of clarifying the derivation.

The most straightforward derivation is done in terms of the vector and electrostatic potentials. Starting from Maxwell's equations and the ion and electron continuity and momentum equations, a set of equations for the vector potentials can be derived.[8] The first equation describes the scattering of a large amplitude light wave (i.e. the pump laser, A_L) by a small amplitude density fluctuation ($\tilde{n}_e$) (Kruer,[8] equation [7.11]);

$$\left(\frac{\partial^2}{\partial t^2} - c^2\nabla^2 + \omega_{pe}^2\right)\tilde{A} = -\frac{4\pi e^2}{m}\tilde{n}_e A_L \quad .$$

(1)

The second equation describes the density fluctuation associated with the beating of the scattered-light wave ($\tilde{A}$) with the pump (Kruer,[8] equation [7.17]);

$$\left(\frac{\partial^2}{\partial t^2} - 3v_t^2\nabla^2 + \omega_{pe}^2\right)\tilde{n}_e = \frac{n_0 e^2}{m^2 c^2}\nabla^2(A_L \cdot \tilde{A}) \quad ,$$

(2)

where v_t is the electron thermal velocity, ω_{pe} the electron plasma frequency, m (e) the electron mass (charge), c the speed of light, and n_0 the background electron density.

The derivation of the coupled-mode equations presented here will be restricted to one dimension, taken to be along the phase gradient, which for SRS is the direction of the density gradient. (The situation is more complicated for stimulated Brillouin scattering where the direction of plasma flow is also important.[23]) The components of the wave numbers along the gradient direction, x, are given by k_j, while the perpendicular components of the wave numbers are denoted by $k_\perp$. It should be noted that all variables are functions of the spatial dimension. With the addition of a resistive Krook term,[8,19] equations (1) and (2) become in 1-D,

$$\left(\frac{\partial^2}{\partial t^2} - c^2\frac{\partial^2}{\partial x^2} + - c^2 k_\perp^2 + \omega_{pe}^2 + 2v_s\frac{\partial}{\partial t}\right)\tilde{A} = -\frac{4\pi e^2}{m}\tilde{n}_e A_L \cos\Psi \quad ,$$

(3)

and

$$\left(\frac{\partial^2}{\partial t^2} - 3v_t^2\frac{\partial^2}{\partial x^2} - 3v_t^2 k_{n,\perp}^2 + \omega_{pe}^2 + 2v_e\frac{\partial}{\partial t}\right)\tilde{n}_e = \frac{n_0 e^2}{m^2 c^2}\frac{\partial^2\left(A_L \bar{A}\cos\Psi\right)}{\partial x^2} \ , \tag{4}$$

where Ψ is the angle between the polarization vectors of the scattered-light and pump waves. The damping rates are defined by

$$v_s = \frac{v_{ei}}{2}\left(\frac{\omega_{pe}}{\omega_{slw}}\right)^2 \ \text{and} \ v_e = v_{ei}/2 + v_L \ , \tag{5}$$

with v_{ei} the electron-ion collision frequency and v_L the Landau damping rate.[24]

A WKB formalism is used to proceed. The wave fluctuations are written in terms of a slowly-varying amplitude and a rapidly-varying phase:

$$A_j(x,t) = 0.5\left[A_j(x,t)e^{i\phi_j} + \text{c.c.}\right] \tag{6}$$

$$\phi_j(x,t) = \int_{x_0,t_0}^{x,t}\left(k_j dx - \omega_j dt\right) \ . \tag{7}$$

Here, $j = 0, 1, 2$ refer to the pump, scattered-light, or plasma wave, respectively. This ansatz is inserted into equations (3) and (4) and the expressions evaluated. A key assumption is that there is no pump depletion so that A_0 is a constant. Equations (3) and (4) become

$$e^{i\phi_1}\left[-i\omega_1\frac{\partial}{\partial t} - \frac{\omega_1^2}{2} - \frac{c^2}{2}\frac{\partial^2}{\partial x^2} - \frac{ic^2}{2}\frac{\partial k_1}{\partial x} - ic^2 k_1\frac{\partial}{\partial x} - \frac{c^2 k_1^2}{2} + \frac{c^2 k_1^2}{2} + \frac{\omega_{pe}^2}{2} - iv_s\omega_1\right]A_1 + \text{c.c.}$$

$$= \frac{-4\pi e^2\cos\Psi}{4m}\left[A_0 A_2 e^{i(\phi_0 + \phi_2)} + A_0 A_2^* e^{i(\phi_0 - \phi_2)}\right] + \text{c.c.} \ , \tag{8}$$

and

$$e^{i\phi_2}\left[-i\,\omega_2\frac{\partial}{\partial t} - \frac{\omega_2^2}{2} - \frac{3v_t^2}{2}\frac{\partial^2}{\partial x^2} - \frac{3\omega_2^2\partial k_2}{2\,\partial x} - 3iv_t^2 k_2\frac{\partial}{\partial x} + \frac{3}{2}v_t^2 k_2^2 - \frac{3v_t^2 k_{n,\perp}^2}{2} + \frac{\omega_{pe}^2}{2} - iv_e\omega_2\right]A_2 + \text{c.c.}$$

$$= \frac{n_0 e^2\cos\Psi}{4m^2 c^2}\left[\frac{\partial^2}{\partial x^2} + i\frac{\partial k_2}{\partial x} + 2ik_2\frac{\partial}{\partial x} - k_2^2\right]\left[A_0 A_1^* e^{i(\phi_0 - \phi_1)} + A_0 A_1 e^{i(\phi_0 + \phi_1)}\right] + \text{c.c.} \tag{9}$$

Because the wave amplitudes are assumed to be slowly varying, several terms are assumed to be small and are dropped:

$$\frac{\partial^2 A_j}{\partial t^2} \ll \omega_j\frac{\partial A_j}{\partial t} \ , \tag{10a}$$

and

$$\frac{\partial^2 A_j}{\partial x^2} \ll k_j \frac{\partial A_j}{\partial x} \ll k_j^2 A_j \ .$$

(10b)

More terms may be eliminated by using the daughter-wave dispersion relations. The electromagnetic-wave dispersion relation,

$$\omega_1^2 - \omega_{pe}^2 - c^2\, k_1^2 - c^2\, k_\perp^2 = 0 \ ,$$

(11)

is used in equation (8), and the electrostatic-wave dispersion relation,

$$\omega_2^2 - \omega_{pe}^2 - 3\, v_t^2\, k_2^2 - 3\, v_t^2\, k_{n,\perp}^2 = 0 \ ,$$

(12)

is used in equation (9).

The coupled equations are now

$$\left[-\, i\omega_1 \frac{\partial}{\partial t} - \frac{ic^2}{2}\frac{\partial k_1}{\partial x} - ic^2 k_1 \frac{\partial}{\partial x} - iv_s\omega_1 \right] A_1 = \frac{-4\pi e^2\, \cos\Psi}{4m}\, A_0 A_2^* e^{i\,\Phi} \ ,$$

(13)

and

$$\left[-i\,\omega_2 \frac{\partial}{\partial t} - \frac{3\omega_2^2 \partial k_2}{2\ \partial x} - 3iv_t^2 k_2 \frac{\partial}{\partial x} - iv_e\omega_2 \right] A_2 = -\frac{n_0 e^2\, \cos\Psi}{4m^2c^2}\, k_2^2 A_0 A_1^* e^{i\Phi} \ .$$

(14)

Only terms which satisfy the resonant phase relation,

$$\Phi \equiv \phi_0 - \phi_1 - \phi_2 = 0 \ ,$$

(15)

where Φ is the total phase, need to be kept. The nonresonant terms in equations (8) through (14) have phases which are proportional to $\exp[i(\phi_0 + \phi_2 - \phi_1)]$ and have been dropped. In addition, the complex conjugate terms may be dropped; they lead to a set of equations which are the complex conjugates of the result presented in equations (17) and (18). In order to consolidate the terms proportional to c^2 in equation (13) and v_t^2 in equation (14), the amplitudes are redefined in terms of a scaled action amplitude:

$$a_j = \sqrt{k_j}\ A_j \ .$$

(16)

The final result is the two coupled-mode equations;

$$\left[\frac{\partial}{\partial t} + v_{g1}\frac{\partial}{\partial x} + v_s \right] a_1 = \gamma_1 a_2^* e^{i\Phi} \ ,$$

(17)

and, after taking the complex conjugate,

$$\left[\frac{\partial}{\partial t} + v_{g2}\frac{\partial}{\partial x} + v_e \right] a_2^* = \gamma_2 a_1 e^{-i\Phi} \ ,$$

(18)

The component of the group velocity of the jth wave along the inhomogeneity is v_{gj}.

The growth rates, γ_1 and γ_2, are defined as

$$\gamma_1 = -i \frac{4\pi e^2}{4m\omega_1} \sqrt{\frac{k_1}{k_0 k_2}} \, a_0 \cos\Psi \, , \tag{19}$$

and

$$\gamma_2 = -i \frac{n_0 e^2}{4m^2 c^2} \frac{k_2^2}{\omega_2} \sqrt{\frac{k_2}{k_0 k_1}} \, a_0 \cos\Psi \, . \tag{20}$$

The product of the two growth rates is the usual homogeneous growth rate,

$$\gamma_0^2 = \gamma_1 \gamma_2 = \frac{k_2^2 v_{osc}^2}{16} \frac{\omega_{pe}^2}{\omega_1 \omega_2} \cos^2\Psi, \tag{21}$$

where the oscillatory velocity is defined to be

$$|v_{osc}| = \left| \frac{e A_0}{mc} \right| = \left| \frac{e E_0}{m\omega_0} \right| \, . \tag{22}$$

III. THE GAIN COEFFICIENT INCLUDING SPATIAL VARIATIONS

An expression for the steady-state gain coefficient is derived in this section following the method of Picard and Johnston,[25] but now allowing the various wave quantities to vary in space along the pump-laser direction. To simplify the exposition of the results, the intensity of the scattered daughter wave (usually the scattered-light wave), I_{scat}, is assumed to grow exponentially from some noise source, I_{noise}, in the plasma:

$$I_{scat} = I_{noise} \exp(GC) \, . \tag{23}$$

The principal result of this paper is equation (34), an integral expression for the steady-state gain coefficient, GC. A more correct determination of the affect of noise on the scattered-light levels than equation (23) has been given by Berger, Williams, and Simon.[26] A determination of the gain coefficient is still necessary, however, and their result may be used with the GC determined here.

The approximation of one-dimensional inhomogeneity in the theory implies that x is measured along the coupling-phase inhomogeneity direction, i.e. the x-coordinate is aligned with the *direction* of the local density normal (assumed constant). (The assumption here will be that this applies well enough over the gain region of the pump beam.) One can therefore project the integration onto any other locally constant direction. Given that the density contours may not be well known until after the experimental modelling is complete and that the scattering geometry is conveniently defined with respect to the pump, we project the x-coordinate onto a coordinate ζ aligned with the pump, so that we have $\zeta = x/\cos\phi$, where ϕ is the angle between the density normal and the pump axis. Converting x-derivatives to ζ-derivatives also implies converting the velocities, so v_{gj} now becomes $v_{gj} \cos\phi$ (henceforth called v_j).

First, the damping term of equation (17) is absorbed by defining

$$a_1 = g_1 \exp\left[-\int d\zeta \; (\nu_1 / v_1) \right]$$

(24)

and substituting into equation (17). This yields a new equation for g_1;

$$\frac{\partial g_1}{\partial \zeta} = \frac{\gamma_1}{v_1} \; g_2^* \exp(D) \; ,$$

(25)

where g_2 is defined analogously to equation (24). Also,

$$\Phi \equiv \int d\zeta \; \kappa(\zeta) \quad \text{and} \; D \equiv \int d\zeta \left(\frac{\nu_1}{v_1} - \frac{\nu_2}{v_2} + i\Phi' \right) \; .$$

(26)

Here Φ is the phase mismatch, κ is the wave-number mismatch defined as

$$\kappa(\zeta) = k_0(\zeta) - k_1(\zeta) - k_2(\zeta) \; ,$$

(27)

$$\Phi' \equiv \kappa(\zeta) \; ,$$

(28)

and D is a damping term. Equation (25) will be put into Schrödinger-equation form in order to use a WKB solution. A similar Schrödinger equation may be derived for the second daughter wave, g_2. Thus, the second derivative of g_1 is obtained by taking the derivative of equation (25) and substituting for $\partial g_2 / \partial \zeta$:

$$\frac{\partial^2 g_1}{\partial \zeta^2} - \left\{ \frac{\partial}{\partial \zeta} \left[\ln\left(\frac{\gamma_1}{v_1} \right) + D \right] \right\} \frac{\partial g_1}{\partial \zeta} - \frac{\gamma_1 \gamma_2}{v_1 v_2} g_1 = 0 \; .$$

(29)

Now the first derivative term will be eliminated by defining

$$g_1 = f_1 \exp\left[\frac{1}{2} \int d\zeta \; \left\{ \frac{\partial}{\partial \zeta} \left[\ln\left(\frac{\gamma_1}{v_1} \right) + D \right] \right\} \right]$$

(30)

and substituting into equation (29). This reduces to a Schrödinger equation for f_1:

$$\frac{\partial^2 f_1}{\partial \zeta^2} + Q_1 f_1 = 0 \; ,$$

(31)

with Q_1 defined in equation (35), whose WKB solution is well-known to be

$$f_1 = \frac{f_{10}}{(-Q_1)^{1/4}} \exp\left[\int d\zeta \sqrt{-Q_1}\right] .$$

(32)

Integration is between the turning points of Q_1 (the region where $Q_1 < 0$). Boundary conditions determine f_{10}.

The spatial amplification of the wave amplitude a_1, is

$$a_1 = f_1 \sqrt{\frac{\gamma_1}{v_1}} \exp\left[\frac{i\Phi}{2} - \frac{1}{2}\int d\zeta \left(\frac{v_1}{v_1} + \frac{v_2}{v_2}\right)\right]$$

(33)

where f_1 is given by equation (32). Examination of equations (32) and (33) shows that the growth of the instability is dominated by the exponential terms. Therefore, only these terms are retained. Because the intensity gain is twice the amplitude gain,

$$GC = 2\,RE \int d\zeta \left[\sqrt{-Q_1} - \frac{1}{2}\left(\frac{v_1}{v_1} + \frac{v_2}{v_2}\right)\right] \equiv 2\,RE \int d\zeta\,[Int] .$$

(34)

Integration of equation (34) is over the range of ζ where the integrand, Int, has a positive real part (i.e. where there is amplification). The new physics of this result is contained in the integrand, which, in terms of fundamental wave quantities, is

$$Int \equiv \left\{ \begin{array}{c} \dfrac{\gamma_1 \gamma_2}{v_1 v_2} + \dfrac{1}{4}\left[\dfrac{v_1}{\gamma_1}\dfrac{\partial}{\partial\zeta}\left(\dfrac{\gamma_1}{v_1}\right) + \dfrac{v_1}{v_1} - \dfrac{v_2}{v_2} + i\Phi'\right]^2 \\[4mm] -\dfrac{1}{2}\left[-\left[\dfrac{\partial}{\partial\zeta}\ln\left(\dfrac{v_1}{\gamma_1}\right)\right]^2 + \dfrac{v_1}{\gamma_1}\dfrac{\partial^2}{\partial\zeta^2}\left(\dfrac{\gamma_1}{v_1}\right) + \dfrac{\partial}{\partial\zeta}\left(\dfrac{v_1}{v_1} - \dfrac{v_2}{v_2}\right) + i\Phi''\right] \end{array} \right\}^{1/2}$$
$$- \frac{1}{2}\left(\frac{v_1}{v_1} + \frac{v_2}{v_2}\right) .$$

(35)

The expression used in the analytic formulation can be obtained from equations (34) and (35) in three steps: (1) setting Φ' equal to the appropriate Taylor expansion, (2) setting all other derivative terms equal to zero, and (3) assuming that each quantity is equal to its value at the three-wave matching point, defined to be where $\Phi = 0$. Again, the velocities represent the components of the group velocities along the direction of the gradient of Φ'. The main result of this work is the evaluation of the gain coefficient from equations (34) and (35).

It will be shown in the next section how these changes affect the scaling of the gain coefficient (equation [34]) with the plasma parameters.

IV. APPLICATION TO SRFS

In this section, the numerical evaluation of equations (34) and (35) for SRFS in a Gaussian density profile are presented. In applying equation (34) to the SRFS problem, the scattered-light wave is identified by subscript 1 and the epw by 2. The coupling of SRFS to the weak anti-Stokes mode has been ignored. The baseline plasma conditions are typical of an exploding-foil plasma and are given in table 1. The analytic results are based on the following expression, from work by Williams and coworkers.[27-29]

$$GC = A \left[\frac{\gamma_0}{(\kappa'')^{1/3} (|v_1 v_2|)^{1/2}} \right]^{3/2} \tag{36}$$

where

$$A = \min \left\{ 7, \ 2\pi \left[\frac{\gamma_0}{v_2} \left(\left| \frac{v_2}{v_1} \right| \right)^{1/2} \right]^{1/2} \right\} . \tag{37}$$

In calculating either the analytic or numeric results, three assumptions have been made. First, the electron temperature was assumed to be constant throughout the plasma. This is a reasonable assumption based on the predictions of a hydrodynamic simulation.[30] The second assumption is that the plasma density profile is Gaussian,

$$n(\zeta) = n_{peak} \exp \left[-\frac{1}{2} \zeta^2 \left(\frac{1}{L_x^2} + \frac{1}{L_y^2} \right) \right] . \tag{38}$$

where ζ is along the direction of the laser, and L_x and L_y are the axial and transverse scale lengths, respectively. The third assumption is that the laser intensity remains constant as it traverses the plasma. This may not be a reasonable assumption because it ignores pump depletion (as assumed in the derivation of [17] and [18]), and the focussing and defocusing of the interaction beam. These two effects were not taken into account in the example scalings presented here.

First, the differences between the analytic and numeric results are shown in figure 2. In this figure, the integrand of equations (35) and (36) are plotted assuming that the laser is incident from the left, so that the ordinate is in units of gain coefficient per length while the abscissa is in units of length traversed by the pump wave. Thus twice the area under the curve is the gain coefficient. The analytic integrand is symmetric about the matching point (where $\Phi' = \kappa = 0$), has a peak at the matching point, and decreases to zero at $\pm 70 \ \mu m$. The numeric result also peaks at the matching point. At each point in the plasma, the numeric integrand is larger than the analytic integrand, meaning that the more exact treatment predicts a higher gain than the spatially-uniform treatment.

Table 1. Typical plasma values used for the calculated scalings.

n_{peak}	0.05	peak density normalized to $4 \times 10^{21} \ cm^{-3}$
intensity	$4 \times 10^{15} \ W/cm^2$	pump intensity
wavelength	527 nm	pump laser wavelength
T_e	1 keV	electron temperature
L_x	0.1 cm	Gaussian scale length along target normal
L_y	100,000 cm	Gaussian scale length along target face
θ_{att}	0°	angle of pump w.r.t. target normal
θ_{scat}	0°	angle of scattering w.r.t. pump laser
θ_{pol}	0°	angle between laser and slw polarization vectors
Z	3.5	plasma Z for CH

The numeric integrand is asymmetric, having a long, low-gain tail on the positive distance side of the matching point. In general, the tail may be on either side of the matching point and is due to the derivative terms in equation (35). That is, if the $\partial/\partial\zeta$ terms are set equal to zero in equation (35), then the integrand is symmetric about the matching point. The derivative terms do little to the gain at other points in the plasma. They do not increase the peak value of the integrand nor do they change the shape of the curve, except for the tail.

The scaling of the gain coefficient with peak density, intensity, and axial scale length are presented for a planar plasma in figures 3 - 8. Figure 3 shows that the numerically-calculated gain coefficient is up to 40% larger than the analytic calculation. The significance of this work becomes apparent when it is recalled that the instability gain scales as the exponent of the gain coefficient. Figure 4 presents the scaling of the gain coefficient with the axial scale length, L_x. Again, the numeric result has a faster scaling with scale length.

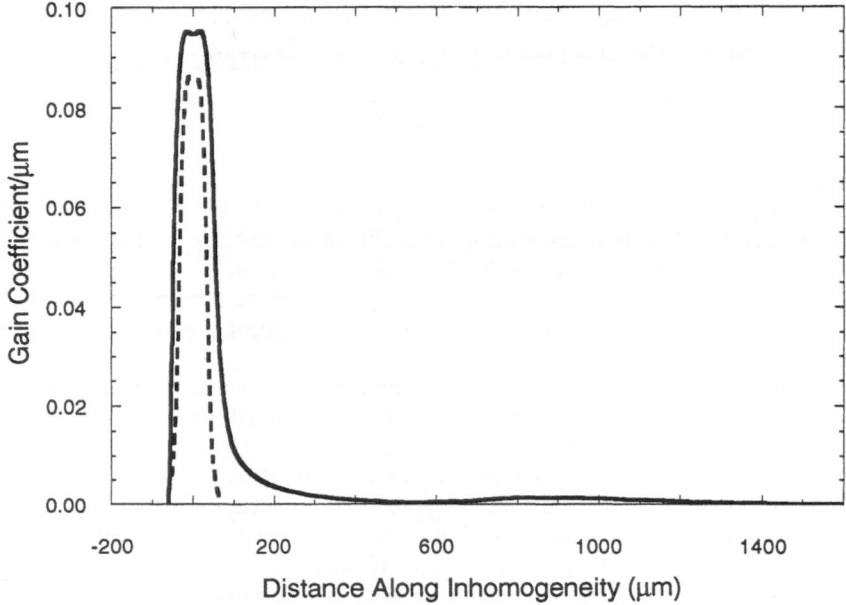

Figure 2. The spatial variation (solid line) of equation (34) compared to the spatial variation of the analytic integrand[27] (dotted line) for parameters given in table 1.

The numerical result also produces higher gain at each intensity, as shown in figure 5. The gain coefficient is up to five times larger than the analytic result. Increasing the pump intensity excites the instability in regions of the plasma that were stable, increasing the gain coefficient in two ways: by increasing the value of the integrand at each point in the plasma and by increasing the width of the gain region. This is shown in figure 6, where the width of the gain region (excluding the low-gain tail) has been plotted. Also plotted is the much shorter analytic gain length, defined to be

$$L_{analytic} = GC_{analytic} * \sqrt{v_1 v_2} / \gamma_0 \; . \tag{39}$$

Two-dimensional plasma effects are quantified by the inclusion of a transverse Gaussian scale length, L_y. To examine these effects, the laser was assumed to strike the plasma at a 50° angle. As can be seen in figure 7, variations of the transverse scale length have little effect on either the analytic or the numeric result until the transverse scale length becomes nearly equal to the axial scale length. At that point, the epw begins to travel perpendicularly to the density gradient. Consequently, v_{g2} becomes small and the gain coefficient becomes large. There is further evidence for this interpretation in figure 8, where the gain-coefficient integrand has been plotted for three different values of L_y. There is no

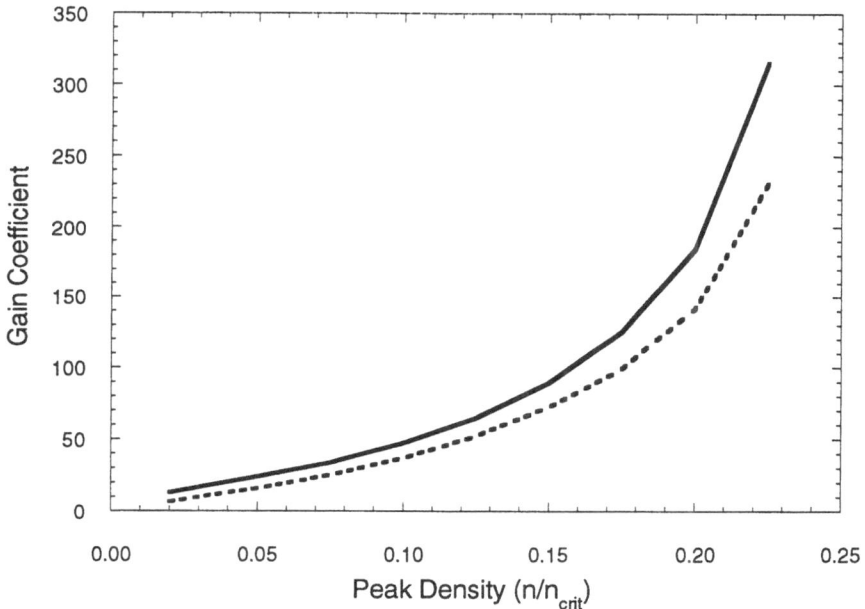

Figure 3. Scaling of the gain coefficient with peak density (normalized to a critical density of 4×10^{21} cm^{-3}). The solid line is equation (34) and the dotted line the analytic result. Parameters for this and subsequent figures are given in table 1 except where explicitly noted in the figure caption or on the abscissa.

difference for the two larger values of L_y where the plasma is still effectively planar. At the smallest scale length, however, the gain region becomes larger and the peak value of the integrand becomes larger.

In summary, it is found that inclusion of spatially-varying group velocities, damping rates, growth rates, and mismatch parameter can have a large affect on the calculated gain coefficient. It is found that for SRFS, this generalization increases the gain coefficient by up to a factor of five.

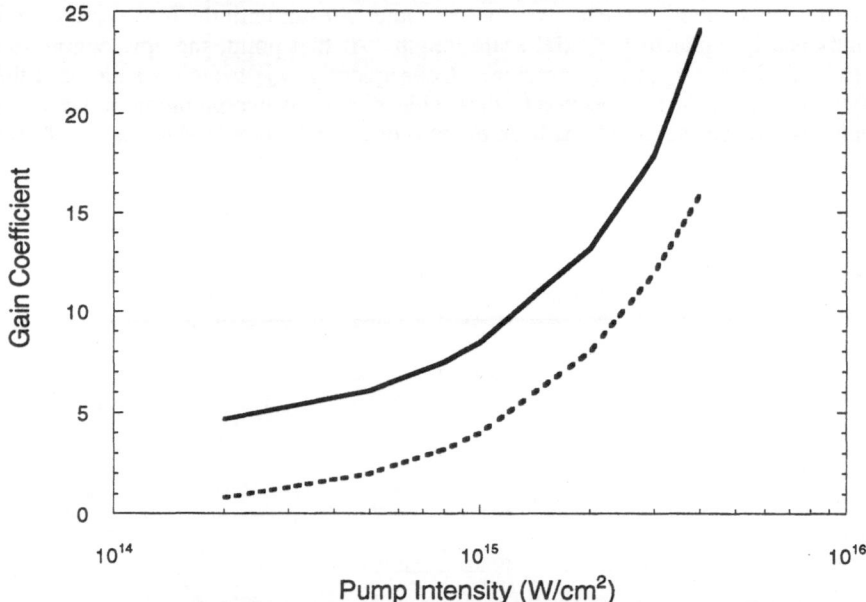

Figure 4. Scaling of the gain coefficient with axial scale length, L_x. The solid line is equation (34) and the dotted line the analytic result.

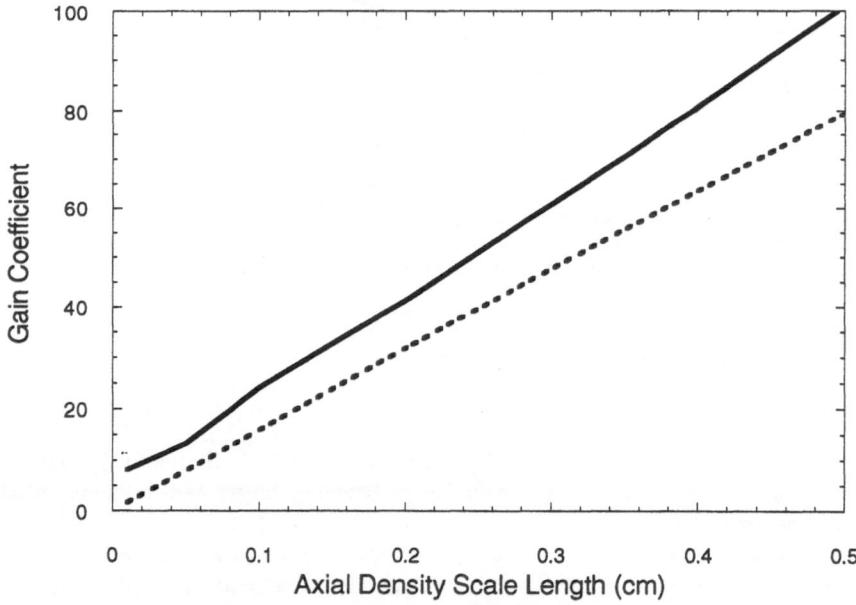

Figure 5. Scaling of the gain coefficient with pump laser intensity. The solid line is equation (34) and the dotted line the analytic result.

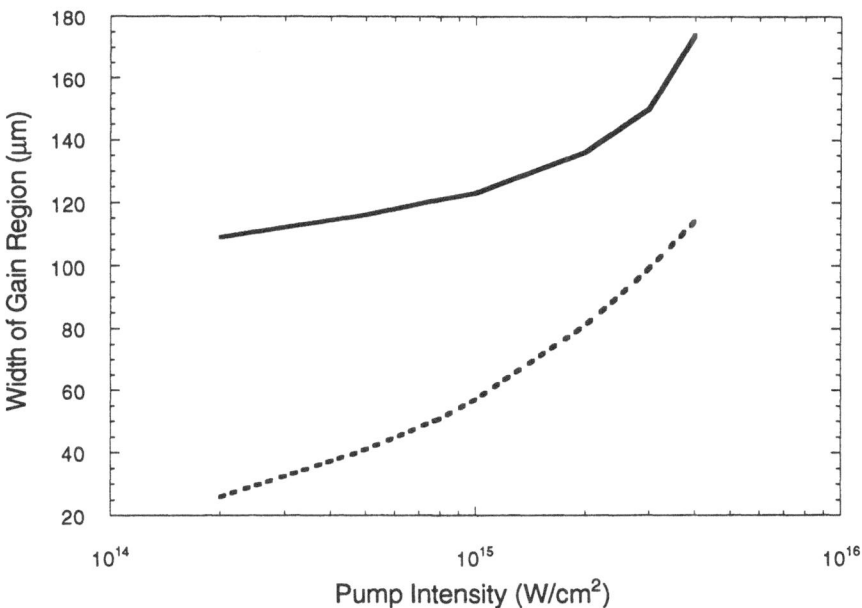

Figure 6. The size of the gain region increases as the pump laser intensity is increased. The solid line is from equation (34) excluding the low gain tail, and the dotted line from equation (39).

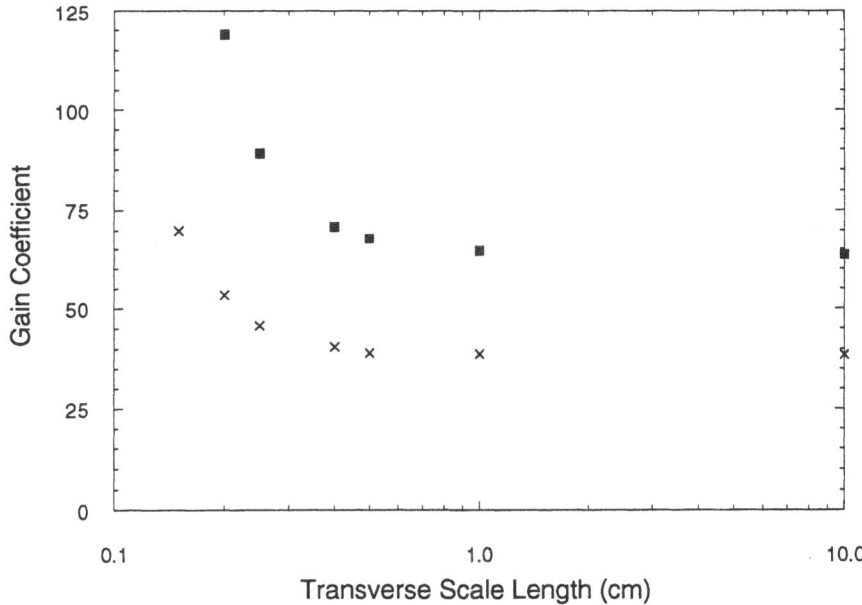

Figure 7. Scaling of the gain coefficient with transverse scale length, L_y. The laser is incident on the plasma at a 50° angle from the target normal. The ■'s are from equation (34) and the x's from the analytic result.

317

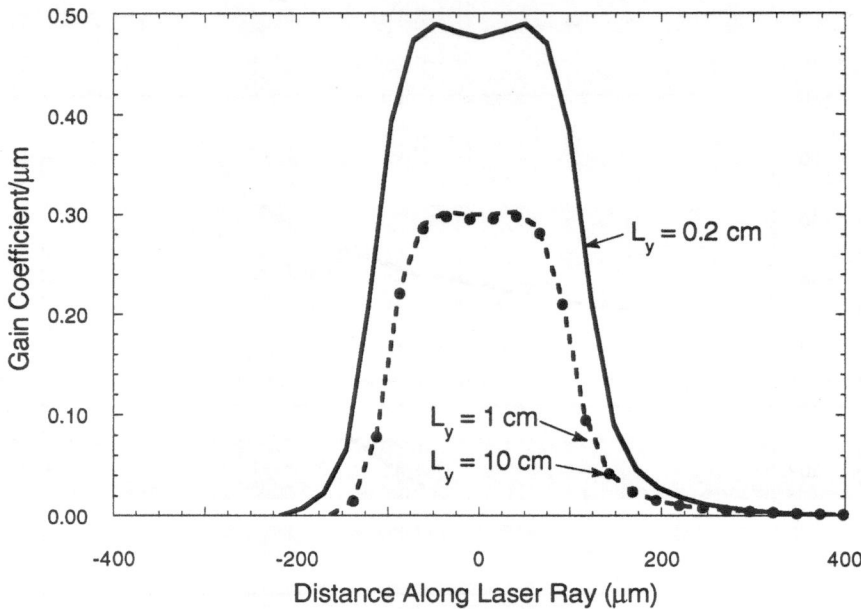

Figure 8. The size of the SRFS gain region increases as the transverse scale length is decreased. The laser is incident on the plasma at a 50° angle from the target normal. The curves are for L_y = 0.2 cm (solid line), L_y = 1 cm (dashed line), and L_y = 10 cm (•).

V. APPLICATION TO AN SRFS EXPERIMENT

An experiment was performed using the NOVA laser facility to measure the convective scaling of SRFS in a low-Z plasma.[31] A preformed plasma[12] was prepared by irradiating a 2 μm thick CH foil, 2.0 mm in diameter, with 15 kJ from seven (preform) beams of the NOVA laser at a wavelength of 351 nm. The preform beams were overlapped to produce a 950 μm diameter spot in a square pulse of 2.0 nsec duration. The plasma was allowed to expand and the peak density to decrease. At 2.2 nsec, a high-intensity, 527 nm beam, also with a 2.0 nsec square pulse, struck the target.

The SRS-scattered light excited by the interaction beam was measured in both the forward and backward directions. Very little backward SRS was measured because the epw for backward SRS was Landau damped due to the low density. The wave number of the SRFS epw was much shorter, however, and it was not significantly damped. Thus, no feedback between backward and forward SRS was possible and the SRFS was genuinely convective. The scattered light was measured with an absolutely calibrated spectrometer/streak camera combination. The data under discussion was measured at 3.6 nsec, when the peak density was 0.02 n_c. At that time, the plasma scale length was estimated to be 0.23 cm by the hydrodynamic code *LASNEX*. The interaction beam was calculated to heat the electrons to a temperature between 0.3 and 2.0 keV, depending on the energy in the interaction beam.

The gain coefficient corresponding to the measured fluence of scattered light, at a density of 0.02 n_c and 27° from the forward direction, is presented in figure 9. The SRFS was first found to be amplified above the background plasma emission at an intensity of 2 x 10^{14} W/cm². As the interaction beam intensity was increased, the gain coefficient increased

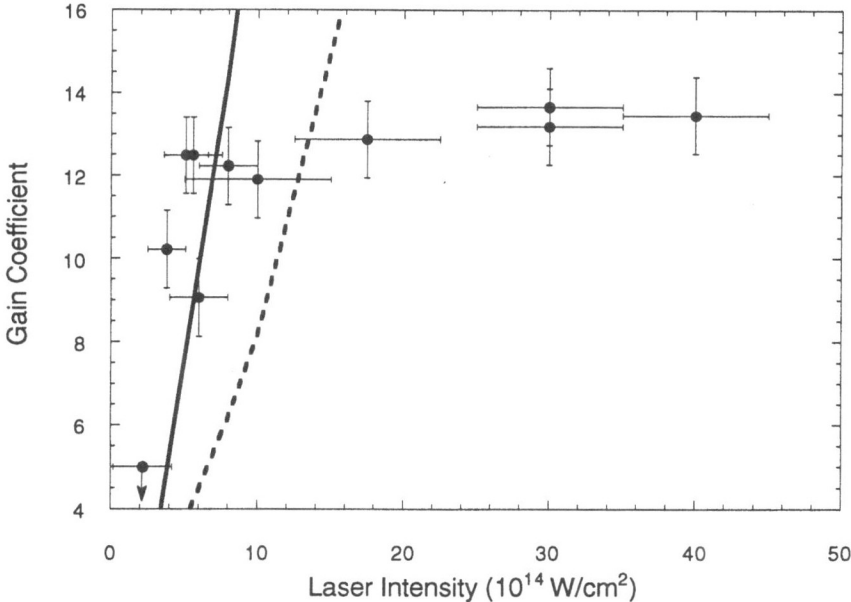

Figure 9. Numeric (solid line), analytic (dashed line), and experimental (•) results compared. The plasma parameters were: $L_x = 0.23$ cm, $L_y = 0.5$ cm, $T_e =$ function of intensity (ranged from 0.4 to 1.1 keV), $\theta_{att} = 50°$, $\theta_{scat} = 0°$, and $\theta_{pol} = 21°$.

dramatically. The signal saturated at a gain coefficient of ≈ 13 for intensities above 1×10^{15} W/cm^2.

The analytic and numerical gain coefficients were evaluated using experimental parameters, except that the scattering was assumed to be direct forward scatter (the parameters are given in the figure caption). This was done because the combination of scale lengths, attack angle, and scattering angle of the experiment meant that the components of the group velocities of the daughter waves along the density gradient were antiparallel. The presence of opposed group velocities significantly complicates the analysis, making complex integration necessary to get the analytic result.[14,25] In principle, the numerical integration should be straightforward, but this is still under study. With that *caveat*, the numeric and analytic results are also compared to the data in figure 9 and both are found to be consistent with the observations. The numeric result yields a larger gain coefficient at each intensity, but both results are consistent with the measured scaling and both are within a factor of 2 of the measured threshold intensity.

At intensities above 10^{15} W/cm^2, the instability has saturated. Usually, the instability itself is assumed to have changed the plasma via some nonlinear coupling and to have thereby caused its own saturation, but this need not be the case. Consider for a moment whether the plasma in the absence in the pump could cause saturation. A linear saturation could happen under the following scenario. At high intensities, the gain is high and the gain length islarge. For example, plotted in figure 10 are the analytic and numeric gain lengths needed to produce the gain coefficient from figure 9. The analytic gain length (equation [39]) increases from zero at low intensity to greater than 200 µm at high intensity. The dip in the curve is related to the dependence of the electron temperature on the interaction beam intensity. The numeric gain length is estimated from graphs like figure 2 (the low-gain tail is included) and increases from 600 µm at threshold to more than 1300 µm at high intensity. Recall that the laser

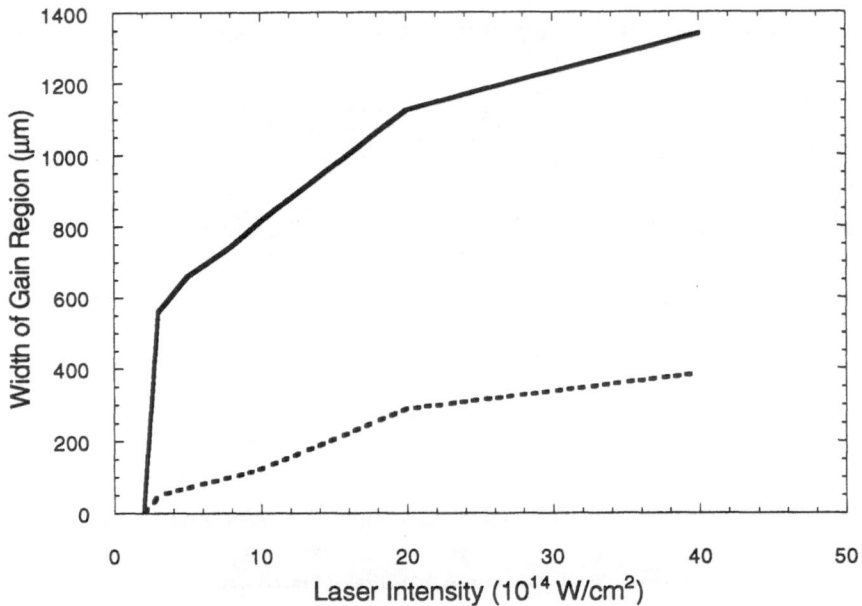

Figure 10. Size of the gain region for figure 9. The solid line is from equation (34) excluding the low gain tail, and the dotted line from equation (39). Plasma parameters are the same as in figure 9.

wavelength was 527 nm, and one sees that to attain the predicted gain, the waves must retain the proper phase relation for over 2000 laser wavelengths! If the plasma were too turbulent to retain this phase relation over such a long distance, the instability would be saturated by convecting out of the gain region in the same way the derived convective gain is a saturated value. Under this premise, figure 10 implies that the plasma provided a gain medium for less than 800 μm.

VI. SUMMARY

In this article, the need for a more accurate treatment of convective-instability gain in large plasmas was discussed. The coupled-mode equations describing a three-wave interaction within a plasma were derived. A reexamination of these equations produced a general, non-analytic expression for the gain coefficient, applicable to any of the parametric instabilities. This work in progress was used to illustrate the effect plasma nonuniformity has on convective gain, both the analytic and numeric gain was calculated for stimulated Raman forward scattering in a Gaussian density profile. The numerically-calculated gain coefficient was found to be up to a factor of 5 greater than the analytic result. The scalings of the gain coefficient with peak electron density, axial scale length, laser intensity, and transverse scale length were presented and were found to be largely consistent with the analytic results. Finally, the numeric results were applied to a convective scaling experiment and found to be in reasonable agreement with experimentally-determined values. A possible linear saturation mechanism was also proposed.

ACKNOWLEDGEMENTS

The author would like to thank R. P. Drake and T. W. Johnston for encouragement to pursue this topic and for a careful critique of this manuscript. Many useful conversations with R. L. Berger, W. L. Kruer, M. V. Goldman, C. J. McKinstrie, and D. D. Meyerhofer are acknowledged, as is the support of H. A. Baldis. The experiment reported in section V was performed with the assistance of D. S. Montgomery, K. S. Bradley, R. P. Drake, and Kent Estabrook. This work was performed under the auspices of the U. S. Department of Energy by Lawrence Livermore National Laboratory under contract number W-7405-ENG-48.

REFERENCES

1. Stephen E. Bodner, Critical elements of high gain laser fusion, *J. Fusion Energy* **1**, 221 (1981).
2. R. L. Keck, L. M. Goldman, M. C. Richardson, W. Seka, and K. Tanaka, Observations of high-energy electron distributions in laser plasmas, *Phys. Fluids* **27**, 2762 (1984).
3. K. Mizuno, P. E. Young, W. Seka, R. Bahr, J. S. DeGroot, R. P. Drake, and K. G. Estabrook, Investigation of ion-acoustic-decay instability thresholds in laser-plasma interactions, *Phys. Rev. Lett.* **65**, 428 (1990).
4. R. P. Drake, R. E. Turner, B. F. Lasinski, K. G. Estabrook, E. M. Campbell, C. L. Wang, D. W. Phillion, E. A. Williams, and W. L. Kruer, Efficient Raman sidescatter and hot-electron production in laser-plasma interaction experiments, *Phys. Rev. Lett.* **53**, 1739 (1984).
5. William L. Kruer, Intense laser plasma interactions: From Janus to Nova, *Phys. Fluids B* **3**, 2356 (1991).
6. R. P. Drake, Three-wave parametric instabilities in long-scale-length, somewhat-planar, laser-produced plasmas, *submitted to Laser and Particle Beams* (1991).
7. R. Paul Drake, Trends in laser-plasma-instability experiments for laser fusion, in: "International Topical Conference on Research Trends in Inertial Confinement Fusion," American Institute of Physics, New York (1991).
8. William L. Kruer, "The Physics of Laser Plasma Interactions," Addison-Wesley Publishing Company, Inc., Redwood City, California (1988).
9. S. H. Batha, D. S. Montgomery, K. S. Bradley, R. P. Drake, Kent Estabrook, and B. A. Remington, Intensity scaling of stimulated Raman forward scattering in laser-produced plasmas, *Phys. Rev. Lett.* **66**, 2324 (1991).
10. R. P. Drake, E. A. Williams, P. E. Young, Kent Estabrook, W. L. Kruer, H. A. Baldis, and T. W. Johnston, Evidence that stimulated Raman scattering in laser-produced plasmas is an absolute instability, *Phys. Rev. Lett.* **60**, 1018 (1988).
11. H. Lowdermilk, C. Yamanaka, S. Nakai, and M. Sluyter, presented at this conference.
12. S. H. Batha, D. S. Montgomery, R. P. Drake, Kent Estabrook, and B. A. Remington, Production of very-large, sub-tenth-critical plasmas for laser-fusion research, *Phys. Fluids B* **3**, 2898 (1991).
13. C. S. Liu, M. N. Rosenbluth, and R. B. White, Raman and Brillouin scattering of electromagnetic waves in inhomogeneous plasmas, *Phys. Fluids* **17**, 1211 (1974).
14. Marshall N. Rosenbluth, Parametric instabilities in inhomogeneous media, *Phys. Rev. Lett.* **29**, 565 (1972).
15. S. H. Batha, R. P. Drake, and T. W. Johnston, Parametric-instability convective-gain coefficient in nonuniform plasmas, *submitted to Phys. Fluids B* (1992).
16. Kyoji Nishikawa, Parametric excitation of coupled waves. II. Parametric plasmon-photon interaction, *J. Phys. Soc. Japan* **24**, 1152 (1968).
17. D. W. Forslund, J. M. Kindel, and E. L. Lindman, Theory of stimulated scattering processes in laser-irradiated plasmas, *Phys. Fluids* **18**, 1002 (1975).
18. C. S. Liu, Parametric instabilities in an inhomogeneous unmagnetized plasma, in: "Advances in Plasma Physics," A. Simon and W. B. Thompson, eds., Wiley, New York (1976).
19. C. J. McKinstrie, A. Simon, and E. A. Williams, Nonlinear saturation of stimulated Raman scattering in an homogeneous plasma, *Phys. Fluids* **27**, 2738 (1984).
20. C. J. McKinstrie and A. Simon, Nonlinear saturation of stimulated Raman scattering in a collisional homogeneous plasma, *Phys. Fluids* **28**, 2602 (1985).
21. Klaus Baumgärtel and Konrad Sauer, "Topics on Nonlinear Wave-Plasma Interaction," Birkhäuser Verlag, Basel (1987).
22. C. J. McKinstrie and D. F. DuBois, A covariant formalism for wave propagation applied to stimulated Raman scattering, *Phys. Fluids* **31**, 278 (1988).
23. R. P. Drake (private communication).
24. R. P. Drake, E. A. Williams, P. E. Young, Kent Estabrook, W. L. Kruer, and D. S. Montgomery, Narrow Raman spectra: The competition between collisional and Landau damping, *Phys. Fluids B* **1**, 2217 (1989).

25.   G. Picard and T. W. Johnston, Decay instabilities for inhomogeneous plasmas: WKB analysis and absolute instability, *Phys. Fluids* **28**, 859 (1985).

26.   R. L. Berger, E. A. Williams, and A. Simon, Effect of plasma noise spectrum on stimulated scattering in inhomogeneous plasma, *Phys. Fluids B* **1**, 414 (1989).

27.   E. A. Williams, in: "Laser Program Annual Report 85," Lawrence Livermore National Laboratory, Livermore, CA (1985).

28.   E. A. Williams and T. W. Johnston, Phase-inflection parametric instability behavior near threshold with application to laser-plasma stimulated Raman scattering (SRS) instabilities in exploding foils, *Phys. Fluids B* **1**, 188 (1989).

29.   A. Simon (private communication).

30.   G. B. Zimmerman and W. L. Kruer, Numerical simulation of laser-initiated fusion, *Comments Plasma Phys. Controlled Fusion* **2**, 51 (1975).

31.   D. S. Montgomery, S. H. Batha, K. S. Bradley, R. P. Drake, K. G. Estabrook, and B. A. Remington, Intensity scaling of stimulated Raman forward scattering, *Lawrence Livermore National Laboratory ICF Quarterly Report* **1**, 13 (1990).

THE NOVA UPGRADE FACILITY FOR ICF IGNITION AND GAIN

W. H. Lowdermilk, E. M. Campbell, J. T. Hunt,
J. R. Murray, E. Storm, M. T. Tobin, J. B. Trenholme

Lawrence Livermore National Laboratory
Livermore, CA

INTRODUCTION

Research on Inertial Confinement Fusion (ICF) is motivated by its potential defense and civilian applications, including ultimately the generation of electric power. The U. S. ICF Program was reviewed recently by the National Academy of Sciences (NAS) and the Fusion Policy Advisory Committee (FPAC). Both committees issued final reports in 1991[1,2] which recommended that first priority in the ICF program be placed on demonstrating fusion ignition and modest gain (G<10). The U. S. Department of Energy and Lawrence Livermore National Laboratory (LLNL) have proposed an upgrade of the existing Nova Laser Facility at LLNL to accomplish these goals. Both the NAS and FPAC have endorsed the upgrade of Nova as the optimal path to achieving ignition and gain.

Results from Nova Upgrade experiments will be used to define requirements for driver and target technology both for future high-yield military applications, such as the Laboratory Microfusion Facility (LMF) proposed by the Department of Energy, and for high-gain energy applications leading to an ICF engineering test facility. The central role which Nova Upgrade would play in the national ICF strategy is illustrated in Figure 1.

In the past 15 years, LLNL has built a series of neodymium-glass lasers of increasing power and energy and used them to explore the physics of ICF target performance. Nova, the latest in this series, was completed on time

Laser Interaction and Related Plasma Phenomena, Vol. 10
Edited by G.H. Miley and H. Hora, Plenum Press, New York, 1992

323

and within budget and has exceeded its performance milestones. Since its commissioning in 1985, Nova has been the world's premier ICF and high-energy-density research facility, used by both the fusion and weapons communities. Figure 2 summarizes the performance of Nova and previous LLNL laser systems, as well as projected performance for the Nova Upgrade.

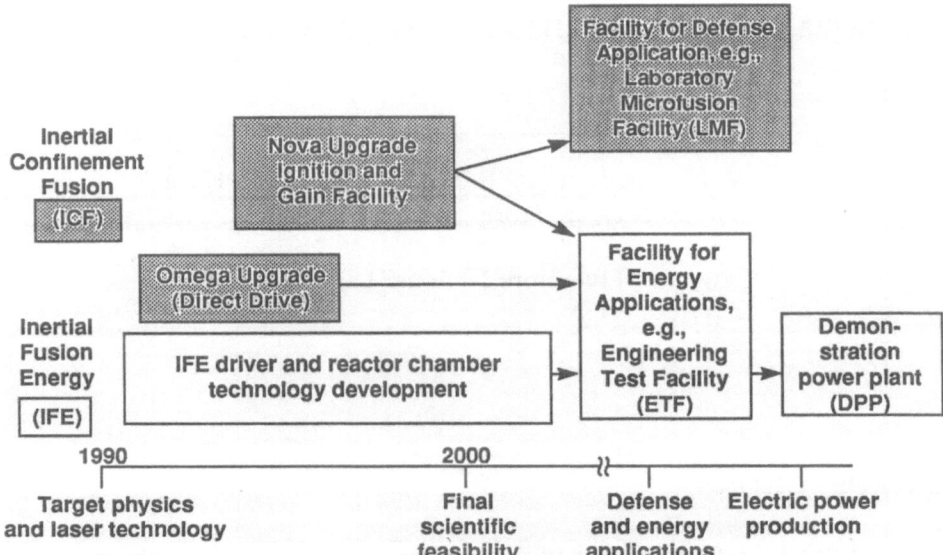

Fig. 1. The Nova Upgrade Facility plays a central role in the national ICF Program strategy, culminating in defense and energy applications. In addition to providing significant near-term weapons physics and high-energy-density experimental capabilities, the Nova Upgrade will be used to define driver, target, and technology requirements for long-range, high-yield military applications, such as the Laboratory Microfusion Facility, and for high-gain energy applications, such as the Engineering Test Facility.

Nova Upgrade Facility

The proposed Nova Upgrade Facility consists of an 18-beamline, high-power, neodymium-glass laser of advanced, cost-effective design and a target experiment area capable of safely containing the nuclear yield generation in the proposed experiments. This facility would be built in the existing Nova building at LLNL. Layout of the facility within the Nova building is illustrated in Fig. 3. Facility specifications are determined by the ignition target requirements, shot schedule, and safety requirements summarized along with proposed cost and schedule, in Table 1.

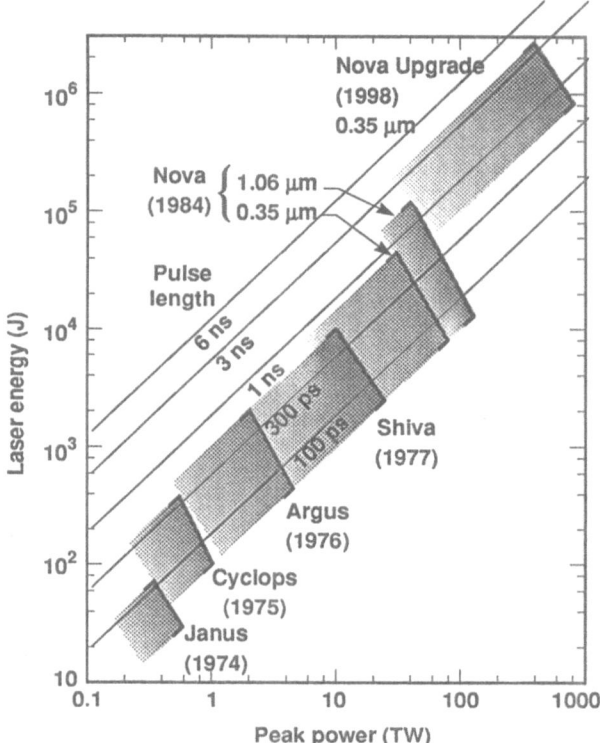

Fig. 2. The performance capabilities of neodymium-glass lasers built by LLNL for ICF research have increased by orders of magnitude. Nova Upgrade's projected capability is an advance in performance of comparable magnitude to those achieved by previous systems.

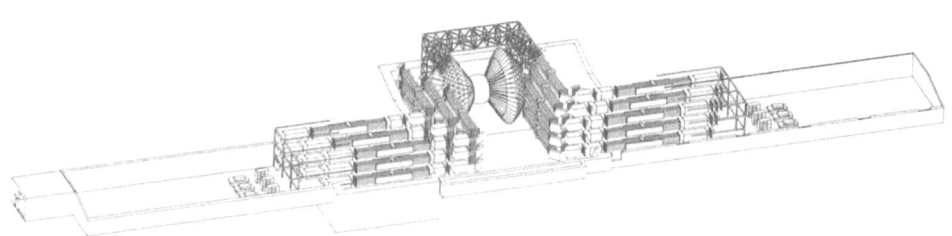

Fig. 3. The Nova Upgrade Facility wil be constructed within the existing Nova building at LLNL. Approximately half of this building was built in 1976 for the Shiva laser fusion system, and the remainder was constructed in 1982 to house the larger Nova laser and experiment area. The ability to upgrade Nova to 1 to 2 MJ of energy within the existing facility results from significant advances in solid-state laser technology made in recent years at LLNL.

Table 1. Specifications for the Nova Upgrade Facility

<u>Laser</u>	
Energy	1-2 MJ at 0.35 μm
Peak power	500 TW (4-ns pulse)
	700 TW (2-ns pulse)
Pulse shape	Continuous or picket fence
Overall duration	10–15 ns
Dynamic range	100–200:1 (continuous)
	10-40:1 (picket fence)
Power balance	5–10% rms
Pointing accuracy	10–30 μrad
<u>Target Area</u>	
Annual experiment yield	100 kJ, 100 shots
	5 MJ, 35 shots
	20 MJ, 10 shots
Re-entry time (unlimited access)	24 hr after 100-kJ shot
Radiation release	<1% of regulatory standard
<u>Cost and Schedule</u>	
Total estimated cost	$400 million
(actual year dollars)	
Construction period	FY1995-99

Construction of the Nova Upgrade Facility would follow the accomplishment of milestones that validate target physics and complete the technical and cost basis for the facility. A technical and financial plan to complete this effort in 3 to 4 years has been presented to and endorsed by the NAS and FPAC. Construction beginning in FY 1995 would lead to experimental demonstration of fusion ignition and gain in the laboratory early in the next decade.

Nova Upgrade Laser

Lasers for ICF research are composed of multiple beams to provide sufficiently uniform illumination of the fusion target. Nova has ten beams, each capable of delivering on target approximately 6 kJ of 0.35-μm radiation in a 1-3 ns pulse. The energy requirement of 1 to 2 MJ for ignition and propagating burn could be delivered by increasing the number of Nova-type beamlines, but the resulting size and cost of the facility would be prohibitive. Therefore, the design challenge for the Nova Upgrade was to make the laser much more compact and lower in cost. The proposed Nova Upgrade design results from substantial progress in recent years in both areas.

Neodymium-glass lasers for ICF are composed of three main sections. In the first section, the pulse is generated by a laser oscillator, given the appropriate shape in time, space, and frequency, and preamplified to modest energy. In the second section, the pulse is amplified to the required energy. In the third section, the pulse is transported to the target area by reflection from mirrors, converted in frequency from the 1.06-μm fundamental wavelength to the third harmonic at 0.35 μm, and focused onto the target.

The major advances in laser compactness and cost reduction leading to the Nova Upgrade have been made principally in the architecture and components of the amplifier section, and in increased damage thresholds of optics in the beam transport section. On Nova and earlier neodymium-glass lasers for ICF, the pulse was amplified by making a single pass through a series of amplifiers of increasing diameter, as illustrated in Fig. 4. A Nova beamline has amplifiers of seven different aperture sizes, from 1 to 46 cm diameter. The Nova Upgrade laser, in contrast, has only the final, largest size amplifier. The laser pulse makes multiple passes (four in the baseline design) through this amplifier, reflecting back and forth between mirrors on either side of the amplifier, as illustrated in Fig. 4. The multipass final amplifier lowers the laser's cost by eliminating the smaller amplifiers that precede it in the Nova-style design. Compactness of the Nova Upgrade laser is achieved by the use of "multisegment" optical components throughout the laser. The multisegment design allows multiple laser beams to be packed close together. An individual Nova Upgrade amplifier, shown in Fig. 5, amplifies 16 separate beams arranged in a four-by-four array. All other components are also segmented, so each beamline consists of 16 optically-independent laser "beamlets." The beamlets have an aperture of approximately 30 by 30 cm. Significantly larger apertures require larger and more costly individual components and have increased amplified spontaneous emission losses, while significantly smaller apertures suffer proportionately larger edge losses. and require larger numbers of components to manufacture, assemble and align. This compact design allows the entire 18-beamline laser to fit in the existing Nova building while delivering 25 to 50 times Nova's 0.35-μm energy.

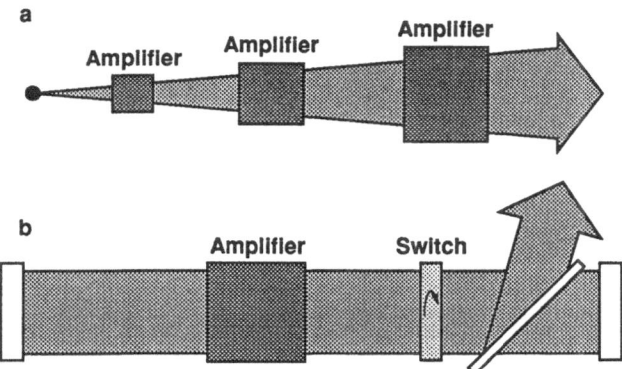

Fig. 4. Nova and earlier neodymium-glass lasers at LLNL used (a) the master-oscillator-power-amplifier architecture. Nova Upgrade uses (b) the multipass geometry to reduce the cost and size of the laser and to improve system performance.

The multisegment amplifier (MSA) further reduces cost by achieving higher efficiency than Nova-style amplifiers.[3] These improvements are a result of the vertical banks of flashlamps, used to pump the neodymium-glass, which are placed between the columns of glass plates, as shown in Fig. 5. Less radiation from these lamps is absorbed in the amplifier's outer reflecting wall than from the lamps placed along the wall. Performance is further improved, in comparison to Nova designs, by reducing the density of flashlamps in the amplifier. A prototype two-by-two multisegment amplifier, shown in Fig. 6, demonstrated energy-storage efficiency of 3.7% at 0.25-J/cm^3 energy-storage density, approximately twice the efficiency of Nova's 31-cm

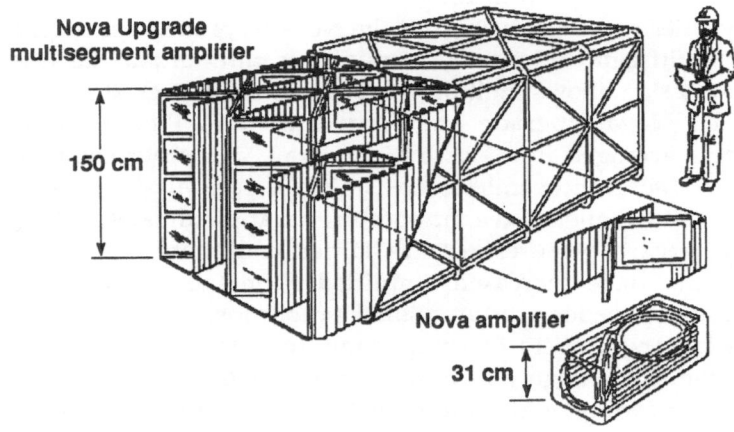

Nova Upgrade multisegment amplifier

150 cm

Nova amplifier

31 cm

Fig. 5. The Nova Upgrade multisegment amplifier (MSA) has four columns and four rows of neodymium-glass plates. The cut-away section shows that the MSA is essentially an array of the proven Nova 31-cm amplifiers, thus giving high confidence in its performance.

amplifier at this storage density. Furthermore, the multisegment amplifier has many fewer mechanical components, flashlamps, and connections than Nova amplifiers delivering the same energy. This feature also contributes to lower cost. Improvements in the efficiency of energy extraction from the amplifier, beam transport and frequency conversion will give Nova Upgrade an overall efficiency of approximately 1% compared to Nova's 0.1% efficiency. This increase in efficiency reduces the size and cost of the capacitor energy storage system.

After amplification, the laser pulse must be switched out of the optical resonator formed by the mirrors on either side of the amplifier and sent on to the beam transport section, as indicated schematically in Fig. 4. Work at LLNL[4] has demonstrated the feasibility of a large-aperture, electro-optic switch at reasonable cost. Figure 7 shows the prototype Pockels cell switch that demonstrated Nova Upgrade performance requirements on switching speed and efficiency. The Pockels cell contains a 1-cm-thick plate of the electro-optic crystal KD_2PO_4 (KD*P) with an aperture of 27 by 27 cm. When

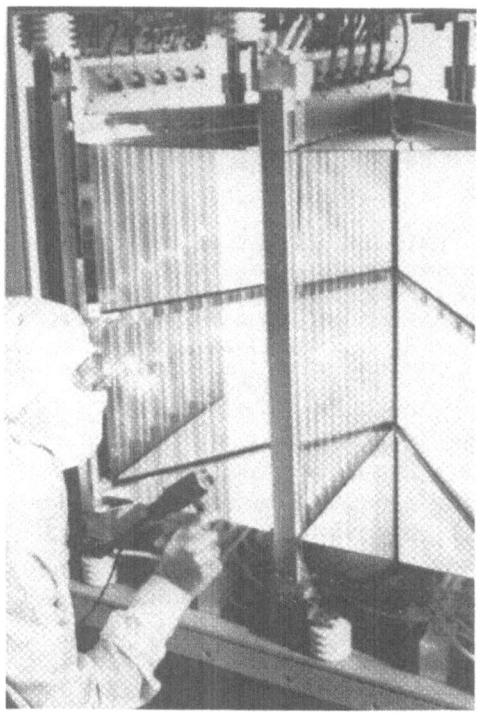

Fig. 6. A prototype two-by-two multisegment amplifier has been built and tested at LLNL. This prototype is one-fourth of a Nova Upgrade amplifier module, and it has operated at more than twice the efficiency (flashlamp energy to stored laser energy) of Nova's 31-cm amplifier.

voltage is applied to the crystal along the direction of light propagation, it rotates the laser pulse's polarization. The laser's propagation direction can then be changed by a polarizer. To apply voltage to the crystal requires a transparent electrode with a high damage threshold. In the prototype Pockels cell, the electrodes are formed on either side of the crystal by a glow-discharge plasma in He gas at approximately 1 Torr pressure.

Fig. 7. This prototype Pockels cell switch, with 27-by-27-cm aperture, demonstrated the feasibility of a high-fluence, large-aperture electro-optic switch at reasonable cost, further validating the multipass design for the Nova Upgrade.

The cost of laser and optical components is proportional to the total beam area. Therefore, to minimize cost, the laser should operate at the highest practical average fluence and intensity. As the beam propagates, the spatial intensity profile becomes less uniform because of diffraction and effects of the nonlinear refractive index, processes which are well understood and can be reliably modeled by computer calculations.[5] The allowable peak fluence is set by the damage threshold of optical components. To minimize system cost, then, the ratio of peak fluence to average fluence should be as small as possible, and optical components with high damage threshold should be developed. LLNL has placed major effort on development of high-damage-threshold optical materials and components. Advances since Nova's construction are summarized in Table 2.

Table 2. Comparison of damage thresholds available for Nova and the Nova Upgrade, expressed in joules per square centimeter for 3-ns pulses

	Nova	Nova Upgrade
Antireflection coating		
(1.06 μm)	8.5	30
(0.35 μm)	4	16
Mirror coating (1.06 μm)	7	26
Polarizer coating (1.06 μm)	5	19 (p-polarization)
		22 (s-polarization)
KDP crystals		
(1.06 μm)	12	35
(0.35 μm)	5	15

To minimize the peak-to-average fluence ratio, the technique of image relaying and spatial filtering was developed at LLNL[6] and has been used on all neodymium-glass ICF lasers since the mid-1970s. Image relaying occurs when two lenses with a common focal point are placed between two amplifier stages, with the separation between the lenses such that an image of the first amplifier lies in the second. An aperture placed in the common focal plane between the lenses will remove intensity modulations on the beam that have high spatial frequencies (short wavelengths), thus reducing the peak-to-average fluence ratio. Analysis and experience with Nova have shown that the peak-to-average ratio can be maintained at less than 2:1. This low-intensity modulation, coupled with the increase in damage threshold noted in Table 2, results in a three-to four-fold reduction in the optical aperture area required per unit energy delivered on target by the Nova Upgrade compared to Nova.

A Nova Upgrade beamlet is shown schematically in Fig. 8. The laser pulse, with energy of 1 to 10 J per beamlet, produced by the pulse-generation section enters the amplifier section by reflection from a small mirror near the image-relay focal plane. The image-relay lenses are in the center of the optical resonator, with an amplifier on one side and the Pockels cell optical switch and polarizer on the other. In the baseline design, this amplifier has nine neodymium-glass plates in each beamlet. The distance between the resonator mirrors is twice the separation between the lenses for optimal relaying. After the pulse makes four passes through the amplifier, it is switched out of the resonator by reflection from the polarizer. It then passes through another amplifier containing five plates in each beamlet. At this point, depending on pulse length, the pulse energy is 5.5 to 11 kJ per beamlet. The pulse then enters the beam transport section, which consists of mirrors and a second image relay that forms an image of the final amplifier near the frequency converter crystals mounted on the target chamber. The frequency converter is similar to that used on Nova, and consists of two KDP crystals in series. The first crystal converts two-thirds of the input light to the second harmonic, and the second crystal mixes that radiation with the remaining fundamental radiation to generate the third harmonic at 0.35 μm. The Upgrade design specifies a 70% conversion efficiency to the third harmonic for square pulses of 1.5- to 4-ns duration. Third harmonic frequency conversion efficiencies above 80% under conditions appropriate for the Upgrade have been demonstrated. After frequency conversion, the pulse is focused through a debris shield onto the target. The final 0.35-μm pulse energy delivered to the target is 3.5 to 7 kJ, depending on pulse duration, per beamlet in an appropriately shaped pulse. The Upgrade design will incorporate necessary temporal and/or spatial coherence control, the requirements of which are the subject of ongoing target experiments.

The Upgrade design is based on individual component performance specifications demonstrated in the laboratory but not as an integrated system. An integrated test of an entire beamlet will be completed, as the NAS report recommended, before Nova Upgrade construction. The prototype beamlet will have sufficient flexibility to explore design alternatives to the baseline that may offer enhanced performance and/or reduced cost for the Nova Upgrade. The beamlet demonstration project is scheduled for completion in 1993.[7]

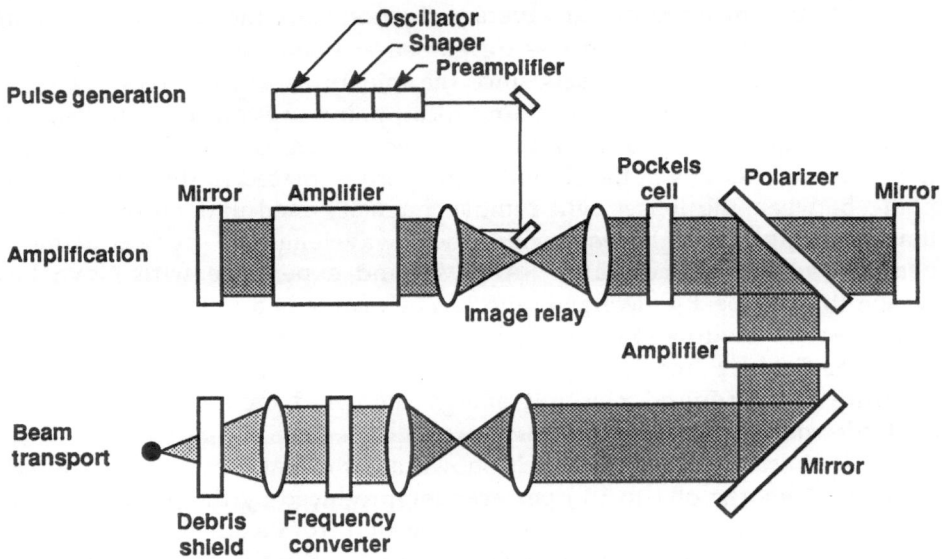

Fig. 8. Each Nova Upgrade beamlet is composed of three sections: pulse generation, amplification, and beam transport. The laser pulse is produced, shaped in space, time, and frequency, and preamplified in the first section. It is injected into the optical resonator and after four passes is switched out by the Pockels cell and polarizer. After final amplification, the pulse is transported to the target area, frequency converted and focused on the target. The laser has 288 beamlets, grouped in 18 beamlines.

Nova Upgrade Target Area

The target area houses the target chamber, diagnostics, and final focusing optics. The existing Nova target area requires no major modification to meet Nova Upgrade scientific and safety requirements. The baseline design for the target chamber is a 5-cm-thick spherical aluminum shell of 4-m radius surrounded by a 70-cm-thick polyethylene neutron moderation. The aluminum first wall provides the vacuum barrier, soft x-ray shield, and shrapnel protection. The 288 laser beamlets enter the target chamber in two clusters, one from the east laser bay and one from the west laser bay. For irradiation symmetry, each cluster of 144 beamlets is divided into a set of concentric cones of beams. The current baseline design has four cones of beams on each side, as shown in Fig. 9.

The total optical-path length of each beam is adjusted to control pulse arrival times at the target, allowing the x-ray drive symmetry to be optimized. The final turning mirrors, KDP frequency converters, focus lenses, and laser diagnostic packages are supported by a spaceframe independent from the chamber. The spaceframe also supports the target alignment viewers, target positioner, and the cryogenic support system. Because the spaceframe, not the target chamber, supports these loads, the mass of the chamber and its resultant activation can be minimized.

The layout of beam optics and protection devices is given in Fig. 10. Protection against soft x-ray damage to the vacuum windows is provided by fused-silica debris shields. With debris shields placed 5.5 m from the target chamber center, single-shot damage from x-rays is not expected for target yields below 20 MJ. Calculations of the shock wave formed in the debris shields by the sudden x-ray heating indicate that tensile stress levels at the rear surface of a 1-cm-thick shield are well below the level that would cause spallation from a 45-MJ-yield shot, which is the maximum credible yield.

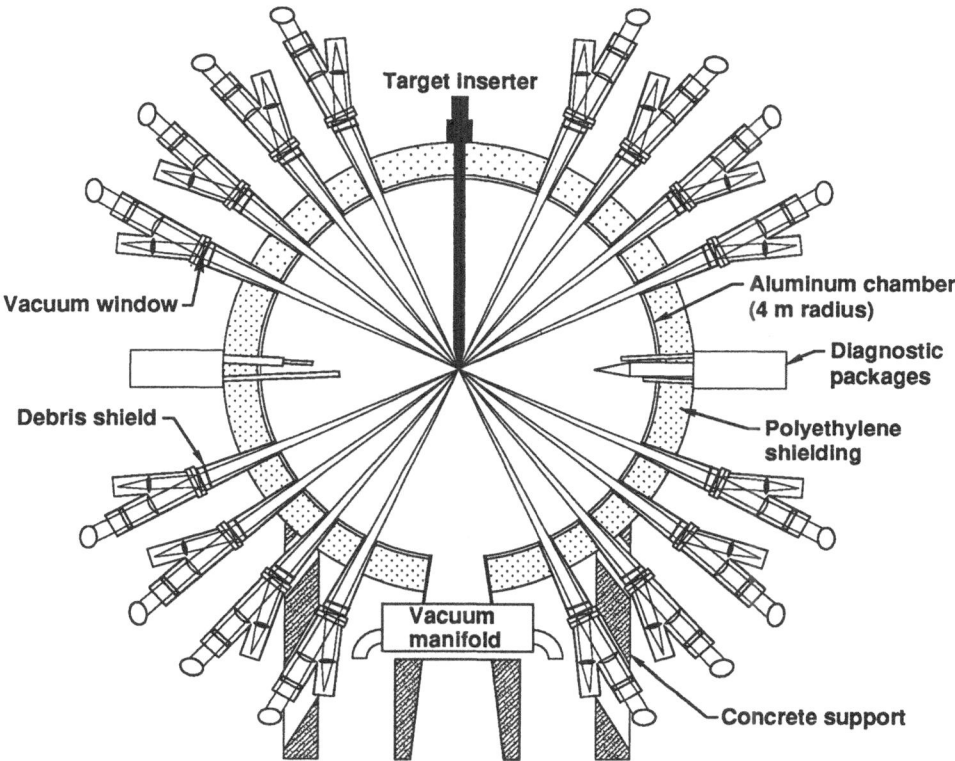

Fig. 9. The 288 laser beamlets enter the target chamber in two clusters: one from the east laser bay and one from the west laser bay. Each cluster is divided into four concentric cones to provide required irradiation symmetry.

Condensation of target debris and ablated aluminum on the debris shields can be removed by periodic cleaning or refinishing. The shields are set at Brewster's angle to eliminate need for an anti-reflective coating, which experiments have shown are unlikely to withstand x-ray emission even from shots with no fusion yield.

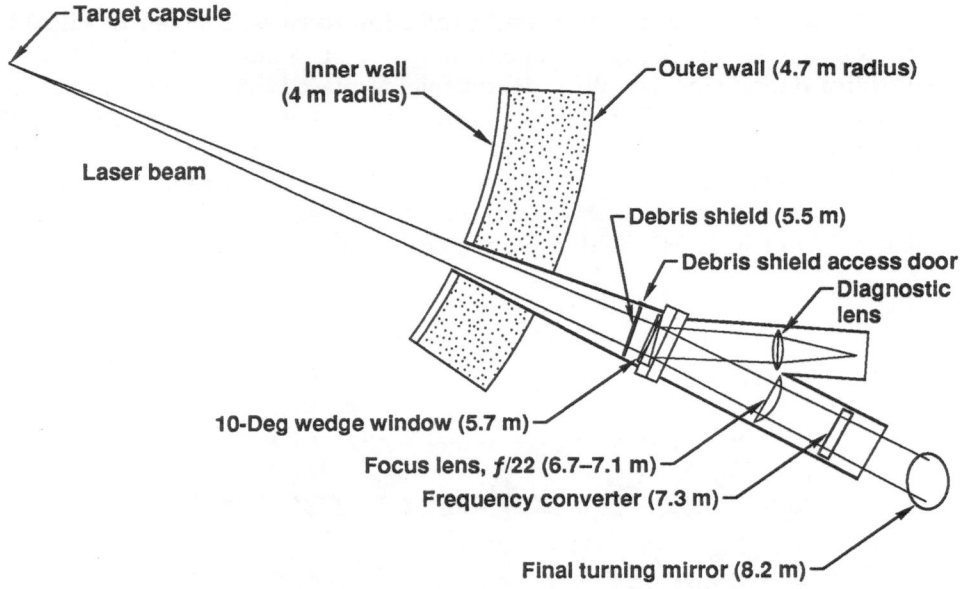

Fig. 10 Beamlet optics and protection devices are designed to maintain optics integrity for yields of up to 20 MJ. Dimensions indicate distance from the center of the target chamber.

Conclusion

The achievement of fusion ignition in the laboratory is the highest priority of the national ICF program, and an upgrade of Nova was recommended by the NAS and FPAC as the optimum path to achieve this goal. The Nova Upgrade facility design emphasizes both low cost and low technical risk. Key components and materials have already been demonstrated individually in the laboratory. The prototype beamlet, which is due to be completed in 1993, will serve as an integrated technology testbed for the Nova Upgrade. Construction of Nova Upgrade beginning in 1995 will allow completion of the project in 1999 with an as-spent cost of approximately $400 million.

References

1. Committee for a Second Review of DOE's ICF Program, "Final Report - Second Review of the Department of Energy's Inertial Confinement Fusion Program, "National Academy of Sciences, Commission of Physical Sciences, Mathematics, and Applications, National Research Council (National Academy Press, Washington, D. C., September 1990).
2. Fusion Policy Advisory Committee, "Final Report - Fusion Policy Advisory Committee" (U. S. Department of Energy, September 1990).
3. Powell, H. T., Erlandson, A. C., Jancaitis, K. S., and Murray, J. E., "Flashlamp pumping of Nd:glass disk amplifiers, " SPIE Proceedings, Vol. **1277**, High-Power Solid State Lasers and Applications (1990) 103-120.

4. Goldhar, J., and Henesian, M. A., "Large-aperture electro-optical switches with plasma electrodes," IEEE J. Quantum Electron. QE-**22** (1986) 1137-1744.

5. Simmons, W. W., Hunt, J. T., and Warren, W. E., "Light propagation through large laser systems, "IEEE J. Quantum Electron. QE-**17** (1981) 1727-1744.

6. Hunt, J. T., Glaze, J. A., Simmons, W. W., and Renard, P. A., "Suppression of self-focusing through low-pass spatial filtering and image relaying," Appl. Opt. **17** (1978) 2053-2057.

7. Murray, J. R., Campbell, J. H., Frank, D. N., Hunt, J. T. and Trenholme, J. B., "The Nova Upgrade Beamlet Demonstration Project," UCRL-LR-105821-91-3, Inertial Confinement Fusion Quarterly Report, April - June 1991, Vol. 1, No. 3 (1991) 89-107.

*This work was performed under the auspices of the U.S. Department of Energy by Lawrence Livermore National Laboratory under contract No. W-7405-Eng-48.

ADVANCEMENT OF INERTIAL FUSION RESEARCH

C. Yamanaka

Institute for Laser Technology
2-6 Yamada-oka Suita Osaka 565
and
Himeji Institute of Technology
2167 Shoshya Himeji 671-22
Japan

ABSTRACT

The laser fusion has made a great progress in the last 20 years. The initiation of the laser fusion was to demonstrate the neutron yield by laser irradiation. A new implosion concept of fuel compression was proposed which tended to reduce the necessary energy of fusion drivers. The laser interaction with plasmas were comprehensively studied to clarify the implosion physics. The several types of drivers had been developed which enabled us to perform the effective fusion experiments. Pellet fabrications had made a great stride in progress to produce fine uniform fuel targets for implosion experiments. The theoretical research as well as the computer simulation methods were widely developed which can predict the processes of the laser fusion in detail.

Recent inertial fusion experiments on the direct drive targets have attained the high neutron yield 10^{13} and the high density compression 600 times liquid density respectively. The electron degeneracy of core plasma was also observed. For the indirect drive target experiments, the radiation confinement was measured to keep the illumination uniformity. The ablation pressure of 100M bar is generated by soft X-ray radiation of 3×10^{14} W/cm^2 over 1 nsec which produces the implosion velocity of 3×10^7 cm/sec. The development of the reactor driver of a few M joule is the most important issue for the inertial fusion energy program.

Now the inertial fusion research has come to the second stage of the development. The ignition and breakeven are in a scope of our program. Several schemes of inertial fusion reactor designs have been investigated. By 2025, a demonstration of inertial fusion energy will be performed.

I. INTRODUCTION

The idea of inertial fusion energy (IFE) is to use drivers to implode a fuel pellet containing deuterium and tritium. The implosion is to compress the fuel and to ignite the controlled thermonuclear reaction. The pellet gain of energy is expected more than 100 which is introduced by the self heating of alpha particle. The goal of plasma

Laser Interaction and Related Plasma Phenomena, Vol. 10
Edited by G.H. Miley and H. Hora, Plenum Press, New York, 1992

337

parameters are estimated to be $T \sim 5\text{keV}$, $\rho/\rho_S \sim 500\text{-}1000$ and $\rho R > 0.3\text{g/cm}^2$ where T is the temperature, ρ is the density of fuel, R is the compressed core radius and ρ_S is the solid density of the D-T mixture.

These data were almost attained individually by laser implosion experiments. However the self sustaining thermonuclear reaction is a target to study. The key issue is to achieve the uniform compression suppressing the instability of implosion. The development of drivers is also very important.

Several concepts on inertial fusion reactor are contemplated. The inertial fusion

FIG.1. GEKKO XII glass laser system

has the potential to provide environmentally attractive and acceptably safe energy. However the cost competitive IFE reactor will require pushing the state of the art in several areas such as the target performance, reactor drivers as well as the reactor technology. The international collaboration is highly expected to resolve these issues.

II. PRESENT STATUS OF IFE

Two types of IFE targets exist: direct drive targets which absorb the energy of the driver directly on the fuel capsule, and indirect drive targets which first convert the driver energy to X-rays which are then absorbed by the fuel capsule. Efficient energy delivery to the fuel capsule is anticipated for the direct drive method, but a greater uniformity of the driver irradiation is required. With the indirect drive method, a larger non-uniformity of the driver irradiation can be tolerated, but at the expense of efficient coupling of energy to the fuel capsule.

(1) Direct drive target

For direct drive, high temperatures and high densities were attained in separate modes. In the high temperature mode the fuel was heated to an ion temperature of 10keV, but it was compressed to D-T densities of only 0.2 g·cm^{-3}. In experiments on the GEKKO XII laser of the LHART target, a Gaussian laser pulse was matched to the development in time of the capsule shell radius in such a way that the fuel compression was at its maximum shortly after the shell was hit by the shock wave moving ahead and reflecting at the target center.[1] D-T neutron yields of up to 1×10^{13} and pellet gains of 0.2% were achieved in these 'stagnation free' conditions.[2,3]

In the High density mode, significant improvement in irradiation uniformity is required by direct drive implosions. Recent implosion experiments used random phase plates (RPPs)[4] to improve the irradiation uniformity. The RPPs are components which divide large laser beams into many beamlets, producing a focal pattern with a smooth envelope and small scale spikes due to interference between the beamlets.

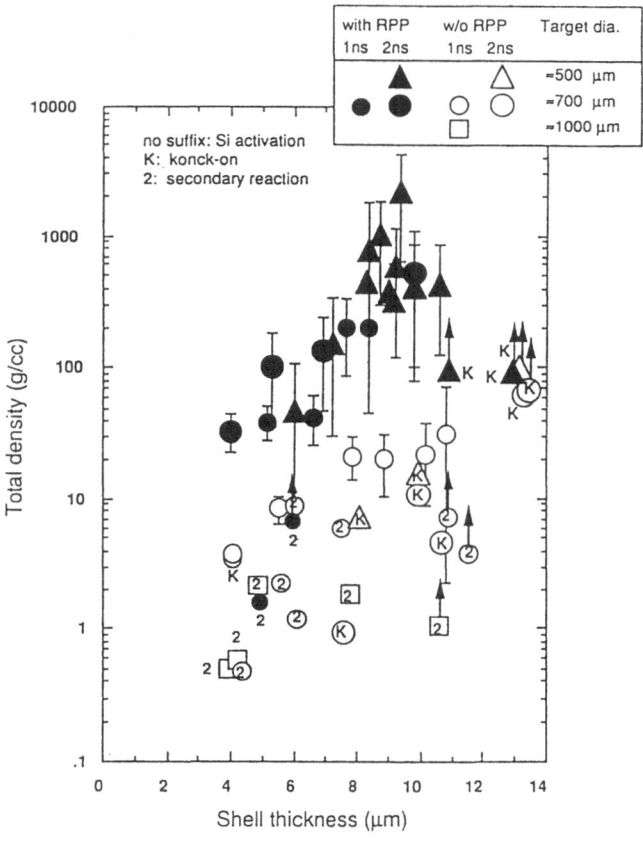

FIG. 2. Fuel density as a function of the initial shell thickness. A density of 1g·cm^{-3} corresponds to the initial density of the pellet material. The average density of nine identical shots (closed triangles at a shell thickness of 8-11 μm) was 600g·cm^{-3} or, equivalently, 600 times LD.[5] .

Direct drive implosion experiments on the GEKKO XII laser (9kJ, 0.5μm, 2ns), using hollow shell targets for which the hydrogen in the plastic molecules was replaced by D and T, demonstrated the achievement of a fuel ρR value of ~ 0.3g·cm^{-2} and fuel densities of ~ 600 times of liquid density.[5] The observed density is plotted in Fig. 2.

The vertical axis gives the total density in g·cm^{-3}; the values can also be converted to times liquid density, since a density of 1g·cm^{-3} corresponds to the initial density of the pellet material. The irradiation non-uniformity in these experiments was significantly reduced to a level of 5% (rms) by introducing RPPs. The target irregularity was controlled to a level of 1%. The fuel ρR value was directly measured by neutron activation of silicon, which was originally compounded in the plastic targets. The fuel densities were estimated from the ρR values using the mass conservation relation, where the ablated mass was separately measured using the time dependent X-ray emission from multilayer targets. The observed densities were in agreement with the results of one-dimensional calculations with convergence ratios of 25-30.

Compressed densities of ~600g/cm^3 were demonstrated on the GEKKO XII laser system with direct-drive implosion of the plastic hollow shell targets. The estimated plasma parameters of the compressed core, which are the Coulomb-coupling constant for ions, Γ, of ~5 and the degree of electron degeneracy θ of ~0.3, indicate that strongly coupled and partially degenerate plasma was obtained. The ratio of the average Coulomb energy of ions to the thermal energy is given by

$$\Gamma \equiv \frac{Z_{eff}^2 e^2 / a}{k_B T}$$

$$Z_{eff}^2 \equiv \langle Z^{5/3} \rangle \langle Z \rangle^{1/3}$$

where $Z_{eff}e$ is an "effective" electric charge in the ion-sphere scaling, k_B is the Boltzmann constant, and a is the ion-sphere radius defined as

$$a \equiv \left(\frac{4\pi n}{3} \right)^{-1/3}$$

The electron system in the compressed core plasma is characterized by the degeneracy parameter θ, which is the ratio of the thermal energy to the Fermi energy of electrons:

$$\theta \equiv \frac{k_B T}{E_F}$$

where E_F is the Fermi energy of the electrons. This highly compressed core plasma may provide a tool for high-density physics study. The yield ratio of the secondary DT neutrons to the primary DD neutrons in such degenerate and strongly coupled plasma was calculated with inclusion of the strong Coulomb-coupling effects, the varied degrees of the electron degeneracy, and the electronic shielding effects. The increase of the secondary nuclear fusion reactions was shown to be effective evidence of the electron degeneracy in such a highly compressed fusion fuel as shown in Figure 3. Furthermore, the density of the compressed core plasma with a known temperature and a known mass was shown to be determined by the yield ratio of the secondary DT neutrons to the primary DD neutrons. The results of the implosion experiments with the deuterated plastic hollow shell targets suggested that the enhancement of the yield ratio provided evidence of the electron degeneracy in the uniformly compressed fuel.[6]

340

(2) Indirect drive target

For indirect drive of cannonball target, a fuel capsule is placed inside a cavity and is illuminated by X-rays generated in the cavity by irradiation with high intensity laser beams. The X-ray flux circulating in the cavity is enhanced due to radiation confinement arising from reemission of X-rays at the high-Z cavity wall.[7] The circulating X-rays are absorbed efficiently by the fuel capsule when the ablator is made of low~medium Z materials, which have low X-ray reemission coefficients. When the fractional area of openings on the cavity is reduced, radiation confinement becomes more effective and the fraction of the X-rays coupled to the capsule increases. The illumination uniformity on the capsule is also improved by the cavity due to transverse radiation diffusion.[8]

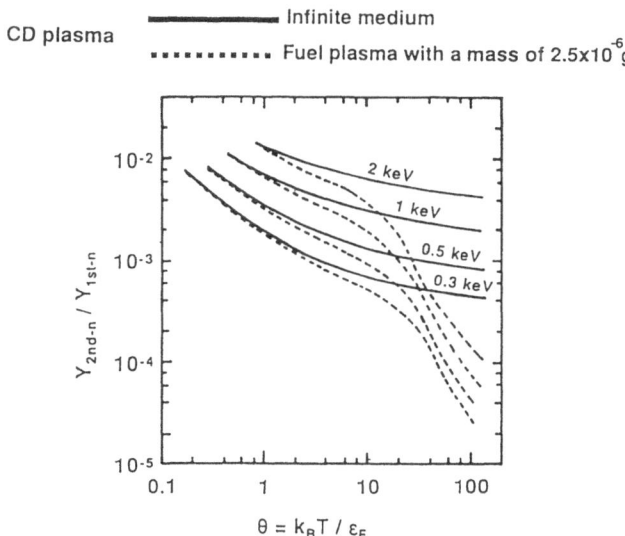

FIG. 3. *Yield ratio of the secondary DT neutrons to the primary DD neutrons, $Y_{2nd\text{-}n}/Y_{1st\text{-}n}$, versus the degree of the electron degeneracy θ in the CD plasma with temperature as a parameter, The solid lines correspond to the yield atio for the infinite medium. The broken lines correspond to the yield ratio for the homogeneous fuel plasma with a mass of 1×10^{-5}g. Requirements for verification of Fermi degeneracy are as follows:*
(i) $(Y_{2nd}/Y_{1st})_{EXP.} \gg (Y_{2nd}/Y_{1st})$ classical
(ii) at the peak temperature, Te is given by the measurement of neutron time of flight.

By irradiation of high-Z materials with short wavelength laser beams, high laser to X-ray conversion efficiency is obtained and generation of high energy electrons and X-rays is reduced. Quantitative understandings on radiation driven ablation, X-ray generation in high-Z materials, and radiation confinement in high-Z cavities have been obtained by the third harmonic light 351nm of GEKKO XII glass laser.

Radiation-driven implosion has been tested using several types of targets with different configurations and parameters. In the recent experiments, a fuel capsule is placed in a cylindrical cavity made of Au which is irradiated by 10 laser beams of 351 nm wavelength at a 5.5kJ total energy. Reproducible implosion was obtained with this target. The implosion velocity of 2.5×10^7 cm/sec has been achieved. Nonuniformities of low order modes were observed during implosion and at the maximum compression. The irradiation uniformity was controlled by careful design of the target structure and the irradiation configuration.

The experimental results were analyzed with a model that we have developed to treat soft X-ray driven ablation in low~medium Z plasmas.[9] Figure 4 shows dependences of the ablation pressure and the maximum acceleration distance on the flux and duration of the incident radiation calculated fo Al using this model. We find that the ablation pressure of 100 Mbar is generated by soft X-ray radiation at ~ 3×10^{14} W/cm² over 1 nsec. This pressure is necessary to implode a fuel capsule to a velocity of ~ 3×10^7 cm/sec.

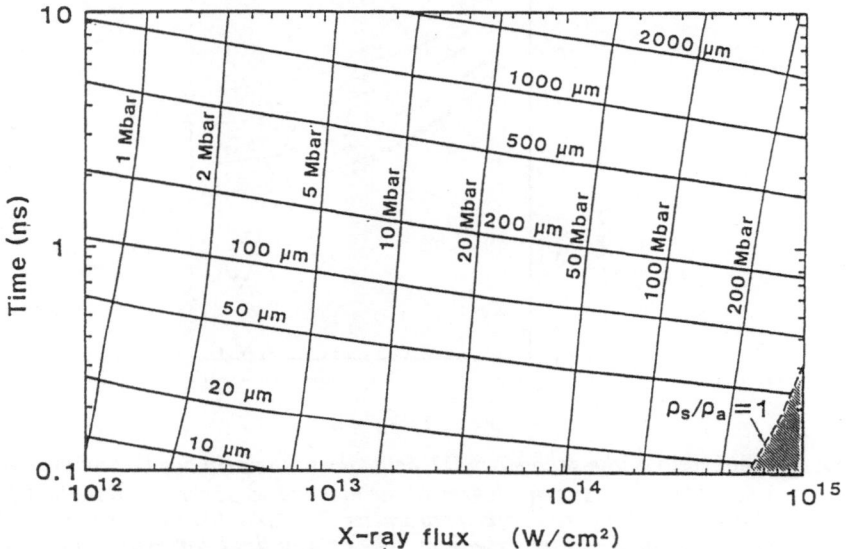

FIG. 4. *Dependences of the ablation pressure and the maximum acceleration distance on the radiation intensity and duration calculated for Al.*[10]

With a DT-filled GMB capsule with a fill pressure of 11 atm, a neutron yield 1.6×10^9 was obtained. The ion temperature determined from the neutron velocity broadening is 4 keV. With a D_2-filled plastic capsule with a fill pressure of 13 atm, the primary neutron yield of 5×10^7 was obtained. The fuel ρR estimated from the secondary neutron measurement is 10mg/cm². The size of the compressed fuel was evaluated from the framing camera images of DT-filled glass microballoon capsules. The convergence ratio of $R_i/R_f = 18$ was obtained for the initial fill pressure of 11 atm, where R_i and R_f are initial and final radius of the fuel. From this convergence ratio, we obtain compressed fuel density of $\rho = 11$g/cm³ (60 × liquid density) and the fuel areal density of 10mg/cm². The latter value is consistent with the secondary neutron measurement. From the framing camera images of the compressed pusher shell, the flux uniformity on the capsule is estimated to be better than 3%.

(3) Simulation analysis

A variety of implosion experiments have been carried out with gas-filled targets, cryogenic targets or deuterized polyethylene shell targets. Results of these experiments have been compared in detail with predictions from one-dimensional fluid codes. The neutron yield, for example, differs by about a factor 2 to 10^3 for the case of gas targets,[2] the difference arising mainly from the nonuniformity of the implosion dynamics.

Two-dimensional simulations have been carried out to explain the above discrepancy.[10] For example, a level of nonuniformity that produces a reduction of a factor two in neutron yield has been studied. In this simulation, a nonuniformity of the $l=6$ mode is assumed in laser absorption. In Fig. 5 snapshots of deformed Lagrange meshes and flow velocity vectors near maximum compression are shown. It is difficult to reduce the neutron yield by larger factors of 10^2 to 10^3 compared to a one-dimensional simulation, if only the nonuniformity of relatively longer wavelength modes ($l<24$) is included. It may be necessary to take into account the role of the instability for relatively shorter wavelength (e.g., $l \approx 100$ to 500).

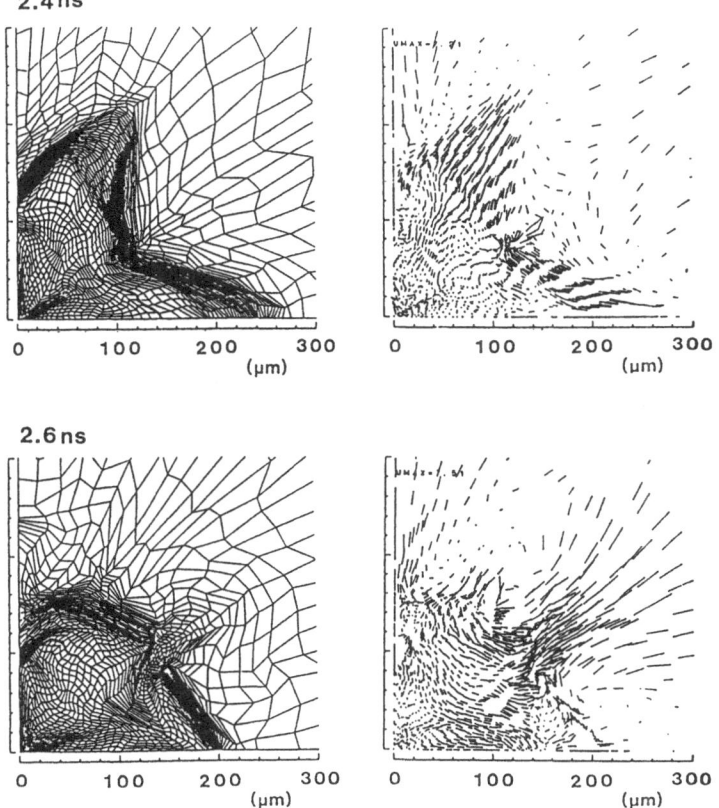

FIG. 5. *Snapshots of deformed meshes and velocity vectors of 2-D simulation for high neutron yield experiment*

The linear and nonlinear phenomena of the Rayleigh-Taylor (R-T) instabilities in the acceleration and stagnation phases are introduced. As for the acceleration phase the turbulent mixing triggered by the nonlinear growth of relatively shorter wavelength perturbations is studied consequently emphasizing the reduced growth of the mixing layer due to the ablative stabilization effect compared to the classical case. As for the stagnation phase, results of three dimensional simulations are to clarify the linear growth rate and the critical amplitude with which the perturbations go into the nonlinear free-falling phase. It is concluded that the speed of the free-falling spikes is enhanced compared to that in 2-D simulation.

III. FUTURE PROSPECT

As the implosion physics has been clarified to improve the implosion parameters, it has come to the second stage of inertial fusion research, the design phase of the inertial fusion reactor.

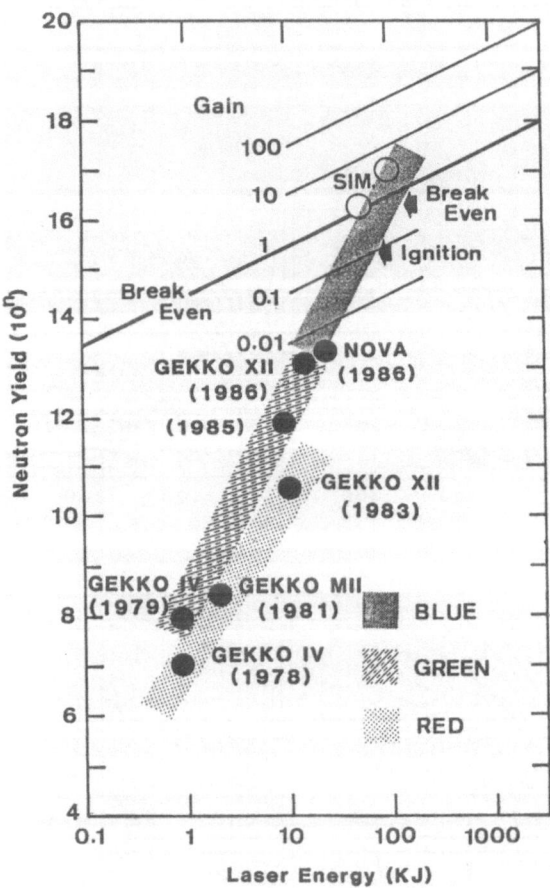

FIG. 6. *Progress and prospect of laser driven fusion. The solid circles are from experiments with GEKKO lasers and NOVA laser, while the open circles are from the present simulation. This figure suggests the possibility of breakeven experiment by 100 kJ class blue laser system.*

To get the ignition, the selfheating of the compressed fuel is performed by the α particles of fusion reactor.

Now the compressed fuel area density has attained a larger value than the α particle range, $\rho R \geq 0.3 \text{g/cm}^2$ and also compressed fuel density reaches 120gr/cm^3. However the hot core has not been observed yet. There might be due to the imperfect uniformity of the implosion process and also to the insufficient laser energy. In any way we are very close to the ignition stage. In Fig. 6 the present status and the future prospect of the neutron yield and the pellet gain estimated by the simulation are shown. The GEKKO up grade system of 100kJ blue light output is expected to demonstrate the ignition with fairly uniform irradiation condition.

In Table 1, the development of IFE reactor scenario is shown. The first phase is the ignition which shows the scientific feasibility. The second phase is the burning test for the engineering feasibility which requests the 500kJ driver and the pellet gain about 10. The following phases are the engineering test reactor and the demonstration reactor respectively. The third phase is hopefully expected in early 2000.

TABLE 1. Development scenario to ICF reactor

	I Ignition Exp	II L F C X	III L F E R	IV L F P R (D to C)
Mission	Ignition (Scientific feasibility)	Burning (Engineering feasibility · High gain pellet design · Reactor engineering · Intence neutron source	Reactor engineering test	Demonstration of power plant
Laser	Nd-glass 100 kJ Single shot	Solid KrF 500 kJ Single-1 shot/min.	500 KJ 1 Hz	4 MJ 1 Hz
Pellet	$Q=1$ 10^{14} N/shot $\rho \geqq 1000 \rho_\circ$ (200g/cm³) $\rho R \geqq 0.3$ g/cm²	$Q=10\sim100$ $10^{14}\sim10^{15}$ N/shot $5\sim50$ MJ/shot $\rho \geqq 2000 \rho_\circ$ (400g/cm³) $\rho R \geqq 1$ g/cm²	$Q=10\sim100$ $10^{18}\sim10^{19}$ N/sec $5\sim50$ MW th	$Q=50\sim500$ $10^3\sim10^3$ MWE $\rho\sim2000\sim4000 \rho_\circ$ (400$\sim$800 g/cm³) $\rho R\sim3\sim5$ g/cm³

IV. CONCLUSION

Various results of the implosion research with high power lasers and the informations for a compressed plasma enable us to design an IFE program aiming the ignition.

The key issue for the next step is to set short wavelength lasers of megajule level as drivers to attain the high pellet gain required for a fusion reactor.

Then the development of the reactor driver is the most crusial theme for the realization of an IFE reactor.

In relevant regions, such as the pellet fabrication, the diagnostics, simulation and the reactor designs, remarkable progress has been performed. Developments in these areas will make it possible to set an adanced IFE programe with the international collaboration.

REFERENCES

1. C. YAMANAKA, S. NAKAI, T. YABE, et al., "Laser implosion of high-aspect ratio targets produces thermo nuclear neutron yield exceeding 10^{12} by use of shock multiplexing" *Phys. Rev. Lett.* **56** 1575 (1986);
   C. YAMANAKA, S. NAKAI, T. YAMANAKA, et al., *Necl. Fusion* **27** 19 (1987);
   C. YAMANAKA, K. MIMA, S. NAKAI, et al., in *Plasma Physics and Controlled Nuclear Fusion Research 1986 (Proc. 11th Int. Conf. Kyoto, 1986),* Vol. 3, IAEA, Vienna **33** (1987).

2. H. TAKABE, C. YAMANAKA, K. MIMA, et al., "Scalings of implosion experiments for high neutron yield" *Phys. Fulids* **31** 2884 (1988).

3. E. STORM, in *Plasma Physics and Controlled Nuclear Fusion Research 1986 (Proc. 11th Int. Conf. Kyoto, 1986),* Vol. 3, IAEA, Vienna 15 (1987).

4. Y. KATO, K. MIMA, N. MIYANAGA, et al., "Random phasing of high power laser for uniform target acceleration and plasma instability suppression" *Phys. Rev. Lett* **53** 1057 (1984).

5. C. Yamanaka, "Advances in inertial confinement fusion" *(Proc. 5th Emerging Nuclear Energy Systems, World Scientific)* P.125 (1989).

6. Y. Setsuhara, H. Azechi, et al., "Secondary nuclear fusion reactions as evidence of electron degeneracy in highly compressed fusion fuel" *Laser and Particle Beams* **8**, 609 (1990).

7. R. PAKULA, and R. SIGEL, "Generation of intense black-body radiation in a cavity by a laboratory pulsed power source", *Z. Naturforsch.* **41a** 463 (1986).

8. A. CARUSO, "High gain radiation-driven targets (I)", *Inertial Confinement Fusion (Proc. Int. Sch. Plasma Phys., Varenna 1988),* CARUSO, A. Ed., Italian Physical Society 139 (1988).

9. T. ENDO, H. SHIRAGA, and Y. KATO, "Quasistationary model for determination of ablation parameters in soft-x-ray driven low-to medium-Z plasma ablation", *Phys. Rev.* **A42** 918 (1990).

10. H. TAKABE, in *Short Wavelength Lasers and their Application,* ed. by C. YAMANAKA, Springer-Verlag, Berlin P307 (1988).

VOLUME IGNITION FOR INERTIAL CONFINEMENT FUSION

Robert J. Stening, Rasol Khoda-Bakhsh[1], Peter Pieruschka,
Gregory Kasotakis, Edgar Kuhn, George H. Miley[2], and
Heinrich Hora[3]

Department of Theoretical Physics, University of New South Wales
Kensington 2033, Australia

Central spark ignition of a pellet or capsule for ICF requires a highly complex compression mechanism where the parameters of ignition must fit within a very small range of values. It is only within these constraints that ICF fusion reactors with 2 to 10 MJ driver energy are possible. In order to relax these restrictions we studied volume ignition where high gains due to self-heating and considerable lowering of the initial temperature are possible, together with a much broader range of parameters and a reduction of Rayleigh-Taylor instabilities. Gains within a factor two compared to spark ignition are achieved. After a detailed study of the DT reaction, results for DD, for DH(3) and for HB(11) are calculated in order to evaluate a clean fuel ICF for the distant future.

INTRODUCTION

There is some doubt whether inertial fusion energy by lasers[1] or heavy[2] or light[3] ion beams will reach the necessary conditions for ignition of the compressed fusion fuel and the Director of the inertial confinement fusion office of the US Department of Energy described the achievement of ignition like "Landing a man on the moon"[4]. And this doubt is expressed even in the face of the marvellous experimental results that inertial fusion is indeed possible with quite high gains for energy inputs of several tens of Megajoules of energy from the Centurion-Halite experiment[5]. Inertial fusion - in contrast to magnetic confinement fusion - is in the most favourable position since this experiment has proven that ICF works with high input energies. This, then, has paved the way to the technological realization of energy generation by fusion.

If the question of ignition is still a concern, then it has to be underlined that this refers only to the very complex spark ignition[1] in laser fusion. But volume ignition[6-12] occurs under much more simplified and much less sensitive conditions than spark ignition and can provide

---

[1]On sabbatical leave from Urmia University, Urmia, Iran.
[2]From Fusion Studies Lab., Univ. Illinois, Urbana, Ill. 61801, USA.
[3]Present address: CERN, CH1211 Geneva 23, Switzerland.

Laser Interaction and Related Plasma Phenomena, Vol. 10
Edited by G.H. Miley and H. Hora, Plenum Press, New York, 1992

gains nearly as high as spark ignition[8,10]. It may well be possible that the experiments of the underground tests of the Centurion Halite experiment[5] were so successful because they fulfilled the conditions for volume ignition.

One excellent success of the volume compression mechanism (yet before ignition could occur) was demonstrated by the direct drive DT laser fusion experiments at the ILE in Osaka[14,15] where the conditions designed for the highest neutron gains gave rise to stagnation - free and shock-free adiabatic compression as is required for volume compression. It was only under these conditions that the x-ray emission resulted in a very homogeneous spatial dependence. The expected volume ignition for the next generation of laser pulses in the 100 KJ range has been computed[16].

Volume ignition is favoured also by one of the earliest pioneers in laser fusion, Stirling A. Colgate[17,18] following many years of complex developments of the spark ignition concept. The motivation for spark ignition was clear: the fusion gain G given by the input of an energy E_0 into a sphere of radius R_0 and uniform initial density n results in a ratio of generated fusion energy to E_0 of[10,19,20]

$$(1) \qquad G = \left(\frac{E_0}{E_{BE}}\right)^{1/3} \left(\frac{n_0}{n_s}\right)^{2/3} = \text{const.} \ R_0 n_0$$

where n_s is the solid state density of a 50:50 DT mixture. The break-even energy E_{BE} is the energy input E_0 where a solid state density DT sphere would produce a gain G= 1. Eq. (1) was derived from numerical calculations of the adiabatic expansion of the plasma and corresponds to a selection of the initial volume (or radius) such that an optimum initial temperature of 17 keV is produced. The break even energy for DT was rather high at 6 MeV. The Rn formulation by Kidder[20] is algebraically identical to the earlier one using the break-even energy[19].

The disadvantage of (1) is that - taking into account a partial burn of the fuel - low gains of values around 50 could only be expected. It was therefore imperative to look for other schemes. That proposed by John Nuckolls[21], using an indirect drive, consists in directing the laser or driving beam energy into a very small central part of the DT fuel with very high temperature and moderate density. This ignites a self sustained fusion combustion wave which propagates into the low temperature but very high density surrounding fuel. It is quite difficult to prepare such temperature and density profiles by spherical irradiation of the pellets, but it seems that this is now under control[1]. In practice it is found that any little change of the profiles will prevent the ignition of the fusion wave and this is the reason why some are cautious concerning the ease of first reaching ignition.

Volume compression was rejuvenated by the numerical discovery[6] of volume ignition where the initial optimum temperature dropped to 4.5 keV from 17 keV, if the plasma was transparent to bremsstrahlung, and dropped to such low initial values as only 1.5 keV if larger spheres of the fuel re-absorb the bremsstrahlung. Gains, with the definition of Eq. (1), of up to 1200 could be reached with densities in the vicinity of 1000 times the solid state - and these densities have been achieved experimentally just recently[22].

This increase of the fusion gain was due to the self-heat of the fuel by the fusion reaction products of the charged alpha particles only[23] (the expected strong heating effect of the neutrons has not been included and may improve all the past and the following results to even better values[24]. This is a kind of self-maintenance of the fusion reaction and explains why the very transparent and easily reproducible computation of the volume ignition obtains values comparable to those reported from spark ignition.

We reproduce here numbers similar to the preceding cases[23]. The result of spark ignition[1] (see Fig. 7 of Ref. [24]) is that a 10 MJ laser pulse produces 1 GJ Fusion energy by indirect drive. Our numbers for volume ignition refer to the gains G of Eq. (1) as shown in Fig. 1 of Ref. [24]. Since the gains G are related to the energy which went into the compressed pellet and no energy loss for the ablation is involved in Eq. (1), one has to add the correction of the hydrodynamic efficiency for the compression by ablation. Taking the 10 MJ laser pulse energy and assuming a 10% hydroefficiency, the Energy E_0 is then 1 MJ. Taking the compression to about 2200 times the solid state density of the case of the spark ignition[1], the gain G of volume ignition in Fig. 1 of Ref. [24] is 850. This means that the 10 MJ laser pulse, after 90% energy loss by the ablation, produced 0.86 GJ fusion energy, which is quite comparable with the value of 1 GJ calculated for the case of spark ignition. It should be mentioned that Basko[8] calculated volume ignition gains which were up to 50% of that of spark ignition.

After this introductory clarification about the advantages of volume ignition, we will now report several results using volume ignition. Our work was directed to cover more than one species of fusion reaction products for the re-heat. This is much more complicated than the earlier treated cases with one species of charged reaction products as is the case for DT or H-^{11}B [10,24].

The new cases should be considered only for some estimation of what possibilities may be expected in the years after the DT fusion reaction has reached the scale of technological energy production.

This first phase of a DT reactor for producing fusion energy, at a low cost similar to that of the light water reactors, should be achieved by a $30 Bill project within 10 to 15 years[10] compared to the $150 Bill needed for a 50 years project for the tokamak reactor whose problem of wall erosion (at least 5 cm per year "and this is too much"[25]) may prevent the success of the concept at all. Furthermore the cost of energy from the Tokamak may be ten times higher than that from light water reactors[26]. Knowledge of the physics of inertial confinement fusion leads us to expect that, within 10 years after the first commercial power station has been produced, the technology may produce inertial fusion energy at a cost 3 to 5 times below that of the light water reactors. This would then really be the beginning of the golden age[10].

DD FUSION REACTIONS INCLUDING ALL CHARGED REACTION PRODUCTS FOR REHEAT

Large scale economic fusion energy production can be based on the nuclear fusion reaction[27] of the heavy hydrogen isotope deuterium, D, with the super heavy radioactive hydrogen isotope tritium, T.

$$(2) \qquad D + T \rightarrow n(14.1 \text{ MeV}) + {}^4He(3.5 \text{ MeV})$$

In order to avoid the use of radioactive tritium and the undesired radioactivity generated by the neutrons (most of which will be used for breeding in lithium surrounding the reaction which will absorb most of the neutrons) an ideal reaction is[28]

$$(3) \qquad H + {}^{11}B \rightarrow 3{}^4He(2.88 \text{ MeV})$$

where less radioactivity is emitted per unit of energy produced[28,29] than the ppm of uranium released when burning coal. It is very difficult, however, to make reaction(3) occur and the technological difficulties associated with a reactor with a hydrogen and boron-11 fuel cannot be overcome within less than about 40 years from now[30]. Another 'clean' reaction which can be considered is the deuterium, D, helium-3, ${}^3He$, reaction[31].

$$(4) \qquad D + {}^3He \rightarrow H(14.7 \text{ MeV}) + {}^4He(3.6 \text{ MeV})$$

This fuel may be an attractive candidate for producing clean electrical power. This is mainly due to the fact that in the fusion reaction the fuel and the main reaction products are charged particles. There remains the problem of the competing reactions (each 50%)

$$(5a) \qquad D + D \rightarrow n(2.4 \text{ MeV}) + {}^3He(0.8 \text{ MeV})$$

$$(5b) \qquad D + D \rightarrow H(3.0 \text{ MeV}) + T(1.0 \text{ MeV})$$

which represent a source of radioactivity.

The deuterium-deuterium (D-D) reaction can be suppressed by nuclear spin polarization[32] and suppression may reach a factor of 95%. However, some plasma parameters for the D^3He reactor differ for the cases with and without the D-D reaction suppression[32] and a calculation of the influence of the D-D reaction on different parameters and on the fusion gain will obviously be interesting. But the first step to calculate the fusion gain associated with the D-${}^3He$ reaction including the DD reactions, within this Section is to carry out calculations on the DD reactions alone. Once these calculations have been done, then inclusion of the DD reactions in the D^3He calculation will be the next step in the following section.

We will give in the following results of our simulations on DD reactions based on volume ignition[5-13,33,34].

The fusion gain calculation for ideal adiabatic compression and expansion of the plasma according to the self similarity model[10,35,36] takes as initial conditions that an energy E_0 is deposited into a fully ionized (50:50) deuterium tritium plasma mixture of uniform density n_0 and spherical volume V_0 (having a radius R_0) resulting in a temperature T_0 and a self-similarity velocity v_0 of the pellet radius with a linear velocity profile into the interior of the pellet. Then the gain G is calculated[6,10,37,38] as the ratio of generation of fusion energy and the input energy E_0 within the compressed fuel during the subsequent adiabatic self-similarity dynamics [eq.(1)]. The initial expansion velocity is taken as zero (see Appendix 1).

If the energy produced per fusion reaction of one deuterium with another deuterium nucleus produces an energy of $\varepsilon_{DD} = 3.2$ or 4.0 MeV, the very simplified fusion reaction gain is defined by the ratio

$$(6) \qquad G = \frac{\text{Reaction energy}}{\text{Input energy } E_0} = \frac{\varepsilon_{DD}}{E_0} \int dt \int dr^3 \frac{n_i^2}{A} <\sigma v>$$

where n_i is the ion density and $<\sigma v>$ is the velocity averaged fusion cross-section (reaction 5a or 5b) with a constant A=4 for binary reactions, otherwise A=2 for the DD case. The velocity averaged fusion cross-section $<\sigma v>$ for a thermalized plasma of temperature T and an averaged mass m_i of the nuclei is given by

$$(7) \qquad <\sigma v> = \frac{\sqrt{m_i}}{\sqrt{\pi}(KT)^{3/2}} \int_0^\infty \frac{m_i}{2} v^2 \sigma(v) \exp\left(\frac{m_i v^2}{KT}\right) dv^2$$

Using Equation (6) with inclusion of fuel depletion, bremsstrahlung loss and reheat from charged reaction products, the gain for DT, D^{3}He and H^{11}B reactions [6-13,33,34] has been calculated.

A special feature of DD fusion is the appearance of a second and third charged reaction product. Therefore, in order to calculate fusion gains for DD reactions we enlarged our previous code to fully include the reheat of all three branches of nuclear reactions products (see Appendix A). In general our model algorithm includes five direct temperature changing elements, the adiabatic cooling, the α particle reheat, the proton reheat, the tritium reheat and the temperature change due to bremsstrahlung losses. As in the cases of DT, D^{3}He and H^{11}B, the collective model for stopping power[10] was adopted in calculating reheat by the charged fusion products. To include the reabsorption of bremsstrahlung, the classical collision frequency was used and not the anomalous collision frequency[10]. The whole detail of the spectral distribution of the bremsstrahlung and its density and temperature dependence was used to calculate the mean free path of the photons for each time step of the plasma dynamics.

The inclusion of self-heat by the fusion neutrons[24] is not yet included in the following discussion since there is not yet a convincing proof available for the proposed models. While these do not result in any improvement for spark ignition of the DT fuel it was estimated that self heat by neutrons may improve the volume compression results by a factor 2 according to Goel[24].

The results of computation of fusion gains G as a function of different input parameters (input laser energy, initial volume, initial pellet density) for DD reactions using our extended code are summarized in table (1). We found that the initial adiabatic burning during compression and expansion of the DD pellet (for 1000 times solid state density), requires an optimum temperature of about 35 KeV and a break even energy, E_{BE}, of 5.5×10^{15} J. These values are very much higher than those values of DT (see preceding section) than for D^3He [30].

Table 1. DD Fusion energy gain yield for some selected input parameter sets of volume ignition model.

$n_0[n_s]$	$V_0[cm^{-3}]$	$E_0[MJ]$	$T_0[keV]$	Gain	Depletion[%]
1000	10^{-1}	14.15×10^3	4.99	1.1	0.9
2000	10^{-1}	24.66×10^3	4.35	5.2	3.7
3000	10^{-1}	29.67×10^3	3.49	13.4	7.7
4000	5×10^{-2}	17.03×10^3	3.00	15.5	7.6
6000	5×10^{-3}	3.21×10^3	3.78	9.4	5.8
8000	10^{-3}	0.96×10^3	4.26	6.6	4.7
10000	5×10^{-4}	0.60×10^3	4.29	6.3	4.5
30000	10^{-6}	6.51	7.67	1.9	2.3
30000	10^{-5}	41.50	4.88	4.7	3.7
30000	10^{-4}	4.60×10^2	5.42	19.9	17.8
30000	10^{-3}	2.43×10^3	2.87	81.1	38.3
30000	10^{-2}	15.52×10^3	1.82	200.1	60.2
30000	10^{-1}	19.0×10^3	1.40	339.4	78.2
50000	10^{-6}	8.60	6.07	3.6	3.6
100000	10^{-7}	4.10	14.49	3.9	9.4
100000	10^{-6}	41.50	14.66	10.1	24.5
100000	10^{-5}	95.48	3.37	56.1	31.1
100000	10^{-4}	6.08×10^{-2}	2.14	156.6	55.3
100000	10^{-3}	4.24×10^3	1.50	303.1	74.8
100000	10^{-2}	32.55×10^3	1.55	463.9	87.8

Energy dependence on the fusion gain of the DD reactions for initial pellet volumes in the range of 10^{-2} - 10^{-7} cm^3 with two different plasma densities, ie, 3.0×10^4 n_s and 10^5 n_s, are shown in figures 1 and 2 respectively. From figures 1 and 2 it is clear that the gain variation of the DD reactions is the same as that of the DT and D^3He reactions except for a slight difference at smaller volumes. There is a pronounced shoulder in the gain-energy curve of the DD reaction at small volumes and this shoulder can be attributed to the two different reactions in the DD plasma which take place with nearly the same probabilities. At higher

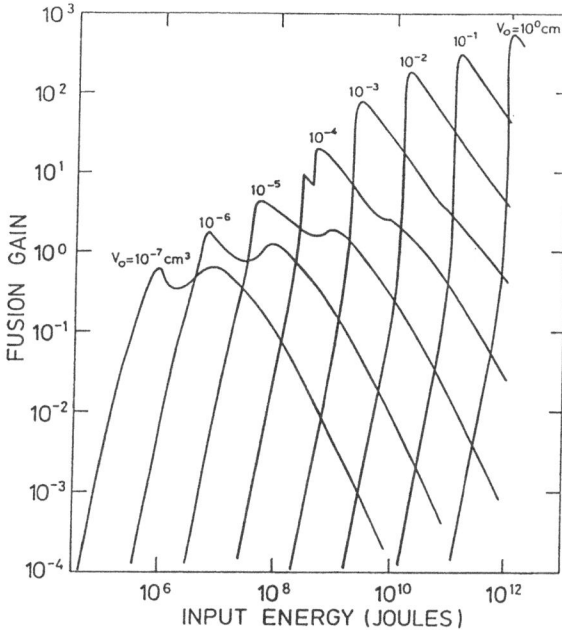

Figure 1. Energy dependence of the fusion gain of the DD reactions for a compression to 30,000 times the solid state density and initial pellet volumes in the range 10^{-7}-1 cm^3. X-axis: input beam energy in joule, Y-axis: fusion gains. Temperature-time analysis of the $V = 10^{-3}$ - 10^{-1} cm^3 cases for selected energies are plotted in figures 3-5 respectively.

volumes and higher input laser energies these shoulders disappear, since under these conditions considerable fusion ignition takes place in the DD pellet.

Figures 3-5 show the time dependence of the plasma temperature T for the case of an initial density of 30,000 times the solid state with initial volumes of 10^{-3} - 10^{-1} cm^3 close to the conditions of volume ignition. The time dependences of the plasma temperature T for the case of an initial density 10^5 the solid state density with initial volumes of 10^4 - 10^3 cm^3 are shown in figure 6-7 respectively. In all of figures 3-7 three cases are given where the input laser energy for each case was slightly increased resulting in increased fusion gain G. The ignition process similar to the DT and D^{3}He cases, can be seen clearly, not only from the strong increase of the gain G when there is little increase of the input energy, but from the temporal increase of the temperature. In the first cases, the temperatures are nearly constant for a long time where the temperature losses by bremsstrahlung and by the adiabatic expansion are compensated by the alpha, proton and tritium self-heat until a strong drop of the temperatures occurs, with fast expansion resulting in a so called 'simple burn'. Except for the case of figure 3 where a sixteen percent higher energy input results in the appearance of ignition instead of the

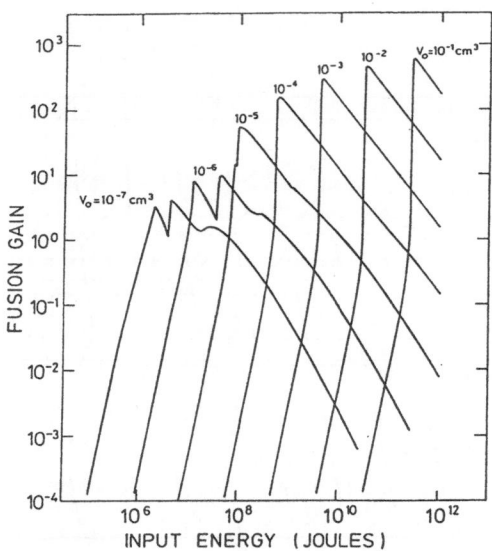

Figure 2. Energy dependence of the fusion gain of the DD reactions for initial pellet volumes in the range of 10^{-7}-10^{-1} cm^{-3}. The initial plasma density was 10^5 times the solid state density. Temperature-time analysis of the V = 10^{-4}-10^{-2} cm^3 cases for selected energies are plotted in figures 6-7 respectively.

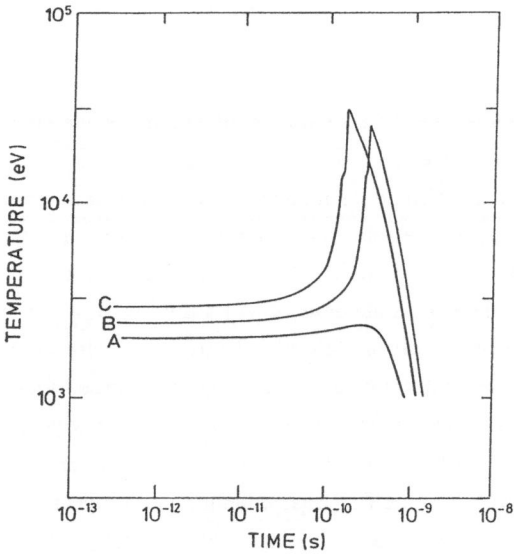

Figure 3. Dependence of the temperature of a DD pellet on time at initial compression to 30,000 times the solid state and initial volume of 10^{-3} cm^3. The input energies E_o were A:1.68 GJ, B: 2.02 GJ and C: 2.43 GJ, the fusion gains G were A:0.9, B:70.3 and C:81.1, the deuterium fuel depletion of A:0.3%, B:27.6% and C:38.3%, and initial temperature of A:1.98 KeV, B: 2.38 KeV and C:2.86 KeV. The maximum temperature of plasma in case B and C were 26.30 KeV, 31.19 KeV respectively.

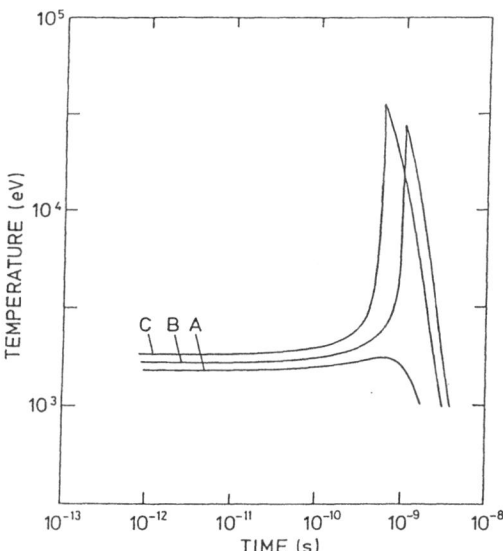

Figure 4. Same as in Figure 3 for a compression to 30,000 times the solid state density and initial volume of 10^{-2} cm^3. Input energy [GJ]: A:12.90, B:14.15 and C:15.52, Gain: A:0.9, B:157.3 and C:200.1, Fuel Depletion [%]: A:0.2, B:43.1 and C: 60.2, Initial Temperature [KeV]: A:1.51, B:1.66 and C:1.82, Maximum Temperature [KeV]: B:28.26 and C:36.01.

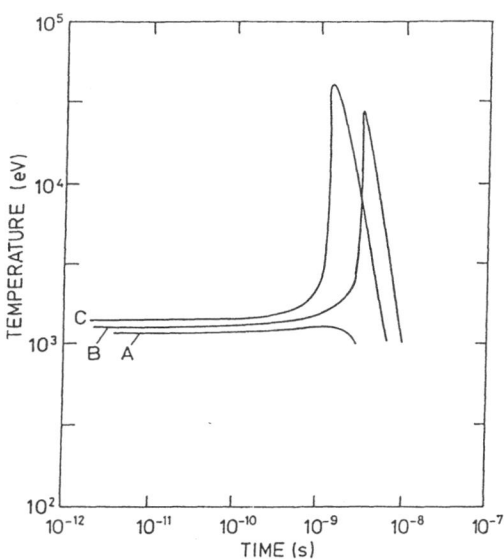

Figure 5. Same as in Figure 3 for a compression to 30,000 times the solid state density and initial volume of 10^{-1} cm^3. Input energy [GJ]: A:98.85, B:108.4 and C:119.0, Gain: A:0.6, B:274.9 and C:339.4, Burn [%]: A:0.2, B: 57.8 and C:78.2, Initial Temperature [KeV]: A:1.16, B: 1.27 and C:1.4, Maximum Temperature [KeV]: B:27.57 and C:41.08.

simple burn, in all other cases less than a ten percent higher energy input is sufficient to result in ignition. All three charged reaction products reheat the plasma more than the losses and the temperature of the pellet increases after the input of the drive energy in each case. The initial conditions of figures 3-7 with maximum temperature (between 31 to 55 keV) where plasma ignition is reached before the fast expansion, and later adiabatic cooling, are shown on each figure.

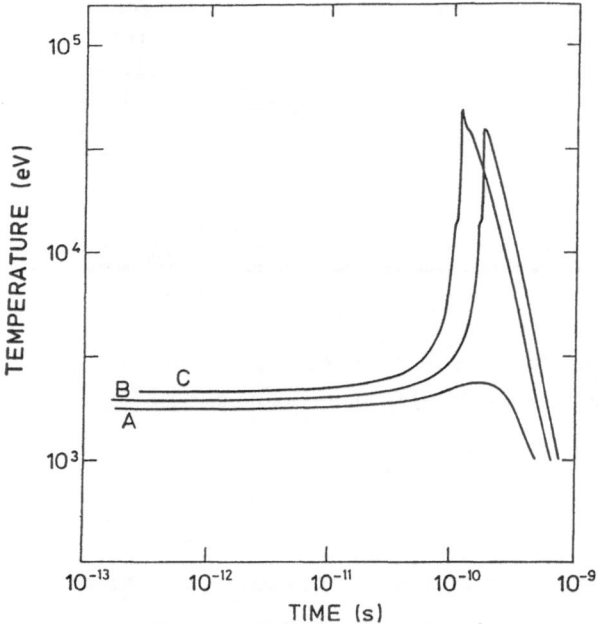

Figure 6. Temperature-time dependence of DD pellets of 10^5 times solid state density, initial volume of 10^{-4} cm^3 and with intake of following energies E_0[GJ]: A:0.50, B:0.55 and C:0.60, resulting in gains of: A: 1.7, B:147.0 and C:156.6 with fuel burn of [%]:A:0.5, B:47.3 and C: 55.3, and initial temperatures of [KeV]: A:1.78, B: 1.95 and C:2.14, maximum temperature of [KeV]: B:39.33 and C:49.10.

Comparison of our results with those of D^{3}He[30,39] shows that, for some initial D^{3}He fusion ignition conditions (see Table 1), the DD plasma also gives fusion ignition or considerable fuel burn up, suggesting the importance of the inclusion of the DD reactions in the D^{3}He calculations. The gains from these reactions are very sensitive to input energies and so inclusion of the DD reactions will give big changes in the detail of the D^{3}He results. Another point of interest is that DD fuel pellets can be used as a middle-term solution of future energy sources if higher compression can be achieved. Assuming a rapid development of an advanced fusion technology the H-^{11}B pellet was suggested[10,39] as an ideal and clean fuel for ICF operation. As a second generation of the nuclear fusion fuel cycle, advanced fuel fusion ignited by the DT burn such as DT/DD or DT/D³He hybrid fuel, has already received considerable attention in the inertial confinement scheme[10,39]. A compression to 10^5 times solid density and an initial volume of 10^{-4} cm³ requires a driver energy of about 0.6 GJ which is far beyond the present technology, but results in a fusion gain, much higher than the H-¹¹B fuel pellet, of about 156. As an example, if these parameters are compared with those of H-¹¹B fuel[10,30,39] with an initial plasma density of $10^5 n_s$ and an initial volume of 10^{-1} cm³, the 2 GJ energy required for ignition of the fuel results in a gain of about 25 [10,40].

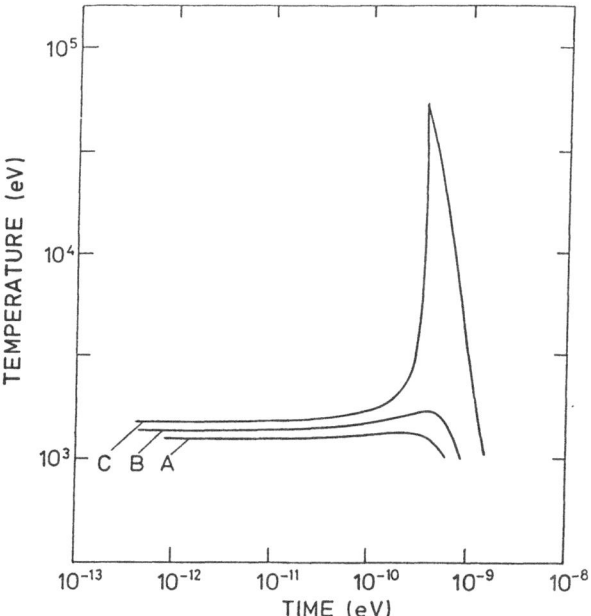

Figure 7. Same as in Figure 6 for a compression to 10^5 times solid state density initial volume of 10^{-3} cm³ and with intake of following energies Eo[GJ]:A:3.53, B:3.87 and C:4.24, Gain:A:0.5, B:1.3 and C:303.1, Burn [%]; A:0.1, B:0.3 and C:74.8, Initial Temperature [KeV]:A:1.20, B:1.36 and C:1.50, Maximum Temperature [KeV]: C:54.92.

Furthermore, our calculation shows that an initial temperature of only 3 KeV is sufficient to burn the DD fuel. This is mainly due to three different sources of reheat with separate stopping power mechanisms and to the models adopted in calculating reheat by fusion products and reabsorption of the bremsstrahlung. We acknowledge here that some of the parameters listed in Table 1 are very far beyond the present technology, but for a sake of comparison and for quite futuristic designs they are included in the Table.

NUMERICAL RESULTS FOR D³He VOLUME IGNITION CALCULATIONS

It has been verified that the fusion gain formula (1)[10,19,20] for optimum gains in simple burn volume compression fusion holds in the D³He case too, as long as the initial conditions of the simulation are 'moderate', i.e. do not lead to volume ignition reactions.

Figure 8 illustrates this behaviour. The lowest curves in this graph correspond to simulations with fuel densities close to solid state density. They are just straight lines with the predicted slope of $\frac{1}{3}$. The right-hand side of Equation (1) gives the relation to the well-known confinement parameter.

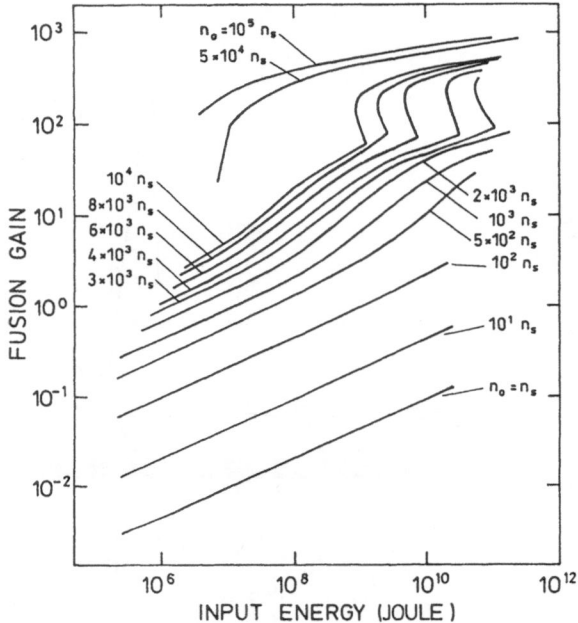

Figure 8. An Energy-Gain scheme for D³He fusion. X-axis: input beam energy in Joule, Y-axis: optimum fusion gains for initial fuel volume $10^{-7} \leq V_o \leq 10^{-1}$ cm³. Each of the curves is assigned to a particular initial target density, labelled at the right end of any curve. The upper gain limit due to saturation by fuel depletion is at about 10^3.

Figure 9 provides the break-even energy (for solid state density) as $18 \cdot 10^{12}$ J (remarkably less than for DD, see preceding Section) at an optimum temperature for ignitionless burn of 52 keV (which is much larger than for DD). For higher initial volumes we already approach conditions under which even for initial density of only solid state density and gains of the magnitude of 1 (this is much lower than for DT or DD) a deviation from the linear law of Equation (1), similar to the DT case[6], can be observed. For DT fuel this rising influence of volume ignition started at gains of about $8 \cdot 10^{6,9,10}$, or 500 times higher input energy than for D³He.

Table (2) summarises gain predictions for different input parameters. The parameters were chosen mainly around the 'edge zone' in Figure (8), just after the tremendous increase of the gain predictions. It is difficult to define a measure for the technological feasibility of those parameter sets. For a particular gain, a decrease in initial compression n_0 happens always at the cost of an increase in input energy E_0 and some sort of weight is required to decide which combination is most likely to be achieved first. But the vast difference of the output parameters for a range of 2000 to 3000 units of initial compression, together with the difficulty to achieve initial densities of more than 10^5 solid state density units or beam energies in the Gigajoule

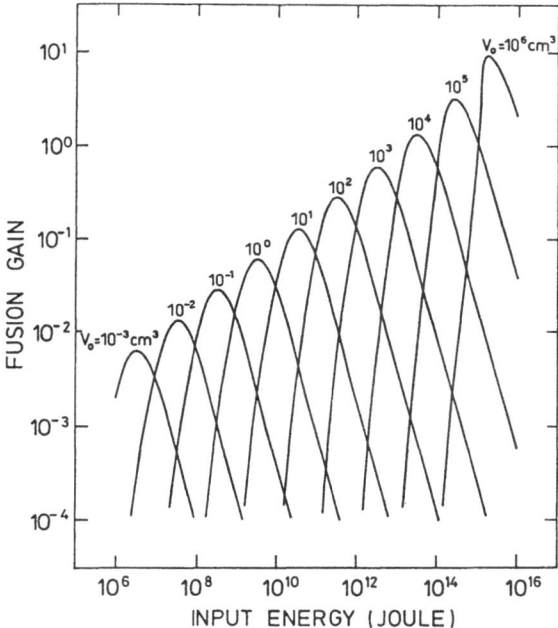

Figure 9. The lowest curve in Figure (8) extended for higher input energies and volumes is constructed as the envelope around energy-gain parabolas for particular initial fuel volumes, here in the range from 10^{-3} - 10^6 cm^3. Note the superlinear deviation from the cubic root law Eq. (1) when approaching large initial volumes.

range certainly imposes restrictions on the area of interest, at least in the near future. Figure 10 is taken as an example of the behaviour of the gain envelope in this 'edge zone' and leads the way to the understanding of the appearance of the edge. There are apparently at least two components involved which contribute to two separate ignition processes. They overlap under certain conditions and hence we find a jumplike increase in energy gain. All analogous sets of gain curves in the interesting energy/density range show basically the same pattern.

Table 2. D^{3}He fusion energy gain yields for some selected input parameter sets mainly in the 'edge' zone of the ignition area of Figure (1).

n_0 [n_S]	V_0 [cm^3]	E_0 [MJ]	T_0 [keV]	Gain	Depletion [%]
500	10^{-1}	$59 \cdot 10^3$	33.9	27.9	38.8
1000	10^{-1}	$106 \cdot 10^3$	48.7	30.4	60.6
2000	10^{-1}	$174 \cdot 10^3$	25.0	70.7	72.5
3000	10^{-1}	$64 \cdot 10^3$	6.1	299.5	75.3
4000	$5 \cdot 10^{-2}$	$39 \cdot 10^3$	5.6	330.4	75.7
6000	$5 \cdot 10^{-3}$	$6.8 \cdot 10^3$	6.5	266.0	70.7
8000	10^{-3}	$2.1 \cdot 10^3$	7.6	212.1	66.0
10000	$5 \cdot 10^{-4}$	$1.3 \cdot 10^3$	7.7	219.0	66.2
50000	10^{-6}	17	9.7	138.0	55.0
10^5	10^{-7}	3.5	10.1	125.0	51.2

The reason for this effect is not as straightforward as suspected earlier[30] and deserves some detailed attention due to the very interesting consequences for the gain efficiency. We will start a preliminary discussion in the next section.

The results presented above are of course model dependent - with the usual shortcomings of any model. Earlier investigations of DT volume compression fusion were remarkable consistent with the results of other authors thus proving the utility of our numerical model in principle.

Problems applying to the particularities of D^{3}He fusion remain to be discussed. The yet unknown solid state density of a symmetric D^{3}He mixture was assumed to be similar to the one used for DT volume compression i.e. $5.8 \cdot 10^{22}$ cm^{-3}. The additional reactions (5a) and (5b) were not included in the above calculations, but the estimated 'losses' can be as low as 5% and consequently are of no major impact. A detailed discussion of these complex influences for the sake of completeness is shown in Appendix B.

The reheat contributions of α particles and protons were simulated neglecting interactions among them. The following results will further elaborate the role of the second charged reaction product.

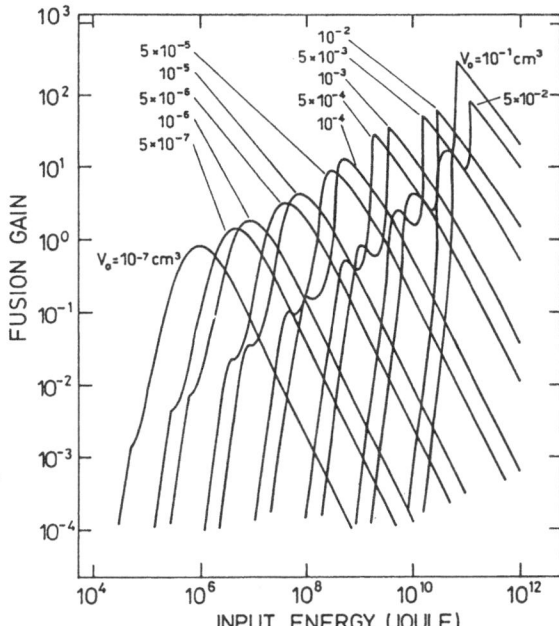

Figure 10. Energy dependence of the fusion gain of the D^3He reaction for initial pellet volumes in the range of 10^{-7} - 10^{-1} cm^3. The initial plasma density was $4.0 \cdot 10^3$ n_s. A temperature - time analysis of the $V = 10^{-2}$ cm^3 case for selected energies is plotted in Fig. (12).

The inclusion of a second source of reheat remains to be investigated. Implicitly it has been assumed throughout the literature that proton reheat should have small effects in comparison with α reheat and was hence neglected. We present in the following a critical collection of data that contradicts this view.

In general our model algorithm includes four direct temperature changing elements, the adiabatic cooling T_{ad} (due to the thermodynamic expansion process), the α particle reheat T_α and the proton reheat T_p (both due to Coulomb collisions with the electrons of the plasma) and the temperature change due to bremsstrahlung T_{br}. The change in temperature during a time interval of the simulated dynamics follows hence as

$$(8) \qquad \Delta T = -T_{AD} + T\nabla A + T_p - T_{br}$$

and the initial temperature for the following time step is then just

$$(9) \qquad T_{n+1} = T_n + \Delta T, \quad T_O = T_O(E_O) = \frac{E_O}{2\, n_o\, V_o\, k_B}$$

361

This is a very interrelated and highly non-linear mathematical description. How much the proton reheat would change the process is therefore a non-trivial question.

The inclusion of the proton was taken here into account by including a second source of collision reheat, where the two reheat contributions were given independent roles in the temperature dynamics. It can be stated in principle that the reheat contributions are always positive, they never really can interfere negatively and the added proton reheat is therefore not likely to significantly change the qualitative appearance of the process but only its quantitative patterns.

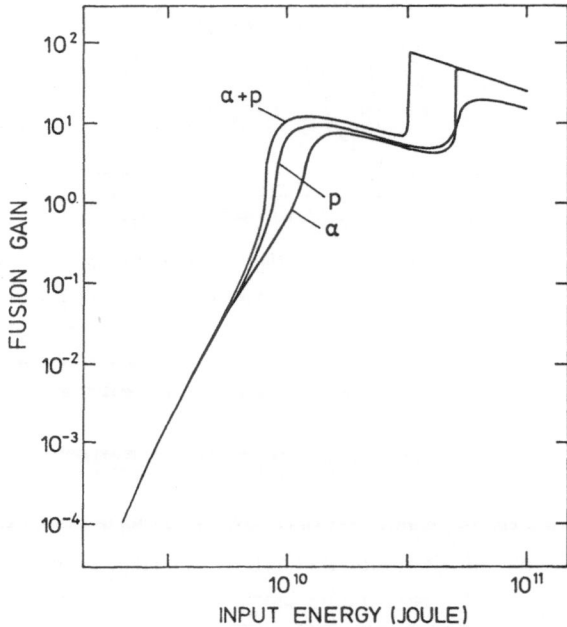

Figure 11. A 'zoom' into the $V = 10^{-2}$ cm^3 curve of Fig. (10). The uppermost curve contains both α and proton reheat contributions (labelled as $\alpha + p$), the middle one just α (labelled α) reheat and the lowest one only proton reheat (labelled p).

Figure (11) gives numerical evidence for this fact, and Table (3) gives a range of gain values under different initial conditions, and compares the result with and without the inclusion of the proton reheat. It is obvious that it is a doubtful step to sacrifice the proton reheat for the sake of mathematical simplicity. In some sense our 'simple' model can thus catch up with this much more complex computation[16] as they are forced to neglect such contributions. Table (3) also indicates that proton reheat can indeed be neglected when no ignition takes place and its

Table 3. The list compares gain efficiencies for D^3He volume compression fusion under selected initial conditions for simulations with and without the inclusion of proton reheat, where the upper row in the gain/depletion boxes represents proton inclusion and the lower row α reheat only.

n_0 [ns]	V_0 [cm^3]	E_0 [J]	T_0 [keV]	Gain	Depletion [%]
10^2	10^{-7}	$3.33 \cdot 10^4$	95.7	0.0287	0.113
				0.0287	0.113
	10^{-1}	$2.7 \cdot 10^{10}$	77.7	3.058	9.746
				3.021	9.628
$4 \cdot 10^3$	10^{-7}	$1.23 \cdot 10^6$	88.5	1.092	3.960
				1.092	3.960
	10^{-1}	$6.8 \cdot 10^{10}$	4.9	384.2	77.0
				26.41	0.053
10^5	10^{-7}	$3.4 \cdot 10^6$	9.8	124.9	50.1
				1.214	0.49

significance rises the more the process converges towards ignition conditions. This is not surprising, as ignitionless burn is always dominated by simple and continuous adiabatic dynamics where even α reheat can be neglected. But ignition is the process that is of interest to us and its behaviour is practically discontinuous[41]. To neglect proton reheat or any other obvious contributions cannot be supported.

It is still to be understood which temperature dynamics takes place behind the peculiar 'nose' of Figure 11. Figure 12 depicts the temperature dynamics of the typical scenario. Figure 13 allows comparison with the common type of ignition that is very similar to the temperature dynamics in the DT and $H^{11}B$ cases[10].

The series of Figures (14) - (16) depicts the temperature dynamics of the typical scenario. Figure (13) allows comparison with the common type of ignition that is very similar to the temperature dynamics in the DT and $H^{11}B$ cases[10].

The series of Figures (14)-(16) reveals more information about the temperature development of Fig.(12). Here the single components of Eq. (8), T_{ad}, $T\nabla a$, T_p, T_{br} are observed in the relevant time period, for the initial conditions of Figure (12). The positive contributions are the two reheats T_α and T_p. As explained before they never can interfere in a

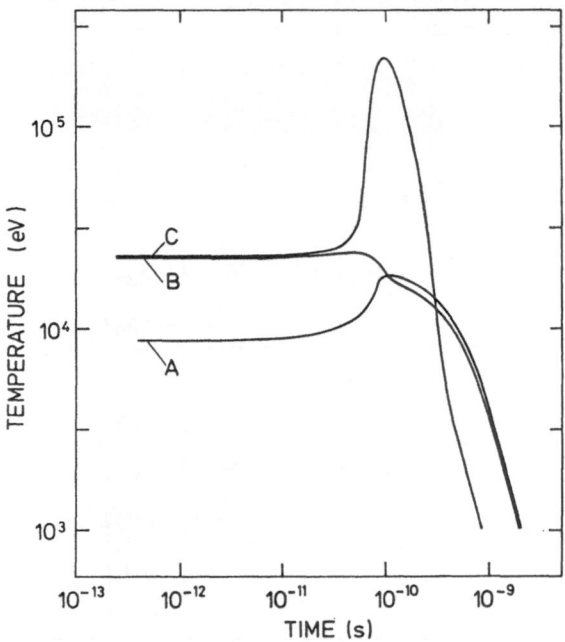

Figure 12. Temperature - time dependence of D^3He pellets ($n_o = 4.0 \cdot 10^3 \, n_s$, $V_o = 10^{-2} \, cm^3$) with E_o of A: 12.12 GJ, B: 31.50 GJ and C: 32.00 GJ, gains of A: 11.96, B: 74.9 and C: 75.11, depletion of A: 4.26%, B: 7.52% and C: 70.7 %

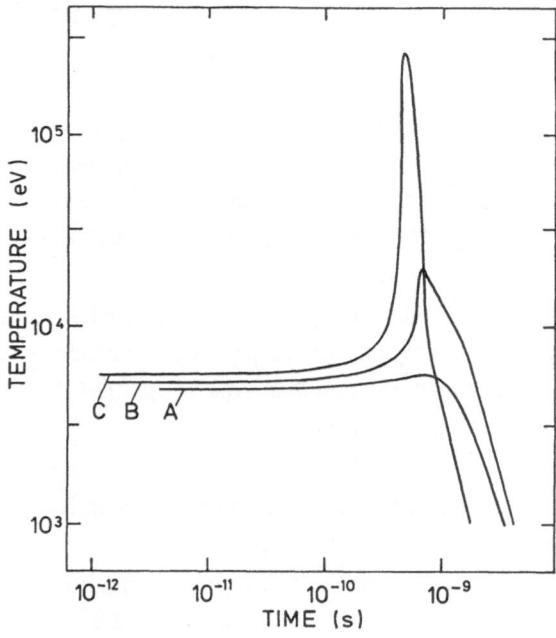

Figure 13. Comparison to Fig. (12): Volume increase of $5 \cdot 10^{-2} \, cm^3$ and same n_o lead to standard ignition dynamics. Input Energy [GJ]: A: 32.97, B: 35.83 and C: 38.94, Gain: A: 0.503, B: 24.03 and C: 330.4, Initial Temperature [keV]: A: 4.77, B: 5.15 and C: 5.59, Burn [%]: A: 0.098, B: 5.07 and C: 75.68.

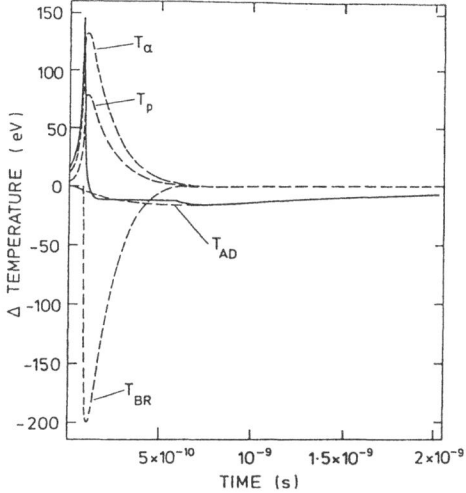

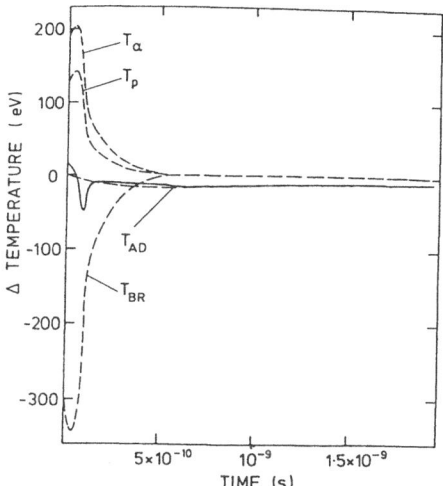

Figure 14. The temperature dynamics analysis for the $E_o = 12.12$ GJ case of Figure (12). Alpha reheat, proton reheat, adiabatic cooling and bremsstrahlung loss are labelled $T\alpha$, T_P, T_{AD} and T_{BR} and are plotted as dotted lines. The solid line represents the superposition of those contributions. (Note: the seeming edge in the bremsstrahlung curve is due to graphical resolution only.)

Figure 15. The temperature dynamics analysis for the $E_o = 31.50$ GJ case of Fig. (12). Alpha reheat, proton reheat, adiabatic cooling and bremsstrahlung loss are labelled $T\alpha$, T_P, T_{AD} and T_{BR} and are plotted as dotted lines. The solid line represents the superposition of those contributions.

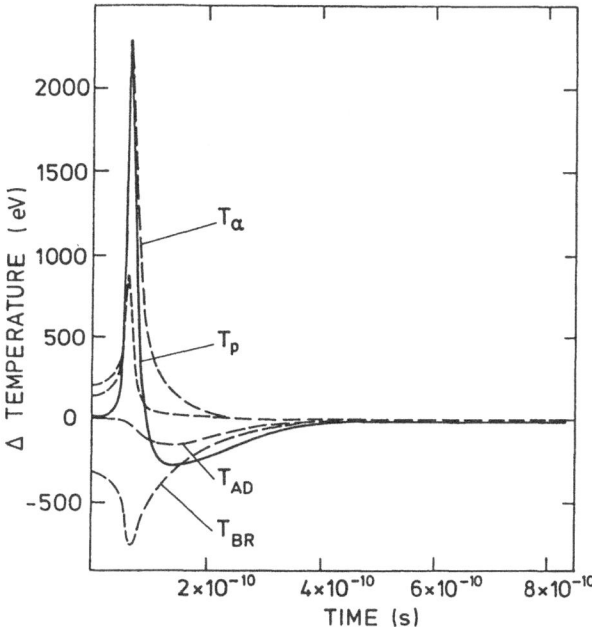

Figure 16. The temperature dynamics analysis for the $E_o = 32.00$ GJ case of Fig. (12). Alpha reheat, proton reheat, adiabatic cooling and bremsstrahlung loss are labelled T_α, T_P, T_{AD} and T_{BR} and are plotted as dotted lines. The solid line represents the superposition of those contributions.

major way. The negative contributions come from T_{ad} which obviously dictates the asymptotic behaviour, and more important from T_{br}.

It becomes evident that the strong non-linearity of the reheat and the bremsstrahlung in dependence of the input energy and of the time leads to the minimal "nose". With rising input energy E_0 the extremum of the value of the reheat T_r^{peak} increases strongly in relation to the minimum of the bremsstrahlung T_r^{peak}, where $T_r = T_p + T_\alpha$. At the same time the times of occurrence of the peaks of reheat and bremsstrahlung (t_r^{peak} and t_{br}^{peak}) shift (on the time axis) against each other, cf Table (4).

Table 4. Retarded volume ignition of D^3He fuel.

Figure	T_r^{peak} / T_{br}^{peak}	t_r^{peak} / t_{br}^{peak}	
7	$T_r^{peak} < T_{br}^{peak}$	$t_r^{peak} \leq t_{br}^{peak}$	First Ignition
8	$T_r^{peak} < T_{br}^{peak}$	$t_r^{peak} \approx t_{br}^{peak}$	'Aborted' Ignition
9	$T_r^{peak} > T_{br}^{peak}$	$t_r^{peak} \geq t_{br}^{peak}$	Second (retarded) Ignition

In other words starting with Fig. 14 the bremsstrahlung loss begins later than the reheat gain, but then plays the major role and thus prevents a first ignition from beginning. In Fig. 15 at slightly higher E_0 practically simultaneous appearance of comparable reheat and strong bremsstrahlung occurs, and hence ignition is still inhibited. A very small rise in input energy (about 1.5%) leads to the scenario of Fig. (16). There the reheat vastly dominates the bremsstrahlung, and we observe volume ignition. The relative influence of bremsstrahlung and reheat is also volume dependent and at a certain limiting volume $V(n_0, E_0)$ bremsstrahlung ceases to disturb ignition - graphically an overlap occurs, i.e. the volume ignition is not retarded any longer but happens in the way with which we are familiar from the study of DT and $H^{11}B$ fuels - though we do not want to exclude that retarded volume ignition took place unnoticed in unrealistic energy/volume regions.

We also show the case of standard volume ignition in Figure (17) - reheat is doubtlessly dominant and drives the volume ignition. A summary of optimum gains G for $D^{-3}He$ is shown in Fig. 18.

After these numerical investigations, we can state that there are no grounds on which proton reheat could be neglected. Though the unusual appearance of a minimum in the energy gain curves invites its interpretation as a pure proton reheat effect, we were able to show that

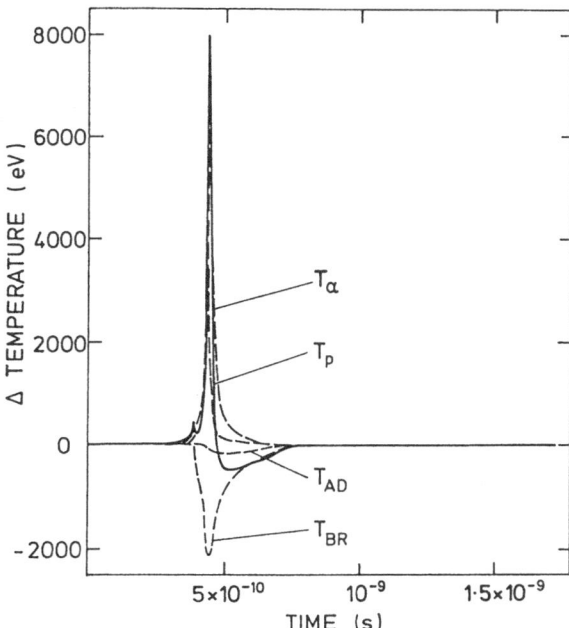

Figure 17. As for Figure 16. The temperature dynamics analysis for the $E_o = 38.94$ GJ case of Figure 13.

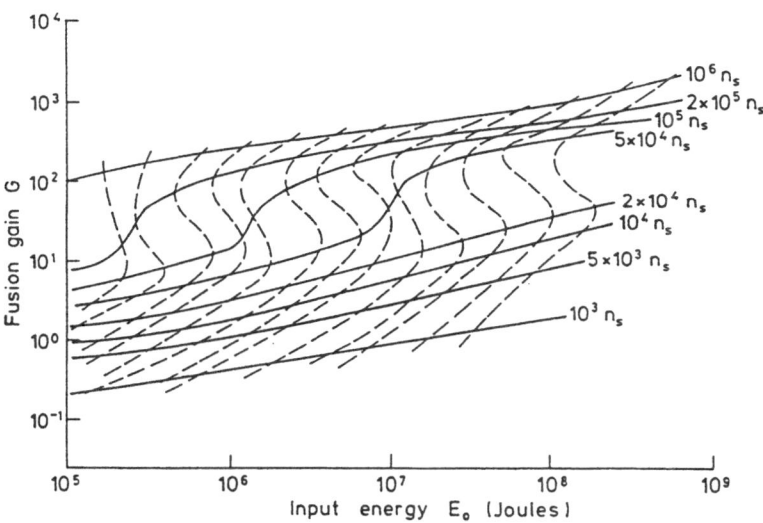

Figure 18. Optimum gains for D-He(3) reactions versus input energy E_o at different compressions.

this is not the case in a direct manner but finds its reason primarily in the complex non-linearity of the competition of reheat and bremsstrahlung.

One point of further discussion concerns the negative slope of the tangent of the edges in Figure 8 that is directly related to the above mentioned retarded volume ignition. A slight increase in initial volume causes a reduction of input energy and at the same time a significant rise in gain output - a challenging effect, likely to open a new round of discussion about the most promising ignition concept.

The examples that we picked for our theoretical considerations above served well as an object of mathematical study, since the gain minimum stretches through the whole range of input volumes, proving thus its consistent and 'non-statistical' character. For higher initial density we found that the minimum starts to vanish for high initial volumes until, at about 10^5 times solid state density, it leaves the considered energy range. Another point of interest is that the density required has already been experimentally achieved by Nakai et al[22], who obtained a DT plasma 1000 times solid state density.

There is still the problem of reaching Gigajoule input energies. But, for an initial density of 5.10^4 n_s, only a 17 Megajoule energy intake is required for a gain of 138. Hence we are close to the 10 MJ used in the ATHENA project[1].

It follows that our calculations are not too far removed from practical feasibility. Further understanding of the interesting underlying phenomena can be expected to lead to further developments.

VOLUME IGNITION OF HYDROGEN-BORON(11) PELLETS

When our volume ignition code was extended to the case of H-B(11), we found that the initial adiabatic burning during compression and expansion required an optimum temperature of about 110 keV. This can be reduced by self-heating of an exceedingly high density plasma (1 x 10^5 times solid state) such that the optimum initial temperature for volume ignition due to reheat and self-absorption of bremsstrahlung is only about 20 keV.

We performed an extensive series of computations similar to those for the case of the DT reactions. Parameters were examined where a drastic decrease of the initial temperature for good gains with reasonable energy input of up to GJ was possible. In the following reported cases the reabsorption of bremsstrahlung was not strong. Indeed, we found in such cases much higher gains with very high fuel depletion and much lower initial optimum temperatures for ignition. However, input energies were a little beyond our limit of input energy.

The results of figure 19 immediately show the conditions required for volume ignition. The time dependence of the plasma temperature during the reaction is shown. As before for DT, we have also illustrated the conditions of a slightly lower initial temperature that did not permit ignition, so that the temperature was nearly constant initially and then dropped monotonically with time. With a little higher input energy the temperature rose to a sharp maximum owing to volume ignition. The gain achieved was 25.3 with respect to the input energy. This is indeed a low value and can be considered only in the far future if much

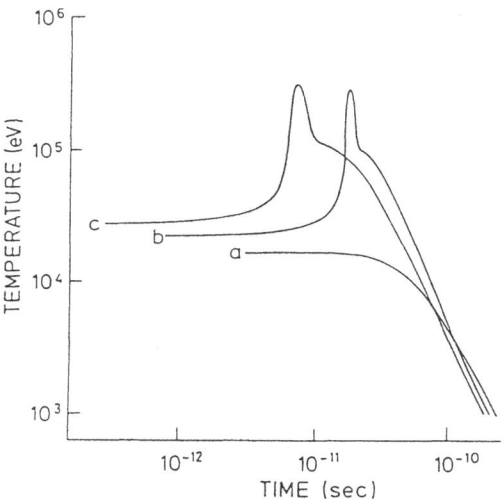

Figure 19. Time dependence of the temperature H-B(11) pellets of 10^5 times solid-state density and 10^{-5} -cm^3 initial volume with intake energies (a) 1.5 GJ, (b) 2.0 GJ, and (c) 2.5 GJ, resulting in gains of 0.09, 25.3 and 21.0, respectively, with fuel burns of 0.21%, 77.7%, and 80.7% and initial temperatures of 16.6, 22.2, and 27.7 keV, respectively.

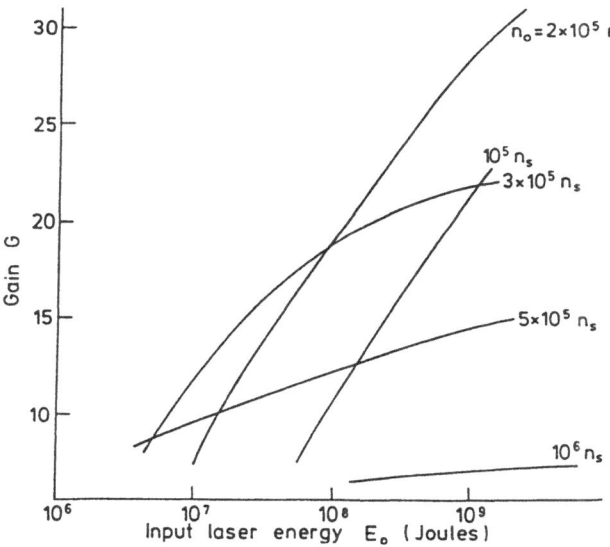

Figure 20. Optimum fusion gains for H-B(11) at compressions above 10^5 n_s versus input energy E_o

improved drivers with high efficiency are developed. At the same time, the methods used to obtain uniform irradiation must be drastically improved if there is to be any hope of achieving the enormous compressed density implied here, in the range of 100,000 times solid-state density. Case c of figure 19 shows that a gain of 21.0 is achieved. The conditions for the future would then be that a laser of at least 30% efficiency is needed and that ways must be found to achieve an ultrahigh compression of 10^5 times solid state. The conversion of the output energy into electricity might be performed with low thermal pollution, possibly with the use of high-efficiency Moseley-type direct conversion.

The predicted compression of the H-B(11) laser fusion requires ten times the maximum estimate for compression of DT (to 10,000 times solid-state density[21]). One thousand times solid-state density has been achieved experimentally[22]. Thus the H-B(11) requirement may not be too difficult to reach within the next 50 years if intensive technological activity on laser fusion with the present or next-step DT reactions is initiated. Compared with the predicted scales of DT reactors, the size of the reactors would be indeed about one order of magnitude larger with respect to the energy production. This one order of magnitude should be insurmountable. It would also very much simplify the Cascade concept[42], since no neutron absorber and its reshuffling are needed, and the Moseley energy conversion can be accomplished in a very clean and static way.

If one goes a little step further in compression, the results of figure 20 are derived. For a density between 10^5 and 10^6 n_s the optimum gains are interesting values for laser energy input down to 10MJ. We observed a decrease of gain on the density above $E_0 = 100$ MJ if the density exceeds 2×10^5 n_s . This retrograde property of the gain is basically different to the monotonous increase for D-^{3}He (Fig. 19) and has to be analysed separately. For an input energy of 100 MJ a gain of 20 can be achieved. For the final case this would mean that if a hydroefficiency of 20% could be reached, a 500-MJ laser pulse would produce 2-GJ fusion energy, which after direct conversion might be 1.8 GJ electrical energy. If one uses the cluster injection laser (as an example) with 80% efficiency, an overall net gain of 3 would be achieved.

In principle, it is known for laser amplifiers that an efficiency of 80% or more can be expected, with simultaneous high gains in the visible or far-UV range from cluster injection FEL[43]. The kinetic energy of condensed speckles with Mach 300 velocity is converted into optical energy. An alternative to the present-day 10-MJ glass laser system[1] is KrF lasers, with which compact 1-Hz repetition rate systems with 25-KJ, 3-ns output of 12-m^2 cross section have been reported[10].

SUMMARY

The preceding evaluations of volume ignition of inertial confinement fusion of DT permitted safe predictions for achieving ignition which were free from the uncertainties of the sensitive parameters of spark ignition. The prediction of a fusion energy power station producing electricity at the same low cost as light water reactors should be possible within 10 to 15 years using an estimated $30 Bill budget. For the use of lasers as drivers, the problems of the pulsation of the interaction have been solved by smoothing techniques[10,40]. Based on the possibilities of improved understanding of the physics involved, it is predicted that, following 10 years more research, the energy costs may be decreased by a factor of 3 to 5 using DT fuel.

Detailed numerical evaluations for volume ignition of D-^{3}He and H-^{11}B fusion were performed where, compared to DT, up to 100 times higher compressions and handling driver energies of about 100 times higher values are necessary. Total energy gains of up to 20 or 3 (Table 5) for the mentioned fuels respectively can be guaranteed. If the DT inertial fusion energy technology is developed now, the dream reaction of hydrogen-boron producing economic nuclear energy without radioactivity should easily be available within 50 years.

Table 5. Total fusion gains G_{total} for D-^{3}He and H-^{11}B reactions using volume ignition at given hydroefficiency, input laser energy E_{laser}, transferred energy E_0 to an initial density n_0 (in multiples of solid state density), with gain of Eq. (1), and gain G_{laser} based on the laser energy using laser and energy conversion efficiencies as shown.

	D-^{3}He		H-^{11}B
Hydroeff.	15%	20%	20%
E_{laser}	100MJ	100MJ	2.5GJ
E_0	15MJ	20MJ	0.5GJ
n_0	$5 \times 10^4 n_s$	$2 \times 10^5 n_s$	$2 \times 10^5 n_s$
G	100	500	26
G_{laser}	15	125	5.1
Laser eff.	40%	40%	25%
E-Conversion	40%	40%	80%
G_{total}	2.4	20	3.08

REFERENCES

1. E. Storm, J. Lindl, E.M. Campbell, T.D. Bernat, L.W. Coleman, J.L. Emmett, W.J. Hogan, Y.T. Horst, W.F. Krupke, and W.H. Lowermilk, Progress in Laboratory High Gain ICF: Prospects for the Future (LLNL Livermore, 1988) Report No. 47312 (August 1988).

2. C. Rubbia, <u>On Heavy Ions Accelerators for Inertial Confinement Fusion</u> IAEA Technical Committee meeting on IDF, April 15-19, 1991, CERN Report PPE/91-117 (14 July 1991).

3. T. Mehlhorn, D. Johnson, J. Bailey, D. Cook, G. Chandler, M. Derzon, M. Desjarlais, T. Haill, T. Hussei, R. Leeper, T. Lockner, M. Matzen, D. McDaniel, L. Mix, R. Olson, J. Quintenz, G. Rochau, C.L. Ruiz, S. Slutz, R. Stinnett, and W. Stugar, <u>Laser Interaction and Related Plasma Phenomena</u>, H. Hora, and G.H. Miley eds. (Plenum, New York 1992) Vol. 10, p.

4. M. Sluyter, <u>Laser Interaction and Related Plasma Phenomena</u>, H. Hora and G.H. Miley eds. (Plenum, New York 1992) Vol. 10, p.

5. W.J. Broad, New York Times, <u>137</u> (No. 47, 451, March 21, 1988).

6. H. Hora and P.S. Ray, <u>Z. Naturforsch</u>. 33A, 890 (1978); H. Hora, <u>Z. Naturforsch.</u> 42A, 1239 (1987).

7. G. Kasotakis, L. Cicchitelli, H. Hora, and R.J. Stening, Laser and Particle Beams, <u>7</u>, 511 (1989).

8. M.M. Basko, Nuclear Fusion <u>30</u>, 2443 (1990).

9. Tan Weihan, and Rehong Liu, Chin. J. Lasers, <u>17</u>, 658 (1990).

10. H. Hora, <u>Plasmas at High Temperature and Density</u> (Springer, Heidelberg, 1991).

11. G.H. Miley, L. Cicchitelli, H. Hora, G. Kasotakis, P.W. Pieruschka, R.J. Stening, and Tan Weihan, <u>13th Int. Conf. Plasma Physics and Controlled Nuclear Fusion Research</u> 1-6 Oct 1990 (IAEA, Vienna 1991) paper IAEA-CN-53/B-111-7, Vol. III, p.177.

12. G.H. Miley, H. Hora, L. Cicchitelli, G.V. Kasotakis, and R.J. Stening, Fusion Technology <u>19</u>, 43 (1991).

13. P. Pieruschka, L. Cicchitelli, R. Khoda-Bakhsh, E. Kuhn, G.H. Miley, and H. Hora, Laser and Part. Beams <u>10</u>, 145 (1992).

14. C. Yamanaka, S. Nakai, T. Yamanaka, Y. Izawa, K. Mima, K. Nishihara, Y. Kato, T. Mochinzuki, M. Yamanaka, M. Nakatsuka, and T. Yabe, <u>Laser Interaction and Related Plasma Phenomena</u>, H. Hora and G.H. Miley eds, (Plenum, New York 1986) Vol. 7, p.395.

15. S. Nakai, Laser and Particle Beams <u>7</u>, 467 (1989).

16. K. Mima, H. Takabe, and S. Nakai, Laser and Particle Beams <u>7</u>, 249 (1989).

17. S.A. Colgate, ICNES Conf. Monterey 1991.

18. Editorial, Newsline (LLNL, Livermore) <u>16</u> (No. 84, 12 Nov. 1991) p.1&4.

19. H. Hora, D. Pfirsch, <u>6th Int. Quant. Electro. Conference, Kyoto, Sept. 1970</u>, Conf. Digest p.10.

20. R.E. Kidder, Nuclear Fusion <u>14</u>, 797 (1974).

21. J.H. Nuckolls, Physics Today <u>35</u> (No. 9) 24 (1982).

22. H. Azechi et al, Laser and Particle Beams 19, 193 (1991).

23. H. Hora, L. Cicchitelli, Gu Min, G.H. Miley, G. Kasotakis, and R.J. Stening, Laser Interactions and Related Plasma Phenomena, H. Hora and G.H. Miley eds. (Plenum, New York, 1991) Vol. 9, p.95.

24. B. Goel, and D. Henderson, A Simple Method to Calculate Neutron Deposition in ICF Targets (Kerforschungszentrum Karlsruhe, Germany) KfK 4142 (October 1986).

25. P.H. Rebut, CERN Colloquium, 13 May, 1991. (CERN Bull. 20/91 page 3).

26. D. Pfirsch and K.H. Schmitter, Fusion Technology 15, 1471 (1989).

27. M.L.E. Oliphant, P. Harteck, and Lord Rutherford, Proc. Roy. Soc. A144, 692 (1934).

28. M.L.E. Oliphant, and Lord Rutherford, Proc. Roy. Soc. A141, 259 (1933).

29. T.A. Waver, Lawrence Livermore Laboratory, Laser Report 9 (No. 12), 1 (1973).

30. R.W.B. Best, Nucl. Instr. Meth. 144, 210 (1977).

31. G.H. Miley, in Laser Interaction and Related Phenomena, Vol. 5 (H. Schwarz et al, Eds.), Plenum Press, New York, p.313 (1981).

32. A.E. Dabiri, Nucl. Instr. Meth. A271, 71 (1988); L. Cicchitelli et al, Laser and Part. Beams, 2, 469 (1984).

33. L. Cicchitelli, S. Eliezer, M.P. Goldsworthy, F. Green, H. Hora, P.S. Ray, R.J. Stening and H. Schizman, Laser and Particle Beams (1988) Vol. 6, 163 (1988).

34. P.S. Ray and H. Hora, in Laser Interaction and Related Plasma Phenomena, Vol. 4B (H. Schwarz, et al., Eds), Plenum Press, New York, p.108 (1977).

35. J.M. Dawson, Phys. Fluids 7, 981 (1969); H. Hora, Physics of Laser Driven Plasmas (Wiley, New York) 1981.

36. R.F. Schmalz, Phys. Fluids 29, 1389 (1986).

37. P.S. Ray, PhD Thesis University of New South Wales (1977).

38. G. Kasotakis, L. Cicchitelli, H. Hora and R.J. Stening, Nucl. Instr. and Methods A278, 110 (1989); R. Khoda-Bakhsh et al, Fusion Technol. 46 (1992).

39. G.H. Miley, H. Hora, L. Cicchitelli, G.V. Kasotakis and R.J. Stening, Fusion Technology, Vol. 19, 43 (1991).

40. M. Aydin, L. Cicchitelli, H. Hora, G. Kasotakis, R. Khoda-Bakhsh, E. Kuhn, G.H. Miley, P. Pieruschka and R.J. Stening, Research Trends in Inertial Confinement Fusion, AIP Proceedings, (Am. Inst. Phys. New York, 4-6 February (1991).

41. L. Cicchitelli, S. Eliezer, M.P. Goldsworthy, F. Green, H. Hora, P.S. Ray, R.J. Stening, and H. Szichman, Laser and Part. Beams 6, 163 (1988).

42. I. Maya, and K.R. Schulz et al, General Atomics, Technol. Dept., San Diego, Rept (GA-A 17842 Oct. 1985).

43. H. Hora, J.-C. Wang, P.J. Clark, and R.J. Stening, Laser and Part. Beams, 4, 83 (1986).

APPENDIX A

The computer code for the volume ignition calculations is presented in the following pages of this Appendix A. It will be immediately understood by the expert in the field. A more detailed introduction for the newcomer in the field is being prepared for a computer oriented publication.

```
1            Program Fusion
2       c    (The file FUSION HELP A1 contains imprtant structural information)
3
4
5       c    ||||||||||||||||||||| COPYRIGHT NOTICE |||||||||||||||||||||||||
6       c    ||                                                          ||
7       c    || (C) 1990 by Prof. H. Hora, Dept. of Theo. Physics,       ||
8       c    ||               University of New South Wales,             ||
9       c    ||               Kensington, N.S.W. 2033, Australia         ||
10      c    ||                                                          ||
11      c    ||  improved by Edgar Kuhn, FH Regensburg, Germany          ||
12      c    ||                 in 1990/91                               ||
13      c    ||                                                          ||
14      c    |||||||||||||||||||||||||||||||||||||||||||||||||||||||||||||||
15
16      c    (Nuclear fusion calculations for the reaction D + 3He -> 1H + 4He)
17      c                                          (and H + 11B -> 3He)
18      c    (The restriction to 2 cases is subject to being changed soon)
19      c    (The interaction between alpha and proton is neglected)
20      c    (cross section data from Greene (USAEC report UCRL-70522 (1967)))
21
22           Parameter(MaxSW=250)
23           Parameter(MaxEn=500)
24           Parameter(pstep=0.04)
25
26
27           Dimension T(MaxSW),SW(MaxSW)
28           Dimension V0(50)
29           Dimension E0(MaxEn),AEX(MaxEn),E1(MaxEn)
30           Dimension DNS(10),DNSNS(10)
31
32      c    (Switches for some optional processes)
33           Logical DoBrems
34           Logical WriteGain,WriteTemp
35           Logical IsTwoComp,IsDHe,IsHB
36
37      c    E0(1)=4.37E+08
38      c    E0(2)=0.00E+00
39      c    E0(3)=0.00E+00
40
41      C****  THIS PROGRAM CALCULATES FUSION ENERGY EFFICIENCY WITH ALPHA REHEAT
42      C****  UNITS EMPLOYED CGS
43      C****  FUSION OF D-3He PLASMA, with consideration of both alpha and proto
44      C****  SV DENOTES REACTION CROSS-SECTION AVERAGED OVER MAXWELL DISTRIBUTION.
45      C****  INPUT DATA FOR TEMPERATURE AND AVERAGED CROSS-SECTION ARE T(I),SW(I).
46      C****  E0 INPUT LASER ENERGY IN JOULES. V0 INITIAL PELLET VOLUME.
47      C****  T0 INITIAL PLASMA TEMP IN EV.
48      C****  ZIA AVERAGE ION CHARGE IN PLASMA.
49      C****  IONIC DEPLETION HAS BEEN TAKEN INTO ACCOUNT.
50      C****  DNSION REACTING ION DENSITY AT A PARTICULAR INSTANT.
51      C****  DNS(I)=INITIAL ION DENSITY
52      C****  XM=AVERAGE IONIC MASS IN TERMS OF PROTON'S MASS
53      C****  EM=MASS OF THE ELECTRON IN TERMS OF PROTON'S MASS.
54      C****  AM=ALPHA'S PARTICLE MASS IN TERMS OF PROTON'S MASS.
55      C****  EF=FUSION ENERGY RELEASED PER REACTION
56      C****  FR=FRACTION OF EF CARRIED BY CHARGED PRODUCTS OF REACTION.
57      C****  Z=CHARGE OF THE REACTION PRODUCT.
58      C****  RRATE=REACTIONS RATE
59      C****   BK=BOLTZMANN'S CONSTANT=1.38E-16 ERG
60      C****  PROTON'S CHARGE=4.8029E-10 ESU
61      C****  PROTON'S MASS=1.6724E-24 G=938.232E6 EV
62      C****  1EV=1.6021E-12 ERG
63      c     DEPTOT=total depletion
64      c     DEPIONS=differential depletion
65
66           Write(6,*) ' '
67           Write(6,*) '               |||||||||||||||||||||||||||||||||||'
68           Write(6,*) '               ||                               ||'
69           Write(6,*) '               || Nuclear fusion calculations   ||'
70           Write(6,*) '               ||                               ||'
71           Write(6,*) '               |||||||||||||||||||||||||||||||||||'
```

```
72          Write(6,*) ' '
73
74    c     {Whether or not the FUSION GAIN/INPUT ENERGY
75    c                  and the TIME/TEMPERATURE relations are plotted}
76          WriteGain=.true.
77          WriteTemp=.false.
78          IsTwoComp=.false.
79          IsDHe=.false.
80          IsHB=.true.
81    c     {Whether or not the subroutine BREMS are applied}
82          DoBrems=.True.
83
84          Write(6,*) 'Status of this fusion run...'
85          If (IsDHe) Then
86            Write(6,*) 'D-He FUSION REACTION.'
87          EndIf
88          If (IsHB) Then
89            Write(6,*) 'H-B  FUSION REACTION.'
90          EndIf
91          Write(6,*) 'Temperature/Time  : ',WriteTemp
92          Write(6,*) 'Energy/Gain       : ',WriteGain
93          Write(6,*) 'BremsStrahlung    : ',DoBrems
94          Write(6,*) '2 Reaction Prods? : ',IsTwoComp
95
96
97
98
99
100   c     File defintions; cf. FUSDHE EXEC file
101   c     Open (unit=8,status='old',file='fusion.input')
102   c     Open (unit=1,status='new',file='fusion.gain')
103   c     {Header of GAIN/ENERGY output file}
104   c     If (WriteGain) Then
105   c       Write(1,*) 'This file is the FUSION GAIN output of FUSDHE.'
106   c       Write(1,19)
107   c  19   FORMAT(/,'LASER ENERGY',1X,'INITIAL TEMP',1X,'SV',5X,
108   c     1'FUSION GAIN',3X,'DEPTOT',4X,'DNSION',6X,'RNIONS',4X,'DEPFR',
109   c     27X,'DEPG',6X,'V0(I) ',6X,'RDOT',5X,'NUM',1x,'LOGFG',/,
110   c     3 'INPUT joule',5X,'E-VOLTS',/)
111   c     EndIf
112   c     {Open (Unit=2,Status='new',file='TEMP')}
113   c     {Header of TEMP/TIME output file}
114   c     If (WriteTemp) Then
115   c       Write(2,*) 'This file is the TEMP/TIME output of FUSDHE.'
116   c       Write(2,181)
117   c  181 FORMAT(///,1X,'NUM',5X,'TIME',6X,'RRATE',6X,'TEMP',9X,'SV',
118   c     1 6X,'DNSION',6X,'FG',8X,'DEPIONS',4X,'DEPTOT',6X,'DEPFR',
119   c     2 5X,'RNIONS',6X,'RDOT',7X,//)
120   c     EndIf
121
122   c     ==> Reading in general data
123         G=0.0
124   c     {Read NP exp. cross sections}
125         NP=0
126         READ (8,1)NP
127       1 FORMAT(I4)
128         write (6,3) NP
129       3 FORMAT(//,10X,'NUMBER OF EXPERIMENTAL INPUT CROSS-SECTION (NP) ='
130        1,I4,/)
131         If (NP.gt.MaxSW) Then
132           Write(6,*) 'Warning: Too many cross section values... STOP.'
133           STOP
134         EndIf
135         READ (8,2)(T(I),SW(I),I=1,NP)
136       2 FORMAT(E10.2,1X,E10.2)
137         write (6,4) (I,T(I),I,SW(I),I=1,NP)
138       4 FORMAT(       10X,'TEMP(',I3,')=',E12.4,8X,'SIGMAV(',I3,')='
139        1,E12.4)
140   c     {Read NL energies}
141         READ (8,1)NL
142         READ (8,5)(E1(J),J=1,NL)
```

```
143        5 FORMAT(E11.4)
144    c      WRITE (1,555)
145    c 555 FORMAT(//,2x,'E0 INPUT LASER ENERGY IN JOULES',9X,'LOG10 OF E0'/)
146    c      WRITE (6,8)(J,E0(J),J,AEX(J),J=1,NL)
147        8 FORMAT(10X,'E0(',I3,')=',E11.4,9X,'LOG10(E0)=AEX(',I3,')=',F 8.4)
148    c      {Number of initial volumes}
149           READ (8,1)NV
150    c      {Read NV initial volumes}
151           READ (8,5)(V0(I),I=1,nv)
152    c      {Number of densities}
153           READ (8,1)ND
154    c      {Number of intervals in the computational integral}
155           READ (8,1)NINT
156    c      <== Stop reading general data
157           write (6,*) 'NV=',NV,' ND=',ND,' NINT=',NINT
158
159
160
161    c      ==> Start reading data SPECIFIC FOR THE PARTICULAR FUSION REACTION
162    c      In this case (cf. above) D+3He => P(14681)+A(3670)
163    c      {Fusion energy released per reaction (total)}
164           If (IsDHe) Then
165             EF=(18.351E+06)*1.6E-19
166           EndIf
167           If (IsHB) Then
168             EF=(8.681E+06)*1.6E-19
169           EndIf
170    c      {Fractions of EF carried by charged products of reaction (total)}
171           If (IsDHe) Then
172             FR=3.670E+06
173             FR2=14.681E+06
174           EndIf
175           If (IsHB) Then
176             FR=8.681E+06
177           EndIf
178    c      {Charge of the reaction products (average)}
179           If (IsDHe) Then
180             Z=2.
181             Z2=1.
182           EndIf
183           If (IsHB) Then
184             Z=2.
185           EndIf
186    c      {Mass of reaction products in terms of electron mass (average)}
187           If (IsDHe) Then
188             MH=4.*1836.
189             MH2=1836.
190           EndIf
191           If (IsHB) Then
192             MH=4.*1836.
193           EndIf
194    c      {Energy of reaction products (average)}
195           If (IsDHe) Then
196             EA=3.670E+06
197             EA2=14.681E+06
198           EndIf
199           If (IsHB) Then
200             EA=2.893667E+06
201           EndIf
202    c      {average mass of charged reaction products (average)}
203           If (IsTwoComp) Then
204             AM=2.5
205           Else
206             If (IsDHe) Then
207               AM=4.0
208             EndIf
209             If (IsHB) Then
210               AM=4.0
211             EndIf
212           EndIf
213
```

```fortran
214   c       {average values of INITIAL plasma}
215   c       {average ionic mass in plasma: here (2+3)/2=2.5}
216           If (IsDHe) Then
217             XM=2.5
218           EndIf
219           If (IsHB) Then
220             XM=6.0
221           EndIf
222   c       {average ionic charge in plasma: here (1+2)/2=1.5}
223           If (IsDHe) Then
224             ZIA=1.5
225           EndIf
226           If (IsHB) Then
227             ZIA=3.0
228           EndIf
229   c       {Solid State Density and its multiples. Caution: Here actually DT}
230   c                                           {solid sate density used}
231           If (IsDHe) Then
232             DNS(1)=5.8E22
233           EndIf
234           If (IsHB) Then
235             DNS(1)=9.4E22
236           EndIf
237           DNS(2)=DNS(1)*1.0E+03
238           DNS(3)=DNS(1)*5.0E+03
239           DNS(4)=DNS(1)*1.0E+05
240           DNS(5)=DNS(1)*2.0E+05
241           DNS(6)=DNS(1)*3.0E+05
242           DNS(7)=DNS(1)*5.0E+05
243           DNSNS(1)=1.0E+00
244           DNSNS(2)=1.0E+03
245           DNSNS(3)=5.0E+03
246           DNSNS(4)=1.0E+05
247           DNSNS(5)=2.0E+05
248           DNSNS(6)=3.0E+05
249           DNSNS(7)=5.0E+05
250
251   c       {mass of electron}
252           EM=9.11E-31/1.67E-27
253   c       {Boltzmann Konstante}
254           BK=1.38E-16
255   c       {50/50 input isotope mix}
256           A=4.
257           pi=ACOS(-1.0)
258   c       {integration upper boundary and interval size; see below}
259           END=pi/2.
260           H=END/NINT
261   c       <== End of reading.
262
263           If (WriteGain) then
264             If (IsDHe) Then
265               Write(1,*) 'DHe fusion reaction      '
266             EndIf
267             If (IsHB) Then
268               Write(1,*) 'HB fusion reaction       '
269             EndIf
270             If (DoBrems) Then
271               Write(1,*) 'with bremsstrahlung      '
272             Else
273               Write(1,*) 'without bremsstrahlung   '
274             EndIf
275             If (IsTwoComp) Then
276               Write(1,*) '2 reaction products      '
277             Else
278               Write(1,*) '1 reaction product       '
279             EndIf
280           EndIf
281           If (WriteTemp) then
282             If (IsDHe) Then
283               Write(2,*) 'DHe fusion reaction      '
284             EndIf
```

377

```
285            If (IsHB) Then
286               Write(2,*) 'HB fusion reaction        '
287            EndIf
288            If (DoBrems) Then
289               Write(2,*) 'with bremsstrahlung       '
290            Else
291               Write(2,*) 'without bremsstrahlung    '
292            EndIf
293            If (IsTwoComp) Then
294               Write(2,*) '2 reaction products       '
295            Else
296               Write(2,*) '1 reaction product        '
297            EndIf
298               Write(2,39) E0(1)
299       39     Format(E11.4)
300         EndIf
301
302         n=1
303    c     { define energie range and steps }
304         If (WriteGain) Then
305         DO 199 d=7,10      ,pstep
306            E0(n)=10**d
307            n=n+1
308       199 Continue
309            NL=n-1
310
311          DO 7 J=1,NL
312               AEX(J)=ALOG10(E0(J))
313        7  CONTINUE
314         Else
315            NL=1
316         EndIf
317
318    c     (Start main double loop (i,j pick certain IonDensity and Init Vol)
319         DO 10 I=4,4
320           DO 10 J=1,13
321            R0=(V0(J)*0.75/PI)**0.333333
322            RIONS0=DNS(I)*V0(J)
323            write (6,9) DNS(I),V0(J),RIONS0
324       9  FORMAT(////,7X,'INITIAL ION DENSITY =',E10.4,7X,'INITIAL PELLET
325       @        VOLUME = ',E10.4,5X,'INITIAL NUMBER OF IONS=',E10.4////)
326
327            If (WriteGain) then
328               Write(1,92) DNSNS(i),V0(j)
329            EndIf .
330    c        {E(k) picks the initial energy value E0}
331
332            Do 10 k=1,NL
333    c          {Output file headers}
334       92      Format (2E11.4)
335       93      Format (3E11.4)
336            Write(6,*) 'The laser energy is ',E0(k),' eV.'
337    c        DEPLET=0.0
338    C        {1 JOULE=6.2419E18 eV}
339            T0=E0(K)*6.25E 18*(2./3.)
340            T0=T0/((1.+ZIA)  *DNS(I)*V0(J))
341            If (WriteTemp) then
342               Write(2,93) DNSNS(i),V0(j),T0
343            EndIf
344    c        {1.0E+01 < T0 < 1.0E+08}
345            IF(T0.LT.1.E1) GO TO 10
346            IF (T0.GT.1.E8) GO TO 10
347            TEMP=T0
348            SV=0.0
349            SUM=0.0
350            ANGLE=0.0
351            C=1.0
352            RDOT=0.0
353            AN=0.0
354            DADOT=0.0
355            TIME=0.0
```

```
356                     RNIONS=RIONS0
357                     DEPTOT=0.0
358                     FAC=SQRT((XM*938.E6)/(5.*T0))/3.0E 10
359                     FAC=FAC/SQRT(1.+ZIA)
360                     ADOT0=1./FAC
361                     ADOT0=ADOT0/R0
362                     DNSION=DNS(I)
363                     NUMBER=0
364
365     c               {Start of integration loop; From 0 to pi/2; Ray (4.17)}
366         15          S=SIN(ANGLE)
367                     C=COS(ANGLE)
368                     TN=TAN(ANGLE)
369                     NUMBER=NUMBER+1
370     c               {Adiabatic law for temperature decrease Ray(4.18)}
371                     TINC=-2.*TEMP*TN*H
372                     EDNS=ZIA*DNS(I)*C**3
373     C               {EDNS ELECTRON DENSITY AT CORRESPONDING PLASMA RADIUS.}
374                     VOL=(4./3.)*PI*(R0/C)**3
375                     DVOL=(4./3.)*PI*R0**3*(1./(COS(ANGLE+H))**3-1./C**3)
376     c               {Ray (4.15) <slightly modified by other authors>}
377                     RDOT=(RDOT**2+5.*TEMP*(RNIONS+ZIA*RIONS0)*(1.-((COS(ANGLE+H))
378         @           /C)**2)/((XM*RNIONS+AM*AN+EDNS*VOL*EM)*938.E6/9.E20))**0.5
379     c               {Ray p.71}
380                     DTIME=R0*(1./COS(ANGLE+H)-1./C)/RDOT
381                     time=time+dtime
382                     IF (ANGLE.EQ.0.0) GO TO 16
383                     ADOT=RDOT*C**2/(R0*S)
384                     GO TO 17
385         16          ADOT=ADOT0
386         17          CONTINUE
387     c               Write(6,*) 'tinc = ',tinc
388     c               {Reheating process of the alpha particle}
389                     CALL RANGE (EA,Z,TEMP,MH,EDNS,RA)
390                     IF (RA.LE.0.0) GO TO 333
391                     IF (TEMP.LE.200) RA=RA*1.5
392                     IF (TEMP.GT.200) RA=RA*(1.49935E-3)*TEMP**1.3441
393     c               {P=absorption probability Ray(4.19)}
394                     P=1.0/(1.+RA*C/R0)
395     c               {TRHT= Temp of reheat}
396                     TRHT=P*(FR/A)*DNSION**2*SV*H/((1.+ZIA)*DNS(I)*
397         @               ADOT*C**3)
398                     TINC=TINC+TRHT
399     c               Write(6,*) 'TRHT = ',TRHT
400     c               Write(6,*) 'tinc = ',tinc
401         333         CONTINUE
402                     If (IsTwoComp) Then
403     c                   {Reheating process of the second charged reaction product}
404                         CALL RANGE (EA2,Z2,TEMP,MH2,EDNS,RA2)
405                         IF (RA2.LE.0.0) GO TO 666
406                         IF (TEMP.LE.200) RA2=RA2*1.5
407                         IF (TEMP.GT.200) RA2=RA2*(1.49935E-3)*TEMP**1.3441
408     c                   {P2=absorption probability Ray(4.19)}
409                         P2=1.0/(1.+RA2*C/R0)
410     c                   {TRHT2= Temp of reheat}
411                         TRHT2=P2*(FR2/A)*DNSION**2*SV*H/((1.+ZIA)*DNS(I)*
412         @                   ADOT*C**3)
413                         TINC=TINC+TRHT2
414     c                   Write(6,*) 'TRHT2 = ',TRHT2
415     c                   Write(6,*) 'tinc = ',tinc
416         666         CONTINUE
417                     EndIf
418                     TEMP=TEMP+TINC
419     c               Write(6,*) 'TEMP = ',temp
420                     If (DoBrems) then
421                         CALL BREM (NUMBER,TEMP,EDNS,ZIA,R0,C,RLOSS,DTIME)
422                         TEMP=TEMP-RLOSS
423                     EndIf
424     c               {Call SEMP taken out from here, cf. end of prog}
425     c               {SV interpolation}
426                     IF(TEMP.LT.0.0) GO TO 50
```

```
427                    IF(TEMP.LT.T( 4)) GO TO 30
428                    IF(TEMP.GT.T(NP)) GO TO 35
429                    DO 25 IT=1,NP
430                       IF(TEMP. GT.T(IT)) GO TO 25
431                       IU=IT
432                       GO TO 28
433        25         CONTINUE
434        28         IL=IU-1
435                    SV=SW(IL)+(TEMP-T(IL))*(SW(IU)-SW(IL))/(T(IU)-T(IL))
436                    GO TO 40
437   c* 30           SV=10**(18.1946*ALOG10(TEMP)-92.5838)
438        30         COEF1=5.653622037
439                    COEF2=-43.65956061
440                    sv=10**(COEF1*ALOG10(TEMP)+COEF2)
441                    GO TO 40
442   C* 35           SV=10**(-2.47723*ALOG10(TEMP)-1.52038)
443        35         SV=SW(NP)
444   c              {End SV interpolation}
445   c              {Ray (4.17)}
446        40         TERM=DNSION**2*SV*H/(ADOT*C**3)
447   c              write(6,701)number,sw(I),sv,T(np),temp,E0(K)
448       701    format('num=',I5,1x,'sw=',e10.4,1x,'sv=',e10.4,1x,'T=',e10.4,1x,
449         1            'TEMP=',e10.4,1x,'E0=',E10.4)
450                    SUM=SUM+TERM
451   C          IF(ANGLE.EQ.0.0)THEN
452   C              DE=0.0
453   C            ELSE
454   C             DE= DNSION*(1.-(COS(ANGLE+H)/C)**3)
455   C            DE=3.*TN*DNSION*H
456   C          ENDIF
457   C        DN= DL+DE
458   C        DNSION=DNSION-DN
459   c              {Depletion}
460                    PRIONS=DNSION*VOL
461                    IF (DNSION.LE.0.0) GO TO 50
462                    RRATE=DNSION**2*SV/A
463                    DNA=RRATE*(2.*PI/3.)*R0**3*DTIME*(1./C**3+1./(COS(ANGLE+H))
464         @              **3)
465                    AN=AN+DNA
466                    DNSHE=AN/VOL
467                    FG= (EF/E0(K))*V0(J)*SUM/A
468                    DEPFG=(E0(K)/EF)*FG*A/2.
469                    DNSION=RNIONS/VOL
470                    DPIONS=2.*RRATE*(2.*PI/3.)*R0**3*(1./ C**3+1./(COS(ANGLE+H))
471         @              **3)*DTIME
472                    DEPTOT=DEPTOT+DPIONS
473                    DEPFR=DEPTOT/RIONS0
474   c              Write(6,*) 'DEPFR = ',DEPFR
475   C        RNIONS=RIONS0-DEPTOT
476   c              {subtract burnt ions from the remaining ones, caution when -
477                    RNIONS=RNIONS-DPIONS
478   c              {End depletion}
479                    GLIM=1.E-2
480                    TLIM=1.E+3
481                    IF (FG.LT.GLIM)   GO TO 47
482                    IF (TEMP.LT.TLIM) GO TO 47
483                    If (WriteTemp) then
484                    WRITE(2,78)NUMBER,TIME,RRATE,TEMP,SV,DNSION,FG,DPIONS,DEPTOT,
485         1            DEPFR,RNIONS,RDOT
486        78         FORMAT(1X,I4,11E11.4)
487                    EndIf
488                    IF (TN.EQ.0.0) GO TO 46
489                    GO TO 47
490        46         TN=1.0E-20
491        47         CONTINUE
492                    ANGLE=ANGLE+H
493                    IF(ANGLE.GT.END) GO TO 50
494                    GO TO 15
495   c              {End of integration loop}
496
497   c              {Fusion gain Ray(4.17)}
```

```
498      50         G=(EF/E0(K))*V0(J)*SUM/A
499                 IF(G.LT.1E-4) GO TO 10
500                 AGY=ALOG10(G)
501                 DEPG=(E0(K)/EF)*G*A/2.
502                 If (WriteGain) then
503                   Write(6,*) 'E0 of ',E0(k),' => ',G,' gain.'
504                   write (1,29) E0(K),T0,SV,G,DEPTOT,DNSION,RNIONS,DEPFR
505        1              ,DEPG,V0(J),RDOT,NUMBER
506      29        FORMAT(   11E11.4,i5)
507                 EndIf
508      10         CONTINUE
509                 STOP
510   c          close(1)
511   c          close(8)
512                 END
513   c      End main double loop and main program
514
515
516
517   c      ***************
518   c      * SUBROUTINES *
519   c      ***************
520
521          SUBROUTINE BREM (NUMBER,TEMP,EDNS,ZIA,R0,C,RLOSS,DTIME)
522   C**** ZSQAV IS AVERAGE VALUE OF ZIA SQUARED USED IN RLOSS
523          ZSQAV=1.0
524   C      EPHOT=1149*TEMP
525          EPHOT=TEMP/6.2419E11
526   C**** 1ERG=6.2419E11 EV
527          WMAX=2*3.1416*EPHOT/6.67E-27
528          WPLAS=5.65E4*EDNS**0.5
529   C**** VEI=ZIA*EDNS*3.1416*4.803E-10**4/(0.9109E-27**0.5*
530   C****@      (3*1.602E-12*TEMP)**1.5)
531          IF(TEMP.LT.0.0) GO TO 320
532          TTMP=TEMP**1.5
533          VEI=5.2576E-7*ZIA*EDNS/TEMP**1.5
534   C      v1=(3.*temp/(6.2419e11*9.1085e-28))**0.5
535   C      rBohr=(6.67e-27)**2/(9.1085e-28*(4.8029e-10)**2)
536   C      Tast=4.176e5*zia**2
537   C      v2=4.*zia**2*((1.+4.*temp/tast)**0.5-1)**2
538   C      Vei=v1*pi*rBohr**2*edns/v2
539          WCRIT=0.2*WMAX
540          RAD=R0/C
541          IF(WPLAS.GT.WCRIT) GO TO 310
542          IF(VEI.GT.WCRIT) GO TO 310
543          ABSL=3.0E10*WMAX**2/(VEI*WPLAS**2)
544          IF(ABSL.LE.RAD) GO TO 310
545          IF(C.EQ.1.0) GO TO 310
546   C      TIMINC=FAC*R0*TN*H/(C*(1.-(R0/RAD)**2)**0.5)
547          RLOSS=1.57E-27*(TEMP*1.602/1.38E-4)**0.5*EDNS*ZSQAV*DTIME*
548        @      2./((1.+ZIA)*3.*1.6E-12)
549          RLOSS=RLOSS*(1.-(RAD/ABSL))
550   C        PRINT 83,VEI,ABSL,RAD,DTIME,TEMP,RLOSS
551       83 FORMAT(7X,'VEI=',E10.4,3X,'ABSL=',E10.4,3X,
552        &'RAD=',E10.4,3X,'DTIME=',E10.4,3X,'TEMP=',E10.4,3X,
553        &'RLOSS=',E10.4,//)
554          GO TO 320
555      310 RLOSS=0.0
556   C        GO TO 319
557   C 316 PRINT 317,TEMP,NUMBER
558      317 FORMAT(10X,'NEGATIVE TEMP=',E12.7,10X,'NUMBER WITH NEGATIVE
559        @          TEMP=',I5)
560   C 319 CONTINUE
561      320 RETURN
562          END
563
564          SUBROUTINE RANGE(E,Z,T,M,D,R)
565   c      {Ray(3.50)}
566          PI=ACOS(-1.0)
567          CU=6.94E6
568          X=CU*T/(M*M)
```

```
569          X=X*(E*CU)**2/(PI*Z*Z*D)
570          TT=PI*D*M*M*Z*Z/(E*CU*E*CU)
571          TT=TT/CU
572          IF(T.LT.TT) GO TO 1
573          X=ALOG(X)
574          CALL EI(X,Y)
575          R=.5*M*Y/(CU*T)
576          GO TO 5
577        1 R=-10.
578        5 RETURN
579          END
580
581          SUBROUTINE EI(X,Y)
582     c    {Evaluates an integral needed by RANGE (cf. Rays thesis)}
583          SUM=0.5772156649
584          SUM=SUM+ALOG(X)
585          EPS=1.E-4
586          N=1
587          TERM=X
588          SUM=SUM+TERM
589        1 N=N+1
590          TERM=TERM*X/N
591          SUM=SUM+TERM/N
592          IF (TERM.LT.EPS) GO TO 2
593          GO TO 1
594        2 CONTINUE
595          Y=SUM
596          RETURN
597          END
598
599     c    SUBROUTINE SEMP(G,TEMP,TRHT,RLOSS,ADOT,H)
600     c    IF(G.LT.1E-1) GO TO 15
601     c    RTIME=H/ADOT
602     c    PRINT 10,RTIME,TEMP,TRHT,RLOSS
603     c 10 FORMAT(10X,E12.4,10X,E12.4,20X,E12.4,20X,E12.4)
604     c 15 RETURN
605     c    END
606
607     C    SUBROUTINE GRAPH(E0(K),G)
608     C    DIMENSION E0(100),AX(100),AY(100)
609     C     DO 77 K=1,NL
610     C          AX(K)=ALOG10(E0(K))
611     C          AY(K)=ALOG10(G)
612     C    IS=JSCALE(AY(AX(K)))
613     C    LINE(IS:IS)='*'
614     C 77  CONTINUE
615     C    RETURN
616     C          If (DoSemp) then
617     C            CALL SEMP (G,TEMP,TRHT,RLOSS,ADOT,H)
618     C          EndIf
619     C    END
```

382

APPENDIX B

This Appendix B reports the small changes of the D-^{3}He fusion gains of the last but one Section where the D-D reactions were ignored. Here we include these and show the differences between the two cases.

The results of computation of fusion gains G as a function of different input parameters for D^{3}He and D^{3}He+DD reactions using our extended code are summarized for various input energies E_0 for a D^{3}He pellet with and without the DD reaction included. At densities less than $10^3 n_s$ we derived a break-even energy of 1.25 x 10^{13} J and an optimum temperature for (ignitionless) burn of 51 KeV, exactly the same as was derived for the fuel pellet with no DD reaction included. The reason for similar values of break-even energy and optimum temperature in both cases is that under conditions where the fuel pellet burns without ignition, the inclusion of the DD reactions makes no difference. This can be further seen from comparison of Figure B1 and B2 where fusion gains are plotted versus input laser energy for both cases.

Energy dependence of the fusion gain of the D^{3}He pellets for initial pellet volumes in the range of 10^{-7} to 10^{-1} cm^3 with two different plasma densities, ie, 4 x 10^3 n_s and 3 x $10^4 n_s$ are shown in figures B3 and B4 respectively, where the upper curves in the figures have DD included and the lower curves have the D^{3}He reaction only. From figures B1 and B2 it is clear that there is no distinguishable difference between gains with and without the inclusion of the DD reactions for lower pellet densities up to $10^3 n_s$. As the pellet density is increased beyond this limit, the difference in the gain for two cases depend on density, pellet volume and input energy. As the values of the input parameters (input laser energy, initial pellet volume, initial pellet density) increases, the gain difference between the two cases becomes larger. Figures B3 and B4 give numerical evidence for this fact, and enable a comparison of the results with and without the inclusion of the DD reactions. It is obvious that it is a doubtful step to sacrifice the DD reaction for the sake of mathematical simplicity. The results indicate that the inclusion of the DD reactions in the D^{3}He pellet gives higher fusion gains than the D^{3}He reaction alone and also the input energy, E_0, required for the D^{3}He pellet for maximum gain is reduced when the DD reaction is included. Considering an extreme case of our calculations with pellet density 3 x 10^4 n_s and pellet volume of 10^{-1} cm^3, inclusion of the DD reactions in the calculations gives rise to a 52.6% increase in the fusion gain and a 40.0% reduction in the input laser energy for maximum gain. This is an interesting result for the design of the driver capacities. This indicates that the inclusion of DD reactions makes less difference as the density and volume decrease.

The inclusion of the DD reactions in the D^{3}He pellet has also an appreciable affect on the plasma temperature. Time dependence of the plasma temperature T of D^{3}He and D^{3}He + DD reactions for the cases of initial densities of 4 x 10^3 and 3 x 10^4 times the solid state with the initial volumes of 10^{-7} - 10^{-1} cm^3 are shown in Figures B4 and B6 as two examples for the behaviour of the temperature. The plasma temperature dependence of the two different cases is not the same. For cases with the D^{3}He reaction alone the fuel does not ignite. The exact values of the input parameters and the results of calculations of gains, maximum plasma temperatures

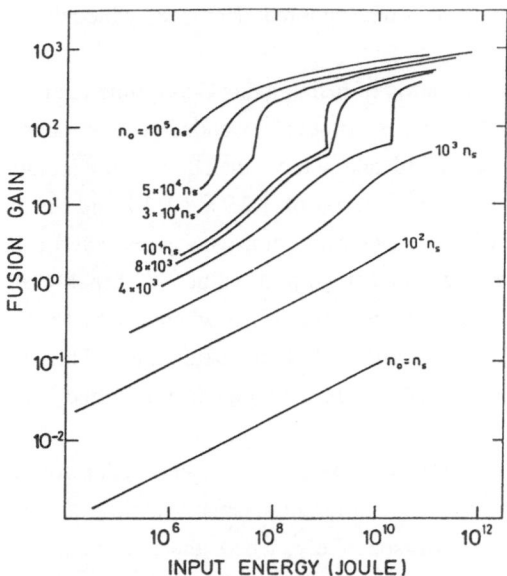

Figure B1. An Energy-Gain Scheme for D³He Fusion. X-axis: Input beam energy in Joule, Y-axis: Optimum fusion gains for initial volume $10^{-7} \leq V_o \leq 10^{-1}$ cm³. Each of the curves refers to a particular initial target density as shown. The largest gain due to saturation by fuel depletion is about 10^3.

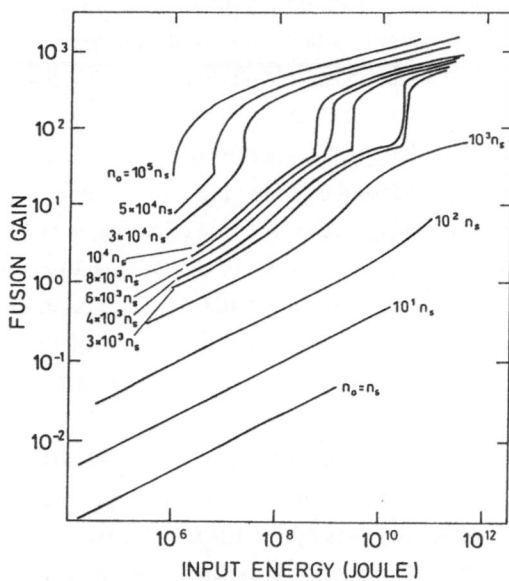

Figure B2. Same as in Figure B1, but with the inclusion of the DD reactions. The largest gain in this case due to saturation by fuel depletion is about 1.5×10^3.

384

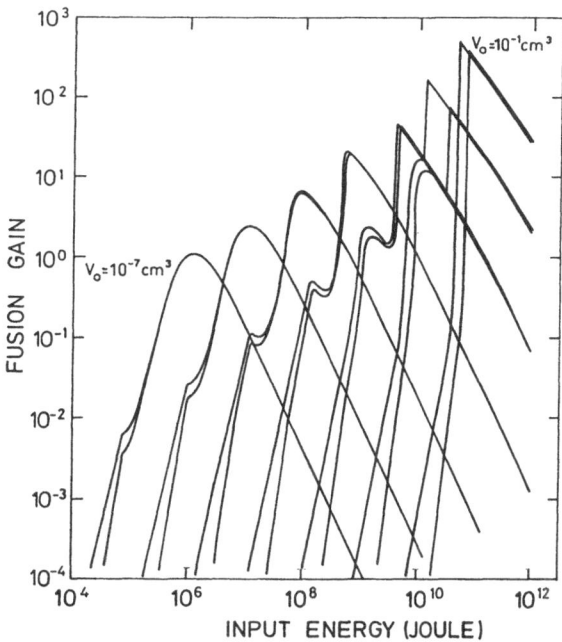

Figure B3. Energy dependence of the fusion gain of the D^3He and $D^3He + DD$ reactions for a compression to 4×10^3 times the solid state density and initial pellet volumes in the range $10^{-7} - 10^{-1}$ cm^3. The gain variation for the D^3He (lower curves) and $D^3He + DD$ reactions (upper curves) is not the same.

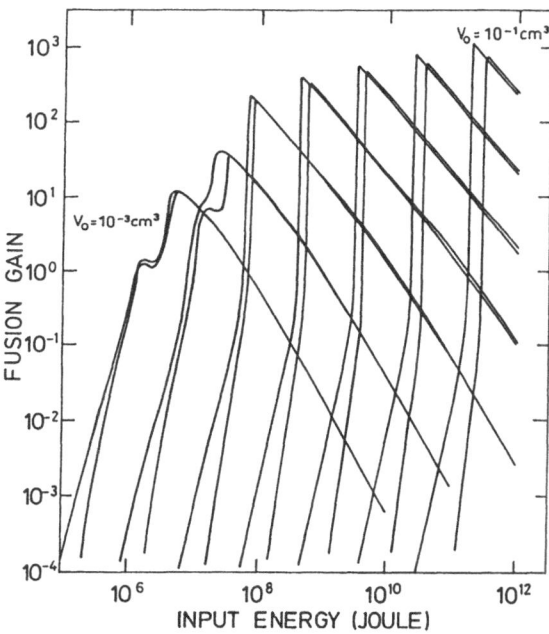

Figure B4:. Same as in Figure B3 with $n_o = 3 \times 10^4 \, n_s$ and initial pellet volumes in the range $10^{-7} - 10^{-1}$ cm^3. The upper curves in the figure represent DD inclusion and the lower curves D^3He reactions only.

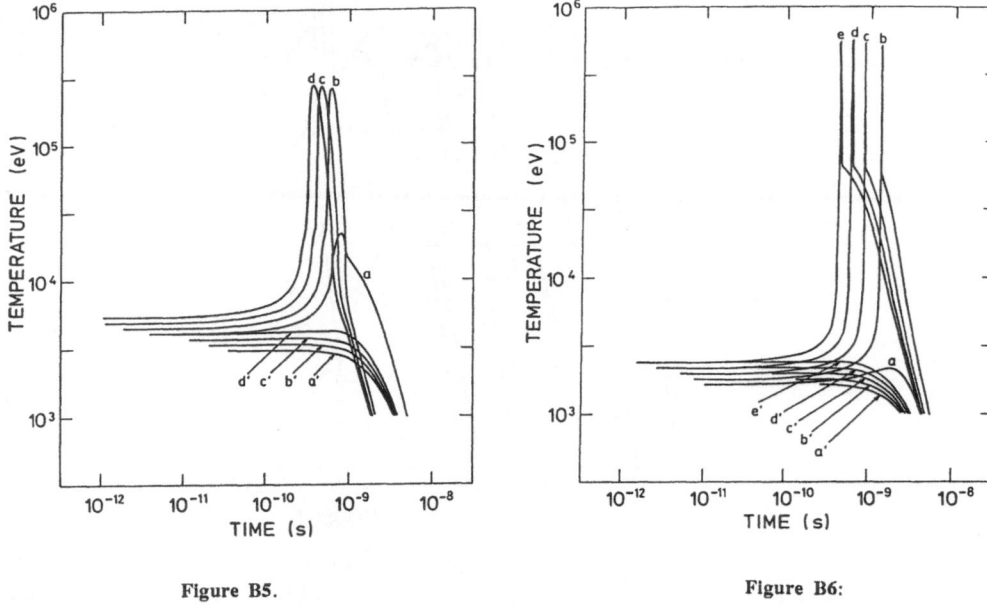

Figure B5.   **Figure B6:**

Figure B5. Dependence of the temperature of a D^3He pellet on time at initial compression to 4×10^3 times the solid state and initial volume of 10^{-1} cm^3. The input energies Eo [GJ] were (a, a'): 42.97, (b,b'): 47.14, (c, c'): 51.71 and (d,d'):56.72 and initial temperature [keV] of (a,a'): 3.087 (b,b'): 3.386, (c,c'): 3.715, (d,d'): 4.075. With inclusion of the DD reactions (a,b,c,d), the fusion gain G were a: 1.16, b: 512.1, c: 472.7, d: 433.0, the deuterium-helium fuel depletion [%] of a:0.709, b: 78.1, c: 79.2, d: 79.5 and the maximum temperature [keV] of plasma a: 22.35, b: 270.6, c: 276.3, d; 279.5. With exclusion of the DD reactions (a',b',c',d'), Gain, a: 0.02, b: 0.045, c': 0.083, d': 0.167, Fuel Depletion [%]; a': 0.003, b': 0.006, c': 0.012, d': 0.028, maximum temperature [keV] (a', b', c', d'; no ignition); a': 3.094, b': 3.408, c': 3.779, d': 4.272.

Figure B6:. Same as in Figure B5 with $n_o = 3 \times 10^4$ n_s and $V_o = 10^{-1}$ cm^3. Input energy [GJ]; (a, a'): 172.3, (b,b'): 189.0, (c,c'): 207.3, (d,d'): 227.4, (e,e'): 249.5, Initial Temperature [keV]; (a,a'): 1.650, (b,b'): 1.810, (c,c'): 1.98, (d,d'): 2.178, (e,e'): 2.389. With inclusion of the DD reactions (a,b,c,d,e), Gain; a: 0.414, b: 1167, c: 1077, d: 986.3, e: 898.1, Fuel Depletion [%]; a:0.265, b: 94.0, c: 95.0, d: 95.4, e: 95.5, maximum temperature [keV]; a: 2.137, b: 512.3, c:534.1, d: 557.2, e: 545.6. With exclusion of the DD reactions (a',b',c',d',e'), Gain a': 0.003, b': 0.007, c' 0.013, d': 0.026, e': 0.051, Fuel Depletion [%]; a' 0.0002, b': 0.0005, C': 0.0011, d': 0.0023, e': 0.0050, Maximum Temperature [keV] (a',b',c',d',e', no ignition): a':1.640, b': 1.890, c':1.987, d': 2.184, e': 2.412.

and fuel depletions of the D^3He and $D^3He + DD$ reactions are given in the figure captions of figures B5-B6.

Another advantage of our extended model is that it enables us to calculate the number of emitted neutrons per D^3He reaction in the fuel pellet. We call this number the neutron ratio factor, "Ratio", which is the ratio of the number of neutrons emitted per D^3He fusion (Ratio=n/D^3He).

This is an important factor in any reactor design and we are now able to accurately take into account the neutron induced damage to the reactor material and the radioactivity generated due to neutron flux. Figures B7-B9 report for the first time the results of the n/D^3He ratio

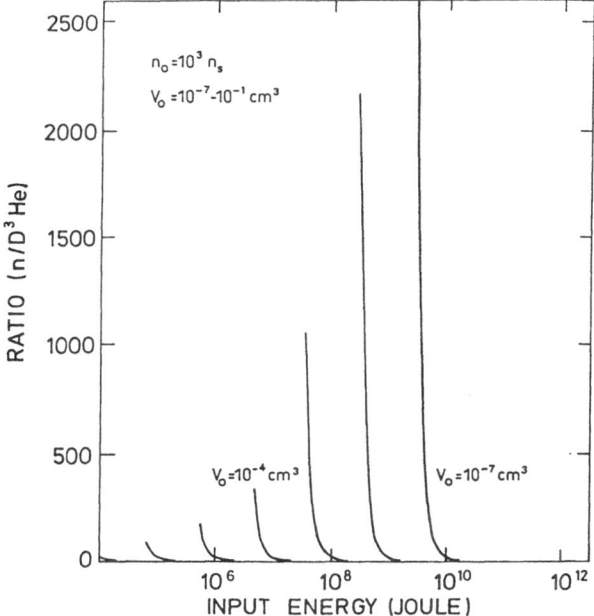

Figure B7:. Dependence of the neutron ratio factor, Ratio = n/D^3He of the D^3He pellet on input laser energy at initial compression 10^3n_s and initial pellet volumes in the range of 10^{-7} - 10^{-1} cm^3.

calculation for the cases of initial densities 10^3n_s, 4×10^3n_s and 10^4n_s with initial volumes of 10^{-7} - 10^{-1} cm^3. As can be seen from the figures and as one might expect, this ratio decreases with increasing input energy E_0, reaching to a very small value when volume ignition occurs for the D^3He pellet. There is a minimum (too small to see in the figures) in the lower tail of the

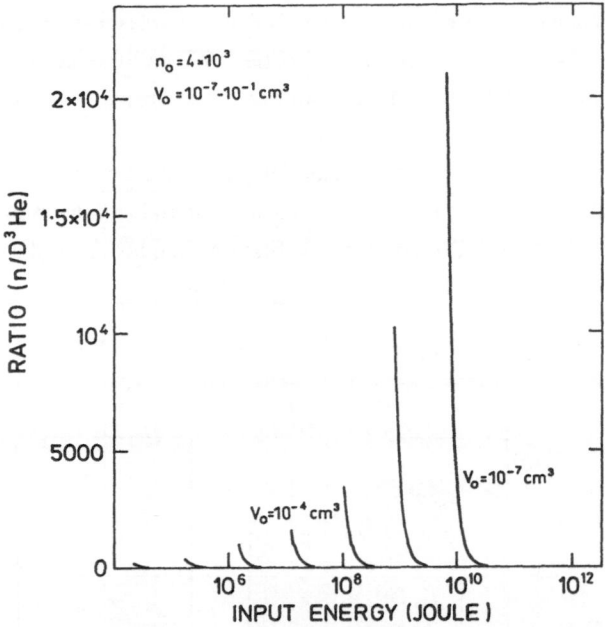

Figure B8:. Same as in Figure B7 for a compression to 4×10^3 times the solid state density, 4×10^3 n_s, and initial volumes in the range of 10^{-7} - 10^{-1} cm^3.

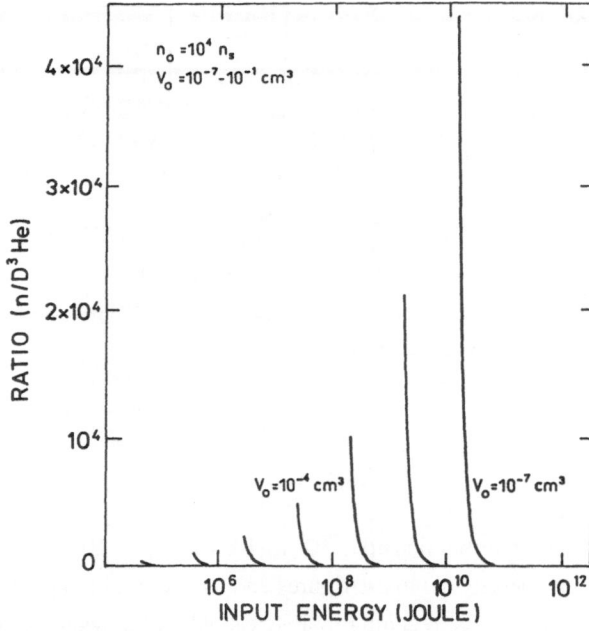

Figure B9. Same as in Figure B7 for a compression to 10^4 n_s and V_0 in the range of 10^{-7} - 10^{-1} cm^3.

neutron ratio curve which corresponds to parameters for which maximum fusion gain is obtained. The ratio not only has an input energy dependence but also has a pellet volume and density dependence. Moreover, the ratio at minimum, corresponding to maximum fusion gain, increases slightly with pellet volume and density. The neutron ratio factor varies from 4% to 6% and the reaction ratio DD/D^{3}He varies from 8% to 12%. As an example, for a compression of 3×10^4 times solid state density, an initial volume of 10^{-5} cm^3 and initial laser energy of 65.93 MJ, the gain is 219.2, the total fuel depletion 61.6, the neutron ratio 4.8% and so the reaction ratio is 9.6%.

Our conclusion is that the problem of the secondary reaction due to two different DD nuclear reactions, taking place at the same time in a D^{3}He pellet, has now been solved. The main advantages of inclusion of the DD reaction in the D^{3}He pellet are as follows.

1) The model can now be considered as complete with respect to different nuclear reactions, reaction charged products and fuel depletion.

2) The exact value of neutron ratio factor (n/D^{3}He) and reaction ratio DD/D^{3}He can be calculated. Since the fuel depletion and the neutron ratio factor is now known, the true value of the neutron flux can be used in any D^{3}He reactor design.

3) The inclusion of DD reactions in D^{3}He pellet results in a significant reduction in the ICF driver energy requirement by a factor depending on the initial condition of calculation.

4) A significantly higher fusion gain is obtained by the inclusion of the secondary reactions.

COMMENTS ABOUT NEUTRON FEEDBACK NPL DRIVEN ICF

G. H. Miley, M. Petra and Y. Shaban

Fusion Studies Laboratory
University of Illinois
Urbana, Illinois 61801 USA

ABSTRACT

The use of a neutron-feedback-coupled Nuclear-Pumped Laser (NPL) driver for ICF would greatly enhance the overall system efficiency. In this case neutrons from the target implosion create nuclear reactions in the laser medium, providing the energy input to the laser. Thus the recycle energy to the laser does not go through an electrical conversion cycle, greatly increasing the system efficiency. Electrical energy is generated in a conventional fashion from energy deposited in the blanket as the energetic neutrons are thermalized.

Here we consider a variation on the conventional DT reactor driven by an NPL; namely, an advanced DT-seeded, D-^{3}He-fueled target. Then, in addition to the efficiency gain due to the neutron coupling, the target allows an added gain through use of a direct energy conversion technique for the large fraction of energy carried by charged particles. While the neutrons produced are reduced compared to a D-T target, they are still adequate to pump the NPL. At the same time, neutron-induced activation and radioactive waste production are decreased.

INTRODUCTION

The possibility of using advanced fuels (AFs) for fusion is gaining increased attention with the realization of the importance of improved environmental and safety factors as well as the need for high efficiency power plants.[1-3] It appears that a D-T spark-ignited target (called AFLINT [4]) represents one of the best **near-term** approaches to burning AFs. Reactor designs employing AF's have generally assumed the use of a heavy-ion-beam driver due to its high energy capability plus its high efficiency [5]. On the other hand, a nuclear pumped laser (NPL) could also provide a high efficiency, and this route is also explored here.

PRIOR DEVELOPMENTS

The potential advantage of using AFs for fusion has been recognized for a number of years following its initial elaboration in 1975 in the book Fusion Energy Conversion[1]. At that time the use of AFs was viewed as a long range goal which would enable fusion power to achieve its ultimate potential as a very clean and efficient energy supply. However, the increasing emphasis on safety and environmental aspects [2,3] of future power sources places even more urgency on the AF approach for fusion.

Two types of AFs have been recognized: Deuterium based reactions, for example, D-D, Catalyzed-D, Semicatalyzed-D, D-^{3}He; and proton based reactions such as, p-^{6}Li, p-^{11}B [1,4,6-7]. Indeed, the possibility of burning p-^{11}B was first discussed in connection with its possible use in inertial confinement fusion (ICF) targets.[6] The plan then was to capitalize on the fact that cyclotron radiation losses are avoided in ICF (vs. magnetic confinement fusion). However, the energy balance required for a reasonable gain is difficult to achieve. Earlier investigations showed that the energetics for such a burn were marginal. More recently it was shown[8,9] that a modest gain is possible with p-^{11}B targets if ultra high compressions (> 10^4) can ultimately be obtained.

Another fuel which verges on being aneutronic is D-^{3}He. A major advantage of D-^{3}He is that its cross section is much more favorable than p-^{11}B.[10] In this case, though, neutrons from D-D reactions occur in the mixture and result in about 5-10% of the fusion energy going into the neutrons.

Studies of these fuel cycles were carried out between 1975-1985 by several groups[1,6,7,11]. While good progress has been made in understanding the physics of proton based fuels, it appears that, until new confinement techniques are discovered, burning p-^{11}B remains beyond our reach. D-^{3}He, on the other hand, offers many advantages and is generally viewed as quite attractive except for the lack of a "natural" ^{3}He source. In 1987, however, the situation was radically changed with the revelation that ^{3}He is implanted in the lunar soil due to bombardment by the solar wind[12]. Subsequent studies indicated that mining of the ^{3}He on a relative near-term basis should be taken as a serious option[13-14]. It should be stressed, however, that three other sources of ^{3}He still remain feasible; namely breeding ^{3}He in a semicatalyzed deuterium reactor, in a special beam-target facility, or by breeding excess tritium and allowing it to decay to ^{3}He[15]. Breeding, e.g. by proton bombardment of a Li plasma target, is generally viewed as complex and expensive due to the cost of electricity needed for the accelerator. However, how breeding compares with lunar mining from an economic point of view has yet to be studied.

D-^{3}He FUSION REACTORS

Almost simultaneously with the rising interest in D-^{3}He, several studies began to stress the environmental impact of fusion. These included the ESECOM study[2] and a contemporary review of the European Fusion Program[3]. A broad generalization of the conclusions from these reviews is that for fusion to be competitive, it must achieve a "reasonable cost" for electricity and simultaneously provide <u>significant</u> advances (compared to other advanced energy sources) in environmental and safety aspects. This conclusion is strongly reinforced by the impact that the Chernobyl accident has had on nuclear power development. As a consequence, there has been an increasing emphasis on combining AFs, particularly $D-^3He$, with near-term goals in the fusion program. For example, a recent study has suggested ways to use $D-^3He$ in next-step magnetic confinement fusion experiments[14].

ANEUTRONIC POSSIBILITIES

For ICF there appear to be three routes for burning AFs. One is to use a D-T spark ignited target of the AFLINT type to burn propagate into a region containing deuterium[4,5]. The second uses spark ignition but would arrange the target (cf Fig. 1) to effectively achieve $D-^3He$ operation[5]. The third would be to employ a homogenous $D-^3He$ mixture in the target which would be ignited by volume compression[8-9].

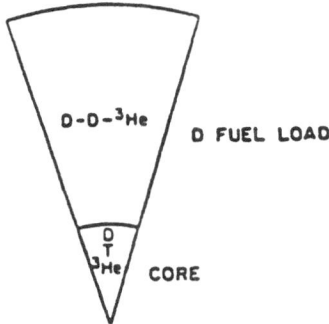

Figure 1. $D-^3He$ pellet with internal tritium and ^3He breeding. Core and fuel load imploded geometry.

In the AFLINT type targets, the original objective was to eliminate the need to breed tritium in the blanket. Tritium used in the spark core is ultimately replaced by unburned tritium (generated in D-D reactions) recovered from the outer region of the target. Similarly, such a target could be designed to be self-sufficient in ^3He. In contrast, the volume compression concept would require a ^3He supply. However, the physics of spark ignition relative to symmetry, etc., is very demanding. Consequently, volume ignition could turn out to be the most practical approach. The manufacture of such targets would also be easier since the internal structure required by separate layers is avoided.

The internal breeding AFLINT targets still result in a non-negligible neutron flux however. While some activation of structure still results, this approach has a significant advantage in reducing the tritium inventory and also in reducing neutron damage to components. Indeed, in the present context of using an NPL driver, the neutron component from the target serves a key role, i.e., to pump the laser via neutron induced reactions. Further, burning the AFLINT type target (with a deuterium outer layer) appears to be feasible with some modest extension of present ICF concepts. Thus this approach potentially offers a relatively "near-term" option.

To illustrate these concepts, we will consider some results from an earlier conceptual reactor study named the LOTRIT reactor[16-19]. LOTRIT used AFLINT targets with a heavy-ion-beam drive. A similar study of a self-breeding ICF target reactor design called "TAKNAWA-I" has been carried out in Japan[20].

Parameters for the AFLINT target used in LOTRIT are compared to a D-T reactor target (cf HIBALL) in Table 1. As seen, an important characteristic of the AFLINT design is the large ρR required, typical of AF targets. The D-T spark plays an essential role in that it offsets the large energy input required to ignite such a target, although the ignition input energy requirement is still a factor of 4 or more above that for a equivalent D-T design. As expected, (compared to the D-T HIBALL target), the AFLINT charged-particle and X-ray output fractions increase significantly with a corresponding reduction in the neutron energy fraction.

Table 1. Comparison of Energy Splits for AFLINT and Conventional DT Targets.

	Fusion target type[a]	
	T and ^{3}He internally bred (AFLINT)	DT
CONCEPTUAL DESIGN	AFLINT	HIBALL
Fusion yield [MJ/pulse]	1000	400
Repetition rate [Hz]	4	5
Cavity power [MW$_{fusion}$]	4000	2000
Number of cavities	5	4
Total power [MW$_{fusion}$]	20000	8000
Pellet data[a]		
Pellet gain	50	83
D[mg]	13.60	1.60
T[mg]	0.10	2.40
^{3}He[mg]	0.72	--
Energy splits [%}		
Neutrons and gamma rays	10	72
X-rays	56	22
Charged particles	34	6

[a]Core density-radius product is $\rho R = 8.9$ g/cm^2 at a core radius R = 100 μm for 2500 times solid density.

It should be emphasized that the AFLINT design was a "point" design which had not been optimized. Indeed there are a number of variations of the design that might be considered, some of which are illustrated in Ref. 5.

As the name LOTRIT implies, the most visible feature of this reactor design is the reduced tritium inventory. Tritium safety represents one of the main challenges for future fusion power plants. The current objective for D-T plants is to obtain total release rates of less than 10 curies per day. Also, possible accident consequences must be minimized by reducing tritium inventories below the levels (typically 10's of kilograms) stored in current design studies. These are very difficult goals to achieve. Further, future regulations could be more strict. Indeed the radiological limitations on tritium are already quite demanding (Table 2). Regulations associated with fission power plants are under constant pressure from various groups who would like to see them made even tighter. If that occurs, development of a D-T fusion power economy could be extremely difficult since there appears to be little margin over present restrictions.

Table 2. Radiobiological data for T_2 and HTO or T_2O (RBE = 1.7).

	HTO, T_2O		T_2	
	Body tissue	Total body	Body tissue	Total body
Maximum permissible concentrations (40 h week) [$\mu Ci/cm^2$]				
Air	5×10^{-6}	8×10^{-6}	2×10^{-3}	4×10^{-4}
Water	0.1	0.2	–	–
Maximum permissible body burden [μCi]	1×10^{-3}	2×10^{-3}		
Critical organ		Body tissue		Skin
Biological half-life		12 days		–

The tritium inventory estimated for LOTRIT is compared with that for the earlier D-T HIBALL reactor design in Table 3. Significant tritium is stored in the target "factory" plus an inventory occurs throughout the handling/manufacturing process. Other inventories occur in the fuel cycle components and in the blanket and coolant. As shown in Table 3, both the total inventory and the "active" inventory for LOTRIT turn out to be roughly two orders of magnitude lower than for HIBALL. This is a very important result which could more than offset technical difficulties in AFLINT target development.

Table 3. Tritium Inventories in the HIBALL and the LOTRIT Concepts (kg).

	LOTRIT 20000 MW$_f$	HIBALL 8000 MW$_f$
Fuel cycle		
cryopumps	0.0370	0.370
fuel cleanup	0.0041	0.041
isotopic separation	0.0083	0.083
	0.0494	0.494
Blanket		
coolant	0.0001	0.013
structure	0.0120	0.012
	0.0121	0.025
Target manufacture	0.1700	4.100
Storage		
targets	0.1700	4.100
uranium beds	0.1700	4.100
	0.3400	8.200
Total inventory	0.5715 kg	12.819 kg
	2.86x10^{-5} kg/MJ	1.60x10^{-3} kg/MJ
Active inventory	0.0615 kg	0.519 kg
(fuel cycle+blanket)	3.08x10^{-6} kg/MG	6.49x10^{-5} kg/MJ

The importance of a reduction in tritium inventory such as achieved in LOTRIT is illustrated further in Table 4. There the tritium released per day is assumed to scale in proportion to the inventory. The results for LOTRIT are compared to that for a D-T plant in terms of the "personnel risk," i.e. in terms of "man-days lost per year" due to the respective plants. The latter is indicated in man-days lost per year due to the health hazard of tritium. From this table we see that the D-T reactor falls less than a factor of 10 below the "allowable" risk levels currently recognized as acceptable. On the other hand, LOTRIT is roughly a factor of 1,000 below this level. In view of the significant possibility that the allowable tritium risk level will be reduced in the future, the large margin available with the LOTRIT concept is a major advantage.

NUCLEAR-PUMPED-LASER DRIVER WITH D-^{3}He TARGETS

In addition to use of a heavy-ion-beam driver, another important route that AF ICF could follow is the use of a NPL driver[21-23]. Such a driver could easily be employed with neutron "rich" D-T targets. However, Beller, et al.[22] recently proposed a NPL combined with a D-T seeded, D-^{3}He AFLINT type target. Such a system would retain the advantages already discussed for the LOTRIT design while offering a high overall efficiency due to the direct neutron coupling to the driver combined with the direct conversion of the increased charged particle energy for the D-^{3}He target.

Table 4. Risk Reduction from Lowering of Tritium Releases by Two Orders of Magnitude. (Single plant estimates)

Risk $\left[\dfrac{\text{man-days lost}}{\text{year}}\right]$	Tritium release per plant [Ci/day]	
	0.1 (LOTRIT)	10.0 (D-T plant)
Global risk	0.000018–0.001800	0.0018–0.1800
Local risk	0.000060–0.002400	0.0060–0.2400
Maximum individual risk	0.000006–0.000060	0.0006–0.0060

Maximum allowable individual risk: 0.0030 man-days lost per year.
Average population allowable risk: 0.0006 man-days lost per year.

In their study, Beller, et al.[24] assumed use of an oxygen-iodine type NPL[25,26]. An energy flow diagram for this system is shown in Fig. 2. Neutrons from the micro-explosion enter a region containing a mix of oxygen and uranium. The latter is in form of a thin coating of UO_2 or uranium metal. Alternately, micron-sized particles can be used. The charged fission fragments from neutron-induced fissions ionize and excite the oxygen atoms. The free electrons produced then cascade down in energy to excite the oxygen to the O_2 ($^1\Delta$) state.

The efficiency of escape of fission fragments from the fuel element into the oxygen gas (η_{ff}) is about 30 to 40% for thin coatings and about 80% for micron-sized particles. Up to 50% of the fission fragment energy (η_{ex}) is captured in the excitation of oxygen.[23] The $O_2(^1\Delta)$ can then be mixed with iodine for use in a high-energy iodine laser, where the excited state of oxygen induces a population inversion in iodine with a chemical efficiency (η_{ch}) of about 50 to 70 percent.[23-26] The energetic oxygen-iodine mixture then lases in an optical cavity with an efficiency, η_{opt}, of 40 to 50 percent.[27] The total efficiency of this NPL system, η_l, is then the product of these four factors ($\eta_{ff}\,\eta_{ex}\,\eta_{ch}\,\eta_{opt}$). Its value will range from 3 to possibly 14%--a very attractive prospective for an ICF driver. (The total efficiency is normalized by the total thermal energy deposited in the fission blanket, not just the energy of the fission fragments.)

Fission fragments carry about 168 of 200 MeV per fission reaction. Thus a fission rate of one per 14.7-MeV fusion neutron is equivalent to an energy multiplication greater than 10 in the fission blanket. A D^3He-fueled target can be ignited with a 1-percent-by-mass DT seed.[5] With correct isotopic ratios in the seed and fuel, such a target can give energy fractions of 1% DT and 4% DD neutrons (f_n), 25% thermal or x-rays (f_x), and 70% charged particles (f_c). Because of the contribution of 2.45-MeV DD neutrons, the average energy of the neutrons from this particular D^3He target pellet is 9.4 MeV. With $\eta_l = 8\%$, $f_n = 5\%$, and only 0.3 fissions per fusion neutron, a 250-MJ target will produce a 5.5-MJ laser pulse. With a target gain of less than 50, this pulse input will produce 250 MJ in the next target. Additionally, the reduced requirement for fusion neutrons increases the fraction of fusion yield available as charged particles.

The large charged-particle fraction obtained from these targets enables the use of direct electric conversion. Indeed a key feature of this NPL-ICF reactor is that a major fraction (880 MW out of 1150 MW--see Fig. 2) of the net electricity produced is from the direct conversion of charged particle energy. Equally important, electrical energy from the power plant is not re-circulated (and stored) to drive the laser.

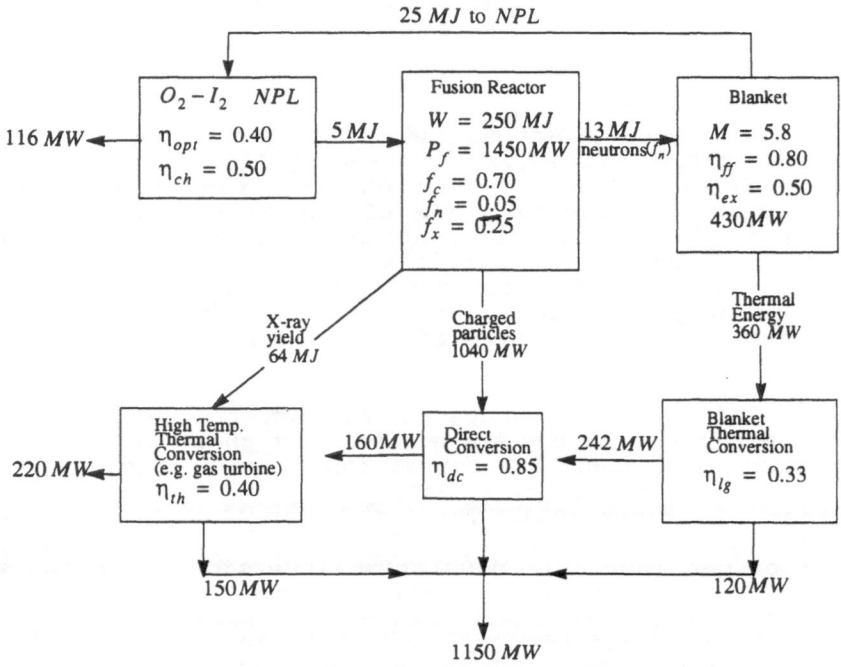

Figure 2. Energy Flow Diagram of the NPL-ICF Power Reactor. Nominal power is 1000 MW_e, and net efficiency with the values shown is 53 percent. Laser efficiency, η_l, with the values in this figure is 8 percent.

The neutron coupling plus energy storage in the $O_2(^1\Delta)$ state serve these purposes. This then avoids the requirement for storing and recirculating large quantities of electric energy which represents a major obstacle for an attractive economic performance in ICF power plants.

The energy flow diagram in Fig. 2 is for a nominal 1000-MWe power plant (with 150 MWe recycled to run auxiliaries). Values for energy conversion factors, energy and power output, and rejected thermal energy are indicated on the figure. (This diagram was used in Ref. 24 to develop energy balance equations. Possible ranges of the various parameters were also discussed and evaluated in that paper.) The net thermal efficiency of the cycle is 53%, assuming 5.8 targets injected per second. A gain of 50 is suggested from a 5-MJ NPL driver having an 8% efficiency, plus the other efficiency values indicated in the figure.

CONCLUSION

In summary, the NPL-ICF reactor concept offers a path to fusion power production with reduced requirements for technological advances in reactor components such as driver systems, injection rate, reactor size, and power conversion equipment[24]. For example, it could operate at an injection frequency of about 6 targets per second, and a gain of only 50. Yet the net efficiency predicted of 53% is quite attractive.

These results suggest that the NPL-ICF could provide economical electrical energy in the future, and its technology may be more easily achieved than anticipated. However, much work remains to evaluate critical issues, including target design, laser coupling, evaluation of auxiliary equipment and power requirements, experimental verification of large-scale nuclear-pumped lasing, demonstration of direct conversion of ICF plasmas, the design of laser optics and cavities, and a detailed economic evaluation.

The goals for the development of a successful fusion power have come to focus solidly on improvements in the safety and environmental aspects of such a plant. The goal for fusion is to achieve a significant advantage in these aspects while at least remaining competitive in terms of economics. This strategy brings to the forefront the importance of developing advanced fuel fusion. Consequently, a fusion confinement approach that can burn AFs (if not in the first generation, in subsequent generations) will gain strong support. The LOTRIT conceptual study results indicate that an AFLINT target leads to a significant reduction in the tritium inventory. However, this advantage must be balanced against the higher ignition energy requirement and the modest gain from AFLINT targets. This requirement can be best handled by selection of an efficient driver such as a heavy-ion beam accelerator, or as discussed here, a nuclear pumped laser. In addition, the energy fraction going to charged particles in AFLINT offers an opportunity for improving the plant efficiency by employing an advanced energy conversion technique.

The NPL-ICF concept also offers other advantages. Not only does this system potentially offer a higher efficiency, but the technology required appears to be a reasonable extrapolation of present experiments. Thus this provides an intriguing route for advanced fuel ICF development.

Acknowledgments

Contributions from various colleagues, especially Drs. H. Hora (U. New South Wales) and D. Beller (Air Force Inst. of Tech.) are gratefully acknowledged. The present manuscript relies heavily on sections of Refs. 5, 18, and 24 in order to consolidate information on advanced fuels for ICF here.

References

1. George H. Miley, Fusion Energy Conversion, American Nuclear Society, 1976.
2. John P. Holdren, et al., "Exploring the Competitive Potential of Magnetic Fusion Energy: The Interaction of Economics with Safety and Environmental Characteristics," Fusion Tech., 13, 7 (1988).
3. David Dickson, "Climate Turns Chilly for European Fusion Program, Science, 241, 154, 1988.
4. G. H. Miley, "The Potential Role of Advanced Fuels in ICF," Laser Inter. and Related Plasma Phenomena, Vol. 5, Plenum Press, 313 (1981).
5. George H. Miley, "Advanced-Fuel Targets for Beam Fusion," Proc. of the 6th Intern. Conf. on High Power Particle Beams (Beams '86), pp. 309-312, Kobe, Japan (1986).
6. T. Weaver, G. Zimmerman, and L. Wood, "Exotic CTR Fuels; Non-Thermal Effects and Laser Fusion Applications," UCRL-4938, Lawrence Livermore National Laboratory (1973).
7. J. D. Gordon, et al., "Evaluation of Proton-Based Fuels for Fusion Power Plants," Report TRW-FRE-007, 1981.
8. H. Hora, G. H. Miley, L. Cicchitelli, G. Kasotakis, P. Pieruschka, R. Stening and T. Weihan, "Limitations of the Pellet Fusion Conditions for Clean Fuel," 13th IAEA Conf. on Plasmas Phys. and Control. Nucl. Fusion Research, Washington, DC, Oct. 1990, Paper IAEA-CN-53/B-3-7.
9. G. H. Miley, et al., "An Advanced Fuel Laser Fusion and Volume Compression of p-^{11}B Laser-Driven Targets," Fusion Tech., 19, 42 (1991).
10. R. Feldbacher, "Nuclear Reaction Cross Sections and Reactivity Parameter Library and Files," INDC(AUS)-12/6, IAEA, Vienna (1987).
11. Bogdan Maglich, "High-energy Fusion: A Quest for a Simple, Small and Environmentally Acceptable Colliding-beam Fusion Power Source," Intern. Conf. on Emerging Concepts in Advanced Nuclear Energy Systems," Graz, Austria, March 29-31, 1978.
12. L. J. Wittenberg, et al., "Lunar Source of ^{3}He for Commercial Fusion Power," Fusion Tech., 10 (1986).
13. Proceedings, Lunar ^{3}He/Fusion Power Workshop, NASA-Lewis, Cleveland, Ohio, April 25-26, 1988. Summary published Fusion Tech., 15, 67 (1989).
14. G. Kulcinski and J. Santarius, "Development Strategy for D-^{3}He Fusion," Proc. Lunar ^{3}He/Fusion Power Workshop, NASA-Lewis, Cleveland, Ohio, April 25-26, 1988.
15. George H. Miley, "^{3}He Sources for D-^{3}He Fusion Power," Rev. of Sci. Instruments, in press.
16. M. Ragheb, et al., "Alternate Approach to ICF with Low Tritium Inventories and High Power Densities," J. of Fusion Energy, 4, 339 (1985).
17. M. Ragheb and G. H. Miley, "Safety Aspects of Tritium in ICF Reactor With Internally-Breeding Targets", Fusion Technology, 8, 2081 (1985).

18. G. H. Miley, "Advanced Fuels for Heavy Ion Beam Fusion," <u>Nucl. Instruments and Methods in Phys. Res.</u>, <u>A278</u>, 281-287 (1989).
19. M. Ragheb, G. H. Miley, & J. Stubbins, "Environmental Considerations of LOTRIT," <u>Trans. ANS.</u>, <u>45</u>, 17 (1983).
20. T. Tazima, "An advanced ICF Reactor 'TAKANAWA-I' With Low Radioactivity and Tritium Self-Breeding in the Pellet," <u>11th International Conference on Plasma and Controlled Nuclear Fusion Research</u>, <u>IAEA</u>, (1986).
21. G. H. Miley, et al., "Fission Reactor Pumped Lasers: History and Prospects," <u>Proc. ANS/NBS Conf.: Fifty Years with Nuclear Fission</u>, pp. 333-342, Gaithersburg, MD, April 26-28, 1989.
22. G. H. Miley, "Neutron Feedback ICF," <u>Atomkernenergie</u>, <u>45</u>, 14 (1984).
23. G. H. Miley, "Nuclear-Pumped Lasers, a Candidate ICF Driver," Fusion Studies Laboratory Report <u>FSL-239</u>, University of Illinois, Urbana, IL (1988).
24. D. E. Beller, F. Whitworth, and G. H. Miley, "Parametric Design Space and Nuclear Analysis of a Nuclear-Pumped-Laser-Driven ICF Reactor," <u>Fusion Tech.</u> <u>15</u>, 772-777 (1989).
25. G. H. Miley and M. S. Zediker, "Nuclear Pumping Oxygen ($^1\Delta$) for an O_2-I_2 Laser," <u>Proc. of the 12th Int. Conf. on Quantum Electronics</u>, June 1982, Munich, Germany (1982).
26. G. H. Miley and M. S. Zediker, "The Nuclear Pumped O_2 ($^1\Delta$)-I_2 Laser," <u>SPIE 335, Advanced Laser Techniques and Applications</u>, 109 (1982).
27. N. G. Basov, M. V. Zagidullin, V. I. Igoshin, V. A. Katulin, N. L. Kupriyanov, "A Theoretical Analysis of Oxygen-Iodine Chemical Lasers," in <u>Proc. of the Lebedev Physics Institute, 171, Research on Laser Theory</u>, A. N. Orayevskiy, ed., Nova Science Publishers, NY (1988), p. 37-71.

DIAMOND PHOTOVOLTAIC CELLS AS A FIRST WALL MATERIAL AND ENERGY CONVERSION SYSTEM FOR INERTIAL CONFINEMENT FUSION

Mark A. Prelas[1,2]

College of Engineering
University of Missouri-Columbia
Columbia, MO 65211

[1]Fusion Research Laboratory, Nuclear Engineering Program
[2] High Bandgap Semiconductor Group, College of Engineering

1.0 Introduction

Diamond technology is a major area of worldwide semiconductor research. It has been said that the current status of diamond semiconductor technology is similar to that of silicon technology in 1960. Most of the research on diamond is in high quality film production (e.g., purity, and single crystalline versus polycrystalline). A few groups are concentrating on the development of diamond electronic devices. In this endeavor, both p-type and n-type diamond films have been produced. The p-type diamond has excellent properties while the n-type diamond is very high resistance. A primitive p-n junction has been demonstrated. Several groups have demonstrated Schottky diodes including the high bandgap semiconductor group at the University of Missouri-Columbia.

Diamond technology has some major advantages over other materials. It is the hardest material known, it has the strongest bonds of any know compound, and it is the most radiation resistant material known. As a film, diamond would be a superior material because of it resistance to damage. For use in fusion devices, diamond would be an excellent first wall material. It will have the lowest sputtering rate of any known material.

This paper introduces a basic concept in which diamond photovoltaic cells serve as both a first wall material and photovoltaic cells for direct energy conversion of light generated by the interaction of fusion products with fluorescer materials. The products from fusion reactions (i.e., ions and neutrons) are used to excite fluorescers then the fluorescence is used to drive the first wall photoelectric cells (See Prelas, 1981).

Several methods of efficiently generating fluorescence with inertial confinement fusion are possible however, for simplicity, this paper will focus on the use of excimers. As will be described, it is possible to achieve energy conversion efficiencies with narrow band ultraviolet fluorescers of 50% (see Prelas, Boody, Kunze, and Miley 1988, Prelas and Charlson 1989, and Prelas, Boody, Charlson, and Kunze 1990). The advantage of high energy conversion efficiencies is significant. It will substantially reduce the requirements on laser efficiency, and on laser size. It will also increase the efficiency of the power plant.

2.0 Energy Conversion

Nuclear technology has been in search of a technologically/economically feasible method of converting energy of nuclear reaction products directly into electricity for many years (Prelas, Boody, Kunze, and Miley 1988). The energy resulting from fusion reactions for example starts out as very high grade energy: the multi-MeV, charged particles and neutrons.

Thermoelectric generation of electricity using the Seebeck (thermocouple) effect has worked successfully for a number of power packages for satellites. Thermionic production of electricity directly inside reactor cores is again being pursued for space power applications. Fluorescence produced from the charged particles from the decay of radioactive isotopes has been used to produce low intensity lights for remote applications, such as runway lights. However, all of these applications involve relatively low power output.

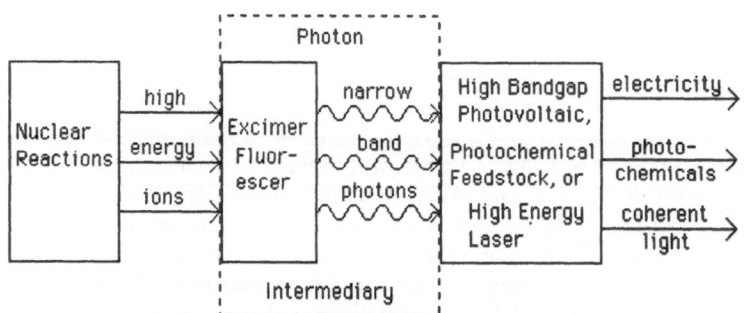

FIGURE 1.   Schematic Diagram of the Photon-Intermediate Direct Energy Conversion (PIDEC) Process.

2.1 PIDEC

The Photon-Intermediate Direct Energy Conversion (PIDEC) process, Figure 1, was first conceived for advanced-fuel high temperature (plasma) fusion (Prelas 1981) which, unlike conventional D-T high temperature fusion, releases its energy as ions, rather than as neutrons. Advanced-fuel high temperature fusion, however, will not be available in the foreseeable future. Narrow band light generation through interactions with both energetic ions and neutrons can be applied to D-T fusion however (Prelas, Romero, and Pierson 1982, and Pappas 1988).

Fusion reactions produce ions and neutrons as shown below:

$$D(T,n)He^4 + 17.6 \text{ MeV} \qquad \text{(eq. 1)}$$
$$\{n(14.08 \text{ MeV}) \ \& \ He^4 \ (3.52 \text{ MeV})\}$$

$$D(D,n)He^3 + 3.27 \text{ MeV (50\% branching)} \qquad \text{(eq. 2)}$$
$$\{n(2.45 \text{ MeV}) \ \& \ He^3 \ (0.82 \text{ MeV})\}$$

$$D(D,p)T + 4.03 \text{ MeV (50\% branching)} \qquad \text{(eq. 3)}$$
$$\{p(3.02 \text{ MeV}) \ \& \ T \ (1.01 \text{ MeV})\}$$

The products from fusion reactions can be used to excite laser media. This paper will focus on one of the many possible methods of exciting a fluorescer with fusion products; excimer flashlamps (Prelas, Boody, Woodall, Speziale, and Wills 1987, Prelas, and Charlson 1987, and Prelas and Boody 1991). Excimers which produce fluorescence within the absorption bands of advanced high bandgap photovoltaic materials can be excited by charged particles (Prelas, 1981, Prelas, Boody, Kunze, and Miley 1988). Neutrons can also be used to transfer energy to liquid or solid media through elastic scattering, inelastic scattering, and neutron capture -with subsequent charged particle, gamma ray, or neutron emission- (Prelas, Romero, and Pierson, 1982). Elastic scattering was examined in detail by Pappas, 1988, as a method of exciting liquid xenon excimers.

The steps in the excimer energy conversion process to be considered are: 1) η_t the efficiency of the transport of the ion and neutron energy from the fuel into the fluorescer gas, 2) η_f the fluorescence generation efficiency, 3) η_c the photon coupling efficiency to material used in the convertor, 4) η_p the useful product (electricity, chemicals, or laser photons) generation efficiency from the converter material per photon absorbed, and 5) η_{ex} the extraction efficiency i.e. the extractable electrical energy (or other desirable energy form). (Processes 1, 2 and 3 represent the nuclear driven flashlamp and processes 4 and 5 represent the photon energy conversion module.) Hence, the system efficiency will be (Prelas 1981, Prelas, and Charlson 1987, Prelas, Boody, Kunze, and Miley 1988, and Prelas, Boody, Charlson and Miley 1991):

$$\eta_s = \eta_t \, \eta_f \, \eta_c \, \eta_p \, \eta_{ex} \qquad \text{(eq. 4)}$$

and the power density deposited in the photon energy conversion material is:

$$Pd_p = Pd_f \, \eta_f \, \eta_c \, V_f/V_p \qquad \text{(eq. 5)}$$

where Pd_f is the fission fragment power deposition in the fluorescer region (W/cm³), V_f is the volume of the fluorescer (cm³), and V_p is the volume of the converter material (cm³).

Similarly the fluorescence intensity on the surface of the photon energy converter material is:

$$I_p = Pd_f \, \eta_f \, \eta_c{'} \, V_f/A_p \qquad \text{(eq. 6)}$$

where $\eta_c{'}$ is the efficiency for coupling the fluorescence to the converter material, and A_p is the area of the converter material (cm²). (Note, the volume coupling efficiency, η_c, and the surface coupling efficiency, $\eta_c{'}$, will differ, with $\eta_c{'} > \eta_c$.)

There are may advantages in using the PIDEC processes over many step processes such as the conventional steam cycle. These advantages are: 1) that it is a direct process producing a useful energy form from high grade energy and thus avoiding the Carnot cycle efficiency limits imposed by thermalization and 2) that it is much simpler, potentially leading to more compact, more reliable, and less expensive energy conversion systems.

The advantage of the PIDEC process over a one-step direct energy conversion process is that of feasibility. The scale length for the transport of the primary high-grade energy must match the geometrical scale of the energy converter (see Figure 2). Energetic ions have a transport length of micrometers while useful energy converters, on the other hand, have a scale length of fractions of meters. For this reason direct conversion of nuclear energy has not previously been possible. What was required was the concept of an intermediate high-level energy converter that can be intermingled with nuclear material on a micrometer scale-

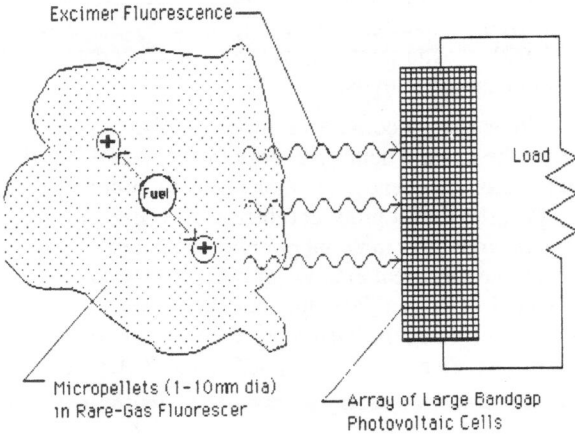

FIGURE 2. Schematic Diagram of the Photon Intermediate Direct Energy Conversion System which uses fusion generated ions.

length but produces an energy form that can be transported to meter scale-length direct converters producing useful output – a sort of "impedance matching" for scale length of energy forms. With PIDEC that scale length matching medium is a fluorescing gas, the nuclear-driven fluorescer. The photons it produces can be transported great distances, making it possible to couple them to various energy conversion processes. Also, some conversion processes require greater power densities than the primary energy sources can provide. A PIDEC's intermediate photon flux can be absorbed in a smaller volume than that in which it is produced, effectively concentrating the flux, enabling achievement of the high threshold power density for such conversion processes.

2.2 Nuclear-Driven Fluorescers

2.2.1 Excimer Fluorescers

Excimer fluorescers are the most efficient radiators known and, because of their unbound lower levels, do not self absorb. They radiate in the single, relatively narrow, band of wavelengths required for efficient photovoltaic energy conversion (see Section 2.3). The intrinsic fluorescence efficiencies of rare-gas and rare-gas halide excimers, based on standard W-value theory, are listed in Table 1. Achievable efficiencies should be near the intrinsic values at the power and electron densities characteristic of nuclear reactions.

Table 1. Theoretical Maximum Excimer Fluorescence Efficiencies.

Excimer	λ (nm)	η_{max}	Excimer	λ (nm)	η_{max}
NeF^*	108	0.43	$XeCl^*$	308	0.27
Ar_2	129	0.50	I_2	343	0.24
Kr_2	147	0.47	XeF^*	346	0.24
F_2	158	0.44	Kr_2F^*	415	0.17
Xe_2	172	0.48	Na_2^*	437	0.46
$ArCl^*$	175	0.48	HgI^*	443	0.19
KrI^*	185	0.37	Li_2^*	459	0.42
ArF^*	193	0.35	$HgBr^*$	502	0.17
$KrBr^*$	206	0.33	XeO^*	547	0.15
$KrCl^*$	222	0.31	$HgCl^*$	558	0.15
KrF^*	249	0.28	K_2^*	575	0.42
XeI^*	253	0.37	Rb_2^*	605	0.41
Cl_2	258	0.32	CdI^*	655	0.13
$XeBr^*$	282	0.29	Cs_2^*	713	0.37
Br_2	290	0.29	$CdBr^*$	811	0.10

In fact one group has reported measuring a nuclear-driven rare-gas excimer fluorescence efficiency higher than that predicted by W-value theory (Prelas, Boody, Kunze, and Miley 1988). Measurements of actual fluorescence efficiencies at various laboratories, including Lawrence Livermore National Laboratory, have demonstrated high fluorescence efficiencies for excimers. Experiments with a variety of excitation sources (e.g. electrons, fission fragments, protons) and particle densities have given fluorescence efficiency values ranging from a few percent to as high as 68% (see review paper by Prelas, Boody, Kunze, and Miley 1988). The most efficient excimer fluorescers are the rare-gas excimers.

2.3 The Photon Energy Converter

The key to the feasibility of this concept is the photovoltaic Photon Energy Converter. The common impression of photovoltaics is that they cannot be very efficient. This misunderstanding comes from the fact that photovoltaics are most commonly employed as "solar cells." And solar cells are not very efficient, ranging from 10-20% for commercial units and reaching as high as about 25% for laboratory cells. However the low efficiency is more due to the characteristics of the solar spectrum than to the photovoltaics devices themselves, especially for the laboratory units with efficiencies of ~25%. The problem with the solar spectrum is that it is very broadband — its ratio of the average photon energy to the width (FWHM) of the spectrum ($E_{mean}/\Delta E$) is about 1. This is good for color vision but quite bad for efficient energy conversion. For excimers, however, this ratio is greater than 10. Under these conditions photovoltaics have intrinsic efficiencies of 75-95%.

Photovoltaic cells for use in photon-intermediate direct energy conversion of electricity will require the development of a doped semiconductor material with a bandgap that matches the UV radiation. With such photovoltaic cells, a system efficiency of 56% for fusion ion driven fluorescence has been projected (Prelas 1981). Studies of fission ion driven fluorescence indicate that system efficiencies of about 40% are possible (Prelas, Boody, Charlson, and Miley 1990).

2.3.1 Photovoltaic Conversion of Narrowband Fluorescence

For a given spectrum, the efficiency of conversion is basically determined by the variation of the irradiance with photon energy and by the substrate bandgap energy, E_g, of the photovoltaic converter. Complete conversion (100%) is not possible because of the width of the solar spectrum. This leads to two competing effects on the efficiency. The first effect is that the energy of all photons with quantum energy $h\nu < E_g$ is lost because they do not have sufficient energy to excite electrons from the valence band to the conduction band. The power density lost in this case is given by

$$P_{lost} = \int_0^{E_g} W(E)\, dE \qquad \text{(eq. 7)}$$

where W(E) is the irradiance in W/cm²/eV. Thus, the lower the bandgap of the photovoltaic converter, the larger the fraction of the total spectrum converted. Competing with this effect however is the fact that, for the photons with quantum energy $h\nu > E_g$ that do contribute, the photon energy in excess of the bandgap energy is lost. Thus the maximum intrinsic efficiency for photovoltaic conversion is

$$\eta_{in} = \frac{\int_{E_g}^{\infty} (W(E)\, \frac{E_g}{E}\, dE)}{\int_{E_g}^{\infty} W(E)\, dE} \qquad \text{(eq. 8)}$$

assuming an ideal collection device.

These two effects are shown graphically in Figure 3 for broadband (solar) and in Figure 4 for narrowband (Xe$_2$*) radiation. In Figure 3 it can be seen that for a broadband spectrum, such as the solar spectrum, there is no value of E$_g$ that results in most of the energy being converted. In Figure 4, on the other hand, it can be seen that, for E$_g$ just less than E$_{min}$, efficient energy conversion takes place.

Typically the effect of the details of the solar radiation spectrum on calculating overall conversion efficiency is translated into a photon flux density, which then relates to an ideal short circuit current density. This is convenient because it is a good assumption that each photon absorbed and collected effectively causes one electron to move around the circuit. Also after each electron thermalizes, that is, gives off energy in excess of E$_g$ to the lattice, it contributes maximally a constant E$_g$ in energy to the overall process.

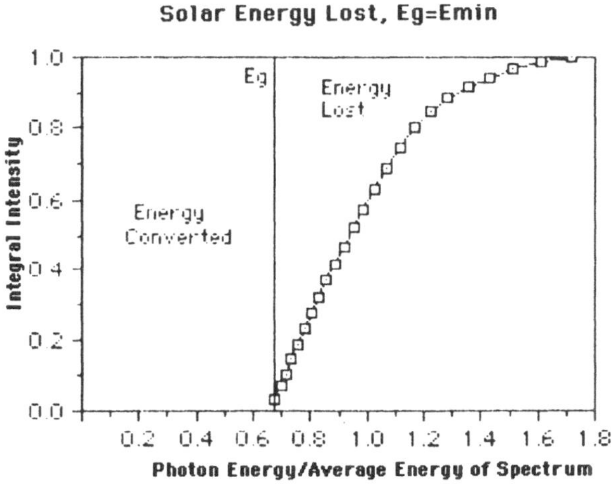

Solar Energy Lost, Eg=Emin

FIGURE 3.   Efficiency of Photovoltaic Conversion of Broadband Solar Radiation Using Photovoltaics with E$_g$=E$_{min}$. Data points are fraction of total number of photons with an energy greater than E versus photon energy normalized to the average photon energy.

This contribution is conveniently modeled in the photovoltaic device using the ideal Shockley model for the p-n junction. Using these concepts the intrinsic conversion efficiency can be written in the following terms,

$$\eta_{in} = \frac{E_g \int_{E_g} N_{ph}(E)\, dE}{\int_{E_g} N_{ph}(E)\, E\, dE} = \frac{E_g}{E_{mean}} \frac{N(E > E_g)}{N_{tot}} = \frac{E_g}{E_{mean}} \eta_{eg} \qquad \text{(eq. 9)}$$

where N$_{ph}$ is the photon flux density, in #/s-cm^2-eV, N(E>E$_g$) is the photon flux in the interval E>E$_g$, N$_{tot}$ is the total photon flux, and η_{eg} is the fraction of photons with E>E$_g$.

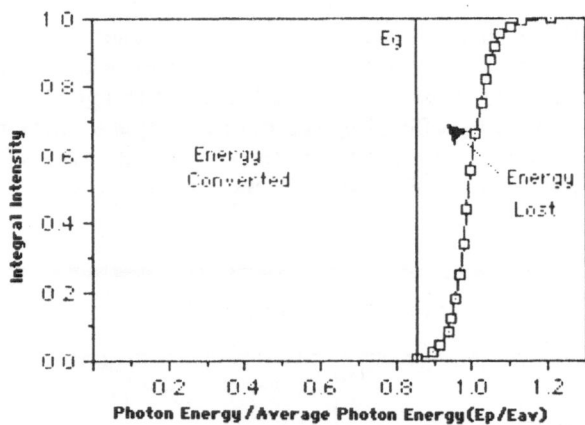

Xe2* Energy Lost, Eg=Emin

FIGURE 4. Efficiency of Photovoltaic Conversion of Narrowband Excimer Radiation Using High Bandgap Photovoltaics. Data points are fraction of total number of photons with an energy greater than E versus photon energy normalized to the average photon energy.

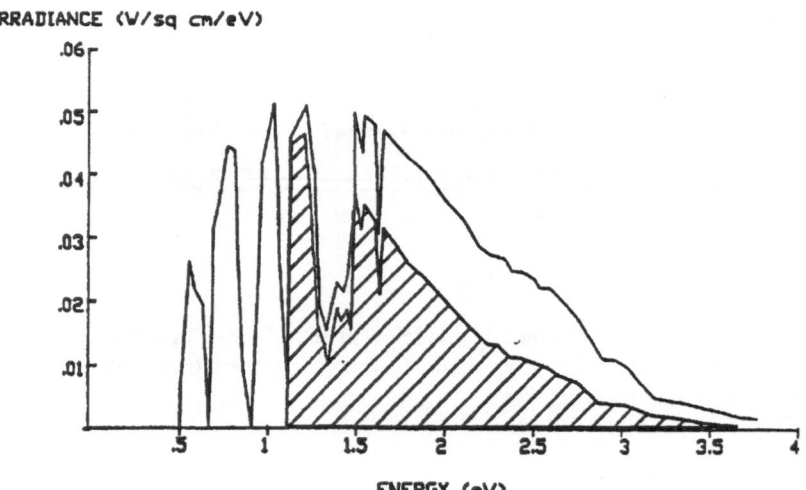

FIGURE 5. The spectral irradiance of the solar spectrum (AM2). The cross hatched area under the curve is the fraction of energy which is usable with a 1.1 eV bandgap photovoltaic cell (taken from Prelas & Charlson 1989).

410

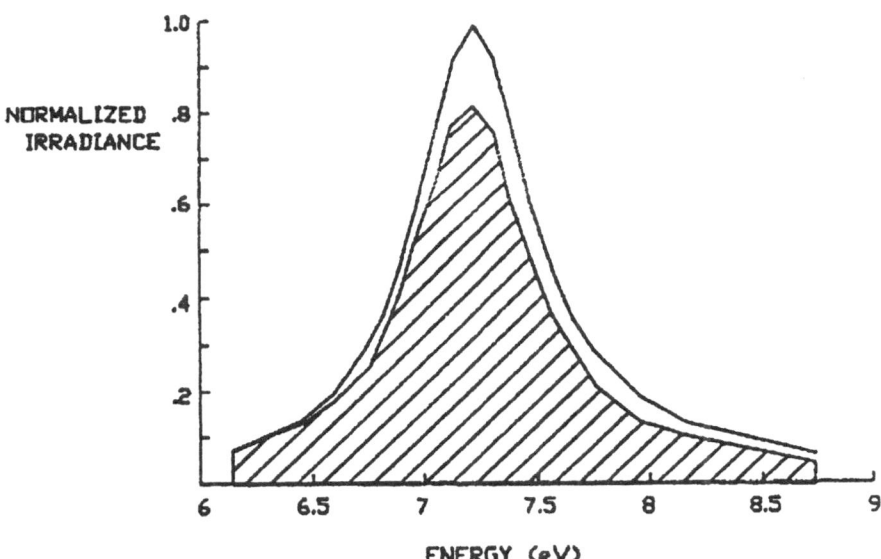

FIGURE 6.   The normalized irradiance of Xe$_2$* excimer source. The cross hatched area under the curve is the fraction of energy which is usable with a 6 eV bandgap photovoltaic cell (taken from Prelas & Charlson 1989).

The solid lines in Figure 5 and Figure 6 show, respectively, the air mass 2 (AM2) irradiance of the solar spectrum (W/cm^2/eV) (Brandhorst et. al. 1975)and the irradiance for Xe$_2$*, a narrowband fluorescer. For Xe$_2$*, E$_{mean}$/ΔE = 14, compared to a corresponding value of 1.3 for the AM2 solar spectrum. For a narrow distribution one can have E$_g$/E$_{mean}$~1 and still have η_{EG}~1. A narrowband spectrum will consequently have the highest intrinsic efficiency. The crosshatched areas in Figures 5 and 6 represent, respectively, the power density converted for the solar spectrum and for the Xe$_2$* spectrum. The ratio of the crosshatched area to the total area in both cases gives the intrinsic efficiency. The white area under each curve corresponds to unconverted power. This figure illustrates that a large portion of the energy in a broad band source, such as a solar source, is wasted while a much smaller fraction of the energy in a narrow band source, Xe$_2$*, is wasted. Figure.7 is a plot of maximum efficiencies for a p-n junction converter versus bandgap energy of the converter substrate material. Two plots are shown, one for an AM2 solar spectrum and one for a Xe$_2$* spectrum. The previously derived equations were used to calculate these curves. Also shown are vertical lines representing the bandgap energies of the two materials theoretically predicted to maximally convert these two spectra. Lines representing Si and diamond are shown for comparison. The curve for the solar spectrum lies in between the two best known curves of this type (Loferski,= 1956, Wysocki and Rappaport 1960). The approximately 30% maximum is thought to be an upper bound on the ability of a single material junction to convert the solar spectrum. The highest known conversion efficiency for silicon, to date, has been 26% obtained with a highly optimized MIS solar cell (Green et. al. 1984).

In contrast to the relatively low values for conversion of the solar spectrum, it can be seen from Figure 7 that efficiencies as high as 80% can theoretically be obtained using a p-n junction and converting the Xe_2^* spectrum. Although it is still speculative about whether or not high quality p-n junctions can be made in materials with bandgaps above 4 eV, it is encouraging to note that high conversion efficiencies are possible.

1.3.2 High Bandgap Photovoltaic Materials

Table 2 lists several potential high bandgap materials. None of them has been studied for photovoltaic applications. One material, diamond (5.5 eV), has recently received intense study as a microelectronic device substrate. However, the most efficient excimer UV radiation sources, the rare-gas excimers, have larger photon energies (>7 eV) and hence the development of higher bandgap photocells for these excimers would be desirable. Table 3 matches the more efficient and desirable fluorescers from Table 1 to materials with appro-

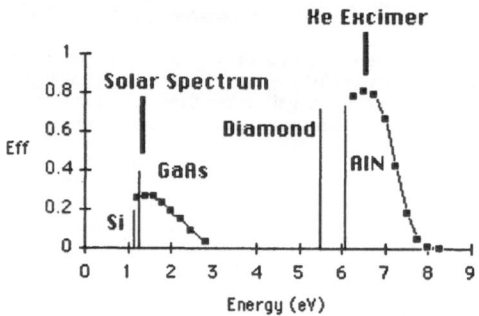

Figure 7.   The theoretical maximum conversion efficiency of photovoltaic cells with various bandgap energies using either a solar spectrum or a Xe_2^* spectrum. Bandgap energies associated with Si, GaAs, diamond, and AlN are shown (taken from Prelas & Charlson 1989).

priate bandgaps from Table 2. The theoretical maximum intrinsic photovoltaic efficiency (the ratio of the bandgap to the mean photon energy) ranges from 75% to 95%) while the corresponding theoretical maximum efficiency for conversion of ion energy to electrical energy (the product of the photovoltaic efficiency and the fluorescence efficiency) ranges from 30% to 45%. If the most optimistic reported values of the fluorescence efficiency were used, the maximum ion-to-electric efficiency would increase to 56%. The outlook for such cells is hopeful since solid-state physicists have focused primarily on materials with bandgaps in the few eV range and have basically ignored the larger bandgap energy ranges. Rare-gas halide excimers have lower photon energies (3.5 eV for XeF*, 5.0 eV for KrF*, and 6.4 eV for ArF*) and, while their fluorescence efficiency may be lower than that of the rare-gas excimers, their photon energy falls in the range of known semiconductor materials.

Radiation damage to the photovoltaics from X-rays, gama rays, charged particles, and neutrons is a concern. High bandgap materials are the most radiation resistant materials known (specifically diamond). Additionally, if a radiation damaged crystal is thermally annealed, the damage disappears. In an operating fusion reactor, the cells must operate a temperatures high enough to cause annealing.

Table 2. Bandgap energies for various semiconductors and insulators.

Material	Bandgap energy (eV)	Melting Temp. (K)
Si	1.1	1685
CdSe	1.44	1370
ZnSe	2.26	1510
ZnS	3.54	1920
SrS	4.1	3000
Ga_2O_3	4.6	1750
UO_2	5.2	3150
C (diamond)	5.4	4300
AlN	6.02	2500
NaCl	7.3	1074
Al_2O_3	7.4	2302
MgO	8	3173
CaF_2	10	1633
LiF	12	1143

* Graphitization Temp.

The question, then, is whether materials can be found with the appropriate bandgap that can be made into photovoltaic cells. Specific work has been done at the University of Missouri to develop high bandgap cells. For example, work has been done which has demonstrated a Schottky barrier diode with diamond and various metal coatings (Zhao, et. al., 1992). The diode has responded to light with frequencies less than 233 nm.

Energy Conversion Cycle for Diamond Walled ICF Reactor

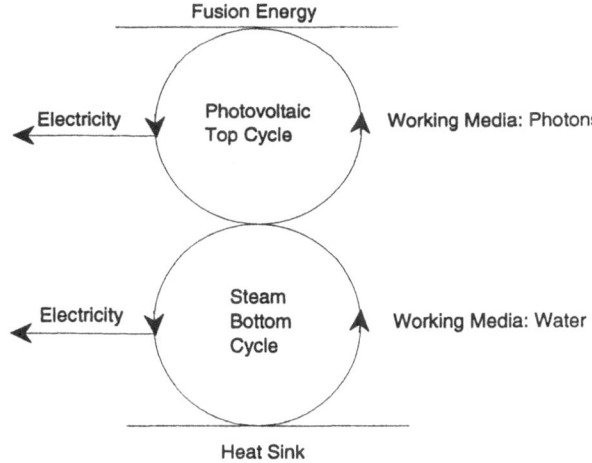

FIGURE 8.   Energy conversion cycles in an ICF reactor using a diamond wall with photovoltaic capability and steam cycle capability. This notation for the cycle is taken from the paper from Prelas, Romero, and Peirson 1982.

Table 3. Theoretical Maximum Intrinsic Photovoltaic, η_{pv}, and Ion-to-Electric, η_{ie}, Efficiencies for Selected Rare-Gas and Rare-Gas Halide Excimer Fluorescers with Matched High-Bandgap Photovoltaic Materials.

Excimer	η_f Fluorescence Efficiency	Energy E_I (eV)	Photovoltaic Material	Bandgap Energy E_g (eV)	$\eta_{pv} = E_g/E_I$	$\eta_{ie} = \eta_{pv} \times \eta_f$
Ar_2^*	0.50	9.6	MgO	8.0	0.83	0.42
			Al_2O_3	7.4	0.77	0.39
			NaCl	7.3	0.76	0.38
Kr_2^*	0.47	8.4	MgO	8.0	0.95	0.45
			Al_2O_3	7.4	0.88	0.41
			NaCl	7.3	0.87	0.41
			AlN	6.0	0.72	0.34
			Diamond	5.5	0.65	0.31
F_2^*	0.44	7.8	Al_2O_3	7.4	0.95	0.42
			NaCl	7.3	0.94	0.41
			AlN	6.0	0.77	0.34
			Diamond	5.5	0.71	0.31
Xe_2^*	0.48	7.2	AlN	6.0	0.83	0.40
			Diamond	5.5	0.76	0.37
ArF*	0.35	6.4	AlN	6.0	0.94	0.33
			Diamond	5.5	0.86	0.30
KrBr*	0.33	6.0	Diamond	5.5	0.92	0.30
KrCl*	0.31	5.6	Diamond	5.5	0.98	0.30
Na_2^*	0.46	2.84	ZnSe	2.7	0.95	0.44
			SiC	2.4	0.845	0.39
Li_2^*	0.42	2.7	$CuAlSe_2$	2.6	0.96	0.40
			SiC	2.4	0.89	0.37
Hg_2^*	0.21	2.58	GaS	2.5	0.97	0.20
			SiC	2.4	0.93	0.19
ArO*	0.11	2.27	GaP	2.2	0.97	0.105
			GaAlAs	2.2	0.97	0.105

3.0 Diamond as a First Wall Material and Energy Conversion System

A conceptual inertial confinement fusion system in which diamond is both a first wall material and energy conversion system is described. Lasers initiate the fusion event in which charged particles and neutrons are generated. The fusion products interact with a high density fluorescer media and creates a narrow band spectra of light. The light and the remaining radiation energy interacts with the wall. Electricity is produced through the interaction of light with the photovoltaic material. In addition, the remainder of the energy heats the wall surface. If a system employing diamond and a fluorescer such as Xe_2^*, the conversion process from fusion to electricity can be 37% efficient as shown in Table 3. The remainder of the energy goes into heating the wall. Temperatures of the wall can be as high as 700 $^\circ$K with a material such as diamond; work on diamond Schottky barrier diodes at the University of Mis-

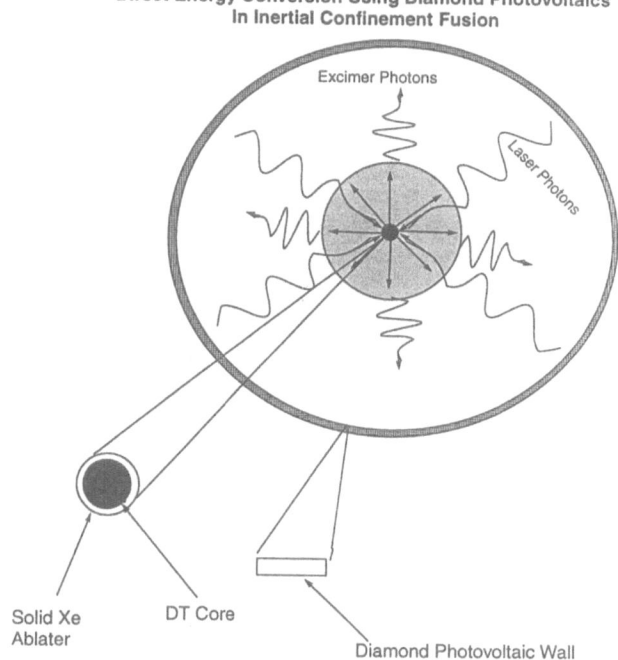

Direct Energy Conversion Using Diamond Photovoltaics in Inertial Confinement Fusion

FIGURE 9.   Illustration of the diamond walled ICF concept.

Excimer	η_f Fluorescence Efficiency	Energy E_l (eV)	Photovoltaic Material	Bandgap Energy E_g (eV)	$\eta_{pv}= E_g/E_l$	$\eta_{ie}=\eta_{pv}x\eta_f$
KrO*	0.13	2.27	GaP	2.2	0.97	0.125
			GaAlAs	2.2	0.97	0.125
XeO*	0.15	2.27	GaP	2.2	0.97	0.145
			GaAlAs	2.2	0.97	0.145

souri-Columbia has demonstrated that the diode performs best at temperatures of about 700 °K (Zhao, et. al. 1992). Thus, a steam cycle or thermoelectric cycle can be used in conjunction with the photovoltaic cycle (see Figure 8).

The ICF concept using diamond photovoltaics is shown in Figure 9. Here it is assumed that the fluorescer media is an excimer. The excimer is placed into the system by developing a special ICF target (Prelas 1988 and Prelas and Charlson 1989). Other methods of generating narrow band light efficiently in the ICF target are possible but not discussed in this paper (Wallace 1991).

In using a topping and bottoming energy conversion cycle, it is possible to produce electricity at efficiencies of over 50%:

$$\eta_{plant} = \eta_{photovoltaic} + \eta_{steam} \times (1 - \eta_{photovoltaic}) \qquad \text{(eq. 10)}$$

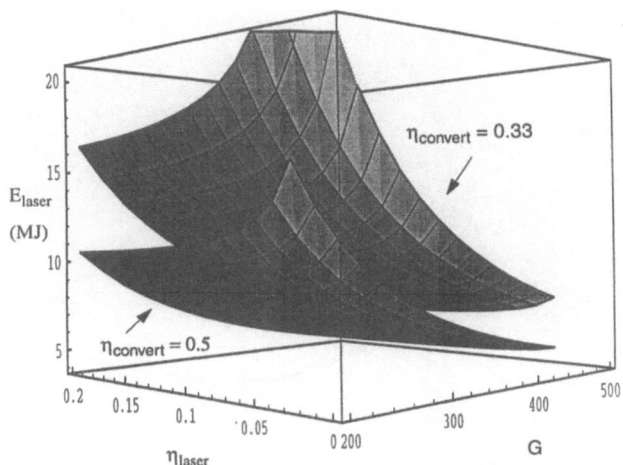

FIGURE 10.   As can be seen in the figure, the required laser energy for an ICF plant capable of producing 1 Giga Watt electric using either a 33% or 50% energy conversion system is shown as a function of the laser efficiency and pellet gain. (See companion paper by Prelas and Boody in this volume of Laser Interactions and Related Plasma Phenomena for description of the model.)

For a photovoltaic conversion efficiency of 37% and a steam cycle conversion efficiency of 33% gives a plant efficiency of 58% from equation 10. A plant efficiency greater than the standard steam cycle efficiency of 33% makes a significant difference in the laser parameters and the plant efficiency of an inertial confinement fusion plant (see Figures 10 and 11). As can be seen in the figures, with a pellet gain of 200 and a laser efficiency of 5%, the required laser energy for a plant with an energy conversion efficiency of 33% is 19.5 MJ as compared to a plant with an energy conversion efficiency of 50% having a required laser energy of 11 MJ. Conversely, the plant efficiency for the two cases is 31% and 48% respectively (where plant efficiency is electrical power/fusion power).

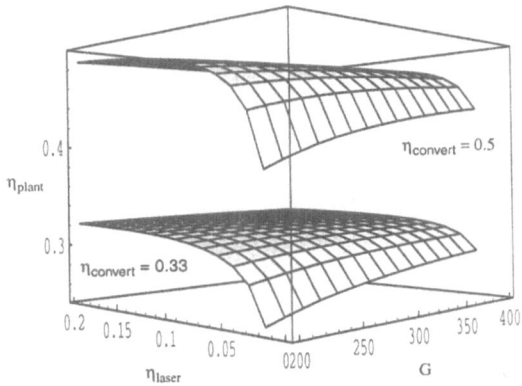

FIGURE 11.   As can be seen in the figure, the plant efficiency for an ICF plant capable of producing 1 Giga Watt electric using either a 33% or 50% energy conversion system is shown as a function of the laser efficiency and pellet gain. Plant efficiency is defined as electric power/fusion power. (See companion paper by Prelas and Boody in this volume of Laser Interactions and Related Plasma Phenomena for description of the model.)

4.0 Conclusions

In conclusion, a diamond first wall offers substantial advantages for inertial confinement fusion systems. Diamond is the most radiation resistant material known and thus would offer substantial advantages as a first wall material. Second, diamond can be made into photovoltaic cells (Prelas 1981, Prelas, 1988, Prelas & Charlson 1989, Prelas, Boody, Charlson, and Miley 1990, and Zhao, et. al. 1992) and could also serve as a photovoltaic energy conversion system. If a narrow band fluorescer could be excited by the fusion products, then very high energy conversion efficiencies are possible (up to 58%). Energy conversion efficiencies greater than that of the standard steam cycle are vary favorable for the scaling of inertial confinement fusion power plants (Prelas & Boody, 1991). As seen in Figures 10 and 11, the scaling of required laser power and power plant efficiencies is favorable. When the energy conversion efficiency is high, the dependence upon laser efficiency is substantially reduced. This is very significant since driver efficiency is one of the limiting factors in inertial confinement fusion.

5.0 Acknowledgments

The author wishes to thank his colleagues Prof. Joe Charlson, Prof. Elaine Charlson, Prof. Jon Meese, and Professor Galina Povovici of the University of Missouri and Dr. Fred Boody of Ion Light Corp. for many fruitful discussions. Additionally, the author is grateful to the Department of Energy's Advanced Energy Program for support of this work.

6.0 References

F. P. Boody, "Remote Pumping of Solid-State Lasers Pumped by Remotely-Located Nuclear-Driven Fluorescers," Ph.D. Dissertation, Nuclear Engineering Program, University of Missouri-Columbia, August 1991.

F. P. Boody, and M. A. Prelas, "Very High Average Power Solid-State Lasers Pumped by Remotely-Located Nuclear-Driven Fluorescers", *Advanced Solid-State Lasers*, Optical Society of America (To be Published 1991)

Brandhorst H., Hickey J., Curtis H., Ralph E. (1975), "Interim Solar Cell Testing Procedures for Terrestrial Applications," NASA TM X-71771.

E. J. Caine and E. J. Charlson, "Junction solar cells made with molecular beam glow discharge bombardment," J. Elect. Mat'ls., 13, (2), 1984.

F. G. Celii, P. E. Pehrsson, H. Wang and J. E. Butler, "Infrared detection of gaseous species during the filament-assisted growth of diamond," Appl. Phys. Lett., 52 (24), June 1988.

C. K. Chen, B. Nechay and B. Tsaur, "Ultraviolet, Visible, and Infrared Response of PtSi Schottky Barrier Detectors Operated in the Front Illuminated Mode", IEEE Trans. on Electron Dev. 38, 1094 (1991).

R. F. Davis, Z. Sitar, B. E. Williams, H. S. Kong, H. J. Kim, J. W. Palmour, J. A. Edmond, J. Ry u, J. T. Glass and C. H. Carter, Jr., "Critical Evaluation of the Status of the Areas for Future Research Regarding the Wide Band Gap Semiconductors Diamond, Gallium Nitride and Silicon Carbide", Materials Science and Engineering B1, 77 (1988).

Green M., Blakers A., Shi J., Keller E., Wenham S. (1984), "19.1% Efficient Silicon Solar Cell," Appl. Phys. Lett. 44, (12).

Gu G., Kunze J. F., Boody F. P., and Prelas M. A., "A UF_6 fueled Visible Nuclear-Pumped Flashlamp", Space Nuclear Power Systems 1988, M. S. El-Genk and M. Hoover, editors, Orbit Book Company, 153-160 (1989);

B. C. Johnson, J. M. Meese, G. W. Zajac, J. O. Schreiner, J. A. Kaduk and T. H. Fleisch, "Characterization and Growth of SiC Epilayers on Si Substrates", J. Superlattices and Microstructures 2, 223 (1986).

Y. Koide, N. Itoh, K. Itoh, N. Sawaki and I. Akasaki, "Effect of AlN Buffer Layer 'teroepitaxial Growth by Metalorganic Vapor Phase Epitaxy", Jap. J. Appl. Phys. 27, 1156 (1988).

W. F. Kosonocky, F. N. Shallcross, T. S. Villani and J. V. Groppe, "160 x 244 Element PtSi Schottky Barrier IR CCD Image Sensor", IEEE Trans. on Electron Dev. ED™32, 1564 (1985).

K. Kurihara, K. Sasaki, M. Kawarada and N. Koshino, "High rate synthesis of diamond by DC plasma jet chemical vapor deposition," Appl. Phys. Lett., 52 (6), Feb. 1988.

Loferski J. (1956), J. Appl. Physics, 27, p 777.

M. R. Melloch, S. P. Tobin, T. B. Stellwag, C. Bajgar, A. Keshavarzi, M. S. Lundstrom, K. Emery, "High efficiency GaAs solar cells grown by molecular©beam epitaxy", J. Vac. Sci. Technology B8, 379 (1990).

G. H. Miley, Direct Conversion of Nuclear Radiation Energy, American Nuclear Society, Hinsdale (Ill.)1970.

K. Okano, H. Kiyota, et. al., "Fabrication of a Diamond p-n junction diode using the Chemical Vapour Deposition Technique," Solid-State Electronics, 34(2), 139 (1991).

D. Pappas, Seminar on Fusion Neutrons As an Excitation source For a Xe Excimer Laser, US-Japan Seminar on Laser Fusion, Honolulu Hawaii, August, 1988.

M. A. Prelas, "A Potential Fusion Light Bulb For Energy Conversion," Poster Paper APS Meeting on Plasma Physics October 11-16, 1981, Bull. Am. Phys. Soc., 26(7), 1045 (1981); APS Press Release picked up by AP and story reported by various newspapers worldwide (See for example Washington Missourian, Washington, Missouri, Wednesday. January 6, 1982, page 7 or Inside R & D Vol. 10, Number 41, October 14, 1981).

M. A. Prelas, J. B. Romero, and E. F. Pearson, "A Critical Review of Fusion Systems For Radiolytic Conversion of inorganics to Gaseous Fuels," Nuclear Technology/Fusion, Vol. 2, 143-164 (1982)

M. A. Prelas, F. P. Boody, J. F. Kunze, and G. H. Miley, "Nuclear-Driven Flashlamps", Lasers and Particle Beams, 6(1), 25,(1988).

M. A. Prelas, "Synergism in Inertial Confinement Fusion: A Total Direct Energy Conversion Package," US-Japan Seminar on Laser Fusion, Honolulu Hawaii, August, 1988.

M. A. Prelas and E. J. Charlson, "Synergism in Inertial Confinement Fusion: A Total Direct Energy Conversion Package," Lasers and Particle Beams, Vol. 7 (3), 449-466 (Aug 1989); .

M. A. Prelas, and F. P. Boody, "Nuclear-Driven Solid-State Lasers for Inertial Confinement Fusion", Laser Interactions & Related Plasma Phenomena, Vol 9, Plenum Press, New York, 197-210 (1991);

M. A. Prelas, E. J. Charlson, F. P. Boody, and G. H. Miley, "Advanced Nuclear Energy Conversion Using a Two Step Photon Intermediate Technique", Prog. In Nuclear Energy, 23 (3), pp. 223-240 (1990);.

Prelas M. A. and Boody F. P., "A Comparison of Electrical, Fusion- Generated-Ion, and Fission-Generated-Ion ICF Drivers," Laser Interactions and Related Plasma Phenomena, Vol. 10 (1992).

Prinz J. (1982), "Bipolar Transistor Action in Ion Implanted Diamond," Appl. Phys. Ltrs., $\underline{41}$ (10)

B. Singh, Y. Arie, A. W. Levine and O. R. Mesker, "Effects of filament and reactor wall materials in low-pressure chemical vapor deposition synthesis of diamond," Appl. Phys. Lett., $\underline{52}$ (6), Feb. 1988.

B. Tsaur, C. K. Chen and J. Mattia, "PtSi Schottky Barrier Focal Plane Arrays for Multispectral Imaging in Ultraviolet, Visible, and Infrared Spectral Bands", IEEE Electron Device Lett. II, 162 (1990).

Wallace C. B., BDM Corp., Albuquerque, NM, Personal Communication, August 1991.

D. E. Wessol, M. A. Prelas, B. J. Merrill, and T. Speziale, "Feasibility Study of a Nuclear Driven $O_2(^1\Delta)$ Generator to Power an 18 MW Average Power Iodine Laser for Inertial Confinement Fusion," *Laser Interactions and Related Plasma Phenomena*, **Vol. 8**, Plenum Press, New York (1988).

Wysocki J., Rappaport P. (1960), J. Appl. Physics, $\underline{31}$, p.571.

H. K. Yasuda, Cascade Arc Plasma Torch, Industrial Contract, 1988-1989.

W. M. Yim, E. J. Stofko, P. J. Zanzucchi, H. I. Pankove, M. Ettenberg, and S. L. Gilbert, "Epitaxially grown AlN and its optical band gap", J. Appl. Phys. 44*ff*, 292 (1973).

S. Yoshida, S. Misawa, Y. Fujii, S. Takada, H. Hayakawa, S. Gonda, and A. Itoh, "Reactive molecular beam epitaxy of aluminium nitride", J. Vac. Sci. Technol., 16(4), 990 (1979).

S. Yoshida, S. Misawa, and S. Gonda, "Properties of $Al_{-x}Ga_{1-x}N$ films prepared by reactive molecular beam epitaxy", J. Appl. Phys. 53, 6844 (1982).

G. Zhao, C. H. Chao, E. G. Charlson, E. M. Charlson, J. Meese, G. Popovica, M. Prelas, T. Stacy, "Silver Diamond Schottky Diodes Formed on Boron Doped HFCVD Grown Diamond," Thrid Annual Diamond Technology Workshop, Organized by Wayne State University, March 17-19, 1992.

ADVANCED ENERGY FOR GREENHOUSE EFFECT

C. Yamanaka

Himeji Institute of Technology
Himeji 671-22
Institute for Laser Technology
Yamadaoka Suita Osaka 565
Japan

NEED FOR ADVANCED ENERGY OPTIONS

The world population explosion began 200 years ago. By 1960 the growth rate was about 2%, so it was growing fast enough to double every 35 years. It is now five billion and the World Commission on Environmental of the United Nations predicts that 60 billion by the end of the next century is a conservative number. Combining a world population rate that stabilizes at 10 billion with the estimated per-capita energy consumption rate, 40 barrels of oil equivalent per person a year yields the world energy consumption rate shown in Fig. 1. The obvious fact from this figure is that we must develop alternative energy sources.

We are facing the prospect of not being able to use fossil fuels in the next century due to the lack of resources. Growing concern about the environment will make it highly desirable to rely on nonfossil fuel before the middle of the 21st century. The problems with burning fossil fuels are such as acid rain, emission of carbon dioxide, ozone hole in the upper atmosphere and so on. Each year, we put about 13 million tons of carbon to the air and an area about double to that of Hokkaido is deforested. We are changing fabric of nature.

The greenhouse effect due to the build up of carbon dioxide in the atmosphere could raise the temperature of the earth's surface a few degrees in the next century. This would produce dramatic changes in world climate. Current observation suggests that the warming trend in beginning.

TECHNOLOGY FOR MITIGATING GREENHOUSE EFFECT

Japan has issued the action program[1] for methods for reducing carbon dioxide emission. This plan aims to recover the earth's environmental destruction done by the industrial activities. Table 1 shows the technological development for this issue. According to this program. artificial treatment of CO_2 will be provided by 2030. The usable technology for mitigating CO_2 emission is still in basic studies which have a wide variety. But the treatment process shall be more effective and at hand than the advanced new energy development such as large scale solar energy and nuclear fusion.

421

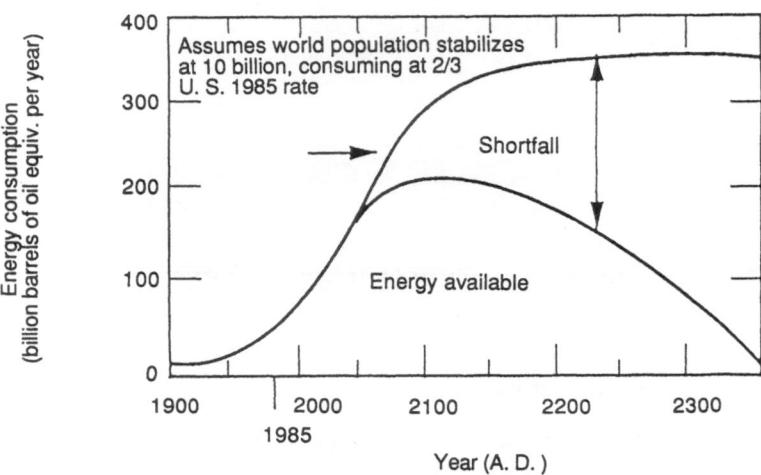

Fig. 1. *Estimated energy consumption*

The following research works are promoted by the sponsorship of the Research Institute of Innovative Technology for the Earth in Japan.

1. CO_2 fixation by artificial photosynthesis
2. CO_2 fixation by Algae
3. Coral reefs as sink of atmospheric CO_2
4. Recovery technology of CO_2

Evaluation of effectiveness of these researches is too early to set, but they shall be useful and promising in future.

Table 1. *Main Technology Developed According to Year*
(from Action Program "The New Earth 21" by MITI)

Year	Main Subject	Example of Technology
2000	Energy Conservation & Efficiency Improvement	Co-Generation System Combined Cycle Generation Fuel Battery, Heat Pump
2010	Nuclear & Renewable Energy	Solar Battery, Geothermal Biomass, Solar Heat
2020	Innovative & Technology	Fixation & Recycling of CO_2 Artificial Photo-Synthesis Hydrogenation of CO_2
2030	Activation of Natural Absorbing Force of CO_2	Biotechnology
2040	Advanced Energy Technology	Nuclear Fusion, Space Solar Power Generation

THE GLOBAL ENERGY SYSTEM

The global energy system can be characterized by three ratios that are important for determining environmental effects.

(1) Carbon emitted per unit energy consumed:
a measure of the share of fossil fuels
(2) Energy used to produce a unit of GNP:
a measure of the efficiency to used energy
(3) GNP per capita:
a measure of human welfare

According to the PNL global energy/economic model originally developed by Edmonds and Reilley, the integrated global model was used to produce scenarios of future greenhouse gas emission. From the report[2] of J. F. Clarke at the IEA International Conference on Technology Responses to Global Environmental Challenges at Kyoto, 1991, Figure 2 shows the evolution of these ratio[3] under the business as usual assumptions which mean population and economic growth continue along familiar paths defined by historic patterns. The carbon intensity of the global system falls until 2050 and then begins to increase. This is due to the progressive exhaustion of the oil

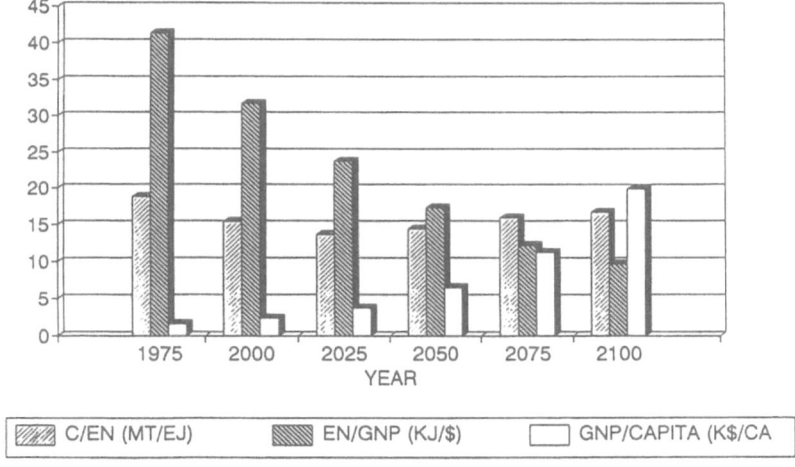

Fig. 2.   *Global Energy System Changes*
            (Business as usual)

and gas later in the century. The mix of energy sources is also shown in Figure 3.
When added to the business as usual assumptions[3], the magnitude and timing of the global carbon tax defines the alternative scenario.
In Figure 4, the carbon tax[3] is imposed to a "moderate" level which would be sufficient to stabilize carbon emissions at roughly their current values throughout the next century. It produces strong changes and reduces the carbon intensity of energy. As for the cost estimation for fusion energy, it has significant safety and environmental characteristics relative to fission energy which are important in determining public attitudes towards fusion. This supports the basic assumption of this scenario that the cost of fusion energy will be determined by its intrinsic technology cost like solar energy.

The timing of Fusion is also very important issue. The ILE Osaka[4] has performed a successful implosion experiment to get approaching high compressed fuel density a kg/cm^3. The ignition and breakeven shall be performed by 2000. Then the fusion demonstration reactor would be by about 2025. Allowing time for construction and operation of demonstration reactors, commercial fusion reactors could become available by 2050.

The analysis of the global energy system is very useful to reveals the mix of technologies which produces the energy demanded by the economy. From these results, it will be impossible to eliminate fossil fuels for a long time in future.

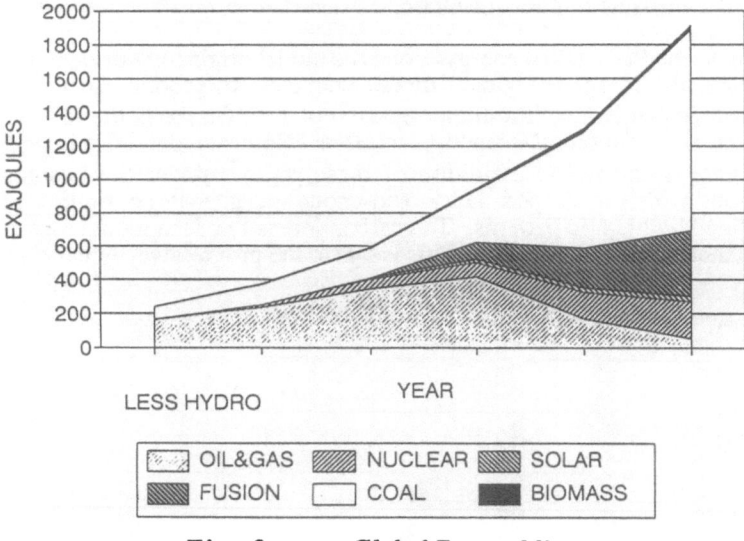

Fig. 3.   *Global Energy Mix*
(Business as usual)

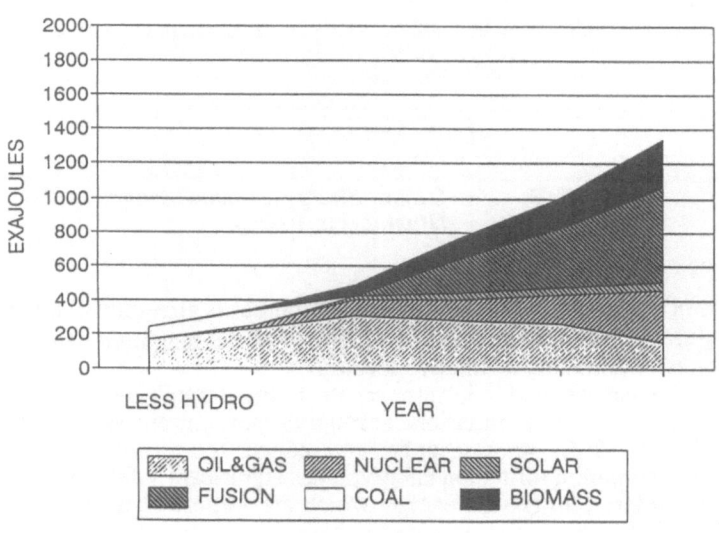

Fig. 4.   *Global Energy Mix*
(Global Carbon Tax)

CONCLUSION

If we are to envision a world in which the standard of living enjoyed today in the developed country is shared by the people of the developing country, we have only three energy options:

(1) Develop ultrasafe nuclear reactors and develop effective safeguards and waste storage systems.
(2) Develop large scale solar energy systems
(3) Develop fusion energy reactors

Nuclear energy from fission is the most developed energy sources. But it has been in decline for more than 2 decades in most of the world by public objections. Fission can be the bridge that connects the time gap when fusion power plants will be available. It is important to develop ultrasafe fission reactor concepts which are developing around the world.

Solar energy is also important, but the diffuse nature will limit its direct use to specialized applications. A concerted, global effort exploiting not only solar energy but other renewable energy sources, hydro, wind, geothermal, tidal, biomass and so on could provide up to 50% of the world's energy needs by about 2150.

The another 50% shall be covered by fusion energy. The most important decision is to undertake the development of non-fossil energy systems in a large scale today.

REFERENCES

(1) O. Yokoyama: Overview of Innovative Technology for Mitigating Greenhouse Effect, IEA International Conference on Technology Responses to Global Environmental Challenges, Kyoto, 1991
(2) J. F. Clarke: The Cost and Benefit of Energy Technology in the Global Context, ibid.
(3) J. Edmonds, J. Reilly: Global Energy, Assessing the Future, Oxford Press (1985)
(4) C. Yamanaka: Introduction to Laser Fusion, harwood academic pub. (1991)

PANEL ON FUSION AS A SOLUTION FOR THE GREENHOUSE EFFECT
OR GLOBAL WARMING, AND, TO HELP FURTHER THAT GOAL,
INERTIAL FUSION TARGETS FOR DEEPLY PENETRATING BEAMS

James Wai-Kee Mark

MT-Enterprises
P.O. Box 2217, Princeton, NJ 08543-2217
P.O. Box 20120, Piedmont, CA 94620-0120

ABSTRACT

The U.S. and other countries are expected to spend
Trillion dollar sums to build hundreds of new electric power
plants in the next few decades. Just in the coming decade,
the U.S. oil and natural gas productions are expected to
decline, and the whole world supply of sililar vital
commodities will also be depleting rapidly. We do have
abundant supplies of coal, but because of concerns raised by
meteorological and geophysical scientists, a panel was
convened to emphasize the potential relation to our fusion
community. After presentations by Prof. G. Miley and Prof.
C. Yamanaka as well as other distinguished panelists,
including himself, Prof. N. G. Basov concluded that the
relation between Greenhouse Effect or Global Warming and the
need for fusion energy is sufficiently important that a
full session should be convened at the next conference.
    In the spirit of contributing to inertial fusion
energy as a solution to these issues, we also added, as part
of our discussion, an outline of a series of targets that use
for advantage deeply penetrating (intense) beams (DPB) for
Hybrid-drive and direct-drive. By DPB we refer to intense
beams that can penetrate through 0.15-0.20 g / cm ^{3} of target
material, or deeper. Particularly for Hybrid-drive (Refs.
1-4), it was earlier announced in Ref. 1 that the concept can
provide advantages for both lasers and particle beams, in
combination or separately (see Section III).
    We suggest a new strategy for making a less costly
inertial fusion reactor by using smaller drivers and reactor
chambers. Towards this goal, we propose to use the lower
beam energy and power, more spread out illumination geometry,
and lower target yields of Hybrid-drive or direct drive to
facilitate smaller drivers and chambers. We also suggest a
new conservative programmatic approach to usage of direct
drive advantages through Hybrid-drive , assisted by a new
approach to direct drive symmetry through judicious use of
Sufficiently Short Pulses (SSP) of beam energy on target. We
can start with Hybrid of indirect and direct drives (Ref.6).

I. INTRODUCTION

A panel discussion was convened and moderated by the author which featured as distinguished panelists, listed here in alphabetical order, Profs. N. G. Basov (Lebedev Physical Institute of the Academy of Sciences, Moscow, Russia), S. Eliezer (Plasma Physics Dept., Soreq N.R.C., Yavne, Israel), H. Hora (CERN, Geneva, Switzerland), G. Miley (Fusion Studies Lab., Univ. Illinois, Urbana, IL, USA), G. Velarde (Instituto de Fusion Nuclear, Madrid, Spain), C. Yamanaka (Institute for Laser Technology, Osaka, Japan).

Prof. G. Miley initiated the discussions on the Greenhouse effect and the urgent need for fusion energy. Prof. C. Yamanaka gave a detailed discussion of the Japanese view. Many informal comments were made by all the distinguished panelists, and many members of the audience, representing to various degrees the views of the individual countries and their respective regional concerns. At various interludes in the discussion and in the final summary panel of the conference, the author J. Mark, added to this view, as well as provided a summary of some targets he designed to help further this goal. Prof. N. G. Basov in particular made the final motion that this interesting and important topic of "Greenhouse Effect and its Relation to Fusion Energy" should be given a more detailed coverage in a dedicated session at the next conference.

Since we are expecting a dedicated session at the next conference, the author and panel chairman shall limit his discussion of the Greenhouse effect to some brief comments in the next section of this paper. But he will elaborate on a few more details of the target concepts he developed to help further that goal of resolving the Greenhouse Effect. (Target discussions are of course more in the style of the remainder of this present conference).

Much of the development cost for inertial confinement fusion (ICF) is related to the cost of the first reactor-scale driver. Driver and reactor chamber size are also major factors in final net cost of electricity. The beam energy and peak power outputs required of a driver are directly related to its size and cost. Reactor chamber cost scales rapidly with size, a parameter influenced by required target energy yield, illumination geometry, etc. We discuss here a series of targets that use for advantage deeply penetrating (intense) beams (DPB) for Hybrid-drive and direct-drive to allow smaller drivers and chambers. By DPB we refer to intense beams that can penetrate through 0.15-0.20 g / cm^2 of target material. Particle beams are a natural candidate, but particularly for the Hybrid-drive concept (Refs. 1-4), we have claimed (Ref. 1, see quotes, Section IIIc), that lasers could eventually access such target regimes. Recent emphases on ultra high power lasers may lead to such developments.

In this paper, we emphasize that Hybrid-drive can be used with Deeply Penetrating Beams (DPB) of either lasers or ion beams alone. We also emphasize some important differences between our generic Hybrid-drive concept of Ref. 1 and our version of tamped direct drive (Ref. 4), as well as our versions of Hybrid-drive where we made specialisations (Section VII). Finally we point out cost-savings advantages for reactor design (Sections III,IV), as well as generic aspects of our ideas on direct drive symmetry which are applicable to choice of illumination geometry

428

(Section IV), as well as to Sufficiently Short Pulses (SSP) of energetic beams of particles or photons (Section V). Comparisons with the earlier literature is given in Ref. 4.

II. THE GREENHOUSE EFFECT AND THE NEED FOR FUSION ENERGY: SOME RESTRICTED COMMENTS

In the panel, we raised the issue of whether there is a comprehensive alternate to energy by a renewable resource. Recent crises have reemphasized again the reality of diminishing world oil supplies, and the accompanying dependence of both industrialized and emerging nations on events occuring in mostly one part of the world (i.e. the middle-east). We have witnessed again the large fluctuations in world oil markets and another period of at least financial concern by many nations. For some major industrialized nations, this concern could be particularly acute since they have limited energy resources of their own. Some emerging nations might well have been wondering whether it is more profitable to purchase more armaments than to pay the increasingly exorbitant energy-cost for industrial development or suffer infinitely delayed development.

Even for the U.S., which is blessed with some 200-year supply of coal, and plenty of fuel for fission power (if that were acceptable by our people), there are major concerns such as increased hydrocarbon emmisions due to burning coal, part of the so-called "Greenhouse effect". Our other problems include interminable delays to license any requisit fission development, due to other concerns raised by increasingly complex litigations and environmental worries. For the U.S., we have in addition the fact that our water for agriculture and population and our industrial base for heavy industries are usually in regions far from our coal reserves and other new resources such as clear sky for solar energy. The regions with greatest amount of heavy industries and related populations are also likely most concerned about the environmental impact of burning coal.

Ice-cap melting due to the Greenhouse effect might make "Little Hollands" out of many low-lying coastal areas presently absorbed by population growth. For areas of the world presently enjoying particularly favorable climate conditions for agriculture, some believers in "Murphy's Law of Applied Statistics" could suspect that the Greenhouse Effect would sometimes result in less desirable agricultural conditions. Selected numerical simulations are making just such charges.

Making the soon to be mandatory transition from dependence on oil alone for liquid fuels will have direct economic consequences even apart from the economic cost for ameliorating the potential environmental concerns for other forms of energy. No less important, but much more difficult to predict, are the economic consequences when global climate change were to occur. Many of these effects are so nonlinear that our discussion only barely touched on some boundaries. For example, if the excess food production capacity of the U.S. were to be affected, world wide food costs would of course rise. Equally likely are the rise in the costs of much of our imports, first directly from food purchasing countries passing along their cost for "survival". But clearly there will be price speculations on worldwide food purchases that will eventually have a dent on most costs of doing business. Defense costs would also go up in most countries as we all face a more unstable world! And as many countries work harder just to attempt to maintain previous standards of living, more energy consumption is a must. So we have a classic "instability situation". only now it involves international economics and politics as well as normal considerations of science and technology.

A possible resource driven economic and political instability
aggravated by environmental feedback factors

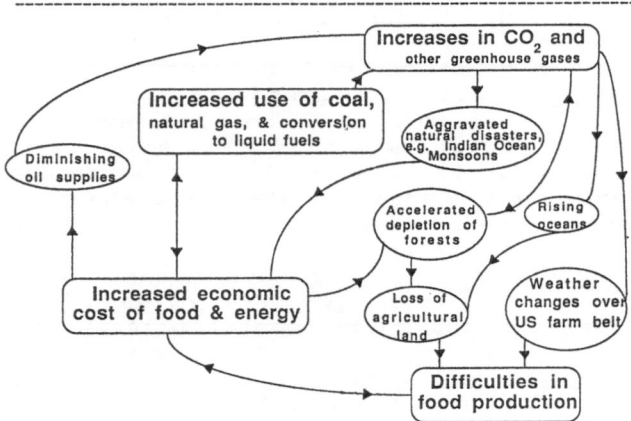

Figure 1. This illustrates some of the feedback loops in the Greenhouse or Global Warming Effect which relates to A Resource Driven Economic and Political Instability Aggravated by Environmental Feedback Factors.

The stakes are so high and possible consequences so worldwide that it is important to at least place some good estimates on the likely worst case scenarios before our colleagues and eventually to a wider audience. Since there is only one patient and we are but all minute cells on that one patient, piecemeal or local solutions by any one participant would be viewed with suspicion by others. Perhaps one acceptable global solution is the development of cleaner energy technology. Hopefully international stability would be enhanced if that were made available to much of the world at reasonable cost.

Fig. 1 illustrates our discussion of "A Resource Driven Economic and Political Instability Aggravated by Environmental Feedback Factors". These are some issues to consider as the U.S. and other countries prepare to spend

Trillion dollar sums to build hundreds of new electric power plants in the next few decades. Just in the coming decade, the U.S. oil and natural gas productions are expected to decline, and the whole world supply of sililar vital commodities will also be depleting rapidly. We mentioned that we do have abundant supplies of coal, but because of the "Global Warming" or "Greenhouse Effect" raised by meteorological and geophysical scientists, there are many discussions of methods of amelioration or even adaptation. There is even an increasing discussion of an urgent need for the U.S. to fully develop and start implementing soon electric power technologies parallel to, and eventually, substituting for sizable fractions of hydrocarbon-based technologies. Easily usable oil and natural gas should eventually be reserved for the petrochemical industries.

Since about half of the global warming effect is due to energy activities, a fraction that might well grow as the effect of CFC's are brought under control by international legislation. We can solve major global energy and environmental problems while contributing to industries by achieving fusion in the laboratory, and then pushing for reactor development. The panelists and many of the audience have provided extensive formal and informal discussions on this issue. Since we are expecting to submit proposals for further detailed papers in the upcoming conference, we return here to the main issue of this particular conference, namely target physics.

III. REACTOR ADVANTAGES FOR TARGETS USING HYBRID-DRIVE AND DIRECT DRIVE WITH DEEPLY PENETRATING BEAMS (DPB)

IIIa. Smaller Drivers and Reactor Chambers means Smaller Cost for Fusion Generated Electricity

Driver and reactor chamber size are major factors in final net cost of electricity. The beam energy and peak power outputs required of a driver are the major factors directly related to its size and cost. Reactor chamber cost scales rapidly with size, a parameter influenced by required target energy yield, illumination geometry, etc. A much more compact fusion reactor chamber is a major factor in reducing electricity costs in ICF.

IIIb. Smaller Drivers and Reactor Chambers are in Agreement with Use of Direct Drive

Once symmetry of implosion is properly addressed, directly driven targets offer substantial advantages in requiring much lower energy and peak power requirements, and lower energy yield for same target gain. Its use of a more spread out geometry are all advantages in the development of a smaller driver and smaller reactor chamber. (For example, more spread out illumination allows more compact arrangement of final focus or pulse-power diodes around reactor chamber). Concurrently, the lower beam energy and peak power requirements for same target gain-driver efficiency product allows further reductions in requirements on final focus and beam propagation through fusion chamber, etc. (much less neutron loading on final optics helps lasers). These factors help in allowing smaller fusion chambers and lower cost for electricity production. Lower heights of pumping of liquid metals in a compact reactor further lower costs. Even dry wall chambers could be accomodated by targets yields below the threshold of a few hundred megajoules, achievable by direct drive targets with proper gain-efficiency products.

Direct drive implosion requires the resolution of a number of issues. Beam illumination geometry is an issue relating the requirements of reactor chamber design, driver final focus and bending, as well as target physics of smoothness of beam energy deposition necessary for adequate implosion symmetry. At least at the initial stages of implosion of most direct-drive targets, ion energy is deposited deep in the target, essentially right on top of the payload mass. There is therefore less smoothing of nonuniformities in deposition in this case than in indirect (radiation) drive. For direct drive targets to work, it is necessary to use some of several new methods for controlling asymmetries. Particularly for ion beam direct drive, until our recent series of efforts (Refs. 2-3), these symmetry issues have never been addressed in regards to showing in 2D hydrodynamics that the targets survive DT burn.

To ameliorate this situation and promote one means for a smaller reactor, we point out now that our "Gaussian Quadrature Beam (Illumination) Geometry (GQBG)" described in the next section is an axisymmetric illumination geometry that enhances the small reactor concept by conforming to the desires of reactor designers (see Section IV).

After that, to further control the evolution or " detuning of optimal symmetry " achieved by any good geometry such as GQBG or others, we point out that the concept of " Sufficiently Short (direct drive) Pulses (SSP) " is crucial (Section V). An application of SSP to the tamped direct drive target, together with GQBG and 32 beams was sufficient to control asymmetries as shown in 2D simulations (Section VI). Now we propose some heretofore unknown applications of the Hybrid-drive target generic concept.

IIIc. Hybrid-drive, the Generic Concept

In the recording of the Hybrid-drive concept first as a U.S. Statutory invention, we emphasized (quote, Ref. 1) :

"The implosion of an ICF target generally occurs in two phases: an initial compression phase, during which the target rapidly and very significantly diminishes in size, and a final peak power phase, during which the reduced dimensions of the target remain relatively static. The initial compression phase has especially high energy deposition uniformity requirements, and the final peak power phase is especially sensitive with respect to energy deposition efficiency, even though the two parameters are always of critical concern during the full course of the implosion.

"The present invention comprises the realization that a vast improvement in gain, at fixed total driver energy, may be achieved by separately and optimally driving the two phases of ICF target implosion. The invention has six potential embodiments: laser-ion beam; ion beam-laser; laser-laser; ion beam-ion beam; single laser; and single ion beam ...." (end quote, Ref. 1).

Hybrid-drive distinguishes itself from our other target concepts by explicitly allowing combined use of say a laser or particle beam for target compression and an unshaped final

<u>pulse of laser or particle beam for ignition, with individual</u> <u>advantages for the drivers as well as for total beam energy</u> <u>and peak power.</u> The central theme of the concept, however, is one or more pulses for compression phase with a final peak power pulse for ignition that is added on after a time delay. By separating these two phases, the target concept allows the driver parameters to be most directly matched to the performance requirements of target physics. It specifically uses to advantage the properties of Deeply Penetrating Beams (DPB) to deliver energy directly, with high hydrodynamic efficiency, for a final ignition pulse on a precompressed shell. A final peak power phase, during which the reduced dimensions of the target remain relatively static.

IIId. <u>New Applications of Hybrid-drive</u>

The <u>generic advanced target concept of Hybrid-drive as</u> <u>conceived can fill numerous interesting roles</u> such as also allowing combined use, to advantage, of several different types of drivers to together drive a target in several "embodiments" or scenarios. (Of course, if two or more drivers were used, to coordinate pulse timing it would help if the drivers have similar pulse power structures, such as a Free Electron Laser and a particle beam).
Second, we point out that the innovative advanced target physics idea of Hybrid-drive (Section VII) allows a second version or test of the use of SSP to address direct drive symmetry (see Sections V,VII).
Even more importantly, we now propose another major role for this target concept. <u>Hybrid-drive also provides a</u> <u>programmatically conservative route to direct drive usage.</u> Because of its introduction of the smaller or even near zero convergence ratio for the bulk of the drive power and energy, it further enhances the embodiment of direct drive concepts. This usage, in one embodiment, results in the so called compromise set of parameters of Section VIIa, where the power of the generic Hybrid-drive advanced target concept allows :
(i) Conservative route to direct drive advantages, including eventually fewer beams because of small convergence ratios.
(ii) Smallest peak power for similar target energy inputs.
(iii)No pulse shaping required for peak power pulse.

IV.   GAUSSIAN QUADRATURE BEAM (ILLUMINATION) GEOMETRY (GQBG)

This illunination geometry espoused by Mark [1,2] has been described in the literature. We emphasise here its characteristic performance, particularly in light of our newly espoused goal of smaller reactor chambers:
1.   It was designed to be the first in principle proof
(Ref. 2) that direct drive could be achieved with an illumination geometry that is based on an axisymmetrically placed set of beams rather than the then usual spherically symmetric ones. The symmetry axis is determined by reactor chamber design.
2.   We now underline that it <u>satisfies the requirements of</u> <u>reactor chamber design particularly for the role of "more</u> <u>spread-out illumination for smaller reactor"</u> (Section IIIa).
3.   We emphasize further freedoms GQBG allows for our smaller reactor in that it can also be used to limit the number of verticle planes through reactor axis which must be occupied by beam ports, a consideration particularly

important to reactor designs using liquid metal wall
protection. The beams are situated on the rims of cones
through reactor axis with cone angles the zeroes of the
Legendre polynomials in sperical harmonic expansion. On these
cones, placement of beams can have some flexibility, for
example to limit the number of verticle planes with beams, or
to do otherwise.
4.   In a harmonic expansion of the asymmetries of target
beam energy deposition profiles, this method systematically
eliminates more and more coefficients of the lowest spherical
harmonics as more and more number of beams are used. By
contrast, previous methods with spherically symmetric
equally-spaced beams only reduces the size of these
coefficients by a least squares method as beam number is
increased.
5.   Thus previous methods simultaneously uses the effects of
increasing beam number to reduce all coefficients of
spherical harmonics, whilst GQBG concentrates its power on
the lowest uneliminated coefficient, leaving the effect of
higher coefficients to other target physics considerations.
6.   The goal is to harmonise target physics with engineering
of reactor chambers using mathematical methods of analytical
symmetry and numerical quadrature. By explicitly
incorporating analytical symmetry considerations, the method
is more amenable to analytical studies.
7.   Within GQBG, choice of beam energy fluxes and beam
tranverse profiles are auxiliaries in additional smoothing
with less beams or of higher harmonics, but not always
necessary. Further smoothing of higher harmonics can use the
effects of internal dispersion of beam angles for beam
particles, or dispersions of photon or particle energies,
thermal effects of target plasmas, or focal shift relative to
the target center; and for laser drivers alone, additional
techniques include refraction smoothing and smoothing due to
decoupling of absorption and ablation layers.
8.   In the ideal limit of ultra thin deposition layers, at
most six beamlets are needed for target symmetry.

V.   SHORT PULSES AS SOLUTION FOR TARGET DETUNING OF
         INITIALLY GOOD ILLUMINATION SYMMETRY

      Even a good illumination geometry usually provides
optimal deposition symmetry only near one target radius which
approximates the beam focal radius. Maintaining adequate
implosion symmetry averaged over the entire implosion relates
to the physics of "detuning of optimal target implosion
symmetry" due to the changes in target radius relative to
beam radius during the implosion. Detuning thus relates to
the choice of beam focal spot achieved and to the proper
addressing of target implosion symmetry. During the drive-
pulse of a typical high-gain ICF target which is driven
directly by laser or particle beams, the target radius
changes by a substantial amount. A good multibeam
illumination scheme in direct drive (e.g. Ref. 1) has
residual asymmetries which can be reduced systematically by
optimal choice of focusing geometry and average beam
intensity profile. But these beam parameters usually remain
fixed during the target implosion, so it is difficult to have
both optimal illumination symmetry and high coupling
efficiency throughout the drive-pulse with a limited number
of beams.

434

However, our recent advances in direct drive offers several recipes for controlling the effects of detuning which is firstly generic to addressing all direct drive, as well as further specialised to the case of DPB (deeply penetrating beams). The method of Ref. 4 and present Section VI uses a judicious amount of tamper mass over the ablator to limit the detuning of optimal residual asymmetries referred to above. The other arises in hybrid drive (Refs. 4 and Section VII), in which the thin, dense, pre-compressed ablator does not move appreciably during the short direct drive pulse of DPB.

The key is often sufficiently short pulses so that the target radius does not change appreciably during the pulse. Concepts of "sufficiently short pulses" (SSP) and "appreciable changes of target radius" are further refined in the following and in specific examples of application below in Sections VI-VII. We now also emphasize that these applications of our generally espoused principle of sufficiently short pulses (SSP) also applicable to really short pulses of other driver situations, such as ultra high intensity laser beams, for example.

Let us now more precisely define the concept of target detuning and SSP. On the level of reducing the asymmetries in a fixed target radius, we can use GQBG as sketched above, or some other scheme. But for a given illumination geometry, deposition uniformity can typically be optimised only in the neighborhood of a single target radius, which we expect to be determined by the radius R_{beam} of the beamlet focal spot. As the target implodes, this optimal target radius for symmetry eventually deviates significantly from the radius R_{dep} of peak ion energy deposition, so that symmetry becomes significantly degraded after some characteristic time t_s for detuning of optimal symmetry. As a consequence, an initially satisfactory symmetry will be degraded during the implosion.

This degradation (for fixed number of beamlets and fixed R_{beam}) occurs because for most target designs, the symmetry detuning time t_s is less than the duration t_d of the direct drive pulse (i.e., $t_s < t_d$). However, residual asymmetry of direct drive detuning can be controlled by numerous target effects. In the next section, we study targets (such as those shown in Fig. 2) with an external high-Z tamper (requires penetration through > 0.15-0.2 g / cm $^=$ of target material, easiest for particle beams). Here we can limit the motion of the deposition region to less than approximately 10% of the initial radius. This is much less than the typical motion of a factor of 2 in radius that takes place in the abscence of a tamper. It is one method of ensuring that t_s > t_d because judicious tamping sufficiently lengthens t_s .

On the other hand, we can omit modifying the existing symmetry detuning time t_s but use sufficiently short bursts of direct drive ion or laser pulses of small duration t_d so that we again satisfy $t_s > t_d$. One of the interesting features of Hybrid-drive with what we called the compromise choice of parameters is that we can accomplish this goal without a t_d so shortened that we end up paying a penalty of drastically increased peak power. (See section VII to illustrate our present specialisation of our generic hybrid-drive target concept of Section IIIc). In fact, the peak power is decreased relative to the requirements for radiation drive, leading to substantial advantages, as well as removing most of the rationale behind use of double-shell targets.

VI. CONTROLLING SYMMETRY DETUNING BY CONTROLLING THE MOTION OF THE BEAM ENERGY DEPOSITION RADIUS

By varying the ratio of tamper and ablator inertias we can make t_s great enough that the entire t_d = 62 ns direct-drive pulse train delivered to our target satisfies $t_d < t_s$. The target in question has the structure illustrated in Fig. 2. Further control over R_{dep} might be possible with more careful tuning of target parameters (it should be noted, however, that varations in tamper mass should eventually involve changes in beam depth of penetration in order to maintain energy efficiency).

Before considering the effects of asymmetries in two or three dimensions, we simulated the full hydrodynamics and DT burn of the target using the spherically averaged ion energy deposition. In this one-dimensional simulation, the beamlets were first assigned a realistic distribution of incident angles relative to the target surface, and the ion energy deposited at each target radius r was then averaged over the sphere at that radius. In effect, this simulates the radial distribution of the illumination by an extremely large number of beamlets. Detailed 2D symmetry results of a 32 beam illumination with GQBG scheme are given in Figures 3a-c. Numerical methods are in Reference 4.

The target of Fig. 2 gives a one-dimensional gain of 16 at 1.8 MJ of input beam energy with R_{beam}= 2.6 mm. This input energy is 40% greater than the minimum energy required to ignite this target in one dimension; the excess corresponds to the 40% safety factor included in all our radiation driven designs to cover three dimensional deviations from the ideal implosion conditions of one dimensional calculations. The drive pulse of Fig. 2 can be generated in the final parts of the induction heavy ion accelerator as the pulse is compressed in length to reach peak power. (In Ref. 5, for example, we show that such a pulse shape could be obtained by relatively simple beam manipulations).

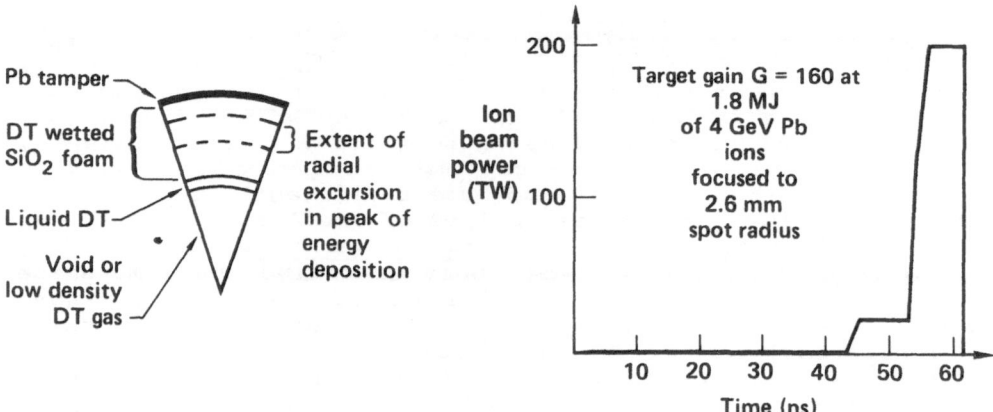

Figure 2. Direct-drive ion beam target with deposition radius motion controlled for symmetry considerations; and drive pulse used.

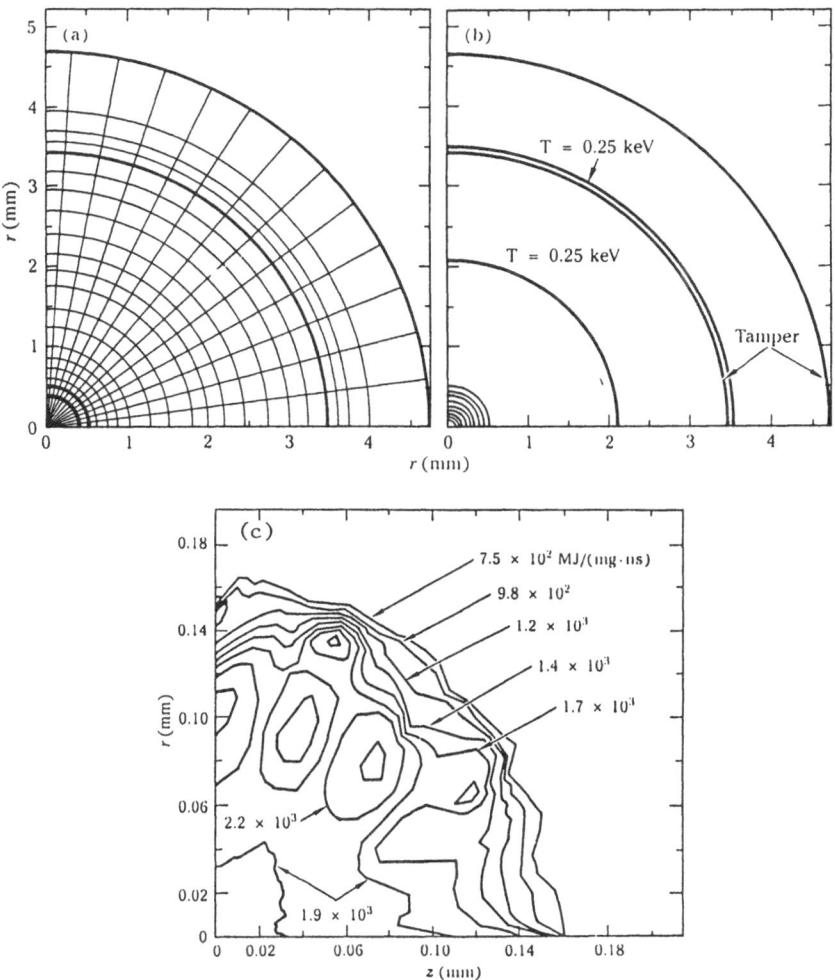

Figure 3.(a) grid and (b) ion-temperature contours of the two-dimensional simulation of the tamped direct-drive target just after the ion pulse is turned off in the one-dimensional simulation; and (c) the DT energy production rate in fuel region near time of peak burn. (GQBG with 32 beams).

VII. HYBRID-DRIVE

Hybrid-drive distinguishes itself from our other target concepts by explicitly allowing combined use of say a laser or particle beam for target compression and an unshaped final pulse of laser or particle beam for ignition, with individual advantages for the drivers as well as for total beam energy and peak power. The central theme of the concept (see generic Hybrid-drive Section IIIc), however, is one or more pulses for compression phase with a final peak power pulse for ignition that is added on after a time delay. By separating these two phases, the target concept allows the driver parameters to be most directly matched to the performance requirements of target physics. It specifically uses to advantage the properties of <u>Deeply Penetrating Beams (DPB) to deliver energy directly, with high hydrodynamic efficiency, for a final ignition pulse on a precompressed shell</u>.

VIIa. <u>Hybrid-drive, further specialisations of the generic concept</u>

We divide the implosion of a high-gain ICF target into two phases (see Fig. 4) : (1) At $t = 0$, a laser or ion beam pulse compresses the initial target shell to high density; (2) At some later time $t = t_h$, a DPB driver pulse delivers energy into a dense shell to provide ignition. Let us now emphasize a special feature of Mark's hybrid-drive (Ref. 1). Since it is true that near adiabatic <u>compression of a target shell uses very little energy and peak power</u>, we can wait for a <u>very dense final shell</u> to provide a hydrodynamically <u>very efficient coupling of the final pulse</u> to the target, as well as smaller and smaller, or even <u>near zero convergence ratio for this final ignition pulse</u>. What we choose to do for a specific example depends on the individual characteristics of the DPB available. Of course, with ion beams, choice of DPB's with very deep penetration is relatively easy. However, in our examples shown below, we choose a compromise which only reduces convergence ratios by factors of several because this compromise choice is more accessible to lasers. For this choice we only require DPB's which can penetrate more than $0.15-0.2$ g / cm^2 of precompressed target material.

VIIb. <u>Hybrid-drive for DPB's that penetrate $0.15-0.2$ g / cm^2</u>

We now concentrate on this compromise choice of parameters. The physical requirements during the first phase are similar to those of all high-gain ICF targets. Careful temporal pulse shaping of the driver power (see Fig. 4b) is necessary to set the DT fuel onto a low adiabat. Concurrently, the ablator is symmetrically compressed to higher density (see Fig. 4a). A high hydrodynamic efficiency is not crucial during this first phase because only about 20% of the total input energy is involved.

In the second phase, little or no temporal power pulse shaping is required (Fig. 4b). With the appropriate ion species and kinetic energy, we can obtain very good coupling of the ion energy to the ablator and thereby maximize the hydrodynamic efficiency. Since about 80% of the input energy is supplied during this phase, an increase in the hydrodynamic efficiency directly reduces required input energy and peak power. We can further improve the hydrodynamic efficiency by <u>dynamically tamping</u> the implosion, for example, by generating an ablation front on the outside of the ion deposition region, as shown in Fig. 4a. Morover, the ablation front can be made to move inward as a function of time to provide dynamic tamping for the ion deposition region. We note, however, that the tamping effect can only provide a modest additional improvement in the hydrodynamic efficiency because the pressure at the ablation front is relatively low. The major improvement in the hydrodynamic efficiency is due to the direct ion deposition into the precompressed shell.

In hybrid-drive, the thin, dense, compressed ablator does not move appreciably during the short heavy-ion pulse. For example, in one of our simulations the radius of peak ion deposition R_{dep} varies from its average value by only 13% during the ion pulse. Our two-dimensional simulations indicate that this motion is small enough that adequate illumination symmetry is maintained.

Figure 5 shows the results of detailed calculations of single shell targets using the hybrid-implosion concept as compared to some of our other targets. Using unpolarized DT fuel, a target gain of about 90 is obtained at 2.5 MJ input energy. In our actual calculations, we used 3 GeV Cs (or 5 GeV Pb) ions with a peak power of 150 TW focused to 3.4 mm spot diameter, D, ($r^{3/2}$ R = 0.004). At higher input energy, we obtained a target gain of 140 at 6.6 MJ. This hybrid drive target used D = 5.1 mm with 4 GeV Cs (or 7 GeV Pb ions) at a peak power of 250 TW ($r^{3/2}$ R = 0.01, unpolarized DT). The peak power and input energy at constant target gain and ($r^{3/2}$ R) are about a factor of two lower than those used in the unpolarized DT targets (Fig. 5).

Many proposed targets use direct drive ions (for early examples see references cited in Ref. 4) to generate efficient target drive. The difference of Hybrid-drive lies primarily in important details of the generic concept of Section IIIc, such as energy deposition in precompressed shells, and only secondarily in the treatment of any residual direct-drive asymmetries and convergence ratios. Other secondary or tertiary differences are target materials, in flight aspect ratios, illumination schemes, etc.

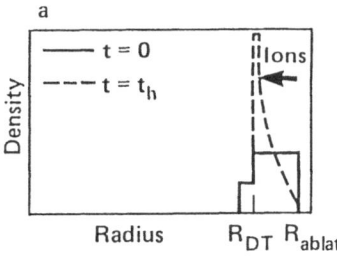 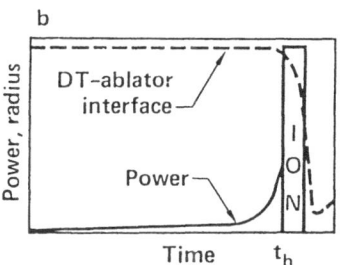

Figure 4. Two-step hybrid-drive implosion. (a) Density versus radius. Ions deposit energy in compressed ablator; optional outer ablation front tamps ion deposition region, increasing hydrodynamic efficiency. (b) Power and ablator radius versus time.

Compared to the target described in Section VI at fixed input energy, this hybrid-implosion target with compromise specialisation of parameters uses a smaller beam spot size and has smaller energy gain. This smaller spot size would place a tighter phase-space constraint on the ion beams. However, this tighter constraint is in part a design trade-off of the compromise parameter choice. When the same ablator material is used in the two targets, the hybrid-implosion target compromise used here has about a factor of two smaller convergence ratio for the ion pulse. The reduction in the convergence ratio is the direct consequence of precompressing the ablator material to higher density and smaller radius before depositing the direct drive ion energy into the ablator material (for the same amount of DT fuel). Convergence ratio is defined here as the radius of peak ion energy deposition, R_{dep}, at the start of the ion pulse divided by the radius of the hot spot at ignition. The two targets also deal with the effects of the residual ion beam asymmetries differently. The same ion beam placement scheme is used in the two cases. The hybrid-implosion target tolerates the effects of the residual asymmetries by : (1) reducing the radial excursion in R_{dep} by choosing the proper duration, t_h, for the direct drive final pulse (too short a t_h would sacrifice the advantage of lower peak power); (2) reducing the effective convergence ratio. Further reductions in convergence ratio is possible within the the generic Hybrid-drive concept, but would reduce focal spot some more. (2D simulation results and numerical methods are summarized in Ref. 4). As more deeply penetrating lasers are developed, this is a promising application of Hybrid-drive. In contrast, the target described in Section VI controls the residual asymmetries by selecting the apporpriate excursion in R_{dep}. This is done through choosing the proper amount of high-Z tamper material (convergence ratio has typically not been reduced).

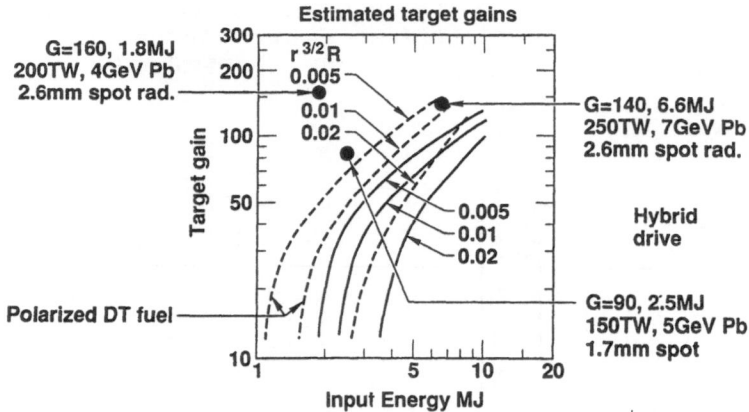

Figure 5. Gain curves of the single-shell targets for several ($r^{3/2} R$) values. The two hybrid-drive points are to be compared with the solid gain curves (unpolarized DT fuel). The dashed gain curves are for targets using 100% spin-polarized DT fuel. Input energy is equal to the total energy deposited in the target.

The stability of an ICF high gain target is a very complex issue and is dependant on the details of target physics.For the typical ablatively driven target, perturbations can grow at two (or more) interfaces during the implosion. One is the ablation front and the second is the fuel-ablator interface when the shock reaches the fuel (Richtmyer-Meshkov). These two unstable interfaces are physically separated by the ablator region. The ions delivered during the second phase of the hybrid-implosion concept deposit their energy in this region at or near the region of peak density and temperature. Since the ablation front is outside of the ion deposition region and this region is exploding during the dominant phase of the target implosion, we expect the deleterious effects due to the perturbations generated at the ablation front to be reduced by the ion depostion.

In this specialisation of the generic Hybrid-drive concept of Section IIIc, the hybrid-implosion concept is a new way to reduce the peak power and energy requirements of the DPB driver. A number of issues generic to direct drive symmetry and stability requirements need to be explored more fully. Ultimately, the utility of this concept will be determined by factors such as cost and other system issues.

VIII. COMMENTS

In this paper, we emphasize that Hybrid-drive can be used with Deeply Penetrating Beams (DPB) of either lasers or ion beams alone. We also emphasize some important differences between our generic Hybrid-drive concept of Ref. 1 and our version of tamped direct drive (Ref. 4), as well as our versions of Hybrid-drive where we made specialisations (Section VII). Finally we point out cost-savings advantages for reactor design (Sections IIIa,b,IV), as well as generic aspects of our ideas on direct drive symmetry which are applicable to Sufficiently Short Pulses (SSP) of energetic beams of particles or photons (Section V). Comparisons with the earlier literature is given in Ref. 4. Some of the physics of using our GQBG illumination (Ref. 2) for direct drive and Hybrid-drive required the invention of new numerical methods such as parallel hydrocode simulations of two 2D plus one 1D runs (see Ref. 4).

REFERENCES

1. James Wai-Kee Mark, "Hybrid-Drive Implosion System for ICF Targets", U.S. Statutory Invention Registration No. H508 (1988), U.S. Department of Energy Case No. S-64,524 (RL-9821).
2. James W-K. Mark (same as James Wai-Kee Mark), "Near Spherical Illumination of Ion-Beam and Laser Targets", <u>Phys. Lett.</u> 114A:458-464 (1986).
3. James W-K. Mark (same as James Wai-Kee Mark), "Recent Livermore Research on Ion Beam Fusion Targets", in proceedings Beams '88 Conference held in Karlsruhe,

Germany, July 4-8, 1988 eds. W. Bauer and W. Schmidt, pp. 785-790 (publ. by Kernforschungszentrum Karlsruhe GmbH, 1988).

4. James W-K. Mark (same as James Wai-Kee Mark), "Recent Livermore Research on Ion Beam Fusion Targets: Utilization of Direct-drive Efficiency During Optimization of Symmetry and Utilization of Polarized DT Fuel, _Laser and Particle Beams_ 9:713-723 (1991).

5. James W-K. Mark (same as James Wai-Kee Mark) et al., in " _Heavy Ion Inertial Fusion_ ", AIP Conference Proceedings No. 152 (American Institute of Physics, New York), pp. 227-235 (1986).

6. James W-K. Mark (same as James Wai-Kee Mark), "Can the Cost of ICF Drivers Be Reduced by Targets Using a Hybrid Indirect-Direct Drive Configuration?", in proc. 1984 Topical Conference on Radiatively Driven ICF-Targets, Title U, December 5, 1984, pp. 181-186 of X-1-85-22 (LA-CP-85-138 Los Alamos, 1986).

LASER FUSION – HIGH DENSITY COMPRESSION EXPERIMENT AND IGNITION PROGRAM WITH GEKKO XII

S. Nakai, K. Mima, H. Azechi, N. Miyanaga, A. Nishiguchi, T. Yamanaka,
K. Nishihara, K. A. Tanaka, M. Nakai, R. Kodama, M. Katayama, M. Kado,
M. Tsukamoto, Y. Kato, H. Takabe, H. Nishimura, H. Shiraga, T. Endo,
M. Nakatsuka, T. Sasaki, T. Jitsuno, K. Yoshida, T. Kanabe*, H. Nakano,
K. Tsubakimoto, M. Yamanaka, K. Naito, T. Norimatsu, M. Takagi, T. Chen,
Y. Izawa and C. Yamanaka*

Institute of Laser Engineering, Osaka University, Osaka 565, Japan
* Institute for Laser Technology, Osaka 565, Japan

I. INTRODUCTION

High density compression of main fuel and stable formation of hot spark at the center of imploded core have been investigated to obtain the scaling and the requirements for fusion ignition and high gain. For this purpose, the experimental data of the high density compression up to 600 times solid density with hollow shell pellet have been analyzed and compared with simulations.

The achieved maximum density with known experimental conditions of targets and laser shows the existence and contribution of some stabilizing mechanisms such as thermal smoothing and ablation stabilization of R-T instability at ablation surface. The reductions of achieved density and neutron yield below the 1D value at higher convergence ratio of implosion give us the technical requirements for ignition and burn. They are (1) pulse tailoring, (2) irradiation uniformity, (3) laser wavelength, and (4) target uniformity and sphericity with a specific designs of fuel pellet such as direct or indirect (cannon ball), and optimized structure of pellet. The high density implosion experiment with deuterized plastic shells is briefly summarized with detail discussions on the reliability of diagnostics and the stability of implosion dynamics (Section II).

Prospects from the present high density implosion experiment towards ignition and high gain is studied numerically. Critical elements to demonstrate fuel ignition and high gain are discussed associated with the uniformity and stability of implosion and laser pulse tailoring (Section III).

The implosion of cryogenic fuel pellet should be investigated because it responds differently compared to GMB or plastic microballoon target due to the low density and low atomic number of hydrogenic material. Cryogenic hollow shell fuel pellets have been developed to provide the quantitatively qualified target for a precision experiment. The target interaction and implosion experiments have been performed to investigate the absorption and energy transport in low Z target, shock formation and preheating on low density target, and implosion of low aspect ratio pellet (Section IV).

Investigation on radiation-driven cannonball targets has made good progress during past two years due to the understanding of the radiative processes taking place in a cavity. Better performance of implosion has been achieved and quantitative analysis has become possible with reasonable accuracy (Section V).

For the improvement of irradiation uniformity, several optical schemes have been developed and tested. They are (1) optical fiber system with random phase plate (RPP), (2) ASE front-end with RPP, and (3) spectral dispersion by grating with ASE front-end and RPP. New technologies such as multi-lens system with broad band front-end and application of liquid crystal for better speckle smoothing are also progressing (Section VI).

Recent progresses of new solid state laser technologies with the laser diode pumping of laser material made the laser fusion scheme a realistic and feasible candidate of the fusion energy development. A reactor driver which has the specification of an out put energy of MJ/pulse, efficiency of $5 \sim 10$ %, short wavelength of $0.3 \sim 0.5$ µm, repetition of $10 \sim 100$ Hz with high controllability of pulse shape and focusing, has become in scope for design and construction. The developments of a new high power solid state laser for industry are providing us the data base for a reactor laser (Section VII).

II. HIGH DENSITY IMPLOSION EXPERIMENT WITH DEUTERIZED PLASTIC SHELLS

Direct-drive implosion experiments at the ILE are addressing two key issues in critical elements of high gain design. One is compressing cold main fuel with relatively low adiabat. Another is obtaining a hot spark plug at the target center. Major efforts have been focussed on the high density issues, using tritiated-deuterated polystyrene shells as a "deputy" of a solid/liquid DT fuel. The conclusion of these experiments [1-3] were that (a) the cold fuel densities almost meet one of the ignition criteria, that of 500-1000 times liquid density (XLD); whereas (b) the hot spark plug failed to perform. We will briefly review these experiments as well as the recently performed cross examination of the high-density compressions, and address a question whether these results are consistent with the present understanding of Rayleigh-Taylor (RT) instabilities.

II-1. Summary of High Density Implosion Experiment

The experiments were performed on the GEKKO XII Nd:glass laser (0.53 µm wavelength, 12 beams) with random-phase-plates implemented. The laser energy was 8-10 kJ with a nominally 1.7-ns (full width at half-maximum, FWHM) flat-top pulse with a rise time equivalent to a 1-1.3-ns Gaussian pulse (FWHM). To improve the laser irradiation uniformity the laser was focused through f/3 lenses behind the target center by five target radii. Due to this irradiation condition 65% of the laser energy was irradiated on the target. The laser irradiation nonuniformity was calculated from the measurements on the target plane to be ~ 20% in root-mean-square (rms) which is a quadrature sum of all l mode considered (l=1-180). No thermal smoothing is assumed here. This value has been changed from the previously quoted value of 14-15% [2,3] because of the higher maximum l mode considered here and the careful re-calibration of the data. The beam energy imbalance was controlled to be ~1% in standard deviation. The power imbalance was not actually measured but estimated as follows. The gain of the each laser chain was initially adjusted so that the energy imbalance became a 1% level. Then a low power pulse (its power was about 1/100 of the full power pulse after the amplification and the harmonic conversion) was injected into the laser chain. The resultant energy imbalance in this case was 4-5% in standard deviation. This value may represent the power imbalance in the early period of the laser pulse.

The targets were deuterated and tritiated plastic shells which have an 1-g/cc initial density and a fuel part of 15-17 wt %. The highest density was observed at 500-µm-diameter and 8-10-µm-thickness. The target material was a co-polymer of deuterated styrene and deuterated *para* trimethylsilylstyrene [$(C_8D_8)_n$- $(C_{11}D_{16}Si)_m$]. Si atoms were included as a chemical composition for compressed density measurements. The Si amount of 5-15 wt% was controlled by changing the n/m ratio from 2 to 0. Small amount of the deuterium in the targets was exchanged with tritium, giving tritium-to-deuterium ratios of 3-10%.

The areal density ,$<\rho R>$, measurements in these experiments were made with the Si activation technique[4]. The ^{28}Si atoms uniformly distributed in the shell were activated by 14 MeV fusion neutrons through the $^{28}Si(n,p)^{28}Al$ reaction. The ^{28}Al atom decays back to ^{28}Si, producing β and γ rays simultaneously with a half - life time of 2.24 min and a distinct γ - ray energy of 1.78 MeV. Some of the created ^{28}Al atoms were collected with the debris collector made of Nb covering 85% of the 4π solid angle, and then transported to a β–and γ–ray counting system. The possible uncertainties included in the Si activation measurements are 1) debris collection efficiency, 2) backgrounds in the measurements, 3) Si amount and distribution in the target, 4) neutron yield measurements, and 5) statistical uncertainty. Intensive examination [5] showed that systematic uncertainties are insignificant, and uncertainties are mostly due to the statistics.

1) The debris collection efficiency is one of the most crucial elements in the <ρR> measurements. The efficiency has been routinely calibrated by using a target containing a radioactive ^{24}Na tracer. This Na calibration is assumed to apply also to ^{28}Al. This assumption was validated from the measured atomic number dependence of the efficiency. Finally, the efficiency has been cross-calibrated in simultaneous <ρR> measurements by using knock-on technique. The reproducibility of the debris collection efficiency was examined and found to be than ±7.3% (fractional) for the 4π-type collector (Fig. 1). In contrast, the collection efficiency of a collector with a much smaller solid angle (3% of 4π) varied by more than a factor of two on a shot-by-shot basis.

2) If aluminum has accumulated in the collector, it could be activated by neutrons or by energetic ions. The resulting ^{28}Al is indistinguishable from the ^{28}Al produced in the imploded target material. Similarly, if contamination is accumulated or contained in the collector, it could produce indistinguishable background. These possible backgrounds were examined under the same experimental conditions as those in the previous high-density compressions except only for no Si containment in the target material. Data taken in this particular condition showed no ^{28}Al signature as shown in Fig. 2 (sum of 8 identical shots), being contrast to the clear ^{28}Al decay observed from the Si contained targets. Therefore, these possible background are insignificant.

Neutron-activation of the collector could produce another background. This effect was also examined in the experiment where the collector was shielded with a steel housing from the target debris and charged particles and then irradiated by DT neutrons with a flux level higher than that at the high-density experiments by more than three orders of magnitude. Even with this extremely high neutron flux, no background was observed in the detection window of the β coincident γ-ray. It is concluded that the background due to the neutron activation of the collector itself is negligible.

The only background we found was a low level background with a half-life of several hours, and easily distinguished from the ^{28}Al signal. This is seen in Fig. 2 after the ^{28}Al signal well decayed. The background level was determined by counting it overnight for many of the

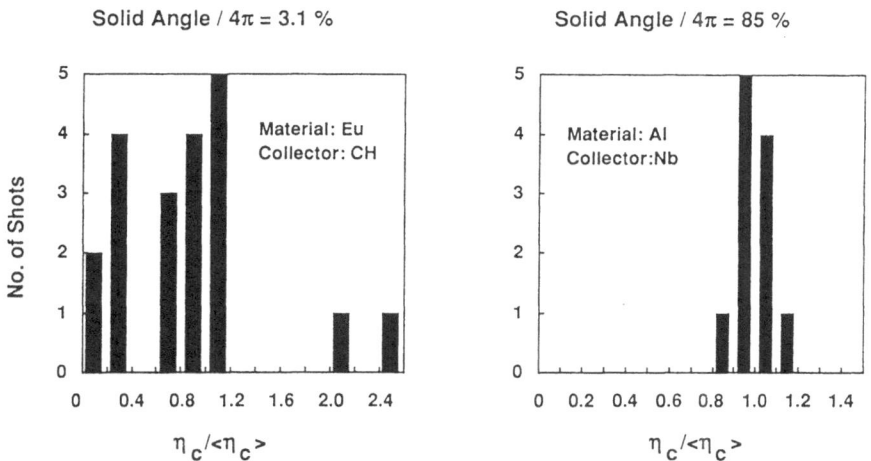

Figure 1 Reproducibility of the debris collection Efficiency η_c for the small solid angle and the 4π-type collectors.

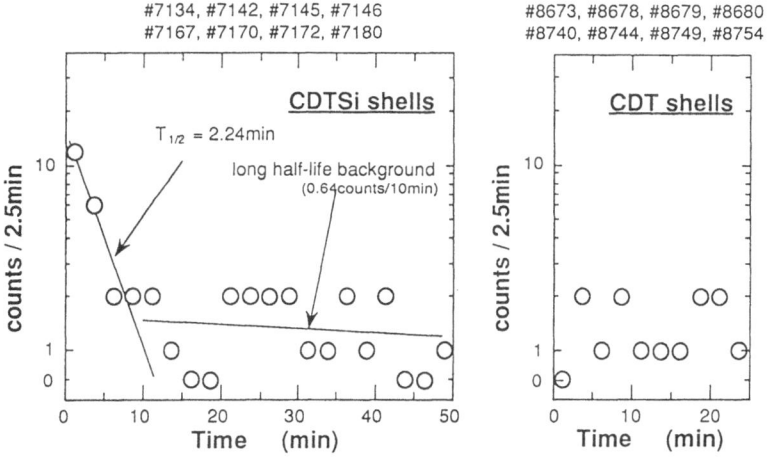

Figure 2 Time decay of 1.6-2.0-MeV γ-rays observed from the targets with and no Si containment.

target shots to be 0.6 counts/10min, including the natural background. This background was not taken into account in the previous reports [1-3] and the <ρR> value determined here will be slightly lower than previously quoted value. [The source of this background is probably ^{93m}Mo produced by deuteron bombardment from the deuterated target on the Nb collector via ^{93}Nb(d,2n) reactions. The decay time of 6.85 h is consistent with the observation. Several γ-ray peaks observed in the spectra are consistent with ^{93m}Mo and two of these lines (1.48 and 0.26 MeV) have a energy sum coincident with the γ-ray energy from ^{28}Al.]

3) The Si amount and distribution in the target could be affected either by polymerization or by tritiation processes. The amount and distribution were examined by measuring Si Kα lines from the targets excited by 15keV- electron beam. The targets used here are identical to those used in the high-density experiments for the size, nominal Si amount, and the tritiation level. Since x-ray reabsorption by Si atoms in the target is negligible, the intensity of Si Kα x rays should be directly proportional to the Si density in the target. The x-ray intensity from the tritiated plastic shell was compared with that from a "pilot" sample target, which was made of pure *para*-trimetylsilylstyrene and therefore Si amount is exactly known. The measured-to-expected ratio of the Kα intensity at several points of the target was 0.98±0.07. It appears, therefore, that the Si amount is almost equal to the nominal value and the distribution in the surface is uniform. Since the electron range of 4 μm in the plastic is shorter than the 10-μm shell thickness, the radial distribution of Si may be evaluated by measuring the intensity ratio of the inner to the outer surface of the target. The measured ratio was 1.15±0.16, indicating that the radial distribution is likely uniform.

4) The neutron yields were measured with two time-of-flight (TOF) detectors, each consists of a plastic scintillator coupled to a photomultiplier. In the present target containing a small amount of tritium, DD neutron yield is comparable to DT yield. Figure 3 shows example where DT neutrons are distinguished from DD neutrons by the TOF difference. These TOF detectors were absolutely and routinely calibrated via Al and Cu activation techniques. We estimate the possible uncertainty of the yield to be less than ±15%. Figure 3 also shows the neutron yield deviation from the mean value.

5) The statistical number of the signal in these experiments were only 2-4/shot and, therefore, the data were averaged over 9 identical shots (26 total counts). Other several shots were excluded from the analysis, because in these shots target had a long wavelength nonuniformity (l <5-10). The long wavelength nonuniformity was estimated from the deformation of interference fringe of the target taken with an interference microscope INTERPHAKO with illumination replaced with He-Ne laser. Figure 4(a) displays examples of "good" and "bad" targets. Figure 4 (b) shows the trend of the <ρR> degradation with increasing target nonuniformity.

We used two neutron time-of-flight detectors.
Detector 1: PilotU + channel plate PMT
Detector 2: NE102 + dynode PMT

• Absolutely calibrated with Al and Cu activation techniques
• ~500 recoil events in each detector for 10^6 yields
• DD neutrons eliminated with TOF

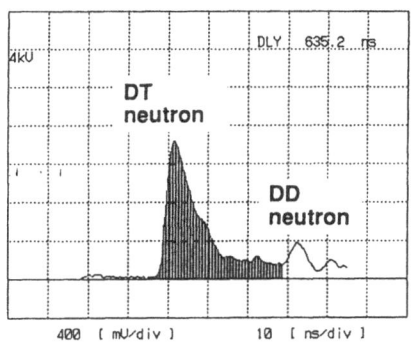

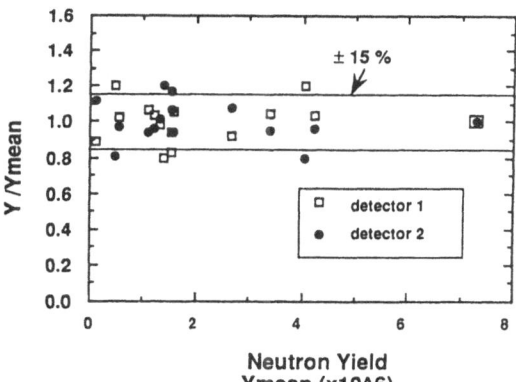

Neutron Yield
Ymean (x10^6)

Figure 3 Example of the neutron time-of-flight signal for the neutron yield measurements. Also shown is the fluctuation of the measured yields around the mean value.

The areal density was then determined to be 0.51 ±0.14 g/cm^2. The uncertainty includes all possible uncertainties mentioned above (statistics, collection efficiency, and neutron yield). The slight reduction of this value from previously quoted one is due to the background discussed above. As shown in Fig. 5 the areal densities agree with predictions of the one-dimensional (1D) simulation, HIMICO, up to a convergence ratio (initial target radius divided by the final radius) of about 30. From this <ρR> and the measured ablated mass, the density was calculated to be 540^{+240}_{-210} g/cc (deuterium density of 500 XLD). At the same time the hydrodynamic efficiency from the absorbed to the core plasma energy (estimated from the core mass and temperature) was 4 - 8 %, which also meets one of the ignition criteria.

We also performed recently a cross-examination of the high density compressions, using an independent technique based on Fermi degeneracy [6]. The electron density in the high-density plasma is calculated from the achieved density to be 2×10^{26} / cc. Since the corresponding Fermi temperature of 1 keV is higher than the plasma temperature of ≈ 0.3 keV, the electrons are expected to be partially degenerate. Charged particle stopping power in such degenerate plasma is reduced from its classical value. One measure of the stopping power reduction is a ratio of the secondary DT to the primary DD fusion neutron yield. This ratio is proportional to the stopping range of 1-MeV DD tritons, if the range is shorter than the fuel <ρR> as in the present case. For plastic targets (Z = 3.5) and T_e = 0.3 keV, the secondary-to-primary yield ratio increases from 2×10^{-3} to 8×10^{-3} with increasing density from 100 to 1000 g/cc. The experimental results are consistent with the previously observed density.

449

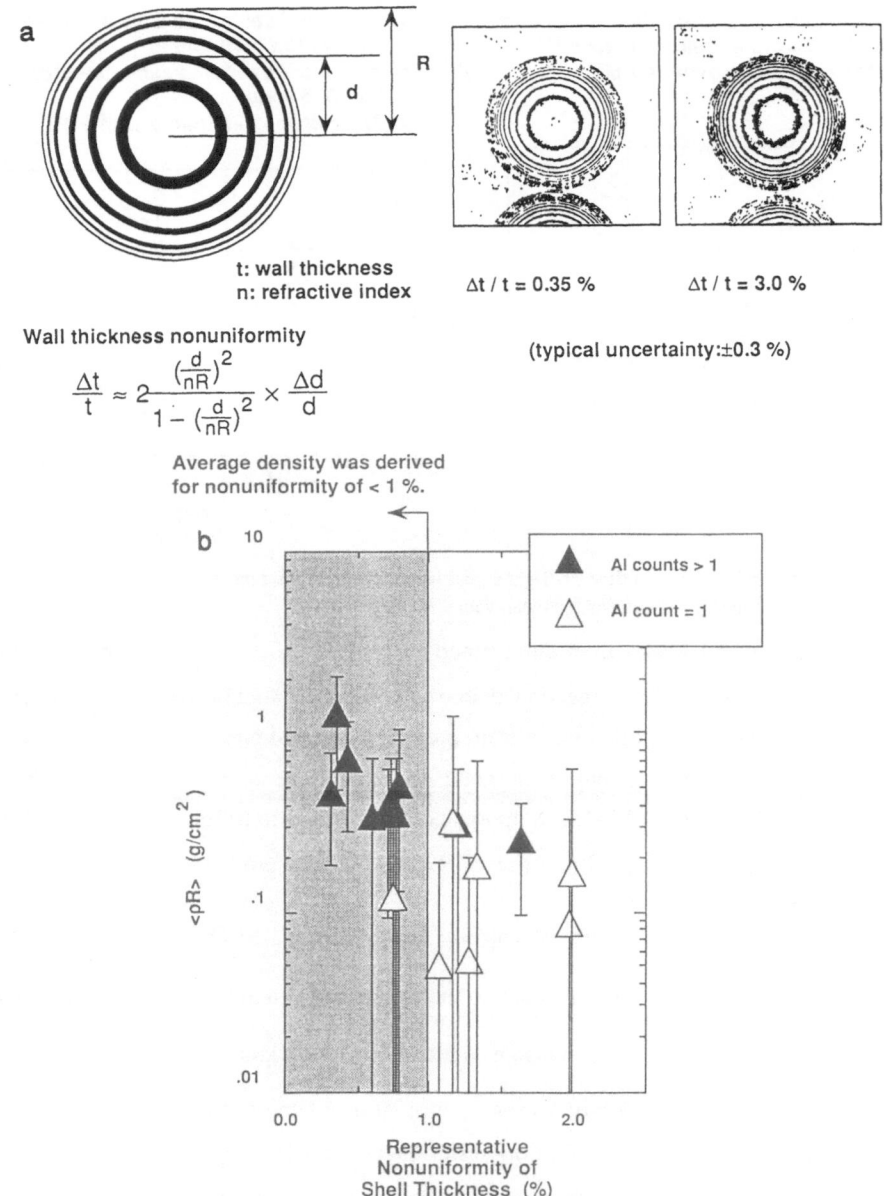

Figure 4 (a) Sample photographs of plastic targets taken with the interference microscope INTERPHAKO and a simple formula to estimate the target nonuniformity. (b) Measured <ρR> vs. target nonuniformity.

The neutron yield was, however, about 10^{-3} of the 1D calculations at a convergence ration of 30 (Fig. 6), where the high density was achieved. We tentatively explained that the remaining irradiation nonuniformity might cause a collapse of the hot spark from which most neutrons are generated in the 1D simulations. This explanation of the hot spark disappearance is also supported by the fact that the ion temperature of ~ 0.5 keV, recently measured with a neutron time-of-flight technique, is consistent with the cold fuel temperature of 0.4 keV rather than the hot spark temperature of 1.2 keV predicted by the 1D simulation.

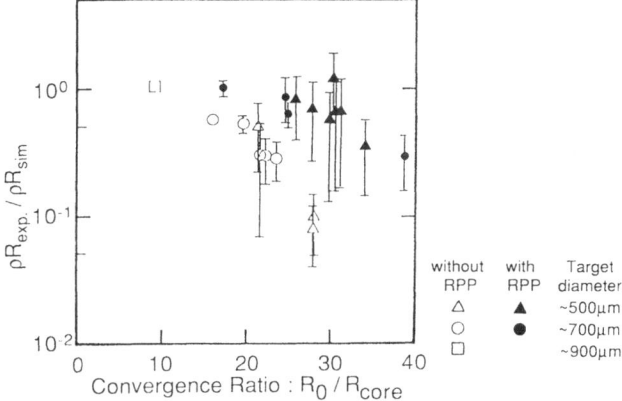

Figure 5 Target areal densities normalized by 1D predictions as a function of the convergence ratio defined as the initial target radius divided by the compressed core radius. Nearly 1D performance is achieved when the convergence ratio is less than 30.

II-2. Analysis of Rayleigh-Taylor Instabilities

Are the success of high density compression and the failure of the spark formation consistent with the present understanding of Rayleigh-Taylor (RT) instabilities? In order for the target to be compressed to the high density, the shell has to survive under the instabilities at least in the acceleration phase. We will compare in this section the mixing width at the ablation front to the shell thickness, using Haan's mix model [7]. The RT instability at the inner surface will be discussed in Sec. II-3.

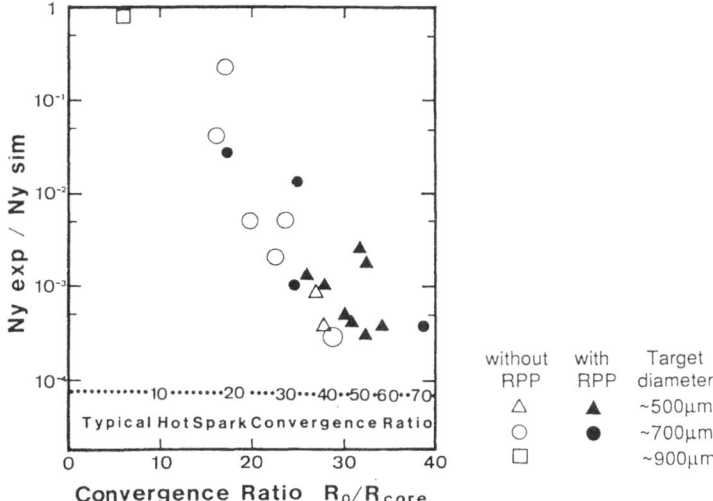

Figure 6 DT neutron yields normalized by 1D prediction as a function of the convergence ratio. Also shown as a horizontal axis is hot spark convergence ratio defined as the initial target radius divided by the hot spark radius.

There are two sources of shell distortions which might be caused by irradiation nonuniformities. The first is RT instability resulting in the exponential growth in time. The second is "secular" growth of distortion in square of time if there is no instability. These two classes of distortion must be coupled and therefore we used a forced RT equation for the shell distortion ξ.

$$\frac{d^2\xi_k}{dt^2} = \gamma_k^2\xi_k + \delta g_k,$$

(1)

where δg_k is the gravity perturbation, γ_k is the RT growth rate, and k is the wave number of the perturbation, which is related to the mode number l and the radius of ablation front R_a by $kR_a = [l(l+1)]^{1/2}$. The gravity perturbation may be given as $\delta g_k/g \approx \delta p_k/p$, where p is the pressure applied on the surface considered. We also used Takabe formula [8] for the RT growth rate including ablative stabilization effect,

$$\gamma_k = 0.9\sqrt{kg} - 3kv_a,$$

(2)

where v_a, the ablation velocity in the frame moving with the shell, is a ratio of mass ablation rate $\dot{m}$ divided by the density ρ_a at the ablation front, This formula has been validated by the excellent agreement with 2D simulation results in various laboratories [9]. We assume that the perturbation generated at the absorption surface decays exponentially away from this surface as exp(-kD), where D is the stand-off distance between the ablation front and the absorption surface at which the maximum absorption takes place. A detail analysis shows significant smoothing [10], but we use the above formula as a conservative evaluation. The pressure perturbation δp_k is therefore related to the perturbation of the absorbed laser intensity δI_k as,

$$\delta p_k/p = (2/3)\delta I_k/I \cdot exp(-kD).$$

(3)

Two classes of initial distortions were included in Eq. (1). The first, commonly referred to as initial imprint, comes from ablated mass and pressure nonuniformity before the target shell substantially moves. We treated the initial imprint so that if there is no instability the ξ_k is equal to the time integral of the perturbation in the ablation front velocity $\int\delta u_k$ dt. The ablation front velocity v should be proportional to $p^{1/2}$ during the time before the first shock reaches the inner surface of the shell, then $\delta u_k/u \approx 0.5\delta p_k/p$. The second is of course the target finish. The surface finish of a plastic shell (fabricated in the same process but has a 1mm diameter) was measured with a phase-shift interferometry (WYKO) for the wavelength range from 40 to 1 μm (corresponding to $l \approx 80-3000$) to be 40 Å in rms with mode dependence of $l^{-1.4}$. Assuming the same l dependence down to the low modes and the same modal amplitude

a_l, the rms roughness of the target used in the compression experiments are calculated from $\xi^{target}(t = 0) = [\Sigma(2l+1)a_l^2]^{1/2}$ to be 0.03 μm. We used this amplitude and the mode dependence in the present model. The RT instability growing from the surface finish may be described by Eq. (1) but without gravity perturbation of the second term in the right hand side. Since perturbations originated from the target finish and laser nonuniformity will have random phase, the combined amplitude will be the quadrature sum,

$$|\xi_k(t)| = [\xi_k^{target}(t)^2 + \xi_k^{laser}(t)^2]^{1/2} \tag{4}$$

The fluid compressibility was phenomenologically taken into account by considering that the areal mass perturbation grows instead of the displacement, so that for a solution ξ_k^0 of eqs. (1)-(3) the perturbation amplitude is given by $\xi_k(t) = [\rho_0/\rho_a(t)]\xi_k^0(t)$, where ρ_0 is the initial target density. The 1D trajectories of the ablation front R_{out}, inside surface R_{in}, the maximum absorption radius R_{abs}, critical density radius R_c (as a reference), and the gravity g are shown in Fig. 7. These were calculated with HIMICO with a flux limiter of 0.1 in the Spitzer-Harm electron thermal conduction and with the Thomas-Fermi equation-of-states where a bonding correction is included. The R_{out} and R_{in} are defined as the radius at which the density becomes 1/3 of the maximum. To eliminate numerical divergence in determining the gravity and the velocity of the ablation front, the R_{abs} was fitted with polynomial regression and then differentiated. The mass ablation ratio and the maximum density histories which are used to determine the ablation velocity are shown in Fig. 8. Here again for eliminating the numerical divergence in the mass ablation rate, the target mass integrated from the target center to R_{out} was fitted by polynomials and differentiated to give the mass ablation rate.

Figure 9 (a) shows the l mode spectra of the laser irradiation nonuniformity, whereas Fig. 9 (b) shows the target finish $\xi_l^{target}(0)$, and the target surface perturbations $\xi_l(t) = (\Sigma_m \xi_{l,m}(t)^2)^{1/2}$ at a time t_s of the first shock traverse as well as at a time $t_{max}=1.9$ ns near which the rms amplitude becomes its maximum. For the perturbation amplitude at t_{max}, those with and without nonlinear saturation effect [7], $\xi_l(t_{max},)$ and $\xi_l^s(t_{max})$ respectively, are shown . It was found by artificially turning off the gravity perturbation in eq. (1) that the low l modes (l <20-30) are caused by the forced RT instability, whereas higher l modes are due to the initial imprint and the target surface roughness.

The mix width at the ablation surface was evaluated as $\sqrt{2}\xi_{rms}$, where ξ_{rms} is the rms perturbation of all modes given as $\xi_{rms}=(\Sigma\xi_l^2)^{1/2}$. This is shown in Fig. 10 with the dotted line as well as the shell thickness (solid line). The shell breaks early in time at 1.6 ns at which the

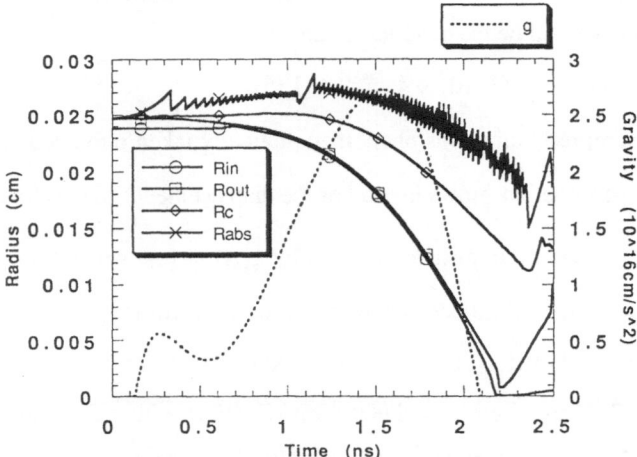

Figure 7 Radius-time trajectories for the typical high density target. The gravity calculated from the ablation front radius Rout is also shown.

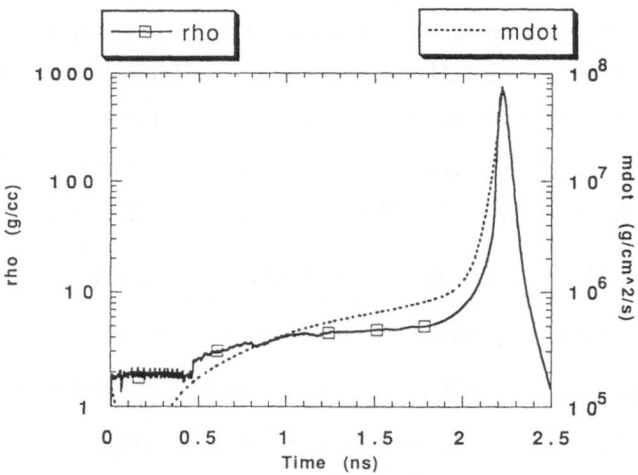

Figure 8 Mass ablation rate and the maximum density as functions of time.

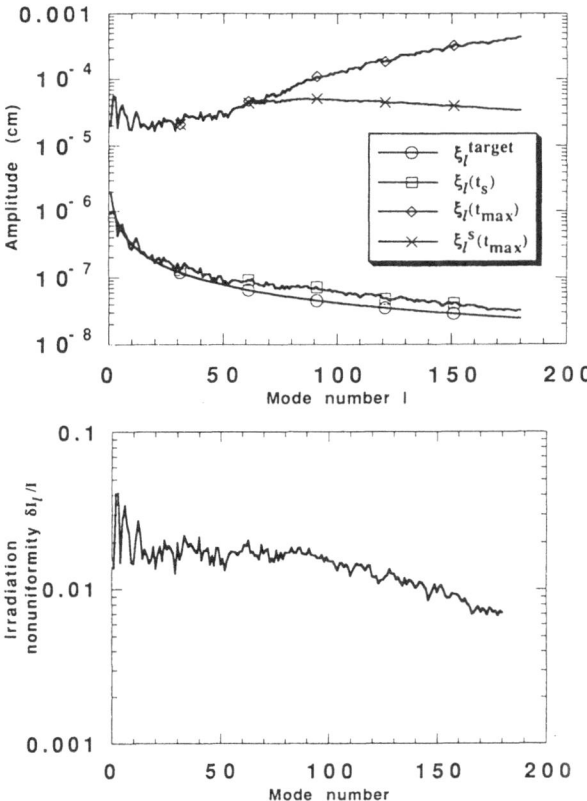

Figure 9 (a) Laser irradiation nonuniformity $\delta I_l / I$. No thermal smoothing is assumed. (b) Modal perturbation amplitude $\xi_l(t_s)$ at the time t_s of the first shock traverse through the shell and that at the time t_{max} of the maximum growth without and with saturation effect, $\xi_l(t_{max})$ and $\xi_l^s(t_{max})$, respectively. The target surface finish $\xi_l^{target} = (2l+1)^{1/2} a_l$ is also shown.

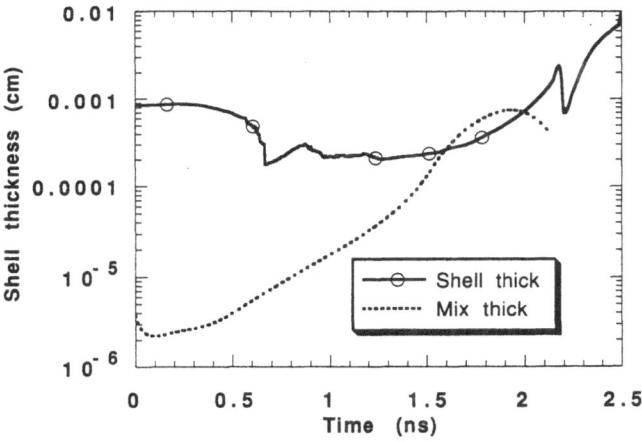

Figure 10 Shell thickness and mixing layer thickness vs. time. The shell brakes at an early time of 1.5 ns.

shell has moved only 1/3 of its initial radius. At present we do not have any interpretation consistent with the experimental observation of the high density. If we add hot-electrons in the simulation by an amount of 1% of the absorbed laser energy, then the mixing thickness becomes less than about half of the shell thickness during the entire time of the implosion. However, the high density is never obtained in this case and the maximum density of 200 g/cc is inconsistent even with the experimental lower limit (330 g/cc). It is unlikely that the collapse of the hot spark due to the RT instabilities in the deceleration phase would increase the density by about a factor of three. Indeed, another model which includes the mixing by anomalous diffusion (details will be discussed in the following sub-section) could not explain the high density compression if the shell is preheated in the acceleration phase.

There are several possible mechanisms which are not included in the present model but have a potential to relax the RT constraints: a) As the target moves, a certain point of the surface "feels" oscillating pressure perturbation in time (dynamic smoothing). A similar effect is also expected to occur for the absorption when the critical density layer moves; b) As Manheimer analyzed[10], the thermal smoothing can be larger than that would be suggested by simple argument of the exp(-kD) form; and c) The mixing itself may decrease the density gradient at the ablation front, thereby reducing the RT growth rate. Perhaps, some or all of these mechanisms might possibly explain the disagreement with the experiments.

II-3. Simulation with Turbulent Diffusion Model

In order to simulate the high density implosion experiment by including the effect of mixing due to the Rayleigh-Taylor instability, we have developed a simple, nonlinear diffusion model. The model is based on the quasi-linear diffusion theory usually used in the kinetic theory of plasmas. The essence of the model is as follows.

Let us assume that ξ represents the displacement due to the instability and f_0 is some physical quantity in case without the instability. In this case, the perturbation f_1 is roughly given to be

$$f_1 = -\xi \cdot \nabla f_0 \tag{5}$$

The convection term including the lowest nonlinear effect is given to be

$$u \cdot \nabla f = u_0 \nabla f_0 + \langle u_1 \cdot \nabla f_1 \rangle \tag{6}$$

where u is the flow velocity and u_0 and u_1 are its 0-th and 1-st order quantities respectively. In Eq. (6), the symbol < > means to take average over the macroscopic scale. Then, the time variation of f_0 is governed by the form;

$$\frac{df_0}{dt} = \nabla(D\nabla f_0) + (\text{other terms}) \tag{7}$$

where

$$D = \left\langle \xi \cdot \frac{d}{dt} \xi \right\rangle \tag{8}$$

In deriving Eqs. (7) and (8), we have assumed the incompressibility condition, $\nabla u_1 = 0$, and used the relation $u_1 = d\xi / dt$. In the spherical geometry, the perturbation ξ is decomposed with the spherical harmonics and the diffusion coefficient D is given to be

$$D = \sum_{l,m} \xi_{l,m} \cdot \frac{d}{dt} \xi_{l,m} \tag{9}$$

In the simulation, the diffusion of Eq. (9) is included in the equation of energy of 1-D implosion code to model an anomalous diffusion by the turbulent mixing, which is driven by the nonuniformity of laser irradiation and the growth of the Rayleigh-Taylor instability.

The target with which the highest density has been achieved is made of C (42.8 %), D (53.5 %), T (1.3 %) and Si (2.4 %) in number fraction. The laser of 8.14 k Joule in 0.53 µm wavelength is irradiated in the form of two Gaussian with 1.3 and 1.0 nsec FHWM's with the separation of 750 psec from peak to peak. Since the random phase plates are used, about 65 % of the irradiated energy is focused on the target.

The one dimensional implosion code ILESTA-1D [11] has been used. In the code, a fitting formula for the equation of state of matters described in Ref. [12] is installed. In the simulation of the high density implosion experiment, not only the mixing but also the radiation preheat are key element, because silicon doped in a target for diagnostic purpose may preheat the target, consequently reducing the achieved density and neutron yield. The radiation transport has been treated with the multi-group flux-limited diffusion model and a non-LTE atomic model has been used.

When a relatively wide spectral grouping is used, the silicon K-line radiation is substantially generated and preheats the shell. Consequently, the both of ρR (density radius product) and Y_N (neutron yield) are reduced due to less compression. This is shown with the data point B in Fig. 11. In this case, the numerical neutron yield coincides with the experimental one which is described by the data E with error bar in Fig. 11, while the ρR product is much less than the experimental one.

In order to improve the line radiation transport, more detail zoning has been done near the K-line radiation (~2 keV) and the zoning has been adjusted so that the amount of hard x-ray coincides with that obtained in the experiment. The result of this case is shown with the point C in Fig. 11. In this case, less preheat allows higher density and the ρR product comes near the edge of the error bar of the experimental one. A hot spot of the radius about 5 µm with the density 100 g / cc and temperature 2-3 keV is generated. The hot spot is surrounded by high

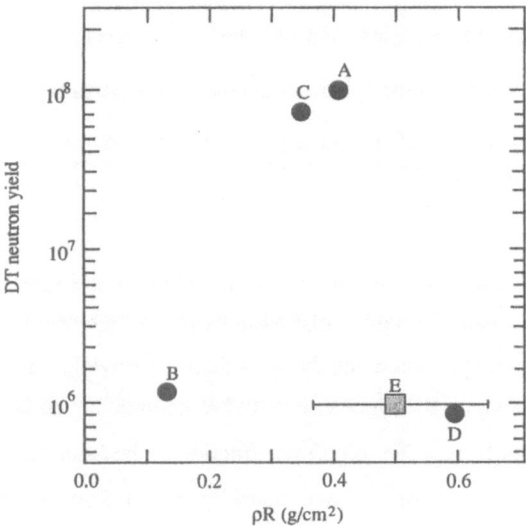

Figure 11 Experimental and numerical yields of ρR and neutrons. The point A (w/o mixing) ; without radiation. The point B (w/o mixing) ; including radiation transport with relatively wide spectral grouping. The point C (w/o mixing) ; after improvement of radiation spectral grouping. The point D ; with the improved spectral grouping plus turbulent mixing effect. The point E ; experimental data with error bar.

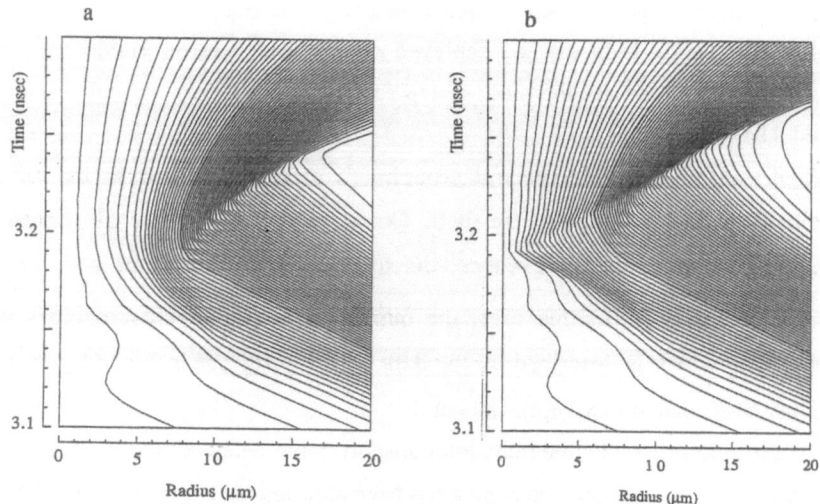

Figure 12 The r-t diagram of the high density implosion experiment near the maximum compression.
(a) Nominal one dimensional simulation.
(b) Turbulent mixing effect is installed in the case (a) as an anomalous energy diffusion.

density ($\sim$ 500 g / cc, $\rho R \sim$ 0.3 g / cm^2) and low temperature ($\sim$ 300 eV) region. This is a typical structure seen in a hollow-shell-implosion with less preheating and no mixing. As the result, higher neutron yield is obtained in the hot spot region and the value is found to be two order of magnitude higher than the experimental one. For reference, the case without radiation transport is plotted with the data point A in Fig. 11.

It is clear that the improvement of the radiation preheat alone can not reproduce the experimental results. As next step, we installed the quasi-linear diffusion model in the ILESTA-1D code. In the code, each Fourier component of Eq. (9) is given by solving the time development of linearized equations to the fluid dynamics. When we assumed 15 % absorption nonuniformity with constant amplitude in $l = 1 \sim 100$, it is found that the ablation front diffuses through the accelerating shell and the burn-through due to the mixing is seen. This is consistent with the result of the model analysis described in Sec. II-2.

For the purpose to reproduce the experimental result, therefore, we have assumed that the acceleration phase is stable because of some reason which is still in open question. Then, we have seen the effect of the mixing in the final compression phase which has been pointed out to be unstable to the Rayleigh-Taylor instability [13]. In the simulation, we start to include the nonlinear diffusion term when the void closure takes place. The total 15 % displacement of $l = 1 - 100$ is imposed. The r-t diagrams near the final compression phase are shown for the cases without mixing [Fig. 12 (a)] and with mixing [Fig. 12 (b)]. We have used the model for the radiation transport with which the data point C in Fig. 11 has been obtained.

In the case without mixing, the hot spot region has high pressure enough to stagnate the surrounding high density region. Since any conduction and energy loss processes can not cool the hot spot, excess neutrons are produced in this region. In contrast, the inclusion of the mixing process allows rapid cooling of the central region due to the anomalous energy diffusion. As the result, the central region can not sustain the surrounding material and no stagnation process is seen. It is noted that the density and temperature structures near the maximum compression are rather flat for the case with the mixing. The resultant neutron yield and ρR values are potted in Fig. 11 with the point D. It is concluded that no formation of the hot spot region due to the anomalous energy diffusion allows to reproduce the experimental implosion parameters

III. PROSPECTS TOWARDS FUEL IGNITION AND HIGH GAIN

It has been demonstrated that high neutron yield up to 10^{13}/shot can be obtained with a thin shell glass microballoon filled with DT gas fuel [14] and high density up to $300 \sim 800$ times solid density can be obtained with a deuterized polyethylene hollow shell target as described in Sec. II. As the next step, simultaneous achievement of high neutron yield and high density is expected to be demonstrated with precision laser systems with better irradiation uniformity and controlled pulse shape. In order to demonstrate fuel ignition, high density and high temperature are required with the parameters such as $\rho R = 0.3$ g/cm^2 and $T_i = 5 - 10$ keV of a compressed D-T fuel. Then, the alpha-particles generated in the central core are substantially stopped by depositing their energies into the fuel, and consequently substantial enhancement of neutron yield can be expected because of the self-heating process.

The required driver energy for fuel ignition strongly depends upon the achievable density. With 1000 times liquid density ($\times$ LD), a compressed spherical DT plasma of $\rho R = 0.3$ g/cm^2 has the total energy of 3 kJoule for the case of $T_i = 10$ keV. This means that with the hydrodynamic efficiency (η_h) of 10 % the required driver energy is roughly 30 kJoule. If the density drops down to 500 $\times$ LD, the required driver energy increases to 100 kJoule with the assumption of $\eta_h = 12$ %. From such estimation, it is reasonable to explore a possibility of fuel ignition demonstration with a driver with output energy of about 100 kJoule.

It has been said that the low isentrope compression is essential to achieve the high density compression and high gain. for the low isentrope compression, we can enumerate the followings as critical elements and/or requirements for laser drivers;

    (1) Pulse shaping,

    (2) Irradiation uniformity,

    (3) Laser wavelength.

The pulse shape is required to precompress the target shell to high density by relatively weak shock waves. After the precompression, the shell is accelerated by a main pulse and finally kinetic energy of the shell is converted to the thermal energy in stagnating at the center. The achievable density is proportional to the accelerated velocity which is controlled by the duration of the main pulse.

The irradiation uniformity is required to keep uniform ablation pressure which drives the shell acceleration. The nonuniformity initially imprinted on the shell through the irradiation nonuniformity starts growing because of the Rayleigh-Taylor instability in the acceleration phase. The spherical effect and ablative stabilization of the Rayleigh-Taylor instability should be studied and used to achieve relatively stable target design. The wavelength of the laser light should be chosen so that less high energy electrons and x-ray are generated through the interaction process and sufficient thermal smoothing can relax the requirement on the laser irradiation uniformity. A combination of long wavelength and short wavelength lasers should be valuable if it can relax the requirement to uniformity without degrading the laser-target coupling.

In the present section, we report the one-dimensional target design of high gain and ignition with the ILESTA-1D code. We report a high gain target design with low isentrope compression mode. However, such implosion mode requires higher irradiation uniformity. It is shown that a high isentrope implosion is better in stability and it can provide a model experiment of a spark region of a high gain target. We present a model simulation of 100 kJoule ignition by high velocity and high isentrope implosion mode. Uniformity requirement is estimate with the forced Rayleigh-Taylor instability model used in Sec. II-2.

III-1. High Gain Target Design by Tailored Pulse

As seen in Sec. II, at the present time it seems to be hard to realize a structured implosion core characterized by central hot spot plus surrounding high density region. It is well known, however, that the structured core is essential to increase the target gain high enough. In the present sub-section, we study a target design for high gain without discussing its stability.

For realizing a structured core plasma with a spark in the center surrounded by a low isentrope main fuel, we use a target consisting of a frozen DT layer contained by a plastic shell. The thickness of the plastic layer is chosen so that almost all the plastic ablates away during the acceleration phase. A target finally used for the high gain design with 1.2 MJoule laser, the detail of which will be shown below, has parameters; DT layer of inner radius 1606 μm and outer radius 1880 μm, and a plastic layer of outer radius 1917.5 μm. The initial aspect ratio of the DT layer is about 9.

A tailored pulse with linearly ramped prepulse followed by a main pulse is irradiated on the target. Short wavelength (0.25 μm) and f/10 lens with d/R = -10 focusing are assumed. The intensity of the prepulse is kept 10^{13} W/cm^2 on target surface with the pulse duration τ_1, while the peak power of the main pulse which has a pulse duration of τ_2 is about 100 times that of the prepulse.

In Fig. 13, the density, temperature and pressure profiles at three different times near the maximum compression are shown for the case without the alpha particle heating, where the inclusion of the alpha particle heating leads to the maximum gain of 140 at driver energy of 1.2 MJoule. It is seen in the figure that the central region with the density 500 × LD and the temperature 3-4 keV forms a spark with $\rho R = 0.3$ g/cm^2 surrounded by a main fuel of density 3000 × LD and temperature of 400 eV. The isentrope parameter of the main fuel is roughly 2.5. It is noted that the entropy increase is mainly due to the heating by a reflected shock generated after the void closure at the center.

The target gain depends on the pulse durations. In Fig. 14-(a), the gain dependence on the main pulse duration τ_2 is shown for the case with an optimized initial pulse duration τ_1 (=

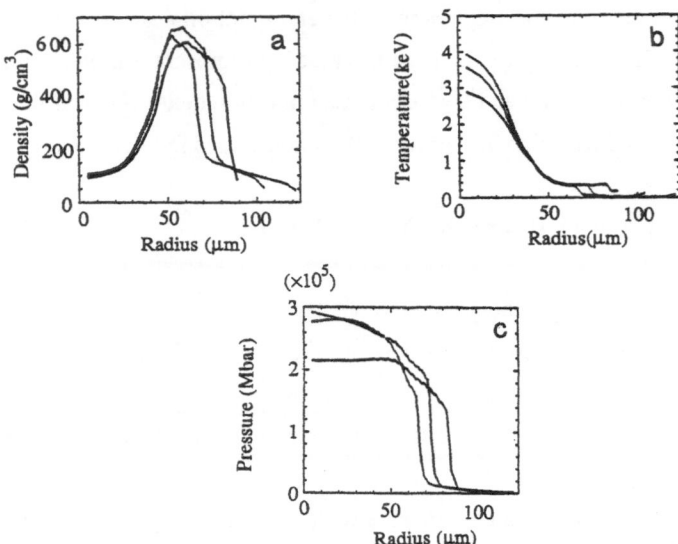

Figure 13 Spatial profiles of density, temperature and pressure at three different times near maximum compression. The implosion mode is the low isentrope one with laser energy of 1.2 MJoule irradiated with a tailored pulse. The spark and main fuel structure is clearly formed.

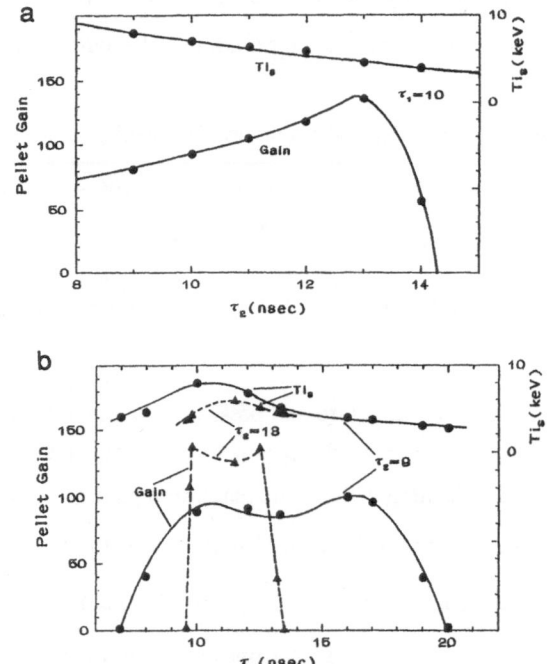

Figure 14 The gain curves as functions of laser pulse timing. The maximum ion temperature near the maximum compression is also shown for the case without the alpha particle heating.

(a) Gain dependence on the pulse duration of the main pulse, where the prepulse duration is kept 10 nsec.

(b) Gain dependence on the pulse duration of the prepulse, where the main pulse duration is kept to be 9 nsec (solid line) and 13 nsec (dotted line).

462

10 nsec). In the figure, the spark temperature for the case without the alpha particle heating is also shown. By decreasing the main pulse duration with keeping the total laser energy, the implosion velocity increases. Then, it is seen in the simulation that the density and temperature at the maximum compression increases as (density) $\propto$ (velocity)^3. Therefore, by reducing τ_2, the resultant increase of the spark temperature triggers the self-heating to abruptly increase the gain near $\tau_2 = 13$ nsec. After the peak gain, the gain gradually decreases with the decrease of τ_2. This is because the spark and main fuel are heated unnecessarily.

On the other hand, the gain dependence on the initial pulse duration shows a flat-peaked profile as seen in Fig. 14-(b). In Fig. 14-(b), the cases with $\tau_2 = 9$ and 13 nsecs are shown with the solid and dotted lines, respectively. When the initial pulse is to short, the second shock wave generated by the main pulse overtakes the first shock which is generated by the initial pulse, before the initial shock passes through the target shell. As the result, the strong second shock wave increases the entropy and avoids the low isentropy implosion. In contrast, if the initial pulse is too long, the second shock wave comes to the rear side of the shell when the rarefaction wave generated after the arrival of the first shock has sufficiently developed. Since the entropy increase is higher at lower density when a given shock wave passes through with the same strength, a relatively low density spark with less ρR product is generated at the maximum compression time. It is noted that the flat peaked gain region decreases as the peaked gain increases [compare the dashed and solid lines in Fig. 13-(b)].

After the same sort of the gain optimization for different laser energies, we finally obtain the driver energy scalings of the gain. In Fig. 15, the resultant gain curve is shown with the shaded region. The upper curve represents the peak gain in varying τ_1 and τ_2 for a given driver energy, while the lower curve is obtained for the case with moderately optimized pulse timing. In the latter case, the gain is rather insensitive to the variation of the timing. It is seen in Fig. 15 that such low isentrope design with predicted gain of about 200 at several MJoule driver requires the achievement of the gain 10 to 30 at 100 kJoule driver. However, such low isentrope implosion requires highly uniform targets and laser irradiations.

When we limit our design for 100 kJoule driver to the ignition demonstration, we can find a design in which the uniformity requirement is more relaxed than that for the design of Fig. 15. Actually, if we design a reactor of the pellet gain 100 at the driver energy of several MJoule, the required core plasma consists of the main fuel with a relatively high isentrope $\alpha = 3$-5 and a moderately compressed spark with density less than 50 g/cm^3 as predicted by Mayer-ter-Vehn's isobaric model [14]. In the ignition demonstration design at 100 kJoule driver, we try to find an implosion in which only the spark plasma modeling the spark of high gain target is produced. Then, relatively high isentrope implosion is allowed.

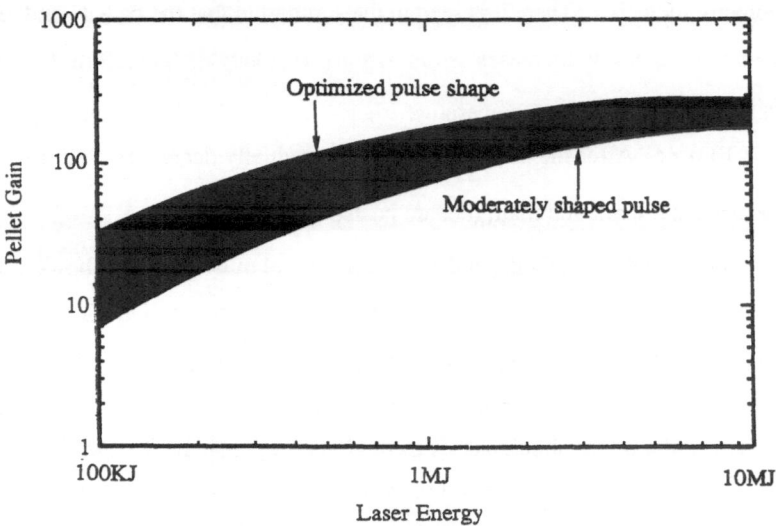

Figure 15 Laser energy scalings of the pellet gain for the low isentrope implosion mode. The upper curve of the shaded area is obtained by optimizing the laser pulse shape, while the lower curve of the shaded area is taken for the case with moderately optimized pulse shape. In the latter case, the pellet gain is rather insensitive to the variation of the pulse timing.

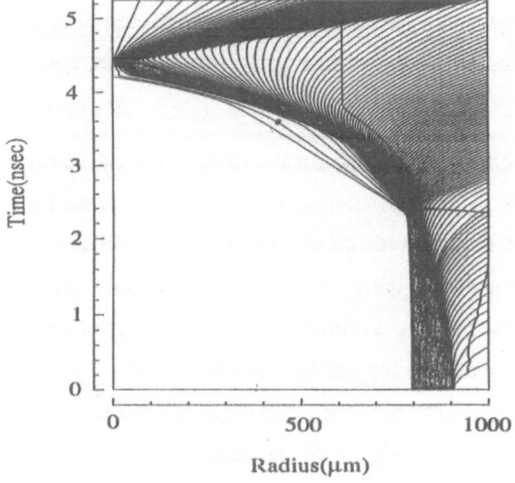

Figure 16 The r-t diagram for the high isentrope implosion. A tailored pulse with a prepulse of 1.06 μm and a main pulse of 0.35 μm is irradiated until the time when the ablation front comes to a half of its initial radius. In the figure the position of the cut-off density is also shown. The longer wavelength prepulse is used to relax the initial imprint due to irradiation nonuniformity via the thermal smoothing process. The shell is accelerated with the isentrope parameter equal to about 10 up to the velocity of 6×10^7 cm/sec.

III-2. High Isentrope Compression

In order to relax the uniformity requirement to drivers and targets, it is required to design a target so that the growth of the nonuniformities due to the Rayleigh-Taylor instability is sufficiently suppressed. In laser driven shell acceleration, the ablative stabilization can reduce the growth rate of the Rayleigh-Taylor instability and the self-consistent growth rate is given by Eq. (2). In order to design the target so that the Rayleigh-Taylor instability is reduced, we have to take account of the following two points;

    (1) Keep the wavenumber k small for a given l mode,

    (2) Enhance the mass ablation velocity.

The point (1) is required to reduce $(kg)^{1/2}$ term in Eq. (2). Since the wavenumber varies with time as $k = l/R_{AF}(t)$, R_{AF} should be large enough in the acceleration phase, where R_{AF} is the radius of the ablation front. This requires that the laser acceleration should terminate when the ablation front comes to roughly a half of the initial radius. The point (2) is required because of enhancing the ablative stabilization. Since V_a = (mass ablation rate)/(ablation front density) in Eq. (2), V_a can be enhanced by (a) increasing the mass ablation rate or (b) reducing the ablation front density. The use of shorter wavelength laser leads to enhancing the mass ablation rate. By using low density shell or increasing the entropy of the shell, we can reduce the ablation front density. We can enumerate at least two ways for reducing the ablation front density,

    (a) Heating by strong shock wave,

    (b) Heating by controlled radiation preheat by doping some radiator material in the shell.

Use of a picket-fence pulse [15] is also useful to reduce the ablation front density.

In the present report, we propose to use a strong shock heating by a tailored pulse. In addition, we propose to accelerate the shell up to the velocity of $6\text{-}7 \times 10^7$ cm/sec according to the idea described in Ref. 16. When the accelerating shell is heated up initially by a strong shock wave, implosion mode becomes the stagnation-free mode and it is hard to compress and heat the fuel using the stagnation dynamics. Then, it is required to heat the fuel up to the ignition temperature of 5-10 keV without invoking to the stagnation dynamics. This requires the high velocity acceleration.

In Fig. 16, the flow diagram of the high velocity implosion is shown. The target consists of 110 µm DT layer and 2 µm over-coated polyethylene layer with the inner radius of 790 µm. The irradiated laser is totally 106 kJoule. The pulse shape consists of a linearly ramped 1.06 µm wavelength prepulse until 2.5 nsec (18 TW at t = 2.5 nsec) and tailored main pulse of 0.35 µm wavelength until 3.8 nsec (130 TW at t = 3.8 nsec). In order to reduce the initial imprint of the laser nonuniformity on the target by the thermal smoothing process, we have used the prepulse of a long wavelength laser. The laser intensity on the target surface is

10^{14} W/cm^2 at t = 2.5 nsec and a strong shock wave is generated by the prepulse. As the result, the target is accelerated with the isentrope parameter $\alpha = 10$. In the simulation, the laser light is designed to be turned off when the radius of the ablation front becomes a half of its initial radius. At the time when the laser light is turned off, the coupling efficiency is 6 %, while it increases up to 12 % at the maximum compression. This is because the pressure work by the high pressure atmosphere near the ablation front continues to accelerate the imploding shell. The maximum velocity of the ablation front is 6×10^7 cm/sec.

In order to study the growth of nonuniformity of the shell dynamics driven by the irradiation nonuniformity, we have used the forced Rayleigh -Taylor instability equation given in Eq. (1). Then, we solve Eq. (1) numerically with the initial condition $\xi_l = d\xi_l/dt = 0$. We have related the pressure nonuniformity $\delta P_l/P_0$ to the laser nonuniformity $\delta I_l/I_0$ with the relation of Eq. (3), where $-kD$ is replaced by $-\Delta R/R_0 l$. In the calculation, we assumed $\Delta R/R_0 = 0.1$. In Fig. 17-(a), the amplitude of each mode is plotted as a function of time for the case where the total nonuniformity of laser irradiation is 2 % (σ_{rms}) with constant amplitude from $l = 1$ to 100.

In the calculation, we also take account of nonlinear saturation by including Haan's model [7]. In Fig. 17-(b), the total amplitude of the displacement at the ablation front (ξ_{AF}^t) is shown and the in-flight distance (X_F) and thickness (ΔX_s) of the shell are also shown. Within the present model the condition $\Delta X_s > \xi_{AF}^t$ may provide the condition that no shell break-up occurs. It is noted that in the present calculation, the mode with $l = 20$ has a peak in amplitude because of less thermal smoothing and moderately large Rayleigh-Taylor growth rate.

III-3. High Density Implosion, Ignition and High Gain

As described in Sec. II, structured core consisting of central spark and surrounding high density is not measured in the present experiment. To our surprise, however, the ablative acceleration process seems to be very stable and more stable than expected with conventional stability analysis based on the thermal smoothing and ablative stabilization. From such standing point, we concluded that if we can design an ignition demonstration with non-structured implosion core, it would be a reliable design.

We plan to perform the ignition demonstration with relatively high isentrope and stagnation-free target design as described in Sec. III-2. This design requires moderate pulse tailoring with combination of 1.06 μm prepulse and 0.35 μm main pulse. By increasing the implosion velocity up to 6×10^7 cm/sec, the core plasma is heated up to 5-10 keV without stagnation heating. A technical innovation is required to fabricate a D-T cryogenic target with

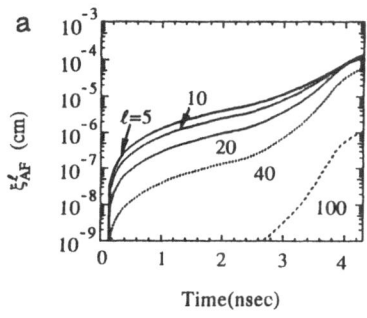

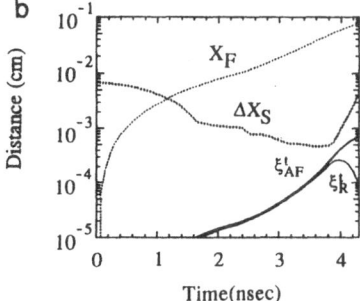

Figure 17 Growth of perturbations at the ablation front calculated with a simple forced Rayleigh-Taylor instability model. In the calculation, the total nonuniformity of laser irradiation is assumed to be 2 % rms with constant amplitude from $l = 1$ to 100.

(a) The amplitude of each mode with $l = 5$, 10, 20, 40 and 100 as a function of time.

(b) The total rms amplitude of the perturbation at the ablation front (ξ_{AF}^{t}) and at the rear side of the

shell (ξ_{R}^{t}). The in-flight distance of the shell (X_F) and the temporal shell thickness (ΔX_s) are also

shown.

the same level of uniformity as the present CD shell target. In addition, better irradiation uniformity of 2 % (compare 20 % in the high density experiment) is required to satisfy the no-shell-break-up condition as seen in Sec. III-2.

In order to scientifically proof the potentiality of such ignition scenario, it would be required to carry out a high velocity implosion experiment with the present CD shell target. If we are successful to achieve high density simultaneously with the high temperature, this can be a direct model experiment of the ignition demonstration based on the present scenario.

From ignition toward high gain, we have to demonstrate the stable formation of structured core. It is essential to realize a compressed core with hot spark and cold main fuel for achieving high gain. In Sec III-1, we have shown a predicted gain curve based on the low isentrope implosion of DT fuel contained by plastic ablator. Optimization of the tailored pulse shape can provide the gain larger than 100 at driver energy of 1 MJoule. However, the peak gain is sensitive to the pulse shape. In such low isentrope implosion, the Rayleigh-Taylor instability is classical and the ablative stabilization can not expected sufficiently. Then, the requirement to laser and target uniformities becomes much severe compared to the case of the ignition demonstration. It is, of cause, necessary to explore some new concept to relax the uniformity requirement in the high gain target design. However, we would be required to improve the uniformity even after the ignition demonstration. There could be a choice of alternative implosion scheme such as the indirect drive implosion. The present status of the indirect drive implosion is presented in Sec. V.

Finally we briefly summarize the issues to be solved.

(A) The ignition demonstration requires

    1) precision laser system with better uniformity,

    2) moderate pulse tailoring,

    3) high velocity implosion demonstration,

    4) cryogenic target fabrication.

(B) The high gain scenario requires

    1) demonstration of the structured core,

    2) control of optimized pulse shape,

    3) much better driving pressure uniformity.

The present status of the cryogenic target fabrication and experimental results are described in the next section.

IV. CRYOGENIC DEUTERIUM EXPERIMENTS IN PLANAR AND SPHERICAL GEOMETRIES

IV-1. Experimental Condition and Target Fabrication

The GEKKO XII laser was used with the parameters of energy up to 8 kJ with a 1.8 nsec full width at half maximum at $\lambda = 527$ nm. The energy balance among the beams could be as good as $\sigma_{rms} = 1$ %. To improve the spatial uniformity of the focused beams, random phase plates (RPP) were installed at each beam. The focusing condition was D/R = -5 for spherical targets, whose diameters can vary from 500 μm to 800 μm. Here D and R are the distance from the target center to the best focus and the target radius. The spot size was typically 600×700 μm with an oval shape for planar target irradiation. As spherical targets, we have used bare foam shells with liquid or solid D_2, plastic shells with solid D_2 layer inside, and foam shells with liquid or solid D_2 overcoated with ablator plastic layers. Foam plane targets with or without plastic layers have been also used.

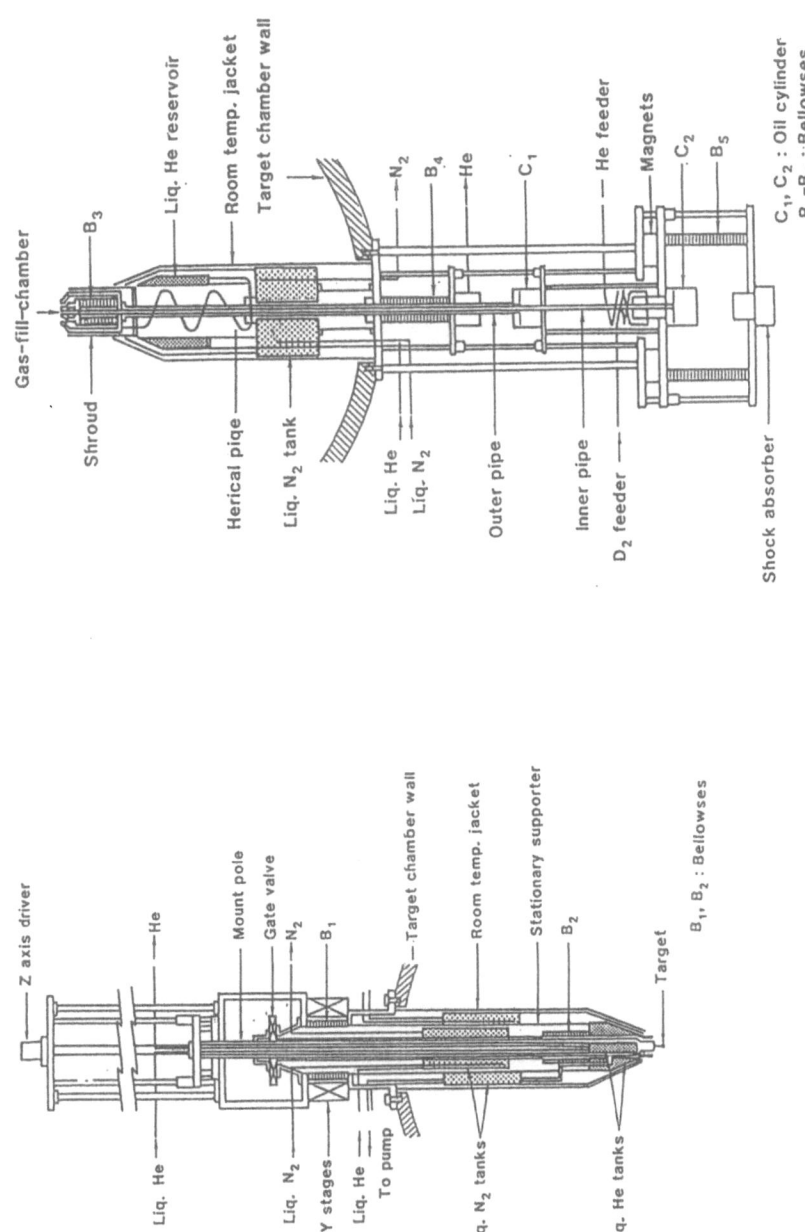

Figure 18 Schematic of cryogenic system.

469

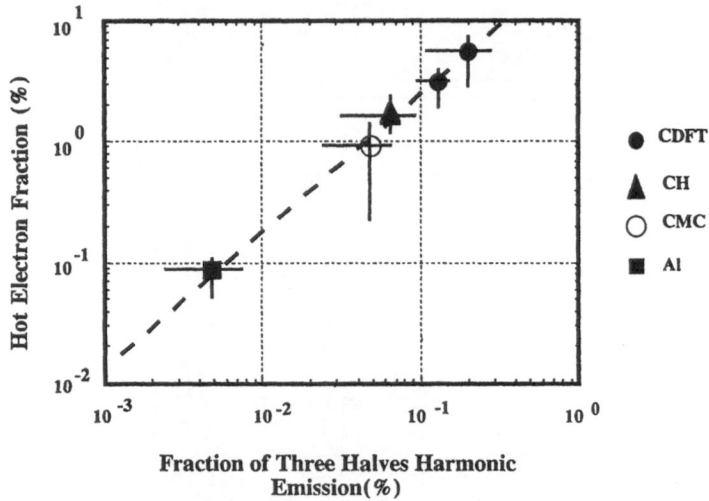

Figure 19 Hard x-ray ($T_h = 14$ keV) vs. three-half harmonic signals.

Figure 18 shows one of the cryogenic systems [17] employed in the experiments. The upper cryostat [Fig. 18 (a)] consists of target positioner, target mount pole, stationary supporter, and a vacuum system, which is activated to evacuate the residual helium gas for cooling the target. Two helium tanks are placed in both the target mount pole and in the stationary supporter. The lower cryostat assembly includes a gas fill chamber with four windows, cryogenic shroud, oil cylinders for actuating the gas fill chamber and the shroud, and a shroud retraction system for the laser irradiation as shown in Fig. 18 (b). Both upper and

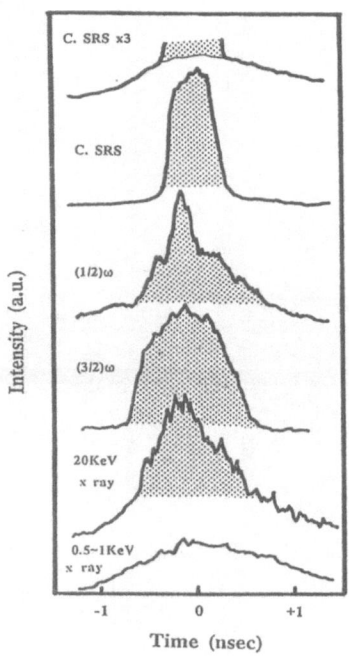

Figure 20 Temporal histories of convective SRS, $\omega/2$, $3\omega/2$, 20 keV hard x-ray, 0.5-1 keV soft x-ray signals.

470

lower parts of the cryogenic system are covered with jackets for thermal shielding. Interferometers are equipped on the same optical axes with the target monitor systems [18]. The laser system is protected with a target detection system [19] against any target misplacement till 500 μsec before the laser irradiation.

IV-2. Experimental Results

a. Corona Instabilities and Preheat

Using low atomic number targets (D_2 or plastic), the study of coronal instabilities is of importance to suppress or control the preheat level in the imploding shell and fuel via such as two plasmon decay instability (TPD) and stimulated Raman scattering (SRS). SRS and TPD have been measured on these targets. Convective SRS has been reduced to more than 1/20 by inserting RPP with D_2+foam targets, while the level of TPD ($3\omega/2$ harmonic light) has not been affected by the presence of RPP. Hard x-ray spectrum has been monitored by the filter fluorescer x-ray (FFX) detector. Hard x-ray signal shows a good correlation with the TPD signals with a typical $T_h = 14$ keV and the fraction of up to several % at $I_L = 3.4 \times 10^{14}$ W/cm^2 as seen in Fig. 19 [20]. Another component for hot electrons could be due to the high energy tail of thermal Maxwellian electrons. Shown in Fig. 20 is the temporal profiles of convective SRS, $\omega/2$ $3\omega/2$, hard x-rays (10-20 keV), and soft x-rays (0.5-1 keV). The hard x-ray and TPD ($3\omega/2$) signals are fairly close. There is another component in the hard x-ray signal whose intensity is low but starts much early and similar to the soft x-ray signal. This component may be attributed to the high energy tail of the thermal electron distribution. The effect of the preheat appears on the shock temporal profile measurement on D_2+foam plane targets prior to the shock appearance. The preheat levels are 13 eV and 35 eV due to the hot electrons and shock, respectively for the 70 μm D_2+foam target at $I_L = 3.4 \times 10^{14}$ W/cm^2.

b. Shock Properties

The shock pressure may be estimated using either the relation of

$$P_{shock} = \frac{\rho_1 - \rho_0}{\rho_1} \rho_0 V^2_{shock}$$

(10)

or the hydro-code simulation [21]. Here ρ_0 and ρ_1 are the initial and the shock compressed densities. V_{shock} is the shock speed and can be measured experimentally. When P_{shock} was calculated using Eq. (10), the density part was assumed at the strong shock limit of ideal fluid. The values from Eq. (10) and the hydro-code results were same within 10 % at 3.4×10^{14} /w/cm^2 and 20 Mbar (2TPa) is obtained at this intensity. This shock pressure reflects the

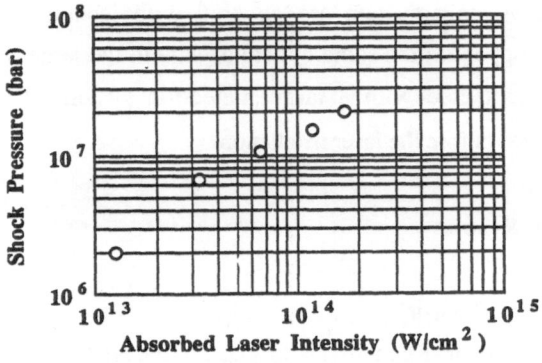

Figure 21 Shock pressure vs. absorbed laser intensity.

ablation pressure if the temperature behind the shock is much higher than the energy required for the ionization and dissociation (ID) of deuterium molecules. The simulation took into account the pressure ionization, Fermi degeneracy, and Coulomb interaction. The pressure is plotted as a function of absorbed laser intensity (I_a) in Fig. 21 using Eq. (10). The absorption rate of the 527 nm laser light onto D_2+foam target has been measured separately and this rate at each laser intensity (I_L) is multiplied to calculate the absorbed laser intensity. The ablation pressure P_a is proportional to $I_a^{2/3}$ as expected from the scaling.

Shock temperature and speed Hugoniot can be obtained for D_2+foam targets, using the independent results of temperature and shock speed as shown in Fig. 21. According to the shock relation, the temperature behind the shock T_1 is proportional to the square of the shock speed, V_{shock}^2 in an ideal plasma gas. Below 5×10^6 cm/sec, the shock temperature deviates and becomes below the above scaling. Two possible sources could contribute to this drop. At the lowest shock speed very small amount of preheat was observed. The small preheat can bring down the shock temperature at a given shock speed. Another possibility may be due to the ionization and/or dissociation. Log T_1 becomes small if γ (specific heat ratio) approaches to

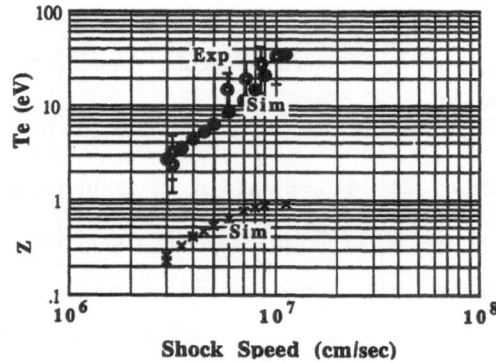

Figure 22 Hugoniot relation in D_2+foam plane targets irradiated by 527 nm.

1 or the ionization and/or dissociation alters the characteristics of the heat capacity of deuterium. Also shown are the results from HISHO, which took into account T_0 and V_{shock} of the observed values to calculate T_1. This region of interest corresponds to the ablation pressure of 1 to 10 Mbar.

c. Implosion

Typical spherical target structure is shown in Fig. 23. The overcoated layer is CH plastic with the density of 1.3 g/cm^3. The foam layer has the density of $30 \sim 100$ mg/cm^3 with the pickcell size of submicron. The thickness of the foam layer is determined by the upper limit of the pressure applicable in the cryostat system. The target diameter can be controlled from 500 to 800 µm. The sphericity of the targets is about 98 %.

Figure 24 shows framing pictures of the cryogenic target implosion. During the implosion of the target, x-ray emissions have been observed at the center prior to the maximum compression on both 2-D x-ray framing and 1-D x-ray streak camera pictures. As the physical mechanisms for the preemission, there are several possibilities such as laser light shine through at early time, rarefaction or shock wave due to the preheat, Rayleigh-Taylor instability, and self-focusing or filamentation. Neutron yield (N_y) versus obtained fuel density-radius (ρR) product shows a dependence similar to the ones obtained for the CD shell implosions, where 600 times the solid density was obtained. However neutron yield is almost order of magnitude higher than the CD shell case for a given ρR. Maximum value of ρR = 20 mg/cm^2 was estimated using the ratio of secondary to the primary neutrons. Using the 1-D HISHO simulation code, ρR of about 100 mg/cm^2 and N_y of 10^9 are predicted when the preheat is taken into account.

V. CANNONBALL TARGET IMPLOSION EXPERIMENT

Radiation driven cannonball targets have been extensively studied as an alternative approach of inertial confinement fusion. In these targets, a fuel capsule is placed inside a cavity made of high-Z material and is heated by incoherent thermal x rays in the cavity created with intense laser light injected through holes on the wall of the cavity. Owing to the geometrical smoothing effect, irradiation non-uniformities on the fuel capsule at high-spatial frequencies arising from small structures inherently involved in the laser beam profile are considerably suppressed. Thus we can devote ourselves to the consequences of lower-mode irradiation non-uniformities resulted from laser illumination configuration and power imbalance among multiple laser beams.

X-ray confinement inside a cavity has very important roles on enhancing radiation

CH Plastic Thickness 4.3 μm

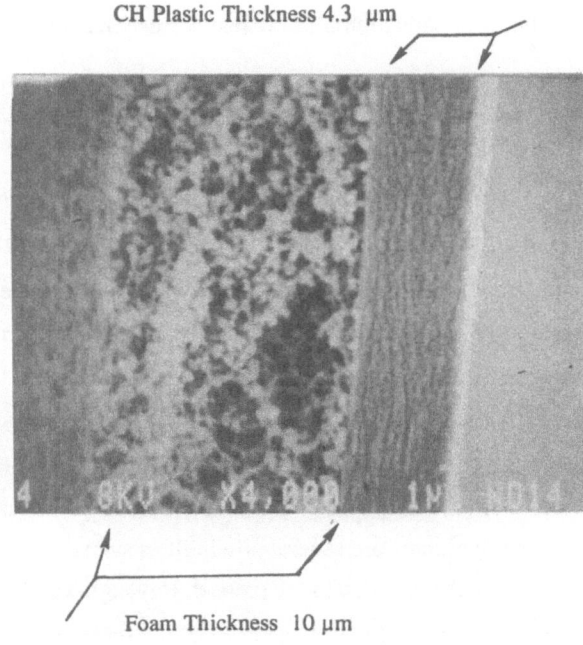

Foam Thickness 10 μm

Figure 23 Cross section of target shell. A 6 μm overcoated plastic layer and 10 μm foam layer can be seen.

Cryotarget with Plastic Overcoat

Framing interval: 100psec

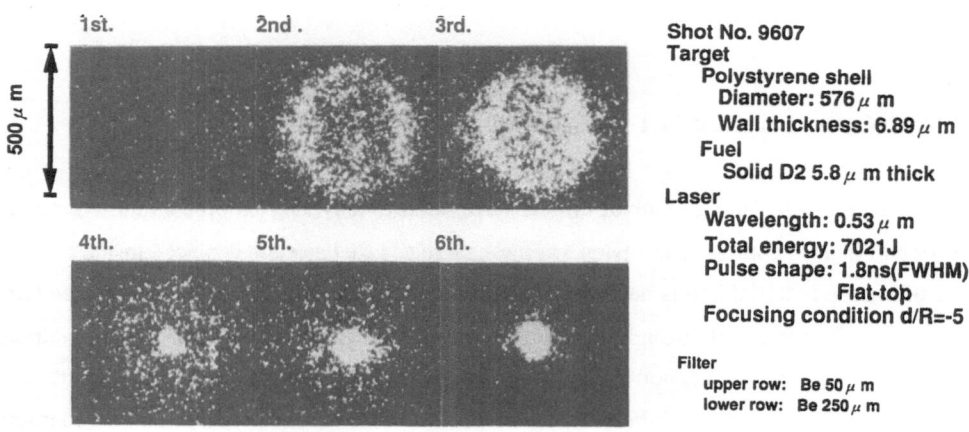

1st. 2nd. 3rd.

500 μ m

4th. 5th. 6th.

Shot No. 9607
Target
  Polystyrene shell
    Diameter: 576 μ m
    Wall thickness: 6.89 μ m
  Fuel
    Solid D2 5.8 μ m thick
Laser
  Wavelength: 0.53 μ m
  Total energy: 7021J
  Pulse shape: 1.8ns(FWHM)
    Flat-top
  Focusing condition d/R=-5

Filter
  upper row: Be 50 μ m
  lower row: Be 250 μ m

Figure 24 X-ray framing picture of imploding cryogenic target. Formed is D_2 solid layer(6 μm) inside a plastic (6.89 μm) shell (576 μm diameter).

fluxes, improving energy transfer efficiency [22] from laser to the capsule, and improving irradiation uniformity [23]. Assuming the quasi-stationary condition for flux balances at surfaces of the cavity wall and the fuel capsule, the power transfer efficiency, which is defined as the net heat power transferred to the inner capsule divided by the source radiation power, is approximately given as [23,24]

$$\eta_{trans} = m^{-1} \cdot \frac{(1 - r_t)}{(N^{-1} + n^{-1} + m^{-1} - m^{-1}r_t)} ,$$

where n^{-1} is the fractional hole area on the cavity wall, m^{-1} the fractional capsule area with respect to the total (including holes) area of the cavity, r_t the reemission coefficient of the capsule, and N the quality factor of confinement [equal to $r/(1-r_c)$ where r_c is the reemission coefficient of the cavity wall]. This relation states that the transfer efficiency is given by the geometrical coupling factor m^{-1} multiplied by the x-ray enhancement factor. Thus it is important to quantitatively evaluate the reemission coefficient and its spectrum of various materials heated by x-ray radiation.

A large mass ablation rate brought by x-ray driven ablation makes it possible to adopt very low aspect ratio targets. It also relaxes requirements for stable implosion via ablative stabilization [8].

A series of experiments has been carried out by using intense blue laser light (351 nm wavelength) from Gekko XII Nd:glass laser system. Quantitative understandings on x-ray confinement and radiation redistribution inside a cavity, propagation of radiative heating wave in medium-Z material, and reemission properties of x-ray heated matters have been obtained.

A fuel capsule placed in a cylindrical gold cavity was imploded by x rays generated by ten laser beams at 5.5 kJ total energy. Good agreement of implosion parameters such as the fuel ρR (density-radius product) was obtained between the experiment and numerical simulations, assuming perfectly spherical symmetry, up to the radial convergence ratio of around 10.

V-1. X-ray Redistribution and Confinement

Fundamental studies on x-ray confinement and radiative energy transport in medium- to high-Z materials have been made through a joint research between the Max-Planck-Institute für Quantenoptik at Garching (MPQ) and Institute of Laser Engineering, Osaka University (ILE) [25]. Evidence of x-ray confinement in a gold cavity heated by laser light (wavelength 351 nm, energy 5 kJ, duration 0.9 ns) was demonstrated, and a maximum brightness temperature of 240 eV and a quality factor of confinement of N=5.3 (corresponding to a reemission coefficient of r=0.84) has been achieved [26]. The propagation of a radiation heat wave through a thin foil of solid gold driven by the intense thermal radiation (which was generated in the same cavity as was used in the study of x-ray confinement) was studied. The propagating radiation heat wave was clearly observed from the delayed onset of intense thermal x-ray emission from a diagnostic foil [27]. The results of these experiments are in satisfactory agreement with theoretical predications [28] based on the self-similar ablative heat wave driven by thermal radiation and with numerical simulations including detailed treatment of atomic physics [11,29].

A three-dimensional model to calculate x-ray intensity distributions on a radiation driven target was made [30]. The model includes energy transfer processes such as conversion of laser light to x rays, radiation reemission from x-ray heated wall of a cavity and influence of a fuel capsule on radiation redistribution. To confirm the validity of the model, intensity distributions of x-ray inside a cylindrical cavity heated by laser light was investigated by measuring a burn-through signal from a diagnostic foil integrated onto the cavity. As is shown in Fig.25 the experimental result is well replicated by the model calculation when the x-ray reemission from the x-ray heated wall is taken into account. By using this model, optimum conditions for uniform irradiation of a fusion capsule by x-ray radiation has been assessed. Calculation was made for a cylindrical cavity heated with the frequency-tripled Gekko XII laser whose details are described in Sec. V-3. Figure 26 shows the nonuniformity σ_{rms} and the energy transfer efficiency E_{abs} / E_x given as a function of the inner capsule diameter D for which the off-set distance L (see the inset figure of Fig.26) is adjusted to give the minimum σ_{rms} value. The size of the cavity and laser conditions are fixed. Due to radiation circulation inside the cavity, the range of the optimum distance L giving the lowest σ_{rms} value becomes wider and the uniformity is significantly improved. The average intensity increases by a factor of 1.5 due to radiation confinement. The model calculation shows that nonuniformity of 2% (mainly due to the mode number of 4), energy transfer efficiency of 11% and x-ray intensity on the capsule of 1×10^{14} W/cm² can be expected under the optimum conditions for the Gekko XII facility.

V-2. Radiation Driven Ablation

When low- to medium-Z materials are heated by x-ray radiation, they become quasi-transparent to the incident radiation mainly because of ionization. The propagating radiation wave is supported by both the external source radiation and the self-emanating radiation from the heated material. The propagation of soft x rays in aluminum was investigated as a part of the ILE/MPQ collaborative experiment. The propagation velocity of the burn-through wave was determined from the measured time-resolved x-ray emission through an aluminum foil covering a diagnostic window made on the wall of the cavity in which intense thermal radiation is confined. The experiment was carried out up to a radiation intensity of 2×10^{14} W/cm² as

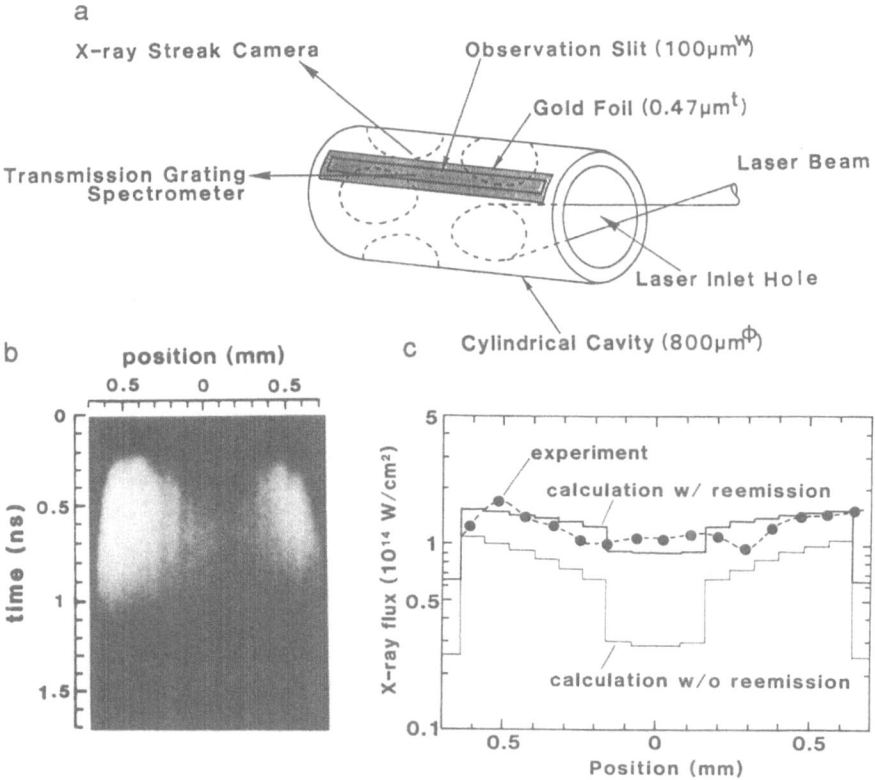

a

X-ray Streak Camera

Observation Slit (100μm^W)

Gold Foil (0.47μm^t)

Laser Beam

Transmission Grating Spectrometer

Laser Inlet Hole

Cylindrical Cavity (800μm$^\phi$)

b

position (mm)

c

Figure 25 Experimental set up for measuring redistribution of radiation inside a cylindrical cavity and an experimental result. (a) Burn-though signal from a diagnostic foil (0.47 μm gold) heated by radiation confined in the cavity was detected by an x-ray streak camera and a transmission grating spectrometer. (b) The output image from an x-ray streak camera. Intense burn-through signals at both ends of the foil appeared in early time followed by weak burn-through radiation. The former signals originated from the directly-heated region with laser and the latter from the indirectly-heated region with thermal radiation. (c) Comparison of model calculation with an experimental result which is indicated by closed circles. The thick and thin solid lines respectively show calculation results including and excluding the effect of reemission on the radiation redistribution.

477

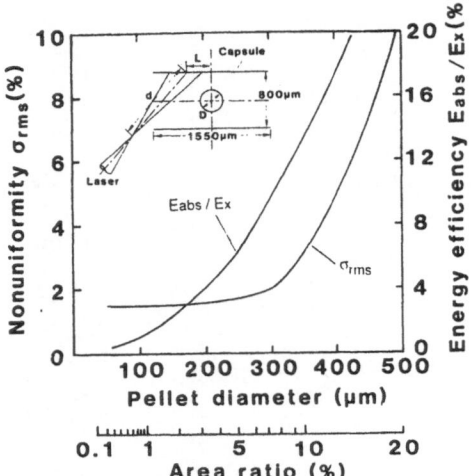

Figure 26 The non-uniformity σ_{rms} and the energy transfer efficiency E_{abs}/E_x given as a function of the inner capsule diameter D for which the distance L is adjusted to give the minimum σ_{rms}. The non-uniformity shows little difference for smaller diameters. The area ratio of the inner capsule to the outer cavity is also given for comparison with the calculated energy transfer efficiency, showing improvement of the efficiency by a factor of 1.5 due to radiation confinement.

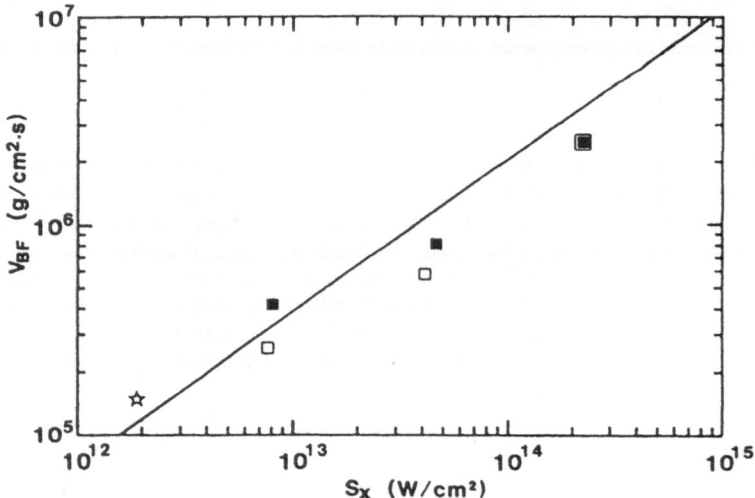

Figure 27 Dependence of the propagation velocity of the burn-through front in aluminum on the flux of the irradiating x rays. Experimental results obtained in the ILE/MPQ joint collaboration (open and closed squares for different observation wavelengths) and in our previous work [star symbol : T. Endo et al., Phys. Rev. Lett. 60 (1988) 1022-1025] are shown. The solid line is the theoretical prediction (Ref.31).

478

shown in Fig.27. The experimental results were analyzed with a model to treat the ablation driven by a radiation heating wave [31] in which contribution of the self emission to the ablation is assumed to be small. The solid line in Fig.27 is the theoretical prediction.

Properties of the x-ray transmission and self-emission were investigated in detail for various materials heated by thermal x-ray radiation. Dependence of the reemission coefficient on the atomic number ranging from 6 to 79 has been measured at the irradiance of 2×10^{13} W/cm^2. The experimental result is well replicated by a numerical calculation with ILESTA [11,32] in which the screened hydrogenic, average ion model is used.

V-3. Cannonball Implosion with Gekko XII Blue Laser

After testing several types of cannonball targets with different configurations and parameters, we are making systematic study of a cylindrical type cavity target [33]. The target structure and the irradiation configurations are schematically shown in Fig. 28. A cylindrical cavity of 800 μm in diameter and 1550 μm in length was irradiated with ten laser beams. Target illumination configuration consists of two bundles of five laser beams each from opposite sides. The bundles have a cone half angle of 50°. Typically the laser energy was 5.5 kJ with a pulse duration of 0.7 ns. Energy balance among the laser beams was within 5%. The fuel capsule was either a glass microballoon containing DT or a plastic microballoon containing D$_2$ and argon gases. The diameter of the capsule was 300 to 320 μm. A CF$_x$ (x=1.4 ~ 1.7) ablator was overcoated to give a total shell thickness of typically, 1.8 mg/cm^2. Both ends of the cavity cylinder were partially closed with ring plates of 600 μm inner diameter. It is estimated that a radiation temperature of T_R ~ 200 eV (which corresponds to the radiation flux of 1.6×10^{14} W/cm^2) and the confinement factor of N ~ 3 are attained under the irradiation conditions described above.

Reproducible results on implosion were obtained. The implosion velocity and the geometrical shape of the capsule in an acceleration stage were measured by means of x-ray backlighting method utilizing an eight-channel x-ray framing camera [34]. The measured implosion velocity of 2×10^7 cm/sec is consistent with a theoretical prediction of radiation driven ablation with the estimated radiation temperature of 200 eV.

The temporal behavior of the imploded core shape was observed by the x-ray emission image from an Ar filled target. The core shape had non-uniformities of low order modes, in particular 2 and 4. These were controlled by changing laser irradiation conditions such as the position and the size of the laser spots on the cavity wall. The best geometrical condition for uniform implosion determined through the observation was found to be in good agreement with the prediction from the model calculations of capsule illumination by x rays. Figure 29 shows

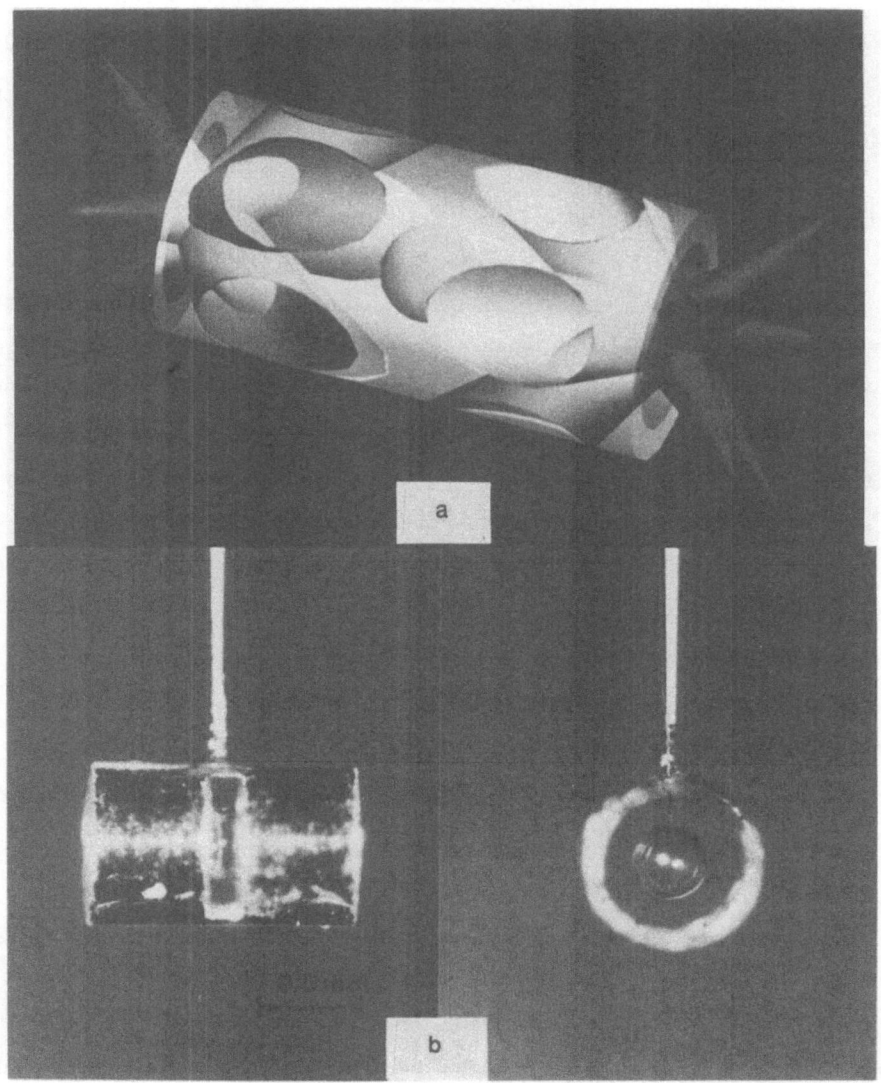

Figure 28 (a) A schematic view of a radiation-driven cylindrical cannonball target irradiated by frequency-tripled (351 nm wavelength) 10 laser beams from Gekko XII system. (b) An example of a cylindrical cannonball for implosion experiments. Observation slits made on the side wall of the cavity and both open-ends of the cylinder are, in later time, covered or closed with thin gold foils and end-rings to enhance radiation fluxes.

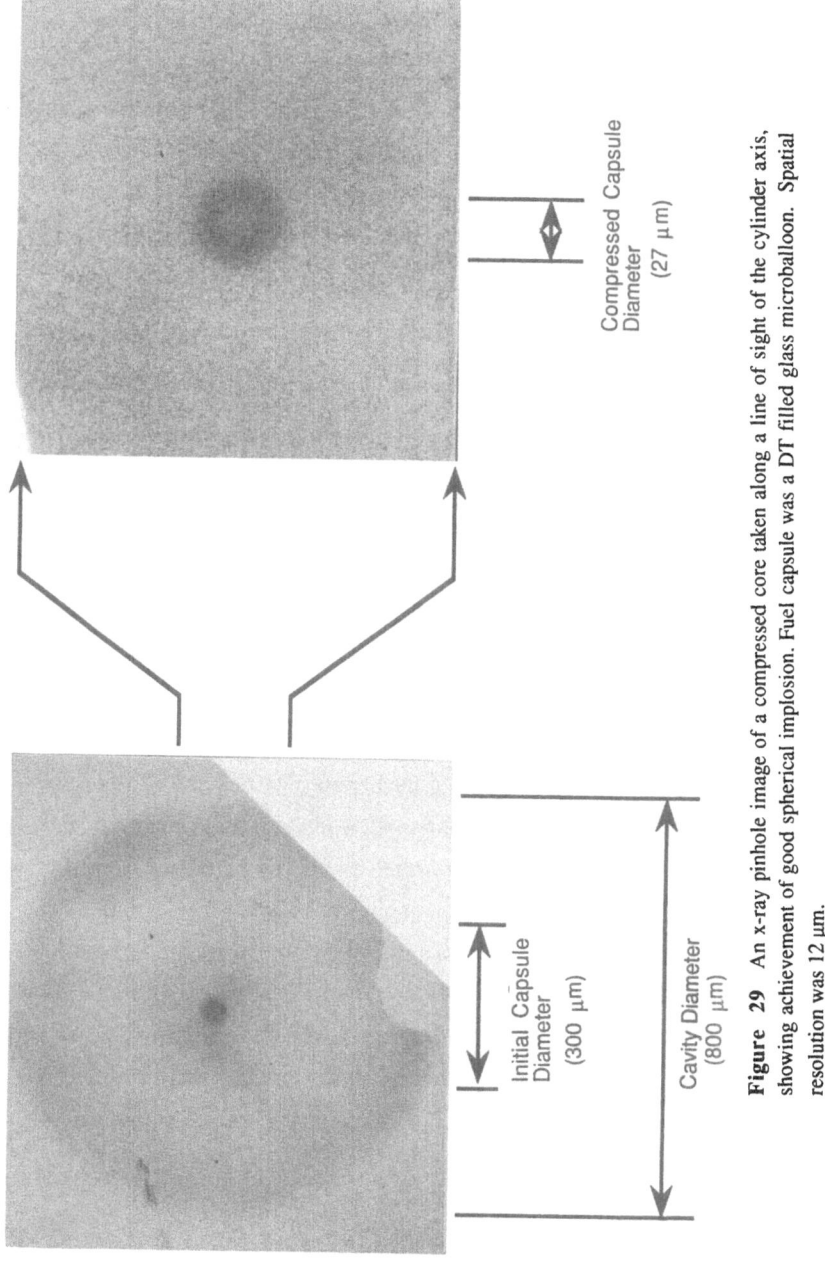

Figure 29 An x-ray pinhole image of a compressed core taken along a line of sight of the cylinder axis, showing achievement of good spherical implosion. Fuel capsule was a DT filled glass microballoon. Spatial resolution was 12 μm.

Compressed Capsule Diameter (27 μm)

Initial Capsule Diameter (300 μm)

Cavity Diameter (800 μm)

481

an x-ray pinhole image (at the photon energy of 3 ~ 5 keV) of the compressed core observed along a line of sight of the cavity cylinder axis. Similar pictures were obtained at the perpendicular direction of the axis, showing achievement of good spherical implosion.

The convergence ratio of R_i /R_f (where R_i and R_f are the initial and final radii) was varied by changing filling pressure of DT gas from 1 to 20 atm. Implosion time measured with an x-ray streak camera and ion temperatures (4 ~ 5 keV) determined from the velocity spread of neutrons were weakly dependent on the initial fill pressure. The size of the compressed fuel was evaluated from the x-ray images of the D-T filled glass microballoon capsules taken with an x-ray framing camera and pinhole cameras. Convergence ratios of 4.5 to 20 were observed corresponding respectively to the initial pressure of 20 to 1 atm. The fuel ρR values derived from this convergence ratio, assuming absence of collapse of spherical convergence and pusher/fuel mixing, were 0.5 to 2 mg/cm^2 [35]. The smallest value (i.e., 0.5 mg/cm^2) was obtained for the medium pressure around 5 atm. But the fuel ρR estimated from the deuteron knock-on method monotonically increases from 0.1 to 2 mg/cm^2 with increasing the fuel pressure. The ρR values provided by the two independent methods agreed at the initial pressure exceeding 2 atm (corresponding to a convergence ratio less than 10).

Numerical simulations with one-dimensional fluid code ILESTA reproduce the experimental ρR value by the knock-on measurement well. The fuel temperature is also closely consistent with the simulation but excluding the cases below 2 atm.. The ratio of the experimental neutron yield to the simulation falls significantly below unity (10^{-1} ~ 10^{-2}) for all pressures. The simulation shows that, due to temperature difference, fusion reaction mainly occurs at a compressed fuel layer adjacent to the pusher so that majority of the neutrons are yielded there. Then the disagreement regarding the neutron yield may suggest occurrence of pusher/fuel mixing at their contact surface. Further details of implosion stability and its influences on the implosion parameters are not convincingly clarified yet.

Systematic studies on radiation hydrodynamics in an x-ray confining cavity and a fuel capsule have attained remarkable progress in these few years. This makes it possible to analyze quantitatively the energy transfer processes from laser to the fusion capsule and to find uniform irradiation conditions of the fusion capsule driven by x rays. Based on these results, driver and target design, aiming ignition and high gain in fusion reaction, is being undertaken by using the simulation code ILESTA.

VI. IMPROVEMENT OF IRRADIATION UNIFORMITY

Recently, in a direct drive fusion experiment, the efforts for a uniform compression of the spherical target become key issues, such concept as a uniform irradiation architecture, a completely spherical shell target with a uniform shell thickness and the investigation of the multi dimensional hydrodynamic behavior of the imploded pellet.

VI-1. Present Status of 12 Beam Irradiation

At ILE, Osaka University, the random phase plate (RPP) [36] was adopted on the Gekko XII system to avoid the lower order mode nonuniformity of the irradiation intensity on the target, which was caused by the irregularities of the laser beam intensity and phase aberration. RPP works as the beam segmentator of the 35 cm$^\phi$ beam to 25000, 2 mm-square beamlets each of which would have a uniform pattern. The RPP reduced the root mean square deviation, σ of the nonuniformity from 36.2 % for the direct laser irradiation to 19.7 % for a case with the RPPs, shown in Fig.30. This nonuniformity has the large components of lower order modes ($l \leq 6$) which were resulted by the large mismatching of the coating thickness of the RPPs. This phase mismatch causes the zero phase central component of the diffracted pattern even with using the focusing condition of a overlapped quasi farfield illumination. We are now preparing new RPPs with a precise coating thickness and the hexagonal element pattern.

The RPP has a disadvantage to generate the speckle structure on the target surface illuminated with the coherent laser light. The thermal smoothing effect is, however, expected to deplete the higher order spatial modes over the mode number of 30, so that σ_{th} was reduced from 17 % to 7 % by using the RPPs.

For more improvements of the irradiation uniformity, especially as to not only the lower modes depletion but also the higher order modes, we have proceeded the partially coherent light generation and the envelope function control of the each beam pattern on the target.

VI-2. Intensity Profile Control on a Target Surface [37]

A number of beamlets segmented by the RPP are usually projected to the target located at the focus position of the main lens. The Rayleigh length of the small segmented beamlet is so long (e.g. 80 mm for F/500 at Gekko XII) that the envelope profile of the irradiated beam is easily controlled by changing the target position. The flat top shape distribution at the plane tangential to the spherical target surface is not suitable to realize the uniform intensity on the target because the multibeam superposition results in the beam overlap to generate the lower order mode nonuniformity, especially $l=6$ for 12 beam system. The bell-shaped distribution is favorable to deplete the lower order mode with a spot size corresponding to the target size. The RPP element with a hexagonal pattern can generate the circularly contoured diffraction pattern which distribution is defined by Bessel sinc function, Besinc^2(r) = $(2J_1(\pi r) / \pi r)^2$. Fig. 31

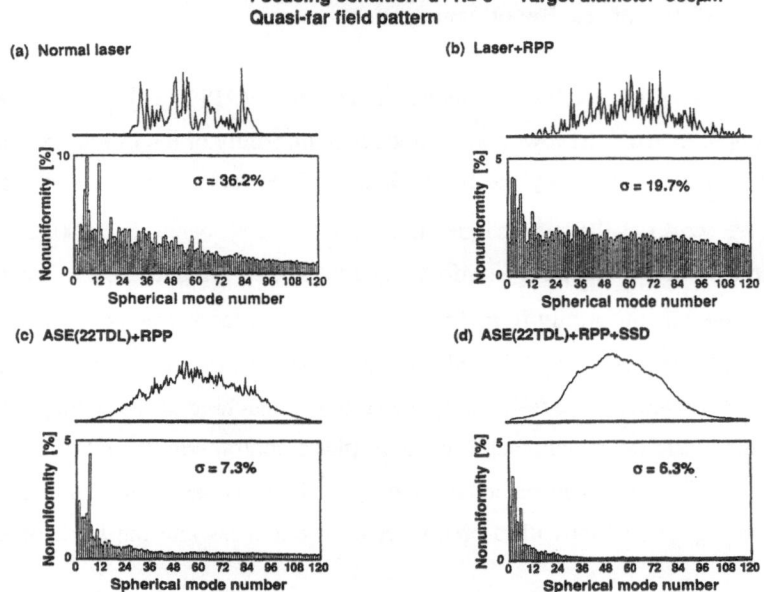

Figure 30 Nonuniformity of the spherical target irradiation on Gekko XII by (a) normal laser, (b) laser with RPPs, (c) angularly diverged ASE with RPPs and (d) ASE having an angular divergence (22 xDL) and spectral dispersion (239 μrad/nm) with RPPs. All lights are frequency-converted to second harmonics.

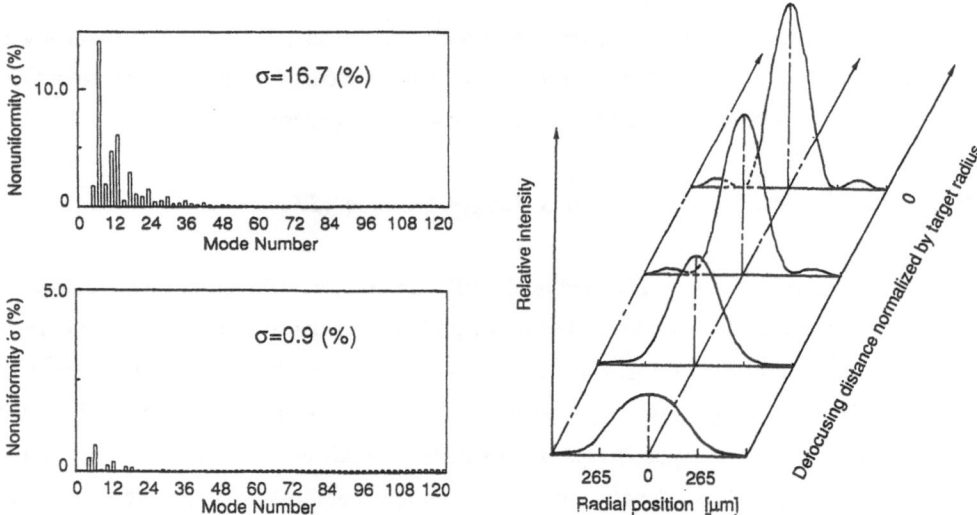

Figure 31 Envelope function of the focused pattern of the segmented beamlets can be controlled by changing the focusing condition on the target.

shows the spherical harmonics distribution of the irradiation nonuniformity for the existing Gekko XII system by changing the target position defined by d/R, where d is a distance between a target center and the focus point, and R is a target radius. It is assumed that the incident light is perfectly incoherent so that the phase interference effect can be ignored. A large number of the center-shifted beamlets, Σ_i Besinc^2 (r-r$_i$) can generate the very uniform irradiation pattern even for twelve beam system. The nonuniformity, σ approaches 1 % level without an assumption of the thermal smoothing.

This scheme can be expected to decrease the nonuniformity of the lower or medium mode components even for a small numbers beam system like Gekko XII, and also easily control the envelope function by slightly changing the target position from the far field irradiation with small energy margin.

VI-3. Partially Coherent Light Sources for Gekko XII

The RPP scheme requires essentially the interference-free superposition of the segmented beamlets which conclude the usage of the incoherent / partially coherent light. We are now developing the two optical systems to generate and control the partially coherent light (PCL) sources.

a. General Consideration of PCL Source

The PCL has special features on both temporal and spatial coherencies. The temporal coherency is controlled by the spectral width of the light and the phase relation of these modes in frequency region. The spatial coherency is governed by the spatial mode structure generated at an emitter. The partially spatial coherent light has the angular divergence much larger than the diffraction limited value.

Considering the propagation of the PCL through an amplifier system from PCL generator, the complete image relaying is absolutely required not to degrade the incoherency of the PCL. Figure 32 shows the schematic of the PCL source and the entrance pupil which is a starting point of the image relaying. The complex degree of coherence, μ_{AB} can be used as a measure of the PCL. μ_{AB} is given by the system parameters as [38]

$$\mu_{AB} = \frac{2 J_1(v)}{v} e^{i\varphi} ,$$

(11)

where $v = \overline{k}(\rho_0 d / R)$ and $\varphi = \overline{k}(r_1^2 - r_2^2)/R$, $\overline{k}$ is a mean wave number of the PCL, ρ_0 is the source radius and R is the distance between the source and the pupil. d is a distance between two points on a pupil surface whose coordinates are given by r_1 and r_2. J_1 (v) denotes the first

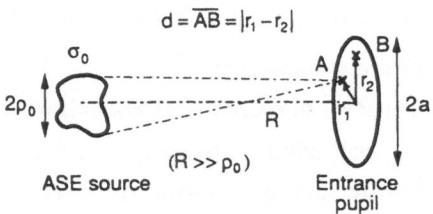

$$d = \overline{AB} = |r_1 - r_2|$$

Figure 32 Spatial coherency is defined by the optical arrangement and source size of the PCL.

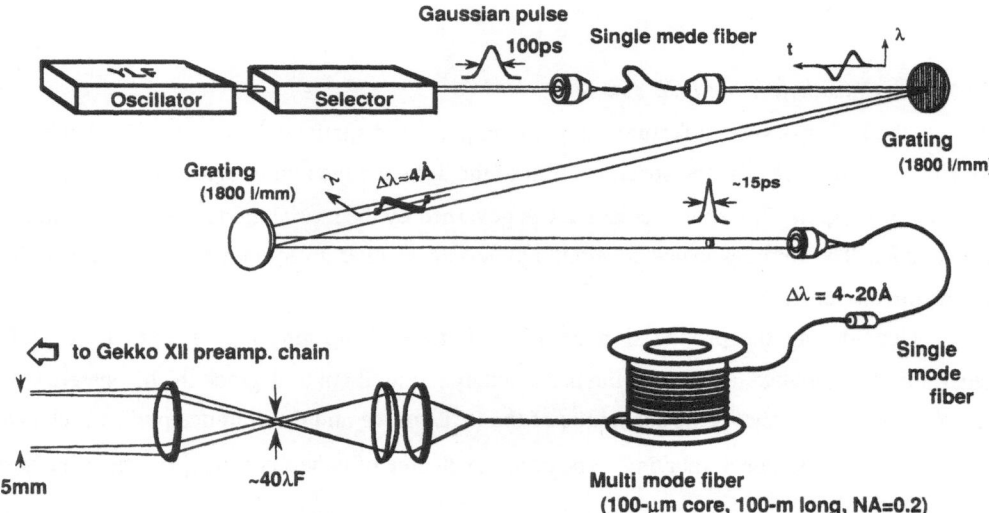

Figure 33 Optical fiber system to generate the PCL. The two stages of the single mode fiber can serve the frequency chirped light, which is mode-mixed by the multimode fiber to destroy the spatial coherency.

order Bessel function which has a first zero point at $v = 3.8$. The coherent zone size, d_c on the pupil is given by $d_c = 0.61 \text{ R } \overline{\lambda} / \rho_0$, which corresponds to the first Airy disk diffracting from the PCL source to the pupil.

Under the assumption that the total system has a sufficiently large aperture in the fully image relaying to keep the beam divergence, the ratio of the coherent zone size, d_c to the diameter of the pupil, 2a is kept constant even on the RPP at downstream of optic system. The incoherent illumination of the target means this ratio larger than that of the element size of the RPP to the whole beam size at a position of the RPP. In our existing system, the element size of 2 mm on the whole beam of 320 mm in diameter requires the ratio, $d_c / 2a \geq 160$.

Before an installation of the PCL front end system, a staging of the preamplifier chain of Gekko XII has been changed to keep the fully image relaying with the tolerance of the beam divergence up to 40 xDL (times Diffraction Limit) by the new installation of many spatial filters. All optical components were carefully arranged to prevent from the optical damage due to beam focusing when the usual coherent laser oscillator is used.

b. Optical Fiber System

The optical fiber system was proposed to generate the spatial incoherency with the broad band glass laser oscillator [39]. We developed the optical fiber system which consists of the two stages of frequency chirping with pulse compression scheme, and the modal conversion by multimode fiber, as shown in Fig. 33, following the modelocked Nd:YLF front end of Gekko XII.

The spectral width of a short pulse output, 100 ps / 10 µJ, is easily broadened by using the self phase modulation in a polarization maintaining silica fiber of 9 µm core diameter. The chirped pulse of 0.4 nm band width limited by the stimulated Raman scattering process can be compressed to 15 ps by a grating (1800 grooves / mm) pair. The second fiber system can control the spectral width up to 0.4 to 2.5 nm by adjusting the incident intensity. The shorter pulse width is required for the fast spatial coherence control in a following multimode fiber system. A typical core diameter (a_c) of a multimode fiber is 100 µm, and the numerical aperture, NA is 0.2 which corresponds to the tolerance of the beam divergence of $40 \times$ DL of Gekko amplifier chain. The length of the multimode fiber is strongly relevant to the temporal coherence of the output beam. The temporal coherence would be reduced by the modal mixing in which the delay time between the adjacent modes is much larger than the coherence time. The length over 100 m is needed for above mentioned fiber where the delay time between the lowest modes is about 20 ps .

The optical fiber smoothed beam is obtained with the repetitive output from the GXII mode locked oscillator and its characteristic is measured with enough energy (1 ~ 10 µJ).

The time integrated farfield patterns (corresponding to the near field image of the output surface of a fiber) measured by a cooled CCD camera. As a reference, the narrow band laser is used with $\Delta\lambda = 0.002$ nm in pulse duration of 100 ps. The beam pattern becomes enough smooth to destroy the speckle structure which is seen for a narrow band laser. The probability density distribution of the intensity is shown in Fig. 34, in which double peak distributions are due to fringes by an interference inside the CCD camera. The standard deviation of the intensity distribution reaches to a few % level. It is noted that the probability at zero intensity is completely vanished so that the modal mixing with broad band spectrum works to deplete the spatial and temporal coherency. The low contrast of the speckle for a narrow band laser can be explained by the incoherent addition of the speckle patterns, but the output pulse width was extended over a few nanosecond through a 100 m long multimode fiber.

The spatial coherency of the PCL by the optical fiber system was measured by Young's interferometer. The coherence area was about one thirtyth of the beam diameter for parameters, $\Delta\lambda = 0.4$ nm, $t_p = 15$ ps and $L = 100$ m. This coherent area size covers 20 elements of the RPP of 320 mm in diameter. The nonuniformity originated from the residual mutual coherence across the beam would be reduced at the target. The statistical property of the fiber PCL is defined by the time averaged band width and the beam divergence. A large scale structure were not observed in both temporal and spatial performances.

The more detail temporal characteristics are measured by the time resolved FFPs to ensure how rapid the speckle structure is time-averaged under the conditions of different spectral widths. The streaked pattern was analyzed by the procedure that the time averaged nonuniformities with several temporal windows are calculated and extended into the spatial modes. Figure 35 shows the typical results for both cases of $\Delta\lambda = 0.76$ nm (a) and 1.59 nm (b). Generally, the smoothing time becomes shorter and the final level of the nonuniformity goes lower for the broader band width. The temporal change of mode structure, however, was not so obvious because of the potential mode scrambling in the multimode fiber. Then a beam pattern can be smoothed in a time scale shorter than several tens picosecond by using such a simple optical fiber system.

c. Amplified Spontaneous Emission of Nd:Glass

The emission from an atomic system has inherently incoherent property when the stimulated process does not occur. The neodymium in glass matrix has the wide spectrum of 20 nm which consists of the homogeneous and inhomogeneous broadened components. Figure 36 shows a recently developed ASE front end system which can be selected instead of the usual Qswitched, mode locked Nd:YLF oscillators of Gekko XII. The ASE system consists of two 25 mm$^\phi$ rod amplifiers (RA25s), one of which works as the ASE generator / amplifier and the other is a four pass regenerative preamplifier with the total small signal gain of

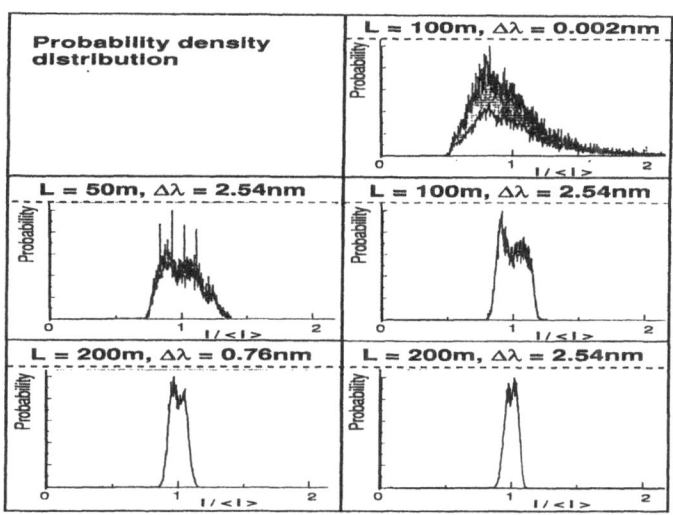

Figure 34 The probability density distribution of the intensity of the time integrated farfield PCL pattern generated by optical fiber.

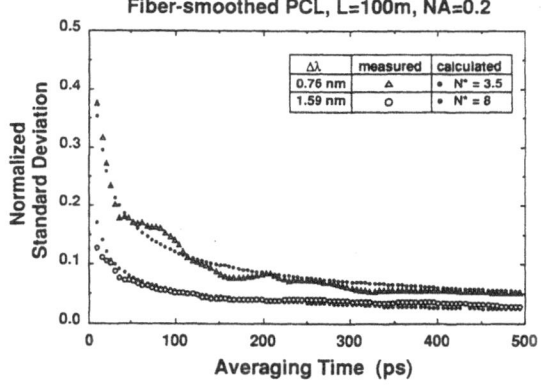

Figure 35 Temporal behavior of the standard deviation of the one dimensional fiber-PCL pattern of $\Delta\lambda = 0.76$ nm (a) and 1.59 nm(b).

$10^6 \sim 10^7$. The RA25 has a phosphate glass rod of 30 mm long and 25 cm in diameter whose pumping energy density is $43 \sim 67$ J / cc corresponding to the single pass small gain of $40 \sim 60$. The Pockels switch having a gating window of 10 ns prevents the parasitic oscillation in the ASE system.

A near field pattern at a front surface of a glass rod is relayed to itself by a rear side spatial filter and a mirror so that the generator takes a role of a double pass amplifier. The beam size is decreased to one fourth and also limitted to 5 mm. The beam divergence is limited by a pinhole at next spatial filter. After four pass amplification an ASE light is introduced to the optics for the spectral angular dispersion by the grating of 1200 grooves / mm where the beam diameter is enlarged to 80 mm. This is effective to improve the efficiency of the second harmonic generation by Type II KDP crystal. The total spectral width can be easily controlled by the iris installed at the far field point in an angular dispersion optics. The wavelength dispersion is adjusted to the wavelength dependence of the matching angle of the KDP crystal of 239 μrad / nm in accordance with the transfer ratio of the beam aperture. The image rotation of 45° is used to align the angle between the polarization direction of the fundamental and an e-axis direction of the KDP crystal. The amplification characteristics of the PCL was measured on the spectral narrowing and the gain saturation at the preamplifier and the main amplifier stages of the Gekko XII [40]. The ASE energy of 10 pJ/ 2 ns was amplified to several μJ in a preamplifier with strong spectral narrowing resulting the spectral width of 2 nm. In a main amplifier chain, there is no significant spectral narrowing and gain reduction in comparison with the narrow band laser light in a case of the spectral width less than 2 nm. These results is well predicted by the numerical calculation with the modified Frantz-Nodvik equation. The final output energy, for example, reaches 15 kJ (1.3 kJ/beam) at 2.2 ns pulse duration and the spectral width of 0.2 to 1.3 nm in spite of the application of the angular dispersion. The beam breakup and other nonlinear effect such as a two photon absorption in a laser glass did not appear in this intensity level of 2 GW/cm^2.

The doubling frequency conversion efficiency is clearly improved with using the angular dispersion scheme on the ASE PCL at an intensity of $0.2 \sim 0.7$ GW/cm^2 in a pulse duration of 2.2 ns. The efficiency near 50 % at 0.25 GW/cm^2 does not change in a region of the beam divergence up to 22 times diffraction limit. A slight degradation of the efficiency is caused due to the beam divergence near 100 μrad. The higher efficiency is expected by increasing the incident intensity over 1 GW/cm^2.

For more improvement of the 2ω conversion efficiency, we are now growing the KTP crystals which have a large nonlinear coefficient and tolerances for angular and spectral bandwidths.

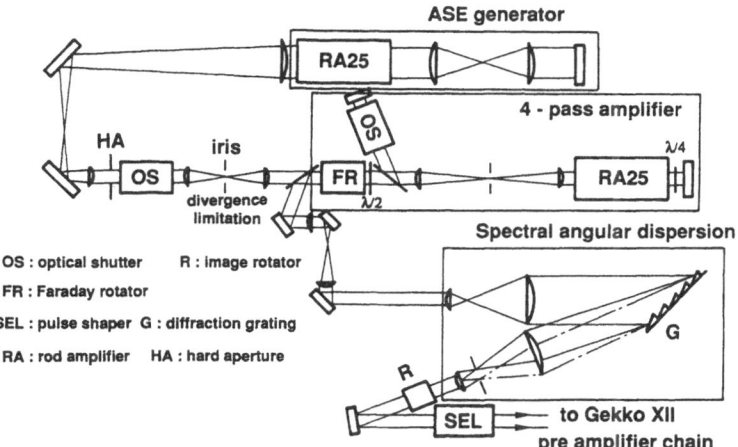

Figure 36 ASE front and system for Gekko XII consists of ASE generator/amplifier and the four pass regenerative pre-amplifier of gain, $10^6 \sim 10^7$.

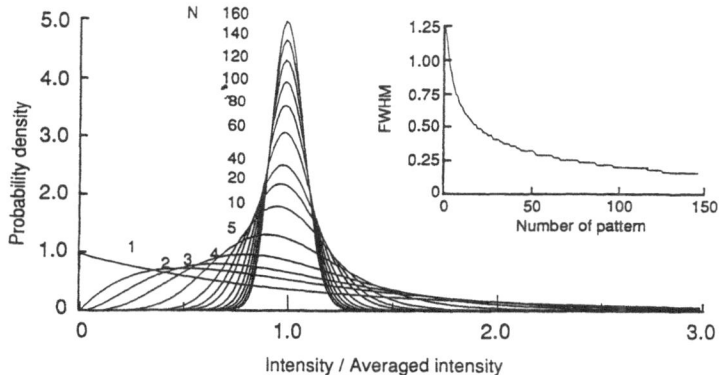

Figure 37 The probability density of the intensity of the incoherent superposition of multi-patterned speckle structures.

VI-4 Evaluation of the Partially Coherent Light Illumination

The PCL is expected to reduce the small scale speckles which appear in a farfield pattern of the randomly phased segmented beamlets. In order to analyse the speckle structure, we introduce the statistical properties of the speckle pattern. The probability density of the intensity for the fully developed speckles (Gaussian speckles) is shown by the negative exponential function.

$$P(i) = \frac{1}{<I>} \exp(-i) \quad , \tag{12}$$

where i is the normalized intensity (= I/<I>) and <I> is the average intensity. If the incident light has a finite beam divergence and the wavelength dispersion, the resultant speckle structure shows the independent (incoherent) superposition of the different pattern structures. As the number of the different speckle patterns is denoted by N, the probability function is given as

$$P_N(i) = \frac{i^{N-1} N^N}{(N-1)! <I>} \exp(-Ni) \quad , \tag{13}$$

The standard deviation, σ of $P_N(i)$ is given by

$$\sigma = N^{-1/2}. \tag{14}$$

This type of function shows in Fig. 37. The number of patterns, N can be used as a measure in an experiment how the speckle structure is destroyed. The nonuniformity less than a few percent needs the incoherent superposition of several thousands patterns. The lower temporal coherency aims also to generate the number of speckle patterns during a given averaging time duration. N should be replaced by t / t_c, where t is the averaging time and t_c is the coherence time of the incident light. These two effects, however, affect complicatedly the resultant performance of the irradiation property. The experimental expression of the temporal evolution of the nonuniformity could be written as

$$\sigma = (1 + t / t_l)^{-1/2} \quad , \tag{15}$$

where t_l is the effective life time of the speckle structure with consideration of both effects of the spatial and temporal incoherency.

The experimental data described below are for ASE with a wavelength dispersion as the PCL.

Figure 38 shows the time integrated (2.2 ns) focused patterns of the frequency converted, green PCL with the RPP on the equivalent surface tangential to the target. The focusing condition is d/R = -5 for R = 250 μm. The experimental parameters are the divergence of the PCL, $\Delta\theta$ and the wavelength width, $\Delta\lambda$ with the wavelength dispersion mentioned above. General feature of the PCL illumination gives the smooth structure without small size speckles. At cases of a lower divergence angle (bottom, left) or a narrower spectral width (upper, right) of the PCL, the lower order mode structures are seen due to the poor effect of the coherency suppression. For large divergfence of 22 times diffraction limit and the 0.8 nm bandwidth, the smoothing effect by the PCL is sufficient to deplete the higher order speckle structure.

The probability density of the intensity of Fig. 38 is calculated and shown in Fig. 39. The measured pattern is first normalized by the locally averaged intensity (1 % area of the whole beam) so that the probability density does not include the envelope shape of the beam pattern. It was found clearly that the partially coherent effect of the incident light is very effective to sharpen of the density distribution through enlarging the beam divergence and the spectral width. The smallest deviation of the measured density is about 2 % for $\Delta\lambda$ = 0.8 nm, $\Delta\theta$ = 22 xDL corresponding to the independent numbers of pattern of 2500. Inserted is the distribution for the coherent laser light with a RPP which has a large deviation.

The focused pattern is also measured by a streak camera to evaluate the temporal behavior of the PCL at a fundamental wavelength. From the streaked images, a temporal evolution of the standard deviation of the intensity distribution was calculated with a various temporal window. Figure 40 shows the results of the smoothing effect of the temporal change of an individual pattern which is assumed to have a finite life time. The solid lines are fitting curves with Eq. (15) at $\Delta\lambda$ = 0.4 nm and the beam divergence as a parameter. The effective life time is estimated by fitting to be t_l = 2.9, 0.9 and 0.7 ps for $\Delta\theta$ = 8, 16 and 32 times diffraction limit, respectively. These values are smaller than 9.2 ps of the coherence time defined by the reciprocal of the whole spectral width, $t_c = (c\Delta\lambda/\lambda^2)^{-1}$ for $\Delta\lambda$ = 0.4 nm. The time integrated standard deviation can be given to be 1.8 % for t = 2.2 ns by extraporation with Eq. (15) for t_l = 0.7 ps ($\Delta\theta$ = 32 xDL) which did agree with the experimental value, 2 % given above.

Finally the spherical irradiation uniformity is calculated by 3D computer simulation with using a measured intensity distribution. Fig. 30 summarizes the calculation results for a normal laser beam, a laser beam with RPP, the ASE having beam divergence of 22 xDL with RPP and the wavelength dispersed ASE with RPP. ASE cases give the much reduction of the higher order modes of the spherically expanded nonuniformity. The residual nonuniformity around mode number less 10 is mainly due to the coating mismatch of the existing RPP. Concludingly, the standard deviation of the nonuniformity would decrease to several percents level with the partially coherent light. Thus it was verified that the PCL with a wavelength dispersion is very effective for the reduction of the small scale ripple structure due to the spatial and temporal coherences.

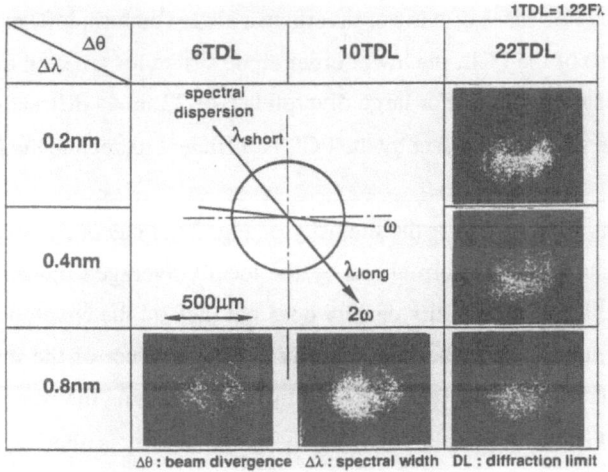

Figure 38 Time integrated focused patterns of the frequency converted PCL.

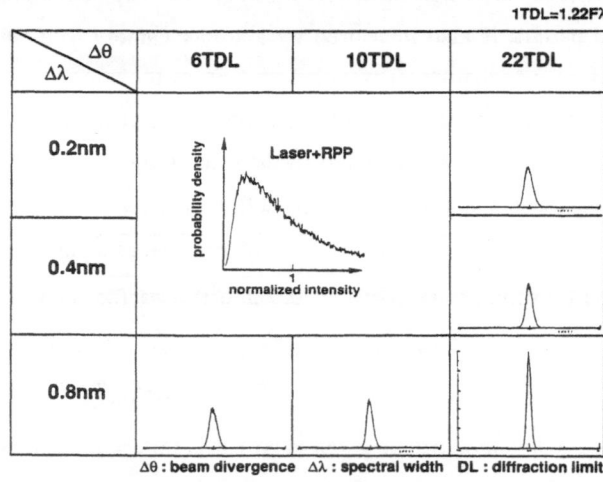

Figure 39 The probability density distributions of the pattern shown in Fig. 38.

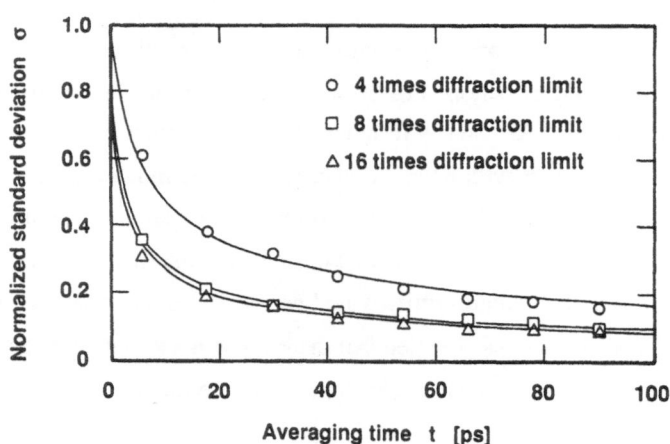

Figure 40 The temporal evolution of the deviation of the intensity distribution as a function of the averaging time.

VI-5. Aspherical Multi-Lens Array for Alternative Approach

Even though RPP is used in an optical smoothing technology, it has same principal problems as follows. The beam pattern with RPP consists of high spatial mode speckles due to the interference of beamlets, and it also consists of low spatial mode modulations originated with neighboring beamlets. These beam modulations are stationary in time. The higher mode modulation and stationary beam pattern are not fatal problems if the dynamic smoothing effect in the target plasma and the thermal smoothing effect are taken into account. The low mode modulation originated by RPP is a difficult problem to solve in a practical manner.

An alternative approach for attaining a uniform beam profile on the target is the multi-lens array proposed by X. Deng et al. [41]. In this scheme, the lowest spatial mode can be controlled by means of choosing the beamlet size. However, the control of the beam profile on the target can be done only with defocusing the beam, and the diffraction due to the edge of the single lens is contained in the individual beamlet pattern.

According to the computer simulation on the target irradiation uniformity, the control of the incident beam profile is vitally important for obtaining a uniform illumination of a spherical target. Hence, here we propose a new scheme using aspherical multi-lens arrays for the beam profile control and beam smoothing. In this scheme, an aspherical single lens coupled with a main focusing lens will be used to generate a beam profile which is suitable for decreasing the irradiation nonuniformity. These segmented beams are overlapped on the target with an aspherical array lens system for eliminating the influence of the intensity modulation in the incident laser beam. The followings are superior points of the aspherical multi-lens array system. 1) A desirable beam profile can be generated with the design of the aspherical lens profile. 2) The lowest mode of the spatial intensity modulation can be determined with the beamlet size. 3) The beam diffraction can be suppressed with shaping the surface at structure edge of the single lens. 4) This scheme is applicable with another beam smoothing technology such as amplified spontaneous emission (ASE) which is effective in higher mode smoothing.

An intensive work to design and fabricate an aspherical multi-lens array have been made in our laboratory.

VII. DEVELOPMENT AND CONSIDERATION FOR REACTOR DRIVER

An ICF power plant requires a driver which has the specifications such as $1 \sim 10$ MJ output in properly shaped pulse at the short wavelength (≤ 500 nm), a repetition rate of ~ 10 Hz, and a total efficiency of ~ 10 % [42,43]. A laser diode (LD) pumped solid state laser has been proposed as a feasible candidate driver which will satisfy these requirements [42,43,44]. LD pumped solid state lasers have greatly advanced to high power systems due to the recent advances of AlGaAs laser diode arrays with power of ~ 4 kW/cm^2 (200 μs pulse duration) [45], efficiency of ~ 60 % [45], and lifetime of 10^{11} shots [46]. Already a one-joule diode pumped Nd:YAG laser system has been constructed which demonstrates an electrical efficiency of 18 %, a repetition rate of 40 Hz, and a peak power of 44 MW (0.75 J at a pulse duration of 17 ns) [47,48]. Relying upon these advancements in the LD pumping, a high potentiality of a solid state laser for the reactor driver has been shown by conceptual designs [42,43,44].

A designing technique has been developed for the conceptual design of a LD pumped solid state laser system having 10 MJ blue output, 10 % overall efficiency, good beam quality, and 10 Hz repetition rate [49]. Here, we took into account the following conditions, and physical and material parameters.

(a)   To prevent the laser damage against optical elements, the fluence of the laser beam in the amplifier was restricted to be 32 J/cm^2 at a pulse width of 10 ns.

(b)   In order to prevent parasitic oscillation in the laser disk material, the product of the largest dimension of the disk and the small signal gain coefficient should be smaller than 4.

(c)   Since solid state laser materials usually have low thermal conductivities, a large thermal load to the disk due to incomplete energy conversion of pumping energy will cause fracture or deterioration of the disk due to large temperature difference between surface and mid-plane of the disk or large temperature rise. The maximum temperature difference and the peak temperature in the disk should not exceed 10 % and 30 % of the transition or melting point of the laser material, respectively.

(d)   The maximum thermal stress in the disk should not exceed 20 % of the fracture strength of the disk to ensure optical quality and long-lifetime operation.

(e)   The number of disks in a single segmented amplifier (Fig. 41) was selected to be 30 because of reducing the losses such as scattering loss at the disk surface and in the He gas coolant. In this case, the single pass transmittance of 90 % will be ensured even by taking into account the transmittance of two polarizers (99 % each) and the two Pockels cells (99 % each) in the regenerative cavity, and scattering loss of 0.1 % per surface of the disk ; then we can realize an extraction efficiency larger than 60 % with the initial small signal gain of larger than 4.

(f)   The aspect ratio of the disk was selected to be 20 in order to ensure the sufficient mechanical strength of the disk to ensure optical quality and long-lifetime operation.

(g)   To achieve an efficient pump absorption, the optical depth, defined as a product of the effective pump light absorption coefficient and the disk thickness, was selected to be 3.5.

(h)   To have an energy storage efficiency larger than 75 %, a rectangular pump pulse width of the laser diode arrays was selected to be half of the fluorescence lifetime of the laser material.

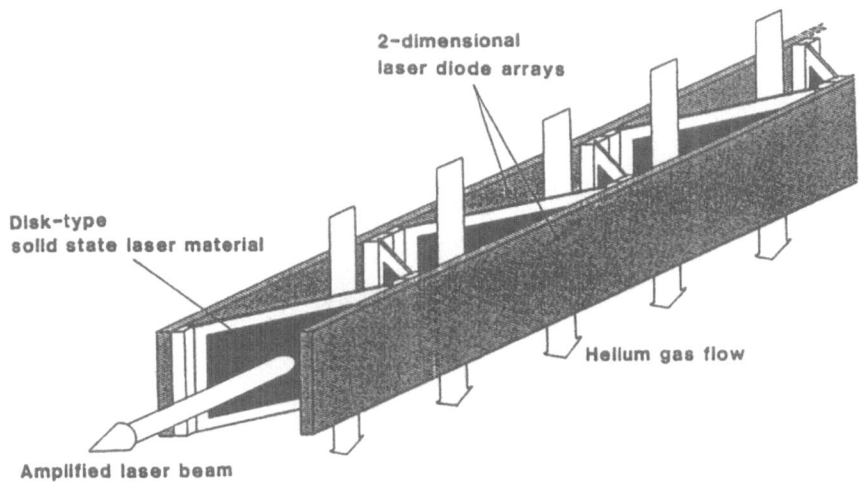

Figure 41 Single segmented amplifier pumped by two dimensional LD arrays. He gas is coolant for the disks.

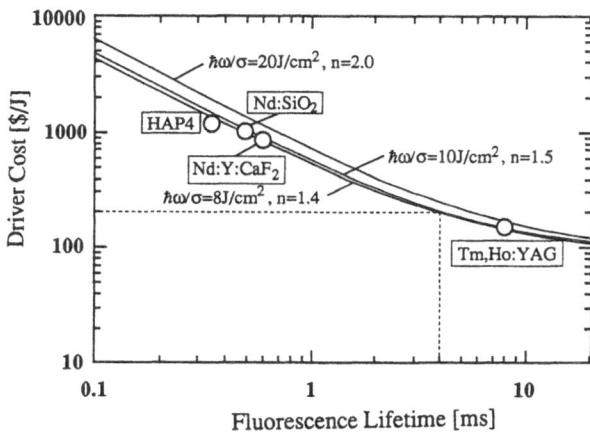

Figure 42 Total cost dependencies on the fluorescence lifetime [49]. The line show the dependencies in the case of $\hbar\omega/\sigma \sim 8$, 10 and 20 J/cm^2 and n $\sim$ 1.4, 1.5 and 2.0, respectively. Open circles are the results of the present designing listed in Table I.

Table I. Design of the laser diode pumped regenerative amplifier with 10 MJ output at 0.35 μm, 10 % overall efficiency, and 10 Hz repetition rate [49].

Laser Material	Glass			Crystal		
	LHG8	HAP4	Nd:SiO$_2$	Nd:YAG	Nd:Y:CaF$_2$	Tm,Ho:YAG
Single SA (30 Disks)						
Disk Size (mm)						
Thickness	5	18	20	20	18	20
Width	100	360	400	400	360	400
Length	183	659	705	834	628	831
Stored Energy (kJ)	1.20	16.8	37.3	1.24	29.8	43.9
2D Laser Diode Array						
Output Power (kW/cm^2)	2.5	2.43	3.06	0.22	2.52	0.23
Radiation Area (m^2)	0.6	7.8	9.6	9.6	7.8	9.6
Total B-integral (radians)	0.339	1.09	2.48	0.202	1.21	4.11
Optimum Pass Numbers	11	11	11	15	13	13
Maximum Extraction Efficiency	0.809	0.759	0.788	0.718	0.754	0.708
10MJ Blue Laser System						
Total Number of SAs	22400	1600	720	21600	900	720

SA : Segmented-Amplifier

(i) Total B integral during the round trips in a segmented amplifier was restricted to be less than 4 radians to have a good beam quality.

Using this conceptual designing technique, we evaluated the feasibility of the systems with six different solid state laser materials for the laser fusion reactor driver with 10 MJ blue output, 10 % overall efficiency, and 10 Hz repetition rate, as shown in Table I. The LD cost was estimated with the present stacked high power LD price and the usual cost reduction factor with mass-product. Here included in the code are the laser material parameters such as lasing wavelength, stimulated emission cross section, fluorescence lifetime, refractive index, nonlinear index of refraction, transition temperature or melting temperature, thermal conductivity and thermal shock parameter. The variable parameters of optimization are the doping density of lasing atom and the thickness of the disk. Although Tm,Ho:YAG crystal has a lasing wavelength of 2.091 μm, a pumping wavelength of 0.8 μm, the system using this material has an overall efficiency of 9.6 % under the assumptions of 85 % frequency conversion efficiency from 2 μm to 1 μm and full cross-relaxation. The systems with the laser materials such as HAP4 glass, Nd:SiO$_2$ glass, Nd:Y:CaF$_2$ crystal, and Tm,Ho:YAG crystal, having, respectively, good thermal shock parameter of 2.0, 14.5, 7.9 and 7.9 W/cm, and reasonable stimulated emission cross section of 3.6, 2.0, 2.0 and 0.9 $\times$ 10^{-20} cm^2, are shown to be an attractive candidates for the reactor driver due to the smaller number of segmented amplifiers. While the systems with such as LHG8 glass (poor thermal shock parameter of 0.4 W/cm) and Nd:YAG crystal (too large stimulated emission cross section of 65 $\times$ 10^{-20} cm^2) will not be selected as the candidates due to too many number of segmented amplifiers.

In Fig. 42, the calculated driver cost dependences on the fluorescence lifetime τ are shown for the solid state materials which have the saturation parameter $\hbar\omega/\sigma$ ($\hbar\omega$ being the laser photon energy, σ being the stimulated emission cross section) of 8, 10 and 20 J/cm^2 and n (refractive index) of 1.4, 1.5 and 2.0, respectively. Plotted are calculated driver costs for HAP4 glass, Nd:SiO$_2$ glass, Nd:Y:CaF$_2$ crystal and Tm,Ho:YAG crystal listed in Table I. It is seen from Fig. 42 that τ longer than 4 ms is desirable to achieve the driver cost per unit blue laser output of lower than \$200/J. In the case of $\tau \sim$ 4 ms, $\hbar\omega/\sigma \sim$ 10 J/cm^2 and n $\sim$ 1.5, the required peak power density for laser diode arrays will be only $\sim$ 0.4 kW/cm^2. From the economical point of view, Tm,Ho:YAG crystal is the most attractive due to its long fluorescence lifetime of 8 ms. With laser system using HAP4 glass, Nd:SiO$_2$ glass and Nd:Y:CaF$_2$ crystal, further reduction in the cost of laser diode arrays should be expected for the economic ICF reactor driver. Through the conceptual design studies with typical solid state laser materials, it is concluded that desirable properties of solid state laser materials for an economical laser fusion reactor driver are found to be $\hbar\omega/\sigma$ of around 10 J/cm^2, a thermal shock parameter of larger than 3 W/cm, a nonlinear index of refraction smaller than 3×10^{-13} esu, and a fluorescence lifetime of longer than 4 ms.

VIII. SUMMARY

The experimental achievement of super high density compression up to more than 600 times solid density gave us the confidence of ICF to reach the ignition and breakeven. The results have been analyzed to obtain the information for experimental requirements to achieve ignition.

Based on the experimental data base, ignition and high gain targets have been designed, which shows the feasibility of ignition and breakeven by ~ 100 kJ blue laser with 1 ~ 2 % nonuniformity of irradiation.

Laser beam control techniques have been developed to get better uniformity of irradiation in direct drive. Cannon ball target experiments also showed the feasibility of good uniformity of driving radiation on the fuel target. coupling efficiency from incident laser to the implosion hydrodynamic energy is the key issue to be investigated.

The progress of high average power laser technology with high efficiency is remarkable and is enough to give the confidence toward the reactor driver of laser fusion.

Laser fusion has now reached the stage to be recognized as an energy development program.

REFERENCES

1   S. Nakai, Bull. Am. Phys. Soc. 3 4, 2040 (1989).

2   S. Nakai et al., in *Laser Interaction and Related Plasma Phenomena Volume 9*, edited by H. Hora and
          G. H. Miley (Plenum Press, New York, 1991) p. 25.

3   H. Azechi et al., Laser and Particle Beams 9, 193 (1991).

4   E. M. Campbell et al., Appl. Phys. Lett. 3 6, 965 (1980)

5   H. Azechi et al. Bull. Am. Phys. Soc. 3 5, 1970 (1990).

6   Y. Setsuhara et al., Laser and Particle Beams 8, 609 (1990).

7   S. W. Haan, Phys. Rev. A 3 9, 5812 (1989).

8   H. Takabe, L. Montierth, and R. L. Morse, Phys. Fluids 2 6, 2299 (1983); H. Takabe, K. Mima, L.
          Montierth, and R. L. Morse, ibid. 2 8, 3676 (1985).

9   M. Tabak, D. H. Munro, and J. D. Lindl, Phys. Fluids B 2, 1007 (1990); J. H. Gardner, S. E. Bodner,
          and J. P. Dahlburg, ibid. 3, 1070 (1991).

10   W. M. Manheimer D. G. Colombant, and J. H. Gardner, Phys. Fluids 2 5, 1644 (1982).

11   H. Takabe et al., Phys. Fluids 3 1, 2884 (1988); H. Nishimura et al., Phys Rev. A 43, 3073 (1991).

12   K. Takami and H. Takabe, Tech. Rep. Osaka Univ. **4 0**, 159 (1990).

13   J. R. Freeman et al., Nucl. Fusion **1 7**, 223 (1977); F. Hattori, H. Takabe and K. Mima, Phys. Fluids **2 9**, 1719 (1986); H. Sakagami and K. Nishihara, Phys. Rev. Lett. **6 5**, 432 (1990), Phys Fluids **B 2**, 2715 (1990).

14   J. Mayer-ter-Vehn, Nucl. Fusion **2 2**, 561 (1982).

15   R. L. McCrory et al., in the proceedings of IAEA-TCM, Osaka, April 14-19 (1991).

16   H. Takabe and K. Mima, ILE Progress Report, ILE 8713P, Osaka University, Dec. 15 (1987).

17   T. Norimatsu, H. Itoh, C. Chen, M. Yasumoto, M. Tsukamoto, K. A. Tanaka, T. Yamanaka, and S. Nakai, submitted to J. Vac. Sic. Tech.

18   M. Tsukamoto, R. Kodama, M. Kado, H. Itoh, M. Yasumoto, T. Norimatsu, M. Nakai, K.A. Tanaka, T. Yamanaka, and S. Nakai, Rev. Laser Engineering, 1 8, 724 (1990).

19   M. Saito, S. Urushihara, K. Suzuki, K. A. Tanaka, T. Yamanaka, and S. Nakai, Rev. Laser Engineering, 1 7, 721 (1989).

20   R. Kodama et al., to be published.

21   K.A. Tanaka et al., to be published.

22   M. Murakami and J. Meyer-ter-Vehn, Nucl. Fusion **3 1**, 1315 (1991).

23   M. Murakami and J. Meyer-ter-Vehn, ibid **3 1**, 1333 (1991).

24   H. Nishimura et al., Kakuyogo Kenkyu **6 3**, 219 (1990) in Japanese.

25   R. Sigel et al. and H. Nishimura et al., 13th International Conference on Plasma Physics and Controlled Nuclear Fusion Research, IAEA-CN-53/ B-2-1, Washington 1-8 October 1991.

26   H. Nishimura et al., to appear in Phys. Rev. **A** (1991).

27   R. Sigel et al., Phys. Rev. Lett. **6 5**, 587 (1990), the longer version is to appear in Phys. Rev. **A** (1991).

28   R. Pakula and R. Sigel, Phys. Fluids **2 8**; 232 (1985), ibid **2 9**, 1340(E) (1986).

29   R. Ramis, R. Schmalz, and J. Meyer-ter-Vehn, Computer Phys. Commun. **4 9**, 475 (1988).

30   M. Nakamura et al., to be published.

31   T. Endo, H. Shiraga and Y. Kato, Phys. Rev. **A 42**, 918 (1990).

32   H. Takabe, "Radiation Transport and Atomic Modeling for Laser Produced Plasmas", ILE Research Report 9008p, Sep. 10 (1990).

33   Y. Kato et al., 13th International Conference on Plasma Physics and Controlled Nuclear Fusion Research, IAEA-CN53/13-2-2, Washington, 1-6 October 1990.

34   M. Katayama, et al., Rev. Sci. Instrum., **6 2**, 124 (1991).

35   The higher ρR value of 10 mg/cm2 was obtained, in the similar way, for a larger diameter ($\sim 380 \, \mu m$) capsule. But the value has not been confirmed yet by the knock-on measurement for the same laser shot.

36   Y. Kato, et al., Phys. Rev. Letters, **5 3**, 1057 (1984).

37   K. Tsubakimoto, Technol. Repts. Osaka Univ., **4 1**, 125 (1991).

38   M. Born and E. Wolf, "Principles of Optics", (1975, Pergamon Press, Oxford) P.514.

39   D. Velon et al., Optics Commun., **6 5**, 42 (1988).

40   H. Nakano et al., Optics Commun., **7 8**, 123 (1990).

41   X. Deng et al, Applied Optics, **2 5**, 377, (1986).

42   K. Naito, M. Yamanaka, T. Kanabe, M. Nakatsuka, K. Mima, and S. Nakai : Rev. Laser Engineering **1 8**, 652 (1990) (in Japanese).

43   W. F. Krupke : Fusion Technology **1 5**, 37 (1989).

44   M. Yamanaka, K. Naito, T. Kanabe, M. Nakatsuka, and S. Nakai : Kakuyugo Kenkyu **6 2**, 79 (1989) (in Japanese).

45   Spectra Diode Laboratories (private communication).

46   R. Beach, D. Mundinger, W. Benett, V. Sperry, B. Comaskey, and R. Solarz : Appl. Phys. Lett. **5 6**, 2065 (1990).

47   R. L. Burnham : Laser & Optronics **7**, 79 (1989).

48   W. Koechner : Rev. Laser Engineering **1 9**, 619 (1991).

49   K. Naito, M. Yamanaka, M. Nakatsuka, T. Kanabe, K. Mima, C. Yamanaka and S. Nakai : Jpn. J. Appl. Phys. (1991) (submitted).

INTERACTION PHYSICS FOR MEGAJOULE LASER FUSION TARGETS

William L. Kruer

Department of Physics
Lawrence Livermore National Laboratory
7000 East Ave., MS L-472
Livermore, CA 94550

ABSTRACT

Some little-explored interaction phenomena for targets irradiated with megajoule lasers are considered. Simple estimates show that the laser plasma interaction then occurs in a hot (multi-keV) plasma with density much less than the critical density. In such plasmas, Raman and Brillouin scattering into the forward hemisphere are potentially significant. A simple model shows that Raman forward scattering can be saturated at low levels by ponderomotive detuning. Calculations also illustrate a suppression of ponderomotive filamentation by plasma-induced beam smoothing.

INTRODUCTION

Laser plasma interactions continue to be a key, high leverage issue for inertial fusion applications. As indicated by the recent evolution[1] in target design strategy, there are significant benefits to increasing the laser light intensity in order to drive hohlraum targets to higher radiation temperature. This could enable more efficient x-ray driven implosions and the demonstration of ignition and gain with a 1-2 megajoule (MJ) laser rather than a 5-10 MJ one. A key issue for this new strategy is the coupling physics upon which more stringent demands are being placed. It is obviously necessary to examine broader parameter regimes; i.e., using laser intensities from near instability thresholds to well above thresholds. It is also important to test various control mechanisms (such as beam smoothing or, say, large bandwidth[2]) which can expand the parameter regime for efficient and benign coupling. This discussion will focus on hohlraum targets[3], since x-ray drive is now the mainline approach to inertial fusion. Figure 1 shows a generic hohlraum target, which is a gold cylinder with holes in the end through which laser light is introduced. The laser light interacts with the Au plasma expanding off the walls to generate x-rays, which are used to implode a fuel capsule.

ESTIMATES OF THE PLASMA CONDITIONS

It is instructive to estimate relevant plasma conditions in a hohlraum for a megajoule size laser driver. In Nova experiments, hohlraums with a characteristic size $L \sim 1$ mm are irradiated with a laser energy of ~ 40 kJ. The size L will scale roughly as the cube root of the laser energy. Hence a hohlraum irradiated with (say) 1.5 MJ will have a characteristic size $L \sim 3$ mm.

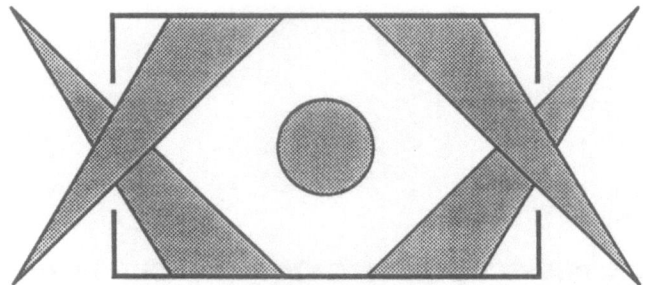

Figure 1.   An illustration of a hohlraum target irradiated with laser light.

In such large high Z plasmas,[4] collisional absorption occurs at a density $n \ll n_{cr}$, the critical density. Consider a light wave impinging on a plasma with density scale length L. The fractional absorption f_{abs} becomes significant ($\gtrsim 1/2$) when

$$\frac{v_{ei}}{c} \frac{n}{n_{cr}} L \cong 1. \tag{1}$$

Here v_{ei} is the electron-ion collision frequency appropriate to inverse bremsstrahlung and c is the velocity of light. A simple flux-limited heat flow model is used to relate the heated electron temperature to the absorbed intensity; i.e.,

$$f_{abs} I \cong 0.2 \, n \, \theta_e \, v_e , \tag{2}$$

where I is the incident laser intensity, θ_e the electron temperature, and v_e the electron thermal velocity. If we take $f_{abs} = 1/2$, Eqs. (1) and (2) give

$$\frac{n}{n_{cr}} \cong \left[\frac{I_{14} \lambda_\mu^4}{16 \, ZL} \right]^{1/3} , \tag{3}$$

$$\theta_e \, (keV) \cong 1.9 \left[I_{14} \lambda_\mu \right]^{4/9} \, (ZL)^{2/9} . \tag{4}$$

Here Z is the ion charge state, L the scale length in cm, I_{14} the laser intensity in units of 10^{14} W/cm^2, and λ_μ the laser wavelength in microns.

As an example, we take L = 0.3, λ_μ = 0.35, I_{14} = 10 and Z = 50. Then n/n_{cr} = 0.08 and θ_e = 6 keV. These crude estimates suffice to show that the laser plasma interaction will effectively be occurring in a large plasma (L ~ 3 mm) with $n \lesssim 0.1 n_{cr}$ and θ_e ~ 5 keV.

STIMULATED BRILLOUIN SCATTERING

First, let's make a few observations concerning Brillouin scattering. Besides reducing the absorption, this process can modify the location of the energy deposition and so is an issue for future targets with their more stringent demands on implosion symmetry. For the projected plasma conditions, this instability is in a different regime than usually investigated. As an example, consider a plasma with θ_e = 5 keV and n = 0.04 n_{cr}. The ion wave associated with backscattering then has a wavenumber $k\lambda_{De} \sim 1$, where λ_{De} is the electron Debye length. The dispersion characteristics of the ion waves change. The modes become more like ion plasma waves than the sound waves usually considered.

One consequence is a change in the threshold intensity. For $k\lambda_{De} \gtrsim 1$, density gradients can have an effect comparable to that of velocity gradients. In particular, the change in the ion wave number k associated with a density gradient with scale length L is

$$\frac{1}{k}\frac{\partial k}{\partial x} \sim \frac{k^2\lambda_{De}^2}{2\left(1 + k^2\lambda_{De}^2\right)}\frac{1}{L} .$$

(5)

Nonlinear frequency shifts might also play a more important role.

It should also be noted that significant equilibration between the electron and ion temperatures can occur in these future plasmas which will be irradiated with multi-nanosecond pulses. An equilibration time is

$$\tau_{eq} \sim \frac{10^{13}\,\mu\theta_e^{3/2}(keV)}{n\,Z^2\,\ell n\Lambda} ,$$

(6)

where τ_{eq} is in ns, μ the ion mass in amu, n the electron density in cm^{-3}, and $\ell n\Lambda$ is a Coulomb logarithm. For $n = 4 \times 10^{20}\ cm^{-3}$ (0.04 n_{cr} for 0.35 μm light), $\theta_e = 5$ keV, $Z = 50$ and $\ell n\ \Lambda = 5$, $\tau_{eq} \sim 2ns$. Hence $\theta_i \sim \theta_e$ when the pulse length exceeds several ns. However, the ion wave is still weakly Landau damped in a high Z plasma.

Several other features of Brillouin scattering are relatively unexplored. Forward scattering at a significant angle can have a growth rate comparable to backscattering. For example, if the incident light wave is scattered by 45°, $|k| \cong 0.8\ \omega_0/c$ and the growth rate is about 0.6 of the growth rate for backscatter. Significant forward scattering at an angle has be observed in simulations by Wilks et al.[6] and in experiments by Batha et al.[7] In addition, Brillouin scattering is enhanced when multiple beams overlap,[8] which obviously happens in the laser entrance hole. The enhancement simply requires that one daughter wave be shared.

STIMULATED RAMAN FORWARD SCATTERING

In hot, low density plasmas the electron plasma wave associated with Raman backscattering is strongly Landau-damped. The resonant instability is transformed to the much weaker process of stimulated Compton scattering on the electrons. However, the electron plasma wave associated with direct forward scattering has a relatively small wavenumber ($k\lambda_{De} \cong v_e/c$) and little Landau damping. Forward scattering at an angle is actually preferred,[6,9] since the growth rate is proportional to $|k|$, which increases with angle. If we neglect transverse variations, the angle for maximum growth is determined by the onset of Landau damping. For $n/n_{cr} = 0.04$ and $\theta_e = 5$ keV, the growth rate maximizes for $\theta \simeq 30°$, where θ is the angle between the wave vectors of the incident and scattered light waves. (For this example, the angle between the wave vectors of the incident light wave and the plasma wave is about 50°.) In practice, transverse inhomogeneity may limit forward scattering to smaller angles.

It is instructive to consider a very simple model for the stabilization of Raman forward scattering by ponderomotive detuning. This nonlinear state can be readily calculated from coupled mode equations which include the density perturbation ponderomotively induced by the growing electron plasma wave. We consider direct forward scattering in an initially uniform plasma with density n_0 irradiated with an intense light wave with vector potential $A_0 \cos(k_0x - \omega_0t)$. The vector potential of the forward scattered light wave ($\tilde{A}s$) and the electron density fluctuation of the associated electron plasma wave ($\tilde{n}$) are expressed as

$$\tilde{A}s = A_c(x) \cos(k_s x - \omega_s t) + A_s(x) \sin(k_s x - \omega_s t),$$

$$\tilde{n} = n_c(x) \cos(kx - \omega t) + n_s(x) \sin(kx - \omega t).$$

Here ω_s and k_s (ω and k) are the frequency and wavenumber of the scattered light (the electron plasma wave). The coefficients are assumed to be slowly varying in x. Equations for these amplitudes are obtained in the usual way:

$$\frac{\partial A_s}{\partial x} = \gamma\, n_c \tag{7}$$

$$\frac{\partial A_c}{\partial x} = \gamma\, n_s$$

$$\frac{\partial n_s}{\partial x} = \alpha\, A_c - \beta\left(n_s^2 + n_c^2\right) n_c \tag{8}$$

$$\frac{\partial n_c}{\partial x} = \alpha\, A_s - \beta\left(n_s^2 + n_c^2\right) n_s \,.$$

The vector potentials have been normalized to A_0, the density fluctuations to n_0, and $x = k_s x$. Here $\alpha = k v_{os}^2/(12 k_s v_e^2)$, $\beta = \left(24\, k_s k^3 \lambda_{De}^4\right)^{-1}$ and $\gamma = \omega_{pe}^2/(4 k_s^2 c^2)$, where k_s is the wavevector of the scattered wave and c is the velocity of light.

The cubic nonlinearity in Eq. (8) is due to the density variation δn_0 induced by the ponderomotive force, i.e.,

$$\frac{\delta n_0}{n_0} = \frac{-v_w^2}{4 v_e^2} \,,$$

where v_w is the oscillatory velocity of an electron in the plasma wave and v_e is the electron thermal velocity. For simplicity the quasi-static approximation has been used. In actuality, the saturated wave amplitudes will overshoot, since there is a finite time for the density to respond.

In terms of the normalized intensity of the scattered light $u = \left(A_s^2 + A_c^2\right)$, the above equations reduce to

$$\frac{\partial^2 u}{\partial x^2} = 4\alpha\gamma u - \frac{\beta^2 \alpha^2 u^3}{2\gamma^2} \,, \tag{9}$$

The solution is trivial. For $u \ll 1$,

$$u = u_0 \exp\left[2\sqrt{\alpha\gamma}\, x\right], \tag{10}$$

which simply describes the convective growth with growth length $k_s \ell_g = (\alpha\gamma)^{-1/2}$. Saturation occurs when

$$u = \left(\frac{16\gamma^3}{\alpha\beta^2}\right)^{1/2} . \tag{11}$$

Since the reflectivity r = u v_{gs}/c, we obtain

$$r \simeq (12)^{3/2} \left(\frac{v_e}{c}\right)^2 k^2 \lambda_{De}^2 \frac{v_e}{v_{os}} \left(\frac{k}{k_s}\right)^{1/2} \frac{\omega_{pe}}{\omega_s} . \qquad (12)$$

Except for numerical factors, Eqs. (11) and (12) can be obtained by simple physical arguments.

We note that the nonlinear levels are quite modest. As an example, consider a plasma with n = 0.04 n_{cr} and an electron temperature of 4 keV irradiated with 0.53 μm laser light with an intensity of 10^{15} W/cm². The reflectivity given by Eq. (12) is then < 1%. Observe also that r scales as $\theta_e^{5/2}/I^{1/2}$. The net dependence on intensity is determined mainly by that of θ_e. (Neglect of the upshifted light wave is a better approximation for scattering at an angle or at higher density.)

It should be noted that other nonlinear effects may be even more efficient in general. For example, when $Z\theta_e/\theta_i \gg 1$ (Z is the ion charge state and θ_i the temperature), the ion acoustic decay instability can be driven by the unstable plasma wave. This instability may clamp the plasma wave amplitude at a small level[10] which corresponds to the threshold for ion acoustic decay.

LASER BEAM SMOOTHING

Let's conclude with some discussion of laser beam smoothing,[11-13] which represents an important control mechanism. The simplest form of beam smoothing is the use of random phase plates to introduce spatial incoherence. An angular spread in wave vectors is created; i.e., the wave vectors are in a cone with a characteristic half-angle $\Delta\theta$ about the original wave vector.

The spatial incoherence can reduce instability generation. Ponderomotive filamentation serves as a good illustration. The angular spread clearly acts to counter nonlinear focussing. A crude estimate is that the instability is suppressed when the correlation length in the transverse direction $\ell_\perp \simeq \dfrac{2\pi}{k_o \Delta\theta}$ is less than the wavelength of the most unstable filament λ_{max}. Within a factor of two, this condition gives an earlier prediction:[14]

$$(\Delta\theta)^2 \gtrsim \frac{1}{2} \frac{n}{n_{cr}} \frac{I_{16} \lambda_\mu^2}{\theta_e(keV)} , \qquad (13)$$

where I_{16} is the intensity in units of 10^{16} W/cm². The same result (again within a numerical factor) is found by requiring that the coherence length in the direction of propagation be less than the growth length.

The addition of temporal incoherence also helps to suppress filamentation. In techniques such as ISI,[12] a small bandwidth smooths the spatial interference pattern on a time scale less than the filament growth time. Simulations[15] have shown a pronounced reduction in filamentation by ISI.

Density fluctuations in the plasma can introduce laser beam incoherence; i.e., broaden the angular spread in the wave vectors. Indeed, small amplitude density modulations with wavelength $\lambda_\perp \ll \lambda_{max}$ in the transverse direction can suppress filamentation. A density fluctuation with amplitude δn couples an incident wave into a light wave at an angle $(\theta \simeq \lambda_\perp/\lambda_0)$ in a distance of $k_o z \sim \left(\dfrac{1}{2} \dfrac{\delta n}{n_{cr}}\right)^{-1}$. This distance is less than a growth length when $\dfrac{\delta n}{n} \gtrsim \dfrac{1}{4}\left(\dfrac{v_{os}}{v_e}\right)^2$, which for typical parameters corresponds to a small amplitude.

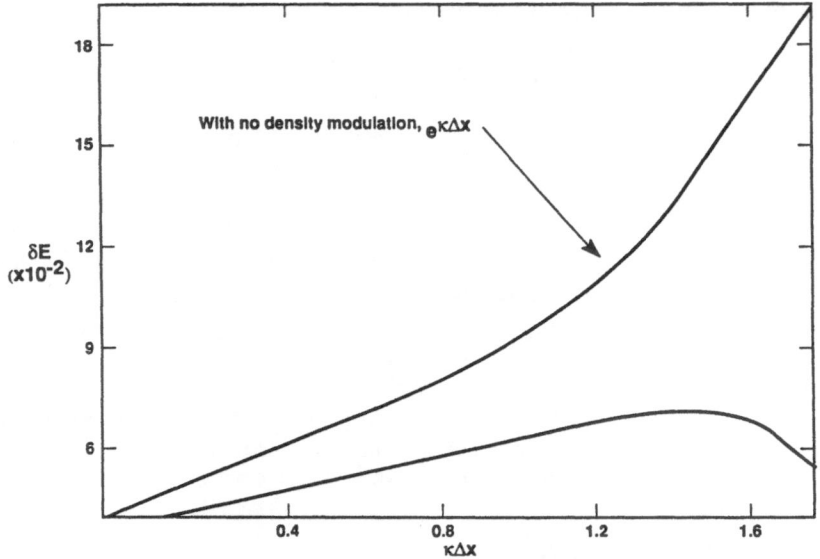

Figure 2.   The calculated growth in space of the amplitude δE of the most unstable filament
with and without an imposed density modulation of amplitude $\delta n/n_{cr} = 0.01$.

This suppression has been observed in simulations[16] of ponderomotive filamentation. A 3D model for light wave propagation including quasi-static plasma response with a saturable nonlinearity was solved numerically. Figure 2 shows the computed spatial growth of the electric field δE associated with the most unstable filament in a plasma with a uniform density of $0.1\ n_{cr}$. With no imposed density modulation, δE grows exponentially as predicted. However, when a small density fluctuation with a wavelength of 0.25 of the most unstable wavelength is imposed, the growth is quenched. Physically, the incident wave is broadened in angle by its interaction with the density fluctuation before filamentation can develop.

These results suggest the possibility of plasma-induced beam smoothing. One scheme is to send along a low intensity seed beam which has an angular spread and a small bandwidth (of order 0.1%). The aim is to induce Brillouin near forward scattering of the main beam, which is thereby smoothed. Generation of a spectrum of density fluctuations which scatter the beam may also be a key feature of the long-term nonlinear evolution of filamentation driven by an initially coherent beam.

SUMMARY

Consideration has been given to some relatively unexplored interaction phenomena of potential interest for targets irradiated with megajoule lasers. Simple estimates show that the laser plasma interaction then takes place in a hot (multi-keV), low density ($n \ll n_{cr}$) plasma. Some novel features of Brillouin scattering as well as stimulated scattering into the forward hemisphere were briefly considered. A simple model for the saturation of Raman forward scattering by ponderomotive detuning was given. Finally, plasma-induced beam smoothing and its potential effect on filamentation were discussed.

Optimizing fusion target designs requires continued improvements in understanding the interaction physics. More detailed measurements of the plasma conditions are needed, particularly in hohlraum targets. Characterization and modelling of the angular distribution of stimulated scattering and the competition with filamentation are important challenges. Finally, an ongoing priority is to better understand the benefits of laser smoothing and how these benefits scale to future targets and irradiation conditions.

ACKNOWLEDGEMENTS

This work was performed under the auspices of the U.S. Department of Energy by the Lawrence Livermore National Laboratory under Contract No. W-7405-ENG-48. This manuscript includes material given in an initial paper at the 22nd Anomalous Conference in April, 1991. I am grateful for discussions with E. M. Campbell, S. C. Wilks, A. B. Langdon, R. L. Berger, B. Lasinski, S. Batha, and R. P. Drake. I thank Gina Cochran for expertly producing the manuscript.

REFERENCES

1. John D. Lindl, in *From Fusion to Light Surfing*, edited by T. Katsouleas (Addison-Wesley, Redwood City, CA 1991) p. 177-190.
2. J. H. Nuckolls, *Phys. Today*, **35**, 24 (1982); R. Sigel. et al., *Phys. Rev. Lett.* **65**, 587 (1990).
3. D. Eimerl, W. L. Kruer and E. M. Campbell, Lawrence Livermore National Laboratory, UCRL-JC-109594 (1992).
4. J. S. DeGroot et al., *Phys. Fluids* **B3**, 1241 (1991); and references therein.
5. C. E. Clayton et al., *Phys. Fluids* **24**, 2312 (1981).
6. S. Wilks et al., Lawrence Livermore National Laboratory, UCRL-JC-108690 (1991).
7. S. Batha et al., *Bull. Am. Phys. Soc.* **36**, 3351 (1991).
8. D. F. Dubois, B. Bezzerides and H. A. Rose, *Phys. Fluids* **B4**, 241 (1992).
9. J. J. Thomson, *Phys. Fluids* **21**, 2082 (1978).
10. R. P. Drake and S. H. Batha, *Phys. Fluids* **B3**, 2936 (1991).
11. R. H. Lehmberg and S. P. Obenschain, *Opt. Commun.* **46**, 27 (1983).
12. S. Skupsky, R. W. Short, T. Kessler, R. S. Craxton, S. Letzring, and J. M. Soures, *J. Appl. Phys.* **66**, 3456 (1989).
13. Y. Kato, K. Mima, N. Miyanaga, S. Arinaga, Y. Kitagawa, M. Nakatsuka, and C. Yamanaka,, *Phys. Rev. Lett.* **53**, 1057 (1984); X. Deng, X. Liang, Z. Chen, W. Yu, and R. Ma, *Chin. J. Lasers* **12**, 257 (1985).
14. A. B. Langdon in Lawrence Livermore National Laboratory, Laser Program Annual Report-83, UCRL-50021-83, pp. 3-35 to 3-361 (August 1984); see also H. A. Rose and D. F. DuBois, *Phys. Fluids* **B4**, 252 (1992).
15. A. J. Schmitt, *Phys. Fluids* **31**, 3079 (1988).
16. W. L. Kruer and P. E. Young, US-USSR Workshop on Optical and Plasma Physics, Los Angeles, CA March 1990.

EXPERIMENTAL STUDIES OF IMPLOSION OF INDIRECTLY

DRIVEN DT-FILLED TARGETS

X.F. Chen, Z.J. Zhen, and H.S. Peng

Southwest Institute of Nuclear
Physics and Chemistry

INTRODUCTION

It is an important subject at present in inertial confinement fusion (ICF) field that balloon targets are irradiated by laser to cause thermonuclear fusion. Laser driven modes may generally be divided into two kinds: indirect drive and direct drive according to process of energy transportation. Direct drive means: laser directly aims at fuel pellet, laser energy is conducted from absorption region to ablation surface by electron conduction. This is a drive mode with high efficiency, but the requirement for irradiation homogeneity is very strict. Indirect drive means: laser energy is first converted to X-ray energy, then fuel pellet contained in hohlraum target is driven by the X-ray. Indirect drive greatly reduces requirements for homogeneity and symmetry of beams, but it is a two-step process, and efficiency is probably low.

Experiments on implosion of directly driven DT-filled targets were conducted on LF-12 laser facility as early as .986. The neutron yields were 5 X 10^6. While experiments on indirectly driven targets were carried out on the same facility in 1990 to research another approach of implosion, the neutron yields were 2 X 10^3~3 X 10^5.

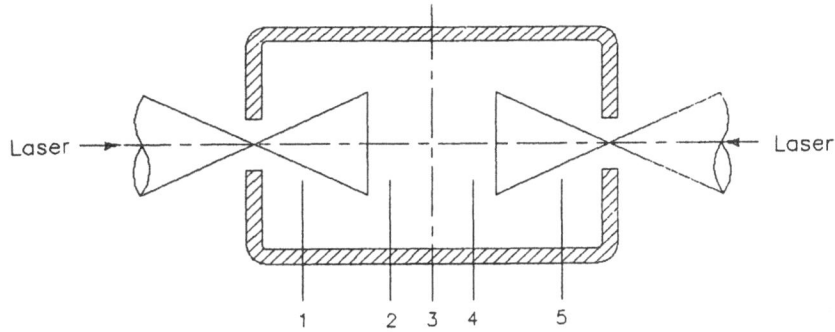

Figure 1. Schematic diagram of the hohlraum target.

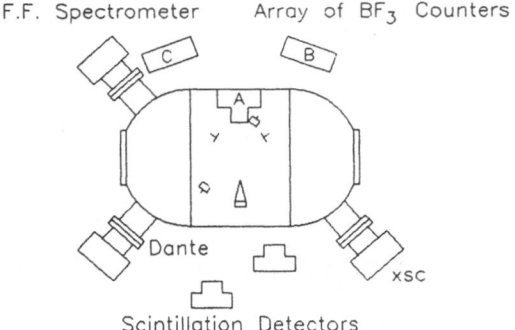

F.F. Spectrometer   Array of BF₃ Counters

Figure 2. Schematic diagram of the experimental arrangement.

EXPERIMENTAL SETUP AND DIAGNOSIS

In the experiments of indirectly-driven implosion, dual laser beams with 500-700J, 0.7-1ns/beam were used, the asynchronization time of the two beams was less than 10ps and their energy asymmetry was ± 10% with energy injection rate 90%.

The target model of indirectly driven implosion was hohlraum to provide a uniform radiation field with high enough temperature. The characteristic of the hohlraum target lies in that its shell is made of high-Z element (Au), DT-filled pellet (the ratio of deuterium to tritium 1:1) is contained in the hohlraum target, and there are two laser-incident holes on the shell (Figure 1). The inside of a hohlraum target may generally be divided into two regions: energy absorption and conversion region (1,5); radiation driven implosion region (3). There are some X-ray transport passages (2, 4) between the two regions.

In the implosion experiments of indirect drive, two laser beams injected into the cavity of the hohlraum through two incident holes are changed to intense X-ray fluid by absorption of the cavity wall, then the X-ray fluid ablate and compresses the DT-filled microballoon contained in the cavity to cause thermonuclear fusion.

The experimental arrangement is shown in Figure 2.

Two sensitive detection systems (arrays of BF₃ counters and scintillation detectors) were used to measure low neutron yields with intense X-rays. Three arrays called A, B and C were used in the experiments, and they were placed at 9. 5, 104 and 129 cm from the target respectively. Array A was put into the target chamber and kept separate from vacuum of the target chamber using a barrel. Nine polyethylene rings were employed as neutron moderators. Delay time and sensitivity of the arrays were calibrated using a

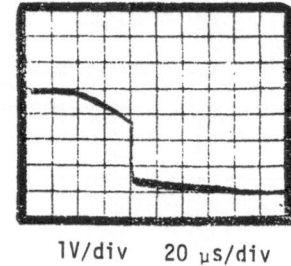

1V/div   20 μs/div

Figure 3. Saturated X-ray pulse.

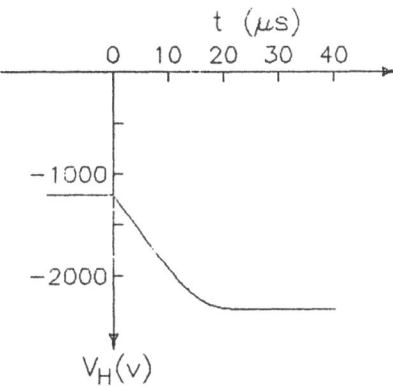

Figure 4. After adding time gate, high voltage of the counter varies with time.

pulse neutron source and an accelerator neutron source.

Delay times of the three arrays were 350, 310 and 210 μs respectively and their sensitivities were about 2 X 10^3, 4 X 10^3 and 7 X 10^3 neutron/count respectively. Pb bricks for shielding (thickness 5-11 cm) were placed around array B and C to reduce the X-ray effect on the detector.

Two scintillation detectors were used in the experiments. They were made up of Φ 180 X 120 mm plastic scintillators and XP 2024 photomultipliers, placed at 1 and 2 m from the target respectively. The thickness of Pb shielding layers was 7-10 cm. Lowest measurable neutron yield was 1 X 10^4.

Two time-resolved X-ray spectrometers and a subkeV streak camera were used in the experiments to monitor radiation temperature and time profiles of X-rays from the hohlraum.

EXPERIMENTAL RESULTS AND DISCUSSIONS

1. Thermonuclear neutron

Comparing with neutron yield measurement of directly driven implosion, neutron yield measurement of indirectly driven implosion would be more difficult because neutron yields are very low and interference from the hohlraum targets is very intense. Figure 3 shows a photograph of hard X-ray pulse obtained by using array A. Saturated X-ray pulse have occupied half of the oscilloscope screen, and the X-ray signal was about 40 times larger than the neutron signal. For indirectly driven implosion experiments where low expected yield required that we place the array close to the targets, the

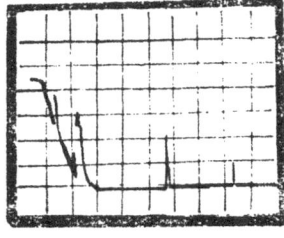

Figure 5. Oscilloscope trace of neutron events measured with array B.

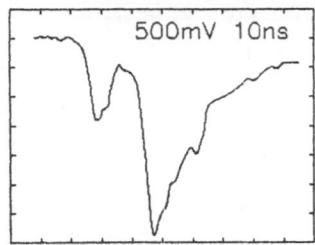

Figure 6. Oscilloscope display of neutron waveform measured.

array was too saturated to be of any use.

We partially solved the array saturation problem by using time gate in negative high voltage version of the arrays and Pb shielding slabs (thickness 5-11 cm).

Figure 4 describes how the high voltage of BF_3 counters varies with time after adding time gate.

In the moment of the X-ray burst, the high voltage of the counter was about 1000V, less than normal operation voltage, which reduced the X-ray effect on the counter.

Figure 5 shows a typical oscilloscope trace of neutron events measured with array B, the initial saturated big pulse corresponds to X-rays, before it disappears, several neutron events overlap on it, then another two neutron pulses are observed.

Figure 6 is typical neutron waveform detected with the scintillation detector. The first peak corresponds to X-ray peak and the second is 14 MeV-neutron. The time interval between the two peaks is in good agreement with flight time of 14 MeV-neutron.

Table 1 shows several typical neutron yields with the laser parameters.

2. Radiation temperature in hohlraum targets

Indirectly driven implosion depends on X-ray transport which is very sensitive to radiation temperature T_R. X-ray fluid

$$F_X = -L_R CVAT_R^4/3.$$

L_R-average free path of X-rays; C-light speed; A-radiation parameter.

So radiation temperature is a key physical value.

For indirectly driven mode, the energy spectrum of subkeV X-rays escaping from

Table 1. Typical neutron yields.

Shot no.	Drive Mode	Laser Parameters				Yn (neutron/shot)	
		E_L (J)		FWHM (ps)		Array	scintillator
90111909	direct	715	636	840	860	5.96 X 10^4	2.8 X 10^4
90111902	direct	706	637	780	840	2.27 X 10^4	7.0 X 10^3
90112002	composite	639	773	810	770	3.15 X 10^5	1.5 X 10^5
90112101	composite	517	709		1100	1.08 X 10^4	
90112202	indirect	600	656	620	690	4.53 X 10^3	
90112603	indirect	680	578	890	700	2.26 X 10^3	

the hohlraum target closely follows a Planck's distribution therefore we employ radiation temperature as characteristic of the spectrum (T_R=Epx/2.8).

Spectra, time profiles and energy of X rays from the targets were observed with two 10-channel time-resolved X-ray spectrometers (0.2-1. 5keV, 200ps), XRDs with absorption foils and a subkeV X-ray streak camera. Radiation temperature in the hohlraums were about 130 ±10 eV.

REFERENCES

1. E. Storm, Progress of Inertial Confinement Fusion at LLNL IAEA 11th conference on Plasma Physics and Controlled Nuclear Fusion Research Kyoco, Japan, 13-20 Nov, 1986.
2. L.R. Veeser, LA-7022-MS, 3, (1978).
3. Yu Ming, CAEP-0001 HL-0001, T, (1988).

LASER PLASMA INTERACTION STUDIES RELEVANT FOR INERTIAL CONFINEMENT FUSION

O. Willi, T. Afshar-rad, M. Desselberger, M. Dunne, J. Edwards, L. Gizzi, F. Khattak, D. Riley, R. Taylor and S. Viana

Imperial College of Science, Technology and Medicine, London, UK

INTRODUCTION

Several basic processes occurring during the interaction of laser irradiation with matter have been investigated by using the high power laser systems of the SERC Central Laser Facility. The main effort of the recent research concentrated on ICF related studies with improved laser illumination uniformity generated by Random Phase Plate (RPP) Arrays and Induced Spatial Incoherence (ISI) techniques or a combination of both. In addition, highly transient plasmas have been produced by a prepulse-free 12 ps high power Raman shifted KrF laser pulse.

An interference code was written to predict the focal spot profiles produced by different smoothing techniques. Simulations were obtained at any time during the laser pulse and were compared to either time integrated or framed (framing time $\simeq$ 140 ps) equivalent focal spot images. The predictions agree well with the experimental observations.

The suppression of instabilities including laser beam filamentation, stimulated Raman and Brillouin scattering was studied in large underdense plasmas with laser beams smoothed by RPP Arrays or ISI. Millimetre-sized cylindrical underdense high temperature plasmas were produced by irradiating thin foil targets with a number of laser beams in a line focus geometry. A separate laser beam interacted axially with the preformed plasma. Significant reductions in both the SRS and SBS levels were seen for both smoothing methods. In addition, time resolved x-ray and optical observations show whole beam self-focusing of the ISI beam in the preformed plasma.

The uniformity of the overdense plasma of laser irradiated targets was investigated by using a novel time resolved x-ray imaging technique with submicron spatial resolution. 2-D images show that laser beam nonuniformities imprint themselves onto the cold target surface at the beginning of the laser pulse generating considerable density perturbations which persist throughout and after the laser pulse with no evidence of smoothing.

Intense, soft x-ray pulses, generated from separate laser irradiated converters, were used to irradiate planar plastic foils. The x-ray heating was investigated by measuring the temperature histories of chlorinated tracer layers buried at different depths in the targets. The temperature diagnostic was a novel time resolved XUV absorption spectroscopy technique using chlorine L-shell transitions. The temporal temperature profiles were reasonably well reproduced by radiation hydrodynamic simulations.

Growth rates of the Rayleigh–Taylor instability were measured in thin foil targets with imposed sinusoidal modulations irradiated by optically smoothed laser beams. A hybrid optical smoothing scheme utilising ISI and RPP was used. The enhancement in the modulation depth during acceleration was observed with time resolved transmission radiography using a soft x–ray backlighting source. The wavenumber dependence and nonlinearity of the RT growth were investigated by using a range of modulation periodicities and depths. The measurements were compared with 2–D hydrocode simulations.

Finally, hot, close to solid density plasmas were produced by irradiating solid targets with a 12 ps 5 Joule Raman shifted KrF laser pulse. The power contrast ratio between the prepulse and the short pulse was less than 10^{-10}. X–ray and XUV observations show that the plasma is highly transient.

MODELLING OF FOCAL SPOT PROFILES BY AN INTERFERENCE CODE AND COMPARISON WITH EXPERIMENTAL RESULTS

In order to model the intensity profile produced by different smoothing techniques an interference code was written. The code solves the 2–dimensional, complex Kirchoff diffraction integral (KDI) in the Fresnel approximation and allows variation of the axial distance parameter to select different planes throughout the focal region. A Fast–Fourier–Transform algorithm is used to solve the integral. RPP and ISI behaviour is modelled using a Gaussian statistics routine to generate appropriate phase fronts. The code was used to predict focal spot intensity profiles at several times during the laser pulse for various smoothing techniques including RPP, ISI or a combination of both. The predicted profiles were compared with experimental equivalent plane images showing good agreement. Figure 1 shows the predicted focal spot profiles for ISI and ISI/RPP after 50 coherence times.

A full account of the results – including analysis of the degree of uniformity achieved and of the mode spectrum present in each case – is published elsewhere.[1] The possibility of employing rectangular–element RPPs for the production of high aspect ratio line–foci has also been investigated both experimentally and with interference code simulations. It is predicted that a square–topped profile can be achieved by adding a small bi–triangular phase–shift to the RPP–system. The method of combining such tilt elements with RPP elements of various shapes can be used to achieve almost any focal spot geometry. A rectangular–element RPP was used to generate a cylindrical plasma using a 12ps KrF laser pulse suitable for x–ray laser research.[2]

STUDY OF INSTABILITIES IN LONG SCALELENGTH PLASMAS WITH AND WITHOUT LASER BEAM SMOOTHING TECHNIQUES

The interaction of intense laser light with large underdense plasmas is of great interest for inertial confinement fusion since fusion pellets will be surrounded by large plasma coronas. Under these conditions various parametric instabilities such as Stimulated Brillouin Scattering (SBS), Stimulated Raman Scattering (SRS) and laser beam filamentation may be very effective, resulting in the reduction of laser plasma coupling, in the production of high energy electrons and in the nonuniform heating of the plasma corona. To simulate fusion conditions, plasmas with scalelengths of about one millimetre were recently produced by focusing four green laser beams of the Rutherford Appleton Laboratory high power glass laser system onto thin foil targets in a line focus configuration.[3] A delayed green laser beam was focused axially into the preformed underdense plasma with an electron temperature and density of about 0.5 keV and 0.1 n_c respectively (n_c is the critical density for

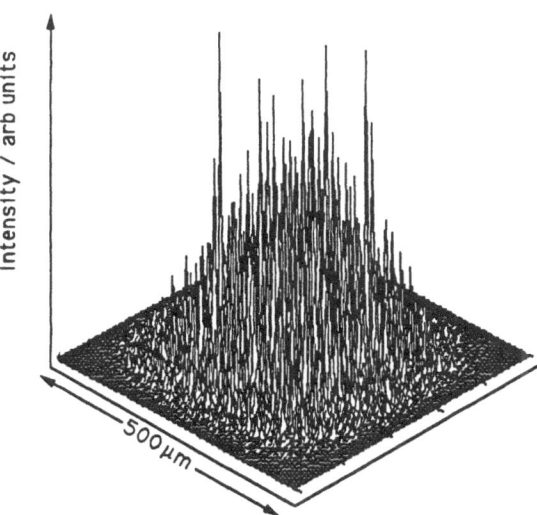

Figure 1a. Predicted intensity profile of an RPP laser beam after 50 coherence times.

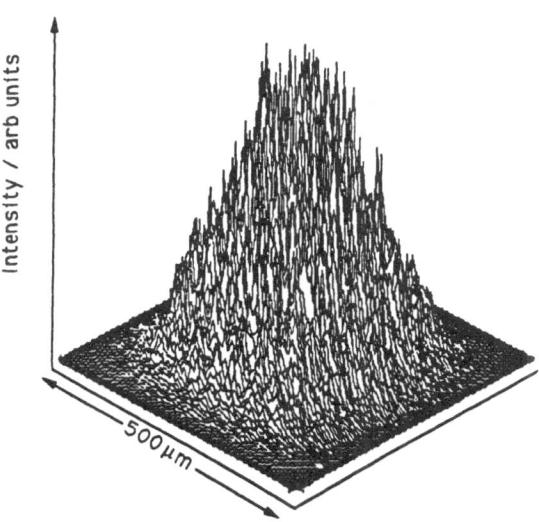

Figure 1b. Predicted intensity profile of a RPP/ISI smoothed laser beam after 50 coherence times.

the interaction beam).[4] Significant levels of SRS were seen and laser beam filamentation[5] and whole beam selfocusing[6] were clearly observed. On some of the interactions, when whole beam self-focusing was seen, anomalously large SBS shifts were observed which are difficult to explain in terms of classical SBS. A simple model of self phase modulation is invoked which in combination with SBS can explain the experimental observations.[7] When the incident laser beam was smoothed either by a RPP or ISI a significant reduction in the absolute levels of these instabilities and the virtual suppression of filamentation were seen.[8,9]

In this paper experimental results of a recent investigation are reported. The preformed plasma was again formed by a line focus configuration using four heating beams. However the heating beams were also smoothed by ISI in contrast to previous measurements in order to produce a more uniform preformed plasma. Either an ISI smoothed infrared (1.05 μm) laser beam or a broadband beam delayed by 2.2 ns was focused axially into the preformed plasma. An intensive set of diagnostics was used to investigate the plasma conditions of the preformed plasma and the nonlinear interaction of the laser beam with the plasma.

The model plasma was generated by irradiating a thin aluminium foil target (700 nm thick, 0.7 mm long, 300 μm wide) which was overcoated on a 100 nm thick formvar substrate. Two pairs of opposing green laser beams smoothed by ISI were superimposed in a line focus configuration as heating beams to form the plasma. Typical irradiances of 10^{14} Wcm^{-2} were used. A separate infrared laser beam typically delayed by 2.2 ns was used to interact with this plasma along its longitudinal axis. Measurements were made either with the broadband beam ($\Delta\omega/\omega \simeq 0.1$ %) or with an ISI laser beam. At the time of interaction the nominal electron density of the preformed plasma was about 0.3 n_c and the electron temperature about 500 eV. The uniformity of the preformed plasma was investigated transversely to the exploding foil target by using optical Moire deflectometry techniques with a probe wavelength of 350 nm. The density profile was measured interferometrically with the 350 nm probe beam propagating along the axis of the preformed plasma. The electron temperature of the plasma was obtained from time resolved x-ray streak spectroscopy. The backscattered Brillouin signal generated by the interaction beam was imaged out via the incident focusing lens onto a calibrated photodiode. In addition, time resolved SBS spectra were recorded with an S1 optical streak camera. A four frame x-ray pinhole camera with a gating time of about 150 ps was used to observe the x-ray emission of the preformed plasma and of the interaction beam.

Figure 2 shows the absolute levels of SBS backscattering for the ISI and broadband interaction beams as a function of the incident irradiance.[10] The focal spot of the interaction beam was 140 μm in diameter and was kept constant for all the data shots. For the broadband laser beam a threshold at an irradiance of about 3×10^{13} Wcm^{-2} is observed with a saturation level between 2 to 6 % of the incident laser energy. For the ISI interaction beam an exponential behaviour is seen with an average SBS value of 0.5 % at an irradiance of 7×10^{14} Wcm^{-2}. The SBS backscattering levels are significantly higher than observed in a previous experiment[8,9] in which a green interaction beam was used, the plasma was less uniform and the electron density was lower (by about a factor of 3) during interaction. However, even if similar electron densities to those obtained previously are used (by irradiating a 500 nm target), the SBS levels observed remain at the same high levels seen with the 700 nm targets. The estimated electron density is consistent with backscattered SRS which was detected by diodes filtered with narrowband interference filters. For the 700 nm targets virtually no SRS backscatter is

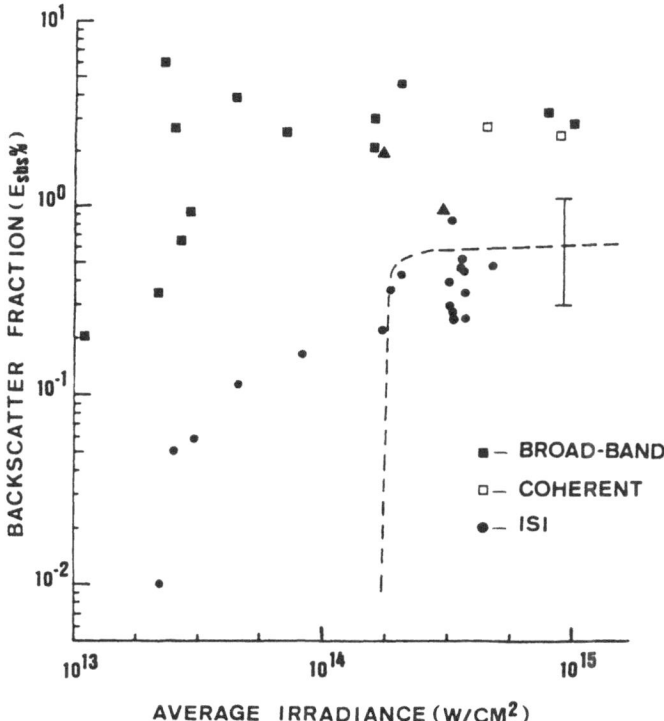

Figure 2. Variation of SBS backscatter fraction with average irradiance for an infrared ISI (illustrated by the solid circles) and broadband (squares) interaction beam. Also shown are the SBS levels for green coherent interactions.

observed. On the other hand, clear SRS signals (at a wavelength of about 1.5 μm) are seen with the 500 nm targets. For comparison, the dashed line in figure 2 represents the SBS levels for green coherent interactions. It is important to note that the SBS levels for the green ISI interactions were below the detection threshold.[8]

For some of the data shots the first experimental evidence of whole beam self-focusing of an ISI laser beam was obtained.[11] For these particular shots the target thickness and "heating" irradiance were adjusted to produce a more collisional interaction. This was achieved by exploding a thinner Al foil target (500nm thick, 0.35mm long, 300μm wide) with a reduced "heating" irradiance $I_H = 3.5 \times 10^{13} \text{Wcm}^{-2}$. The effect was to produce a preformed plasma with a similar peak electron density of $0.3n_c$ but with a reduced electron temperature of 300 eV at $t_D = 1.7$ns. The result of the cooler plasma conditions will be to enhance the strength of the thermal self-focusing mechanism (γ_{TH}) which is strongly temperature dependent, $\gamma_{TH} \sim T_e^{-5}(N_e/N_c)^2$.

Fig.3 shows a sequence of three x-ray framing images (140ps gate-time) of the preformed plasma taken with the gating period of the framing camera ending 1ns before the interaction, 50ps before the peak of the pulse and 150ps after the peak of the pulse respectively. The interaction irradiance was $I_{av}=2.5 \times 10^{14}Wcm^{-2}$. As can be seen in figures 3b and 3c a strong filamentary channel has formed with a transverse scalelength of less than $50\mu m$ in diameter, ie. this is significantly smaller than the beam at the plasma input-plane. At the exit-plane we also observe a strong emission plume extended beyond the exit, which is consistent with hot plasma being expelled from a filament channel by the beam break-out. Further confirmation was provided by optical probing images taken simultaneously with the x-ray records where a density cavity at the input was seen. In addition, the SBS backscattered signal increased by a factor of 8 for these shots compared to the data taken with similar irradiances where no self-focusing was observed.

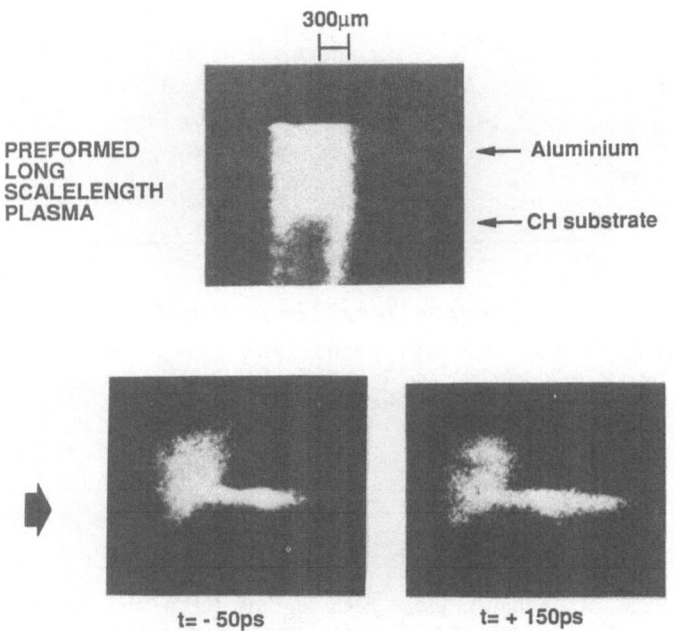

Figure 3. A time sequence of x-ray framing images (140ps gate-time) recording the preformed plasma a) 1ns before the start of the pulse b) 50ps before the peak of the pulse and c) 150ps after the peak of the pulse. The interaction beam delay $t_D=1.7$ns with $N_e \approx 0.3N_c$ and $T_e \approx 300$eV. The average interaction irradiance $I_{av}=2.5 \times 10^{14}Wcm^{-2}$, focal-spot size=$140\mu m$. A strong emission channel is clearly identified in Figs.3b and 3c consistent with beam self-focusing.

INVESTIGATIONS OF THE UNIFORMITY OF OVERDENSE PLASMAS BY XUV PROBING TECHNIQUES

After the first experimental evidence of jet formation in the underdense plasma corona of laser irradiated targets was provided using optical probing techniques,[12,13] considerable theoretical interest was stimulated. A number of mechanisms for the production of the jet-like structures have been proposed in theoretical studies. For a review see Ref. 14.

Until now, no experiment has however clearly identified the responsible mechanism, nor has any evidence been obtained of the effect of these instabilities on the ablation rate or on thermal smoothing. Of particular importance to the ICF programme is the question of whether the jets exist only in the subcritical region (all experimental work until now has provided information only about the subcritical region) or whether they extend into the supercritical region, where their effects are potentially far more detrimental for implosion symmetry. It is of paramount importance to investigate the superdense region for the occurrence of these instabilities with good spatial and temporal resolution.

A novel experimental techniques based on only recently available multilayered mirror technology for use in the XUV spectral region has been developed.[15] As it is well known, optical probing is limited to the sampling of relatively low density regions $n_e < n_c$, since the optical rays are refracted out of the imaging optics by the steep plasma density gradients. Shorter wavelengths are refracted less and simulations show that XUV probe wavelengths approximately of 100 Å will, in combination with the imaging effectively probe the conduction region up to several times critical density.

The imaging system consists of a spherical mulitilayered mirror operating at a wavelength of 130 Å with a bandwidth of about 20 Å. The overall magnification of the system was about 50x. The spatial resolution has been measured by using a zone plate as an object and was found to be about 0.8 μm, limited by the resolution of the microchannel plate intensifier which was used as a detector. The temporal resolution was 150 ps which was the gating time of the microchannel plate detector. A higher spatial resolution was achieved when Kodak 101–01 film was used. The mirror images the self–emission and also the shadow of the expanding plasma generated by either a short pulse (100ps) x–ray backlighter flash (produced by irradiating a separate gold target using another beam) when film was used as a detector or by a long pulse x–ray backlighter (1.5 ns) when the microchannel plate detector was used.

The uniformity of the overdense plasma was investigated by irradiating thin gold and Cu wire targets with coherent, ISI and ISI/RPP smoothed laser beams.[16] Figures 4a and b show two images taken on 101–01 film showing the effect of ISI and ISI+RPP hybrid irradiation (in these experiments, beam energy levels were adjusted to provide constant irradiation at $\approx 5 \times 10^{13}$ Wcm^{-2} for direct comparison). For the target in Fig. 4a, a 20μm gold wire, defocussed ISI irradiation from an aspheric f/10 lens is incident from the left and provides regularly spaced spikes of intensity (corresponding to the individual beamlets). These generate jets of blow–off plasma which persist throughout and after the pulse (the backlighter frame is taken at 2.3ns after the peak of the drive pulse). In the same image, tight–focussed ISI irradiation from an identical lens is incident from the right. In this latter case, all the beamlets are overlapped and the time–integrated profile is smooth. From the image, however, it is evident that the blow–off is not uniform with jets of material similar to those observed in the defocussed case. The structures are no longer regular since, in the case of completely overlapped beamlets, the nonuniformity in the focal spot is of a random nature. From this image, it is evident that laser focal spot nonuniformities – present only over timescales short compared to the laser coherence time – imprint themselves onto the target at these early times. This imprint initiates density nonuniformities which are never smoothed out thereafter. Figure 4b shows an image also taken on 101–01 film, with hybrid RPP+ISI irradiation incident on a 10μm gold wire (again at an irradiance of 5×10^{13} Wcm^{-2}). The small–scale ($\approx 10\mu$m) speckle structure produced by the RPP is smoothed by the ISI action over many coherence times. In Figure 4b, structure on the scale of the speckle is clearly observed and persists throughout the pulse and thereafter as in the case of the ISI–only irradiation (Figure 4a). In this case the structure appears to be regular since it is on the scale of the RPP speckle. As with the ISI–only case one is lead to the conclusion that, at early times, the laser structure imprints itself onto the target, initiating nonuniform ablation.

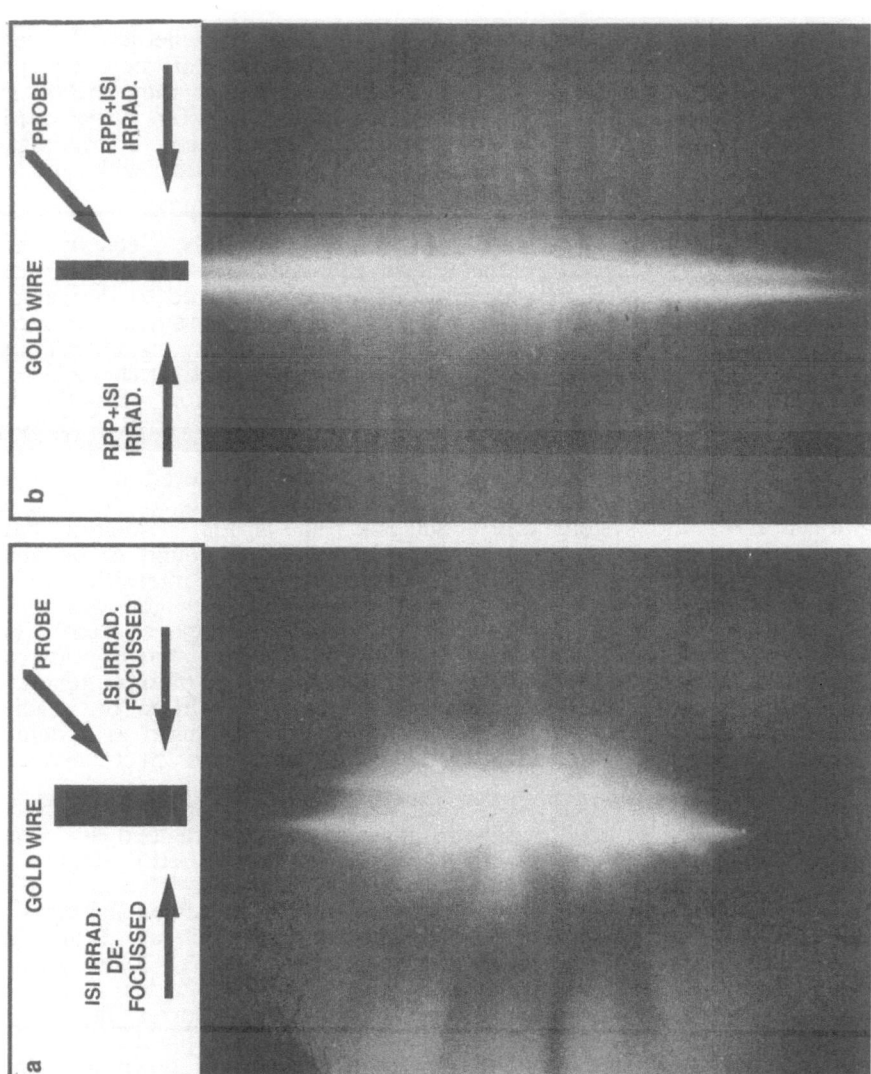

Figure 4. **(a)** A 20μm Gold wire is irradiated from the left by defocussed (≃ 250μm from best focus, into the beam near-field) and from the right by tight-focussed ISI radiation. In both cases f/10 focussing optics was used with λ=0.53μm and I≃5x10¹³Wcm⁻². Regularly spaced nonuniformities (corresponding to the individual beamlets) are clearly observed on the defocussed ISI side (left). On the tight-focussed ISI side (right) nonuniformities are also still clearly visible, although these are now of a random nature since all the beamlets are overlapped.

(b) A 10μm Gold wire irradiated by RPP+ISI hybrid radiation from both sides (I=5x10¹³Wcm⁻², λ=0.53μm). Gross nonuniformities in the self-emission and blow-off plasma are clearly visibvle in the X-ray image. In both Figs. 4 a,b the backlighter frame is taken 2.3ns after the peak of the main pulse. The white arrows indicate the initial left- and right-hand edges of the target.

The effects of radiation imprint on the plasma formation was tested with uniform irradiation. A uniform plasma was observed with an intense source of soft x-rays which was produced by irradiating a separate foil target. This foil consisted of a 1000 Angstrom plastic (parylene, C_8H_8) substrate overcoated with 500 Angstroms of gold. Two green beams (incident at an angle of 30 degrees to the normal of the foil) are used to irradiate the foil at an intensity of about 10^{14} Wcm^{-2}. The x-rays produced by the gold plasma are transported through the thin plastic substrate. This is ablated away during the pulse and, apart from serving as a substrate for the gold deposition, also absorbs the remainder of the optical pulse after this burns through the gold. The foil was placed at a distance of $250\mu m$ from the wire target, providing an estimated irradiance (in x-rays) of 10^{13} Wcm^{-2} under our conditions. The question is whether such a uniform plasma can be used to thermally smooth intensity nonuniformities in an incident laser beam avoiding the initial imprint problem. This question was investigated by irradiating a Cu wire target with a soft x-ray drive. An ISI laser beam delayed by 1.2 ns was used to interact with the preformed plasma. It was evident from the observations that there were no visible jet-like structures, neither in the self-emission nor in the backlit plasma indicating that thermal smoothing of the nonuniformities in the interaction beam had occurred.

MEASUREMENT AND ANALYSIS OF RADIATION TRANSPORT IN RADIATIVELY HEATED FOIL TARGETS

The transport of soft x-ray radiation through thin foil plastic targets, up to $6\mu m$ thick, was recently studied using time resolved XUV spectroscopy in the 10 to 70 Å region. An intense source of x-ray radiation was produced by overcoating the front side of the target with 100nm of gold and irradiating it with green laser light at an intensity between 1 and 5×10^{14}Wcm^{-2}. The soft x-rays transmitted through the target were diagnosed by a time resolved, grazing incidence flat field spectrometer coupled to an XUV streak camera.[17] The experiments were simulated with a multi-group radiation transport model which was coupled to a hydrodynamic code. Overall agreement was obtained between the experimental observations and the simulations. In detail, however, a smaller shift in the carbon K-edge was observed experimentally in transmission during heating than predicted theoretically.[18] These measurements were extended by using a separate burnthrough foil target to generate an intense source of soft x-rays. The soft x-rays transmitted through the burnthrough target were used to heat the thin plastic foils. Absorption spectra observed in the region of the carbon K-edge using a time resolved flat field spectrograph indicated that the smaller shift seen experimentally in the carbon K-edge is due to the presence of bound-bound transitions not included in the initial calculations.[19] In this paper we report on measurements of temperature profiles as a function of time and depth in plastic foil targets heated by an intense soft x-ray source using a novel time resolved XUV absorption spectrocopy technique.[20]

Thin plastic foil targets were heated by an intense Pseudo-Planckian source of soft x-ray radiation. The x-ray radiation was generated by illuminating a separate target consisting of a thin 100 nm gold layer which was supported on a $1\mu m$ plastic substrate with six frequency doubled RPP smoothed beams (800 ps in duration) of the VULCAN laser arranged in a cluster configuration. The focal spot was typically 500 μm (FWHM) in diameter as measured with an x-ray pinhole camera. The soft x-ray radiation transmitted through the burnthrough target was used to heat a separate planar CH foil, positioned parallel to the burnthrough foil and separated from it by approximately $250\mu m$. A thin chlorinated tracer layer ($C_8H_6Cl_2$), $0.3\mu m$ thick, was buried at different depths in the CH foil targets. The soft x-rays transmitted through the target were analysed using a time resolved, grazing incidence flat field spectrometer coupled to an XUV streak camera. The temporal and spectral resolutions were aproximately 50 ps and 0.3 Å respectively. The soft x–ray radiation was reflected into the spectrograph by a pair of highly polished silica mirrors acting as a high frequency cut off filter for x–rays with energies above approximately 600 eV to reduce multiple order effects.

Numerical simulations were carried out using a multi–group radiation transport model coupled to the 1–D Lagrangian hydrodynamics code MEDUSA. The radiation transport code uses 116 groups in the 0–100keV energy region. Group averaged Planck mean opacities are used and are calculated at the material temperature, in–line with the hydrodynamics using an average–atom (AA) screened–hydrogenic approximation in LTE, based on the model XSN. Only bound–free and free–free transitions are considered in the model. The behaviour of the laser irradiated thin gold layer was not calculated in the code because of the marked non–LTE behaviour of the x–ray emitting region of the gold plasma which could not be accurately calculated by our LTE model and so an experimentally measured heating spectrum was used instead. Small inaccuracies in the gold spectral emission do not cause any serious errors in the material heating, with simulations including the M, N and O band emission not predicting any significant differences in the overall results. The x–ray conversion efficiency from the burnthrough target was estimated from time–resolved and time–integrated XUV photodiode measurements taken under similar experimental conditions when a value of 5.5∓2% was found. The value used in the 1D simulations was adjusted to give best agreement with the experimental measurements to account for the geometrical coupling of the source emission to the target and variations in the conversion efficiency.

Detailed spectral analysis of the experimental data has been performed using a spectral analysis package (SAP) containing an extensive atomic data base (oscillator strengths, level and transitions energies) generated by a multi–configuration Dirac–Fock atomic physics code. In the calculation of the transition energies, the initial and final atomic states are optimised separately. Ionic distributions are calculated with the Saha equation. The lineshape is taken to be Lorentzian. Synthetic absorption spectra are generated from the calculated opacities by solving the time–dependent equation of transfer in one dimension for a given plasma density, temperature and thickness. The predicted spectrum is convolved with the instrumental function, in this case taken to be Gaussian with a FWHM given by the instrument resolution. Plasma conditions are inferred by comparing the predicted spectra with those observed experimentally. Calculations were performed for the tracer layer only. The remainder of the target introduces no discontinuous absorption features in the energy range of interest and only affects the slope of the underlying continuum as verified when targets with no tracer layers were used. The analysis is based mainly on 2p–3d transitions of F–like to S–like chlorine ions.

Targets with 0.3μm tracer layers buried 0.2μm and 1.8μm below the surface of a 2μm CH target were irradiated with an x–ray flux of between 2–$5\text{x}10^{12}\text{Wcm}^{-2}$. Several distinct differences between the two target configurations were observed. When the chlorinated layer is placed 0.2μm below the front surface of the target, the 2p–3d transitions shift towards shorter wavelength more rapidly than when the layer is positioned at a depth of 1.8μm (rear layer) indicating a more rapid ionization of the material in the front layer. This can also be seen from the earlier turn on of Na–like and Ne–like absorption features (by ≈200–300ps) in the spectrum from the front layer. In addition, the chlorine becomes more highly ionized in the front layer indicating that a higher temperature is attained here. This can be seen from the appearance of F–like absorption features approximately 500ps after the start of the x–ray pulse. In contrast, little evidence is seen for the presence of significant fractions of F–like ions in the rear layer. Also, the 2p–3d

Na–like absorption feature remains prominent for the layer at the rear of the target whereas it becomes almost unnoticeable after about 900ps when the layer is positioned near the front surface.

The temperature histories of the tracer layers were inferred as a function of time by comparing the experimental data with detailed, synthetic, absorption spectra calculated by SAP. Calculations were performed for the temperatures between 5 and 50eV at 5eV intervals and for densities between 1 and 0.005g/cc with the ratio between successive values of aproximately 2. Figure 5 shows a comparison between a densitometer trace taken approximately 900ps after the start of the emission from the front layer spectrum and an absorption spectrum caclulated with SAP for a density of 0.02g/cc and a temperature of 40eV. For comparison, a trace taken across the spectrum produced by the chlorinated layer positioned near the rear of a plastic target is also shown. Figure 6 shows a comparison between the temperature histories inferred from the absorption spectra and those predicted by the radiation hydrocode using the "best fit" x–ray conversion efficiency of 2.5%. The solid curves show the range of temperatures predicted by the hydrocode to exist in the tracer layers. The vertical error bars on the experimental points include the estimated uncertainty in the inferred temperature both due to the discrete density and temperature values at which the synthetic spectra were calculated and due to spatial gradients in the layers parallel and perpendicular to the target surface. For reference, the x–ray pulse shape is also shown in figure 6.

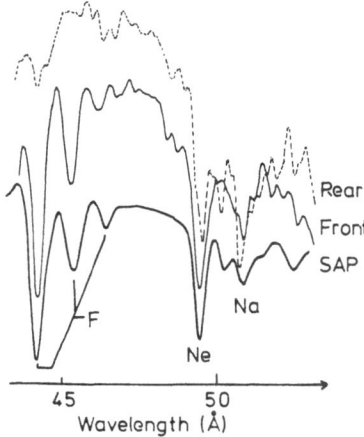

Figure 5. Traces taken near the peak of the emission and a spectrum predicted by SAP for 40eV and 0.02g/cm^3. Some 2p–3d transitions are marked. For clarity the spectra have been displaced vertically.

The overall agreement between the experimentally measured points and the hydrocode simulations is reasonably good. In particular, the different shapes of the temporal profiles for the front and rear layers are reproduced well. In the simulations, the rear layer is "preheated" (0–400ps) to temperatures above that of the ambient CH by radiation mainly in the carbon window below the K–edge where the intervening CH is optically thin. A significant fraction of the x–rays in this region is absorbed in the chlorinated layer due to b–f chlorine L–shell transitions. As the layer heats and ionizes, its opacity in the carbon window region decreases as the chlorine L–shell b–f threshold shifts towards higher energy. As a result the layer heating rate is reduced and a temperature plateau is established (400–800ps). The intervening CH is heated mainly by x–rays in the energy regions just above and well below the carbon K–edge where the photon mean free path is sub–micron. As the material is ionized, the opacity is reduced and, in particular in this case, the K–edge shifts towards higher energy allowing radiation in this region to propagate into the cooler material where it is strongly absorbed. It is the propagation of this "ionization front" that is responsible for the step in the simulated rear layer profile at around 850ps. Similar behaviour is not predicted for the front layer because the 0.2μm of the intervening CH is optically thin to the source over most of the photon energy spectrum and therefore, will have little effect on the layer heating.

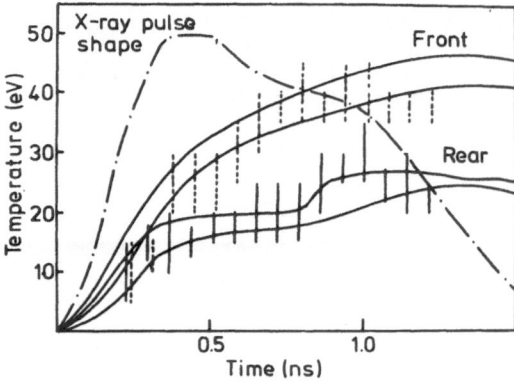

Figure 6. Comparison of experimentally measured temperature histories and those predicted by the radiation hydrocode. The x–ray pulse shape is shown.

STUDY OF RAYLEIGH TAYLOR INSTABILITY USING X–RAY BACKLIGHTING TECHNIQUES

The Rayleigh–Taylor instability occurs in any accelerating fluid system in which the density and pressure gradients are of opposite sign. In the case of ICF experiments, the hot, low density ablating plasma accelerates a cooler, more dense part of the target and is thus succeptible to the RT instability. In an ICF implosion small imperfections in the target manufacture or intensity nonuniformities present in the laser irradiation may initiate RT growth sufficient to disrupt the target symmetry to an unacceptable degree. Here we describe experimental measurements in which the growth rates of the Rayleigh–Taylor instability were measured in thin foil targets with imposed sinusoidal modulations irradiated by optically smoothed laser beams. A hybrid optical smoothing scheme utilising ISI and RPP was used. The enhancement in the modulation depth during acceleration was observed with time resolved transmission radiography using a soft x–ray backlighting source. The wavenumber dependence and nonlinearity of the RT growth were investigated by using a range of modulation periodicities and depths. The measurements were compared with 2–D hydrocode simulations.[1,21]

Six frequency doubled green (0.53μm) beams arranged in a hexagonal cluster with a cone angle of 13 degrees, were focused onto thin foil targets with f/10 lenses. The six superimposed beams generated a smooth, flat–topped spatial intensity profile providing uniform acceleration across the target surface. The total energy delivered on target was approximately 250 Joules with a pulse duration of 1.76 ns (FWHM) resulting in an incident irradiance of about 1.5×10^{14} Wcm^{-2}. This was kept constant throughout the experiment to allow direct comparison of the results for different target specifications. The laser beams were smoothed by a hybrid scheme consisting of Induced Spatial Incoherence (ISI) and Random Phase Plate (RPP) arrays. The ISI was generated using a broad–band oscillator ($\Delta\omega/\omega=0.1\%$) and a 6x6 echelon to produce 36 independent beamlets. The RPP's were placed immediately in front of each of the six focusing lenses. In addition to profile smoothing, the RPP's have the effect of increasing the far–field focal spot size to a large (approx. 335 μm) diameter. An x–ray pinhole camera, filtered for an x–ray energy band between 0.84 keV and 1.63 keV was used to monitor the uniformity of the focal spot produced by the drive beams. Intensity variations of about 1% (calculated with a 2D weighted averaging procedure) were seen over a spatial wavelength of 10μm. The far field focal spot distribution was calculated using an interference code. This showed close agreement to the experimentally determined profile structure and uniformity.

The targets consisted of low density (0.9g/cm^3) (CH_2) plastic approximately 16μm thick with sinusoidal modulations of periodicity 30μm, 50μm, 70μm and 100μm. Modulation depths between 1.8μm and 4.6μm were investigated, with the modulations always facing the drive beams. The accelerated targets were backlit with a Mg backlighter source which was generated by a separate green laser beam 2.5 ns in duration. The transmitted fraction of the x–rays produced by the backlighter source was imaged onto a streak camera by means of a pinhole with an overall spatial resolution of 21 μm. The image was filtered for a spectral wavelength window between 0.84 and 1.63 keV containing the magnesium Heα and Lyα transitions. The backlighter spectrum was measured using a time–integrated crystal spectrometer with filtering identical to that used in front of the streak camera. From this the relative contribution to the image intensity at each wavelength in the backlighter could be determined. Control experiments with only the drive irradiation (to determine the contribution to the image caused by self–emission from the target) and with backlighter illumination only were conducted. In addition to the transmission radiographs, streaked edge–on radiographs were recorded to obtain the target acceleration.

Computer simulations of the experiment were performed using the 2D Eulerian hydrodynamics code POLLUX which has been extensively tested under various conditions. The code was modified to incorporate the sinusoidal target surface modulations and matched to the experimental conditions by varying the absorbed irradiance until the inertial motion measured in the experiment was reproduced in the simulations. The absorbed intensity was chosen as the variable parameter, since it is the least well–characterised experimental quantity due to the effects of lateral energy transport. It was found that approximately 50% of the energy needs to be absorbed to reproduce the experimentally determined inertial motion. This value is comparable to previous measurements under similar conditions.

From the transmission radiographs the ripple amplitude growth rates were determined by microdensitometry of the streak records to give quantitative measures of the observed X–ray contrast levels at different times. Both the initial modulation depth and the instability growth rate were then deduced from exponential curves fitted to the data. The growth rates measured in this way are plotted in figure 7 as fractions of their classically predicted values, the errors representing the spread of the amplitude of modulation evident at late times. The measured initial (t=0) modulation depth were convolved with the modulation transfer function (MTF) of the imaging system and compared to the known initial modulation depths (accurately measured by microscopy before the experiment). No significant discrepancy was found (error < 0.4μm). Further, no growth rate analysis was carried out during the initial period in which the shock passes through the target. In this time, the growth is not due to the RT instability but due to the shock driven Richtmeyer–Meshkov instability.

Also plotted in figure 7 are the measured growth rates of the NRL experiment[22] and those predicted by the POLLUX simulations (for both high and low modulation depths) and the predictions for high modulation depth targets after convolution with the MTF of the imaging system. Several points arise. Firstly, the qualitative wavelength dependence of the observed growth rates is in fair agreement with the dependence predicted by the simulations; growth was observed at all modulation periodicities including 50μm. Secondly, the growth rates predicted by the simulations are (except for two points) outside the error bars and greater than the observed rates. This may not be purely due to the

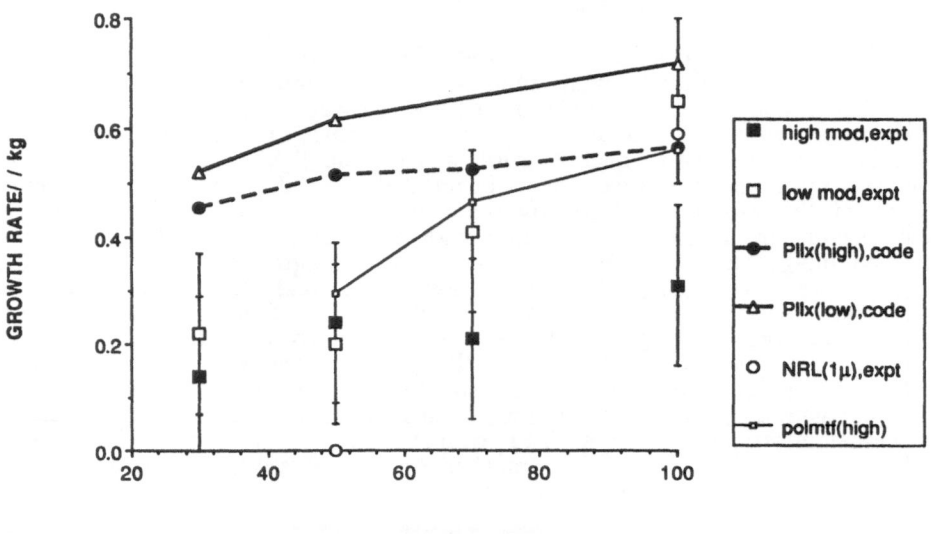

Figure 7. Measured RT growth rates presented as a fraction of the classical growth rate (kg)$^{1/2}$ where k is the modulation wavenumber and g is the acceleration for both high (4.5 μm) and low (2.6 μm for 30 μm and 50 μm periodicity; 1.8 μm for 70 μm and 100 μm periodicity) modulation depth targets. Also shown are POLLUX results for 1.8 μm (low) and 4.5 μm (high) modulation depth targets, MTF–corrected POLLUX results for high modulation depth targets, and experimental results from [Ref. 22].

nonlinearity of the growth, but may in part be due to radiation transport effects not included in the modelling. These would, in principle, reduce the growth rate. The excitation of higher order modes also reduces the growth rate of the fundamental. To investigate the extent of such higher–order contributions to the growth of the instability, the X–ray radiographs were spectrally analysed with a Fourier decomposition package. It was found that small amplitude higher order modes are clearly present at late times ($\approx$1.6ns after the start of the laser drive pulse) for 100μm modulation periodicity targets. Such higher harmonic modes were not clearly observed in the case of the $50 \mu m$ targets, where only the fundamental mode was found to have a significant Fourier component (for the case of $30 \mu m$ targets, the resolution of the imaging system was insufficient to detect higher harmonic growth).

PLASMA PRODUCTION WITH 12 ps HIGH POWER UV LASER PULSES

The interaction of short pulse and wavelength lasers with solid targets is of great interest for various areas in plasma and atomic physics. Since the laser energy is absorbed close to the initial target surface a large thermal conduction wave is launched into the solid material before hydrodynamic expansion occurs resulting in the formation of a hot, close to solid density plasma.[23] Besides fundamental studies of plasma and atomic physics, these plasmas are of great interest as pump media for X-ray lasers.[24] In addition, plasma conditions may be generated in the laboratory which are interesting for astrophysical applications. We have recently carried out studies of plasmas produced by irradiating various solid targets with a 3.5 ps KrF laser pulse.[24-27] Here we report on observations of plasmas produced by irradiating targets with a 12ps KrF laser pulse at an intensity up to 1×10^{17} Wcm^{-2}. The contrast ratio between the prepulse and the short pulse intensity was less than 10^{-10} resulting in an irradiance of the prepulse which is well below the threshold for forming a pre-plasma.

The experiment was carried out at the SERC Central Laser Facility using the short pulse, high power KrF system SPRITE.[28,29] A KrF pulse was amplified in a series of KrF amplifiers and Raman shifted in methane resulting in a wavelength of 268 nm. A total laser energy of about 5J in a pulse 12ps in duration was obtained. The laser beam was focused onto planar targets with an f/4 parabolic mirror. With a focal spot of 10 μm (FWHM) in diameter, irradiances of about 1×10^{17} Wcm^{-2} were obtained on the target surface. Time resolved X-ray spectroscopy was used to diagnose the plasma. A curved TLAP crystal in a Johann configuration was coupled to an x-ray streak camera covering a spectral range between 5 to 7 Å. The temporal resolution was about 15 ps.

Figure 8 shows an x-ray streak record taken on an aluminium target which was irradiated at an intensity of about 1×10^{16} Wcm^{-2}. As can be seen in the spectrum the aluminium transitions are very broad, particularly during the laser pulse. Also, the higher series members such as Heϵ are not observed. This spectral behaviour is characteristic for emission originated from very high density plasmas. The electron density of the emitting region of the plasma was calculated as a function of time by comparing the experimental line widths of Heγ and Heβ from the time resolved data with those predicted by the atomic physics code RATION and SPECTRA. The main broadening mechanism is Stark although it was estimated that about 15% of the line width is caused by opacity broadening. A peak electron density of about 1.5×10^{23} cm^{-3} was obtained for a bare aluminium target. Electron densities up 3×10^{23} cm^{-3} were obtained for targets with an aluminium layer buried below a plastic layer $0.2 \mu m$ in thickness and irradiated at an intensity of about 5×10^{16} Wcm^{-2}. The electron temperature is obtained from the Heβ:Lyβ line ratio of spectra recorded under similar experimental conditions. Because of the short time scales involved, the plasma is highly transient, especially at early times. To check the validity of using a steady state atomic physics code the 1-D hydrocode simulations were post processed with a time-dependent, average atom model. Both the time dependent and steady state ionisation stage populations were calculated. It was seen that the time dependent analysis gives a distinctly different temperature profile compared to the steady state interpretation. Electron temperatures up to 800 eV are obtained with the time dependent analysis whereas a maximum temperature of about 600 eV is estimated with the time independent analysis. The time dependent interpretation seems to be more consistent with the experimental observations

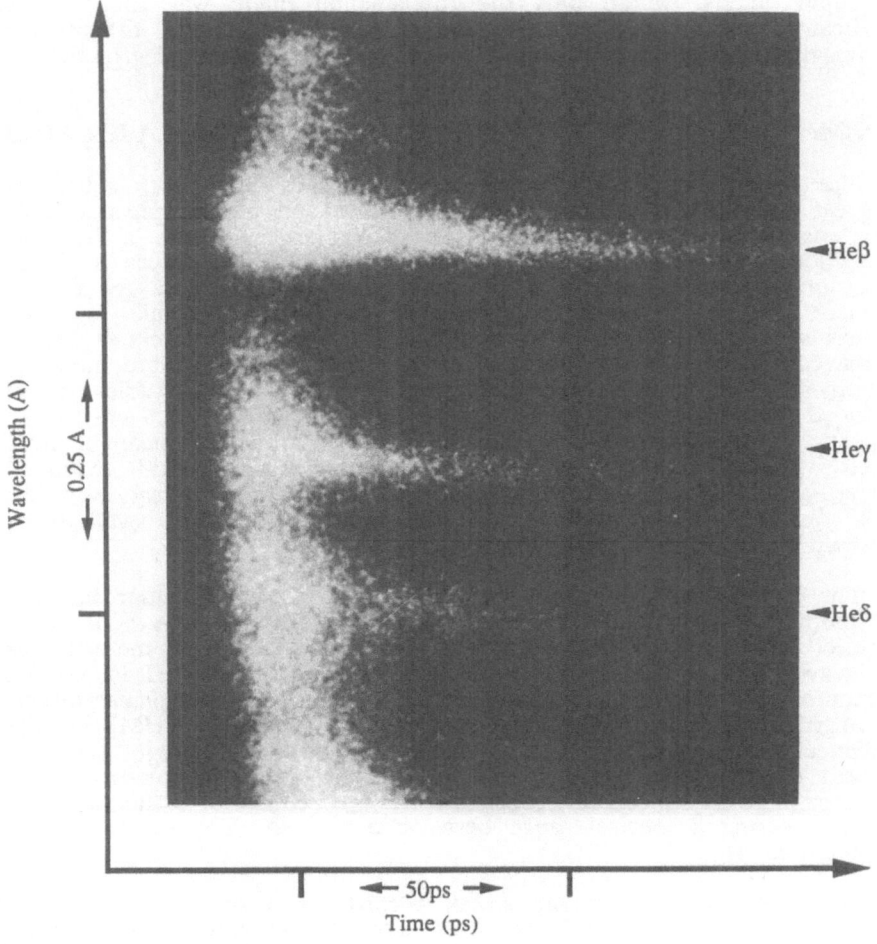

Figure 8. Time resolved spectrum obtained when an aluminium target was irradiated at an intensity of about $1\times10^{16}\mathrm{Wcm^{-2}}$.

since hydrogenic like chlorine emission was seen on KCl targets irradiated at similar intensities.

The experimental conditions were modelled with the 1-D Lagrangian hydrocode MEDUSA. The ions are assumed to have a perfect gas equation of state (EOS), while the electrons are described by a Thomas Fermi EOS with modifications to give the correct solid density. The ionisation balance of the plasma is calculated with a time dependent average atom approximation in NLTE. The laser energy was absorbed via inverse bremsstrahlung and 5% of the remaining laser energy was dumped at the critical density surface to simulate resonance absorption. A flux limiter of 0.1 was used in the simulations since this value agreed best with measurements taken on plastic coated aluminium targets. Simulations of these targets under the same conditions, but with a flux limiter of 0.03, revealed that the temperature would not be hot enough to ionise the target to hydrogenic like aluminium wheras a flux limiter of 0.1 would. For these conditions MEDUSA predicts an absorption of 30%.

CONCLUSIONS

In summary this paper gives a review of our ICF related work carried out during the last two year. The main effort of the research concentrated on studies with improved laser beam uniformity. The interaction of a smoothed laser beam with large underdense homogeneous plasmas and solid targets was investigated. It was seen that the levels of SRS, SBS and filamentation were suppressed when smoothed laser beams were used. It was also observed, however, that the large initial nonuniformity present in smoothed laser beams imprint themselves onto the cold target surface early in the laser pulse. A novel hybrid scheme was proposed where a uniform plasma is formed with an intense soft x-ray source. The preformed plasma is then irradiated with a smoothed laser beam avoiding the initial imprint. The heating of foil targets by soft x-rays was investigated experimentally and computationally. The Rayleigh–Taylor growth was studied in targets driven by combination of ISI and RPP smoothed laser beams. Finally, highly transient plasmas were generated with a prepulse-free 12 ps, 0.268 μm laser pulse.

ACKNOWLEDGEMENTS

The authors thank the staff of the Central Laser Facility for their assistance and technical support. This work was funded by several SERC grants. We would like to acknowledge the contributions made by Dr. A. Giulietti to the long scalelength interaction experiments and by Dr. S. Rose to the radiation transport studies.

REFERENCES

1. M. Desselberger and O. Willi, submitted to Phys. Fluids B.

2. M. Desselberger, L Gizzi, V. Barrow, J. Edwards, F. Khattak, S. Viana, O. Willi and C. Danson, Appl. Optics, March 1992.

3. S. Coe, T. Afshar–rad, D. Bassett, J. Edwards and O. Willi, Opt. Commun. 81, 47 (1991).

4. O. Willi, D. Bassett, A. Giulietti and S. Karttunen, Opt. Commun. 70, 487 (1989).

5. S. Coe, T. Afshar–rad and O. Willi, Opt. Commun. 73, 299 (1989).

6. S. Coe, T. Afshar–rad and O. Willi, Europhys. Lett. 13, 251 (1990).

7. T. Afshar–rad, S. coe, A. Giulietti, D. Giulietti and O. Willi, Europhys. Lett. 15, 745 (1991).

8. S. Coe, T. Afshar–rad, M. Desselberger, F. Khattak, O. Willi, A. Giulietti, Z. Q. Lin, W. Yu and C. Danson, Europhys. Lett. 10, 31 (1989).

9. O. Willi, T. Afshar–rad, S. Coe and A. Giulietti, Phys. Fluids B 2, 1318 (1990).

10. T. Afshar–rad, M. Desselberger, L. Gizzi, O. Willi, F. Khattak and A. Giulietti, submitted to Phys. Rev. Lett.

11. T. Afshar–rad, L. Gizzi, M. Desselberger, F. Khattak, O. Willi and A. Giulietti, accepted in Phys. Rev. Lett.

12. O. Willi, P. Rumsby and S. Sartang, IEEE J. of Quant. Electr. QE–17, 1909 (1981).

13. O. Willi, P. Rumsby, C. Hooker, A. Raven and Z. Lin, Opt. Commun. 41, 110 (1982).

14. M. G. Haines, Can. J. Phys. 64, 914 (1986).

15. M. Desselberger, T. Afshar–rad, F. Khattak, S. Viana and O. Willi, Appl. Optics 30, 2285 (1991).

16. M. Desselberger, T. Afshar–rad, F. Khattak, S. Viana and O. Willi, submitted to Phys. Rev. Lett.

17. G. Kiehn, T. Garvey, R. Smith, O. Willi, A. Damerell and J. West, Proc. SPIE Int. Soc. Opt. Eng. 831, 150 (1987).

18. J. Edwards, V. Barrow, O. Willi and S. Rose, Europhys. Lett. 11, 631 (1990).

19. J. Edwards, M. Dunne, L. Gizzi, O. Willi, S. Rose, C.A. Back and C. Chenais–Popvics, Proceedings of the Sarasota Workshop on the Properties of Hot Dense Matter (World Scientific, Singapore, to be published).

20. J. Edwards, M. Dunne, D. Riley, R. Taylor and O. Willi, Phys. Rev. Lett. 67, 3780 (1991).

21. M. Desselberger, O. Willi, M. Savage and M. Lamb, Phys. Rev. Lett. 65, 2997 (1990).

22. J. Grun et al., Phys. Rev. Lett. 58, 2672 (1987).

23. O. Willi, T. Afshar–rad, V. Barrow, J. Edwards and R. Smith, Proc. of the International Conference on LASERS' 89, edited by D.G. Harris and T. M. Shaw, STS Press, McLean, VA, 40 (1990).

24. R. A. Smith, V. Barrow, J. Edwards, G. Kiehn and O. Willi, Appl. Phys. B 50, 187 (1990).

25. O. Willi, G. Kiehn, J. Edwards, V. Barrow and R. Smith, OSA Proc. on Short Wavelength Coherent Radiation: Generation and Applications, edited By R. Falcone and J. Kirz, Vol. 2, 194 (1988).

26. O. Willi, G. Kiehn, J. Edwards, V. Barrow, R. Smith, J. Wark and E. Turcu, Europhys. Lett. 10, 141 (1989).

27. J. Edwards, V. Barrow, O. Willi and S. Rose, Appl. Phys. Lett. 57, 2086 (1990).

28. M. Steyer, I. Ross and J. Lister, Annual Report to the Laser Facility Commitee 1990, RAL–90–026, 90 (1990).

29. E.C. Harvey, C.J. Hooker, J.M. Lister, I.N. Ross, M.J. Shaw, G.J. Hirst, M.H. Key, P.A. Rodgers and J.E. Andrew, Annual Report to the Laser Facility Committee 1990, RAL–90–026, 79 (1990).

ICF PHYSICS AND NUMERICAL SIMULATION AT DENIM

G. Velarde, J.M. Aragonés, R. Falquina, J.J. Honrubia, J.M.
Martínez-Val, E. Mínguez, E.J. Moralo, J.M. Perlado, M. Piera,
P.M. Velarde

Instituto de Fusión Nuclear (Denim)
Jose Gutierréz Abascal, 2; 28006 Madrid

1. INTRODUCTION

The physics and numerical simulation of ICF at DENIM have progressed
in the areas of numerical modelling and target design. The goal in the first area
has been to improve our computational capability to simulate the key issues of
ICF, including the hydrodynamic stability along the implosion and stagnation
phases, the ignition propagation, and the energy transport. In addition we have
paid some atention to the simulation of current ICF experiments, specially
those related with the X-ray energy conversion, in order to validate and adjust
our numerical models. The efforts have been mainly concentrated in atomic
physics, 1D and 2D multigroup radiation transport and in the development of a
robust two-dimensional (2D) hydrodynamics algorithm.

In the area of target design, we have studied the effect of pulse shaping in
the implosion of a laser target with 1 mg of DT. The goal has been to minimize
the energy necessary in order to obtain a high gain and to reduce the aspect
ratio of the target in order to minimize hydrodynamic instabilities.

2. 2D HYDRODYNAMICS

Accurate 2D simulations of the ignition process to be applied to the design
of high gain targets and to the modelling of near term experiments require

advanced and robust hydrodynamic algorithms. A new finite difference method for shock hydrodynamics using a staggered mesh called RMFCT[1] has been developed. This method uses a flux corrected transport method to control diffusion in the Lagrangian phase. The algorithm is conservative, free streaming invariant and well behaved around shocks. The method has been successfully tested with 1D Lagrangian and Eulerian problems in planar geometry[1].

Recently, we have extended the RMFCT method to 2D hydrodynamics and written a code that is now under testing. We present here some results about the Richtmyer-Meshkov (RM) instability (a Rayleigh-Taylor instability with an impulsive aceleration[2]) as an example to calibrate our present 2D hydrodynamic capability. This instability could produce mixing in the fuel-pusher interface of an ICF target, but only for the case of stagnation-free targets (the only one presently being considered). The action of the instability is restricted to times before the maximum temperature is reached.

A simple test problem with $R = \rho_+/\rho_- = 10.0, 6.48$ (R being the density ratio in contact discontinuity before and after the pasage of the shock wave), is analyzed to test the accuracy of the growth rate with mesh spacing. RM instability grows linearly at early times when $k\zeta < 1$, k being the wave number and ζ the amplitude of the initial perturbation. In the simulations we start with a large amplitude ($k\zeta = 0.42$), which implies that the sinusoidal profile is broken at early times. Values of the wavelength/mesh spacing are 10, 30, 120 (in reference 3 they are 20, 50, 100). We measured the evolution of amplitude with time for different mesh spacings. As seen in figure 1, all cases agree with linear theory[4] (line with $d\zeta / dt = 0.45$) when this approximation can be applied. However, we only get some saturation at final times when a higher number of zones is used (see Figure 2). Damped oscillations of the transmitted shock are clearly observed for all cases. In order to eliminate any uncertainty in the boundary conditions, cases with two wavelengths width have been run on high resolution (120 × 120). The results agree essentially with the previous ones. It is interesting to see that only the lower resolution case follows the linear profile, when no structure is seen in the perturbed surface.

We have also performed simulations with phase reversal (heavy to light shocking, $R = 0.1$), with the final leftward state consisting in a rarefaction wave. The amplitude of the perturbation as a function of time is shown in figure 3 for several mesh spacings and it can be seen to remain in the linear regime (growth rate of 0.66 for $\zeta < 0$) during the simulation. All the cases give a similar behaviour. For several mesh spacing the amplitude of the perturbation remains

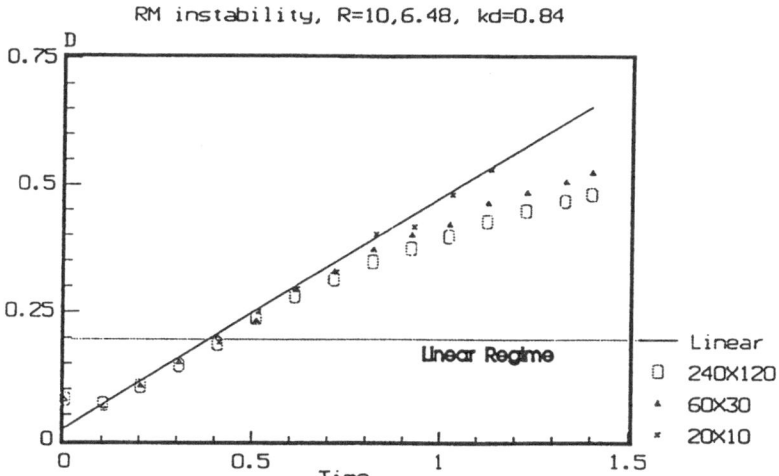

Figure 1.   Evolution of the interface amplitude (ζ) for the case of R > 1.

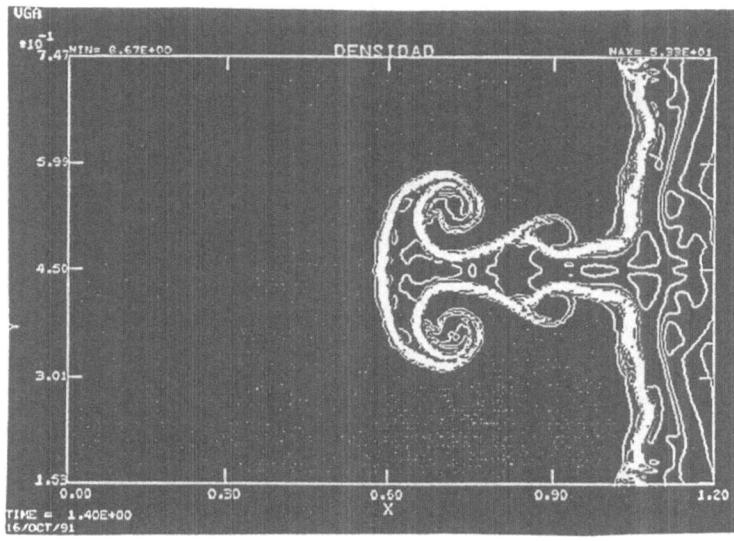

Figure 2.   Density Isocontours for the R > 1 case at the end of the simulation time.

in the linear regime (growth rate of 0.66 for $\zeta < 0$) during the simulation and all of the cases follow similar behaviours.

In a realistic simulation, only a few zones per wavelength (~ 10) are used. Thus, the algorithm should give enough accuracy with this spatial resolution. According to the poor results here presented on the coarser mesh (see figure 1), higher accuracy in contact discontinuities is required in order to reduce the number of mesh intervals and to increase the global efficiency of the algorithm.

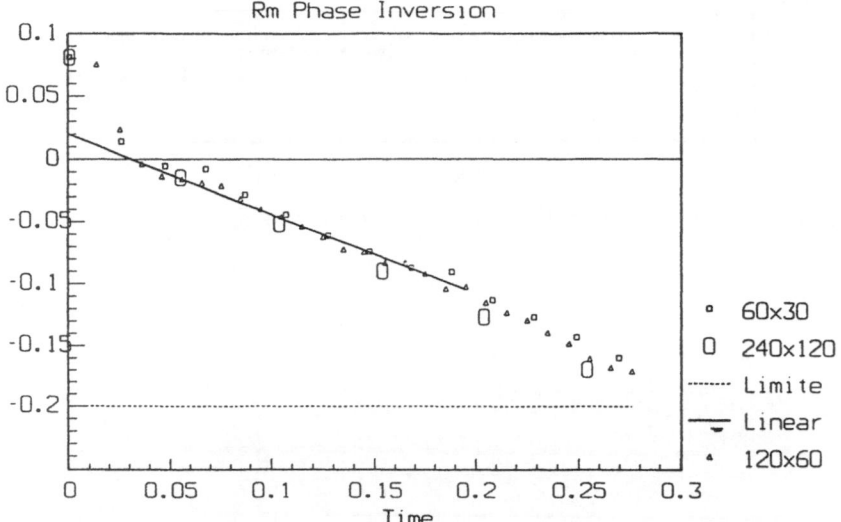

Figure 3.　　Evolution of the interface amplitude (ζ) for the case of $R < 1$ (phase inversion)

3. ATOMIC PHYSICS MODELS

We have developed several atomic models in LTE and non-LTE, which have been used to determine opacities for ICF elements, and have been compared with other models.

Using the average atom (AA) concept, two different algorithms were reported[5,6] for generating atomic data needed for the calculation of the opacities. The first one uses a corrected analytical hydrogenic formula to determine orbital energies[7]. The second one solves the radial Dirac wave equation using a self-consistent potential. A comparison between those two models has been reported[8].

Usually the plasma is composed of ions at different ionization states, whose abundances or probabilities have to be known to evaluate frequency dependent opacities. To simulate these situations, a detailed configuration accounting (DCA) model in the frame of LTE was developed [6,9]. This new model is more expensive than the former ones, but it gives more realistic results.

When opacities for low-Z elements are determined, the DCA model generates good approximations, but for high-Z materials, e.g. gold, the DCA model gives usually poor results because many configurations existing in the plasma are neglected. We have also recently included in our DCA model the unsolved transition array concept (UTA) [10], assuming j-j coupling to calculate the interaction energy. With this concept, lines in each transition array strongly overlap and form an absorption band, in such a manner that the real line profile is of the Voigt type in which we consider the Doppler, collisional and absorption band widths.

In figures 4.a-c the main differences in the frequency dependence of the opacity for a gold plasma-at normal density and a temperature of 750 eV using the above referenced models are shown.

For Non-LTE situations, we have three levels of description: (1) Average Atom (AA) collisional-radiative equilibrium model with populations depending on the principal quantum number, (2) AA with populations depending on nlj[11], and (3) DCA considering the ground state and an arbitrary number of excited levels for each ionic state[12]. The comparison between the first two models for the ionization of a gold plasma is depicted in figure 5, where some differences between both models are shown. These differences arise from the different screening model used, which varies from the simple hydrogenic of the first model to the self-consistent HFS of the second one.

The hydrogenic AA model has been used to compute the Non-LTE ionization, opacities and emissivities in the simulation of X-ray conversion experiments. The coupling of this model with the hydrodynamics is done through tables, which contain the information to construct the opacities and emissivities in a fine frequency mesh into the hydrodynamic code. Then, the fine mesh opacities are collapsed through the Rosseland and Planck means to the multigroup mesh and interpolated in density and temperature.

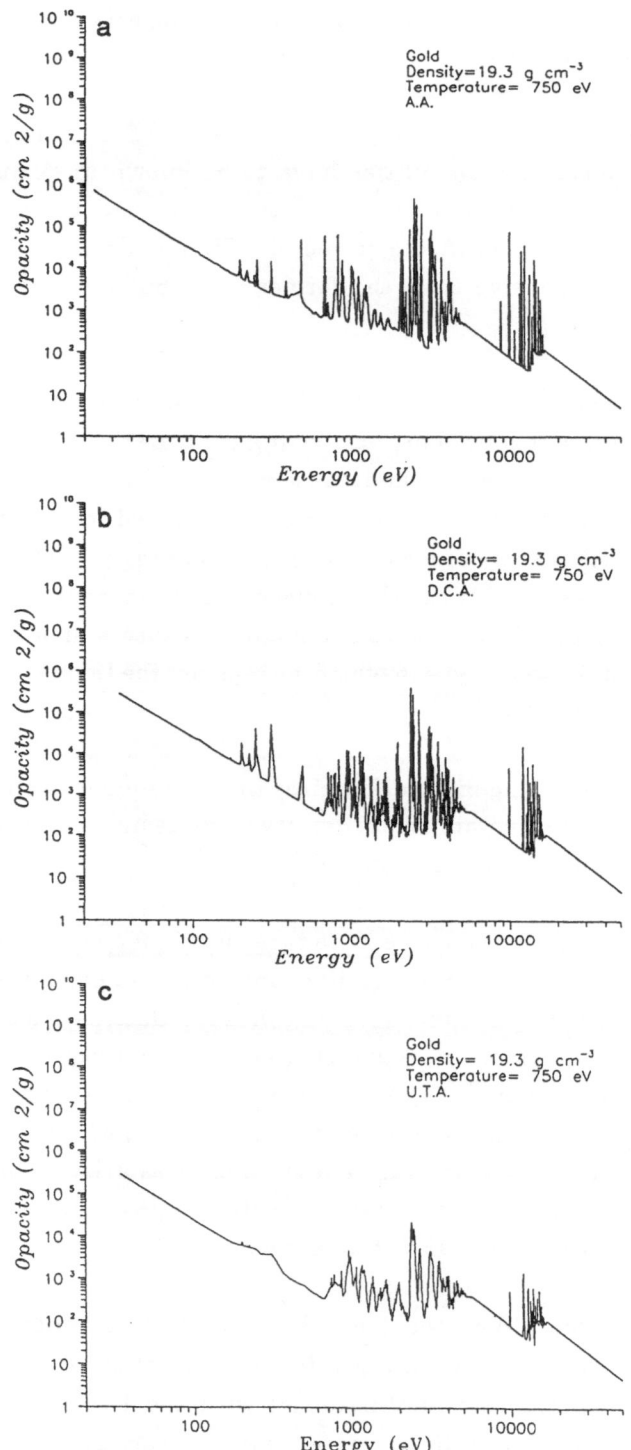

Figure 4.   Gold opacity versus frequency at normal density and a temperature of 75 eV. (a) AA, (b) DCA and (c) DCA with UTA.

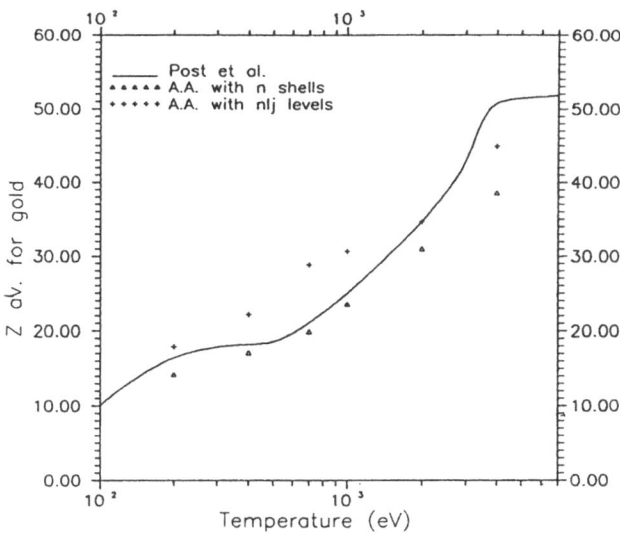

Figure 5.   Comparison of non-LTE ionization of gold using several models (the data labeled by Post et al. have been taken from reference 13)

4. MULTIGROUP RADIATION TRANSPORT

In the last years we have developed a new 1D radiation-hydrodynamics model as a basic tool to simulate the X ray energy conversion of laser and ion beams. The radiation transport algorithm solves the true (S_n) multigroup equations by means of a positive, second order differencing scheme that verifies the diffusion limit for intermediate and optically thick regimes. The multigroup equations are solved implicitly by using multifrequency-grey synthetic acceleration techniques to accelerate the convergence of the radiation source. The details of the algorithms can be found elsewhere[14,15].

Recently, this scheme has been extended to 2D settings including the angular dependence and using the S_2 low order operator to accelerate the convergence of the S_n equations. The difference between the one and two dimensional algoritms is that the matrix of the 2D S_2 equations is too large to invert by direct methods. Thus, we have developed a multigrid scheme to solve iteratively these equations[16]. As it is well known, the multigrid method is particularly effective to accelerate the low frequency error modes in such a manner that for a wide variety of cases we obtain a high computational efficiency with the multigrid solver alone. However, in the cases of extremely thick intervals, the high frequency error modes could converge slowly, resulting in a slow convergence rate of the whole algorithm. In this case, these modes are accelerated by a generalization of the two-cell block inversion method proposed in reference 17 for 1D settings. The final algorithm with multigrid and block inversion is very efficient to accelerate the radiation source when compared with the standard source iteration method, but it is somewhat costly to couple with the hydrodynamics in a production code. Instead, we are using it as a reference code where the radiation transport includes the angular dependence and the multigroup capability, avoiding problems of boundary conditions, boundary layers and flux limiters.

In order to couple radiation transport with hydrodynamics in a production code we have developed a simpler 2D multigroup flux-limited diffusion algorithm that includes the synthetic acceleration of the radiation source.

The 1D radiation-hydrodynamic model has been used to simulate the X-ray conversion experiments performed at Limeil, which consist of the illumination of gold disks with different widths and different laser intensities[18]. The atomic model used for the emissivities and opacities is a Non-LTE AA model based on the radiative-collisional equilibrium with dielectronic recombination. The scaling of the conversion efficiency with the laser intensity is shown in figure 6, where we can see a good agreement between the computed values and the experimental results. A good measure of the radiation transport modeling is the calculation of the energy emitted by the rear side in the form of X-rays. This is shown in figure 7 as a function of the target thickness for a laser intensity of $5 \cdot 10^{14}$ W/cm^2. In that case, we also have a reasonable good agreement with experiment.

However, the spectra emitted by the rear side present some discrepancies with those observed in the experiments and in particular with the attenuation of M-band photons (2-3 KeV), which are not completely damped at thicknesses close to 1 micron. This fact reveals some uncertainty degree in the gold opacities and emissivities used in the simulations which is possibly related to the modelling of the line profiles.

5. HIGH-GAIN TARGET DESIGN

In previous works, we have explored the performance of single-shell laser targets in order to obtain a high gain with pulse energies of the order of a few

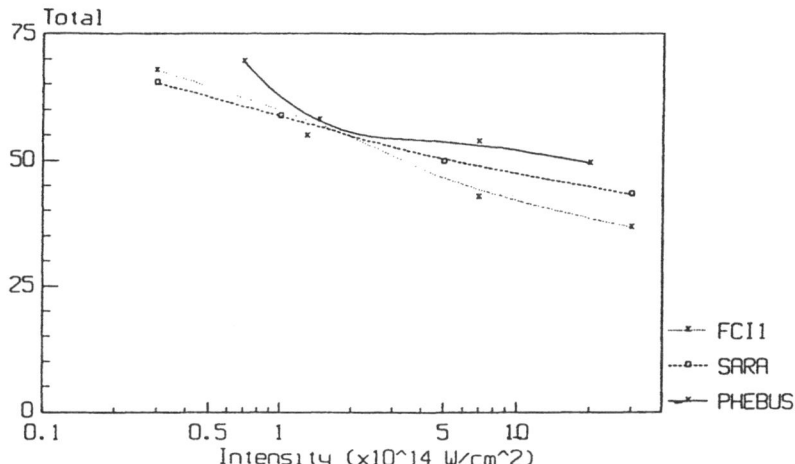

Figure 6.   Total X-ray conversion efficiency as a function of the laser
intensity (PHEBUS results taken from reference 18)

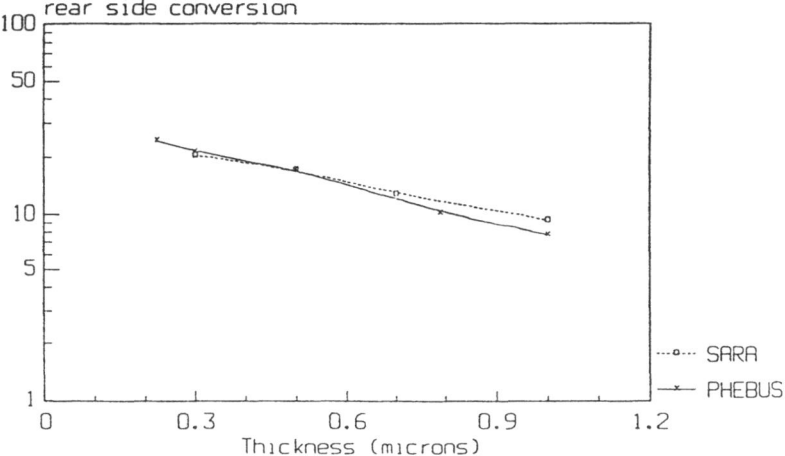

Figure 7.   Rear side X-ray conversion efficiency for several thicknesses of
gold. I= 7 1014 W cm-2 (PHEBUS results taken from
reference 18)

MJ[19]. The targets are formed by a low-Z material layer (CH) to absorb and convert the laser energy into hydro motion, and a layer of 1 mg of cryogenic DT. The low-Z absorber avoids the conversion of the beams into X-rays. The initial aspect ratio of these targets was moderately high in order to obtain a high hydrodynamic efficiency. Numerical simulations showed that Gaussian pulses do not give a high compression of the DT because the fuel is shock heated and expanded before its compression. Thus, we considered a moderate pulse shaping of the Gaussian profile by limiting the maximum laser intensity. We found that, in order to obtain high gain contitions ($\rho R \approx 3$ gcm^{-2} and $T \approx 3$keV) with pulse energies of 2-3 MJ, one should have in-flight aspect ratios (IFAR) greater than 200, which can seriously affect the hydrodynamic stability along the implosion process. If the initial aspect ratio of the target is reduced to limit the maximum IFAR, the hydrodynamic efficiency decreases to unacceptable values that require large pulse energies.

The next step has been to consider a careful pulse shaping to obtain a high compression of the D-T with the minimum pulse energy and with low IFAR values ($<$100) in order to avoid hydrodynamic instabilities. We have used pulses based on Kidder´s theory of homogeneous isentropic compression of hollow shells[20].

The 1D code NORMA[8] has been used to simulate the evolution of targets with 1 mg of DT and different CH masses, chosen to optimize the implosion velocity. The results show that the optimum mass of the CH layer is 2.3 mg, which corresponds to an implosion velocity of $4.6 \cdot 10^7$ cm/s. The initial aspect ratio is roughly 12 with an outer radius of 0.2 cm.

The pulse considered in the simulations is defined as follows

$$P = \left| \begin{array}{ll} CP_o \left[\dfrac{1}{(1 - \tau^2)^{5/2}} - 1 \right] & 0 < \tau < \tau_a \\ P_0 & \tau_a < \tau < 1, \end{array} \right.$$

where P_o is the maximum laser power impinging upon the target, $\tau = t/t_c$, $\tau_a = t_a/t_c$, t_a is the time when the maximum power is reached, t_c is the time of central void closure and C is a constant determined by the continuity of the pulse at $t = t_a$. The maximum power is 830 TW, which corresponds to an intensity of the order of 10^{15} Wcm^{-2}, and the total energy of the pulse is 3 MJ. The gradient scale length of the corona (L) is such that we are in the reactor target regime corresponding to $L/\lambda \approx 10^4$, and $I\lambda^2 \approx 10^{14}$ W.µm^2/cm^2 of reference

21, which is under the threshold of the convective SRS, but may be above the threshold of the absolute SRS.

The evolution of the target is as follows. First, the prepulse produces a moderate shock wave that locates the fuel in a high density and low temperature isentrope. Then, the ablation process begins and the ablation pressure increases monotonically with time, continuously compressing the D-T layer without producing shock waves that could preheat the fuel. At the end of the pulse, almost all of the CH has been ablated. In fact, the DT carries out almost 80% of the implosion kinetic energy. The maximum IFAR is less than 100 and is reached at the end of the pulse. The hydrodynamic efficency is, roughly 6%, as expected from the low IFAR, and the velocity profile of the D-T is almost flat with an average implosion velocity $4.6 \cdot 10^7$ cm/s . The high implosion velocities reached allow to obtain a good ignition propagation and a high compression of the fuel ($\rho R \approx 3.4$ g·cm^{-2}) just before the shock wave produced by the void closure is reflected by the fuel-pusher interface. The convergence ratio of the fuel is 23.

We can conclude from our simulations that a pulse of the Kidder type has a good coupling with the target, which allows to have a high compression of the fuel with reasonable energies and low IFAR. Because of the high sensitivity of the target perfomance to the pulse shaping that we have observed in our simulations, we are currently studying more elaborate pulses. In addition, the stability of the corona for reactor targets should be addressed to assess the feasibility of low aspect ratio targets.

REFERENCES

1. Velarde, P.M., "FCT and Projection Methods for Lagragian Hydrodynamics" DENIM 213, Institute of Nuclear Fusion, submitted to *J. Comput. Phys.* (1990).
2. Richtmyer, R.D., "Taylor Instability in Shock Acceleration of Compressible Fluids", *Comm. Pure. App. Math.*, XIII, 297-319 (1960).
3. Meyer, K.A., and Blewett, P.J., "Numerical Investigation of a Shock-Accelerated Interface between Two Fluids", *Phys. Fluids*, 15, 753-759 (1972).
4. Houas, L., Farhat, A., and Brun, R., "Shock induced Rayleigh-Taylor Instability in the Presence of a Boundary Layer", *Phys. FLuids*, 31, 807-812 (1988).

5. Velarde,G., et al., *European Space Agency Scientific and Technical Publications Branch* **207**, 201 (1984).

6. Mínguez, E., et al., *Laser and Particle Beams* **6**, 265 (1988).

7. More, R.M., Lawrence Livermore National Laboratory. *Report N. UCRL-84991 part I-II.* Preprint (1981).

8. Velarde, G., et al., *Laser and Particle Beams* **4**, 239 (1986).

9. Mínguez, E., and Gámez, M.L., *Laser and Particle Beams* **8**, 103 (1990).

10. Bauche-Arnoult, G., Bauche, J., and Klapisch, M., *Phys. Rev. A* **31**, 2248 (1985).

11. Mínguez, E., and Falquina, R., submitted for publication.

12. Gámez, M.L., and Mínguez, E., accepted for publication in *World Scientific Publishing*, W.H. Goldstein, C. Hooper, J.C. Gauthier, J.F. Seely and R.W. Lee, editiors,

13. Post, D.E., et al. *Atomic Data and Nuclear Data Tables*, Vol. 2D, 5 (1977).

14. Honrubia, J.J., *Laser and Particle Beams* **8**, 1 (1990).

15. Honrubia,J.J. and Morel, J. E., *Nucl. Sci. Eng.* **104**, 91 (1990).

16. González, M.C., and Honrubia J.J., "An S_2 Synthetic Acceleration Method for Two-Dimensional Transport Problems", *Trans. Am. Nucl. Soc.*, **63**, 184-186 (1991).

17. Barnett, A., Morel, J.E., Harris, D.R., *Nucl. Sci. Eng.*, **102**, 1 (1989)

18. Babonneau, D., Bocher, J.L., Bayer, C., Decoster, A., Juraszcek, D., Pemine, J.P., and Thill, G., *Laser and Particle Beams* **9**, 2 (1991)

19. Velarde, G., et al, "High-Gain Direct-Drive Target Design for ICF", *Laser Interaction and Related Plasma Phenomena*, Vol 9, Eds. H. Hora and G. Miley, Plenum Press, New York (1991)

20. Kidder, R.E., *Nucl. Fus.* **16**, 1 (1976)

21. Max, G.E., et al, "Scaling of Laser. Plasma Interactions with Laser Vawelength and Plasma Size", *Laser Interaction and Related Plasma Phenomena*, Vol. 6, Eds. H. Hora & G. Miley, Plenum Press (1984).

NUMERICAL SIMULATIONS AND NUMERICAL ANALYSIS OF REACTOR-SIZE INDIRECTLY DRIVEN INERTIAL FUSION TARGET IMPLOSIONS

N.A. Tahir and C. Deutsch

L.P.G.P.* Bât. 212, Université Paris XI
91405 ORSAY Cedex, France

and

T. Blenski and J. Ligou

Institut de Génie Atomique, Département de Physique
Ecole Polytechnique Fédérale de Lausanne
CH-1015 Lausanne, Switzerland

ABSTRACT

This paper presents numerical simulations of implosion of a reactor-size indirectly driven inertial fusion target which contains 4 mg DT fuel. The spherical fuel shell is followed by a low-Z, low-ρ lithium stabilizer surrounded by a high-Z, high-ρ gold tamper. The gold shell acts as an ablator as well as a radiation shield. The target is irradiated by a uniform radiation field characterized by a temperature 300 eV. The target material is heated by the radiation field and is ablated away. The pressure generated by the material ablation drives the target. We have carried out numerical simultations of target implosion using a three-temperature radiation-hydrodynamic computer code MEDUSA-KAT [Tahir *et al.*, *J. Appl. Phys.*, **60**, 898 (1986)] and detailed analysis of these implosion simulations are presented in this paper.

1. INTRODUCTION

Indirect drive inertial confinement fusion (ICF) has a number of advantages over direct drive ICF inculding uniformily of illumination that leads to symmetry of implosion. Moreover the "Centurion Halite" experiments[1] have demonstrated that indirect ICF reactor-size targets may work in practice whereas no such experience has been made with directly driven targets. On the other hand, extensive theoretical and numerical simulation work has been reported in the literature on directly driven reactor-size targets[2-10], but very little is known about indirectly driven targets[6,11,12]. Reference[6] describes fusion gain curves without giving any details about the target design and the implosion physics. References[12,13] although give all the necessary details about the target physics but these targets are just "breakeven" targets (Fusion energy gain : Input thermal radiation energy). Moreover these targets work as exploding pusher targets and not as ablative driven targets. These targets are designed for use in a Laboratory Microfusion Facility (LMF) driven by heavy ion beams. Consequently these targets are small and they contain a fraction of μg DT fuel whereas in a reactor system similar to that described in[14,15] one would need a much bigger target with a few mg DT fuel.

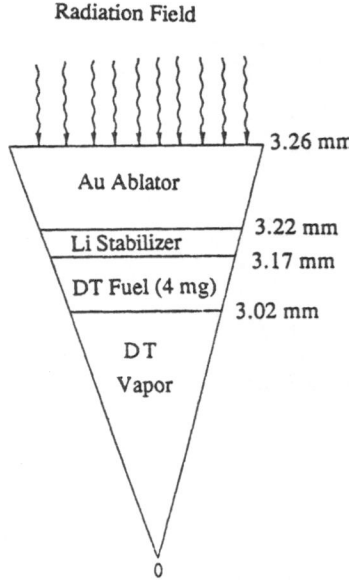

Figure 1. Target initial conditions

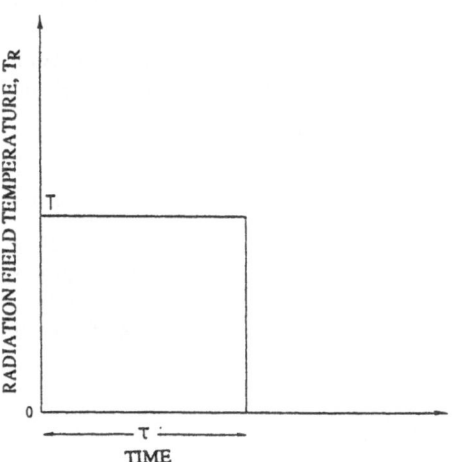

Figure 2. Time history of the input radiation temperature pulse.

We have designed a reactor-size radiation driven ICF target shown in Fig. 1 which is suitable for use in a reactor-system. Such a reactor system may be driven by energetic heavy ion beams and the beam energy can first be converted into thermal radiation which can then be used to drive the target. This target contains 4 mg DT fuel which is surronded by a low-Z and low-ρ lithium stabilizer that is followed by a heavy gold shell. This gold shell acts as an ablator as well as a radiation shield which protects the fuel from radiative preheat. The target is uniformly irradiated by a Planckian radiation field characterized by a radiation temperature, $T_r = 300$ eV which is constant in time. The time history of the imput radiation temperature pulse is shown in Fig. 2. It is an unshaped (Box Pulse) whereas we know that we need a reasonably shaped pulse[2-9] in order to achieve high target gain. In the present calculations we are not primarily interested in fusion gain but we want to study the process of ablation including calculation of mass ablation rate, ablation velocity and the ablation pressure. The maximization of ablation pressure is desirable as it increase the hydrodynamic efficiency which is advantageous for achieving fuel ignition.

These simulations have been carried out using a one-dimensional, three-temperature, Lagrangian, radiation-hydrodynamic computer code MEDUSA-KAI[13] which is an extensively updated version of the two-temperature code MEDUSA[14]. The physics included in the former simulation model is as follows. Atomis physics processes are treated either by a Saha equilibrium model or by a steady state density and temperature dependent model[15]. This latter model assumes that the plasma is in an ionizational equilibrium due to a balance between collisional ionization and radiative decay plus thre-body recombination. Ions are treated as a classical ideal gaz while for the electrons we use analytic fits to the Los Alamos equation of state (EOS) data. Electron thermal conduction and radiation diffusion (a three-temperature modes) are means of energy transfer in the plasma. Los Alamos opacity date is used for DT and Li while for gold we use data calculated by[16]. Local α-particle deposition is considered while the neutrons are allowed to escape freely.

2. TARGET SIMULATION RESULTS

In this section are present simulation results of the implosion of the target shown in Fig. 1. A radiation temperature pulse with $T_R = 300$ eV is applied at the outer target boundary at $t = 0$. The time variation of the radiation temperature is shown in Fig. 2. This radiation temperature pulse has a length $\tau = 10$ nsec. A temperature of 300 eV corresponds to a power density of the radiation field = 830 TW/cm^2 and the total energy absorbed by the target is = 11 MJ. It is well known[2-5,8] that a reasonably shaped input pulse is required to achieve high target gain. It has been shown[8] that use of a box pulse like the one shown in Fig. 2 does not ignite configuration in which the entire fuel is uniformly heated and almost uniformly compressed. One needs to have a temperature = 5 keV in the fuel in order to achieve thermonuclear burn, otherwise the radiation losses will suppress ignition. For details see[8]. Energy required to heat DT to 5 keV, is $E_h = 5.8 \times 10^2 M$ J/g. Total input energy in our cases is $E_i = 11$ MJ And assuming a 10 % energy fuel coupling efficiency, only 1.1 MJ energy is coupled to the fuel whereas to heat 4 mg DT fuel to 5 keV, we need 2.32 MJ. Therefore we do not achieve successful burn of our target fuel. However in the present case we are not interested in achievement of burn, but we wish to study the process of radiation ablation of the target and corresponding ablation pressure generated by the ablating material. Therefore we used a simpler input radiation pulse instead of a shaped pulse.

In Fig. 3 we plot the density, ρ, electron temperature, T_e, radiation temperature, T_r, and pressure P along the target radius at $t = 1.049$ nsec. It is seen that the radiation wave has penetrated into the gold shell and the electrons in this region are heated by the radiation field. The radiation temperature is higher than the electron temperature ahead of the radiation Marshak wave[17-19]. It is also seen that ahead of the ablation front the density is about three times higher than the solid gold density due to compression by ablation pressure. Some variables are plotted in Fig. 4, but at $t = 3.66$ nsec which shows that the Marshak wave has penetrated further into the fold tamper and a significant amount of gold has been ablated away. The input radiation temperature pulse is switched off at $t = 10$ nsec. The ablation pressure drives the target to void closure and the maximum compression is achieved at $t = 17.25$ nsec (see Fig. 5). This figure shows a volume type configuration with a fuel density $\sim 6 \times 10^4$ Kg/m^3. Also the ion temperature in the inner half radius of the fuel is about 2 keV while in the outer half radius is 1 keV. Since a temperature of 5 keV is needed for the α-particle bootstrap heating to take place which is necessary to maintain thermonuclear burn, the ignition process in our simulations is quenched and we do not get a substantial amount of energy output.

It is also seen that the density of the lithium stabilizer at the time of void closure is comparable to the fuel density and this configuration is very advantageous from the stability point of view.

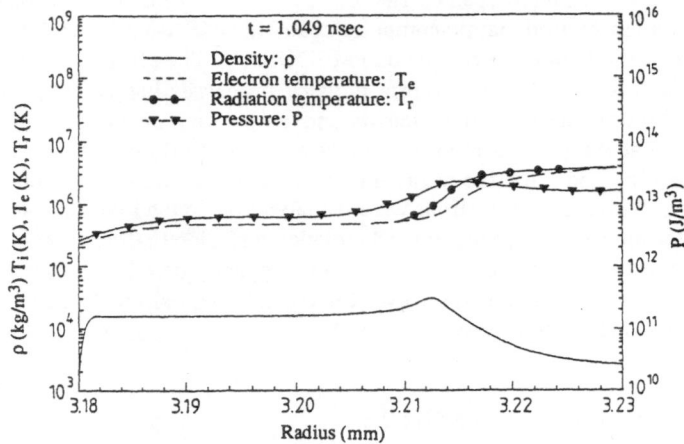

Figure 3. Density ρ, Electron temperature: T_e, Radiation temperature: T_r, and Pressure: P vs. Target radius in the gold ablator shell at t = 1.049 nsec.

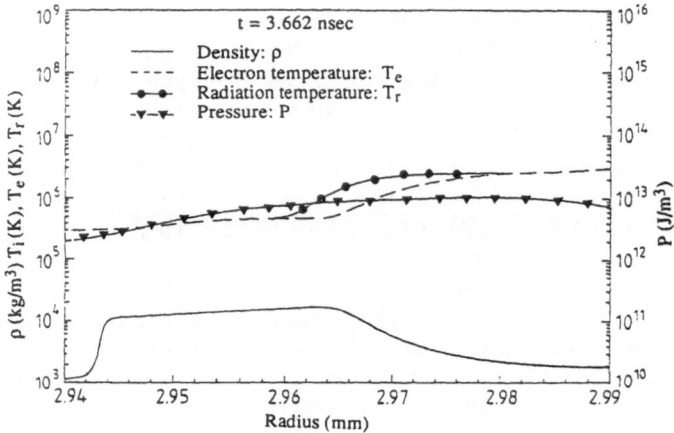

Figure 4. Same as in Fig. 3, but at t = 3.66 nsec.

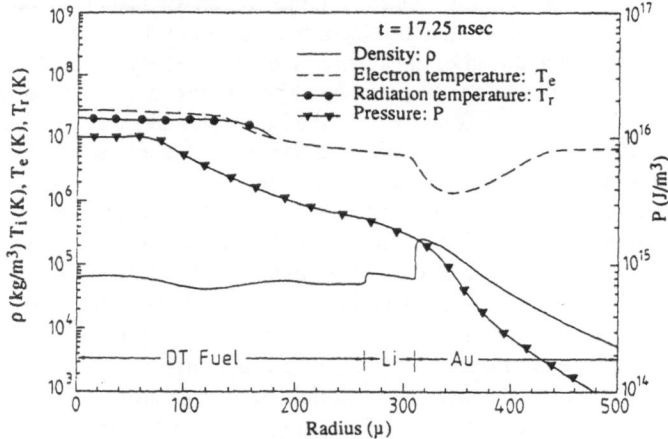

Figure 5. Same as in Fig. 3, but at t = 17.25 nsec. and in the fuel, lithium stabilizer and unablated part of the gold shell.

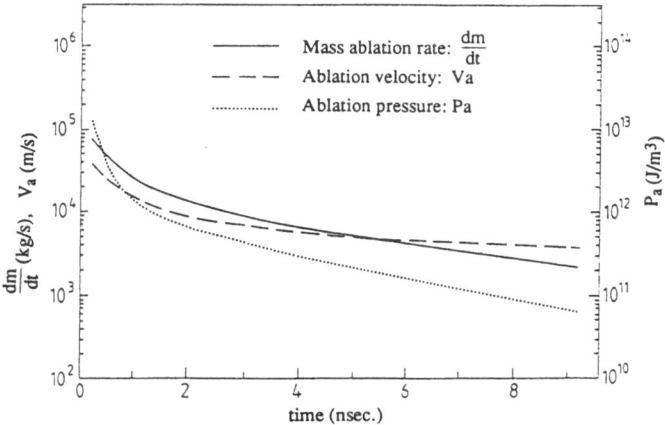

Figure 6. Mass ablation rate, $\frac{dm}{dt}$, Ablation velocity: V_a and Ablation pressure: P_a vs time.

In Fig. 6 we plot the mass ablation rate $\frac{dm}{dt}$ (kg/s) , ablation velocity V_a (m/s) and ablation pressure, P_a(J/m³) vs time. The mass ablation rate is calculated from the simulations and then the ablation velocity and ablation pressure are evaluated as

$$V_a = \frac{1}{\rho_a A} \frac{dm}{dt}$$

and

$$P_a = \frac{1}{2} \rho_a v_a^2$$

where ρ_a is the density at the ablation surface and A is the area of the ablation surface.

It is seen that the above three quantities decrease with time. This is due to the fact that the radiation Marshak wave slows down with time. As a result the mass ablation rate decreases which reduces the ablation pressure. It is seen that the ablation pressure is of the order of 100 MBar at the beginning of the compression phase which is sufficient to drive the implosion.

It is also known[20] that the hydrodynamic stability of these radiation driven targets is much better than that of the directly driven targets. A simple analysis of the calculations have confirmed that this target may be stable during the compression phase.

3. CONCLUSIONS

We present simulations of implosion of a radiation-driven, reactor-size ICF target using a one-dimensional, three-temperature computer code. The radiation field is characterized by a radiation temperature = 300 eV and the intensity of the radiation field is constant in time. It has been shown that like directly driven targets, one needs a reasonably shaped input pulse in order to get ignition and burn in case of indirectly driven targets. In the present calculations we are not interested in the thermonuclear burn of the target because the burn phase of an indirectly driven target is very similar to that of a directly driven target, a problem which has been thoroughly analyzed and well understood[2-10]. The compression phases, on the other hand, are very different in the two type of targets. In this paper we only concentrate on the compression phase and we have analyzed the radiation ablation of the target and ablation pressure has been calculated. It has been found that in a radiation driven target, the ablation pressure dominates the plasma thermal pressure. Therefore the target is driven by the ablation pressure and because of the uniformity of this ablation pressure the implosion is expected to be symmetric.

REFERENCES

1. W.J.Broad, *International Herald Tribune*, March 22 (1988).
2. J. Nuckolls *et al.*, *Nature*, **239**, 139 (1972).
3. J.S. Clarke *et al.*, *Phys. Rev. Lett.*, **30**, 89 (1973).
4. R.E. Kidder, *Nuclear Fusion*, **19**, 223 (1979).
5. S.E. Bodner, *J. Fusion Energy*, **1**, 219 (1981).
6. R.O. Bangerter *et al.*, *Phys. Lett.*, **88 A**, 225 (1982).
7. N.A. Tahir and K.A. Long, *Atomkernenergie*, **40**, 157 (1982).
8. N.A. Tahir and K.A. Long, *Nucl. Fusion*, **23**, 887 (1983).
9. N.A. Tahir and K.A. Long, *Phys. Fluids*, **29**, 1282 (1986).
10. C. Deutsch, *Ann. Phys. Fr.*, **11**, 1 (1986), and
    also *Laser Part. Beams*, **2**, 449 (1984).
11. N.A. Tahir and R.C. Arnold, *Laser Part. Beams*, (1991).
12. N.A. Tahir and R.C. Arnold,*Phys. Fluids*, **B3**, 1717 (1991).
13. N.A. Tahir *et al.*, *J. Appl. Phys.*, **60**, 898 (1986).
14. J.P. Christiansen *et al.*, *Comput. Phys. Comm.*, **7**, 271 (1974).
15. N.A. Tahir, "Simulation Studies of Laser-Compression of Matter, *Ph.D. Thesis,*
    Glasgow University (1978).
16. T. Blenski and J. Ligou, *J. de Physique*, **C7**, 259 (1988).
17. R.E. Marsahak, *Phys. Fluids*, **1**, 24 (1958).
18. N.A. Tahir and K.A. Long, *Laser Part. Beams*, **2**, 371 (1984).
19. K.A. Long and N.A. Tahir, *Laser Part. Beams*, **4**, 287 (1986).
20. H. Takabe, *ILE Osaka Report, ILE-QPR-60*, 28 (1983).

ION BEAM COUPLING AND TARGET PHYSICS EXPERIMENTS AT SANDIA NATIONAL LABORATORIES

Thomas A. Mehlhorn

Target Physics Analysis Division
Sandia National Laboratories
Albuquerque, New Mexico

INTRODUCTION

For many years there has been an international effort aimed at producing the ignition of an ICF capsule in the laboratory. The majority of these efforts utilize lasers of varying wavelengths to deliver the energy to initiate the ablation of the target, the compression and ignition of the fuel, and the propagation of the fusion burn. One alternate to this scheme is to provide the drive energy in the form of a light ion beam produced by an efficient pulse power accelerator. A related method uses beams of heavy ions from high intensity versions of traditional high-energy accelerators. Until recently, only a limited number of target experiments have been performed using ion beams because most of the research has been directed towards developing beams of sufficient intensity and energy to perform significant target experiments. However, Sandia has recently completed a series of target experiments that are contributing to ICF science.

While it has been difficult and expensive to develop lasers with the energy necessary to drive an ignition implosion (≥ 1 MJ), it is easy to focus the laser energy to very high intensities and to heat very small quantities of matter to high temperatures. Therefore, laser-driven target experiments have been performed early in the development process. Furthermore, since lasers are used in many other applications and at many institutions, the technology is developed in a broad way. With light ion beams produced by pulsed power accelerators things are somewhat different. It is relatively inexpensive to build pulse power machines that are capable of delivering large quantities of energy into an ion beam and thence to a macroscopic-sized target. However, it has proved much more difficult to focus these beams to high intensity. Therefore, the application of these beams to target experiments has occurred later in the development process. Furthermore, the intense light-ion beam technology is less mature than the laser technology and does not have as large a pool of institutions working on parallel developments.

Within the past two years, proton beams have been generated by the Particle Beam Fusion Accelerator II (PBFA II) at Sandia National Laboratories that have sufficient intensity and energy to drive interesting target physics experiments on ignition-sized targets. Newly developed lithium beams show promise of even higher target physics performance in the coming year. This article will describe the target physics basis of the light ion fusion program, discuss the present status of our beam generation and focusing capability, and review the status of our understanding of ion beam deposition physics. Some comparison with the related status of the laser-driven ICF will be presented. Then some of the recent proton-driven target physics experiments on PBFA II will be described. Finally, our plans for future target experiments driven by high intensity lithium beams will be discussed.

OVERVIEW OF SANDIA'S TARGET PHYSICS PROGRAM

The goal of the Sandia ICF program is clear; that goal is to make fusion in the laboratory. Although research in driver development and beam focusing has consumed much of the total effort of Sandia's ICF program, this effort has been ultimately directed at the goal of making fusion in the laboratory. The technology of high intensity light ion beams has only been studied since about 1975 with pioneering work at Cornell[1], Sandia[2], and the Naval Research Laboratory (NRL)[3]. It has been necessary to first develop this technology to produce and focus ion beams to the energies and intensities necessary to drive fusion targets. In 1989, a proton beam was focussed to an intensity of 5 TW/cm^2 averaged over the surface of a 6 mm diameter sphere[4]. In 1991, proton beams were used to perform the first series of significant target experiments aimed at contributing to the knowledge base of ICF science. Ongoing research in the understanding, development, and focusing of light ion beams is aimed primarily at the generation of lithium ion beams which will be superior in their target-drive characteristics. At present, two series of target physics experiments using lithium ion beams are planned for 1992.

In response to the recommendations of the National Academy of Sciences review committee report on the Sandia ICF program[5], we have established two clear objectives for the light ion program for 1992. The first goal is to demonstrate the achievement of a radiation-dominated hohlraum driven by a light ion beam. The second is to develop a clear and credible path for the light ion program to ignition, high gain, and energy applications.

THE PHYSICS OF INDIRECTLY-DRIVEN LIGHT-ION TARGETS

It is the energy deposition mechanism, not the capsule implosion physics, which comprises the most significant difference between ion and laser targets. Figures 1a and 1b schematically show the concepts for both laser and light ion targets that are indirectly-driven by thermal radiation. It is important to note that the fuel capsule in both the laser- and ion-driven target could be very similar because they are both driven by a thermal radiation spectrum, and both drivers need to implode and compress the fuel to the same density and temperature conditions to achieve ignition. Those differences that might exist between the fuel capsule designs will be primarily due to differences in driver pulse shaping. Target ablation into a foam, rather than into a vacuum, might possibly lead to other differences. However, the most important distinction between the two approaches lies in the mechanisms of primary energy deposition and transport in the target.

This difference in the deposition mechanism leads to a light ion beam target configuration that is substantially different from either direct- or indirect-drive laser target configurations. With a light ion beam driver the fusion capsule is imbedded within a foam-filled hohlraum as shown schematically in Figure b. The ions penetrate the external shell and volumetrically deposit their energy in the low-density foam, which converts the ion beam energy into x-rays. The radiation, in turn, bathes the fusion capsule and provides the drive for the capsule implosion. In a laser target, the laser beams enter through holes in the external case and deposit their energy on the inner surface of the case. Radiation is produced along the heated surface which irradiates the fusion capsule, and drives the implosion.

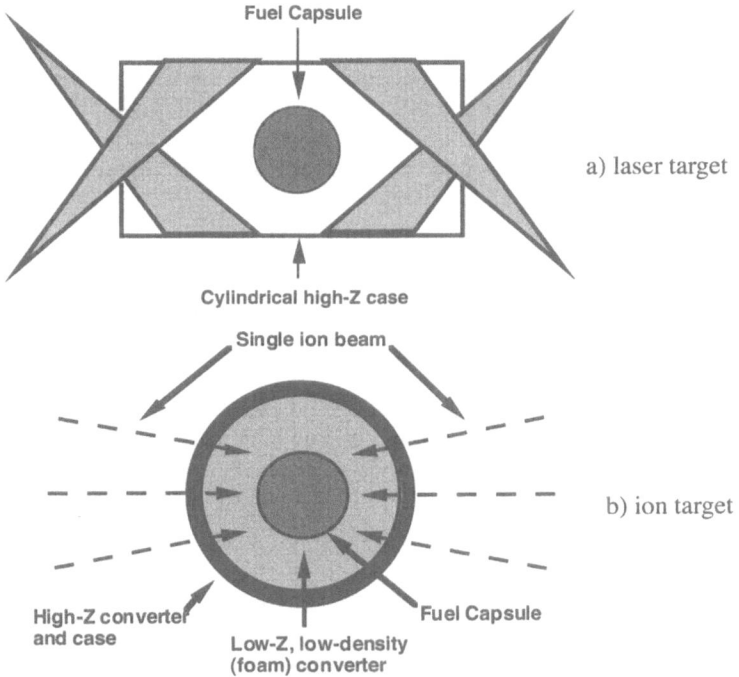

Figure 1. Generic sketches of indirectly-driven inertial confinement fusion targets: a) laser fusion target, b) light ion beam fusion target.

The light ion beam hohlraum has several advantages over its laser counterpart. In a well-designed target, the target absorbs all of the incident ion energy with the low-density foam absorbing about 70% of the energy and producing a thermal electron spectrum. The absence of holes minimizes energy loss and increases the overall target efficiency. The volumetric deposition characteristics of ion beams provide initial smoothing of the ion beam energy and make radiation smoothing more effective. This smoothing during deposition and x-ray generation also makes the time-dependent symmetry of the drive less of an issue. Volumetric deposition minimizes the density and temperature gradients in the target which decreases the opportunity for instabilities to occur in the deposition region. Ion energy spread, which is analogous to bandwidth for lasers, can also contribute to the robust nature of the ion-driven target and the avoidance of instability-generated target coupling and pre-heat problems. (Note, however, that future energy applications involving time-of-flight (TOF) bunching of the ion beam will require a minimal instantaneous energy spread in order to be efficient.) Energy conversion into hydrodynamic motion of the converter material is also minimized in a light-ion hohlraum. The unique and favorable deposition and hohlraum physics of light ions contribute to a more robust and broadly based ICF program.

ENERGY DEPOSITION BY LIGHT-ION BEAMS

One of the strengths of ion-beam-driven fusion is that the beam-target coupling is expected to remain "classical" up through the intensities required to ignite an ICF capsule and achieve high gain. While coupling of laser energy to matter has been found to involve many nonlinear effects (which has dictated the movement of the laser programs to shorter and shorter wavelengths), theory and experiments to date indicate that the ion-beam energy coupling involves only well understood "linear" physical processes. Furthermore, our knowledge of the physics of energy deposition of intense ion beams has been extended and corroborated through the work for the past 15 years of many scientists throughout the world. Moreover, with ions one has the possibility of tailoring both the ionic species and the accelerator voltage to simultaneously optimize beam generation, beam transport, and target coupling efficiencies.

Because the Debye shielding distance in both solid and plasma targets is much less than the interparticle spacing of an intense ion beam, the classical theory of single-particle stopping power applies in ion-driven ICF[6]. Furthermore, studies of gradient-driven[7] and streaming instabilities[8] in plasmas indicate that such instabilities do not play a significant role in the ion stopping process. The energy loss of ions is primarily due to energy transfer to both bound and free electrons in the target material via Coulomb interactions [9]. The energy loss to bound electrons results from both ionization and excitation. The controlling variables of the ion stopping power, dE/dx, of an ion beam in an ICF plasma are shown in Equation 1:

$$\frac{dE}{dx} \propto \frac{Z_{eff}^2}{E} \left[(Z_T - \bar{Z}) \ln\left(\frac{E}{I_{av}}\right) + \bar{Z} \ln\left(\frac{E}{\hbar\omega_p}\right) \right] \qquad (1)$$

where E is the ion energy, Z_{eff} is the effective ion charge, Z_T is the atomic number of the target material, $\bar{Z}$ is the average ionization of the target material, I_{av} is the average ionization potential of the bound electrons, and $\hbar\omega_p$ is the energy of a plasma wave having the electron plasma frequency ω_p. The first term within the brackets represents the stopping power of the $(Z_T - \bar{Z})$ bound electrons, while the second term represents the stopping power of the $\bar{Z}$ free electrons. We note that the stopping power, as given by Equation 1, increases as the denominator of each of the "Coulomb logarithms" decreases; i.e. as I_{av} and $\hbar\omega_p$ are decreased.

The average ionization potential, I_{av}, that is used in the Bethe theory of ion stopping[10] characterizes the minimum allowed energy transfer to the bound electrons of an atom. Thomas-Fermi theory shows that I_{av} scales as 12.5 Z_T (eV). This scaling, coupled with the functional dependence of the stopping power on the average ionization potential in Equation 1, indicates that a low-Z material is more effective in slowing an ion beam than a high-Z material. Therefore, the low-Z foam will absorb energy more efficiently than the high-Z case in a light ion target. Near the end of the ion range where the ion energy is small, the slowing of the ions by elastic Coulomb collisions between the ion and the target nuclei can become significant[11]. For accuracy, our models of ion beam deposition in targets include both these effects, although the primary energy transfer is to thermal electrons.

As the ion beam heats and ionizes the target material, the free plasma electrons participate in the slowing process. The minimum energy transfer (corresponding to the maximum impact parameter in collision theory) to the plasma electrons is characterized by the energy of a plasmon ($\hbar\omega_p$). Free electrons can be much more efficient at stopping the ion beam than bound electrons, and therefore "range shortening" occurs as the target ionizes. Consider a plasma where $n_e \sim 10^{23}/cm^3$. The plasma frequency $\omega_p = (4\pi n_e e^2/m_e)^{1/2} \sim 2 \times 10^{16}/sec$; therefore $\hbar\omega_p \sim 10$ eV. Comparing this value with the Thomas-Fermi scaling for I_{av}, we see that the stopping power for bound and free electrons will be about the same for low-Z material. However, for higher Z materials, the free electrons will have a higher stopping power because their minimum energy transfer limit will be much less[12].

As the electrons are removed from an atom, the average ionization potential of the remaining bound electrons is increased. Therefore, in partially ionized material the enhanced stopping power of the free electrons is partially counterbalanced by a decrease in the stopping power of the remaining bound electrons. Experiments that have been performed to date for protons and deuterons have confirmed that the ion beam stopping power is correctly predicted by our computational models, and that the ion range is shortened by up to a factor of 2 as the target ionization increases[13]. Range shortening further enhances the efficiency of low-Z materials in ion stopping as compared to high-Z materials, thus further increasing the effective specific energy deposition (MJ/g) of a low-Z foam compared to a high-Z case.

The stopping power of an ion depends on the energy, species, and charge state of the ion as well as on the composition, density, and temperature of the target material. The parame-

Table 1. Summary of enhanced ion deposition experiments performed at NRL and Sandia through 1985. The dE/dx enhancement factor gives the stopping power ratio of heated to cold material in these experiments.

Ion	Ion Energy (MeV)	Material	Ion Range (mg/cm^2)	Intensity (TW/cm^2)	Specific Deposition (TW/g)	T_e (eV)	dE/dx Enhancement Factor
d	1.25	aluminum	5.5	0.06	26	4-5	1.3
d	1.25	aluminum	5.5	0.31	82	13-17	1.5
d	1.25	Mylar	3.4	0.06	27	2-4	1.0
d	1.25	Mylar	3.4	0.31	133	9-11	1.6
p	1.8	aluminum	9.8	1.4	144	48	2.0
p	1.8	nickel	13.8	1.4	101	42	1.5

ters for experiments which investigated the enhanced stopping power of intense proton and deuteron beams in Mylar, aluminum, and nickel are shown in Table 1. The first four rows summarize the parameters from the first such experiments performed at NRL[14] using deuteron beams. At the lowest specific depositions, little or no enhancement of stopping power was observed. At the higher depositions, enhanced energy loss was seen for both Mylar and aluminum. The last two rows refer to experiments performed at Sandia in 1985[15] using proton beams. Using the higher intensity proton beams from the PROTO I accelerator, the enhanced stopping power of an intermediate-Z material was studied. Enhanced energy loss was observed for both aluminum and nickel. Further increases in beam intensity will enable researchers to study the enhanced stopping power of high-Z materials. These experiments have verified our stopping power models at specific power depositions of up to 150 TW/g. These models must still be validated at the higher beam intensities, specific deposition rates, and material temperatures required for ignition and gain capsules.

For non-protonic ions the effective charge, Z_{eff}, of the projectile plays an important role because the ion stopping power has a quadratic dependence on effective charge, whereas it only has a logarithmic dependence on average ionization potential. The theory of the evolution of the effective charge of the projectile in ionized targets is also well developed[17,18] and has been experimentally verified for high-Z projectiles in a hydrogen plasma at Gesellschaft fur Schwerionenforschung (GSI) Darmstadt[19]. Stopping power enhancement of up to a factor of 3 for heavy ions occurs when the effect of the reduced recombination of plasmas on the effective charge is combined with the enhanced stopping power of free electrons. Experimental measurements of the effect of enhanced effective charge on ion stopping power in plasmas heated by lithium beams remains to be performed.

Although our understanding of the ion energy deposition at the TW level is experimentally consistent with classical energy transfer primarily to a "thermal electron" spectrum, potential preheat mechanisms in ion-driven targets exist and their significance must be quantified for each specific target design. The electron spectrum generated by ion deposition has a $1/T^2$ dependence, where T is the electron kinetic energy[20]. Kinematically, the maximum electron energy that is produced is $T_{max}=2m_e v_i^2$. Monte Carlo simulations of target preheat using this electron spectrum indicate that electron preheat is not a problem. Hard x-rays can be produced by characteristic inner-shell line emission via ion impact ionization. These x-rays should also produce negligible preheat. The most probable preheat source is elastically-scattered, low-Z "knock-on" ions that have ranges longer than the high-Z ions that produce them[21]. Knock-ons have been shown to have the potential for a few-eV fuel preheat, depending on target design.

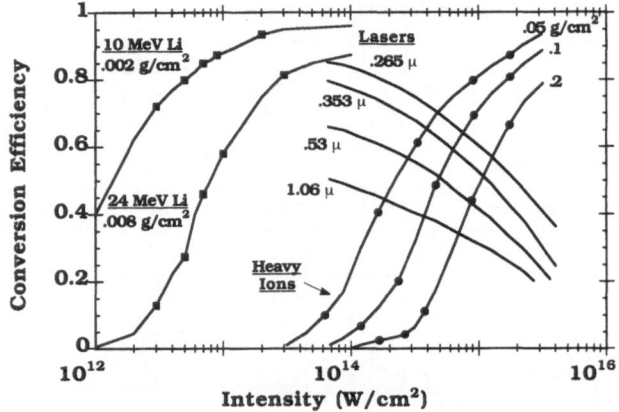

Figure 2. X-ray conversion efficiency as a function of intensity for lasers of varying wavelengths and ions of varying ranges[23].This graph has been modified to include the predicted conversion efficiency of shorter range light ions[16] assuming a 20 ns beam pulse into a carbon target.

DEPOSITION INTENSITIES AND RADIATION CONVERSION

The peak ion beam intensity on target, the total energy delivered to the target within a hydrodynamically acceptable time, and the ion beam range in the target are among the parameters that most directly affect the performance of an ion-driven target. The ion beam intensity (TW/cm^2) divided by the ion beam range (g/cm^2) determines the specific power deposition (TW/g) in the target. The energy density (MJ/cm^2) divided by the range gives the specific energy deposition (MJ/g). The magnitude and uniformity of the specific power and energy provide the proper figures of merit for judging the performance of an ion-driven hohlraum for driving a capsule. At present, the divergence of our ion beams[22] limits the specific power and energy deposition in the target because it limits the amount of the beam that reaches the target.

The process of converting ion beam energy to x-rays is significantly different from the production of x-rays through a laser plasma interaction. Although both drivers rely on converting beam energy into internal energy, ion beams deposit their energy volumetrically at solid density whereas the laser beam deposits its energy at the critical density surface. For ion beams, the component of scattering and reflection is insignificant. The issue then becomes one of heating the material to a temperature that is high enough so that the radiation field intensity is adequate to drive a target ablator. The x-ray conversion efficiency depends on the ion mass and energy, and on the target material.

It is interesting to contrast the predicted x-ray conversion efficiencies of ion beams and lasers. Figure 2 demonstrates that the ion beam conversion efficiency to radiation is predicted to increase as the beam intensity is increased. On the other hand, the conversion efficiency of lasers at all wavelengths is shown to decrease with increasing intensity. Thus, radiation conversion efficiency scaling is another possible advantage of the ion approach to ICF. A simple model of this process[16] predicts that a 10 TW/cm^2 Li beam with an energy of 10 MeV will be ~80% efficient in converting ion energy into x-rays. The conversion efficiency into x-rays is predicted to exceed 90% at 100 TW/cm^2 for 24 MeV lithium.

PRESENT BEAM FOCUSING CAPABILITIES

The specific deposition rates that are consistent with the present beam focusing capabilities of the ion diode on the PBFA II accelerator are summarized in Table 2.

Table 2. Present PBFA-II specific deposition rates into a cold CH foam.

Ion	Ion Energy (MeV)	Range (mg/cm^2)	Intensity (TW/cm^2)	Specific Deposition (TW/g)
Proton	5	32	3.5	110
Proton	5	32	5	157
Lithium	9	2.6	1	380

Proton beams of up to 5 TW/cm^2 have been generated on PBFA II[4]; the 3.5 TW/cm^2 value is perhaps more representative of the average performance on a complex target experiment. A lithium beam of 1 TW/cm^2 has recently been generated on PBFA II. A contour plot showing the size and distribution of the focal intensity as measured by a time-dependent ion movie camera diagnostic is shown in Figure 3. Referring again to Table 2, we can see the coequal role of ion intensity and ion range in determining the target performance. Although the intensity achieved with lithium beams is less than that of protons, the specific power deposition in TW/g is larger for the lithium beams. This means that the lithium beams are actually more effective at heating the target material than the proton beams because their range is so much shorter.

Total beam energy is also important in determining the heating of a target material. Over our seven best analyzed shots, we have also shown that we can reproducibly generate lithium beams with energies of 170±30 KJ. These values are indicative of the energy-rich nature or the light-ion pulsed power technology. Moreover, as will be discussed later, improvements in energy coupling and vertical focusing of these lithium beams that are planned for 1992 should substantially increase the intensity and modestly increase the beam energies on target.

EXPERIMENTS ALONG THE PATH TO IGNITION

As Sandia's beam generation and focussing capability is increasing, the range of significant target experiments that can be performed with these beams multiplies. Figure 4 is a modification of a recent plot from a GSI report[24] that shows anticipated temperatures

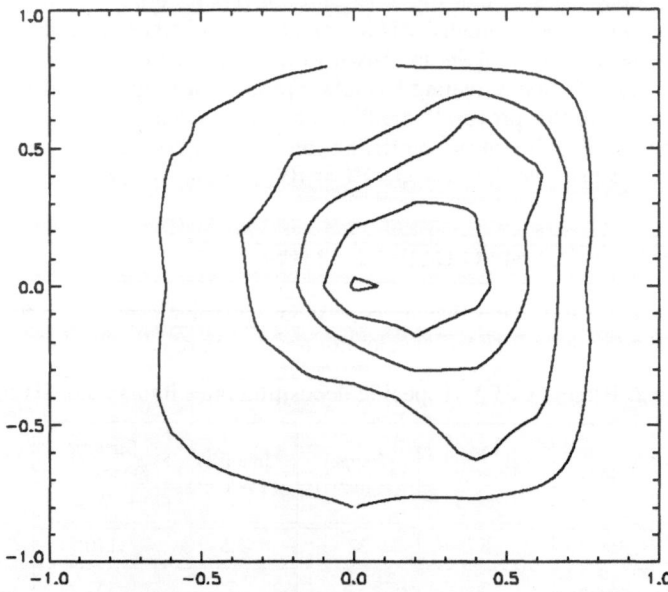

Figure 3. Contour plot of the focal intensity of a lithium beam on target for PBFA-II shot 4443 as viewed by the time-dependent ion movie camera. Each contour represents an increment of.2 TW/cm², with the outer contour having a value of.2 TW/cm². The spatial scale of the plot is in centimeters.

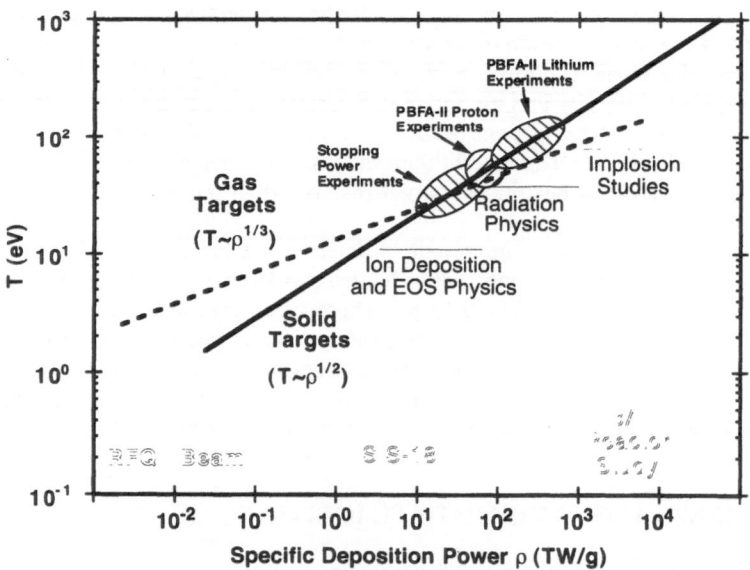

Figure 4. ICF target physics issues that can be studied by ion beams of increasing specific deposition power and associated increasing target temperature. Present and projected capabilities of light- and heavy-ion machines are noted. The original figure, found in Reference 24, has been modified to include the light ion capabilities.

generated in gaseous and solid targets as a function of the specific deposition power of an ion beam. The plot indicates that ion deposition and equation of state (EOS) experiments can be appropriately performed at specific powers of 10-100 TW/g, radiation physics experiments begin between 100 and 1000 TW/g, while implosion studies become possible at 1000 TW/g. The notations along the abscissa indicate the specific power regimes of various existing and planned experimental facilities in the European HI Program. The RFQ beam at GSI already exists and the heavy-ion synchrotron-storage ring (SIS) is anticipated to achieve the listed power by about 1995. Superimposed on this plot are three regions that have already been explored, or will be explored in the near future using light ion beams. The lowest specific deposition region labeled "Stopping Power Experiments" spans the specific depositions listed in Table 1 that were realized in previous stopping power experiments. The next higher region labelled "PBFA-II Proton Experiments" refers to the recent proton target series; those experiments achieved specific deposition powers of approximately 100 TW/g in foam and will be the subject of the next two sections. The highest shaded region labeled "PBFA-II Lithium Experiments" shows the specific powers that should be available using lithium beams in 1992. This figure also demonstrates that target physics experiments on the PBFA II accelerator are generating important data for both the Light and Heavy Ion Programs.

TARGET PHYSICS EXPERIMENTS ON PBFA II

The recent experimental target series on PBFA II reached specific deposition intensities of 100 TW/g and generated the first real data on x-ray production and the radiation physics from targets in a hohlraum-like geometry. Other experiments with direct-drive exploding pusher targets studied the hydrodynamics of ignition-size targets. These experiments are the subject of the next two sections. On these experiments we successfully obtained time-integrated, time-resolved, imaging, and spectral diagnostic information from an extensive array of diagnostics as listed in Table 3:

Table 3. Summary of diagnostics fielded on PBFA-II target experiments.

Time-integrated grazing incidence spectrometer	Time-resolved grazing incidence spectrometer
Bolometer	Transmission grating spectrograph
11-channel filtered XRD array	11-channel filtered PIN array
Elliptical crystal spectrograph	Ion movie camera
Magnetic spectrometer	Energy-resolved 1-D x-ray streak cameras
Time-integrated x-ray pinhole cameras	Time-resolved x-ray pinhole camera

DIRECT-DRIVE IMPLOSION EXPERIMENTS

Although exploding pusher target performance does not scale to ignition conditions, these targets can provide useful target physics data. Furthermore, since exploding pusher targets are less sensitive to drive uniformity than ablative targets, they can be studied using direct-drive by the ion beam. It is important to note the size difference of the targets used in these experiments with those of similar experiments using laser beams. The targets imploded in these experiments were 6 mm in diameter while typical laser-fusion capsules are only ~0.5 mm in diameter. This is directly related to the energy rich, intensity poor status of the light ion technology and the energy poor, intensity rich status of the laser technology. Experiments with our exploding pusher targets served several purposes: 1) they studied the implosion characteristics of ignition-size targets, 2) they helped to push target fabrication capabilities for ignition-size targets, and 3) they provided data with which to compare our radiation-hydrodynamic calculations in order to normalize and validate our simulation models.

The pie diagram in Figure 5a shows a cross sectional view of the targets that were used in these experiments. The targets were comprised of layers of plastic of varying composition. The Parylene-D in the inner wall contained chlorine which emits 2.6 keV x-rays when irradiated by an MeV ion beam. Ion impact ionization creating a hole in the K-shell followed by a 2P-to-1S radiative transition is the mechanism that produces these diagnostically valuable x-rays. These x-rays provided a monitor of the fuel/pusher interface motion for the duration of the beam pulse on target. The PVA layer acted as a gas barrier to seal the target. The outer Parylene-N layer completed the shell thickness that was appropriate for the range of 6 MeV protons.

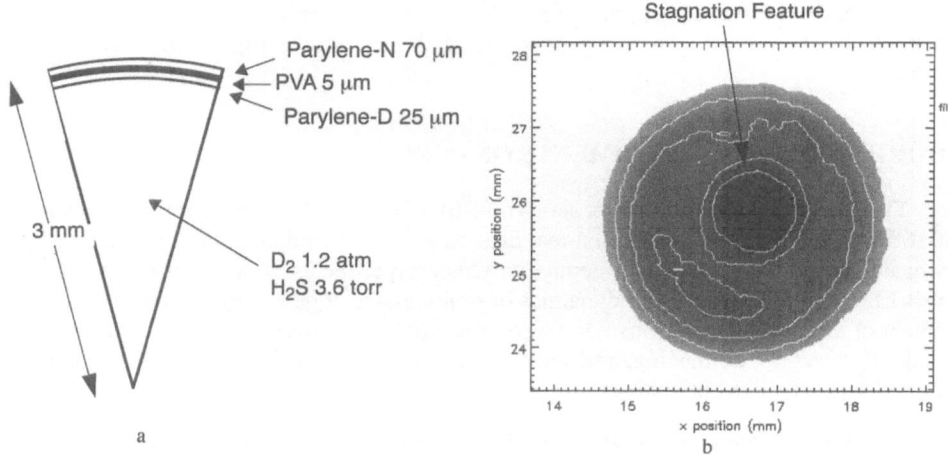

Figure 5. a) Pie diagram of a direct-drive exploding pusher target fielded on PBFA II and b) time-integrated x-ray image of the imploded target as viewed through a ~2 keV x-ray filter in the coordinate system of the film plane. Note the prominent central stagnation feature.

The 1.2 atm deuterium gas fill of these targets contained an additional 3.6 torr of hydrogen sulfide gas (H_2S). The gas fill was not meant to produce measurable neutrons in these experiments. Rather, the H_2S was added in an attempt to observe K-shell emission from thermally stripped sulfur. These spectroscopy experiments are similar to argon-seeded experiments performed in the laser fusion program.

The results of these experiments are still the subject of intensive analysis and review, and will be reported in future articles. A time-integrated image obtained by an x-ray pinhole camera with an ~2 keV filter is shown in Figure 5b. This image shows the bright ring of the initial image of the chlorine layer, the implosion of the chlorine layer with time, and a bright stagnation feature on-axis. The implosion feature is about 1 mm in diameter, in good agreement with simulation predictions.

One of the most important results that we have presently obtained from these exploding pusher target experiments is the demonstration that complex target physics experiments can be fielded on the PBFA II accelerator. This signals a progression of the ion beam program from primarily driver and beam development into an integrated target physics program. These experiments also demonstrated that ion beam energy can be efficiently coupled to ignition-sized spherical targets (these were the first ever spherical targets fielded on PBFA II). The experimental data obtained is consistent with our hydrodynamic simulations in terms of the diameter of the imploded core and the implosion symmetry with respect to the measured beam uniformity.

HOHLRAUM-LIKE TARGET EXPERIMENTS

In these experiments, a cylindrical version of an ion-driven hohlraum was used to study radiation and hohlraum physics issues. Since the specific deposition power of our proton beams have reached the 100 TW/g levels which was noted in Figure 4 as being consistent with the beginnings of radiation physics experiments, and since the indirect-drive approach that the light ion program has embraced is critically dependent on radiation physics, these experiments were extremely important.

Figure 6a shows a cross sectional view of the targets used in these experiments. A low-density hydrocarbon foam was placed within a gold cylinder, approximating a cylindrical hohlraum. The exterior gold cone was part of the beam characterization and energy accounting diagnostics on these experiments while the 5 Torr of argon was part of the gas used to provide space charge and current neutralization of the ion beam during transport. The vacuum barrier excluded the argon from the x-ray diagnostic line-of-sight to allow viewing of the thermal emission from the target.

Initial analysis of streak camera data observing emission from the foam indicates a brightness temperature of about 35 eV on PBFA-II shot #4531. An analysis of filtered XRD and bolometer signals indicate similar brightness temperatures across many spectral cuts. This consistency not only adds credibility to these independent measurements, it also is consistent with the generation of a Planckian x-ray spectrum. If this result is shown to be true this would be highly significant in that it would indicate that at these specific deposition levels the ion beam coupling is indeed into a thermal spectrum. A sample x-ray pinhole camera image of the target as viewed from below is shown in Figure 6b.

Another important conclusion that can be drawn from the data is that the foam was heated and reached the optically thin state necessary for radiation transport in an ion beam hohlraum. This conclusion is based on a comparison of the shape of the experimental XRD traces with those predicted by hydrodynamic simulations. Note that due to the long proton range and the correspondingly thick gold walls used in these experiments, the energy absorption of these hohlraums tended to be wall-dominated rather than foam-dominated. This situation should be reversed in upcoming lithium-driven hohlraum experiments. Note, however, that even though the energy absorption of these hohlraums was wall-dominated, the foam-filled hohlraums did become significantly hotter than the empty hohlraums (e.g. the bolometers observed a four-fold increase in signal for the filled hohlraums). Furthermore, the heated foam did prevent the outer gold cylinder from collapsing as it did for empty hohlraums. Thus, the foam did contribute favorably to the target performance.

Once again, much more analysis of this data is necessary before a full and consistent picture is available. This analysis will include post shot calibration of diagnostics, incorporation of proper diagnostic and film response functions in the simulation codes, and analysis with more accurate beam parameters obtained from beam characterization diagnostics. We can now say that we have made significant technical progress in target design, diagnostics, and experiments on PBFA II. The thermal source target experiments have demonstrated the ability of light ions to heat foam-filled hohlraums. The foam-filled hohlraums appear to have achieved their predicted performance including reaching an optically thin state and producing a near-Planckian spectrum. And finally, high quality ion-driven target data has been obtained PBFA II.

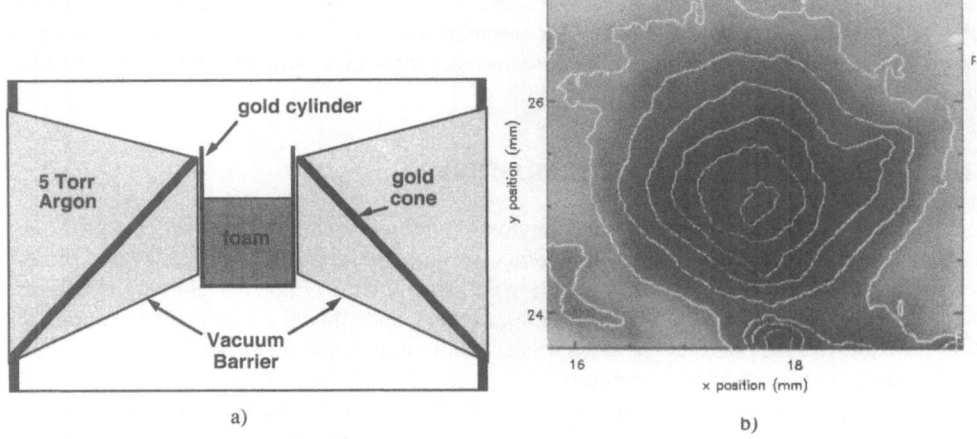

Figure 6. a) Cross sectional view of cylindrical hohlraum-like target and b) time-integrated x-ray image of the thermal emission from the central foam region of the target in the coordinate system of the film plane.

FUTURE BEAM CAPABILITIES AND TARGET EXPERIMENT PLANS

Now that the first significant target experiments aimed at developing a path to ignition for light ion beams have been performed on PBFA II we are planning the future. Table 3 is similar to Table 2 with the addition of a few entries indicating our anticipated capabilities for target experiments with lithium beams in the coming year. The superiority of present and anticipated lithium beams for studying target physics is clear. The specific deposition power of a 9 MeV lithium beam is 10 times that of a 5 MeV proton beam for the same focal intensity. This table indicates that planned lithium-driven hohlraum experiments can fully explore the radiation physics and hohlraum physics regimes, and achieve the conditions necessary to perform initial hydrodynamic experiments. It is interesting to note that present ignition target designs call for 24-30 MeV lithium ions at intensities of between 100 and 120 TW/cm^2. This translates to specific deposition powers of between 5000 and 10000 TW/g. We see, therefore, that a 10 TW/cm^2 lithium beam having a specific deposition rate of almost 4000 TW/g will be testing ion beam deposition and radiation conversion at near-ignition levels. Such experiments will give us an early indication of the validity of our deposition and radiation conversion models and should indicate whether non-linear processes in ion deposition or radiation conversion will be a problem for ignition targets.

Table 4. Present and future PBFA-II specific deposition rates into a cold CH foam.

Ion	Ion Energy (MeV)	Range (mg/cm²)	Intensity (TW/cm²)	Specific Deposition (TW/g)
Proton	5	32	5	157
Lithium	9	2.6	1	380
Lithium	9	2.6	5	1900
Lithium	9	2.6	10	3800

SUMMARY

The first complex ICF target experiments to be performed on an ion accelerator have been successfully fielded on PBFA II. These experiments have successfully obtained time-integrated, time resolved, imaging, and spectral diagnostic information from an extensive array of instruments. The target data has shown that we can efficiently couple ion beam energy to ignition-size spherical targets. The data has also shown that we can couple energy into a foam-filled hohlraum, produce an optically-thin foam, and generate an x-ray yield which preliminary analysis indicates is consistent with a Planckian spectrum.

ACKOWLEDGEMENTS

The target experiments described in this article were the result of a large collaborative effort. The target modelling was performed by T. Hussey, R. Olson, and R. Humphreys. The targets were fabricated by D. Derzon, P. Sawyer, F. McNamara, and J. Aubert at Sandia, by MST-7 at Los Alamos, and by B. Ramer, ICF Section-Rocky Flats. Target diagnostics were modeled, fielded, and analyzed by G. Chandler, M. Derzon, P. Rockett, J. Bailey, A. Carlson, J. Torres, J. Pantuso, J. Hunter, D.J. Johnson, C. Ruiz, A. Schmidlapp, R. Dukart, and A. Moats. Ion beam diagnostics were fielded by G. Rochau and D. Wenger. The PBFA-II accelerator is operated by a large crew within the PBFA-II Accelerator Experiments Division.

The lithium focusing experiments described in this article were performed by R. Stinnett and the PBFA-II operations crew. The ion movie camera data was analyzed by P. Mix and by D.J. Johnson.

Note that much of the data in this article is new and has only been treated here in an overview fashion. Much of this data will be the subject of further detailed analysis and publication in the near future.

REFERENCES

1. S. Humphries, Jr., J.J. Lee, and R.N. Sudan, J. Appl. Phys. **46**, 187 (1975).

2. D.J. Johnson, G. W. Kuswa, A.V. Farnsworth, Jr., J.P. Quintenz, E.J.T. Burns, and S. Humphries, Phys. Rev. Lett. **42**, 610 (1979).

3. S.J. Stephanakis, D. Mosher, G. Cooperstein, J.R. Boller, J. Golden, and S.A. Goldstein, Phys. Rev. Lett. **37**, 1543 (1976).

4. D.J. Johnson, et al. Proc. 7th IEEE Pulsed Power Conference, Monterey, CA, June 11-14, 1989.

5. Final Report of the National Academy of Sciences, "Review of the Department of Energy's Inertial Confinement Fusion Program, (1990).

6. T.A. Mehlhorn, J. Appl. Phys. **54**, 6522 (1980).

7. G.L. Payne and J.D. Perez, Phys. Rev. A **21**, 976 (1980).

8. J.A. Swegle, Comments Plasma Phys. Controlled Fusion **7**, 141 (1982).

9. E. Nardi, E. Peleg, and Z. Zinamon, Phys. Fluids **21**, 574 (1978).

10. H.A. Bethe, Ann. Phys. (Leipzig) **5**, 325 (1930).

11. J. Linhard and M. Scharff, Phys. Rev. **124**, 128 (1964).

12. C. Deutsch, Ann. Phys. Fr. **11**, 1 (1986).

13. T.A. Mehlhorn, J.M. Peek, E.J. McGuire, J.N. Olsen, and F.C. Young, J. Physique Colloq. **44** C8-39 (1983).

14. F. C. Young, D. Mosher, S. J. Stephanakis, S.A. Goldstein, and T.A. Mehlhorn, Phys. Rev. Lett. **49**, 549 (1982).

15. J.N. Olsen, T.A. Mehlhorn, J. Maenchen, and D.J. Johnson, J. Appl. Phys **58**, 2958 (1985).

16. J.K. Rice and T.W. Hussey, Sandia National Laboratories, private communication of the results of calculations for light ion beams based on radiation conversion model discussed in Reference 23.

17. E. Nardi and Z. Zinamon, Phys. Rev. Lett. **49**, 1251 (1982).

18. Th. Peter and J. Meyer-ter-Vehn, Phys. Rev. A **43**, 1998 (1991).

19. D.H.H. Hoffmann, K. Weyrich, H. Wahl, D. Gardes, R. Bimbot, C Fleurier, Phys. Rev. A **42**,2313 (1990).

20. R. O. Bangerter, "Proceedings of Heavy Ion Fusion Workshop", ANL-79-41, 415 (1979).

21. S. A. Slutz, J. Appl. Phys. **53**, 3957 (1982).

22. J. P. Quintenz, et al. IAEA Technical Committee Meeting on Drivers for Inertial Confinement Fusion, Osaka Japan, April 15-19 1991, to be published in IAEA-TECDOC series.

23. Figure from, J.D. Lindl, R.O. Bangerter, J.W-K Mark, and Yu-Li Pan, Heavy Ion Inertial Fusion AIP Conference Proceedings, **152**, 89 (1986).

24. R. Bock, "Status and Perspectives of Heavy Ion Inertial Fusion", GSI-91-13 (1991).

TIME-RESOLVED SPECTROSCOPIC MEASUREMENT OF ABLATION PLASMA PRODUCED BY INTENSE PULSED ION BEAM

Kiyoshi Yatsui, Norihide Yumino and Katsumi Masugata

Laboratory of Beam Technology, and Department of
Electrical Engineering, Faculty of Engineering, Nagaoka
University of Technology, Nagaoka, Niigata 940-21, Japan

ABSTRACT

Characteristics of ablation plasma of target produced by
the irradiation of an intense pulsed light-ion beam on the
solid targets are studied in detail by time-resolved
spectroscopic measurement. From the measurement of Stark
broadening, we have estimated the electron density to be
$(1.2 \sim 1.5) \times 10^{18}$ cm^{-3} at 10 mm downstream from the
aluminium target irradiated by the ion beam with diode
voltage of 720 and 880 kV, and ion-current density of 2 and
3.5 kA/cm^2, beam power density of 1.5 and 3 GW/cm^2,
respectively. Assuming local thermodynamic equilibrium, we
have evaluated the electron temperature to be $(2.1 \sim 2.7)$ eV
at 10 mm downstream from the target from the measurement of
light intensity ratio of successive ionization stages. The
expansion velocity of the ablation plasma has been found to
be $(2.2 \sim 2.8)$ cm/μs.

1. INTRODUCTION

An intense, pulsed, light-ion beam (LIB) can be used to
produce high-density "ablation" plasma of the target when
the LIB is irradiated onto the surface of the solid targets.

Such the plasma has been known to be applied to various material sciences.[1~11]   Since the pulse width is short compared to the thermal conduction time, the phenomenon can be considered as "adiabatic".   In this sense, the above process can be considered as the energy transfer from the kinetic energy of LIB to the thermal energy.   Using the above plasma, we have proposed "Ion-Beam Evaporation" (IBE) and succeeded in quick preparation of various thin films,[4, 5, 7~11] where the films are prepared on the substrate placed nearby the target.   In fact, various thin films have been prepared by IBE, by which we have succeeded in the preparation such as ZnS, ZnS:Mn, B, C, ITO, YBaCuO, apatite, ZrO_2, and h-BN.   In addition, the ablation plasma produces very high pressure of more than 10^5 bar.   A strong shock wave is also produced, by which surface modification takes place such as the drastic change of the resistance against wear.[3]   In the practical applications to various material sciences, it is very important to understand clearly the basic characteristics and properties of the ablation plasma.

To diagnose the ablation plasma, a spectroscopic measurement is known to be very effective.[12]   Previously, we have successfully developed a spectroscopic diagnostic system of anode plasma by the combination of a spectrometer with a streak camera,[13] where spectroscopic data are acquitted and transacted continuously both in time and wavelength.   This system can be possible for various data processings where streak images are stored in a micro-computer with the digital data.   These images are taken by a SIT camera, and then led to a frame memory by A/D converter.   By this technique, we have succeeded in highly time-resolved measurement of the spectra as well as the real time processing.

Extending the same technique to the ablation plasma of the target, we would like to present time-resolved spectroscopic measurement on the density, temperature and expansion velocity of the ablation plasma produced by LIB.

2. EXPERIMENTAL SETUP

Figure 1 shows the outline of the experimental setup. The experiment was carried out in the pulse-power generator, "ETIGO-I", at Nagaoka University of Technology.[1] The output parameter of the generator is as follows: voltage = 1.2 MV, current = 240 kA, power = 0.3 TW, pulse width = 50 ns, and energy = 14.4 kJ. The ion diode used is a magnetically insulated diode (MID), where the anode-cathode gap is 10 mm wide. On the surface of the aluminium anode a polyethylene sheet (160 mm x 160 mm x 1.5 mm thick) has been attached as the flashboard. The ion beam extracted is focused geometrically at 160 mm downstream from MID. The target, made of aluminium, is placed at the focusing point by an angle of 45° with respect to LIB.

The light signal from the ablation plasma is introduced into the monochrometer (focal length = 25 cm, grating = 1800/mm) through a lens of f400, two mirrors, and a lens of f200. The magnification of this optical system is approximately 2. The incident slit width of the monochrometer and the streak tube are 50 μm and 150 μm, respectively. The spatial resolution of this system is 300 μm x 100 μm.

Figure 2 shows the outline of the time-resolved system. The spectra of the ablation plasma from the monochrometer are time-resolved by a streak camera (temporal disperser). The streak images are taken by a SIT camera, and A/D converted, which is recorded into a frame memory (8 bit, 256 channel x 256 channel, 16 frame) as digital data. These data of images memorized are analyzed by a temporal analyzer, giving a profile of spectra both in time and wavelength. The wavelength covered is ~ 8 nm/frame. The resolution of the wavelength and the time of this system is ~ 0.2 nm and ~ 50 ns, respectively.

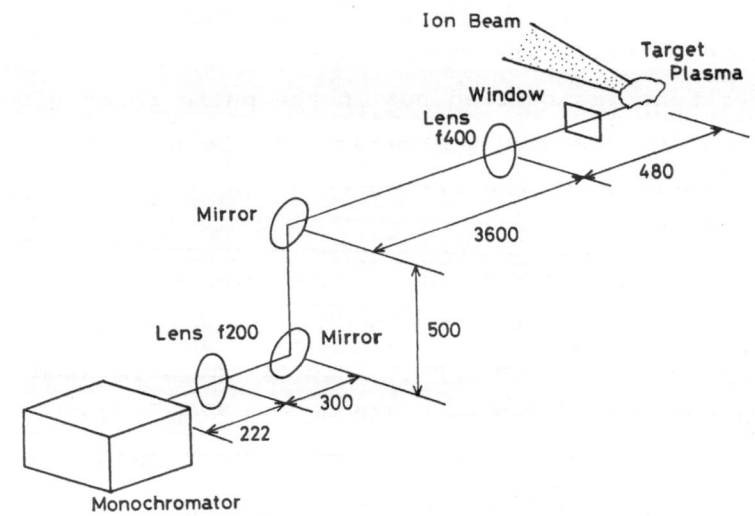

Fig. 1   Outline of the experimental arrangement.

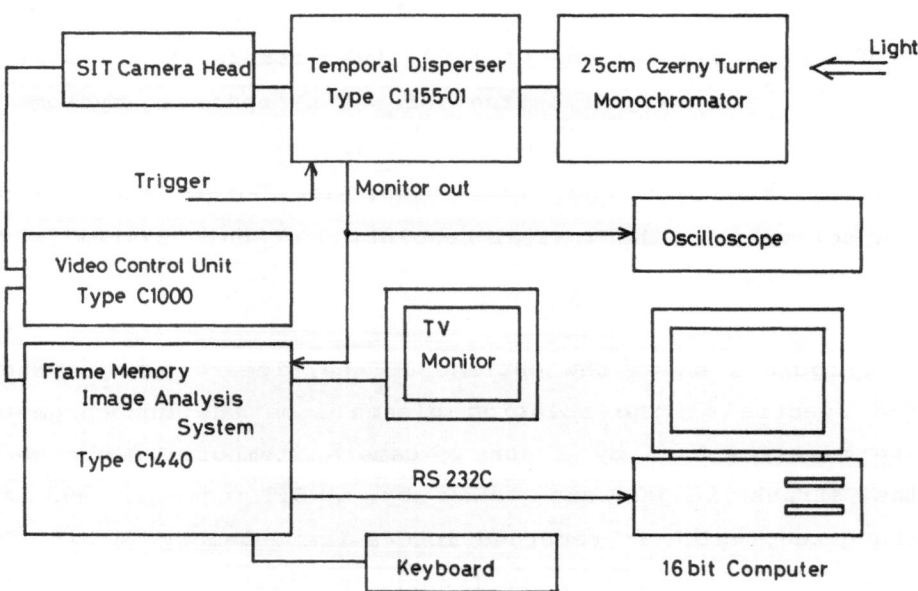

Fig. 2   Schematic diagram of time-resolved
spectroscopic system.

3. EXPERIMENTAL RESULTS

3.1. TYPICAL WAVEFORMS

Figure 3 shows typical waveforms of the diode voltage inductively corrected (V_d), the diode current (I_d), and the ion-current density (J_i) measured by a biased-ion collector, where V_{ch} (charging voltage of Marx generator) = 35 kV. From Fig. 3, we see

$V_d \sim$ 880 kV, $I_d \sim$ 30 kA, P_d (beam power density) = $V_d I_d \sim$ 3 GW/cm², τ (pulse width) $\sim$ 50 ns, $J_i \sim$ 3.5 kA/cm².

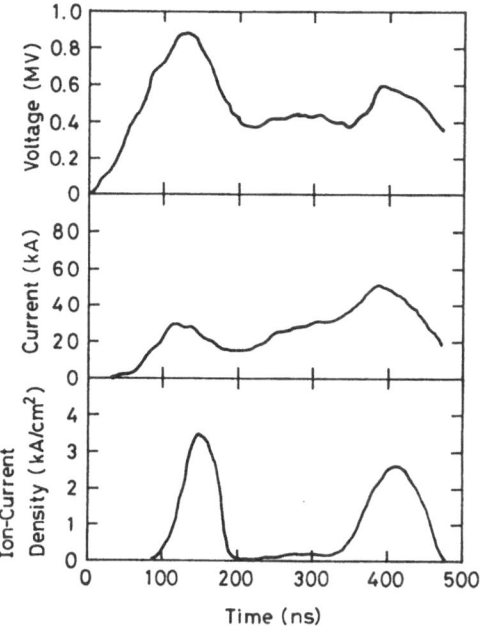

Fig. 3   Typical waveforms of diode voltage (top), diode current (middle) and ion-current density (bottom).

Similarly, at V_{ch} = 30 kV, we have found

$V_d \sim$ 720 kV, $I_d \sim$ 22 kA, $P_d \sim$ 1.5 GW/cm², $J_i \sim$ 2.0 A/cm².

The experiments were carried out under the above two conditions.

3.2. ELECTRON TEMPERATURE

Comparing relative intensities of two lines of AℓII,

$$3p^2-3s4p \quad (466.31 \text{ nm}),$$
$$3s4p-3s5d \quad (559.32 \text{ nm}),$$

we here evaluate the electron temperature with the assumption of local thermodynamic equilibrium (LTE).[12,14]

Figure 4 presents typical traces of the above two lines, where $P_d \sim 3$ GW/cm^2 and z (distance from the target) = 20 mm. The pulse width of the light emission from the plasma is seen to be $\sim$ (300 $\sim$ 500) ns. The sensitivity of the whole system that includes the monochrometer is 2.2 for λ = 446.31 nm if we put it one for λ = 559.32 nm. Taking the above sensitivity into account, we have evaluated the electron temperature, which is summarized in Table I. The electron temperature at z = 10 mm is typically estimated as

$$T_e \sim (2.1 \sim 2.7) \text{ eV}.$$

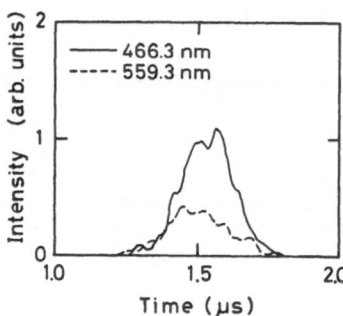

Fig. 4   Time variation of two lines of AℓII (466.31 and 559.32 nm).   z = 20 mm, $P_d \sim 3$ GW/cm^2, shot #5604 and #5605.

3.3. ELECTRON DENSITY

It is well known that the line spectrum of singly ionized ion of aluminium due to the transition of $3p^2-3s4p$ (466.31 nm), and that the broadening is mainly due to the

Table I. Electron temperature of the ablation plasma
estimated by light intensity ratio of successive
ionization stages under the assumption of LTE.

		T_e (eV)	
V_d (kV)	P_d (GW/cm²)	z = 10 mm	z = 20 mm
720	1.5	2.1	1.4
880	3	2.7	1.7

Stark effect.[12]   Since several coefficients due to this
transition have been studied by many authors,[15] we here
diagnose its broadening to evaluate the electron density.

The line broadening ($\Delta\lambda$) due to Stark effect is given
by[16]

$$\Delta\lambda = 2W(n_e/10^{16})$$
$$+ 3.5A(n_e/10^{16})^{1/4}(1 - 1.2n_D^{-1/3})W(n_e/10^{16}). \quad (1)$$

Here, W is the impact width parameter, A the ion-broadening
parameter, n_e the electron density (cm⁻³), and n_D the number
of particles within the Debye sphere.

For AℓII, we know A is quite small, and the second term
in eq. (1) can be neglected.   Furthermore, at T_e = 2 eV, we
estimate

$$W \sim 4.7 \times 10^{-2} \text{ nm.}$$

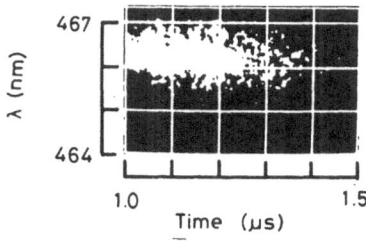

Fig. 5   Streak photograph of AℓII (466.31 nm).
z = 10 mm, $P_d \sim 3$ GW/cm², shot #5508.

Figure 5 shows a typical streak photograph of AℓII
(466.31 nm).   Figure 6 shows the time variation of the
streak photograph of AℓII as shown in Fig. 5.   From Fig. 6,
we see the maximum broadening of the line spectrum,

$$1.3 \sim 1.35 \text{ nm.}$$

Fig. 6   Three-dimensional
    plot of $A\ell$ II (466.31 nm)
    shown in Fig. 5.
    Shot #5508.

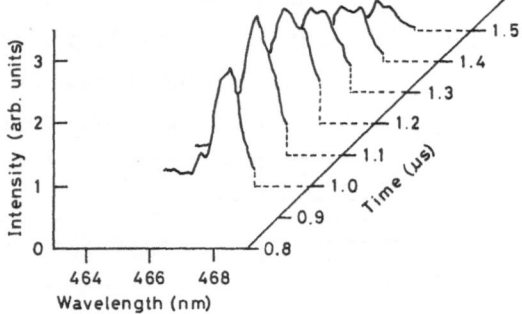

Figure 7 shows the time variation of the electron density, which was evaluated from Fig. 6.   In Fig. 7, t = 0 in the abscissa means the time when the diode voltage starts to rise.   From Fig. 7, we see that the pulse width is about 500 ns, and that the density has the peak of

$$n_e \text{ (peak)} \sim 1.6 \times 10^{18} \text{ cm}^{-3}.$$

The result of n_e thus estimated is summarized at Table II.

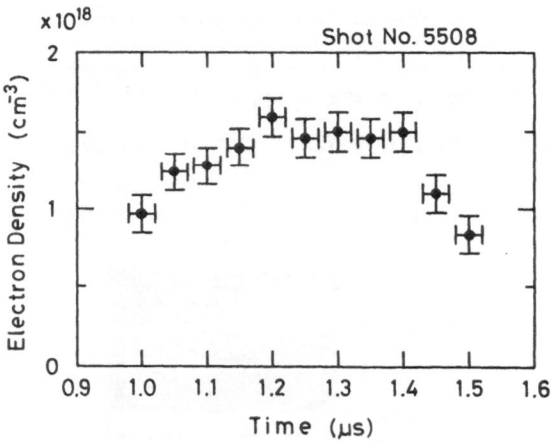

Fig. 7 Time variation of electron density.   Shot #5508.

Table II. Electron density of the ablation plasma
at various positions from the target.

		n_e $(\times 10^{18}$ cm$^{-3})$		
V_d (kV)	P_d (GW/cm^2)	z = 10 mm	z = 20 mm	z = 30 mm
720	1.5	1.2	0.6	-------
880	3	1.5	0.6	0.17

3.4. EXPANSION VELOCITY

Figure 8 shows the time variation of line spectrum of
AℓII of 466.31 nm at P_d ~ 3 GW/cm² at various position from
the target.   From Fig. 8, we see the time difference (Δt)
separated by 10 mm is

$$\Delta t ~ 375 \text{ ns.}$$

We estimate the expansion velocity of the ablation plasma
(v_a) to be

$$v_a ~ 2.8 \text{ cm/}\mu\text{s.}$$

For P_d ~ 1.5 GW/cm², we have also calculated

$$v_a ~ 2.2 \text{ cm/}\mu\text{s.}$$

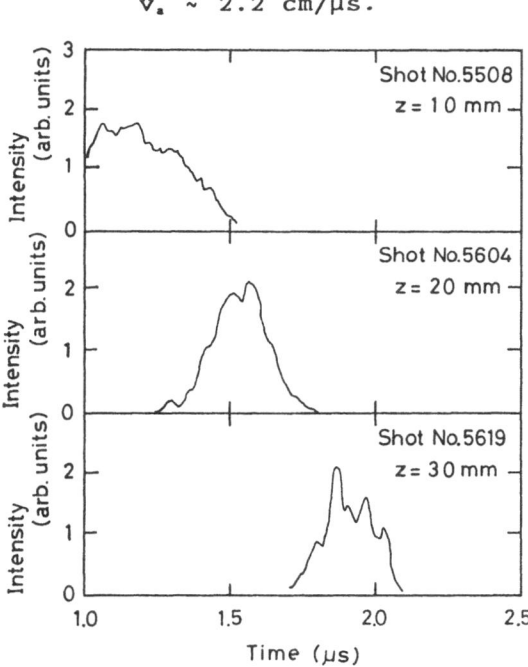

Fig. 8   Time variation of line spectrum of AℓII (466.31 nm)
at z = 10 mm (top; shot #5508), 20 mm (middle; shot
#5604)   and   30 mm (bottom; shot #5619)   at P_d ~ 3
GW/cm² .

4. CONCLUDING REMARKS

By the time-resolvable spectroscopic technique, we have
measured the ablation plasma produced by the irradiation of

LIB on the aluminium target using the Nagaoka "ETIGO-I" pulse-power generator, typically operated at V_d (diode voltage) ~ 720 and 880 kV, J_i (ion current density at the geometrically focusing point) ~ 2 and 3.5 kA/cm², P_d (beam power density) ~ 1.5 and 3 GW/cm², and τ (pulse width) ~ 50 ns, respectively. We have observed, at z = 10 mm downstream from the target, n_e (electron density) ~ (1.2 ~ 1.5) x 10^{18} cm^{-3} from the Stark broadening, T_e (electron temperature) ~ (2.1 ~ 2.7) eV from the light intensity ratio using LTE model, and v_a = (2.2 ~ 2.8) cm/µs from the time variation of line spectra.

Using these basic characteristics and properties, we would like to clarify the quality of the thin films and the deposition rate prepared by IBE, the mechanism of surface modification after being irradiated by LIB, and so on in the connection with the above plasma parameters.

ACKNOWLEDGEMENTS

This work was partly supported from the Grant-in-Aid from the Ministry of Education, Science, and Culture of Japan. The authors would like to express their sincere thanks to Dr. Y. Shimotori, now at Nippon Seiki Co. Ltd., for his useful support and advices in the measurements.

REFERENCES

1) K. Yatsui et al., Laser & Particle Beams 3, 119 (1985).
2) K. Kamata et al., Laser & Particle Beams 5, 495 (1987).
3) A. D. Pogrebnjak et al., Phys. Lett. A129, 259 (1987).
4) Y. Shimotori et al., J. Appl. Phys., 63, 968 (1988).
5) K. Yatsui et al., Proc. Int'l Joint Symp. on Newer Trends of Surface Modification, Welding Res. Inst., Osaka Univ., Japan, 21 (1988).

6) Y. Nakagawa et al., Proc. 7th Int'l Conf. on High-Power Particle Beams, Karlsruhe, Germany, II, 1475 (1988).

7) Y. Shimotori et al., Jpn. J. Appl. Phys., 28, 468 (1989).

8) K. Yatsui et al., Proc. 20th Symp. on Ion Implantation and Submicron Fabrication, Inst. Phys. and Chem., Japan, 21 (1989).

9) K. Yatsui, Laser & Particle Beams 7, 733 (1989).

10) T. Takaai et al., Proc. Int'l Conf. on Surface Engineering -Current Trends and Future Prospects-, Toronto (International Scientific Committee, Canada) (1990).

11) K. Yatsui et al., Proc. 3rd Int'l Conf. on Intense Ion Beam Interaction with Matter, Albuquerque, USA (in press) (1990).

12) H. R. Griem: "Plasma Spectroscopy" (McGraw Hill, N Y, 1964).

13) Y. Kawano et al., Laser & Particle Beams 7, 277 (1989).

14) W. L. Wiese, M. W. Smith and B. M. Glenon: "Atomic Transition Probabilities", Vol. 2, Sodium through Calcium, NBS 22, (U. S. of Commerce, 1969).

15) A. W. Allen et al., Phys. Rev., A11, 477 (1975).

16) J. T. Knudtson et al., J. Appl. Phys., 61, 1771 (1987).

FOCUSING AND PROPAGATION OF PROTON BEAM

Keishiro Niu

Teikyo University of Technology
Uruido, Ichihara, Chiba 290-01, Japan

ABSTRACT

It is the aim of this paper to design a fusion power plant whose electric output power is 1GW, and to find a way for breaking through fusion technically and energy-economically. Proton beams, whose total energy is 12MJ, pulse width is 30ns and beam number is 6, are chosen here as the energy driver. Because of low quality of proton beams, the target should be indirect driven and its radius should be large. The target with the radius of 8.7mm is the spherical cryogenic hollow one, which has double shells and five layers. The reactor has double solid walls. The inner wall rotates around the axis to induce a centrifugal acceleration. Flibe as the coolant protects the solid walls from damage and breezs tritium. The key technology of this power plant is for beam focusing and beam propagation. To suppress the beam divergence by the electrostatic force due to unneutralized proton charge, the simultaneous electron beam launching is proposed. When the excess electron beam current is -50kA, the induced magnetic field in the azimuthal direction confines the beam in the radius of 5mm, provided that the beam path is covered by the metal guide whose radius is 1cm.

INTRODUCTION

PBFA-II in Sandia National Laboratory has been started to be used for beam-target interaction experiment.[1,2] Ion beam has a preferable stopping range in the target. It is expected that fuel compression and fuel heating will be done efficiently by using light ion beam as energy driver in the near future. Light ion beam has a high energy conversion rate from electric energy to beam kinetic energy. Especially in the case of proton beam, the conversion rate with more than 30% is achieved. From the point of view of ion source, proton is easily supplied. If the ion source includes other ion species beside proton, proton is extracted from the ion source first than other species. Always proton hits target first. No preheat of target occurs by other ion species.

Being taken the facts described above into consideration, proton beam is chosen here as energy driver. The optimum particle energy of proton for ICF is 4MeV.[3] In order to launch a large amount of beam energy such as 12MJ to the target, the beam current becomes strong, and the local divergence angle rmains not small. Thus the beam focusing and beam propagation is a most important issues for proton beam. In this paper, the simultaneous electron beam launching with the proton beam launching is proposed. The electron beam carries a excess current of 50kA. This current induced the azimuthal magnetic field, which confines the beam in a small radius. In order to strengthen the beam confinement effect by the magnetic field, beam path is proposed to be covered by a metal guide.

Figure 1 shows the power plant schematically. There are 36 modules of power supply systems numbered by 1 in Fig.1. Each system stores 1.3MJ of electric energy. Since the plant is operated with the frequency of one Hertz, the input electric power to the plant is 46.8MW. As the electric output power from the plant is expected to be 1GW, the net power amplification

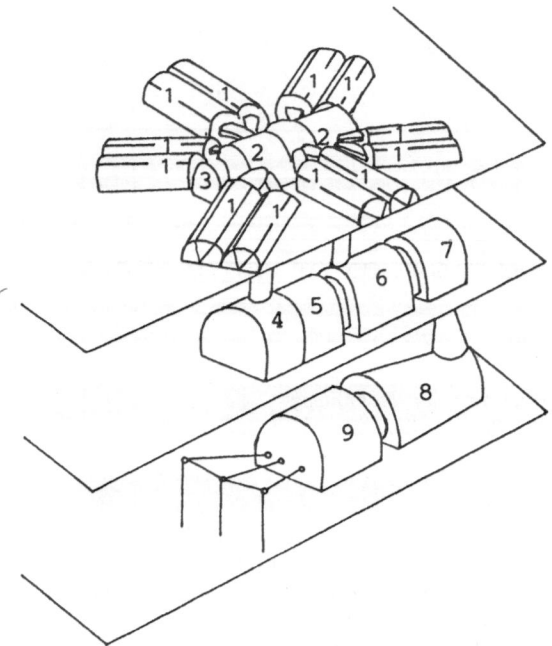

Fig. 1. Overview of power plant.

factor of the plant is about 20. Six modules of power supply system are combined to extract one proton beam. The target at the center of reactor cavity numbered by 2 in Fig.1 is irradiated by six proton beams with the total beam energy of 12MJ. The target releases the fusion thermal energy of 3GJ. The target cavity consists of double solid walls. An electric motor numbered by 3 rotates the inner solid wall of reactor cavity. By the action of centrifuge due to rotation, the coolant, flibe, with the thickness of 50cm flows along the inner solid wall of reactor cavity. This molten salt, flibe, absorbs the fusion thermal energy, especially the neutron kinetic energy, and protects the solid wall from damage. From the reactor cavity 2, the coolant, flibe, and the reactor gas, argon, flow down to chamber 4, where unburned deuterium in the argon gas is separated. In the next chamber 5, unburned tritium and breezed tritium are separated from argon gas and flibe, respectively. From the chamber 6, flibe and argon gas are sent back to the reactor cavity, after heat exchange is carried out from flibe to another molten salt, $NaF-BF_3$. The heat exchanger from $NaF-BF_3$ to water steam is in the chamber 7. A steam turbine and an electric power generator are located in chambers 8 and 9.

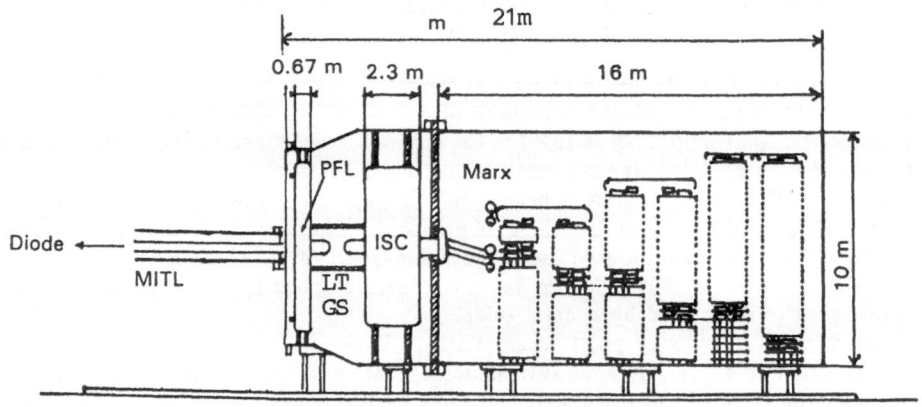

Fig. 2. schematic diagram of power supply system.

POWER SUPPLY SYSTEM

A schematic diagram of power supply system is shown in Fig. 2. The power supply system consists of Marx generator, intermediate storage capacitance, laser trigger gap switch and pulse forming line. The pulse power is sent from pulse forming line to diode through magnetically insulated transmission line. The parameters of Marx generator, intermediate storage capacitor and pulse forming line are summarized in Table 1.

Table 1. Parameters of power supply system.

Marx Generator	36 modules
Charging Voltage	200kV
Capacitance of a Bank	1.7μ F
Number of Capacitor Banks	40
Stored Energy	1.3MJ
Output Voltage	8MV
Cylindrical Intermediate Storage Capacitor	36 modules
Insulator	Water
Inner (Anode) Radius	3m
Outer (Cathode) Radius	4m
Length	2.3m
Charging Time	103ns
Pulse Forming Line	36 modules
Input Voltage	8MV
Output Voltage	4MV
Length	0.67m
Pulse Width	30ns

The total stored energy in the Marx generators of 36 modules is 46.8MJ. The total energy of six proton beams is expected to be 12MJ. The energy conversion rate to beam kinetic energy is 26%. If the ion source is made of water, for example, 1/3 of beam energy is conveyed by oxygen and is ineffective. The total stored energy in the Marx generators must increase to $46.8 \times 2/3$ MJ=70.2 MJ. The proton beams with 12MJ should have high quality, that is, the incident angles at the target surface is within 30 degrees and the spread of particle energies is less than 10%.[4] The proton particles which exceeds the limits described above are inefficient for fuel implosion and are excluded from 12MJ. If the argument is confined in the plasma part apart from reactor, the main facilities for proton beam fusion is capacitor banks. The cost to construct these power supply systems is less by one or two orders of magnitude in comparison with other fusion machines such as Tokamak or glass laser. When the power plant of proton beam fusion is operated with the frequency of 1 Hertz, the plant must have a quick charging system. However the cost to construct increases only twice.

BEAM FOCUSING AND BEAM PROPAGATION

The side view of the diode to extract proton beam and triode to extract electron beam is shown in Fig. 3. Because the proton beam propagates with the high speed of 1/10 of light speed, proton charges at the leading part of the beam is not neutralized by electrons in the background plasma, and induces a strong electrostatic field, which causes the beam divergence during the propagation.[5,6] In order to delete this electrostatic field, it is proposed to launch simultaneously the electron beam from the triode with proton beam as is shown in Fig. 3. The parameters related to diode and triode are given in Table 2.

Since the electron mass is much smaller than proton mass, the stored energy in banks to extract electron current of -16.65MA is order of 5kJ only. At the center axis, the diode has the magnetic coil, which induces the radial magnetic field of $B_r = 2.77 \times 10^{-4}$T. By the Lorentz force due to this field, the proton particle rotates around the axis with the average velocity $v_\theta = 2.77 \times 10^4$m/s during propagation. This proton rotation induces the axial magnetic field B_z, which stabilizes the beam propagation.

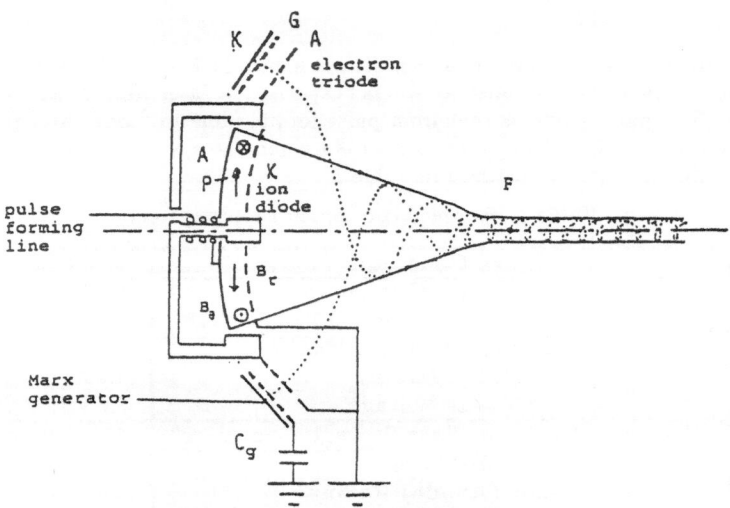

Fig. 3. The side view of diode and triode.

To insulate the electron current in the diode, this week B_r only is not enough. Figure 4 shows the front view of diode. The too large anode surface is divided into moderate size of six pieces. To each anode piece, one cathode ring is set. The pulse power is fed from one power supply system to each couple of an anode piece and a cathode ring. Between the connecting

Table 2. Parameters related to diode and triode.

Bank for Anode of Triode	6 modules
Output Voltage	-6.2kV
Stored Energy	3.93×10^2J
Bank for Grid of Triode	6 modules
Capacitance	354pF
Charging Voltage	-168kV
Stored Energy	5kJ
Diode	6 modules
Anode Inner Radius	r_i=1cm
Anode Outer Radius	r_o=32.5cm
Anode Area	$S_A = 3.32 \times 10^3$cm^2
AK Gap Distance	d_{AK}=9.51mm
Azimuthal Magnetic Field	B_θ=4.55T
Radial Magnetic Field	$B_r = 4.54 \times 10^{-4}$T
Anode Voltage	V_A=4MV
Proton Current Density	$j_p = 5kA/cm^2$
Proton Total Current	$I_p = j_p S_A$=16.60MA
Triode	6 modules
Cathode Inner Radius	r_{in}=35cm
Cathode Outer Radius	r_{out}=43.4cm
Cathode Area	$S_K = 2.08 \times 10^3$cm^2
KA Gap Distance	d_{KA}=1cm
Cathode Voltage	V_K=-6.20kV
KG Gap Distance	d_{KG}=2.0mm
Grid Voltage	V_G=-168kV
Electron Current Density	j_e=-8kA/cm^2
Electron Total Current	$I_e = j_e S_K$=-16.65MA

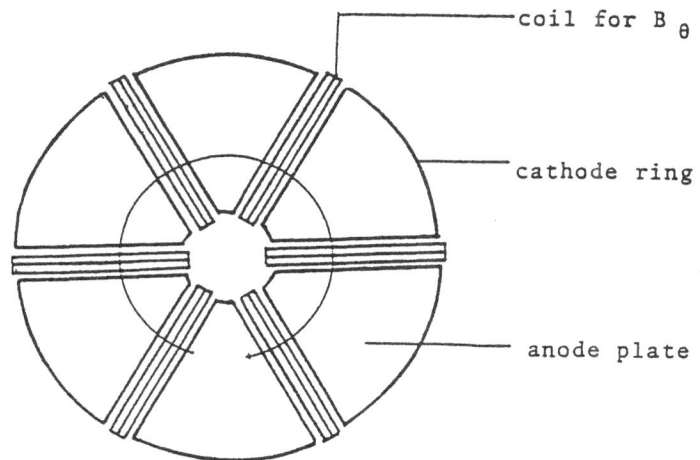

Fig. 4. Front view of diode.

two anode pieces, one magnetic coil is set and supplies the azimuthal magnetic fields B_θ, whose average intensity is 4.55T, in order to insulate the electron current in the diode.

The number density of electron beam during propagation is controlled by grid voltage in triode. When the charging voltage to grid of triode is lower than -168kV, more number of electrons from triode than proton from the diode is extracted. Excess electrons soon run away from the proton beam, forming the negative radial electrostatic field, which shrinks the radius of propagating proton beam. The electron propagation velocity is decided by the anode voltage of triode. When the anode voltage of triode is -6.20kV, the electron propagation velocity is a little larger than that of proton with particle energy of 4MeV. Thus the net current of the combined beam with proton and electron is -50kA. That is, the beam is electron-current-rich and induces a negative azimuthal magnetic field, which confine the beam in a small radius. As Fig. 5 shows, if the beam path is covered by a metal guide, the effect on beam confinement by the azimuthal magnetic field is strengthened. Thus the circumstance of beam propagation proposed here is similar the plasma in Tokamak. The net electron current of 50kA induces the azimuthal

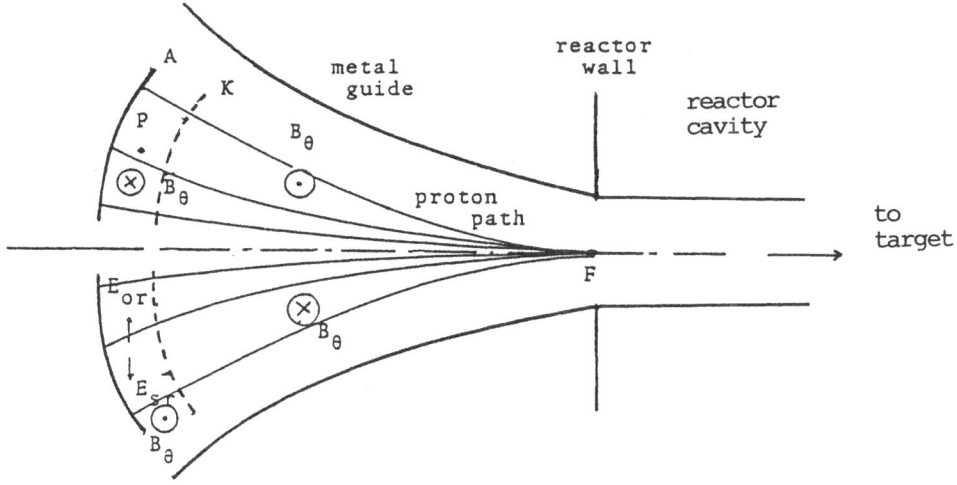

Fig. 5. Metal guide around the beam path.

magnetic field (poloidal field) whose intensity is 30T on the beam surface, whose radius is 5mm, surrounded by the metal guide, whose radius is 1cm. Rotation of proton particles around the propagation axis produces the axial magnetic field (troidal field), which stabilizes the beam propagation. Although the metal tube expands radially by the magnetic pressure, the tube remains unmoved during a short period of beam propagation (beam pulse width is 30ns and beam propagation time in reactor is 100ns).

SIMULATION OF BEAM PROPAGATION

Focusing and propagation of proton beam is simulated. The equation of motion for a super particle (proton of electron) is

$$M \frac{d^2 \mathbf{r}}{dt^2} = Q(\mathbf{E} + \mathbf{v} \times \mathbf{B}) \tag{1}$$

where M is the mass, Q the charge, $\mathbf{v}$ the velocity, $\mathbf{E}$ the electric field, $\mathbf{B}$ the magnetic field, $\mathbf{r}$ the space coordinate and t the time. The Maxwell equations for the electric field $\mathbf{E}$ and the magnetic field $\mathbf{B}$ is in the following integral forms.

$$\mathbf{A}(\mathbf{r}, t) = \frac{\mu}{4\pi} \int d^3 \mathbf{r}' \frac{\mathbf{j}(\mathbf{r}', t')}{|\mathbf{r} - \mathbf{r}'|} \tag{2}$$

$$\phi(\mathbf{r}, t) = \frac{1}{4\pi\epsilon} \int d^3 \mathbf{r}' \frac{\rho(\mathbf{r}', t')}{|\mathbf{r} - \mathbf{r}'|} \tag{3}$$

$$\mathbf{E}(\mathbf{r}, t) = -\frac{\partial \mathbf{A}(\mathbf{r}, t)}{\partial t} - \nabla\phi(\mathbf{r}, t) \tag{4}$$

$$\mathbf{B}(\mathbf{r}, t) = \nabla \times \mathbf{A}(\mathbf{r}, t) \tag{5}$$

where $\mathbf{A}$ is the vector potential, ϕ the scaler potential, $\mathbf{j}$ the current density, ρ the charge density, μ the magnetic permeability in vacuum and ϵ is the dielectric constant in vacuum. The simulation code is 3-dimensional in space. Here the axis-symmetry of the phenomena is assumed to be realized. Rotation motions of protons and electrons are neglected. Other large assumption is done for the charge, $\rho = 0$, in order to simplify the calculation. That is, the beam motion is charge free and no electrostatic field appears. Through each cross-section, the beam is assumed to have a net current of 50kA. Figure 6 shows the side view of the metal guide fro beam path. The radius of the narrow straight part is 1cm. Figure 7 shows the beam profile. The figure indicates that the beam radius once expands a little at $z = 75$cm at $t = 10.8$ns. However, at $t = 21.7$ns, the beam radius remains small by the action of the azimuthal magnetic field at $z = 85$cm.

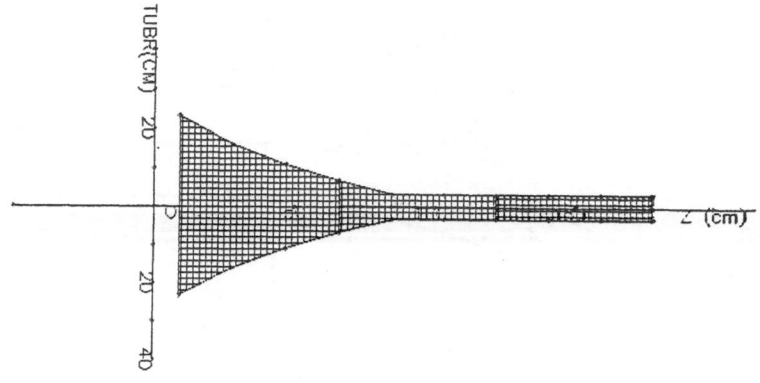

Fig. 6. Side view of metal guide.

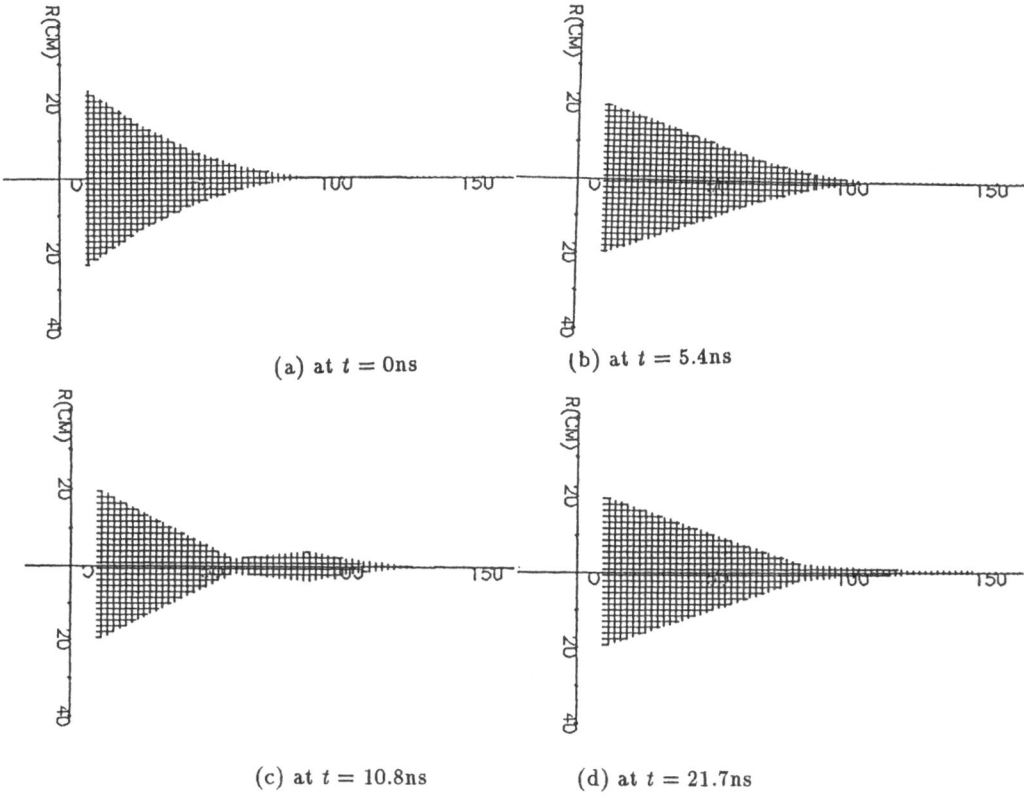

(a) at $t = 0$ns   (b) at $t = 5.4$ns

(c) at $t = 10.8$ns   (d) at $t = 21.7$ns

Fig. 7. Profile of propagating beam.

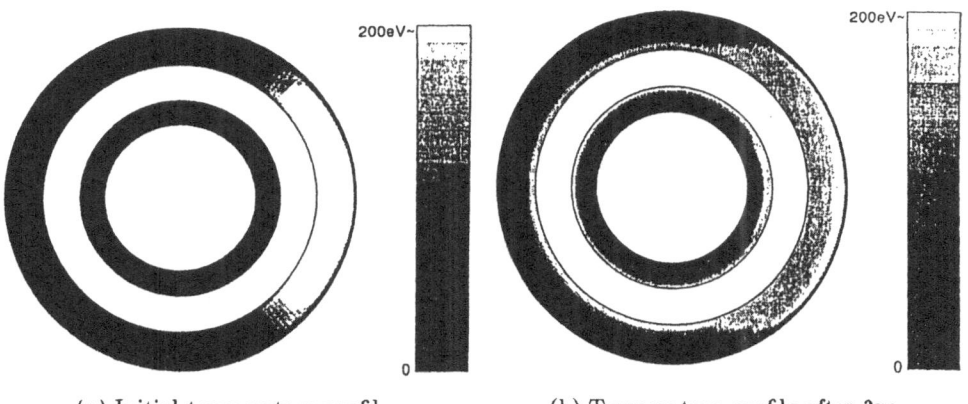

(a) Initial temperature profile.   (b) Temperature profile after 3ns.

Fig. 8. Temperature profiles in a indirect driven target

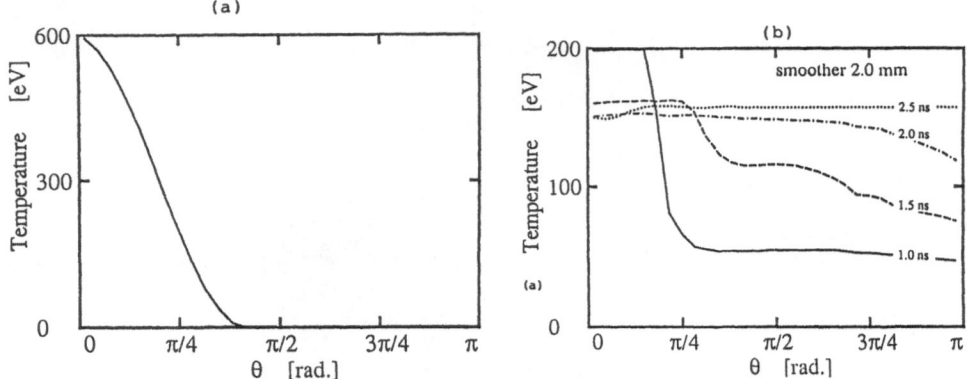

(a) Initial temperature on the inner surface of outer shell.
(b) temperatures on the outer surface of inner shells for several time steps.

Fig. 9. Temperatures along surfaces.

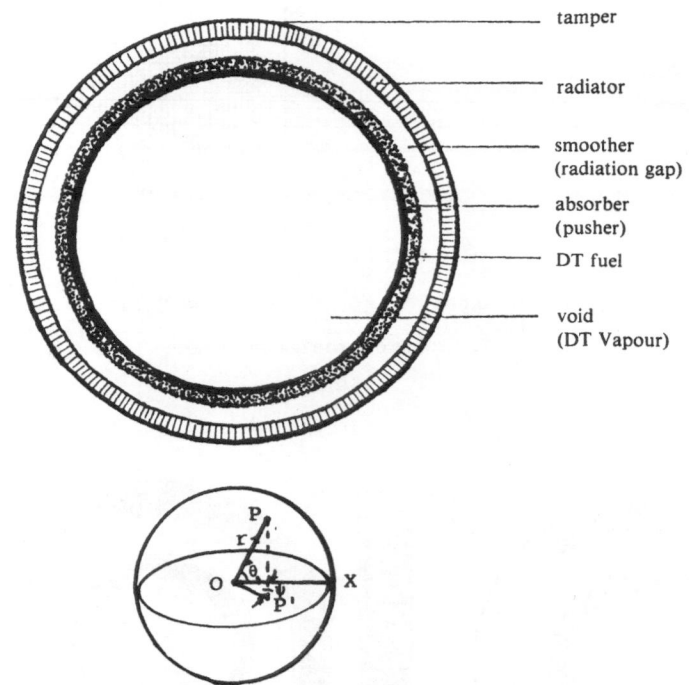

Fig. 10. Indirect driven target

INDIRECT DRIVEN TARGET

Since spherical symmetric beam irradiation on target surface is not expected in the case of proton beam, target should be indirect driven one and target should have a large radius.[7] Indirect driven target is effective for nonuniform beam irradiation. Recently, an analysis has been given for indirect driven reactor target[8], besides the small target.[9-11]

Figure 8 shows a result of 2-D simulation regarding radiative energy transport only without the fluid motion. The figure make clear that the vacuum radiation gap with the thickness of 2mm in the indirect driven target plays a role on radiation mixing. Figure 8 (a) gives the initial temperature profile in the target before the radiative energy transport starts. One proton beam impinges to the target from the right direction, and heats right part of outer shell. Figure 8 (b) shows the temperature profile after 3ns. It turns out that the temperature of outer surface of the inner shell increases uniformly through the radiation mixing in the radiation gap between the outer and inner shells. Figure 9 (a) indicates the initial temperature along the inner surface of outer shell. The temperatures along the outer surface of inner shell are drawn several time steps in (b). At the time step 2.5ns after the start of radiative transport, the temperature on the outer surface of inner shell becomes homogeneous. Although the target is irradiated by 6 proton beams in our case, there are energy unbalance among beams. Instead of direct driven target, indirect driven target is inevitably adopted in proton beam fusion. Our target is a cryogenic spherical hollow target, which has double shells and five layers.

The parameters of indirect target are tabulated in Table 3.

The total radius of the target is $r_t = 8.716$mm while the total thickness of shells is $\delta = 3.216$mm. The aspect ratio of this target is $A = 2.71$. The target is irradiated by the six proton

Table 3. Parameters of indirect driven target.

Indirect Driven Target	$r_t = 8.716$mm, $\delta = 3.216$mm, $A = 2.71$
Lead Tamper	
Density	$\rho_{Pb} = 11.3$g/cm^3
Thickness	$\delta_{Pb} = 23.4\mu$g
Mass	$M_{Pb} = 120$mg
Rate of Beam Energy Deposition	$C_{Pb} = 20\%$
Lead Radiator	
Density	$\rho_{ra} = 2.13$g/cm^3
Thickness	$\delta_{ra} = 690\mu$m
Mass	$M_{ra} = 1.49$g
Rate of Beam Energy Deposition	$C_{ra} = 80\%$
Temperature	$T_{ra} = 8$K$\rightarrow 1.7$KeV$\rightarrow 600$eV$\rightarrow 200$eV
Expansion Velocity	$v_{ra} = 1.61 \times 10^5$m/s
Inward radiation Intensity	$I_{ra} = 4.06 \times 10^{13}$W/cm^2
Radiation Gap (Smoother)	
Thickness	$\delta_{sm} = 2$mm
Radiation Temperature	$T_{sm} = 600$eV$\rightarrow 200$eV
Closing Time	$\tau_{sm} = \delta_{sm}/v_{ra} = 12.4$ns
Aluminum Absorber (Pusher)	
Density	$\rho_{Al} = 2.7$g/cm^3
Thickness	$\delta_{Al} = 217\mu$m
Mass	$M_{Al} = 2649$mg
Temperature	$T_{Al} = 8$K$\rightarrow 600$eV$\rightarrow 200$eV
Pressure	$p_{Al} = 7 \times 10^7$Pa$\rightarrow 10^{12}$Pa
Propagation Velocity of Hot Region	$v_{Al} = 5.42 \times 10^3$m/s
Transparent Time	$\tau_{Al} = \delta_{Al}/v_{Al} = 40$ns
Solid DT Fuel	
Density	$\rho_{DT} = 0.19$g/cm^3
Thickness	$\delta_{DT} = 286\mu$m
Mass	$M_{DT} = 23$mg
Inside Void	
Radius	$r_v = 5.5$mm
Saturated Vapor Pressure	$p_v = 7 \times 10^7$Pa

beams, whose beam energy is 12MJ, whose pulse width is 30ns and whose particle energy is 4MeV. Eighty percent of beam energy deposits in the radiator layer. The temperature of radiator layer increases from 8K to 1.61keV and the radiator layer emits soft x-rays. Twelve percent of x-ray energy escapes through the tamper layer to the outside. Meanwhile, the radiation layer expands in the radiation gap, and radiator temperature decreases to 600eV. The radiation gap is filled by soft x-rays, whose initial radiation temperature is 600eV. The outer surface of aluminum layer absorbs x-rays and the temperature goes up to 600eV. Soon the aluminum surface expands in the radiation gap. This expansion causes the decrease in aluminum, radiation and lead-radiator temperatures to 200eV. The inward radiation intensity across the radiation gap from lead radiator to aluminum absorber is $I_{ra} = 4.06 \times 10^{13}\text{W/cm}^2$ at $T = 200$eV. By the expansion of radiator and absorber, the radiation gap closes 12.4ns after the start of beam irradiation. When the thickness of radiation gap is small, the radiation mixing is not enough and the radiation temperature in the radiation gap is not uniform. But the thickness of radiation gap is too large, then energy loss due to expansion of radiator and absorber becomes large and causes the decrease in radiation temperature. There is the optimum value for the thickness of radiation gap. The thickness of 2mm is optimum. The outer surface of aluminum absorber becomes hot and transparent for x-rays. The propagation velocity of transparent region is $v_{Al} = 5.42 \times 10^3\text{m/s}$. Thus $\tau_{Al} = 40$ns after the start of beam irradiation, whole the aluminum layer becomes transparent for x-rays. This transparent time $\tau_{AL} = 40$ns is longer than the beam pulse width $\tau_b = 30$ns. The pressure of aluminum pusher (absorber) reaches 10^{12}Pa at $T_{Al} = 200$eV. This pusher pressure accelerates the solid fuel toward the target center.

At the 3ns after the acceleration, the implosion velocity of DT fuel arrives at $u = 3 \times 10^5$m/s. With this velocity, the Mach number of fuel is $M = 12.8$. The supersonic flow of fuel in the decreasing cross-section inside the void of target compresses the fuel adiabatically.[12] Finally the fuel has the density $\rho_f = 220\rho_s$ (ρ_s is the solid density), the temperature $T_f = 4$keV and the fusion parameter $< \rho_f R > = 35\text{kg/m}^2$. The burn fraction of fuel arrives at 35%. Thus a target yields fusion output thermal energy of 3GJ.

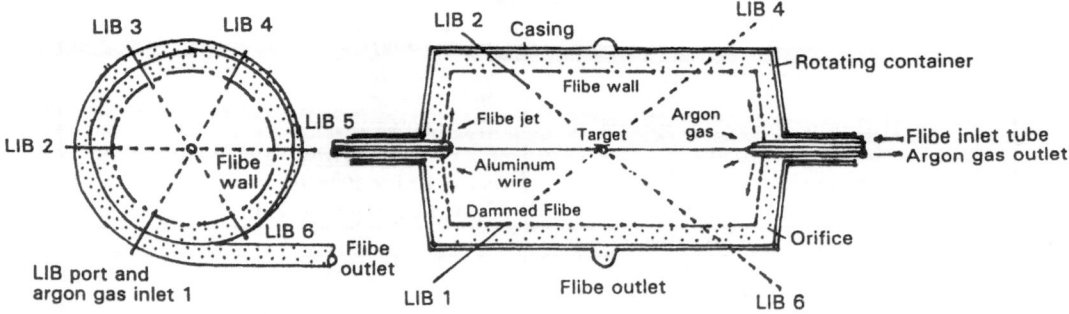

Fig. 11. Fusion reactor

FUSION REACTOR

The schematic diagram of proton beam fusion reactor is given in Fig. 11.[13] The cylindrical reactor cavity is surrounded by double solid walls. The outer wall is fixed on the ground while the inner wall rotates around the cylinder axis. If the radius of inner wall is 5m and the wall rotates one revolution per second, the acceleration of the wall surface due to the centrifuge is 20 times the acceleration due to the gravity. From the inlets at the cylinder axis, molten salt, flibe, flows into the reactor cavity and flows down along the inner solid wall by the action of centrifuge even under the ceiling part. The outer solid wall has six holes, through which proton beams are launched. The inner solid wall has six metal beam-guides, which are connected with the target located at the reactor center. For the repetitive frequency of power plant is one Hertz and the rotation of inner solid wall is one revolution per second, the synchronized six proton

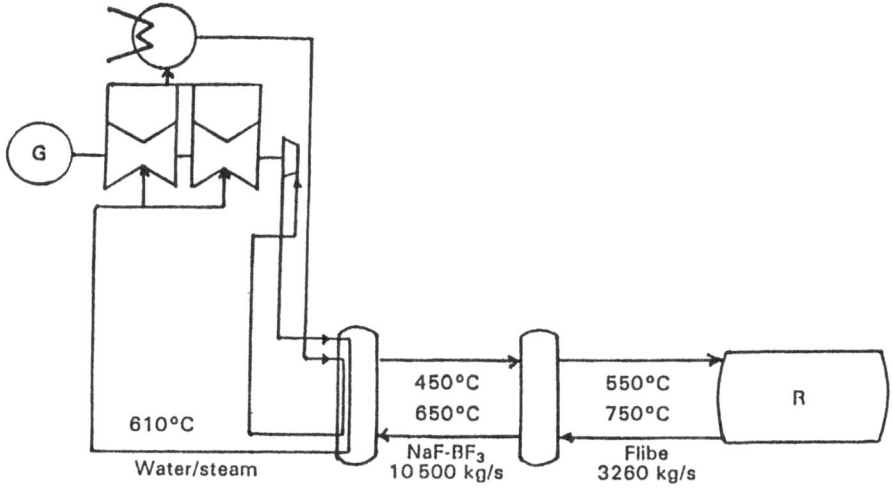

Fig. 12. Heat exchange system.

beams are launched when the holes on outer and inner walls overlap. The target releases the fusion thermal energy of 3GJ. The flibe layer whose thickness is 50cm absorbs the fusion energy including energetic neutrons and increases its temperature to 750°C. The heat exchange system of our fusion reactor is shown in Fig. 12. Flibe is chemically stable and handy as the coolant for proton beam fusion reactor. On the other hand, the operation temperature of flibe should be more than 500°C, in order to flow in the reactor with a high mobility. The generator G in Fig. 12 generates the electric power of 1GW.

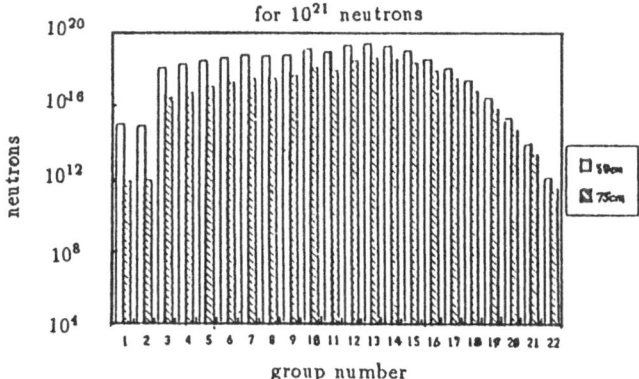

Fig. 13. Neutron spectra on inner surface of inner solid wall.

Figure 13 indicates the neutron spectra on the inner surface of inner solid wall after neutrons passes through the flibe layer. The target yields 10^{21} neutrons with 14MeV. The figure gives tow spectra for the two thicknesses of flibe layer, 50cm and 75cm. The abscissa shows the group number of neutron energy. The relation between group number and neutron energy is given in Table 4.

The number of neutrons impinging into flibe layer with the particle energy of 14MeV (group 1) is 10^{21}. The number of neutrons which leak out from the flibe layer to the inner solid wall with

Table 4. Group number and neutron energy

Group Number	energy range of neutron
1	12-14MeV
2	10-12MeV
3	8-10MeV
4	6-8MeV
5	5-6MeV
6	4-5MeV
7	3-4MeV
8	1.4-3MeV
9	900keV-1.4MeV
10	400-900keV
11	100-400keV
12	17-100keV
13	3-17keV
14	550eV-3keV
15	100-550eV
16	30-100eV
17	10-30eV
18	3-10eV
19	0.1-3eV
20	0.04-0.1eV
21	0.01-0.04eV
22	0.025-0.01eV

the particle energies more than 10keV is 10^{18}. The latter reduces by three order of magnitude from former. The life time of the inner solid wall with thickness of 1/4 inch is estimated to be 420 years with respect to the swelling by this leaking out neutrons.

When the thickness of flibe layer is 50cm, the volume of flibe in a reactor is $314m^3$. The total mass of this flibe is 600 tons, among which the lithium mass is 200 tons. The total lithium mass on the earth is estimated to be 1.8×10^{10} tons. By using this lithium, we can construct 9×10^7 proton beam fusion reactors. This is an enough number of fusion reactors for the human beings.

Figure 14 shows the tritium breezing ration in the flibe layer with the thickness of 50cm. The abscissa is the time after the neutron yield from the target. The real line shows the breezing

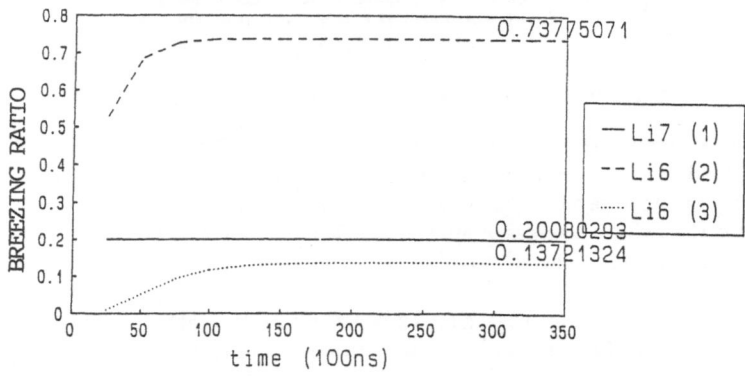

Fig. 14. Tritium breezing ration in flibe layer.

ratio by ^{7}Li in the flibe. The broken line shows the ratio by ^{6}Li collided with neutrons with medium energy, while the dotted line indicates the ratio by ^{6}Li collided with thermal neutrons. At $t = 350$ns, the sum of three ratios does not exceeds unity. Figure 14 shows that main part of tritium breezing comes from ^{6}Li. Flibe is the mixture of LiF and BeF$_2$ with the mol fraction of 2:1. The concentration of ^{6}Li in the natural lithium is only 2.5%. If ^{6}Li is little enriched in flibe, tritium breezing ratio exceeds unity in our proton beam fusion reactor.

SUMMARY

This paper wants to show the high possibility of proton beam as energy driver of inertial confinement fusion, although the beam quality in the case of proton beam is much lower than others at present. In this paper, a new method is proposed for focusing and propagation of intense proton beam. For long term research for fusion, proton beam seems to have a high ability to break through fusion technically and economically. Of course, there are many subject for proton beam fusion. The life times of capacitor banks and gap switches, for instance, for repetitive usage remains uncertain. In order to solve these problem, however, proton beam fusion does not included fatal defect.

REFERENCES

1. D. L. Cook, Bulletin of American Phys. Soc. **36**, 2480 (1991).
2. T. Mehlhorn et al., Abstract of 10th Int. Workshop on Laser Interaction (Monterey), 45 (1991)
3. K. Niu, H. Takeda and T. Aoki, Laser and Particle Beams **6**, 149 (1988).
4. M. Tamba, N. Nagata S. Kawata and K. Niu, Laser and Particle Beams **1**, 121 (1983).
5. T. Kaneda and K. Niu, Laser and Particle Beams **7**, 207 (1989).
6. K. Niu, P. Mulser and L. Drska, Laser and Particle Beams **9**, 149 (1991).
7. K. Niu, T. Aoki, T. Sasagawa and Y. Tanaka, Laser and Particle Beams **9**, 283 (1991).
8. N. A. Tahir and C. Deutsch, Abstract of 10th Int. Workshop on Laser Interaction (Monterey), 71 (1991).
9. M. Murakami and J. Meyer-ter-Vehn. Nuclear Fusion **31**, 1315 (1991).
10. M. Murakami and J. Meyer-ter-Vehn. Nuclear Fusion **31**, 1331 (1991).
11. N. A. Tahir and R. C. Arnols. Phys. Fluids B**3**, 1717 (1991).
12. K. Niu and T. Aoki, Fluid Dyn. Res. **4**, 195 (1988).
13. K. Niu and S Kawata, Fusion Tech. **11** 365 (1987).

FOCUSING AND TRANSPORTATION OF INTENSE LIGHT ION BEAM IN INERTIAL CONFINEMENT FUSION

Takeshi Kaneda[1] and Keishiro Niu[2]

[1]Social Systems Dept. 1
Mitsubishi Research Institute, Inc.
ARCO tower, 8-1, Shimomeguro 1-chome, Meguro-ku,
Tokyo 153, Japan

[2]Physics Laboratory
Teikyo University of Technology
2889-23 Ohtani, Uruido
Ichihara, Chiba 290-01, Japan

INTRODUCTION

The minimum requirement for the light ion beam (LIB) as the energy driver is that the instability must not grow up enough while the propagation time. The LIB will diverge by its own strong Coulomb repulsive force as it is, because the LIB is an ensemble of the charged particles. The complete charge neutralization must be required by all means. One of the way for the charge neutralization of LIB during the propagation is to fill up the reactor cavity with the inert gas of 0.1-1 Torr. The electrons in this inert gas are expected to neutralize the charge of LIB. The self-pinch propagation method has been proposed recently as a new idea. Such a propagation will be achieved by setting the ratio of the number density of background plasma to that of beam particles to be 1:10. As the current density of the beam is very intense, the self-induced magnetic field in the azimuthal direction pinches the beam itself. Because, the self-pinched plasma is not stable, the magnetic field in the axial direction to stabilize the beam is induced by the rotating motion of the beam. This magnetic field corresponds to the toroidal magnetic field in the tokamak machine. In this· paper, the effects of magnetic field on the stabilization of the beam is investigated numerically as an eigenvalue problem. This is an analysis for macroscopic stabilities. The importance of the effect of electric field on the beam propagation was pointed out recently. That is, the leading- and tailing edges diverge by electric fields. These electrostatic fields are estimated on the basis of the kinetic theory and the beam divergence is calculated in the later part of the present paper.

MACROSCOPIC STABILITY FOR ROTATING BEAM

The scheme employed here for stability analysis is of an eigenvalue problem. This scheme is applied to the LIB propagation in the background plasma which fills inside the reactor. In order to stabilize the beam propagation, the ion beam particle rotates around the propagating axis. The beam rotation enduces the magnatic field inthe axial direction. The magnetic field corresponds to the "toroidal field" in tokamak. Propagating beam ions are confined in a small radius by the azimuthal magnetic field (corresponding to the poloidal magnetic field in tokamak) which is induced by the self-current of beam in the axial direction. The toroidal field is necessarily required to exist to stabilize the plasma confined by the poloidal field. The electro-magnetic fields induced around the beam are schematically shown in the following figure.

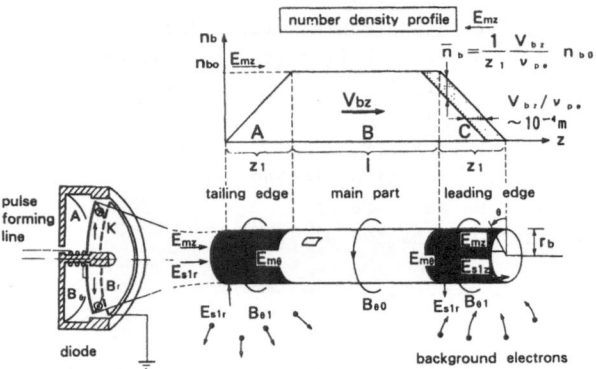

Fig. 1 Beam Configuration

Because of the high particle energy of propagating beam, the beam particle can be considered to be collision-free. The Larmore radii of beam particles and the amplitudes of their betatron oscillations are comparable to the propagating beam radius. Thus the macroscopic description is very poor for the beam. The growth rates of disturbances depend completely on the steady solution. Thus the macroscopic treatment gives very limited information about the stability. Surely the steady solution is easily obtained in the form of a distribution function. (The Maxwellian form for the beam distribution is not preferred, because it is not confined by the magnetic field.) But the stability analysis becomes much more complex in the case of microscopic treatment because of its many independent variables, although the formalism is similar to the macroscopic one. In this paper, for simplicity we choose the macroscopic description for the beam as an example in which the eigenvalue problem is applied. The governing equation for the beam are,

$$\frac{\partial n_b}{\partial t} + \nabla \cdot (n_b \mathbf{v}_b) = 0 \quad , \tag{1}$$

$$m_b n_b \{ \frac{\partial \mathbf{v}_b}{\partial t} + \frac{1}{2} \nabla \mathbf{v}_b^2 - \mathbf{v}_b \times (\nabla \times \mathbf{v}_b) \} = -k_B \nabla (n_b T_b) + e_b n_b (\mathbf{E} + \mathbf{v}_b \times \mathbf{B}) \quad , \tag{2}$$

$$\mathbf{j}_b = e_b n_b \mathbf{v}_b \quad . \tag{3}$$

The election in the background plasma is collision-dominant. The beam current, however, is very strong. Hence the induced magnetic field B is very intense. The mean electron Larmore radius is frequently shorter than the electron mean free path. In such a circumstance, Ohm's law is also a very poor expression. But in accordance with the simple expression for the beam, the electron motion is expressed by

$$j_e = \sigma (E + v_p \times B) \quad . \tag{4}$$

The Maxwell equations for electromagnetic fields are

$$\nabla \times B = \mu_0 (j_b + j_e) + \mu_0 \varepsilon_0 \frac{\partial E}{\partial t} \quad , \tag{5}$$

$$\nabla \times E = -\frac{\partial B}{\partial t} \quad , \tag{6}$$

$$\nabla \cdot B = 0 \quad , \tag{7}$$

$$\varepsilon_0 \nabla \cdot E = \rho \quad . \tag{8}$$

Here n is the number density, v is the velocity, m is the particle mass, e is the particle charge, T is the temperature, k_B is the Boltzmann constant, E is the electric field, B is the magnetic flux density, σ is the electric conductivity, j is the current density, μ_0 is the magnetic permeability in vacuum, ε_0 is the dielectric constant in vacuum and ρ is the electric charge density. The independent variable t is the time and indicates the space gradient. The suffixes b, e and p refer to the beam, the electron in the background plasma and the background plasma itself, respectively. Since the beam energy equation is not solved, the beam temperature T_b is assumed to have a given profile in the radial direction. The velocity v_p of the background plasma is assumed to be zero. The charge neutrality is considered to hold everywhere, andhence the induced electrostatic field is zero. Thus, eq. (8) does not need to besolved. Equation (7) is satisfied automatically. Since the stability of a propagating LIB is insensitive to high-frequency disturbances caused by electrons, the equation fo electron motion used here (Ohm's law eq. (4)) does not include the inertia term of electron. The displacement current in eq. (5) is neglected in accordance with the condition of charge neutrality. The typical parameters of this rotating LIB are given in the following table.
These are enough values to obtain a practical amount of fusion energy from a targat. This is the goal of our fusion study in LIB-fusion.

In eqs. (1)-(8), the dependent variables are divided into two parts as,

$$\phi(r,t) = \phi_0(r) + \delta\phi(r,\theta,z,t) \quad . \tag{9}$$

Then we have the two groups of equations for steady state and equations for disturbances. Here the dependent variables in the steady parts are assumed to be

Table 1 Typical Parameters for LIB Energy Driver

Total beam energy	10 MJ
Number of modules	6 modules
Pulse length	30 ns
Beam number density	10^{22} 1/m^3 (proton)
Background plasma number density	10^{21}-10^{22} 1/m^3 (Argon)
Particle energy	8 MeV
Kinetic energy in z-direction	4 MeV
Kinetic energy in θ-direction	3 MeV
Thermal energy	1 MeV
Beam current in z-direction	10 MA
Beam radius	3-5 mm
Beam velocity	3x10^7 m/s
Transport length	5 m

a function of r only. Then, the equations for steady state used to specify the propagation velocity v_{bz0} and rotation velocity $v_{b\theta}$ of the beam ion disappear. Therefore, v_{bz0} and $v_{b\theta}$ are the arbitrary while v_{br0} must be zero. Here v_{b0} is chosen to be

$$\mathbf{v}_{b0} = \left(0 , \Omega v_{bz0}\frac{r}{r_b} , v_{bz0} \right) \quad . \tag{10}$$

The ratio of rotation Ω is defined by

$$\Omega = \frac{v_{b\theta0}(r_b)}{v_{bz0}} \quad . \tag{11}$$

The equation of continuity for beam ions is lost because $v_{br0}= 0$. If the fundamental equations are linearized with respect to the disturbance (see eq. (9)), the disturbance term can be Fourier-expanded as follows;

$$\delta\phi(r,\theta,z,t) = \sum_k \sum_l \sum_\omega \delta\phi^*_{kl\omega}(r) \exp i(kz + l\theta - \omega t) \quad , \tag{12}$$

where k and l are the wave numbers in the z- and θ-directions respectively. By substituting the above equation into the fundamental equations, the eigenequation is obtained. In general, system of ordinary differential equations for the disturbance as the system of eigenequations for given boundary conditions cannot be solved analytically. The highest order of the given differential equations is second order in this case. The ordinary differential equations are transformed into the difference equations on the basis of the values at mesh points along the r-axis. Therefore, the differential term of the first order in the linearized equation can be expressed as

$$\frac{d\delta\phi^*(r)}{dr}\bigg|_j = \frac{\delta\phi^{**}(r_{j+1}) - \delta\phi^{**}(r_{j-1})}{2\Delta r} \quad . \tag{13}$$

Here the quantities with ** denote the solution $\delta\phi^*$ in the system of
difference equations. By using eq.(13), the system of ordinary differential
equations for the disturbance are transformed into that of difference equations
which are based on the values at the mesh points r_j. In another expression, the
system of difference equations can be written in the following matrix equation
as the eigenvalue equation,

$$L\,\delta\phi^{**}_{kl\omega} = 0 \quad , \tag{14}$$

that is,

$$(M - \omega I)\,\delta\phi^{**}_{kl\omega} = 0 \quad . \tag{15}$$

In the above equation, each row of the coefficient matrix L has a linear term of
ω, and the other terms are independent of ω. In eq.(15), the matrix L is
divided into two parts of M and ωI, where I is the unit matrix. Here M is
obtained as follows, we divide each row of the coefficient matrix L by the
coefficient of the term which ω is proportional to. In L, all terms which
includes ω exist on the diagonal positions. For nonzero amplitude of the
disturbance vector $\delta\phi^{**}_{kl}$, the eigen equation (eq.(14)) derives the following
determinant-free equation,

$$|M - \omega I| = 0 \quad . \tag{16}$$

From this determinant equation, the solutions ω are obtained as eigenvalues of
the matrix M. For the steady solution (when $\Omega = 0$, i.e. $B_z = 0$), the angular
velocities ω_r of disturbances in a sausage mode ($k=100/2\pi$ m, $l=0/2\pi$ rad) are
indicated in the following figure. Here ω_r is the real part of the complex
eigenvalue. One line in the figure corresponds to one solution of ω_r.

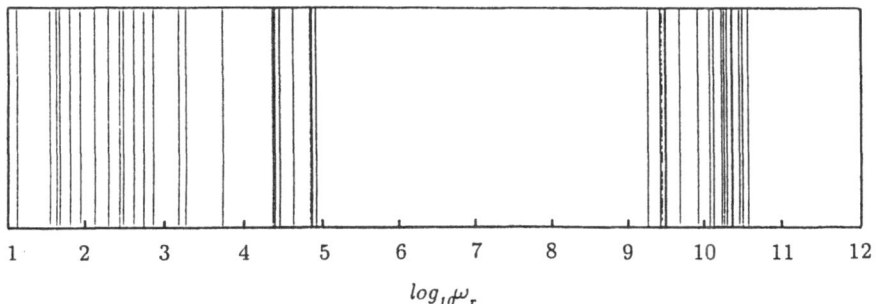

$$log_{10}\omega_r$$

Fig. 2 Eigenfrequency Spectrum of the Disturbances in the Case
of k=100, l=0 and Ω=0.
Each line indicates an eigenfrequency.

In the Fig. 2, many angular velocities concentrate near $\omega_r=3\times10^9$ rad/s.
About 1/3 of the calculated eigenvalues concentrate near $\omega_r=3\times10^9$ rad/s. The
phase velocities $v = \omega_r/k$ of these disturbances are nearly equal to the beam
propagation velocity. The effect of the rotation velocity of the propagating
beam on the growth rate γ is investigated on the basis of the simulation
results. As to the maximum growth rate of the sausage mode is obtained as
$\gamma=2.04\times10^8$ rad/s, in the case of $k=500/2\pi$ m and $\Omega=0.6$. In general, the maximum
growth rate appears in the fundamental mode in the r-direction and in the
sausage mode (or, at most, in the kink mode) regarding the wave number l.

597

In the previous chapter, the analyses are given for the rotating LIB with infinite length, that is, the beam has no leading and tailing edges. The phenomena are steady. Unsteady physical quantities, however, of the beam with finite length, that is, those for the beam which has the leading and tailing edges, remains unsolved. The beam duration time is about 30ns and is determined relating with the implosion time of the target. In the LIB-ICF, 30ns is considered to be the most suitable duration time and this value is slightly changeable dependent on the target structure, but almost fixed. A short-pulse and high-density beam is preferable compared with a long-pulse and low-energy-density beam, because a high temperature and high density plasma shouldbe confined during a short period when the inertia of plasma protects from its expansion. The beam, whose pulse width is 30ns and whose propagation velocity is 3×10^7m/s has a pulse length of about 1m. From a point of view, however, it is almost impossible to make a beam whose number density has the step-like profile. The number density at the leading edge is determined by the rising time of diode voltage.

Ions in the background plasma are assumed here to be at rest completely. The electron in the background plasma is treated as a fluid and the equation of motion is derived as the 1st moment of the Boltzmann equation as follows,

$$m_e n_e \{ \frac{\partial \mathbf{V}_e}{\partial t} + \frac{1}{2} \nabla \mathbf{V}_e^2 - \mathbf{V}_e \times (\nabla \times \mathbf{V}_e) \}$$

$$= e_e n_e (\mathbf{E} + \mathbf{V}_e \times \mathbf{B}) - \nabla p_e - e_e n_e \frac{\tau_L}{\tau_{ei}} \mathbf{E} - m_e n_e \frac{\mathbf{V}}{\tau_{ei}} . \tag{17}$$

The last two terms are derived from the collision term of the Boltzmann equation as an approximation. Here, V is the mean velocity, n is the number density, p is the pressure, τ_{ei} is the mean electron-ion collision interval and τ_L is the electoron Larmor period. Subscript e refers to the electron (the electron charge is denoted by e_e instead of $-e$). Electrons in the background plasma are in strong magnetic fields. The averaged electron Larmor radius is much smaller than the electron mean free path. Thus the main part of electron mean velocity

comes from drift motion . Except the very limited region near the two

edges of the beam, the velocity of background electorns follows this equation. Electron collisions with ions in the background plasma modify this drift velocity. An electron encounters a collision with an ion in the background plasma every mean collision interval τ_{ei} on the average. The electron loses its velocity by the collision. After the collision , the electric field E accelerates the electrons during one Larmor period. The last term in eq. (17) expresses this acceleration. Being the facts described above taken into account, the electron velocity is derived approximately form eq. (17) as follows,

$$\mathbf{V}_e = (1 - \frac{\tau_L}{\tau_{ei}}) \frac{\mathbf{E} \times \mathbf{B}}{B^2} - \frac{\nabla p_e \times \mathbf{B}}{e_e n_e B^2} + \frac{e_e \tau_L^2}{2 m_e \tau_{ei}} \mathbf{E} \quad \text{for } \tau_L < \tau_{ei} , \tag{18}$$

On the other hand, in the region where $\tau_L > \tau_{ei}$, the background electrons are colision-dominant. In such a region the electron velocity is considered to be zero, that is,

$$\mathbf{V}_e = \frac{e_e \tau_L^2}{2\, m_e \tau_{ei}} \mathbf{E} \sim 0 \quad \text{for } \tau_L > \tau_{ei} \; . \tag{19}$$

The electron current is given by

$$\mathbf{j}_e = e_e n_e \mathbf{V}_e \; . \tag{20}$$

The Maxwell equations for electromagnetic fields are given by

$$\nabla \times \mathbf{B} = \mu_0 \mathbf{j} + \mu_0 \varepsilon_0 \frac{\partial \mathbf{E}}{\partial t} \; , \tag{21}$$

$$\nabla \times \mathbf{E}_m = -\frac{\partial \mathbf{B}}{\partial t} \; , \tag{22}$$

$$\nabla \cdot \mathbf{B} = 0 \; , \tag{23}$$

$$\varepsilon_0 \nabla \cdot \mathbf{E}_s = \rho \; , \tag{24}$$

where j is the current and ρ is the charge density. Subscripts m and s denote the electromagnetic and electrostatic field, respectively. The current j is expressed by

$$\mathbf{j} = \int_{-\infty}^{\infty} \sum_s e_s \mathbf{v}_s f_s \, d^3 \mathbf{v} \quad (\; x;\; b,e,i\;) \; . \tag{25}$$

The charge density is expressed by

$$\rho = \int_{-\infty}^{\infty} \sum_s e_s f_s \, d^3 \mathbf{v} \quad (\; x;\; b,e,i\;) \; . \tag{26}$$

The velocity distribution function f_b of propagating beam which has the leading and tailing edges is expressed here in a form as

$$f_b = f_{b0}(r,\mathbf{v})\, g_b(z,t) \; , \tag{27}$$

where subscript 0 refers to the steady state and the first term the right hand side is the velocity distribution function of the steady state and is expressed here by

$$f_{b0}(r,\mathbf{v}) = f_{b0}(H, P_\theta, P_z) \; . \tag{28}$$

599

The second term, $g_b(z,t)$ is an unknown function to be solved. The velocity distribution function for the propagating beam should be modified form that for the steady state, because the beam has two edges.

The general solution is,

$$F(z_0, \xi) = 0 , \qquad (29)$$

$$z_0 = z - V_z t , \qquad (30)$$

$$\xi = g_b \exp\left\{-\int\lambda(z,t)\ dt\right\} , \qquad (31)$$

$$\lambda(z,t) = -\frac{e_b}{m_b}\left\{(E_{s1r} - V_z B_{\theta 1})\frac{\partial f_{b0}}{\partial v_r} + \frac{1}{r}E_\theta \frac{\partial f_{b0}}{\partial \theta} + E_z \frac{\partial f_{b0}}{\partial v_z}\right\}\frac{1}{f_{b0}} . \qquad (32)$$

In eq. (29), F denotes an arbitrary function of the two integrals z_0 and ξ. The right hand side in eq. (32) is a function of z and t as well as r and v through f_{b0}. Now g_b is assumed to be weekly dependent on r and v, and at the first step to solve g_b, g_b is treated as a function of z and t only. From these equations, maximum values of modified electromagnetic fields are given by approximately, in the case of $z_1 = 0.5$ m,

$$B_{\theta 1}^{max} = \frac{1}{2}r_b\mu_0(e_b V_z g n_{b0} + j_{e1z}) \qquad (33)$$

$$E_{m1z}^{max} = \frac{1}{4}V_z\mu_0(e_b n_{b0} V_z \frac{dg}{dz} + \frac{dj_{e1z}}{dz})\frac{r_b}{2} \qquad (34)$$

$$E_{e1z}^{max} = \frac{1}{2}\frac{e_b}{\epsilon_0}\frac{dg}{dz}\frac{V_z}{\nu_{pe}}n_{b0}(r_b+l-\sqrt{l^2+r_b^2}) \qquad (35)$$

$$E_{s1r}^{max} = \frac{e_b V_z n_{b0} r_b}{4\epsilon_0 \nu_{pe}}\frac{dg}{dz} \qquad (36)$$

On the basis of its result, the average electron Larmor period inside the beam is calculated as $\tau_L = 3.6\times10^{-13}$ s. By considering the fact that the electron mean collision interval is $\tau_{ei} = 5.1\times10^{-11}$ s, it is clear that the electrons do the intense Larmor motion around the beam particles and have the drift velocity.

$$E_{m1z}^{max} = 2.26\times10^7\ V/m , \qquad (37)$$

$$E_{e1z}^{max} = 8.98\times10^7\ V/m , \qquad (38)$$

$$E_{s1r}^{max} = 4.52\times10^7\ V/m . \qquad (39)$$

REFERENCES

1) H. Murakami, T. Aoki, S. kawata and K. Niu, Laser & Particle Beams 2(1984) 1.
2) T. Kaneda and K. Niu, Laser & Particle Beams 7(1989), part2, p. 207.
3) K. Niu and S. Kawata, Fusion Tech. 11(1987), p.365.

SYMMETRY ANALYSIS OF PERTURBED ION BEAM-TARGET INTERACTIONS

C.K. Choi
School of Nuclear Engineering
Purdue University

INTRODUCTION

Accurate treatment of the intensity-dependent ion beam depositing its energy fronts on a highly distorted material interface of ion-beam driven target is essential to the symmetry and stability analyses of heavy-ion ICF target. Due to its highly non-linear nature of the distorted material interface caused by beam intensity perturbations, analytic treatment of this symmetry analysis of "perturbed" beam-target interactions is very much limited.

Recently, a fluid-implicit particle (FLIP) method has been developed to study the hydrodynamic behavior of ion beam-plasma interactions[1,2,3]. This method uses finite-sized "particles" in a particle-in-cell representation with a three-dimensional ray trace to represent arbitrary beam illumination and intensity geometries. Plasma motion is modelled by integrating the equations of motion for each particle implicitly in time on an arbitrarily-adaptive computational grid. The physics included in the numerical model comprises 2-d compressible hydrodynamics, 2-T energy transport for electrons and ions, average charge states (Saha relation), Sesame equation of state with equilibrium mix, and the ion beam energy deposition with a unified slowing-down theory. Radiation transport is not included in the model. Neither are atomic-physics processes that are not in local thermodynamic equilibrium (LTE). Burn physics and convergence effects are also neglected, as the hydrodynamic response to perturbations in target material interfaces and beam-intensities is the primary emphasis of this study.

A simplifying planar rather than spherical geometry is adopted which is in many cases justified for the early implosion stage. Convergent effects in spherical geometries are likely to amplify any perturbations in planar geometry. The problem is attacked from a symmetry point of view to study the effects of a non-uniform beam deposition layer on an initially smooth target (Figure 1).

FORMALISM

The hydrodynamics, energy transport, and equations of state for the target plasmas are described in Equations (1) through (7):

$$\frac{d\rho}{dt} = -\rho\nabla\cdot\mathbf{V},$$
(1)

$$\rho\frac{d\mathbf{V}}{dt} = -\nabla p,$$
(2)

$$\rho\frac{d\varepsilon_i}{dt} = \nabla\cdot(\kappa_i\nabla T_i) + \omega_{ei}(T_e - T_i) - p_i\nabla\cdot\mathbf{V} + \mu\Phi + \lambda(\nabla\cdot\mathbf{V})^2 + S_i,$$
(3)

$$\rho\frac{d\varepsilon_e}{dt} = \nabla\cdot(\kappa_e\nabla T_e) - \omega_{ei}(T_e - T_i) - p_e\nabla\cdot\mathbf{V} + S_e,$$
(4)

$$p = p_i + p_e, \qquad p_e = p_e(\rho,\varepsilon_e), \qquad p_i = p_i(\rho,\varepsilon_i),$$
(5)

$$\varepsilon = \varepsilon_i + \varepsilon_e, \qquad T_e = T_e(\rho,\varepsilon_e), \qquad T_i = T_i(\rho,\varepsilon_i),$$
(6)

$$\rho = n_e m_e + n_i m_i = n_i(\overline{Z}m_e + m_i), \qquad n_e = \overline{Z}n_i,$$
(7)

where ρ and $\mathbf{V}$ are the density and velocity of the fluid, respectively. Since discontinuous solutions are not allowed in the transport of momentum, an artificial viscosity is added to the pressure gradient term in Eq. (2) in regions of compression such as shocks in the flow. Eqs. (3) and (4), the specific internal energies for ions and electrons, include thermal diffusion (with diffusivities κ_i and κ_e), collisions (with collision frequency ω_{ei}), pdV work (with ion and electron pressures p_i and p_e), viscous dissipation (with coefficients of a shear viscosity μ and bulk viscosity λ, and a dissipation function Φ). There is also an arbitrary source (beam-plasma energy exchanges S_i and S_e),

$$S_{i(e)}(x) = \frac{I(x)}{E(x)}\left[\frac{dE(x)}{dx}\right]_{i(e)},$$
(8)

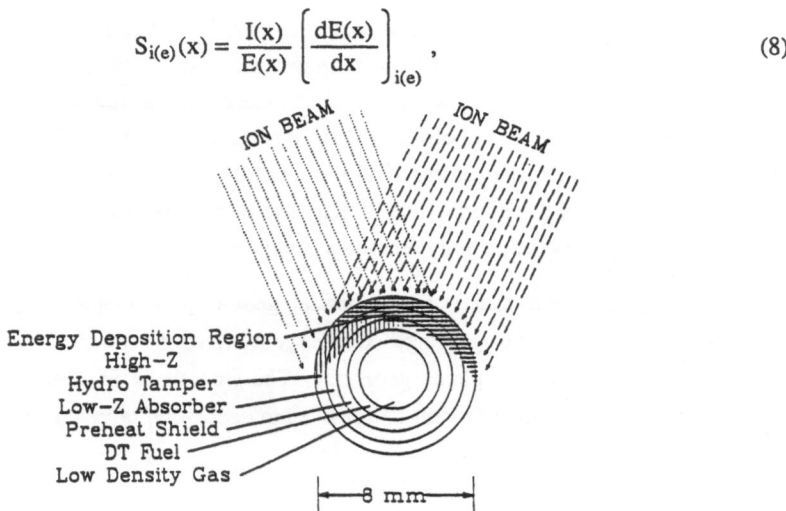

Figure 1. Energy deposition asymmetries expected on a typical direct-driven ion ICF target with multi-beam irradiation.

where I(x) is beam intensity, and E(x) and dE(x)/dx are ion beam energy and energy loss per unit path length, respectively.

Given the conservation equations, one can linearize by expressing the flow variables as sums of zero and first order terms: For example, for the zero-th order,

$$\frac{d\rho_o}{dt} + \rho_o \frac{\partial v_o}{\partial z} = 0; \quad \rho_o \frac{dv_o}{dt} + \frac{\partial p_o}{\partial z} = 0; \tag{9}$$

and, for the first order,

$$\frac{\partial \rho_1}{\partial t} + \frac{\partial v_o}{\partial z}\rho_1 + \rho_o k_x u + \rho_o k_y v + \rho_o \frac{\partial w}{\partial z} + \frac{\partial \rho_o}{\partial z} w = 0;$$

$$\rho_o \frac{\partial u}{\partial t} = k_x p_1; \quad \rho_o \frac{\partial v}{\partial t} = k_y p_1; \text{ and } \frac{\partial v_o}{\partial z}\rho_1 + \rho_o \frac{\partial w}{\partial t} + \frac{\partial v_o}{\partial z} w = -\frac{\partial p_1}{\partial z}; \tag{10}$$

where $\rho = \rho_o + \rho_1$ and $\vec{v} = \vec{v}_o + \delta\vec{v}$ with $\delta\vec{v} = (u, v, w)$.

Separate energy transport for the ions and electrons requires separate ion and electron equations of state (EOS) in order to close the system of conservation equations. The SESAME EOS data library from the Los Alamos National Laboratory[4] is used. Energy loss, dE/dx, to free electrons and plasma ions is described by a unified theory that includes both friction and diffusion in velocity space[5] as

$$\left[\frac{dE}{dx}\right]_{UT} = \frac{8\sqrt{\pi} n_2 \bar{Z}_2^2 e^4 Z_{eff}^2}{v m_2 v_2} \left[\frac{erf(\xi)}{\xi} - \frac{2m_2}{\sqrt{\pi}\mu_2} \exp(-\xi^2)\right]$$

$$\times \left[\ln \frac{\mu_2 T_2^{3/2}}{2.934\sqrt{\pi} m_2 \sqrt{n_2} |Z_{eff} e^3 Z_2^2|} + h_2\right], \tag{11}$$

where h_2 is given by (h_e for electron and h_i for ion stopping medium):

$h_e = 3 \ln(v/v_{eo}) + 0.5,$	$h_i = 2 \ln(v/v_{io}),$	for $v \gg v_{eo}$;
$h_e = \ln 2 - 0.557,$	$h_i = 2 \ln(v/v_{io}),$	for $v_{eo} \gg v \gg v_{io}$;
$h_e = \ln 2 - 0.577,$	$h_i = \ln\left[2(m_e/m_i)^{1/2}\right] - 0.577,$	for $v \ll v_{io}$;
$v_2 \equiv (2 T_2/m_2)^{1/2},$	$v_{20} \equiv (T_2/m_2)^{1/2},$	$\xi \equiv v/v_2.$

(12)

This unified formalism, which combines both the binary collision and collective wave phenomena, is capable of handling an arbitrary stopping medium (electrons or ions) without introducing the Coulomb logarithm, and thus is valid for all interaction ranges.

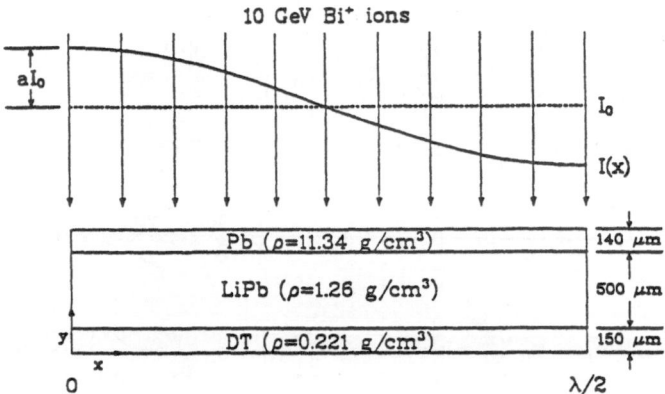

Figure 2. The initial planar target configuration subjected to a normally incident beam having a perturbed single mode intensity I(x). The beam, normally incident and focused at $y = -\infty$, carried an intensity I(x) held constant in time up to 30 ns.

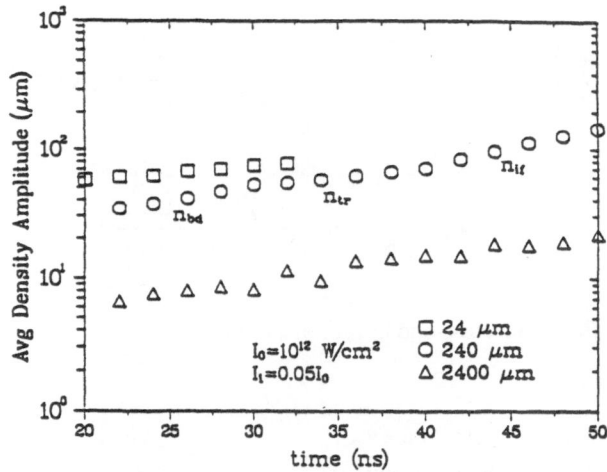

Figure 3. Computed average perturbation amplitudes (one-half of the peak-to-valley value) of density of a uniform planar target for three beam intensity perturbation wavelengths, each with a 5% amplitude about 1 TW/cm^2.

SYMMETRY ANALYSIS

The present fluid model for heavy-ion beam-plasma interactions, FLIP-PHD (Plasma Hydro Dynamics)[1,2] is based on the FLIP model for fluid flow with the following characteristics: (1) Electric and magnetic fields are neglected; (2) plasmas are assumed to be collisional; (3) ions dominate momentum transport and electrons dominate energy transport; (4) ion beam energy is partitioned differently between plasma ions and electrons, forcing $T_e \neq T_i$; and (5) partial ionization and non-ideal gas EOS are included.

Both zero- and one-dimensional simulations have been performed by FLIP with the proton beam on aluminum foil to compare the FLIP with the known results. The pressure profiles as function of time from the zero-dimensional model of Evans[6], and the simulations by FLIP were compared for both 15 μm and 30 μm thick aluminum foils that were irradiated by a 16-TW/cm^2, 1-MeV proton beam with the average range of 2.8 mg/cm[7,8]. The model predicts reasonably well the initial rise in pressure as well as the approximate maximum, calculated by FLIP for 15-μm foil, but breaks down thereafter when a rarefaction propagates back through the foil.

The proton-foil interaction recently studied at the Naval Research Laboratory[7] was compared with the FLIP calculation for 1-D planar implosion. The FLIP calculation used 50 zones with 16 particles per cell and incorporated physics discussed in the previous section. The grid boundary was allowed to move with a semi-Lagrangian in order to follow the expansion of foil.

A planar target that consists of a tamper layer of lead ($\Delta_{Pb} = 140$ μm), an absorber/pusher layer of lithium seeded with lead ($\Delta_{LiPb} = 500$μm), and cryogenic liquid DT fuel layer ($\Delta_{DT} = 150$μm) is considered. Since convergence effects are not modelled here, the void on the inner side of the DT layer will be treated as an infinite vacuum in planar geometry. The 10 GeV B_i^+ ions depositing 4.37 MJ of energy in the target with 4 mg of cryogenic DT over a period of 30 ns are considered to simulate the present planar target. A simple sinusoidal intensity variation within the beam was assumed for this analysis (Figure 2):

$$I(x,t) = I_o(t) \, [1 + a \cos(kx)] \,. \tag{13}$$

This form of intensity is convenient for parameter studies, since by varying the three quantities $I_o(t)$, a, and k one can reasonably assess a target sensitivity to intensity magnitude, perturbation amplitude, and perturbation wavelength. This design grew from a design originally proposed by Bangerter and Meeker[9], which to date remains the generic single-shell design (tamper, absorber/pusher, and fuel layer) for ion-driven ICF targets.

This symmetry analysis corresponds to a uniform planar target driven for 30 ns by a normally incident, 10 GeV B_i^+ "perturbed" beam having an intensity variation $I(x,t) = I_o(t) \, [1 + a \cos(kx)]$, where x is defined in the lateral direction of the target with $0 \leq x \leq 120$μm, $I_o = 1$ TW/cm^2, a = 0.05, and kd = 20. Computational results are illustrated in Figure 3. It shows the pattern of density amplitude growth with three distinct slopes visible through the points denoting growth rates n_{bd} ("beam driven" for 22 to 30 ns), n_t ("transitional" for 30-40 ns), and n_{if} ("interfacial" for 40-50 ns). Density amplitudes for $\lambda = 2400$μm are very small, almost at the level of computational noise. The short-wavelength calculation ($\lambda = 24\mu m$) is terminated at 32 ns because of problems with the finite grid instability, which rendered the results suspicious at later times. It is apparent in Figure 3 that the maximum density perturbation growth rate occurs at $\lambda = 240\mu m$.

CONCLUSIONS AND SUMMARY

Two-dimensional planar ion-beam target has been modelled and the computational results indicate that the target implosion symmetry is found to be sensitive to spatial variations in beam energy deposition. Variations in beam intensity are found to be the limiting factors for target implosion at high beam intensities. The beam intensity perturbations with wavelengths on the order of payload (i.e., DT plus fraction of LiPb absorber/pusher) shell thickness result in higher implosion asymmetries.

REFERENCES

1. D. Kothe, C. Choi, and J. Brackbill, *Laser Interactions and Related Plasma Phenomena*, Vol. 8, pp. 701-721 (1988).

2. D. Kothe, and C. Choi, *Laser Interactions and Related Plasma Phenomena*, Vol. 9, pp. 623-642 (1991).

3. D. Kothe, J. Brackbill, and C. Choi, *Phys. Fluids*, B2(8), pp. 1898-1906 (1990).

4. K.S. Holian, ed., Los Alamos National Laboratory report, LA-10160-MS (1984).

5. C.K. Choi and M. Hsiao, *Nucl. Technol./Fusion 3*, pp. 273-279 (1983).

6. R.G. Evans, *Laser and Particle Beams 1*, pp. 231-239 (1983).

7. J.E. Rogerson, R.W. Clark, and J. Davis, *Phys. Rev. A 31*, pp. 3323-3331 (1985).

8. B.P. Goel, G.A. Moses, and R.R. Peterson, *Laser and Particle Beams 5*, pp. 133-154 (1987).

9. R. Bangerter and D. Meeker, "Ion beam inertial fusion target designs," Lawrence Livermore National Laboratory report, UCRL-78474 (1976).

ACCELERATION OF ELECTRONS BY LASERS IN VACUUM

Thomas Häuser and Werner Scheid

Institut für Theoretische Physik
der Justus-Liebig-Universität
Giessen, Germany

Heinrich Hora*

CERN, CH1211 Geneva 23, Switzerland

ABSTRACT

Acceleration of electrons in vacuum to high energies is possible by laser wave packets with a finite lateral extension of the laser beam. Such an acceleration can be easily understood for electrons which are laterally injected into the beam and, after they are accelerated in a half wave length, are emitted at the other side of the laser beam. Present days lasers should reach electron energies up to about 100 MeV. For electrons with TeV energies laser pulses with a power in the order of 1000 EW (Exa=10^{18}) are needed. A laser focus collider is discussed. A detailed numerical study of the acceleration of the electrons takes the effects of both the transversal and longitudinal components of the laser fields into account. Also an additional transverse static magnetic field is investigated in connection with the deflection of the electron beam out of the laser beam. It is shown that the longitudinal electric field reduces the energy gain by the transversal electric field maximally up to 40 %.

---

* Permanent address: Department of Theoretical Physics, University of New South Wales, Kensington 2033, N.S.W., Australia

INTRODUCTION

A free electron in vacuum can never gain or loose energy (ignoring Compton effect) if it is passed by a plane electromagnetic wave[1]. Nevertheless a net acceleration of electrons by lasers in vacuum can be reached if special pulses of laser light are applied. Any of these special wave fields can be produced by a superposition of plane waves since the electromagnetic field amplitudes can be linearly superimposed, but the conclusion can not be drawn from this linear superposition that the net electron acceleration vanishes. The Lorentz force on the electron is nonlinear[2], and all the rules of linear superposition are no longer valid for the forces.

The concept to produce intensity minima by laser fields and to accelerate electrons by acceleration of these minima[3] was clarified in all details by following up the complete higher order single electron motion in these fields and by proving that the acceleration does occur[4] as globally expected from the ponderomotive force[2] or by using more general conservation theorems[5]. An experiment with microwaves indicates that electrons are gaining energy with such techniques[6].

Other schemes were discussed how asymmetric plane wave packets may accelerate electrons in vacuum by adding weak static electric[7] or magnetic fields[8]. By looking at the action of longitudinal components of the microwave fields in cavities, the use of lasers in linear accelerators[9] was suggested where high laser fields in the boundaries of the cavity may not be too much damaging[10]. The use of the longitudinal fields was stressed[11] when geometrically produced longitudinal components form bent wave fronts in the focus. Also the superposition of two slightly crossed beams[9], or the recently discovered longitudinal components derived from the exact solution of the Maxwellian equations can be used[12].

The following contribution reports mainly on computations of electron acceleration in Maxwellian fields where besides the transversal fields also longitudinal fields, which were considered before in an isolated way only[11], are included. The discussion of these transversal fields for electron acceleation was published before[13,14]. Here, we are presenting the results of the total field interaction for the first time for a laser beam of finite diameter and short pulse duration interacting with a free electron.

In order to bridge the gap to the preceeding computations of the isolated action of the transversal fields, we are first reporting about some extensions of the results gained after the last conference[14]. We shall see at least by basic considerations that laser pulses of very high power are necessary. The result is that for competing with electron accelerators up to the TeV range, such lasers are not yet existing. In this context it is interesting to note that the present developments of lasers are very promising. If very powerful lasers may be developed, one has to think on alternative concepts, as proposed here, what these lasers may offer for accelerators in the field of high energy physics.

In this contribution we first sketch the conceptual ideas on the acceleration of electrons by plane electromagnetic half wave pulses. Then we report the main results of our calculations for the acceleration of electrons by realistic short laser pulses.

HALF WAVE ELECTRON ACCELERATION BY LATERAL INJECTION

Our preceeding results[13,14] were based on the exact solutions of the relativistic electron motion in infinitely spread stationary plane electromagnetic waves. The result was that the initially resting electron receives an acceleration by the transversal electric field. Then the magnetic field vector of the light wave produces the axial motion acting via the Lorentz force in the direction of the wave propagation. The electron gains the energy during a single half wave and would be slowed down during the second half wave fulfilling the result[1] that a symmetric plane wave can not transfer energy to a free electron. The question was then discussed to use a single or a sequence of positive half waves. Apart from the proposal to apply a Duguay shutter[13] with fs laser pulses for this purpose, a very original concept was developed by Bessonov[15] with his "conventionally strange electromagnetic waves". If a wave packet is separated in two halves with one half shifted by 180° in phase and superposed again, the inner parts vanish by interference; a half pulse at the beginning and a half counterpulse at the end will remain.

Using a half pulse wave it was possible to clarify theoretically at what laser intensities one could expect electrons with TeV energies. The acceleration lengths during the half wave interaction are in the range of 30 cm for neodymium glass lasers, which is the distance until the half wave has taken over the electron. During this acceleration the electron has a lateral motion in the order of 500 optical wave lengths for the mentioned TeV case. Another case of laser acceleration with a similar preference for the lateral motion happens with ions[16] when they are accelerated to very high energies in the laser beam after relativistic self-focusing[17].

In order to avoid the difficulties with the generation of half waves, we considered the injection of the electron into a laser beam from the side (Fig.1, point A) such that it reaches the other side (point B) after a half wave acceleration along the acceleration length. Then the electron is emitted from the beam with the energy gained by the laser field. For this

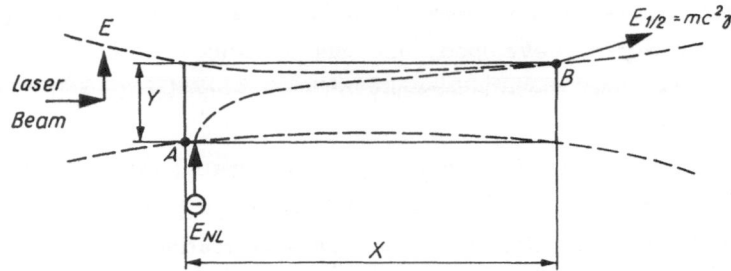

Fig.1   Acceleration of an electron in a laser beam, where the electron is laterally injected at point A with an energy E_{NL} to overcome the nonlinear force. It is then accelerated by a half wave in order to reach the energy $E_{1/2}$ with which the electron is emitted on the other side of the beam at point B.

purpose it was necessary to provide an initial lateral injection energy E_{NL} to the electron. For simplification a nearly rectangular radial intensity profile of the beam is chosen (Fig.2) where the range Δr of increase of the intensity is rather small against the total beam radius. The action of disturbing forces is limited to this small peripheral range only and the whole acceleration of the electron occurs similarly to the case of the plane wave when the electron reachs the intensity in the interior of the beam.

Although the Maxwellian exact solutions of the laser field have a longitudinal component outside the inner plane wave in the peripheral ranges of the beam, their effect to the whole acceleration may be small and probably fully compensated with respect to the counteracting longitudinal forces at A and B. When injecting the electron at point A towards the beam axis one may provide the electron with an initial axial motion with an energy

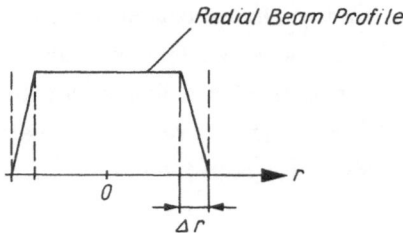

Fig.2   Radial intensity profile of the laser beam as function of the radial coordinate r.

which just compensates the slowing down of the electron by the longitudinal electric field interaction when moving through the outer parts of the beam. The longitudinal fields are in phase of 90° to the transversal fields obtained from the first exact derivation of the longitudinal fields in Chapter 12.3 of Ref. 18 which was achieved for a triangular amplitude profile which followed the paraxial approximative derivation[19], before the Gaussian profile was treated exactly[12].

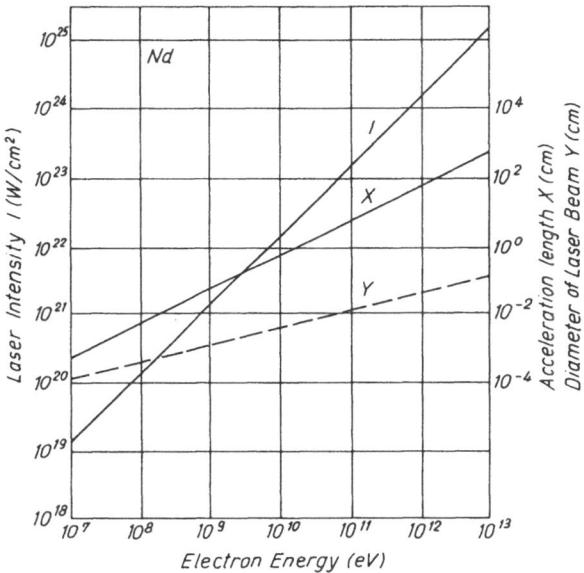

Fig.3   Acceleration of electrons inside a half wave length to energies $E_{1/2}$ for lateral injection into laser beams and the corresponding laser intensities I for a neodymium glass laser. The acceleration length during the half wave interaction is denoted as x and for the lateral motion as y determining the beam diameter[14].

Fig. 3 and Fig. 4 present the results of the necessary intensities for neodymium glass and KrF excimer lasers for achieving a desired electron energy $E_{1/2}$ after the half wave acceleration by the beam. Simultaneously the acceleration length x and the necessary lateral length y which determines the beam diameter are given. Due to this necessary lateral motion, the power of the beam has to arrive at very high values. These values are independent of the wave length (Fig.5) while the lateral nonlinear force injection energy E_{NL} is dependent on the wave length (see Fig.5). The electron energy $E_{1/2}$ is obtained as function of the laser power

$$E_{1/2} = \sqrt{32}(e/\pi^{3/2})(\mu_o/\epsilon_o)^{1/4}\sqrt{P} \approx 20 \ \sqrt{P/Watt} \ eV, \tag{1}$$

where we assumed that the intensity I and P are related as $\pi(y/2)^2I=P$ with the lateral length y.

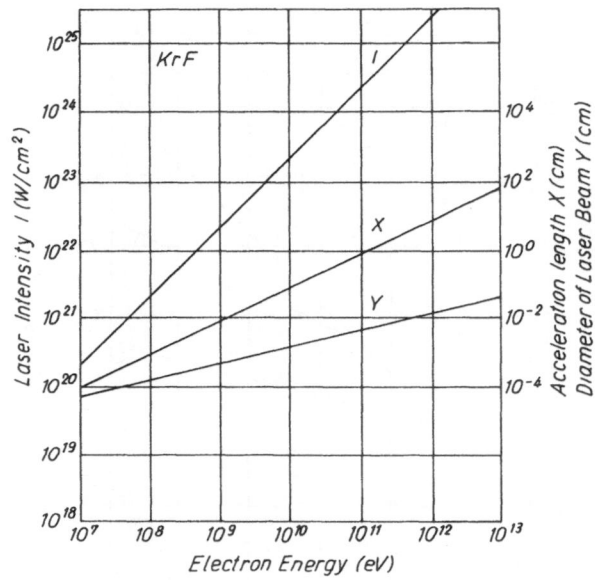

Fig.4 The same as shown in Fig.3 for a KrF excimer laser

The considerable lateral motion has the disadvantage of the necessary high laser power, but has the advantage that the beam has a rather large cross section. Therefore, a large number of electrons can be accelerated per interaction. Assuming that MeV electron beams of MA/cm^2 have been produced, we can estimate that pulses of 10^{13} electrons can be accelerated to the range of TeV energies. It can be predicted with a similar reliability, as predictions for the CLIC and other electron accelerators for the next century are being performed, that high luminosities of 10^{33} s^{-1}cm^{-2} can be reached in connection with colliders.

For a first verification of the acceleration of electrons one may use laser beams which are now available. A neodymium glass laser producing pulses of 20 J energy and 1 ps duration (20 TW) has been achieved[20]. From Fig.5, this results in a maximum energy gain of 89 MeV. Using the optically ideally shaped iodine laser beam[21] with 2 kJ energy in 200 ps pulses corresponding to 10 TW, a maximum energy gain of 63 MeV can be obtained. The neodymium glass laser with pulses of 1 kJ and 0.3 ps should arrive at gains of the electron energy of GeV. In each of these cases the optimum beam diameters, focusing lenghts[14] and injection energies of the electron have to be used as just described. The main difficulty in extending the 20 TW laser to a laser beyond PW [20] consists in the need for optically highly accurate gratings with a diameter of a half meter[22].

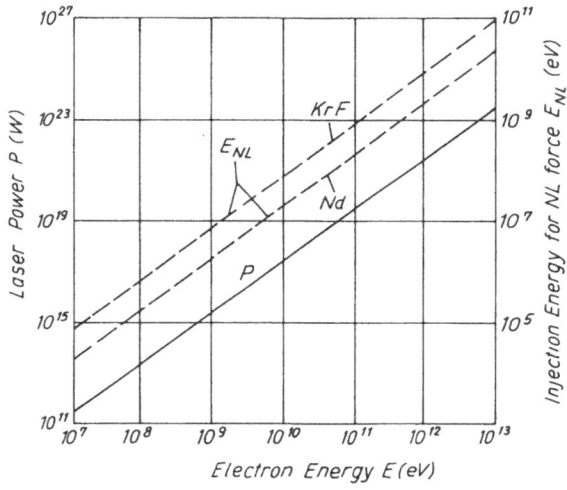

Fig.5   The laser power P for achieving an acceleration in a half wave length to the energy $E_{1/2}$. The corresponding energies for the lateral injection to overcome the nonlinear force E_{NL} are given for neodymium glass and KrF excimer lasers (dashed lines).

ACCELERATION OF ELECTRONS BY LASER PULSES

In the following we discuss the realisation for the acceleration of electrons in vacuum by short laser pulses. Laser beams have a much higher field strength than applied in common linear accelerators. In this chapter we study the question how these laser fields can be used for an effective acceleration of electrons up to energies of 1 TeV.

The exact solution of electron acceleration by plane linearly polarized electromagnetic waves in vacuum was shown in an earlier paper[13]. Starting with an electric and magnetic field propagating in x-direction

$$\vec{E} = E_y(x-ct)\vec{e}_y, \tag{2}$$

$$\vec{B} = B_z(x-ct)\vec{e}_z, \tag{3}$$

the solution of the relativistic equation of motion

$$m\frac{d}{dt}(\gamma\vec{v}) = -e\vec{E} - \frac{e}{c}\vec{v}\times\vec{B} \tag{4}$$

was found for an initially resting electron $(\vec{v}(t=0)=0)$. The solution for the relativistic Lorentz factor γ $(\gamma=1/\sqrt{1-v^2/c^2})$ is given by

$$\gamma = 1 + \frac{1}{2}A^2(u), \tag{5}$$

with $u=x-ct$ and the function

$$A(u) = \frac{e}{mc^2}\int_o^u E_y(u')du', \tag{6}$$

depending on the electric field strength of the laser only. It can be immediately seen that in the case of an oscillating field the electron is only oscillating with no net energy gain. Furthermore, the elongation of the electron was studied and is given by

$$x = -\int_o^u \frac{1}{2}A^2(u')du', \tag{7}$$

$$y = -\int_o^u A(u')du'. \tag{8}$$

The electron receives its maximal energy after the acceleration by a single positive half wavelength, before the following negative half wavelength starts to deaccelerate the electron. This behaviour was discussed in the chapters before. Now the question arises, under which conditions one can use the energy of only one half wavelength for an effective acceleration. For $\gamma = 1.96\cdot10^6$ which corresponds to an electron energy of 1 TeV, an intensity of about 10^{22} W/cm^2 is needed by using a CO_2 laser with $\lambda=10.6$ μm. This intensity is very high and is not reached today for lasers.

The electron with a final energy of 1 TeV has travelled 3.9 m in the CO_2 laser beam parallel to the beam axis (x-direction) during one half wavelength has overtaken it. Perpendicularly to the beam axis in y-direction the electron moves only 0.53 cm. That means the motion of the electron is nearly parallel to the beam axis with only a small deviation in the lateral direction.

The motion of the electron inside a half wavelength gives us an estimation for the necessary laser power P. Because the electron should stay inside of the laser beam during the acceleration, the radius of the beam has to be at least in the dimension of the lateral elongation. Choosing the diameter equal to the elongation y in lateral direction (Eq.(8)), we obtain the energy of the electron as a function of the laser power as given in Eq.(1). Since the energy is no longer a function of the wavelength of the laser, all laser types would need the same power to reach a certain electron energy.

The optimum method for electron acceleration by lasers would be to use short pulses in the femto second range. Since only a single half wavelength of the beam is needed, the aim of development must be powerful short pulses.

The results shown up to here are obtained for infinite plane electromagnetic waves. Now we want to discuss the more realistic case of laser pulses with a finite lateral radius. In the simplest approach we choose the field as:

$$\vec{E} = E_y(x-ct)f(r)\vec{e}_y, \tag{9}$$

$$\vec{B} = B_z(x-ct)f(r)\vec{e}_z. \tag{10}$$

The function $f(r)$ describes the decrease of the laser field in the radial direction ($r=\sqrt{y^2+z^2}$). The above ansatz is only an approximative solution for the laser fields because it does not solve the Maxwellian equations. This fact can be easily seen from $\vec{\nabla}\vec{E} \neq 0$ which is false in the vacuum. By inserting this ansatz into the right hand sides of the equations:

$$\frac{\partial\vec{E}}{\partial t} = -c\vec{v} \times \vec{B}, \tag{11}$$

$$\frac{\partial\vec{B}}{\partial t} = c\vec{v} \times \vec{E}, \tag{12}$$

one obtains longitudinal components for the laser beam. The longitudinal and transversal fields fulfill the equation $\vec{\nabla}\vec{E} = 0$. We assume Eqs.(11) and (12) as equations for an iterative procedure. Continuing this iteration one finds that a second iteration yields correction terms of minor importance.

For simplicity we have applied the simplest possible model. Choosing the transversal laser field with components E_y and H_z of sinus-type in the propagating direction, we can write with a Gaussian decreasing radial function ($r^2=y^2+z^2$)

$$\vec{E}_y = -E_o\sin(kx-\omega t)\exp(-(\frac{r}{R})^2)\vec{e}_y, \tag{13}$$

$$\vec{H}_z = -E_o\sin(kx-\omega t)\exp(-(\frac{r}{R})^2)\vec{e}_z. \tag{14}$$

Inserting this ansatz into Eqs.(11) and (12), we obtain the longitudinal fields (E_x, B_x) as

$$\vec{E}_x = \frac{2y}{kR^2} E_o \cos(kx - \omega t) \exp\left(-\left(\frac{r}{R}\right)^2\right) \vec{e}_x,$$ (15)

$$\vec{H}_x = \frac{2z}{kR^2} E_o \cos(kx - \omega t) \exp\left(-\left(\frac{r}{R}\right)^2\right) \vec{e}_x.$$ (16)

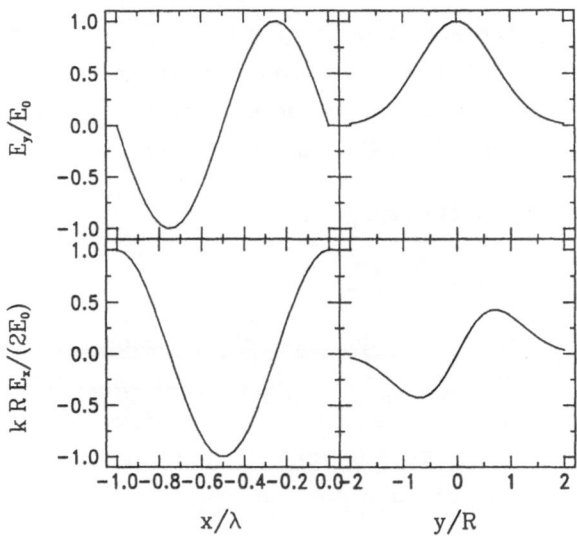

Fig.6 The transversal and logitudinal components of the electric field given in Eqs.(13) and (15) in dependence of the coordinates x and y for fixed times. The upper two curves are E_y/E_o=-sin(kx) and exp(-(y/R)²) for y=0 and x=-1/4 λ, respectively. The lower two curves are $kRE_x/(2E_o)$=coskx and (y/R)exp(-(y/R)²), respectively. The longitudinal component is by a factor of about kR smaller than the transversal component.

Fig.6 shows the components of the electric field. The maximal longitudinal fields are by about a factor kR smaller than the transversal fields. Therefore, in most cases where the laser wavelength λ is small against the radius R of the beam, we find very small longitudinal fields. Nevertheless these fields can be very important during the acceleration of the electrons. The significance of these fields can easily be seen if one transforms the fields into the rest frame of the electron. The Lorentz transformation does not change the longitudinal fields. On the other hand the transversal fields are diminished by a factor of $1/\gamma$. Therefore, at high electron energies one also has to regard the longitudinal field components of the beam.

The sidewise motion can be increased by choosing an additional static magnetic field kicking the electron out of the beam. Fig.7 shows this arrangement. First the electron is accelerated by a positive half wavelength. During this time it travels a distance of 3.9 m in the case of a CO_2 laser pulse. Before the negative half wavelength slows the electron down, it is kicked out of the beam due to its transversal motion and by the additional magnetic field.

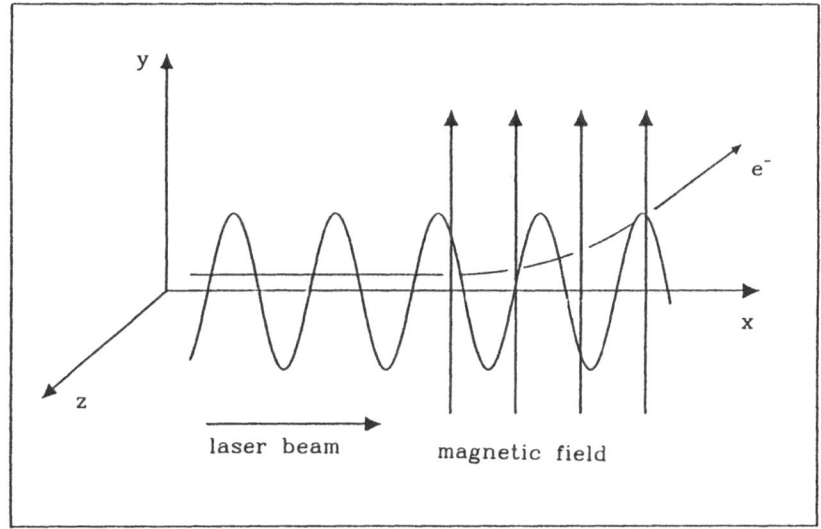

Fig.7   The principle of the laser acceleration. The electron is accelerated by a single half wavelength nearly parallel to the laser beam. The distance is 3.9 m for a CO_2 laser and a final electron energy of 1 TeV. By taking advantage of the transversal motion and an optional additional static magnetic field, the electron leaves the beam before the negative half wavelength will deaccelerate it.

NUMERICAL CALCULATIONS

Choosing the laser field as described in Eqs.(13)-(16), we carried out numerical calculations of the acceleration of electrons. In order to demonstrate the importance of the longitudinal fields, we present results with and without the longitudinal components.

Figure 8 shows results of a calculation without the longitudinal field components and for initially resting electrons. In this figure the Lorentz factor γ is given as a function of the starting position in the yz-plane, after a full wavelength has passed the electron. The results are obtained for a beam of a CO_2 laser (λ=10.6 μm) with a radius of 0.5 cm. The maximal strength E_o of the laser field is $3 \cdot 10^{14}$ V/m which corresponds to a laser power of $9.4 \cdot 10^{21}$ W. An energy gain is received by the electrons, since they enter the zone of lower field strength already after the first half wave length has crossed them.

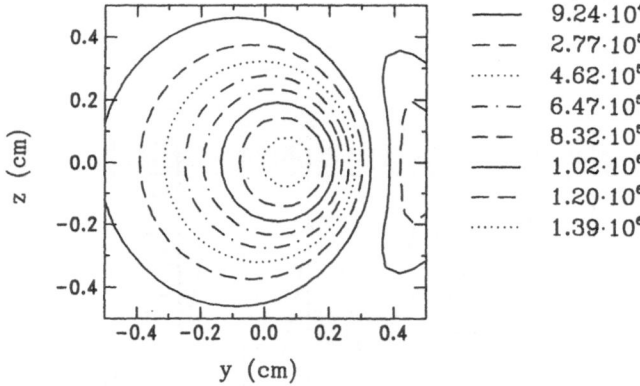

Fig.8   The Lorentz factor γ of the electrons after one wavelength has passed them as a function of their starting position in the yz-plane. The calculation is carried out for a CO_2 laser with a maximal electric field strength of $3 \cdot 10^{14}$ V/m and a radius of 0.5 cm. The longitudinal fields and the static magnetic field are not included in this calculation. Electrons belonging to the circles in the middle of the figure are accelerated out of the beam in negative y-direction. The curves on the right hand side belong to electrons which first move inside the beam and then are pushed out in positive y-direction.

The first half wavelength accelerates the electrons in the negative y-direction. Electrons with initial positions in the middle of the figure have left the laser beam after one half wavelength has passed them. So they can keep their energy. Electrons with initial positions inside the two curves on the right hand side of the figure are pushed by the first half wavelength towards the center of the beam and by the second half wavelength in the positive y-direction out of the beam.

The importance of the very small longitudinal fields ($E_x = E_y/1500$ in our example) is shown by the results in Fig.9 which we obtained with the same parameters as used in Fig.8, but including the longitudinal fields. We find that the energy of the electrons is decreased by about 40% due to the influence of the small longitudinal field components in comparison with the results of Fig.8. With similar other calculations we proved that higher order field components are of minor importance.

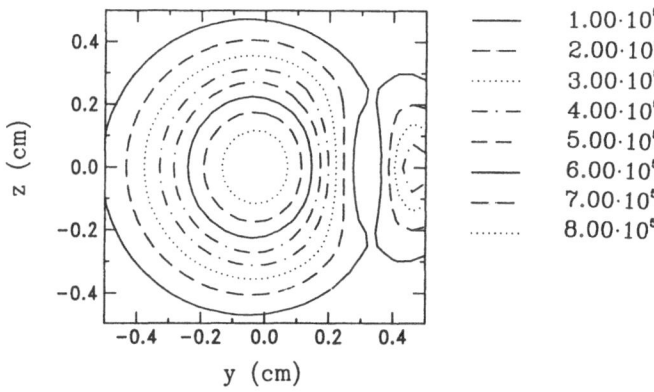

Fig.9   Same as in Fig.8 but with longitudinal field components. The figure shows the importance of the small longitudinal fields.

The influence of the longitudinal field components can be understood in principle with the aid of Fig.6 which shows the longitudinal (E_x) and transversal (E_y) field components as functions of x and y. An electron starting in the center of the beam is accelerated by the first half wavelength. During this time the electron is moving away from the beam axis and undergoes an accelerating and deaccelerating part of the longitudinal field (lower two graphs of Fig.6). When first the accelerating part of the longitudinal field acts, the electron stays inside the beam. In this part the radial function is small and only a small acceleration of the electron occurs by the longitudinal electric field. When the deaccelerating part of the longitudinal field comes into action, the electron has moved into the region where the radial distribution has its maximum. Then in total the electron gets deaccelerated by the longitudinal field.

The acceleration can be optimized by switching on a static magnetic field as shown in Fig.7. Calculations were carried out with a magnetic field of 10 T in y-direction. The results are shown in Fig.10. The parameters of the laser are the same as in Fig.9. We obtain an energy gain of about 40% in comparison with the values of Fig.9. This demonstrates the advantage of the static magnetic field which kicks the electrons out of the beam.

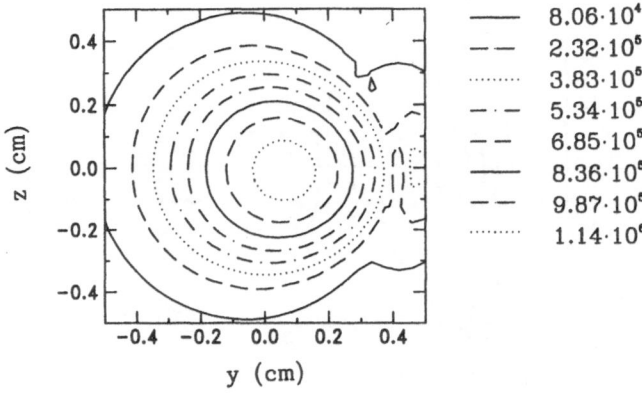

Fig.10   Same as in Fig.9 but including an additional static magnetic field of 10 T. The efficiency of the acceleration is increased by the static magnetic field in comparison with the results of Fig.9 where no static magnetic field is applied.

LASER DEVELOPMENTS AND APPLICATION TO HIGH ENERGY PHYSICS RESEARCH

To reach TeV electrons one needs laser beams with a power of $2.5 \cdot 10^{21}$ W. Apart from the mentioned lasers with short pulses in the PW range (one may add the proposal of F.P. Schäfer for a 100 J - 100 fs KrF excimer laser[23]) the next higher step is under design by building a 10 fs -1kJ XeF_2(A-C) laser of rather moderate costs which should produce 100 PW pulses[24]. With this power the electrons would be accelerated up to 6.3 GeV energy. This laser beam may be extented to a 1 m^2 XeF_2 laser beam with 3 fs -10 kJ pulses (3EW=$3 \cdot 10^{18}$W) which can accelerate electrons to energies of 36 GeV. It comes close to the best existing accelerator LEP with 50 GeV electrons. If one would be able to combine 30x30 of these beams, 2700 EW would be reached resulting in the TeV electrons of high luminosity. Therefore, we suggest to develop lasers of the size of 1000 EW operation which is competitive - but probably less expensive - compared to the about 10 years of development and costs of the present SSC accelerator project.

One may ask what other comparable concepts for colliders could be reached. The following suggestion is to focus a 0.5 ps carbon dioxide laser pulse with a focal diameter of 5 wave lengths and an intensity of $3 \cdot 10^{26}$ W/cm^2 into sulphur of solid state density. Electrons with a density of 10^{24} cm^{-3} are then quivering with an energy of about 100 GeV in between the sulphur nuclei having the same energy but with counterstreaming velocities. Any charged collision products will have similar collisions where the motion of the negative particles is like that of the electrons and that of the positive particles like the motion of the nuclei. Because of the relativistic motion the particles can not be distinguished by their masses. The advantage consists in the fact that the charged collision products take part in the following collider interaction.

This laser focus collider would need a carbon dioxide laser design of the kind of the ANTARES laser of Los Alamos[2] with shortened pulses and a cross section of the laser beam of about 100m x 100m. For the focusing a Fresnel lense-like optics or a mirror with about 100m diameter can be used which should not be damaged by the radiation. If one shot per hour would be performed with 10^{18} nuclei in the focus volume, only a 30 MW average power would be needed for the operation and a luminosity of $6 \cdot 10^{35}$ cm^{-2}s^{-1} can be reached. Operation with one interaction per minute would need 11.8 GW average power and provides a luminosity of $3.6 \cdot 10^{37}$ cm^{-2}s^{-1} [25]. The rather modest particle energy of 100 GeV of the collider with this enormous luminosity and the interesting collisions of leptons and hadrons and of their subsequent products could be of high interest for studying the symmetry breaking in the quark-gluon plasma[26].

The question of the expelling of the particles from the laser focus by the nonlinear forces and/or by electric double layers has been estimated to be sufficiently low for the short time of laser interaction by using comparable computations of the escape process in the case of a laser focus with a quivering motion of 10 GeV energy for the efficient production of antiprotons by lasers[27].

If by this way the development of very big lasers and of such of medium size with high technological performance would be opened, many other applications in high energy physics would be possible with numerous further applications in other fields. One example should illustrate this here. If the Petawatt (1 kJ-1ps) neodymium glass[20] or KrF[22] lasers would be realized with about 1 Hz or higher repetition rate, an ideal source for positrons can be built. The electron-positron pair production with lasers has been discussed since many years[28] and it has been estimated that the efficiency for the positron production can be very high[29]. Even if one would work with an efficiency of 1% only for positrons gained by electrostatic extraction, the PW laser pulses focused into solid targets with intensities of 10^{21} W/cm^2 would result in pulses of the order of 10^{13} positrons with durations down to a few ps. While the present accelerators can use classical positron sources, the conditions of the futural lepton colliders as e.g. CLIC (CERN linear collider) or that under consideration at Serpukhov or DESY may well need sources of positrons created with lasers.

Finally we may mention another interesting diagnostic scheme for high energy physics using the high power lasers. If a PW neodymium glass laser pulse would be focused into a collision area filled with electrons and positrons from the LEP accelerator with a laser focusing length of 1 to 2 meters and a focus diameter of about 10 wave lengths, any charged particle in the focus would receive a quivering motion and an additional energy up to a maximum value similarly to the ionization and emission of electrons from helium in the focus of a laser beam[30]. The breaking of the conservation of the Liouville theorem[31] would happen in the same way as in the case of the helium atoms[32].

Also high power lasers can be used to calibrate the energy measurements of bosonic mesons resulting in LEP experiments to such an accuracy that the details of the decay mechanisms of the bottom mesons can be followed up. By working with laser pulses with a power of 10 PW for the mentioned interaction region in LEP, the maximum energy increase for B-particles would be 0.16 MeV, for the decay products of D mesons 1.47 MeV, for K mesons 1.79 MeV and for pions 6.31 MeV. The statistical evaluation of the energy increases provides detailed information of the decay of the B-paticles which have a life time of 1.18 ps, being just in the time range of the mentioned PW laser systems.

ACKNOWLEDGMENTS

One of the authors (H.H.) is grateful for helpful discussions about this topic with Dr. E.J.N. Wilson and the participants of an Academic Traning Course at CERN initiated by him for the author. Thanks are also given to other colleagues at CERN, especially to Dr. H. Haseroth, Dr. E. Schindl, Dr. F. Caspers, and Dr. E. Jensen.

REFERENCES

1. A. Sessler, Am.J.Phys. 54, 505 (1986)
2. H. Hora, Plasma at High Temperature and Density (Springer, Heidelberg, 1991)

3. H. Hora, Nature $\underline{333}$, 337 (1988); Opt.Comm. $\underline{67}$, 431 (1988)

4. L. Cicchitelli, and H. Hora, IEEE J. Quantum Electr. $\underline{26}$, 1833 (1990)

5. V. Petrzilka, Cz.J. Physics $\underline{41}$, 807 (1991)

6. R.Z. Olshaan, A. Gover, S.Ruschin, and H. Kleinman, Phys.Rev.Lett. $\underline{58}$, 483 (1987)

7. S. Kawata, T. Maruyama, H. Watanabe, and I. Takahashi, Phys.Rev.Lett. $\underline{66}$, 2072 (1991)

8. V.V. Apollonov, A.I. Artemev, Yu. L. Kalachev, A.M. Prokhorov, and M.V. Fedorov, Pisma-Zh.Eksp.Teor.Fiz. $\underline{47}$, 77 (1988) (JETP Lett. $\underline{47}$, 91 (1988))

9. K. Shimoda, Appl.Optics, $\underline{1}$, 33 (1962); R.B. Palmer, Particle Accelerators $\underline{11}$, 81 (1980)

10. R. Meerson, and T. Tashima, A mesoscopic linear accelerator driven by superintense subpicosecond laser pulses, Report Inst.Fusion Studies, Univ.Texas, Austin TX, IFSR No.504 (1991) unpublished

11. M.O. Scully, Appl.Phys. $\underline{B51}$,238 (1990); F. Caspers, and E. Jensen, Laser Interaction and Related Plasma Phenomena, H. Hora and G.H. Miley eds. (Plenum, New York, 1991) Vol.9, p.459

12. L.Cicchitelli, H. Hora, and R. Postle, Phys.Rev. $\underline{A41}$, 3727 (1990)

13. W. Scheid, and H. Hora, Laser and Particle Beams, $\underline{7}$, 315 (1989); L. Cicchitelli, H. Hora, and W. Scheid, Advances Accelerator Concepts, AIP Conf. Proceedings No. 193, C. Joshi ed. (Am.Inst.Phys., New York, 1989) p. 17

14. L. Cicchitelli, H. Hora, A. Scharmann, and W. Scheid, Laser Interaction and Related Plasma Phenomena, H. Hora and G.H. Miley eds. (Plenum, New York, 1991) Vol.9, p. 467

15. E.G. Bessonov, Nuclear Instr.Meth. $\underline{A308}$, 135 (1991)

16. T. Häuser, W. Scheid, and H. Hora, Phys.Rev. $\underline{A45}$, 1278 (1992)

17. T. Häuser, W. Scheid, and H. Hora, J.Opt.Soc.Am. $\underline{B5}$, 2029 (1988)

18. H. Hora, Physics of Laser Driven Plasmas (John Wiley, New York, 1981)

19. M. Lax, W. Louisell, and W. McKnight, Phys.Rev. $\underline{A41}$, 3727 (1975)

20. M. Andre, J. Coutant, R. Dautray, M. Decroisetete, L.A. Lompre, M. Naudy, C. Manus, G. Mainfray, A. Migus, D. Normand, C. Sauteret, and J.P. Watteau, Industrial and Scientific Uses of High Power Lasers, SPIE Proceedings No. 1502, J.P. Billon and E. Fabre eds. (SPIE, Bellingham, 1991)

21. Iodine laser: see reference in Proceedings ECLIM Conference (1991)

22. J. Hermann, and B. Wilhelmi, Laser for Ultrashort Light Pulses (Akademieverlag, Berlin,1987)

23. F.P. Schäfer, ECLIM Conference (Schliersee) Febr. 1990

24. L.D. Mikheev, Laser and Particle Beams $\underline{10}$ No.2 (1992)

25. H. Hora, Lasers for Accelerators, CERN Lecture Series, 1991-1992 Academic Training Programme

26. H. Schopper, Nova Acta Leopoldina NF63, 78 (1990); H. Hora, CERN-PS/DL-Note-91/05, August 1991

27. A.S. Christopoulos, H. Hora, R.J. Stening, H. Loeb,and W. Scheid, in Laser Interaction and Related Plasma Phenomena, H. Hora and G.H. Miley eds. (Plenum, New York,1988) Vol.8, p.245

28. J.W. Shearer, J. Garrison, J. Wong, and J.W. Swain, Laser Interaction and Related Plasma Phenomena, H. Schwarz and H. Hora eds. (Plenum, New York, 1974) Vol.3B, p.803; H. Hora, ibid. p. 819

29. W. Becker, Laser and Particle Beams 9, 603 (1991)

30. B.W. Boreham, and H. Hora, Phys.Rev.Lett. 42, 776 (1979)

31. C. Rubbia, On Heavy Ion Accelerators for Inertial Confinement Fusion, CERN-PPE/91-117 (24 July 1991)

32. B.W. Boreham et al., Laser and Particle Beams 10 No.2 (1992)

THOMAS-FERMI LIKE AND AVERAGE ATOM MODEL EQUATIONS OF STATE FOR HIGHLY COMPRESSED MATTER AT ANY TEMPERATURE.

P. Fromy, C. Deutsch, and G. Maynard

L.P.G.P.* Bât. 212,
Université Paris XI
91405 Orsay, Cedex, France

ABSTRACT

We develop a systematic comparison of equations of state computed within the three (TF, TFD and TFDW) Thomas-Fermi modellings with results obtained from the Average Atom Model.

A special emphasis is laid on using the same analytic approximations for exchange, correlation and gradient corrections. Analogies and discrepancies with respect to temperature behaviour are also stressed out.

1. INTRODUCTION

It is well-aknowledged that the understanting of pellet compression for the purpose of achieving Inertial Confinement Fusion (ICF) requires accurate equations de state (EOS) on a very large range of density and temperature values[1,2]. Heavy as well as light materials have to be considered. As expected, a lot of effort have already been devoted to this problem. Amongst the many theoretical frameworks employed for attacking it, a lion share has been provided by Thomas-Fermi (TF) like and its modern offsprings such as the Density Functional Theory (DFT)[3]. The basic idea is then to combine Poisson equation, accounting for the nucleus-electron interaction, with Fermi statistics for the electron jellium. This scheme may also be extended to include exchange within electron component (Thomas-Fermi-Dirac (TFD) model). Subsequently, the diverging electron density at the nucleus has also been corrected by including gradient corrections (Thomas-Fermi-Dirac-Weizäcker (TFDW) model). These modellings build up essentially a classical picture with electrons treated within a BKW-like and semi-classical setting[4]. More recently, it has been proposed [MORE 1979] to treat quantum-mechanically the bound electrons, while keeping those with a positive energy, the free ones, in a classical approximation. One thus obtains the so-called Average Atom Model (AAM) of frequent use for ICF target studies. All these developments obviously point out the need for a detailed comparison of the respective merits and limitations displayed by these various methods of EOS calculation.

Already, some studies have already been devoted to a critical evaluation of accuracy versus numerical efforts of several approaches[1].

The main goal of this work is to compare at the EOS level, the performances of the above four advocated methods : TF, TFD, TFDW and AAM. In order to make meaningfull such a comparison, we intend to work out identical analytic approximations in the various models for specific physical features: exchange, correlation or gradient corrections. In so doing, we expect a fair evaluation of the characteristics of each approach. One of the by-products of such studies is that an accurate EOS evaluated within a TF frameworks for a given temperature-density range, does not always require the most sophisticated modelling, i.e. TFDW, for instance. Such numerical remarks are of obvious interests for handling in a most economical way, the heavy numerical codes describing ICF target compression in various conditions.

In Sec. 2, the status of TF like approximations is asserted.

A systematic derivation of TF, TFD and TFDW thermodynamics is given in Sec. 3 within DFT formalism. The AAM is detailed in Sec. 4. Numerical results are presented and discussed in Sec. 5.

2. THOMAS-FERMI-LIKE APPROACHES: GENERAL REMARKS

Before proceeding with the formal developments, it is worthwhile to recall briefly the status of Thomas-Fermi like approximations for atoms, molecules, plasmas, solids, etc...

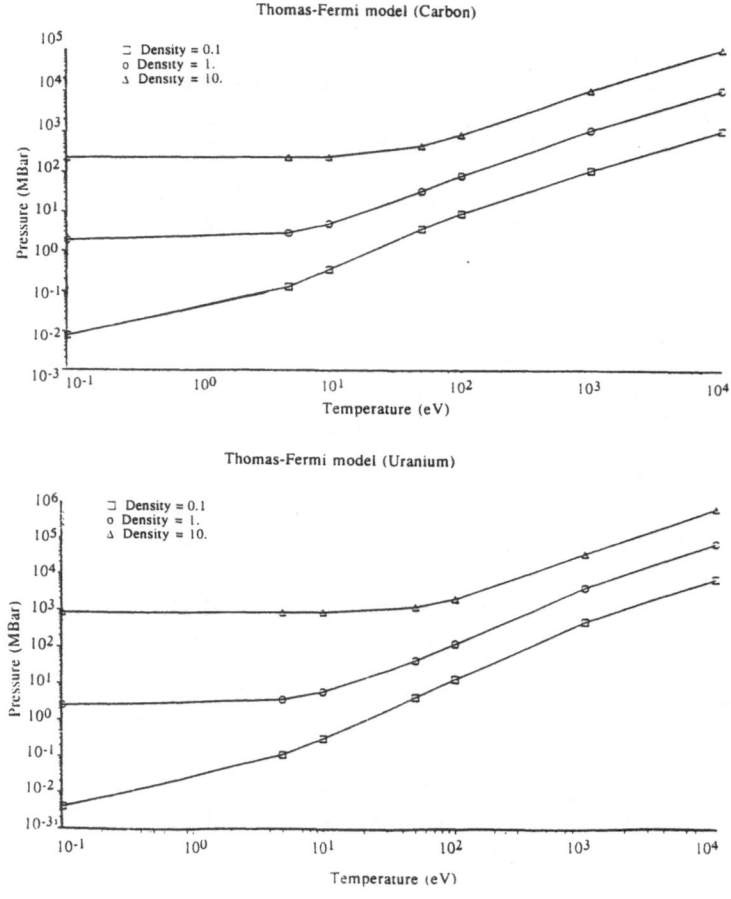

Figure 1. Pressure (Mbar) as a function of Temperature (eV) for Carbon and Uranium computed with Thomas-Fermi approximation for 3 densities measured in solid density (normal conditions).

A lot of rigorous mathematical results[5,6] have been produced which highlight the initial TF model as an exact asymptotic limit when the atomic number Z becomes very large. One thus expects that exchange, correlation, relativistic corrections etc. compensate each other for $Z \rightarrow \infty$. We shall see that those requirement are indeed numerically fullfilled.

In the sequel, compressed matter is represented by a classical fluid of ions of charge Z embedded in a inhomogeneous and degenerate jellium of electrons. Most of the calculations are conducted within the atom sphere of radius $R_0 = (\frac{4}{3} \pi n_I)^{-1/3}$, n_I being the ion density. Suitable boundary conditions have to be selected at atom edge.

DFT DERIVATION OF TF LIKE THERMODYNAMICS

A/- DFT formalism

The Density Functional Formalism (DFT)[3] allows for a unified derivation of the various Thomas-Fermi (TF) like models, at any Temperature T[7].

The basic idea is to minimize the Grand potential quantity $\Omega = PV$ written as:

$$\Omega[n] = F[n] - \mu Z$$

$$\equiv F_e[n] + \int v(\vec{r})n(\vec{r}\,')d(\vec{r})$$

$$+ \frac{1}{2}\iint \frac{n(\vec{r})n(\vec{r})}{|\vec{r} - \vec{r}|} d(\vec{r})d(\vec{r}\,') + F_{xc}[n] - \int \mu n(\vec{r})d(\vec{r}) \tag{1}$$

for a mixture of inhomogeneous electrons and nuclei with charge Z.

$F_e[n]$ denotes the kinetic free energy part of a free electron gas. The 2nd and 3rd term respectively pertain to the potential energy of a inhomogeneous electron fluid within the external potential $v(\vec{r})$ and the electron potential energy. $F_{xc}[n]$ is the exchange-correlation contribution to the free energy. μ is the system chemical potential. $n(\vec{r})$ is the electron density keeping track of Coulomb interactions. The present approach provides a $T \neq 0$ formulation of the response problem for a inhomogeneous electron fluid. The present system is supposed to consist of isolated spherical atoms. We thus consider a pointlike charge Z within a spherical cavity with a positive density of charge ρ_+, for $r > R_0$.

The electron gas inhomogeneity is introduced through a bounded perturbation fulfilling

$$\int |\delta n(\vec{r})| d(\vec{r}) < \infty$$

Putting $n' = n + \delta n$, one obtains

$$\delta\Omega[n] = \Omega[n'] - \Omega[n] = F_e[n'] - F_e[n] + F_{xc}[n'] - F_{xc}[n]$$

$$+ \int v(\vec{r} \, (n'(\vec{r})-n(\vec{r})) \, d(\vec{r} - \int \mu(n'(\vec{r})-n(\vec{r})) \, d(\vec{r})$$

627

$$+ \frac{1}{2} \iint \frac{d\vec{r}\ d\vec{r}'}{|\vec{r} - \vec{r}'|} (n'(\vec{r})\ n'(\vec{r}') - n(\vec{r})\ n(\vec{r}')) \tag{2}$$

Equating to zero, the first order variations, one observes that the exact electron density satisfies

$$\frac{\delta F_e}{\delta n}[n] + \frac{\delta F_{xc}}{\delta n}[n] + v(\vec{r}) + \int n(\vec{r}')/|\vec{r} - \vec{r}'|\ d(\vec{r}') = \mu\ , \tag{3}$$

Where $F(n)$ = free energy density given by

$$F[n] = \int F(n)\ d(\vec{r})$$

Eq. (3) is formally identical to that of a system of independent particles submitted to the external field

$$V(\vec{r}) = v(\vec{r}) + \int n(\vec{r}')/|\vec{r} - \vec{r}'| + \frac{\delta F_{xc}}{\delta n}[n] \tag{4}$$

This analogy is used below, within the Average Atom Model framework.
The minimization of the Grand potential yields a chemical potential under the form

$$\frac{\delta F_e}{\delta n} + \frac{\delta F_{xc}}{\delta n} + v(\vec{r}) + \int n(\vec{r}')/|\vec{r} - \vec{r}'|^{-1}\ d(\vec{r}') = \mu\ , \tag{5}$$

which when combined with Poisson equ. relating $n(\vec{r})$ and $V(\vec{r})$, yields a closed system. Technically speaking, it appears more appropriate to solve these equs. on $V(\vec{r})$. One thus makes use of the simple boundary conditions:
- close to the nucleus, the e-i potential turns Coulombic
- It vanishes at the atom edge ($V(R_0) = 0$).

B/ Thermodynamic quantities

We are now considering the thermodynamics derivation.
We restrict our treatment to the degenerate electron component. The ionic one might be included through the strongly coupled and classical one-component-plasma (OCP) approximation results[8,9]. At best, the ion contribution to thermodynamics would be a few per cents of the electron one considered here.
Within the Grand canonical ensemble formalism, one obtains the thermodynamic derivatives under the form ($\beta = (k_\beta T)^{-1}$)

$$E = \frac{\partial}{\partial \beta}[\beta\ (\Omega + \mu Z)]_{V,n} \tag{6}$$

$$P = \frac{\partial}{\partial V}[\beta\ (\Omega + \mu Z)]_{\beta,n}\ s^3\ , 0 \leq s \leq 1$$

P fulfills electroneutrality when volume V is changing.

From the above $\Omega[n]$ expression, one derives E as

$$E = K + X + W \tag{7}$$

$$K = \frac{\partial}{\partial \beta} [\beta\, F_e[n]]_{V,n} \quad ,$$

$$W = \int v(\vec{r})\, n(\vec{r})\, d(\vec{r})$$

$$+ (\tfrac{1}{2}) \iint n(\vec{r})\, n(\vec{r}') / |\vec{r} - \vec{r}'|\, d(\vec{r})\, d(\vec{r}')$$

$$X = \frac{\partial}{\partial \beta} [\beta\, F_{xc}[n]]_{V,n} \quad , \text{ exchange}$$

P may be similarly displayed. It is also worthwhile to estimate the exchange free energy density up to 1st order in the interaction, i.e. (all T values)

$$F_x(n) = -\frac{1}{2\pi^3 \beta^2} \int_{-\infty}^{\eta} [I_{-1/2}(\eta')]^2\, d\eta' \, , \tag{8}$$

in terms of

$$n = (\frac{\sqrt{2}}{\pi^2})\, \beta^{-3/2}\, I_{1/2}(\eta) \quad , \tag{9}$$

and the Fermi function

$$I_\alpha(z) = \int_0^{\infty} \frac{x^\alpha\, d\,x}{e^{x-z} + 1}$$

Let us also recall that the kinetic free energy may also be explained as:

$$F_k(n) = (\frac{\sqrt{2}}{\pi^2})\, \beta^{-5/2}\, [\eta\, I_{1/2}(\eta) - (2/3)\, I_{3/2}(\eta)\,] \tag{10}$$

THOMAS-FERMI-DIRAC-WEIZÄCKER (TFDW) MODEL

As a most sophisticated representative of the TF class, we detail here the treatment of the TFDW model including exchange (Dirac) within the electron component as well as gradient (Weizäcker) corrections near nucleus.

A/ Zero Temperature (T = 0)

Through the replacements

$$\frac{\partial \mathcal{F}_k}{\partial n} \to K_t \, n^{2/3} \quad , \quad K_t = (3\frac{\pi^2}{2})^{2/3}$$

$$(11)$$

$$\frac{\partial \mathcal{F}_{xc}}{\partial n} \to K_x \, n^{1/3} \quad , \quad K_x = -(\frac{3}{\pi})^{1/3}$$

one recovers the central relationship

$$K_t \, n^{2/3} + K_x \, n^{1/3} + (1/72) \, ((\vec{\nabla} \, n/n)^2 - 2\nabla \, n/n) = U(r) + \mu, \tag{12}$$

where

$$U(r) = \frac{Z}{r} - \int d\vec{r}' \, n(\vec{r}') \, / \, |\vec{r} - \vec{r}'| \quad , \tag{13}$$

All T = 0 TFlike models are formulated within this equation. When the density increases the kinetic term overwhelms the others. As a consequence the Thomas-Fermi limit which restricts to this kinetic term is the right asymptotic limits for all TFlike models.

All lot of attention has already been paid to TFDW at T = 0. For instance, it has been extensively used for investigating isostructural transitions in highly pressured materials[10].

B/ Finite Temperature (T ≠ 0)

Using the expression[11].

$$\frac{h(n)}{n} = -(\frac{\sqrt{2}}{24}) \, \beta^{3/2} \frac{d}{d\eta} \, (\frac{1}{I_{-1/2}(\eta)}), \tag{14}$$

for the inhomogeneity contribution to electron kinetic energy, the free energy density reads

$$\frac{\partial \mathcal{F}_k}{\partial n} - \frac{\partial}{\partial n} \, (\frac{h}{n}) \, (\vec{\nabla} n)^2 - 2h \frac{\Delta n}{n} + \frac{\partial \mathcal{F}_{xc}}{\partial n} = U + \mu \tag{15}$$

Upon introducing ø(r) through

$$\varphi(r) = \frac{Z}{r} \, \text{ø}(r) = \mu + U(r) + \frac{1}{2\pi^2} \quad , \tag{16}$$

with the notations

$$\frac{d^n \text{ø}}{dr^n} = \text{ø}^{(n)} \quad , \quad \frac{d^p n(r)}{dr^p} = n^{(p)}$$

one reaches the T ≠ 0 TFDW extension $(h(1) = \frac{dh(n)}{dn})$

$$\emptyset^{(4)} = \frac{2\pi r n}{Zh} \left[(h - n h^{(1)}) \left(\frac{n^{(1)}}{n}\right)^2 + \frac{\partial \mathcal{F}_k}{\partial n} + \frac{\partial \mathcal{F}_{xc}}{\partial n} - \frac{Z}{r} \emptyset(r) + \frac{1}{2\pi^2}\right], \qquad (17)$$

Where n(r) and ø(r) are connected by the Poisson equation

$$n(r) = \left(\frac{Z}{4\pi r}\right) \emptyset^{(2)}(r)$$

Some partial results have already been obtained for the TFDW model[11,12]. Our present goal is to perform a systematic comparison with TF et TFD models, as well as with the Average Atom Model (AAM) detailed below. Therefore, we emphasize the same approximations for the thermodynamic quantities and energy levels behaviours in the different approaches. Thus, we shall be entitled to perform a meaningfull comparison.

C/ Numerical Procedure

If n(r) denotes an approximate solution of the TFDW model, far from the nucleus it should becomes TFD, so that

$$n_W(r) = \lambda \, n_D(r) \qquad , r > r_1 \quad , \qquad (18)$$

in terms of the TFD density $n_D(r)$. Close to the nucleus, one gets a finite density

$$n_W(r) = \lambda \, \alpha \, (1 - 18 \, Zr + \beta r^2 + \gamma r^3) \qquad r \le r_1, \qquad (19)$$

with

$$r_1 = \frac{1}{12 \, Z}$$

which does not need to be highly accurate.

α, β, γ are continuity determined while γ is fixed though normalization.
$n_D(r)$ is used to derive a cubic spline approximation for ø(r), through Poisson equation.
The previously discussed boundary conditions yield the constraints

$$\emptyset(0) = 1 \text{ and } \emptyset^{(2)}(0) = 0$$

Continuity of U(r) and its gradient, together with the electroneutrality requirement of the atomic cell, provide two additional relationships

$$\emptyset(Ro) \qquad = Ro \, \emptyset^{(1)}(Ro)$$

$$\hspace{10cm} (20)$$

$$\emptyset^{(2)}(Ro) \qquad = Ro \, \emptyset^{(3)}(Ro)$$

The last one derives from $\frac{dn}{dr}(Ro) = 0$.

D/ Thermodynamics

TFD and TFDW fulfill the same relationship between $\phi(r)$ and μ, i.e.

$$\mu = \frac{Z \, \phi(Ro)}{Ro} - 1/2 \, \pi^2 \quad , \tag{21}$$

The potential reads as

$$U(r) = \frac{Z}{r} \phi(r) - 1/2 \, \pi^2 \quad , \tag{22}$$

while the electron density is given by

$$n(0) = (\frac{Z}{4\pi}) \, \phi^{(3)}(0) \quad , \tag{23}$$

At $T = 0$, the pressure is

$$P = \frac{2}{5} K_t \, [n \, (Ro)]^{5/3} - \frac{1}{36} n''(Ro) + \frac{1}{4} K_x \, [n(Ro)]^{4/3} \tag{24}$$

$n''(r)$ is derived from Poisson equation with

$$n''(r) = (\frac{Z}{4 \, \pi}) \frac{[r^3 \, \phi^{(4)} - 2r^2 \, \phi^{(3)} + 2r\phi^{(2)}]}{r^4} \quad , \tag{25}$$

and the final quantity

$$n''(Ro) = \frac{Z}{4 \, \pi} \, [\frac{\phi^{(4)} \, (Ro)}{Ro}] \quad , \tag{26}$$

through the initial differential equation.

At $T \neq 0$, the pressure reads as :

$$P = (\frac{n\partial \mathcal{F}_k}{\partial n} - \mathcal{F}_k)_{r=Ro} - 2 \, (h \, n'')_{r=Ro} + (\frac{n\partial \mathcal{F}_{xc}}{\partial n} - \mathcal{F}_{xc})_{r=Ro} , \tag{27}$$

AVERAGE ATOM MODEL (AAM)

Now, we depart in a significant way from the above classical approaches. Bound electrons are treated in a quantum-mechanical fashion, while maintaining a classical framework for the free ones. If ψ denotes a bound state wave function, the energy of bound electrons is derived through $\psi(R_0) = 0$. Band effects may be taken care of with $\frac{d\psi}{dr}(R_0) = 0$. A given electron is considered free as soon as its energy is > 0.

The Fermi statistics and the central field approximation are maintained, giving an average atom representation. It is generally believed that in the region of high temperature

632

and/or pressure, the TFD model gives a reasonably accurate electron potential. This potential, in turn, can be used as a basis to treat the atom in a more quantum-mechanical way by solving the single-electron wave equation for the bound states.

The basic assumption is that during an elementary stopping process, the target ion is nearly instantaneously neutralized by plasma fluctuations. This hypothesis is particularly relevant to matter composed by intense ion beams, in view of the relatively long pulse time ($\sim$ 20-30 ns) allowing the target species to be considered in local thermodynamic equilibrium (LTE) with comparable electron and ion temperature T and ion number density n_I.

The numerical procedure is initialized by taking each $R_0 = (\frac{3}{4\pi n_I})^{1/3}$. The AAM schema is implemented as follows[13].

(a) Bound electrons are considered independent. They are supposed to move in a spherically symmetric and self-consistent potential $V_{eff}(r)$ (see Eq. (30))., the same for all electrons. However, exchange and correlation are still retained in $V_{eff}(r)$. The total electron density

$$\rho(r) = \rho_b(r) + \rho_f(r) ,$$

fulfills

$$4\pi \int_0^{R_0} dr\, r^2 \rho(r) = Z \quad , \tag{28}$$

with $\rho_b(r)$ in terms of single-electron eigenquantities (ε_i, ψ_i).

(b) The average atom assumption allows us to replace the various excitation states in target with those of a fictitious atom with noninteger occupation numbers for excited orbitals. $V_{eff}(r)$ is taken constant within each subshell.

(c) Conceptually and numerically speaking it is most appropriate to derive (ε_i, ψ_i) from a Dirac equation, which proves especially convenient to tackling the eigenvalue problem through two coupled first-order Eqs (31). In this paper, we do not intend to emphasize relativistic and spin effects of increasing relevance for heavy elements. However, this could be easily achieved within the present framework.

The bound electron radial density $\rho_b(r)$

$$4\pi\rho_b(r) = \sum_{nlj} \frac{(2j + 1)}{\exp[\beta(E_{nlj} - \mu)]+1} \times [A^2_{nlj}(r)+B^2_{nlj}(r)] , \tag{29}$$

with the total electron energy

$$E'_{nlj} = E_{nlj} + V_{eff}(r)$$

and

$$V_{eff}(r) = V_{cb}(r) + V_{exch}(r) + V_{cor}(r) \quad , \tag{30}$$

a sum of Coulomb, exchange, and correlation potentials is readily expressed in terms of radial Dirac wave functions deduced from

$$\frac{d}{dr}\begin{pmatrix}A\\B\end{pmatrix} = \begin{pmatrix} -\dfrac{K}{r} & \dfrac{V_{eff}(r) - E_{nlj}}{cs} \\[2mm] -\dfrac{V_{eff}(r) - E_{nlj}}{cs} & \dfrac{K}{r} \end{pmatrix}\begin{pmatrix}A\\B\end{pmatrix} \tag{31}$$

where $K = -s(j + \frac{1}{2})$, $c = 137.037$ (a.u.), $s = \pm 1$.

633

Thomas-Fermi model (Carbon)

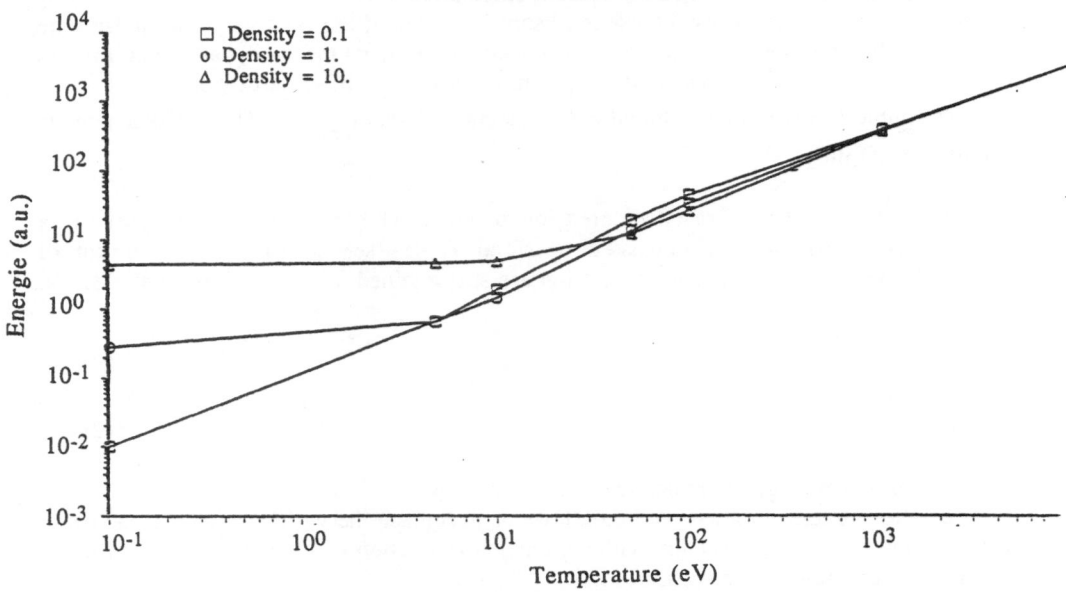

Thomas-Fermi model (Uranium)

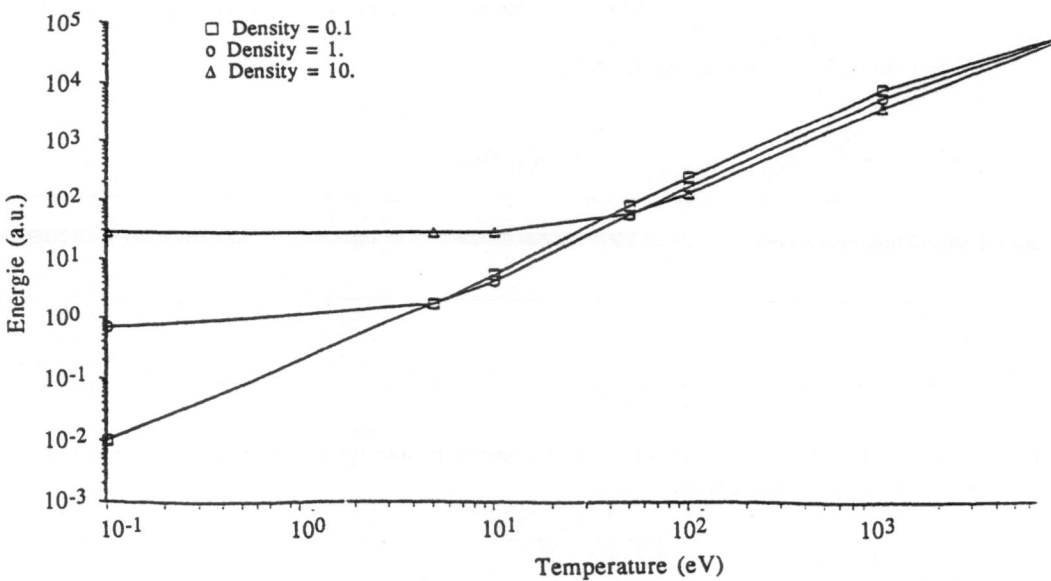

Figure 2. Energy (atomic units) as a function of Temperature for Carbon and Uranium computed with Thomas-Fermi approximation for 3 densities measured in solid density (normal conditions).

TABLE 1. Energy levels (a.u.): Thomas-Fermi-like and the Average Atom Model (AAM) data are obtained for $T = 0$ and $\rho = \dfrac{1}{10}$ solid density.

	HF	TF	discrepancy %	TFD	discrepancy %	TFDW	discrepancy %	A.A.M.	discrepancy %
Carbon									
1 S	11.3436	8.0592	28.95	9.4923	16.32	9.4305	i6.86	9.9533	12.25
2 S	0.712603	0.2883	59.94	0.4525	36.50	0.35803	49.75	0.50838	28.65
Aluminium									
1 S	58.6336	51.807	11.64	55.291	5.70	55.513	5.32	55.207	5.84
2 S	4.9293	3.47118	29.58	4.335	12.05	4.07415	17.34	3.8955	20.97
Fe									
1 S	262.83	248.078	5.61	255.703	2.71	256.721	2.32	55.207	5.84
2 S	32.476	27.4916	15.34	29.503	9.15	29.459	9.28	3.8955	20.97
Ag									
1 S	943.158	916.923	2.78	932.864	1.09	934.527	0.915	929.120	1.49
2 S	141.993	133.601	5.91	138.308	2.59	138.550	2.42	136.770	3.6
Pb									
1 S	3255.875	3231.99	0.73	3279.37	-0.72	3267.81	-0.36	3230.61	0.77
2 S	587.8814	579.343	1.45	592.65	0.81	591.368	-0.59	577.307	1.79
Uranium									
1 S	4279.23	4267.54	0.27	4329.30	-1.17	4309.84	-0.71	4253.535	0.60
2 S	806.159	800.15	0.74	817.533	-1.41	814.922	-1.08	749.281	0.70

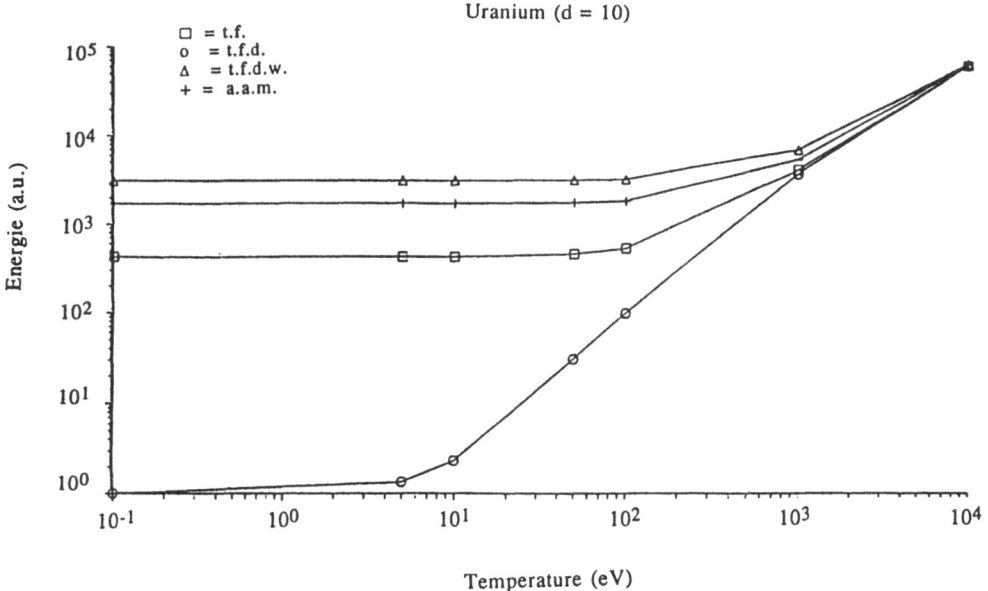

Figure 3. Internal energy (atomic units) as a function of Temperature for Uranium at 10 times solid density, computed through TF, TFD, TFDW and Average Atom Model (AAA).

RESULTS AND DISCUSSION

On table 1, TF, TFD and TFDS models are used through potentials worked out in Schroedinger equation to determine the 1S and 2S levels for a few selected elements. Results are contrasted (left) to isolated atom Hartree-Fock (HF) ones, taken as reference values. On the right, the corresponding AAM data are also plotted. The shrinking of discrepancies between HF and all TF models is obvious when one reads vertically the table, from light to heavy species. Clearly, for Uranium the Thomas-Fermi approximation discloses a status of an exact asymptotic limit, in accord with expectations. TF like and AAM data are obtained for $T = 0$ and $\rho = 10^{-1}$ solid density (normal conditions)

The effect of density and temperature on Pressure and Internal Energy is depicted on Figs 1 and 2 respectively.

Three densities (in unit of normal solid) are considered.

The behaviour of light element (Carbon) and heavy one (Uranium) looks very much alike. However, high temperature displays a much stronger effect on E than on P.

The three Thomas-Fermilike methods (TF, TFD and TFDW) are contrasted to the Average Atom model (AAM) results for Internal energy and Uranium at ten times solid density. Considerable discrepancies are shown for low temperature. The AAM and TFDW data are quite close, the one to the other, and feature the most accurate outputs (Fig. 3)

REFERENCES

1 R.M. More, K.H. Warren, D.A. Young and G.B. Zimmerman, *Phys. Fluids*, **31**, 3059 (1988).
2 C. Deutsch, *Ann. Phys.*, Fr., **111**, 1 (1986).
3 P. Hohenberg, and W. Kohn, *Phys. Rev.*, **136**, B864 (1964).
4 L. Landau, and E.M. Lifschitz, *Quantum Mechanics*, Pergamon, Oxford, p. 259 (1977).
5 E.H. Lieb, and B. SIMON, *Adv. Math.*, **23**, 22 (1977).
6 E.H. Lieb, *Rev. Mod. Phys.*, **53**, 603 (1981).
7 D. Mermin, *Phys. Rev.* **137**, A1441 (1965).
8 C. Deutsch, Y. Furutani, and M.M. Gombert, *Phys. Repts*, **69**, 86 (1981).
9 C. Deutsch, *Phys. Scri.*, **T2**, 192 (1982).
10 J. Meyer-Ter-Vehn, and W. Zittel, *Phys. Rev.*, **B37**, 8674 (1988).
11 F. Perrot, *Phys. Rev.*, **A20**, 586 (1979).
12 Y. Furutani, M. Shigesada, and H. Totsuji, *J. Phys. Soc. Japan*, **55**, 2653 (1986).
13 X. Garbet, C. Deutsch, and G. Maynard, *J. Appl. Phys.*, **61**, 907 (1987).

LASER INDUCED SHOCK WAVES AND DYNAMIC DAMAGE

S. Eliezer[1,2],I. Gilath[2] and Y. Gazit[2]

[1]Instituto de Fusión Nuclear, E.T.S. Ingenieros Industriales
José Gutierrez Abascal, 2;28006 Madrid, Spain
[2]Soreq N.R.C. Yavne 70600, Israel

ABSTRACT

A high irradiance short pulsed Nd: glass laser was used to generate shock waves in alminum (Al) and copper (Cu). Dynamic fracture at hypervelocity impact conditions was observed as a result of reflected shock waves as tensile waves from the back surface of the samples. Damage development from incipient spallation to complete sample perforration were obtained in planar shock wave conditions. Maximun elongation at fracture at ultra high strain rate was measured for several metals. Shock wave attenuation and dynamic strength were evaluated for Al and Cu. The experimental results of threshold for spall and spall width at this energy were compared with numerical simulations.

The laser pulses were also used to generate hemispherical shock waves in planar targets (focal spot smaller than target thickness). A linear experimental relationship was obtanied between laser energy for threshold spall conditions and the cubic target thickness. This relation is equivalent with the progatation of a strong point explosion where the internal energy per unit volume of the shocked material is constant.

INTRODUCTION

Spall is a dynamic fracture of materials, extensively studied in ballistic research[1]. The term spall as used in shock wave research is defined as planar separation of material, parallel to the wave front, as a result of dynamic tensile stress components perpendicular to this plane.[2] Spall in ductile materials is controlled by localized plastic deformation around small voids that grow and coalesce to form the spall plane. Spall in brittle materials takes place by dynamic crack propagation without large-scale plastic deformation.[3-7]

There are different experimental methods to study dynamic fracture and spall: (a) by plate impact,[2, 8, 9] (b) by high explosives[2, 10] and (c) by laser-induced shock waves.[11-20]

The plane parallel plate impact experiments are mostly used because the inputs are well known and controllable, and the results are easily interpreted. Impact velocity and geometry are changed to control the (rectangular) pressure pulse and the strain rate. The spall width is known a priori. Diagnostics usually include the measurement of the free surface velocity-time history or the shock stress-time history during impact. To a good approximation, the explosively produced shock has a triangular shape, but the position of the spall layer and the tensile loading rate are not easily determined in advance.

A review article on the effect of shock waves on materials by Davison and Graham[9] summarized the available information on spall strength up to 1979. This information is extended up to 1983 in another review article by Meyers and Aimone[21]. This article describes the (nonlaser) spall research done mainly at Stanford Research Institute, at Sandia and at Los Alamos laboratories. It highlights the main advances: the concept of damage function, the equation describing nucleation and growth rates, and the generalized formulation for fragment size. It also discusses the theoretical aspects of crack and void growth, the computer codes used, and the metallurgical aspects of spall. The microstatistical fracture mechanics that relate the kinetics of material failure microstructural level to continuum mechanics were reviewed by Curran and Seaman[5].

Shock loads can be generated by high-speed impact, explosives or by intense short-time energy deposition. The advantage of the laser induced damage as compared to most of other impact experiments is the study of thermomechanical damage at strain rates of the order of 10^7 s^{-1}, corresponding to hypervelocity impact condictions.

In laser-matter interaction under consideration the inward momentum is imparted to the target by the ablation of the hot plasma rather than by the momentum of the laser photon themselves. The heated material is blown off the target and the ablation drives a shock wave into the target. At the final stages of the laser pulse the ablation pressure at the surface drops and a rarefaction wave propagates into the target, following the shock wave. This pressure wave is aproximately of a triangluar shape and it propagates into the target (see Fig. 1). After reaching the back surface of the target it is reflected and a tension (negative pressure) is created. There is a minimum pressure value P_s for spall to occur.

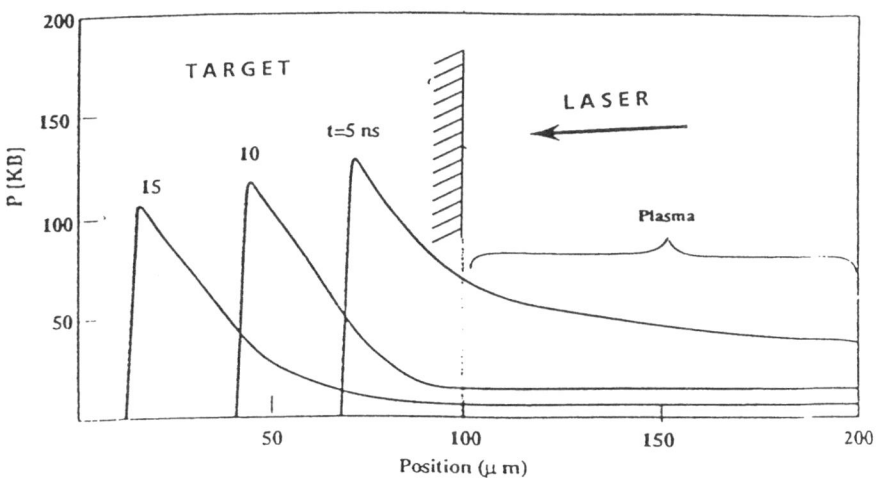

Figure 1.   Shock wave propagation at different times for a 100-μm Al foil. (I = 2 × 10¹¹ W/cm²).

Unsing our laser-induced shock waves, spall in metallic targets was investigated and all stages of damage or dynamic failure were identified. For damage evaluation, metallurgical or scanning electron microscopy was used. For internal damage evaluation, the samples were sectioned, polished, and examined by optical microscopy. The intensities of the 3.5ns Nd:glass laser were in the rage of 10^{10}-10^{12} W/cm², and the foil thickness was in the 100-600μm range. The laser-generated shock wave pressure was in the range of a few hundred kilobars (kb). The shock wave traversed the foils in a few tens of nanoseconds (ns). The controlled stepwise increase in laser energies allowed us to find the stages of damage evolution from incipient to complete perforation of the target foils. The incipient spall for ductile metals was identified from the level of separate voids in aluminum, while brittle metals from the level of cracks. Those voids or cracks coalesce resulting in a continuous spall layer. For higher laser intensities the spall layer breaks away and target penetration is observed for specific high intensities.

COMPUTER SIMULATIONS

Computer simulations of the laser-induced spall were performed[18], including the laser absorption, shock wave travel through the foil, and the spall phenomena. The simulation were done for aluminum (Al) and copper (Cu) targets. The simulations were based on the one-dimensional (1D) MEDUSA computer code, which was expanded to include the spall phenomena, using simple spall criteria. An estimate of the strain rate was drived from simulations and it was shown that the strain rate in these experiments are about 10^7sec^{-1}: A typical shock wave profile at different times is given in Fig. 1, for 100μ Al foil and a laser irradiance of $I = 2 \times 10^{11} \text{ W/cm}^2$.

Table 1. The spall pressure and pressure gradient in Al and Cu.

Material	P_{spall} (kb)		$P_{gradient}$ (kb/mm)
	Simulation*	Experiment	
Copper	25 ± 5	20 ± 7	180 ± 60
Aluminum	25 ± 5	25 ± 8	60 ± 20

*The ± 5 in simulation means: the range of 20 to 30 kb depends on the strain rate ($\dot{\varepsilon}$). For $\dot{\varepsilon} = 1.5 \times 10^7 \text{ sec}^{-1}$ the simulation yields $P_{spall} \approx 20$ kb (for Al and Cu) while for $\dot{\varepsilon} = 4.5 \times 10^7 \text{ sec}^{-1}$ it gives $P_{spall} \approx 30$kb

A two dimensional (2D) code was used[20] in order to describe the spall phenomena (which is by definition at least a 2D effect) for Al and Cu targets. The numerical simulations were compared with our experiments. Good results were obtained between experimental and simulation results for the spall pressure and the spall widths. In table 1 the spall pressure derived in experiments and in simulation is given. For a strain rate $\dot{\varepsilon} = (1.5 \div 4.5) \times 10^7 \text{sec}^{-1}$ the simulations yield a spall pressure in the range of $20 \div 30$ kb. In table 2 the spall widths as a function of foil widths and laser irradiance are given. A good agreement is obtained between simulation and experiments.

Material	Foil width (μm)	Laser intensity (10^{11} W/cm²)	Spall width (μm) experim.	Spall width (μm) simulat.
Al	100	0.72	28 ± 5	25 ± 4
Al	100	2.0		23 ± 4
Al	300	2.5	42 ± 6	47 ± 6
Al	300	5.0		45 ± 6
Al	600	5.0	68 ± 3	76 ± 8
Al	600	9.0		77 ± 8
Cu	100	1.0	15 ± 4	14 ± 4
Cu	600	4.0	30 ± 4	38 ± 6
Cu	600	6.0	56 ± 6	68 ± 8

THEORETICAL ESTIMATION OF STRAIN RATE AND SPALL STRENGTH

The strain rate in ballistic experiments is estimated[22] as $\dot{\varepsilon} = \Delta\varepsilon/\Delta t$, where $\Delta\varepsilon = \varepsilon = 2U_p/C$, where ε is the strain, U_p is the particle velocity, and C is speed of sound. This gives only rough estimate of the strain rate. The exact strain rate associated with spall is difficult to determine accurately. Using the analogy of laser experiments with ballistic experiments for strain-rate calculations, Δt can be substituted by T_L. The strain for laser experiments can be written in the same way, and the rough estimate for the strain rate is accordingly

$$\dot{\varepsilon} = (2U_p/CT_L) \quad s^{-1} \tag{1}$$

By substituting for U_p and C the values compatible with few hundred kb, we obtain the following strain rate:

$$\dot{\varepsilon} = (3-4) \times 10^7 \quad s^{-1} \tag{2}$$

The strain rate can also be calculated from the simulation itself:

$$\dot{\varepsilon} = \frac{-\dot{\rho}}{\rho} = -\frac{1}{\rho}\frac{\partial \rho}{\partial x}C, \tag{3}$$

where $\partial \rho / \partial x$ is the density gradient obtained from the simulations at time and position right before the spall occurs, and C is the speed of sound. From[18] the 1D simulation we deduce $\partial \rho / \partial x$. The density gradient is laser-intensity dependent and in our simulations varied in the range of -5×10^6 to -2×10^7 kg/m^4. Therefore the strain rate will be in the range of $(1-4) \times 10^7 s^{-1}$, in agreement with the strain rate calculated above.

The strain rate in the planar shock wave can be also defined as the gradient of the particle velocity (U_p) in the shock wave

$$\dot{\varepsilon} = \frac{\partial U_p}{\partial x} \tag{4}$$

Using the relation $\partial x = C\partial t$ one can write

$$\dot{\varepsilon} = \frac{1}{C}\frac{\partial U_p}{\partial t} \tag{5}$$

Using the connection of the free surface velocity (U_{FS}) with the particle velocity near the rear surface

$$U_{FS} = 2U_p, \tag{6}$$

one can determine the strain rate in the area. The simulation[20] for U_{FS} and eqs. (5) and (6) yield a strain rate in the range of $(1.5 \div 4.5) \times 10^7 sec^{-1}$ in good agreement with the above estimates. Fig. 2 describes the free surface velocity as a function of time for 100μm Al foil irradiated by a 0.72×10^{11} W/cm^2, 3.5nsec laser pulse[20]. From Fig. (2) one can estimate the spall pressure,

$$\sigma_{small} = \frac{1}{2}\rho C \Delta v \tag{7}$$

where Δv is defined in Fig. 2, ρ is the target density and C is the speed of sound. The values obtained from eq (7) are given in table 1 (the simulation results).

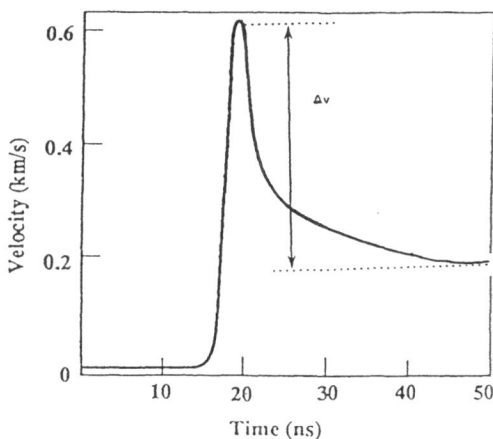

Fig. 2.   Free surface velocity as a function of time for 100 µm Al foil irradiated by a 7.2×10^{11} W/cm², 3.5 nsec.

EXPERIMENTAL ESTIMATION OF SPALL STRENGTH[19]

At threshold spall, we have the same spall pressure and the same shock velocity at the rear surface for different foil thicknesses. Using targets with decreasing thicknesses, the laser induced plasma ablation presure P_{th} represents the real spall stress (P_S) in addition to the value corresponding to pressure decay through the target thickness Δx

$$P_{th} = P_S + \frac{\partial P}{\partial x} \Delta x + \ldots . \tag{8}$$

By extrapolating the experimental values for zero thickness $\Delta x = 0$, the limiting value will represent the dynamic spall pressure. For small values of Δx the Taylor expansion can be approximated by the first two terms given in eq. (8). In this case the slope will represent the pressure gradient. Experiments were performed to evaluate the pressure decay and spall pressure. The threshold pressure necessary to obtain incipient spallation of the samples was calculated from the plasma ablation pressure using our simulations, or alternativbley the following equation[23],

$$P_{kb} = 230 A^{7/16} Z^{-9/16} \lambda^{-1/4} \tau^{-1/8} \left(\frac{I_L}{10^{12}} \right)^{3/4} \tag{9}$$

where, the pressure is given in kilobars, A is the atomic number, Z is the ionization number, λ is the laser wavelenght in μm, τ is the laser pulse time in nanoseconds, and I_L is the laser intensity in W/cm². Equation (9) was compared with our simulation code[18] for the materials under investigation in the irradiance regime of 10^{10}-10^{12} W/cm², and an agreement of $\pm$ 30% was found. The laser-light absorption for the above irradiances is about 80% $\pm$ 10%. The ionization Z was calculated[24] using our PLASMOR computer code. Taking this input into account our estimation of spallation pressure and pressure decay are known within a factor of two. A linear dependence was obtained for the pressure as a function of target thickness for aluminum and copper. These experimental results are given in table 1.

DYNAMIC STRAIN MEASUREMENTS

For low intensities no damage is observed on the back surface of the target. With incrasing intensity an internal crack develops which can be seen as a swelling of the back surface. By carefuffy controling the laser intensities, the elongation (l) and the spall thickness (x_0) can be measured, from incipient fracture up to the breaking of the spall layer (see Fig 3). The strain in Fig 3 is defined as

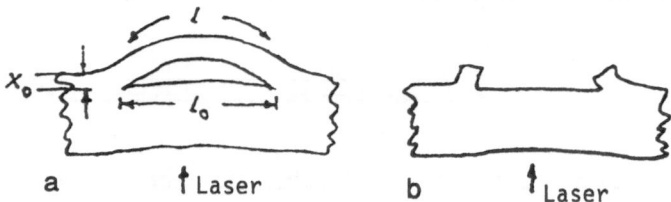

Fig. 3.   (a) Internal fracture. (b) spall layer break-away

$$\epsilon = \frac{l - l_0}{l_0},$$ (10)

and is measured at different stages. In table 3 values of ϵ are given as a function of laser irradiance and energy. The spall x_0 is aso given in table 3. It is interesting to point out that x_0 does not depend on the laser irradiance (see table 3). However x_0 changes (almost linearly) with target thickness (see table 2)

Table 3.   Spall development as a function of laser intensity for Al target

Sample	I (W cm^{-2})	E (kJ cm^{-2})	X_o (μm)	ϵ
1	8.0×10^{11}	2.4	48 ± 5	0.0050
2	9.5×10^{11}	2.8	45 ± 5	0.0476
3	1.1×10^{12}	3.4	45 ± 5	0.0557
4	1.4×10^{12}	4.2	45 ± 5	0.0761
5	1.8×10^{12}	5.3	50 ± 5	0.0783

TWO DIMENSIONAL SHOCK WAVES [25]

For the two dimensional (2D) shock wave experiments the laser light was focused on the target to focal spots of 0.05 mm (radius, R_L) . The target material was pure Al foil of d = (0.5 ÷ 1.3) mm thickness, i.e. d ≫ R_L. The laser is a single beam Nd : glass system capable of delivering up to 120 joules energy in 3.5 nsec. The controlled stepwise change in laser energies allowed to find the threshold energies for incipient spalling. The spall was observed by microscopy on sectioned targets. For each sample thickness about 10 laser shots were performed. In Fig 4 the energy is given as a function of cubic target thickness for threshold spall condition. The following relation is obtaned

$$(\frac{E_L}{Joules}) = 45.3 \, (\frac{d}{mm})^3 + 4.9 \tag{11}$$

At incipient spall, for different target thicknesses, the same shock parameters prevail at the bak surface. Therefore, in thicker targets, higher energies (and thus ablation pressure) are applied on the front surface to make up for the energy loss in the target. Thus equation (11) is equivalent to the propagation of a strong explosion where the pressure decreases with the distance according to the relation

$$P \simeq \eta \, \frac{E_L}{R^3} \tag{12}$$

where R is the distance from the explosion and η is an efficient parameter. Taking for P a pressure threshold damage of (25 ± 5) (kb) (see table 1), eq. (12)

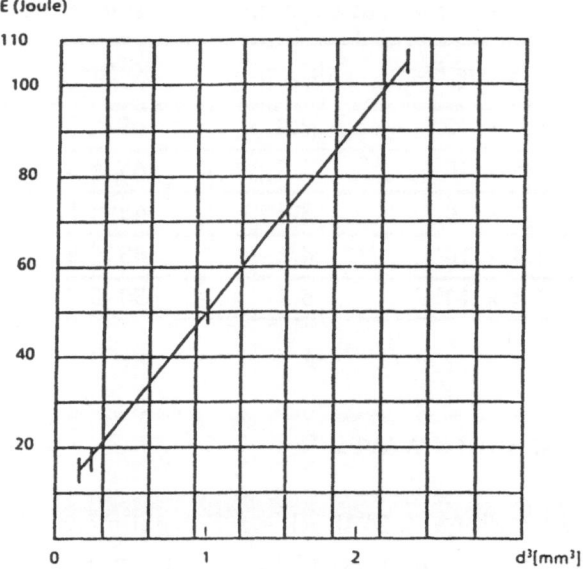

Fig. 4.   Energy as a function of cubic target thickness for threshold spall
conditions. In these experiments the laser focal spot is much smaller
than the target thickness (2D shock waves).

and Fig. 4 give $\eta = (3 \div 5)\%$. In particular for $d = 1mm$, $E_L = 50j$ and $P = 25kb$ one gets $\eta = 5\%$.

The physical content of eq. (11) shows that the energy per unit volume transferred for the shock wave to the target is constant. The experimental results indicate that for a given laser energy the pressure attenuates as the cubic distance (i.e. volume dependent).

CONCLUSIONS

This paper summarizes the experimental and theoretical work [14-20,25] by the authors and collaborators in the subject of laser induced dynamic damage in Al and Cu. This research suggest that the high intensity laser seems to be a good device to measure high strain rate damage in materials.

Experimental techniques supported by 1D and 2D simulations were developed to measure:
a) spall strength (pressure).
b) shock wave decay.
c) dynamic strain.
d) strain rate.

646

Furthermore, 2D wave experiments in material are suggested through the threshold spall damage in materials. This analysis may open the way for 2D shock wave analysis.

Acknowledgement

One of the authors (S.E.) would like to thank Prof. G. Velarde, Prof. J.M. Martínez-Val and the staff of the *Instituto de Fusión Nuclear-Universidad Politécnica de Madrid* for their kind hospitality. The useful and enlightening discussions with the people of the Institute are very much appreciated.

REFERENCES

1. A. Zukas, T. Nicholas, H.F. Swift, L.B. Gresczuk and D.R Curran, *Impact Dynamics* (Wiley, New York, 1982).
2. J. N. Johnson, J. Appl. Phys. **52**, 2812 (1981).
3. D.A. Shockey, L. Seaman and D.R. Curran, *Metallurgical Effects at High Strain Rates* (Plenum, New York, 1973), pp. 473-499.
4. S. Cochram and D. Banner, J. Appl. Phys. **48**, 2729 (1977).
5. D.R. Curran and L. Seaman, Phys. Rep. **147**, 253 (1987).
6. D.R. Curran, L. Seaman and D.A. Shockey, *High Strain Rate Phenomena in Metals* (Plenum, New York, 1981), Chap. 9.
7. D.A. Shockey, L. Seaman and D.R. Curran, Int. J. Fract. **27**, 145 (1985).
8. Z. Rosenberg, G. Luttwak, Y. Yeshurun and Y. Partom, J. Appl. Phys. **54**, 2147 (1983).
9. L.Davison and R.A. Graham, Phys. Rep. **55**, 255 (1979).
10. Y. Maron and B. E. Blaugrund, J. Appl. Phys. **53**, 356 (1982).
11. A. R. Larson, "Calculation of laser induced spall in aluminun targets", Los Alamos Laboratory Report No. LA-5619-MS (1974).
12. J. A. Fox and D.N. Barr, Appl. Phys. Lett **22**, 594 (1973).
13. C.G. Hoffmann, "Some effects of laser irradiation on aluminum, "Los Alamos Laboratory Report No. LA-6189-M (1975).
14. I. Gilath, D. Salzmann, M. P. Dariel, L. Kornblit, and T. Bar-Noy, J. Mater. Sci. **23**, 1825 (1988).
15. I. Gilath, S. Eliezer, M. P. Dariel, and L. Kornblit, in *Proceedings of the International Conference on Impact Loading and Dynamic Behaviour of Materials* (Informationsgesellschaft, Bremen, 1988) pp. 669-676.
16. I. Gilath, S. Eliezer, M. P. Dariel, and L. Kornblit, Appl. Phys. Lett. **52**, 1128 (1988).

17. I. Gilath, S. Eliezer, M. P. Dariel, and L. Kornblit, J. Mater. Sci. Lett. **7**, 915 (1988).
18. S. Eliezer, I. Gilath and T. Bar-Noy, J. Appl. Phys. **67**, 715 (1990).
19. S. Eliezer, Y. Gazit and I. Gilath, J. Appl. Phys. **68**, 356 (1990).
20. V.E. Fortov, V.V. Kostin and S. Eliezer, J.Appl. Phys. (in press).
21. M.A. Meyers and C. T. Aimone, Prog. Mater. Sci.. **28**, 1 (1983).
22. D.E. Grady, J.Mech. Phys. Solids **36**, 353 (1988).
23. C.R. Phipps et. al, J. Appl. Phys. **64**, 1083 (1988).
24. H. Szichman, S. Eliezer and D. Salzmann, J. Quant. Spectrosc. Radiat Transfer, **38**, 281 (1987).
25. I. Gilath and S. Eliezer, "Hemispherical shock wave decay in laser-matter interaction" (to be published).

HOT PLASMA MODEL FOR CLUSTER IMPACT FUSION

Y. E. Kim,[†] M. Rabinowitz,[‡] J.-H. Yoon,[†] and R. A. Rice[†]

[†]Department of Physics, Purdue University, West Lafayette, IN 47907
[‡]Electric Power Research Institute, Palo Alto, CA 94303

ABSTRACT

Line broadening for the proton energy spectrum from the deuterium–deuterium fusion reaction is used to extract the effective deuterium temperature in the recent cluster-impact fusion experiments. The two primary contributions to the proton line width are due to temperature and kinematic broadening. By deconvoluting these contributions, we have determined a temperature ~ 20 keV for the deuterium–deuterium fusion reaction. This is substantially hotter than conventional estimates. Our analysis shows that the proton spectrum is an additional diagnostic tool that can rule out high energy monoenergetic contaminants in explaining the unexpectedly high yield. Furthermore, the proton spectrum indicates that the high temperature results from a one–dimensional rather than a three–dimensional velocity distribution.

I. INTRODUCTION

In recent experiments, Beuhler et al.[1,2] observed unexpectedly high rates of $\sim 1 - 10 s^{-1}/D$ pair for deuterium–deuterium $(D - D)$ fusion when singly charged clusters of D_2O molecules $((D_2O)_n^+, n = 25 - 1300)$ accelerated to 150 to 325 keV (with a beam current of $\sim 1 nA$) and were impacted onto TiD, C_2D_4 and $ZrD_{1.65}$, targets.[1,2] In addition, $D - D$ fusion was observed with $(H_2O)_{115}^+$ clusters impacting upon a C_2D_4 target.[2] A virtual invalidation of the former results[1,2] was reported by Fallavier et al.[3] who carried out similar experiments using pure deuterium clusters $D_{200}^+ - D_{300}^+$; they observed no fusion events upon bombarding TiD and C_2D_4 targets. However, the results of Beuhler et al.[1,2] recently received confirmation by Bae et al.[4] for $(D_2O)_{115}^+$ and $(H_2O)_{115}^+$ cluster beams impacting on a C_2D_4 target. Beuhler et al.[5] have recently shown from their time–of–flight experiments with pulsed cluster beams that light–fragment contaminants cannot be responsible for observed fusion events. The experimental yields for $(D_2O)_{100}^+$ clusters are $\sim 10^{25}$ times higher than that expected[6,7] from single D^+ clusters at $D - D$ center of mass (CM) energies of 150 eV/D, and $\sim 10^{100}$ times higher than at 15 eV/D for $(D_2O)_{1000}^+$. A number of theoretical models[8–12] have been proposed but they underestimate the observed fusion yields[1,2,4] by many orders of magnitude for large clusters with $n > 100$. It has been recently shown[10–12] that heavy atomic partners (such as the oxygen in D_2O) in the molecule play a vital role in explaining the apparently conflicting negative results for $(D)_n^+$ clusters.

Recently, we have proposed a cluster–impact fusion theory[13–16] which explains and reproduces the observed $D - D$ fusion rates[2] for $(D_2O)_n^+$ cluster striking TiD, C_2D_4, and $ZrD_{1.65}$ targets. The theory is expressed in the form of a universal scaling equation with two physically reasonable parameters: (1) the energy enhancement factor or effective temperature $T_e(n)$, and (2) the fraction $\nu(n)$ of projectile cluster and target D atoms reaching high–temperature, T_e. Fitting these parameters to the experimental data for $(D_2O)^+_{100}$ clusters impacting on TiD, C_2D_4, and $ZrD_{1.65}$ targets for cluster energy, $E_t = 150 - 300$ keV results in $\nu(100) = 0.75 \times 10^{-2}$ and $T_e(100) \approx E_t/46$. The effective temperature of $T_e(100) = 3.3 - 6.5$ keV used for the case of $E_t = 150 - 300$ keV is substantially larger than conventional estimates (< 700 eV) based on classical molecular–dynamics simulation calculations.[17-19] In this paper, we describe a theoretical analysis for extracting the effective temperature T_e from the experimentally measured proton energy spectrum, as in the case of $D - D$ fusion in a hot plasma for which the neutron energy spectrum has been used as a diagnostic of the temperature,[20-22] and show that the predicted proton yields calculated with the extracted temperature are in very good agreement with the experimental data for 150–300 keV $(D_2O)^+_{100}$ clusters impacting on C_2D_4, TiD, and $ZrD_{1.65}$ targets.

II. FUSION YIELD

In the cluster–impact fusion (CIF) experiment[2], both the proton yield and the proton energy spectrum from the reaction

$$D + D \rightarrow T(1.01 MeV) + p(3.02 MeV),\tag{1}$$

were measured. For reaction (1), we use the conventional parameterization for the cross–section,

$$\sigma(E_{cm}) = \frac{S(E_{cm})}{E_{cm}}e^{-(E_G/E_{cm})^{1/2}},\tag{2}$$

where E_{cm} is the deuteron CM kinetic energy which is related to the deuteron kinetic energy E in the laboratory (LAB) frame by $E_{cm} = E/2$. E_G is the "Gamow energy" given by $E_G = (2\pi\alpha Z_i Z_t)^2 Mc^2/2$ where Z_i, Z_t, and M are the incident nuclear charge, the target nuclear charge, and the reduced mass, respectively. The numerical values for the Gamow energy are $(E_G)^{1/2} = 31.39$ and 44.40 (keV)$^{1/2}$ in the CM and LAB frames, respectively.

Since the usual two– and three–parameter fits for $S(E_{cm})$[23] are valid only for low energies ($E << 1 MeV$), we use the recent four–parameter fit[15]

$$S(E_{cm}) = (S_O + S_1 E + S_2 E^2)e^{-\beta\sqrt{E}}\tag{3}$$

for the reaction $D(D,p)T$, which fits the experimental values of $\sigma(E_{cm})$ for $E_{cm} \lesssim 10 MeV$ ($E \lesssim 20 MeV$) with $S_O = 54.84$ keV–barns, $S_1 = 0.212$ barns, $S_2 = 0.206 \times 10^{-4}$ barns/keV, and $\beta = 1.943 \times 10^{-2}$ (keV)$^{-1/2}$.

In CIF experiments, $(D_2O)^+_n$ molecular clusters impact on the deuterated target with a cluster velocity of $\sim 10^7 cm/s$ in an impact period of $\Delta t \approx 10^{-14} s$. As shown in the following, a three–dimensional velocity distribution results in an asymmetrical line broadening contrary to the experimental data[2], whereas a one–dimensional (1–dim.) velocity distribution yields a symmetric broadening consistent with the data[2]. We assume that due to the cluster impact a fraction f_D of projectile deuterons in the cluster $(D_2O)^+_n$ and target deuterons will develop a 1–dim. velocity distribution

$$f(v_z) = (\frac{m}{2\pi kT_e})^{1/2}e^{-mv_z^2/2kT_e}\tag{4}$$

with an effective temperature or energy $E_e = kT_e$ (k is the Boltzmann constant). Note that $\int_{-\infty}^{\infty} f(v_z)dv_z = 1$, $\overline{v_z} = \int_{-\infty}^{\infty} |v_z|f(v_z)dv_z = \sqrt{\frac{2kT_e}{m\pi}}$, and $\frac{1}{2}m\overline{v_z^2} = \int_{-\infty}^{\infty} \frac{1}{2}mv_z^2 f(v_z)dv_z = kT_e/2$. m is the deuteron rest mass. Although the total number of atoms is limited to

$3n$ per $(D_2O)_n^+$ cluster, the total number N_t of atoms in a cluster–beam with a current of $\sim 1 nA$ is large, $N_t = 0.625 \times 10^{10}(3n)/s$ ($\approx 2 \times 10^{12}/s$ for $(D_2O)_{100}^+$), so one has a reasonable statistical system impacting the target.

Assuming the deuterons with their velocity distribution described by eq. (4) are moving into the target after the cluster impact, the total fusion proton yield per cluster due to the $D(D,p)T$ reaction is given by[14,15]

$$Y(n, E_t, kT_e) = 2ng f_D I(kT_e) \tag{5}$$

with

$$
\begin{aligned}
I(kT_e) &= \frac{1}{\overline{v_z}} \int_{-\infty}^{\infty} P(|v_z|)|v_z| f(v_z) dv_z = \frac{1}{kT_e} \int_0^{E_t} P(E) e^{-E/kT_e} dE \\
&= n_D^t \int_0^{E_t} \frac{\sigma(E_{cm})}{|dE/dx|} (e^{-E/kT_e} - e^{-E_t/kT_e}) dE
\end{aligned}
\tag{6}
$$

where $P(E_d)$ is the probability for a deuteron with the LAB kinetic energy E_d to undergo $D(D,p)T$ fusion while slowing down in the deuterated target,

$$P(E_d) = n_D^t \int_0^{E_d} \frac{\sigma(E_{cm})}{|dE/dx|} dE. \tag{7}$$

Here the target deuteron number density n_D^t is $5.68 \times 10^{22} cm^{-3}$, $7.05 \times 10^{22} cm^{-3}$, and $8 \times 10^{22} cm^{-3}$ for the targets TiD, $ZrD_{1.65}$, C_2D_4, respectively, and dE/dx is the stopping power of the target for a deuterium projectile.[14,24]

For the cases considered in this paper, E_t in eq. (5) is set to infinity with little loss of accuracy. The factor g (set to 0.5) in eq. (5) is included to account for the fraction of f_D that is lost in the backward direction away from the target.

III. PROTON LINE BROADENING

The proton spectrum measured by Beuhler et al.[2] for the case of a $(D_2O)_{115}^+$ cluster with cluster energy $E_t = 275$ keV impacting the C_2D_4 target has a broad width as shown in Fig. 1. There are several mechanisms which contribute to line broadening of the proton energy spectrum. A dominant one is kinematic broadening due to the finite acceptance angle of the proton detector. Resolution of the silicon surface barrier detector is ~ 10 keV. Since the contribution to the full–width at half–maximum (FWHM) enters in quadrature for a Gaussian (or symmetric) distribution, its contribution would be negligible. For the observed FWHM value of $\Delta E_p \approx 320$ keV of Beuhler et al.[2], the circular proton detector used had a surface area of 3 cm^2 and was placed at a distance of 1 cm from the target perpendicular to the incident beam, corresponding to a detector orientation angle of $\theta_d = 90°$ and a maximum acceptance angle of $\theta_L = 90° \pm 44.34°$ where θ_L is the proton scattering angle with respect to the incident beam direction (chosen to be along the positive z–axis) in the laboratory frame (i.e., $\theta_L = 0°$ for forward scattering). The deuterated targets were oriented at $\theta_L = 45°$.

For an incident deuteron with LAB kinetic energy E_d, the emitted proton energy $E_p(\theta_L)$ for the $D(D,p)T$ reaction is given by $E_p(\theta_L) = \left[a + (a^2 + b)^{1/2}\right]^2$ where $a = (m_d m_p E_d)^{1/2} \cos \theta_L/(m_t + m_p)$ and $b = [(m_t - m_d)E_d + m_t Q]/(m_t + m_p)$ with $Q = 4.033$ MeV. For a circular detector of radius r_0 at a distance l_0 from the target, we define the proton spectral probability function for a monoenergetic (E_d) beam of D^+ (assuming an isotropic angular distribution for protons in the CM frame) as

$$P_{E_p}(E_d) = n_D^t \int_0^{E_d} \frac{\sigma(E_{cm})}{|dE/dx|} F_S(E_p, E(\theta_L), \theta_d)\, dE \qquad (8)$$

where $F_S(E_p, E(\theta_L), \theta_d)$ is a proton spectral distribution function given by

$$F_S(E_p, E(\theta_L), \theta_d) = \frac{1}{2\pi} \cos^{-1}\left\{ \frac{|\ell_0 - R_0 \cos\theta_L \cos\theta_d|}{R_0 \sin\theta_L \sin\theta_d} \right\} \left|\frac{d\Omega_C}{d\Omega_L}\right| \left|\frac{d\cos\theta_L}{dE_p}\right| \qquad (9)$$

where $R_0 = (\ell_0^2 + r_0^2)^{1/2}$, θ_d is the angle between the beam direction and the detector orientation,

$$\left|\frac{d\cos\theta_L}{dE_p}\right| = \frac{(m_t + m_p) + [(m_t - m_d)E_d + m_t Q]/E_p}{4(m_d m_p E_d E_p)^{1/2}}, \qquad (10)$$

and

$$\left|\frac{d\Omega_C}{d\Omega_L}\right| = \frac{(1 + \gamma^2 + 2\gamma\cos\theta_C)^{3/2}}{|1 + \gamma\cos\theta_C|}, \qquad (11)$$

with $\cos\theta_C = \cos\theta_L \left(1 - \gamma^2 \sin^2\theta_L\right)^{1/2} - \gamma\sin^2\theta_L$ and $\gamma = (m_d m_p E_d)^{1/2} [m_t(m_t + m_p)Q + m_t(m_t + m_p - m_d)E_d]^{-1/2}$.

III.1. 1–Dimensional Case

For the case where the deuteron velocity distribution is given by eq. (4), the differential fusion proton spectrum is given by

$$\frac{dY(kT_e)}{dE_p} = N_{kT_e} \frac{1}{kT_e} \int_0^\infty e^{-E/kT_e} P_{E_p}(E)\, dE$$

$$= N_{kT_e} n_D^t \int_0^\infty \frac{S(E_{cm})}{E_{cm}|dE/dx|} e^{-(E/kT_e + \sqrt{E_G}/\sqrt{E})} F_S(E_p, E(\theta_L), \theta_d)\, dE \qquad (12)$$

with N_{kT_e} chosen so that the maximum value of $dY(kT_e)/dE_p$ is normalized to unity. The proton spectrum calculated with eq. (12) and $kT_e = 20$ keV (solid curve in Fig. 1) agrees well with the experimental data[2], and implies that an effective temperature of 20 keV was achieved for the case of 275 keV $(D_2O)_{115}^+$ impacting on the C_2D_4 target.

For the monoenergetic deuteron beam with LAB kinetic energy E_d, the differential fusion proton spectrum as a function of E_p is given by

$$\frac{dY(E_d)}{dE_p} = N_{E_d} P_{E_p}(E_d) \qquad (13)$$

where N_{E_d} is chosen so that the maximum value of $dY(E_d)/dE_p$ is normalized to unity. The proton energy spectra are calculated and plotted in Fig. 1, using eq. (13) with $E_d = 20$ keV (dashed curve), 206 keV (dot–dashed curve), and 275 keV (dot–dot–dashed curve), and eq. (12) with $kT_e = 20$ keV (solid curve).

As can be seen from Fig. 1, the shape and width of the proton spectrum due to the monoenergetic 20 keV D^+ (or 200 keV D_2O^+ or 220 keV D_3O^+) contaminants (dashed curve, representing the kinematic contribution) are different from the data[2] and the result calculated from eq. (12) for a deuteron velocity distribution, eq. (4), with $kT_e = 20$ keV (solid curve) and thus can be ruled out.

Using electrostatic deflection, Beuhler et al.[2] showed that ions of less than about 75% (206 keV) of the full energy (275 keV) could not contribute to the fusion proton yield. The remaining high–energy (206 – 275 keV) D^+ contaminants can be ruled out since the calculated proton spectra (dot–dashed and dot–dot–dashed curves in Fig. 1) for 206-275 keV D^+'s are expected to have much larger values of the FWHM (540-620 keV) than the observed value of $\Delta E_p \approx 320$ keV.

To assess the effect of line broadening due to the straggling of the deuterons as they slow down in the target, we have computed the proton energy spectrum for the case in which the deuteron velocity vectors of eq. (4) are uniformly oriented within a forward ($0° \pm 25°$) cone. The calculated proton spectrum (dotted curve) is plotted in Fig. 1 for comparison with the previous results (solid curve) for the forward ($0°$) deuteron direction. Both results (dotted and solid curves) are nearly identical and imply that line broadening due to deuteron straggling is expected to be small compared to temperature and kinematic line broadening. A one–dimensional velocity distribution confined within the $0° \pm 25°$ cone is consistent with the forward direction of the CM velocity of the incident deuterons.

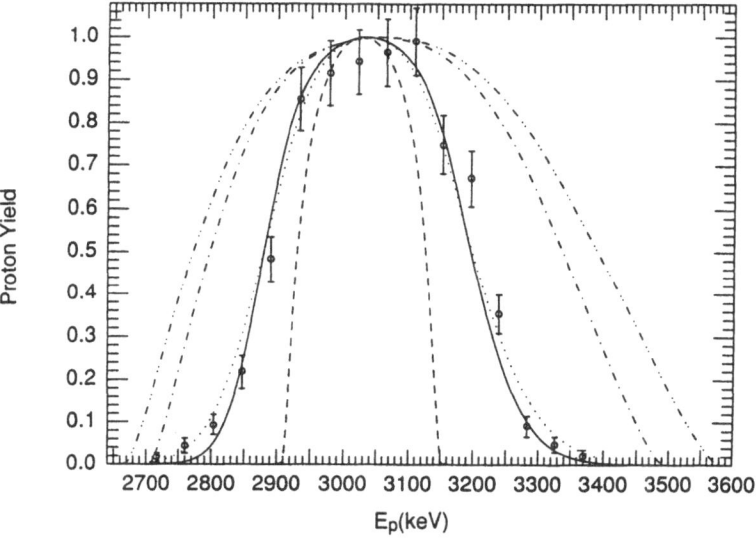

Fig. 1.   Proton energy spectrum calculated from eq. (13) [monoenergetic D^+] with $E_d = 20$ keV (dashed curve), 206 keV (dot–dashed curve), and 275 keV (dot–dot–dashed curve); from eq. (12) using a 1–dim. velocity distribution, eq. (4), with $kT_e = 20$ keV (solid curve), and the case where the deuteron velocity vectors are oriented within a 25° forward cone (dotted curve). The data are from ref. 2.

We have used a more general spectral distribution function than that given by eq. (9) to estimate the effect of beam spreading with a cross–sectional radius r_b on the proton line broadening. For a practical value of $r_b = 0.25$ cm, we find that FWHM increases by less than 10 keV (i.e., $\Delta E_p \lesssim 330$ keV) compared with our previous estimate of $\Delta E_p(kT_e = 20 \text{ keV}) \approx 320$ keV. Therefore, the effect of beam spread on the proton line broadening is much smaller than the experimental value of $\Delta E_p \approx 320$ keV.

III.2. 3–Dimensional Case

For a three–dimensional (3–dim.) MB deuteron velocity distribution,

$$f(\vec{v}) = (\frac{m}{2\pi kT})^{3/2} e^{-mv^2/2kT_e} \tag{14}$$

with the normalization $\int f(\vec{v}) d^3 v = 1$ and $m\overline{v^2}/2 = 3kT_e/2$, the differential fusion proton spectrum for the CIF experiment of Beuhler et al.[2] is given by (assuming the deuteron velocity vectors are uniformly oriented from the impact point into a hemisphere in the target),

$$\frac{dY^{(3)}(kT_e)}{dE_p} = N_{kT_e}^{(3)} n_D^t \int_o^\infty dE \int d\theta_d \frac{\sigma(E_{cm})(1 + E/kT_e)}{|dE/dx|} e^{-E/kT_e} F_S(E_p, E(\theta_L), \theta_d) H(\theta_d) \tag{15}$$

where $N_{kT_e}^{(3)}$ is chosen so that the maximum value of $dY^{(3)}(kT_e)/dE_p$ is normalized to unity and F_S is given by eq. (9). $H(\theta_d)$ is defined as

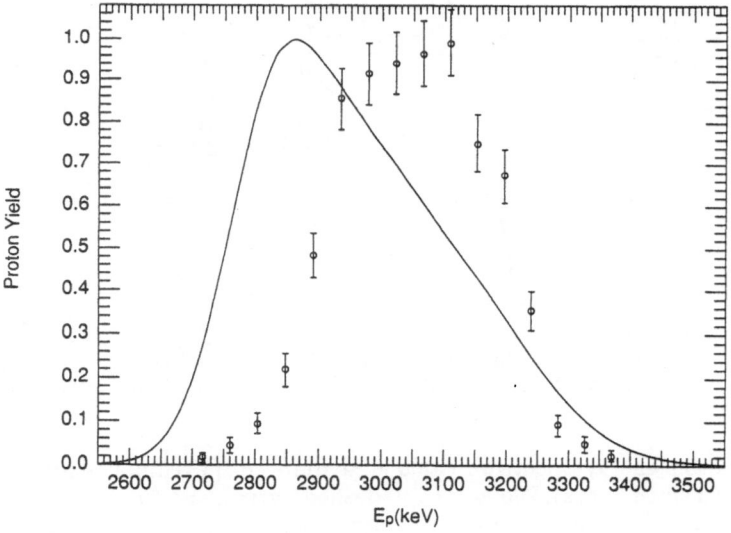

Fig. 2. Proton energy spectrum calculated from eq. (15) using a 3–dim. velocity distribution, eq. (14), with $kT_e = 20$ keV. The data are from ref. 2.

$$H(\theta_d) = \sin\theta_d \cos^{-1}(cot\theta_d)/\pi, \qquad \frac{\pi}{4} \leq \theta_d \leq \frac{\pi}{2}$$

$$= \sin\theta_d[1 - \cos^{-1}(|cot\theta_d|)/\pi], \quad \frac{\pi}{2} \leq \theta_d \leq \frac{3\pi}{4} \qquad (16)$$

$$= \sin\theta_d, \qquad \frac{3\pi}{4} \leq \theta_d \leq \pi$$

The proton energy spectrum calculated with eq. (15) and $kT_e = 20$ keV is plotted as a function of E_p in Fig. 2 for comparison with the experimental data. As can be seen from Fig. 2, the proton energy spectrum calculated with the 3–dim. velocity distribution leads to an asymmetrical line broadening which is inconsistent with the data.[2] A different kT_e value would affect the FWHM but the skewness would still remain.

IV. PREDICTIONS FOR PROTON YIELD

Assuming that kT_e for $(D_2O)^+_{110}$ is the same as that $(kT_e = 20$ keV) for $(D_2O)^+_{115}$ at the same value of $E_t = 275$ keV, the parameter f_D in eq. (5) is determined to be 5×10^{-4} by fitting the total proton yield (calculated from eq. (5)) to the data for the C_2D_4 target.[2] For the same value of $f_D = 5 \times 10^{-4}$, $kT_e \approx 11$ keV is required to fit the experimental value of the proton yield at $E_t = 175$ keV. Assuming that kT_e scales linearly with E_t, i.e., $kT_e = (E_t/A) + B$, the parameters A and B can now be determined to be $A = 11$ and $B = -5$ keV using $kT_e = 10.9$ keV and $kT_e = 20$ keV at $E_t = 175$ keV and 275 keV, respectively. Using $f_D = 5 \times 10^{-4}$, $A = 11$, and $B = -5$ keV, the total proton yield is calculated as a function of E_t, and plotted as a solid curve in Fig. 3(a). The calculated results agree well with the $(D_2O)^+_{100} - C_2D_4$ data of Beuhler et al.[2], which is also shown in Fig. 3(a). Using the same values of f_D, A, and B, theoretical proton yields for $(D_2O)^+_{100}$ on TiD and $ZrD_{1.65}$ are also calculated as our predictions and compared with the corresponding experimental data[2] in Figs. 3(b) and 3(c), respectively. Our predicted results are in fairly good agreement with the experimental data for both the TiD target and the $ZrD_{1.65}$ target. The predicted form of the linear scaling $kT_e = E_t/11 - 5$ keV can be tested by experimental measurements of the proton energy spectra as a function of E_t.

To provide our predictions of the proton line broadening from the same experimental set-up of Beuhler et al.[2] for the case of $(D_2O)^+_{115}$ (and $(D_2O)^+_{110}$) on C_2D_4 at different values of E_t, we have extracted the expected FWHM, $\Delta E_p(kT_e)$, from the proton spectra calculated with eq. (12) using the same parametric values of $f_D = 5 \times 10^{-4}$, $A = 11$, and $B = -5$ keV. The predicted results for ΔE_p are plotted as functions of both kT_e and E_t in Fig. 4, and should be tested in future CIF experiments.

V. ONE–DIMENSIONAL ENERGY ENHANCEMENT

The highly directional nature of cluster–impact fusion (CIF) and the extremely short time of impact ($\sim 10^{-15}$ to 10^{-14} sec) favor an approximately one–dimensional (reduced dimensionality) thermalization process. There may only be sufficient time for high temperature thermalization along the beam direction. As demonstrated in ref. 14 the lateral motion is much less than a mean free path so that there are few collisions to produce thermalization perpendicular to the beam direction. Our results shown in Figs. 1 and 2 from analysis of the 3.025 MeV fusion proton energy spectrum support this view of quasi one–dimensionality since the spectrum is symmetrical. If the Maxwell–Boltzmann velocity distribution were three–dimensional, the proton line shape would be skewed. For

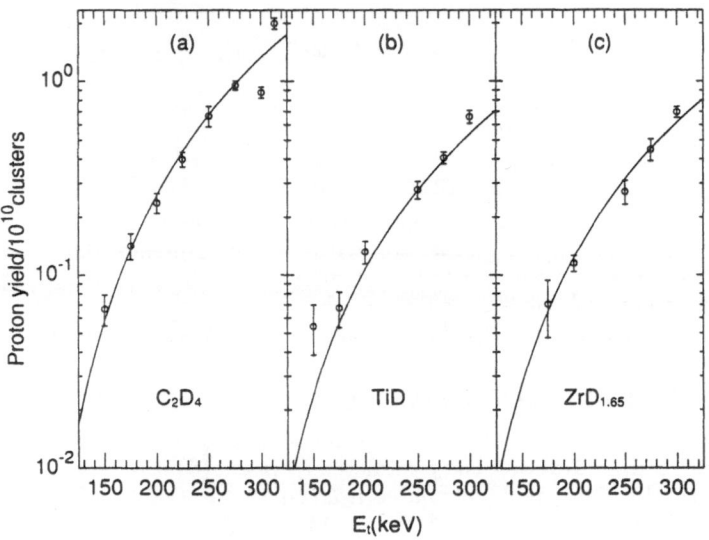

Fig. 3. Calculated fit to the $(D_2O)^+_{100} - C_2D_4$ proton yield of Beuhler et al.[2] and corresponding predictions for the $(D_2O)^+_{100} - TiD$ and $(D_2O)^+_{100} - ZrD_{1.65}$ yields as compared to the experimental data[2].

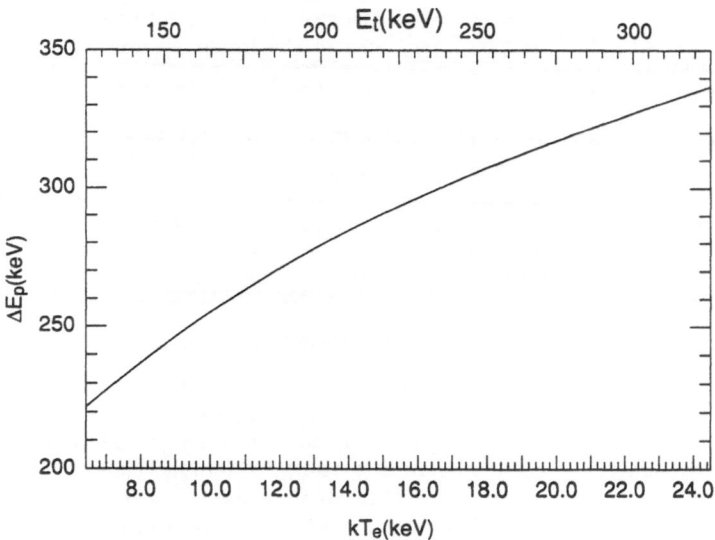

Fig. 4. Expected $(D_2O)^+_{115}$ proton spectrum FWHM, ΔE_p, as a function of kT_e and E_t using the same parameter fit as in Fig. 3.

a given particle energy, E, 1–dim. thermalization produces an effective temperature T that is three times as high as that for the 3–dim. case because $T = 2E/fk$, where f represents the deuteron degrees of freedom.

We note that $(2n)f_D$ in eq. (5) can be less than unity since f_D represents a product of two fractions, $f_D = \nu f_C$ where ν is a fraction of $(2n)$ deuterons in a cluster (i.e., $1 \le 2n\nu < 2n$) and f_C is a fraction of clusters in the incident beam which participate in heating the deuteron plasma to a high average temperature kT_e. Energy conservation for the 1–dim. case requires $2n\nu(kT_e/2) \le E_t$. When the trailing cluster atoms move through this plasma, a high–velocity tail may develop for a fraction $\nu(n)$ of projectile cluster and target deuterons due to mechanisms (yet to be investigated and understood) such as (a) multiple backscattering of deuterons between target and projectile heavy atoms[10–13], (b) pinch instability heating due to magnetic confinement[15,25,26] (which is also conducive to a one–dimensional velocity distribution), (c) focussed converging shock–wave heating, (d) needle–shape cluster impact fusion (e) other collective effects related to electron degrees of freedom, etc.

In the following, we discuss several possible mechanisms and processes such as: converging shock–wave fusion (V.1), possibility of three–deuterium fusion (V.2), and needle–shape cluster impact fusion (V.3).

V.1. Converging Shock–Wave Heating

Shock–wave heating has been previously proposed by us as a possible mechanism for CIF.[13–16] Upon impact, the leading edge of the cluster creates a plasma $\sim$ few Å in thickness consisting of target and cluster ions and electrons which is compressed between the target cavity surface and the trailing cluster atoms. As the trailing cluster atoms begin to plow through the plasma, they produce a shock–wave which initially heats the plasma to temperature T_2.

The plasma can be treated as an ideal gas to good approximation, even at high densities. The sound (or acoustic) velocity v_a in an ideal gas can be written as[27–29]

$$v_a = [\gamma R T_a/m_a]^{1/2} = 1.18 \times 10^4 [T_a(K)/m_a]^{1/2} cm/s \; . \tag{17}$$

where γ is the ratio of the specific heats ($\gamma = C_p/C_v = 5/3$ for an ideal monoatomic gas), T_a is the initial gas temperature in front of the shock wave discontinuity, m_a is an average atomic weight and R is the universal gas constant ($R = 8.314 \times 10^7 erg/K$). For a cluster with total kinetic energy E_t and rest mass nM_m the shock–wave velocity v (4/3 of the cluster velocity[28]) is

$$v = 4(2E_t/nM_m)^{1/2}/3 = 4\sqrt{2} \times 10^{10} \left[E_t/(nM_mc^2)\right]^{1/2} cm/s \; . \tag{18}$$

The Mach number M_1 is then

$$M_1 = v/v_a = 4.79 \times 10^6 \left[E_t(\mathrm{keV})m_a/nT_a(K)M_mc^2(\mathrm{keV})\right]^{1/2} \; . \tag{19}$$

Typically, for a $(D_2O)^+_{100}$ cluster ($M_mc^2 = 1.865 \times 10^7$ keV and $m_a = 20/3$) at $E_t = 300$ keV and $T_a = 300$ K, $M_1 = 2.87 \times 10^2$. Therefore, the Mach number will be very large, thus creating a strong shock wave.

The temperature behind a strong shock–wave front ($M_1 >> 1$) in an ideal gas is given by[27–29]

$$T_2(n) = \frac{2\gamma(\gamma - 1)}{(\gamma + 1)^2} M_1^2 T_a = 7.17 \times 10^{12} \frac{m_a}{n} \frac{E_t}{M_m c^2} K \ . \tag{20}$$

For example, for $(D_2O)_{100}^+$ with $m_a = 20/3$ (average atomic weight of O and $2D$) and $E_t = 300$ keV, we obtain $kT_2 = 0.67$ and 0.067 keV for $n = 100$ and 1000, respectively.

Although the mean free path of deuterons at T_2 can be comparable to or larger than the cluster diameter of ~ 15 Å$-$ 40 Å for $n = 50 - 1000$, most of them will still be in the region (a few Å thick) between the shock–wave front and trailing part of the cluster during the impact period. A novel ingredient suggested previously[13–16] is enhanced heating due to dynamic interaction of the back–flow gas with the leading edge of the remaining cluster. As a result of back–scattering, nearly half of the deuterons in an expanding gas at T_2 will flow into the back hemisphere toward the trailing cluster atoms. This back–flow of deuteron gas will be heated even further by the shock wave generated by the trailing cluster atoms since the cluster velocity relative to the back–flow deuteron gas is greater than that relative to the target. Since the acoustic velocity remains the same, the Mach number and T_2 increase (cf., eqs. (19) and (20)). Such dynamical and continuous shock–wave heating of the plasma will further enhance the temperature during the impact period. This yields the enhanced temperature $T_e(n)$

$$T_e(n) = \tau(n, E_t) T_2(n) \tag{21}$$

where $\tau(n, E_t)$ is expected to be a smooth function of n and E_t. For the case of 300 keV $(D_2O)_{100}^+$, $kT_e \approx 20$ keV $= 30\, T_2$ with $T_2 = 0.67$ keV, i.e., $\tau = 30$.

A very large enhancement factor $\tau \approx 30$ can be also understood in terms of a heating mechanism involving focussed converging shock waves. T_2 given by eq. (20) is for plane shock–wave heating and can be also applied for the case of a reflected shock–wave. Upon impact, a compressive shock wave of velocity v (given by eq. (18)) propagates into both the target (transmitted shock wave) and the cluster (reflected shock wave).[30] T_2 can be interpreted as the temperature behind the reflected shock wave (which lasts for the impact period). T_2 can be made to be substantially larger, if the reflected shock wave is focussed to converge. It has been observed in cluster impact experiments[31] that energetic molecular clusters can produce craters with a diameter comparable to the size of the cluster. As stated by Beuhler et al.,[2] the probability (which may be related to f_C) of subsequent hits of these craters by other incoming molecular clusters is not negligible. The reflected shock wave generated upon cluster impact may be hemispherically converging because of the hemispherical shape of the crater bottom. The converging situation is similar to the case of high velocity jets produced by hollow shaped charge detonation.[32,33]

Another possibility of generating a converging shock wave is that of the cylindrically converging shock wave.[34] When a cluster hits the bottom of a similar size crater, the leading edge of the projectile cluster creates upon impact a plasma consisting of target and cluster atoms (ions) and electrons which is partially trapped in the micro–crater of several Å size between the target cavity surface and the trailing cluster atoms. The trailing part of the cluster can provide a geometry similar to that for the cylindrically converging shock wave[34] which is similar to a spherical implosion. Both the hemispherically reflected converging shock wave and the cylindrically converging shock wave can be regarded as a quasi–spherical implosion and can provide reasonable shock–wave heating mechanisms for achieving the high temperature $kT_e \approx 20$ keV.

V.2. Possibility of Three–Deuterium Fusion

In 3–dim., the collision frequency per particle is $F_3 = n\sigma_c v$, where n is the particle number density, σ_c is the collision cross section, and v is the mean thermal velocity. Decreasing the dimensionality or degrees of freedom decreases the number of ways potentially colliding particles can miss each other. Some very simplified equations illustrate that a significant increase in collision frequency may thus be achieved.[35] In two dimensions the collision frequency is $F_2 = n^{2/3}(\sigma_c)^{1/2}v$. In one dimension the collision frequency is $F_1 = n^{1/3}v$. If CIF increases the probability of a nearly one–dimensional collision with essentially the absence of angular momentum in the final state, then this permits a greatly increased fusion rate for three-body collisions. Thus one should investigate the possibility of reactions such as: $D + D + D \rightarrow T + {}^3He$, $D + D + D \rightarrow {}^6Li + \gamma$, $D + D + D \rightarrow T + p + D$, and $D + D + D \rightarrow {}^4He + D$ which are highly improbable in 3–dim., but which are much more probable for the case of high number density and 1–dim. as suggested by Rabinowitz.[35] The results of 4.5 – 5 MeV tritium production observed in recent experiments[36,37] may be attributable to $D(2D,{}^3\mathrm{He})T(4.76\ \mathrm{MeV})$.[38]

V.3. Needle–Shape Cluster Impact Fusion

There may be a subtle contribution to the extraordinarily high fusion rate also related to reduced dimensionality. It is possible that D_2O clusters may form into dendritic needles, characteristic of H_2O and D_2O during the freezing process, much the same as in snowflakes. If this is the case as previously suggested[11,12], this simple model may have both explanatory and predictive power. Of course nearly spherical clusters may be possible for small n, and for certain cluster sizes corresponding to magic numbers such as found in nuclei. One may expect a significantly higher fusion yield from needle clusters than from spherical clusters.

When the D_2O needles impact end–first upon the target, only the atoms near their tips may participate in the extremely brief thermalization period. Thus a smaller number of D's may become thermalized to an even higher effective temperature by energy transfer due to compressive collision with the trailing O's in the bulk of the cluster. This is effectively a 1–dim. compression like a piston into a cylinder. In 1–dim. the probability for energy enhancement by double-backscattering of D's from heavier atoms like O, Ti, C, Zr etc. is greatly increased. Of course, when the cluster needle does not impact end–on with its major axis roughly parallel to its velocity vector, this model predicts that the fusion rate would be greatly reduced. A corollary to this prediction is that not only do a limited number of D's in a given cluster participate in high temperature thermalization, but only a small fraction (possibly related to f_C) of the incoming clusters can participate due to needle orientation.

There may be a yet more subtle contribution related to an electromagnetic effect. When a positively charged cluster needle strikes a negative target, for an instant it is like a whisker into which electrons from the target move in an extremely short time to charge it up. If the needle has a very high aspect ratio (length to trailing tip radius) to sufficiently enhance the electric field it may even produce field–emission. (This will not occur for a negative cluster striking a positive target.) This may produce a transient azimuthal magnetic field with pinch confinement of the plasma to 1–dim. and pinch instability heating. If the magnetic effect is critical, a possible prediction is that for a reversed polarity experiment in which a negatively charged cluster impacts upon a positively charged target, the fusion rate will be correspondingly less.

VI. CLUSTER IMPACT FUSION WITH $(H_2O)_n$

Since a 275 keV $(H_2O)^+_{115}$ cluster beam upon impact is expected to produce a 1–

dim. proton velocity distribution with $kT_e \approx 20$ keV as is the case for a 275 keV $(D_2O)_{115}$ cluster beam, it is useful to search for fusion reactions involving protons with $kT_e \approx 20$ keV. For light nucleus targets, there are three fusion reactions, ${}^9Be(p,D)^8Be$, ${}^9Be(p,{}^4He)^6Li$, and ${}^{11}B(p,2{}^4He)^4He$, which have sufficiently large values of cross–sections at Gamow peak energies. The integrand in eq. (6) has a maximum value approximately at the Gamow peak energy, $E_{lab}^{max} = [E_G(lab)^{1/2}kT_e/2]^{2/3}$ and $E_{cm}^{max} = E_{lab}^{max}m_t/(m_i + m_t)$ where m_i and m_t are the rest masses of the incident and target particles, respectively. For $D(D,p)T$, ${}^9Be(p,D)^8Be$, ${}^9Be(p,{}^4He)^6Li$, and ${}^{11}B(p,2{}^4He)^4He$, $E_{cm}^{max} = 29.1$, 104.4, 104.4, and 123.8 keV, respectively, and the numerical values of their cross–sections are $\sigma(E_{cm}^{max}) \approx 0.5 \times 10^{-3}b$, $1.20 \times 10^{-3}b$, $1.2 \times 10^{-3}b$, and $2.2 \times 10^{-3}b$, respectively, where we use $kT_e = 20$ keV and eq. (2) for $\sigma(E_{cm})$ with $S(E_{cm})$ given in ref. 23. Since their cross–sections at their Gamow peak energies are similar, the fusion yields with 275 keV $(H_2O)_{115}^+$ for ${}^9Be(p,D)^8Be$, ${}^9Be(p,{}^4He)^6Li$, and ${}^{11}B(p,2{}^4He)^4He$ are expected to be the same order of magnitude as that of the proton yield from $D(D,p)T$ with 275 keV $(D_2O)_{115}^+$. The above predictions can be tested experimentally using a ~ 300 keV $(H_2O)_{\sim100}^+$ cluster beam on ${}^9Be$ and ${}^{11}B$ targets, and bromellite (BeO), phenazite (Be_2SiO_4), boron oxide (B_2O_3), hexa–boron silicide (B_6Si), tri–boron silicide (B_3Si), etc.

VII. SUMMARY AND CONCLUSIONS

Previously, we had shown by means of our theory that CIF is hot fusion due to focussing of the incident cluster energy on a small atomic scale.[13-16] One or more mechanisms such as focussed shock waves, magnetic pinch instability heating, one–dimensionality, double backscattering, etc. may contribute to the energy enhancement of a limited number of deuterons from effective temperatures ~ 100 eV to ~ 20 keV. In this paper, we have analyzed the line broadening for the proton energy spectrum from the $D - D$ fusion reaction to directly determine that the effective deuteron temperature is ~ 20 keV in agreement with our previous theoretical prediction. Additionally, we have demonstrated that the proton spectrum implies that the high temperature is due to a one–dimensional rather than a three–dimensional velocity distribution. The energy enhancement with the temperature ~ 20 keV is in the direction parallel to the beam velocity, and quite small in the perpendicular direction. Quasi one–dimensionality greatly increases the probability of energy enhancement by double backscattering, as well as three–body fusion. The easily achieved high temperature in CIF makes it feasible to consider aneutronic fusion reactions which require such higher temperatures to make them competitive with ordinary $D - D$ fusion reactions.

Since low energy (< 20 keV) resonances are expected to yield much narrower FWHM than the data ($\Delta E_p \approx 320$ keV) for the proton energy spectrum, any theoretical models based on such low–energy resonances can be ruled out. Thus the proton spectral broadening can be used to test the theoretical model for CIF in addition to discriminating the effect of possible contaminants. Therefore, it is very important to measure not only the total proton yield but also the proton energy spectra simultaneously in future CIF experiments.

Finally we note an advantage of CIF over laser induced fusion. CIF as a particle beam method is an efficient way of coupling the input energy into the target nuclei, rather than first coupling into the electronic system and then into the target nuclei as in laser fusion.

ACKNOWLEDGEMENTS

The authors wish to acknowledge valuable comments by Drs. Young K. Bae, Gary S. Chulick, and Robert Vandenbosch. This work has been supported in part by the Purdue Research Foundation and the Electric Power Research Institute.

REFERENCES

1. R. J. Beuhler, G. Friedlander, and L. Friedman, Phys. Rev. Lett. 63, 1292 (1989).

2. R. J. Beuhler, Y. Y. Chu, G. Friedlander, L. Friedman, and W. Kunnmann, J. Phys. Chem. **94**, 7665 (1990).

3. J. Fallavier, R. Kirsch, J. C. Poizat, J. Remillieux, and J. P. Thomas, Phys. Rev. Lett. **65**, 621 (1990).

4. Y. K. Bae, D. C. Lorents, and S. E. Young, SRI Report MP91–023 (February 1, 1991) to be published in Phys. Rev. A.

5. R. J. Beuhler, Y. Y. Chu, G. Friedlander, L. Friedman, A. G. Alessi, V. LoDestro, and J. P. Thomas, Phys. Rev. Lett. **67**, 473 (1991).

6. M. Rabinowitz, Mod. Phys. Lett. **B4**, 665 (1990).

7. Y. E. Kim, Fusion Technology **17**, 507 (1990).

8. C. Carraro, B. Q. Chen, S. Schramm, and S. E. Koonin, Phys. Rev. **A42**, 1379 (1990).

9. P. M. Echenique, J. R. Manson, and R. H. Ritchie, Phys. Rev. Lett. **64**, 1413 (1990).

10. M. Rabinowitz, Y. E. Kim, R. A. Rice, and G. S. Chulick, in AIP Proceedings (No. 228) of Int. Work. on Anom. Nucl. Eff. in Deuterium/Solid Syst., Oct. 22–24, 1990, pp. 846–866.

11. Y. E. Kim, M. Rabinowitz, G. S. Chulick, and R. A. Rice, Mod. Phys. Lett. **B5**, 427 (1991).

12. Y. E. Kim, R. A. Rice, G. S. Chulick, and M. Rabinowitz, Mod. Phys. Lett. **A6**, 2259 (1991).

13. Y. E. Kim, G. S. Chulick, R. A. Rice, M. Rabinowitz, and Y. K. Bae, Chem. Phys. Lett. **184**, 465 (1991).

14. Y. E. Kim, M. Rabinowitz, Y. K. Bae, G. S. Chulick, and R. A. Rice, Mod. Phys. Lett. **B5**, 941 (1991).

15. Y. E. Kim, M. Rabinowitz, Y. K. Bae, G. S. Chulick, and R. A. Rice, "Hot Plasma Shock–Wave Theory of Cluster–Impact Fusion", to be published in AIP Proc. of Int. Symp., Nikko, Japan, June 6–8, 1991.

16. Y. E. Kim, M. Rabinowitz, Y. K. Bae, G. S. Chulick, and R. A. Rice, "Cluster Impact Hot fusion", in Proc. of ICENES '91, to be published in Fusion Technology.

17. M. I. Haftel, "Molecular Dynamics of 300 keV $(D_2O)_{100}$ Clusters on $Ti - D$", submitted to Phys. Rev. A.

18. M. Hautala, Z. Pan, and P. Sigmund, "Accelerated Deuterons in Cluster Fusion Experiments", submitted to Phys. Rev. A.

19. O. H. Crawford, "Cluster–Impact Fusion: Yields from Binary–Collision Sequences", submitted to Radiation Effects and Defects in Solid.

20. G. Lehner and F. Pohl, Z. Physik **207**, 83 (1967).

21. G. Lehner, Z. Physik **232**, 174 (1970).

22. H. Brysk, Plasma Phys. **15**, 611 (1973).

23. W. A. Fowler, G. R. Caughlan, and B. A. Zimmermann, Ann. Rev. Astr. Astrophysics **5**, 525 (1967); **13**, 69 (1975).

24. H. H. Anderson and J. F. Ziegler, *Hydrogen Stopping Powers and Ranges in All Elements*, Pergamon Press, New York (1977).

25. A. Hasegawa et al., Phys. Rev. Lett. **56**, 139 (1986).

26. Y. E. Kim in *Laser Interaction and Related Plasma Phenomena* (International Conference Series), Vol. 9, edited by H. Hora and G. H. Miley, Plenum, New York (1991), pp. 583–592.

27. L. D. Landau and E. M. Lifshitz, *Fluid Mechanics*, Pergamon Press, London (1959).

28. Ya. B. Zel–dovich and Yu. P. Raiser, *Physics of Shock Waves and High Temperature Thermodynamics Phenomena*, 2 vols., Academic Press, New York (1967).

29. F. Winterberg, Z. Naturf. **19a**, 23 (1964).

30. H. Polachek and R. J. Seeger, in *Fundamentals of Gas Dynamics* (edited by H. W. Emmons), Princeton University Press (1958), Section E, pp. 482–522.

31. M. W. Matthew, R. J. Beuhler, M. Ledbetter, and L. Friedman, Nucl. Inst. & Meth. Phys. Res. **B14**, 448 (1986).

32. G. Birkhoff, D. P. MacDougall, E. M. Pugh, and Sir G. Taylor, J. Appl. Phys. **19**, 563 (1948).

33. R. J. Beuhler, G. Friedlander, and L. Friedman, "Fusion Reactions in Dense Hot Atom Assemblies Generated by Cluster Impact," to be published in Accts. Chem. Res. (1991).

34. A. R. Kantrowitz, in *Fundamentals of Gas Dynamics* (edited by H. W. Emmons), Princeton University Press (1958), Section C, pp. 350–413.

35. M. Rabinowitz, Mod. Phys. Lett. **B4**, 233 (1990).

36. G. Chambers, G. Hubler, and K. Grabowski, in AIP Conf. Proc. (No. 228), pp. 383–396.

37. E. Cecil, in AIP Conf. Proc. (No. 228), pp. 375–382.

38. Y. E. Kim, in AIP Conf. Proc. (No. 228), pp. 807–826.

ION CLUSTER STOPPING IN COLD TARGETS FOR ICF

C. Deutsch

CNRS-GDR-918 "Interaction Ions Lourds- plasma dense"
Bât. 212, Université Paris XI
91405 Orsay, Cedex, France

ABSTRACT

Heavy ion clusters are considered for driving an ICF pellet. They are shown to be effective at much lower energy and intensity than required for atomic ions. Direct pressure imparted to a DT pellet is estimated.

The stopping of strongly correlated ion cluster debris flowing in a degenerate electron jellium target is considered for polarized (with respect to velocity) projectiles. Chains and rectangle geometries are considered. Systematic trends are stressed out.

INTRODUCTION

The obvious interest of using the heaviest drivers for compressing ICF targets make desirable the consideration of atomic clusters with arbitrarily large mass as potential drivers. The production of clusters and, their internal properties have been given a certain attention, recently, from the inertial thermonuclear fusion point of view[2]. At Orsay, we are currently investigating the possibility of accelerating from the terminal of a tandem accelerator Au clusters built on 2 up to 7 atoms, with a positive charge equal to 1 or 2.

The considered energy range will be $35 < E/A < 380$ keV. Another ion source aims at producing organic clusters containing several thousands of atoms with one positive charge per one thousand atomic mass. We thus expect to accelerate linear clusters with A = 50000 and Z = 50. In the sequel we briefly examine the potentialities of cluster drivers for ICF. The main emphasis in this work will be on the specific features of the cluster stopping properties in a cold jellium target. The main attention is given to the geometry of the relevant interactions. Polarized clusters with respect to their velocity in the target are successively considered through a few specific examples.

Also, we pay a special attention to dilute and dense homogeneous jellium target at room temperature.

Specific plasma target of interest for ICF will be taken up in another forthcoming work. Finally, it should be appreciated that the crucial fragmentation issues are simplified here by taking for granted the so-called Maximum Entropy Principle (hereafter refered as MEP). This implies that the given cluster projectile will break under impact into the largest number of its smallest building blocks. Namely, the ionized atoms. Atomic units (a.u.) are used in the sequel.

FRAGMENTATION SCENARIOS

Elaborating a little bit further the above given arguments, it has to be appreciated that in the given energy range the cluster projectile is highly likely to experience a partial coulomb explosion. This means that the resulting debris will be at least once ionized. Moreover, their relative velocity is expected to be small compared to the projectile one over most of its quasi-linear range within the target. Therefore, these debris are expected to fly in a highly correlated motion with relative distances of order of a Bohr radius a_0.

The given Coulomb explosion takes place on a femto second scale length, which supports the MEP model.

Other fragmentation scenarios based essentially on combinatorial arguments might also be considered. However, they do retain as an exact asymptotic limit, the MEP scenario.

This corresponds to a sudden projectile-target interaction with a maximum produced disorder compatible with initial conditions.

Such approach might be implemented through a kind of Saha-like approach, provided the fragmentation times (and their inverses) remain much shorter than that pertaining to the thermal expansion of the N-cluster-target interaction volume V. Then, a temperature concept makes sense for describing the cluster evolution.

Of special relevance to the present work is the dynamical parameter

$$x = \frac{V}{\lambda^3} \times e^{-a_B/k_B T} \tag{1}$$

in terms of the cluster binding energy (if any) a_B and the thermal wavelength λ of the monomer subunit given by $\lambda = \hbar/(2\pi m_i k_B T)^{1/2}$.

In practical conditions of interest for ICF, the cluster kinetic energy should be larger than 10 kev/a.m.u. in order to treat the projectile-target interaction within a stopping formalism[1,2].

Then $a_B \sim eV/atom$ is much smaller than $k_B T$.

So $x \sim \dfrac{V}{\lambda^3}$

For $x = 0$, one just gets a fused and unbroken cluster, for larger values one witnesses fragmentation in a few relatively large subunits. Many other regimes are also likely to occur when x increases further. For $x \gg M$, one reaches a particularly interesting situation called multi-fragmentation.

Then the cluster M explodes into M elementary basic units. Such a behaviour can be associated to the Coulomb explosion alluded to previously[1] provided the resulting ions are not too heavily charged. Such a pattern appears highly desirable for optimizing the correlated stopping picture detailed below.

Otherwise stated, in order to have ion debris experiencing Coulomb repulsion, the detached electrons toll should comply with the structure of the cluster projectile and also with the available kinetic energy.

TARGET CONSIDERATIONS

A detailed system analysis of a Cluster Ion Beam (C.I.B.) driven ICF target will be reported elsewhere. The target is likely to display a rather simple structure. It could be made essentially of roughly 1 mg of D+T fuel, in close analogy with Momentum Reach Beam (MRB) targets proposed recently.

The main point we want to emphasize here is the tremendous flexibility introduced into the accelerator constraints by using CIB as heavy ion drivers.

According to obvious scaling laws, the efficiency of various particle drivers may be discussed with a simple relationship namely

$$\frac{n \, I \, M \, (\sum\limits_{i} Z_i)^2 \, \Delta t}{r^2 \, E \, Z'_i} \sim constant \tag{2}$$

where : n = Number of ion beams, I = Current intensity in each beam, M = Ion Cluster Atomic mass, Z_i = Charge of ion debris, E = Ion cluster kinetic energy, Δt = pulse duration, r = neck radius at pellet and Z'_i = Inflight cluster charge.

Taking for instance, a given plausible CIB with $M_i = 50000$, $Z'_i = 50$ and $E_i/M_i \sim 10$ keV, we see that a complete fragmentation into ion debris with unit charge is equivalent to a standard direct drive (HIBALL) HIB, provided $nI \sim 10^7$ ampere.

Other parameters featuring Eq. (1) are expected to retain their HIB values. Such a modest requested beam intensity, 4 orders of magnitude below its HIB counterpart, seems to lie within the reach of existing linear accelerating structures. RFQ are obvious candidates

As far as target compression is concerned, the present MRB concept introduces the possibility of a directly imposed external pressure (hammer effect) on the pellet.

Let us consider, for instance, Au_n^+ metallic clusters with n = 2-7, and assume that we have enough sources for bombarding uniformly the target surface. This is quite feasible if each ion source is followed by a Tandem like accelerating structure. In terms of the projectile kinetic energy E and its linear range R in target, the ablation pressure reads as (A = target atomic number)

$$P(100 \text{ Mbar}) = \frac{N^{1/3} \text{ x } E(keV/a.m.u.) \text{ x } 0.00565 \text{ x } A}{R(\mu m)} \tag{3}$$

Huge pressures may indeed be achieved at reasonable projectile energies. The atom distribution within cluster is taken homogeneous in space

POLARIZED CLUSTER STOPPING

Now, we turn to the most plausible CIB stopping scenario[2]. As already stated in Sec. 2, the debris resulting from the cluster impact on pellet is expected to fly in a highly correlated relative motion. The given target is modeled presently by a fully degenerate electron jellium with a T = 0 Fermi temperature. Such a homogeneous medium is characterized by a dimensionless parameter

$$r_S = (\frac{4}{3} \pi n_e)^{-1/3} a_0^{-1} \tag{4}$$

in terms of the electron number density n_e.

Then, we consider a given cloud of cluster ion debris in target, as a 2-body superposition. The stopping analysis then puts emphasis on the individual ion contribution, and on the correlated one, as well.

The latter is treated here as a superposition of dicluster contributions.

We start our analysis by considering a frozen (polarized) configuration of ion debris flying in target.

Obviously, many geometries can be taken into account. Figs. 1-4 features a polarized cluster structures with respect to velocity in target.

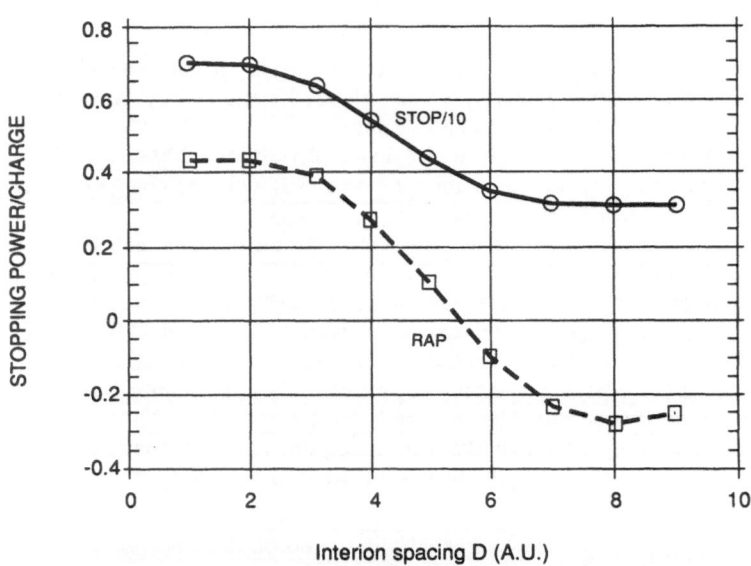

(a) Stopping 10-chain unit charges//V = 1 A.U.
in jellium with R_s = 2.1446

Interion spacing D (A.U.)

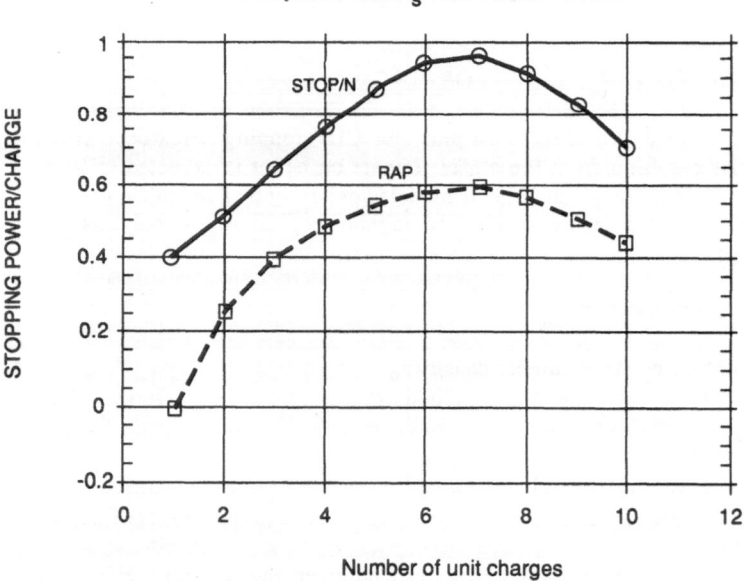

(b) Stopping chains unit charges//V = 1 A.U.
in jellium with R_s = 2.1446, D = 1 A.U.

Number of unit charges

Fig. 1. Stopping chains parallel to velocity in
dense jellium (R_s = 2.1446).

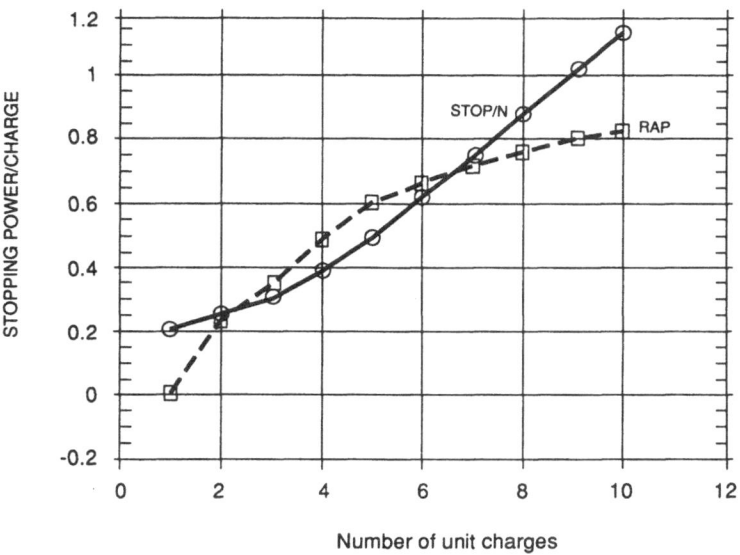

(b) Stopping chains unit charges / N = 2 A.U.
in jellium with R_s = 2.1446, D = 1 A.U.

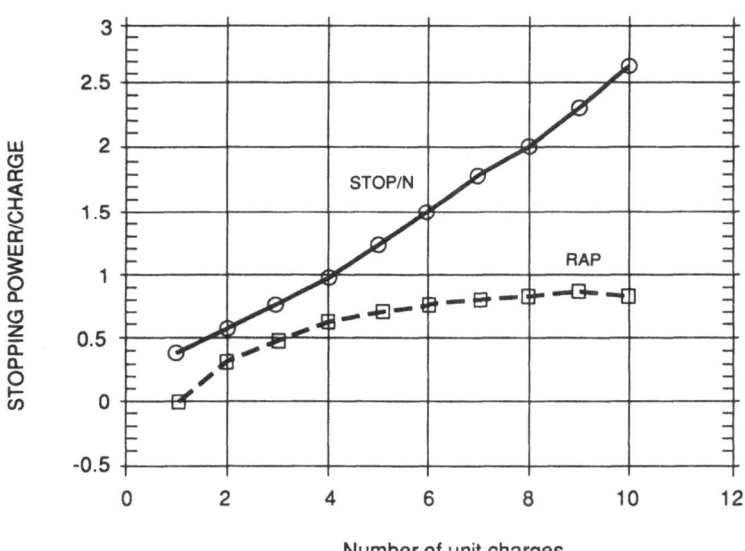

(a) Stopping chains unit charges transverse to V = 1 A.U.
in jellium with R_s = 2.1446, B = 1 A.U.

Fig. 2. Stopping chains (N-Sheild) transverse to velocity
in dense jellium (R_s = 2.1446).

667

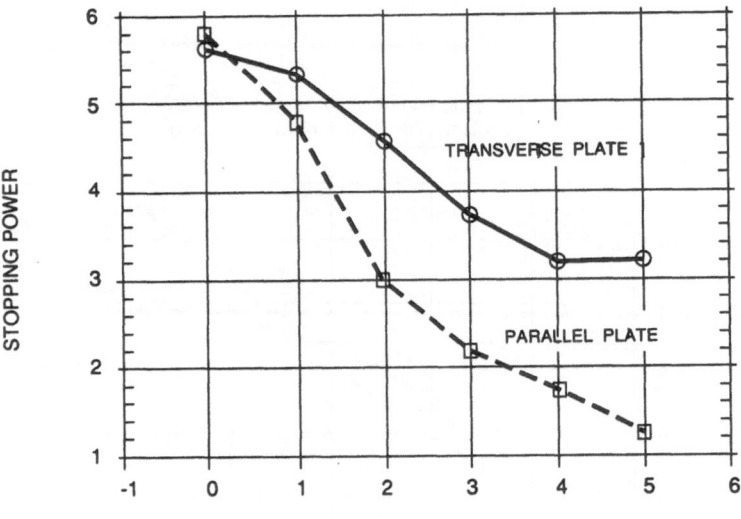

Fig. 3. Stopping 1+1+1+1 cluster resp. paralled and
orthogonal to V = 1 A.U. in jellium
(R_2 = 2.446), B = 1 A.U.

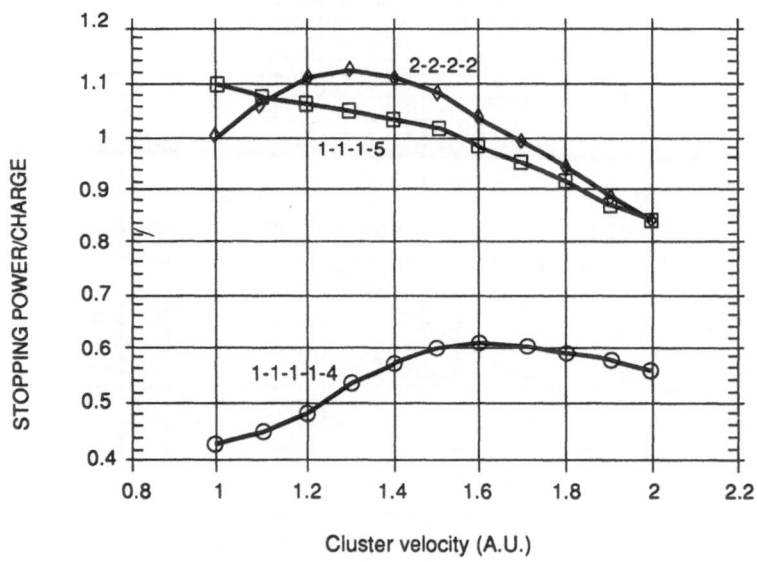

Fig. 4. Stopping quadrangles // V with total charge 8 in
jellium (R_s = 2.1446), B = D = 2 A.U.

DICLUSTER STOPPING

Suppose that two swift point charges z_1e and z_2e proceed with velocity $\vec{v}$ in a medium characterized by the dielectric function $\varepsilon(k,\omega)$. Separation between the charges is specified by the vector $\vec{R}$ with components D and B in directions with, and perpendicular to $\vec{v}$, respectively. In linear-response theory and the first Born approximation, the energy loss of the cluster per unit length to electronic excitation in the medium may be written[1] as ($k^2 = \kappa^2 + \omega^2/v^2$),

$$S_c = \frac{2e^2}{\pi v^2} \int_0^\infty \kappa\,d\kappa \int_0^\infty \frac{\omega\,d\omega}{k^2} \times \mathrm{Im}[\frac{-1}{\varepsilon(k,\omega)}][z_1^2+z_2^2+2z_1z_2J_0(\kappa B)\cos[\frac{\omega D}{v}]]$$

$$\equiv (z_1^2+z_2^2)\,S_p + 2z_1z_2S_v(B,D) \tag{5}$$

In order to get easily surveyed analytical results, we employ a simplified form which exhibits collective and single-particle effects. We take

$$\mathrm{Im}[\frac{-1}{\varepsilon(k,\omega)}] = \frac{\pi\omega_p^2}{2\omega}\,[\delta(\omega-\omega_p)\theta(k_c-k)+\delta[\omega-\frac{\hbar k^2}{2m}]\theta(k-k_c)] \ , \tag{6}$$

where $\hbar\omega_p$ is the plasma energy of the electron gas and the choice $k_c = (2m\omega_p\hbar)^{1/2}$ allows the two δ functions in Eq. (6) to coincide at $k = k_c$ in the k-ω plane.

In this approximation

$$S_p = \frac{e^2\omega_p^2}{v^2}\ln[\frac{2mv^2}{\hbar\omega_p}] \ , \tag{7}$$

Using Eq. (6) in Eq. (5) one finds

$$S_v(B,D) = \frac{e^2\omega_p^2}{v^2}\,[\cos[\frac{\omega_pD}{v}]\int_0^{\kappa_c}\frac{\kappa J_0(\kappa B)}{k^2+\omega_p^2/v^2}\,d\kappa + \int_{\kappa_c}^{\kappa_2}\frac{dk}{k}\cos[\frac{\hbar k^2D}{2mv}]J_0(QB)]$$

$$\equiv \frac{e^2\omega^2}{v^2}*G(B,D)) \ , \tag{8}$$

where $Q^2 = k^2 - (\omega_k/v)^2$, $\kappa_c^2 = k_c^2 - \omega_p^2/v^2$,

$\omega_k = \hbar k^2/2m$, and $k_2 = 2mv/\hbar$

$J_0(x)$ is the usual Bessel function.

N-CLUSTER STOPPING

According to the 2-body superposition principle advocated above, the stopping power of a N-cluster is straight forwardly given by

$$S_G = (\sum_i Z_i^2) S_p + 2 \sum_{1 \le i < j \le N} Z_i Z_j S_v(B_{ij}, D_{ij}) , \qquad (9)$$

$$\equiv \text{point} + \text{CØRR} , \qquad (9a)$$

in terms of Eqs. (7) and (8). It is obvious that S_c is upper bounded by the pointlike (individual) stopping of the coagulated charge $\sum_{i=1}^{N} Z_i$.

Specializing Eq. (9) to a 4-cluster, it reads now

$$S_c = \frac{e^2 \omega_p^2}{v^2} * \{ (\sum_{i=1}^{4} Z_i^2) \ln (\frac{2mv^2}{\hbar \omega_p}) + 2(Z_1 Z_2 + Z_3 Z_4) G(0, D) + 2(Z_1 Z_3 + Z_2 Z_4) G(B, O)$$

$$+ 2(Z_1 Z_4 + Z_2 Z_3) G(B, D) \} \qquad (10)$$

with $G(B, D)$ defined by the 2nd r.h.s. in Eq. (8). This rectangle flowes parallel to its velocity.

In the sequel, we shall estimate quantitatively the total stopping S_c and the correlated part CØRR, through the ratio RAP given as

$$RAP = \frac{\text{CØRR}}{S_c} , \qquad (11)$$

RAP is always upper bounded by the coagulated $\sum_{i=1}^{N} Z_i$ charge limit, so that

$$RAP \le 1 - \frac{\sum_{i=1}^{N} Z_i^2}{(\sum_i Z_i)^2} , \qquad \text{all } Z_i \ge 0 \qquad (12)$$

The upper limit is reached for $B = D = O$, as expected.

Maximum stopping efficiency is achieved when all the debris bear the same charge Z_i. Moreover RAP values are then Z_i independent.

These features are obviously due to the quadratic structure of Eq. (9).

On the other hand, in similar conditions, a cluster geometry mostly orthogonal ($B > D$) to velocity, is likely to experience a strongest enhanced stopping. These remarks take care of the $B\omega_p/v$ and $D\omega_p/v$ dependence specifying a given projectile-target behaviour.

RESULTS

On Fig. 1, we display stopping charge for a chain of unit positive charges flowing parallel to its velocity in a dense jellium target. The RAP quantity (11) is also given. STOP/N shows saturation. This is not the case of similar chains flowing orthogonal to velocity which exhibit a largest stopping effect on Fig. 2.

Rectangles B x D and B x C respt. parallel and orthogonal to velocity V with D//V and B,C$\perp$V are analyzed on Figs. 3 and 4. Fig. 3 shows the behaviour of unit charges on transverse and parallel such plates. As expected, from the chains (Figs. 1 and 2) the transverse geometry experiences a largest stopping contribution. Fig. 4 explains a charge repartition effect for squares flowing parallel to velocity. 2+2+2+2 refers to a square with charge 2 at each vertex 1+1+1+5 denotes a similar asymmetric repartition. 1+1+1+1+4 is a centered square 1+1+1+1 with charge 4 at its center. The evenly distributed cluster (2+2+2+2) gets the largest stopping effect.

REFERENCES

1 G. Basbas and R.H. Ritchie, "Vicinage Effects in Ion-Cluster Collisions with Condenses Matter", *Phys. Rev. A*, **25**, 1943 (1982).
2 C. Deutsch, "Cluster Stopping in Dense Plasmas", *Laser Particle Beam*, **8**, 541 (1990).

ATTENDEES

Christine Bamiere
S.G.D.N.
51 B^d de Latour
75700 Paris, France

N. G. Basov
Levedev Physical Institute
  of the Academy of Sciences
Moscow, USSR

Steven H. Batha
Princeton University
Plasma Physics Laboratory
P. O. Box 451
Princeton, NJ 08543

Irving Bigio
Los Alamos National Laboratory
P. O. Box 1663
Los Alamos, NM 87545

Bruce W. Boreham
Dept. of Applied Physics
UCCQ, Rockhampton
QLD 4700, Australia

Chan K. Choi
Purdue University
School of Nuclear Engineering
West Lafayette, IN 47907

C. B. Collins
The University of Texas at Dallas
Center for Quantum Electronics
P. O. Box 830688, MS NB11
Richardson, TX 75083-0688

John S. De Groot
Department of Applied Science
  and Plasma Research Group
University of California-Davis
Davis, CA 95616

Claude Deutsch
LPGP Bat. 212
Universite de Paris XI
91405 Orsay Cedex, France

Shalom Eliezer
Plasma Physics Dept.
SOREQ N.R.C.
YAVNE 70600, Israel

Leonida Antonio Gizzi
Plasma Physics Group
The Blackett Laboratory
Imperial College
SW7 2BZ London, United Kingdom

Martin Gundersen
University of Southern California
Dept. of EE-Electrophysics
SSC-420, MC 0484
Los Angeles, CA 90089-0484

Charles D. Hendricks
Lawrence Livermore National Lab.
P. O. Box 808
Livermore, CA 94550

J. Javier Honrubia
Instituto de Fusion Nuclear
Jose Gutierrez Abascal, 2
28006 Madrid, Spain

Heinrich Hora
Dept. of Theoretical Physics
The University of N. South Wales
P. O. Box 1
Kensington, New South Wales
Australia 2033

Zhang Jun
Institute of Applied Physics
P. O. Box 8009
Beijing 100088
People's Republic of China

Takeshi Kaneda
Social Systems Department,
Mitsubishi Research Institute, Inc.
3-6, Ohtemachi 2, Chiyodaku
Tokyo 100, Japan

Kim M. Kassabasian
U. S. Navy (Civilian)
Systems Engineering Branch (H11)
Naval Surface Warfare Center
Dahlgren, VA 22448-5000

Yeong E. Kim
Department of Physics
Purdue University
West Lafayette, IN 47907

William Kruer
Lawrence Livermore National Lab.
P. O. Box 808
Livermore, CA 94550

Sergei Kuznetsov
Department of Chemistry
Moscow State University
119899 Moscow, RUSSIA

Yim T. Lee
Lawrence Livermore National Lab.
P. O. Box 5511
Livermore, CA 94550

W. P. Leemans
Lawrence Berkeley Laboratory
1 Cyclotron Rd, MS-71H
Berkeley, CA 94720

W. Howard Lowdermilk
Lawrence Livermore National Lab.
P. O. Box 5508, MS L-490
Livermore, CA 94550

James Mark
MT-Enterprises
P. O. Box 2217
Princeton, NJ 08543-2217

David J. Mayhall
Engineering Research Division
Lawrence Livermore National Lab.
P.O. Box 808, Mail Code L-156
Livermore, CA 94550

Tom A. Mehlhorn
Sandia National Labs
Div. 1262
P. O. Box 5800
Albuquerque, NM 87115

David J. Mencin
University of Missouri-Columbia
Nuclear Engineering Program
Columbia, MO 65211

George H. Miley
University of Illinois
Fusion Studies Laboratory
103 South Goodwin Avenue
Urbana, IL 61801

Noriaki Miyanaga
Institute of Laser Engineering
Osaka University
Yamada-oka 2-6, Suita
Osaka 565, Japan

Katsu Mizuno
Lawrence Livermore National Lab.
P. O. Box 808, L-418
Livermore, CA 94550

Sadao Nakai
Institute of Laser Engineering
Osaka University
2-6 Yamada-oka
Suita Osaka, 565, Japan

Keishiro Niu
Physics Laboratory
Teikyo University of Technology
2289-23 Ohotani, Urido
Ichihara, Chiba 290-01, Japan

674

John Nuckolls
Lawrence Livermore National Lab.
P. O. Box 808
Livermore, CA 94550

Chun-Mou Peng
ABB Systems Control
2550 Walsh Ave.
Santa Clara, CA 95051

Hansheng Peng
China Academy of Engineering Physics
P. O. Box 501
Chengdu, China

Maria Petra
University of Illinois
Fusion Studies Laboratory
103 South Goodwin Avenue
Urbana, IL 61801

Mark A. Prelas
323 Electrical Engineering
University of Missouri-Columbia
Columbia, MO 65211

Andrei Rode
Laser Physics Centre
The Australian National University
P. O. Box 4
Canberra ACT 2601, Australia

Fred Schwirzke
Department of the Navy
Naval Postgraduate School
Monterey, CA 93943-5100

Andrei Shikanov
Lebedev Physical Institute
Acadmey of Science of Russia
Moscow, RUSSIA

Marshall Sluyter
U. S. Department of Energy
Inertial Fusion Division, DP-243
Washington, DC 20545

Edward Teller
Lawrence Livermore National Lab.
P. O. Box 808
Livermore, CA 94550

Guillermo Velarde
Instituto de Fusion Nuclear
Jose Guierrez Abascal, 2
28006 Madrid, Spain

J. Paul Watteau
Commissariat a L'Energie Atomique
Ctr D'Etudes de Limeil-Valenton
B. P. 27
94190 Villeneuve, St. Georges
FRANCE

Oswald Willi
Imperial College
Prince Consort Road
London SW7 2BZ
United Kingdom

Chen Xiaofeng
Southwest Institute of Nuclear
    Physics and Chemistry
P. O. Box 525-77
Chengdu, 610003, China

Chiyoe Yamanaka
Institute for Laser Technology
Yamadaoka 2-6
Suita, Osaka 565 JAPAN

Kiyoshi Yatsui
Laboratory of Beam Technology
Nagaoka University of Technology
Nagaoka, Niigata 940-21, Japan

Jick H. Yee
Lawrence Livermore National Lab.
7000 East Avenue
P. O. Box 808, L-156
Livermore, CA 94550

AUTHOR INDEX

A

Afshar-rad, T.   517-534
Alvarez, R. A.   233-250
Aragonés, J. M.   535-546
Aydin, M.   181-195
Azechi, H.   251-280, 443-502

B

Basov, N. G.   25-28
Batani, D.   171-179
Batha, B. H.   305-322
Batha, S. H.   305-322
Biancalana, V.   171-179
Blenski, T.   547-552
Boody, F. P.   67-78
Borghesi, M.   171-179
Brunner, W.   105-109

C

Cameron, S. M.   197-208
Campbell, E. M.   323-335
Carroll, J. J.   151-166
Chapman, H. N.   111-123
Chen, T.   443-502
Chen, W.   91-97
Chen, X. F.   511-515
Chessa, P.   171-179
Choi, C. K.   601-606
Collins, C. B.   151-166
Cun, Y.   99-104

D

De Groot, J. S.   197-208
Deha, I.   171-179
Desselberger, M.   517-534
Deutsch, C.   547-552, 625-636, 663-671
Drake, R. P.   197-208

E

Dunne, M.   281-288, 517-534

E

Edwards, J.   281-288, 517-534
Eliezer, S.   137-149, 167-170, 637-648
Endo, T.   443-502
Estabrook, K. G.   197-208

F

Falquina, R.   535-546
Fan, D.   91-97
Fromy, P.   625-636
Fruchtman, A.   167-170
Fu, S.   91-97

G

Gazit, Y.   637-648
Gilath, I.   637-648
Giulietti, A.   171-179
Giulietti, D.   171-179
Gizzi, L.   517-534
Gizzi, L. A.   171-179
Gu, Y.   91-97

H

Häuser, T.   607-624
He, X.   91-97
Henis, Z.   137-149
Hong, C.   151-166
Honrubia, J. J.   535-546
Hora, H.   11-21, 181-195, 347-389, 607-624
Hunt, J. T.   323-335

I

Izawa, Y.   443-502

677

SUBJECT INDEX

Irradiation uniformity, 251
    274
    computer simulation of,
        274
    smoothing by, 251

Lasant pressure dependence,
    60

Laser driven hohlraums, 23
    (also see Inertial
        confinement fusion,
        indirect drive
        target)

Laser fusion, 5, 443
    (also see Inertial
        confinement fusion)
    GEKKO XII blue laser, 479
    high density compression
        experiment, 443
    megajoule targets, 503
    progress, 5

Laser induced shock waves,
    637

Laser interaction, 181
    pulsation and stuttering
        of, 181

Laser light, 167
    circularly polarized, 167

Laser plasma interaction,
    517

Lasers, 1, 24, 29, 32, 34
    excimers, 34
    high intensity, 1
    multi-kilojoule, 24
    nulcear flashlamp pumped,
        29, 32
        atomic iodine, 29

Magnetic field, 167
    generation of, 167

Micro-channel plates, 111

Microwave bandwidth
        broadening, 233
    two-dimensional calcu-
        lation of, 233

Multimode optical fiber,
    258

Muon catalyzed fusion, 137

Nova upgrade facility, 323

Nuclear radiation effects
    on lasant, 55, 67

Parallel plate capacitor,
    214
    voltage collapse in, 214

Parametric instabilities,
    305
    experimentalist's view-
        point, 305

Partially coherent light,
    252

Partially coherent light
        source, 259
    characteristics of, 259

Plasma instabilities, 171
    laser driven, 171

Progress, 99
    at CAEP, 99

Propagation, 529
    of ion beam, 529

Pulse generator switch, 217
    voltage collapse in, 217

Pumping, 40, 42, 43, 52
    ^{10}B, 40, 52
    ^{3}He, 42, 52
    XeBr flourescence under
        ^{10}B, 43

Radiation thermal con-
        duction, 126
    self-similar solution of,
        126

Radiation trapping, 105
    its influence on the
        gain, 105

Radiatively-heated foils,
    281
    time-resolved measure-
        ments, 281

Sandia's target physics
        program, 554
    indirectly-driven light-
        ion targets, 554

Second harmonic, 173, 272
    conversion, 272

Shock waves, 637
    laser induced, 637